PRODUCTIVITY AND QUALITY IMPROVE.
IN ELECTRONICS ASSEMBLY

PRODUCTIVITY AND QUALITY IMPROVEMENT IN ELECTRONICS ASSEMBLY

Sponsored by
the Institute of
Industrial Engineers

Johnson A. Edosomwan
& Arvind Ballakur

McGraw-Hill Book Company

New York   St. Louis   San Francisco   Auckland   Bogota
Hamburg   London   Madrid   Mexico   Milan
Montreal   New Delhi   Panama   Paris   Sao Paulo
Singapore   Sydney   Tokyo   Toronto

Library of Congress Cataloging-in-Publication Data

Productivity and quality improvement in electronics assembly.
"Sponsored by Institute of Industrial Engineers."
Bibliography: p.
1. Electronic apparatus and appliances—Production control.
2. Electronics apparatus and appliances—Quality control.
I. Edosomwan, Johnson Aimie, 1953- . II Ballakur, Arvind.
III. Institute of Industrial Engineers (1981-)

TK7836.P75   1988   621.381   88-13376
ISBN 0-07-019026-7

Published in 1988
Printed in the United States of America
94 93 92 91 90 89   5 4 3 2 1
ISBN 0-07-019026-7

CONTENTS

Preface

Electronics assembly processes are constantly changing and relying heavily on the use of sensitive parts and sophisticated technology, tools, and equipment. There is no doubt that electronics products will continue to greatly affect the quality of life now and in years to come. The rate of technological change in electronics has progressed from the invention of the transistor, the development of the chip silicon transistor, and the invention of the integrated circuit or silicon semiconductor chips.

In the past three decades semiconductor technology has advanced by a rapid increase in the density and number of transistors on a chip. In the early 1980s, two increases in the number of transistors on a chip occurred: large-scale integration and the very large-scale integration of chips. The number of transistors per chip could climb to a number ranging from several hundred million to about one billion by the year 2000. Experts describe the next era as ultra large-scale integration.

To improve productivity and quality in a highly sophisticated electronics assembly environment, new techniques, methodologies, and principles are needed. This book assembles useful productivity and quality improvement ideas and techniques that have applications in electronics assembly.

The introductory section presents background information on electronics assembly, productivity, quality, and just-in-time (JIT) management definitions and concepts. This section also examines electronics industry trends. A framework for managing productivity and quality in electronics assembly is presented, and assessment tools and techniques for improving productivity and quality are discussed. A framework and the fundamental concepts of JIT such as housekeeping, set-up reduction, kanban production, vendor partnership, total quality control, etc., are presented. Additionally, some of the management issues in implementing JIT are discussed.

Part I focuses on productivity and quality management in electronics assembly. Total productivity and quality improvement techniques are discussed. This part emphasizes models, techniques, and methodologies to improve productivity and quality in electronics assembly. Part II focuses on JIT application in electronics assembly, and several case studies of JIT applications are presented.

Electronics assembly support systems are covered in Part III. Issues such as materials control, electronics testing, and information systems requirements are examined. Part IV covers automation issues in electronics assembly processes. This part provides case studies and an automation framework for electronics assembly, as well as total system integration efforts in electronics assembly. Part V focuses on product and process design issues in electronics assembly. Design for manufacturability, testability, safety, etc., is emphasized, and designing ergonomically acceptable electronics assembly work stations, products, and processes are examined.

Part VI presents a step-by-step methodology for integrating product and process design in high technology equipment production. A specific application of the methodology to circuit pack design and assembly is presented. Part VII offers a methodology for classifying and coding electronics components. The appendices present glossary of terms and sources of information of electronics assembly.

The material presented in this book is intended to serve the practical needs of engineers, managers, system designers, and analysts directly involved in improving productivity and quality in electronics assembly.

The editors gratefully acknowledge the contributions from the authors of invited papers and publishing organizations that granted permission for reproduced papers. The support of members of the Electronics Industry Division of the Institute of Industrial Engineers, the Industrial Engineering and Management Press Staff, and the McGraw-Hill staff is highly appreciated. Our thanks to Donald L. Morelli and Rob Hooper for the assistance provided during the preparation of this book. We are also grateful to our employers, IBM and AT&T Bell Laboratories, and to our families and friends for their support and encouragement.

User's Guide

As stated in the preface, this book provides useful background information and selected papers on specialized topic areas in electronics assembly. Users of this handbook should follow these steps to enhance their knowledge in the various topics covered.

Step 1: Read the introductory section thoroughly. This section is intended to familiarize the reader with a basic knowledge of productivity, quality, and just-in-time techniques in electronics assembly.

Step 2: Use the productivity and quality improvement matrix (PQIM) presented at the beginning of parts 1-5 to match papers of interest with your specific needs. The PQIM organizes all the papers by electronic assembly processes and focus areas.

Introduction: Productivity, Quality, and Just-In-Time Improvement Concepts in Electronics Assembly

This introduction presents background information on electronics industry products, processes, and growth rate in a competitive world economy. A definition of electronics and important statistics on the electronics industry are presented. Productivity and quality are defined, as well as key concepts involved in comprehensive productivity and quality management in electronics assembly. A framework for implementing just-in-time (JIT) manufacturing to improve productivity and quality in electronics assembly is presented. JIT manufacturing is defined, and its significance in electronics manufacturing is examined. Nine elements of the JIT framework are discussed. The role of JIT in attaining the business objectives of a company is also emphasized. Additionally, the requirements for improving productivity and quality at the source in electronics assembly are presented.

BACKGROUND INFORMATION ON ELECTRONICS INDUSTRY

The electronics industries include various companies and organizations concerned with the design, development, manufacture, assembly, and application of electronic parts, tools, technologies, components, and systems. In the past four decades, electronic technologies and related products have made the most influential impact on the quality of working life, industrial productivity, employment, international trade, and global communications. Micro-electronics technology has revolutionized our way of living and has had a tremendous impact on almost everything we do.

Industry analysts predict that the electronics industry could grow by 10% or more annually. Electronics has grown from a $200 million U.S. industry in 1927 to $201 billion in 1985 and is expected to reach $800 billion by the Year 2000.

For example, in the computer industry alone, the Electronic Industries Association (EIA) reported that some 138,000 mini-computers, defined as those priced from $5,000 to $40,000, were sold in the United States in 1979 alone. The purchases of the computer-aided design (CAD) systems, the first step in integrating computer-aided design and computer-aided manufacturing systems (CAD/CAM), was about $570 million in 1979. CAD system purchases could hit a record of $9 billion by the Year 2000.

Figure 1 presents total U.S. sales of electronics in billions of dollars from 1976 to 1985. The rate of growth in sales has been tremendous, with an average growth rate of 7% or more per year. The growth rate for the industry as a whole

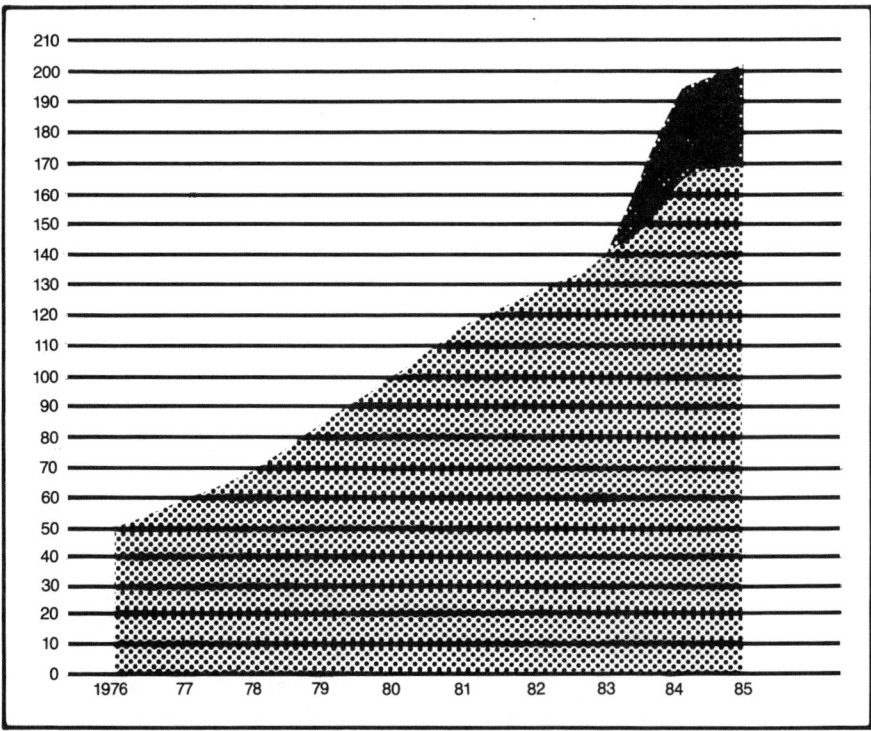

Figure 1. *Total U.S. sales of electronics ($ billion), 1976 - 1985.*
Source: Electronics Industries Associations' 1986 Electronic
Market Data Book, p. 4.

Because of the addition of "other related products and services" to Figure 1, this graph depicts two sets of trend data. The darker portion of the graphic represents the inclusion of "other related products and services." The lighter portion portrays the sales trends of the industry groups that have been traditionally measured (excluding "other related products and services").

in consumer electronics, industrial electronics, electronic components, communications, computers, and other related products and services is shown in Table 1, Figure 2, and Figure 3, respectively. According to an EIA report in 1986, U.S. factory sales of electronic equipment systems and components totaled $209.7 billion, an increase of 5.5% over the 1985 total sales of $198.9 billion. An estimated 27% of the $209.7 billion total is accounted for by sales to the defense electronics market.

In the United States, exports of electronic products increased 7.8% in 1986 to $33.5 billion, whereas imports increased 16.2% to $50.4 billion, resulting in a balance-of-trade deficit of $16.9 billion.

In November 1983, the U.S. Office of Technology Assessment issued a report on international competitiveness in electronics. The following recommendations were made for maintaining the continuing competitiveness of the U.S. electronics industry:

1. High-quality education and training (including retraining) for engineers, technicians, and other skilled workers.
2. A strong technological base stemming from basic research and applied research and development with long-term objectives, including the diffusion of results, in fields such as solid-state electronics, optical devices, communications technologies, computer-aided design of circuits and systems, and computer software.
3. Economic adjustment policies that smooth flows of capital and labor within the economy, aiding growing firms in their efforts to compete while providing well-paying jobs for the domestic labor force.
4. Adequate supply of investment capital for new start-ups as well as rapidly expanding established firms.
5. An international trading environment that places U.S. firms on a generally equal footing with their competitors in other countries, including those that have well-developed industrial policies intended to protect or promote domestic manufacturers.

ELECTRONICS DEFINED

Electronics is a branch of science and technology that involves the study, analysis, application, and control of the conduction of electricity in a vacuum, gas, liquid, semi-conductor, superconductor, or miniconductor. Electronic equipment has the unique ability to influence the flow of electricity. The various electronic products use parts, subassemblies, and materials to provide and perform several electronic functions. Important milestones in the development of electronics technology are schematically shown in Table 2.

TABLE 1 – Factory Sales of Consumer Electronic Products
(Including Imports $ Million; United States, 1977-1986)
Source: Electronics Industries Association's 1987 Electronic Market Data Book, p. 6

CATEGORY	1977	1978	1979	1980	1981	1982	1983	1984	1985	1986
Monochrome TV Receivers	530	549	561	588	505	507	465	419	309	328
Color TV Receivers	3,289	3,674	3,685	4,210	4,349	4,253	5,002	5,538	5,562	6,024
Projection TV					287	236	268	385	488	529
Video Cassette Recorders	180	326	389	621	1,127	1,303	2,162	3,585	4,738	5,258
Color Cameras					147	232	303	355	228	59
Video Disc Players					55	54	81	45e	45e	45e
Audio Systems[1]	606	748	748	809	720	573	630	976	1,372	1,370
Separate Audio Components	1,275	1,143	1,178	1,424	1,363	1,181	1,268	913	1,132	1,358
Home Radio	523	436	436	468	501	530	565	661	379r	408
Portable Audio Tape Equipment[2]	1,208	1,649	1,739	1,403	1,157	971	1,102	1,191	1,140 r	1,389
Car Audio[3]	534	582	623	1,368	2,000	2,100	1,900	2,484	2,300e	2,800e
Blank Audio Cassettes					242	219	249	275	277	304
Blank Video Cassettes						357	580	931	1,285	1,480
Total	8,145	9,107	9,359	10,891	12,453	12,516	14,575	17,758	19,255	21,352
Estimates by Consensus[4]				7e	44e					
GRAND TOTAL	8,145	9,107	9,359	10,898e	12,497e	16,826e	20,550e	23,438e	25,294e	28,721e

[1] Before 1981, data includes console phone.
[2] Before 1980, data includes some tape equipment other than portable.
[3] Before 1980, data reflects factory-installed car audio products only.
[4] Includes home computers, software, programmable video games, video game cartridges, telephones, telephone answering devices, video discs, video cassette players, and satellite earth station systems.
r-Revised.
e-Estimated.
na-Not available.

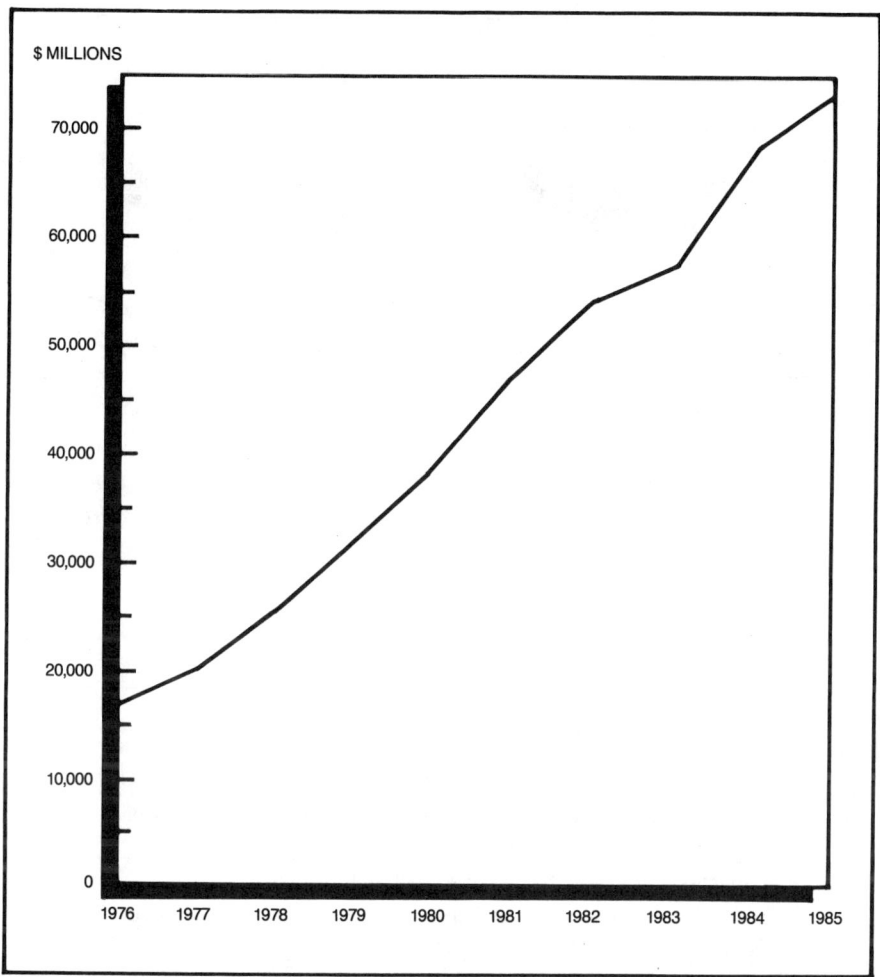

Figure 2. *U.S. factory shipments of industrial electronics equipment, 1976-1985.*
Source: U.S. Department of Commerce.

EMPLOYMENT IN ELECTRONICS INDUSTRIES

Table 3 shows the average annual employment in the electronics industries in the United States from 1981 to 1985. According to the Bureau of Labor Statistics, average annual employment in the United States reached 1,829,000 persons in 1985. Industry analysts predict an employment growth rate of 5% or more, especially in areas of computer manufacturing, communications development, and

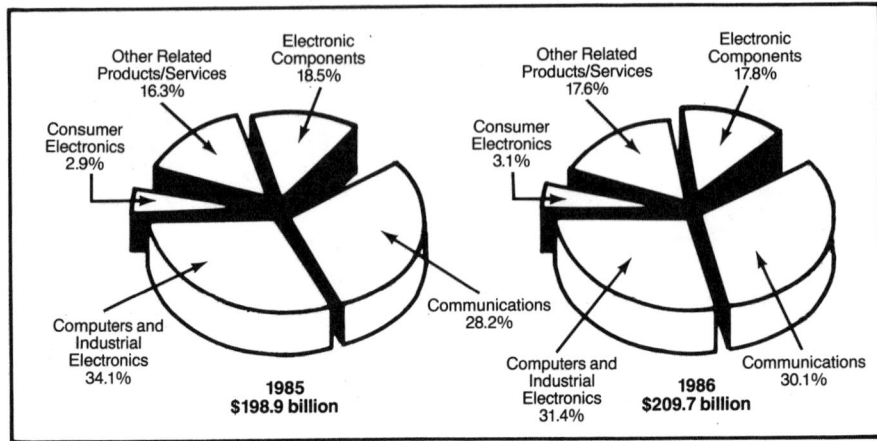

Figure 3. *U.S. factory sales of electronics by industry group (%), 1985-1986. Source: Electronics Industries Association's 1987 Electronic Market Data Book, p. 3.*

TABLE 2 – Milestones in Electronics Technology Development, 1883 - 2000

YEAR	MAJOR ACTIVITY
1883	Control of electrons achieved. Thomas Edison induced electrons to jump across partial vacuum between a carbon filament and a metal plate in his light bulb.
1894	Oliver Lodge developed a system of wireless communications in England. Electronics communication started.
1897	Electron discovered by Sir Joseph John Thompson
1906	Three-element vacuum tube invented by Lee Dee Forest. Radio broadcasting made possible by vacuum tubes.
1920	Radio sets sold to the public for the first time.
1930s	Development of television, radar, oscilloscope voltmeter, and electromechanical analog calculators and computers.
1947	AT&T Bell Laboratories announced the invention of transistors that consist of semi-conducting material such as silicon.
1959 to 1962	Integrated circuit developed and produce in mass quantities.
1970s	Combination of thousands of circuits into a single silicon chip became possible. The technique is known as large-scale integration.
1980s	Very large scale integration of circuits was made possible.
1990s to	Ultra large scale integration of circuits expected. Major advances in superconduction and optical circuitry are also expected.

TABLE 3 – Average Annual Employment in the Electronics Industries (thousands of workers); United States, 1981-1985
Source: U.S. Department of Labor, Bureau of Labor Statistics

	1981	1982	1983[r]	1984[e]	1985
Consumer Electronics (SIC 3651)	81.4	70.8	67.2	71.8	67.7
Communications Equipment (SIC 366)	556.7	569.4	573.1	616.6	659.3
Electronic Components & Accessories (SIC 367)	557.3	558.2	578.7	672.7	657.2
Computers (SIC 3573)	385.3	403.3	421.2	460.9	445.0
TOTAL[1]	1,580.7	1,601.7	1,640.2	1,822.0	1,829.0

[1] Excluded from total are estimates for SIC Codes 3611, 3622, 3643, 3693, 3699, 3811, 3821, 3822, and 3931 because these contain major, non-electronic product categories.
[e]-Revised.

consumer electronics. An estimated 1,776,300 persons were employed in electronics manufacturing and related activities in 1987. Figure 4 shows employment in U. S. electronic industries for 1986 versus 1995. Table 3 shows annual employment in electronics industries from 1981 to 1985. Employment comparison by industry in the United States from 1976 to 1985 is presented in Figure 5.

OVERVIEW OF ELECTRONICS ASSEMBLY

Electronics products have a wide spectrum of applications ranging from household consumer goods (phones, radios, televisions, etc.) to sophisticated industrial and defense systems (computers, communications equipment, space flights, etc). *Electronics assembly,* as referred to in this book, deals with the assembly of a typical electronic product that exploits a card-on-board scheme. In this scheme, various functions are realized on individual printed circuit boards (PCB) and integrated into a system by interconnection via a backplane.

Electronic components (such as transistors, resistors, capacitors, etc.) are mounted on PCBs. A PCB has wiring pathways that are printed on the board. A PCB also has an interconnection device at one of its edges. The electronic device could be a cabinet or a collection of shelves, that holds the PCB. The device has nesting slots that guide the PCBs onto the backplane of the system.

A material flow process for PCB assembly is shown in Figure 6. The process starts with components and PCBs being supplied by the storeroom. For those devices that are surface-mounted, the boards are prepared and stencil-printed. The components are then placed and soldered using a solder technique such as vapor phase reflow or infrared reflow technique. The boards are then prepared for through-hole components. These components are inserted using dual in-line package (DIP), axial, radial, or variable center distance (VCD) insertion

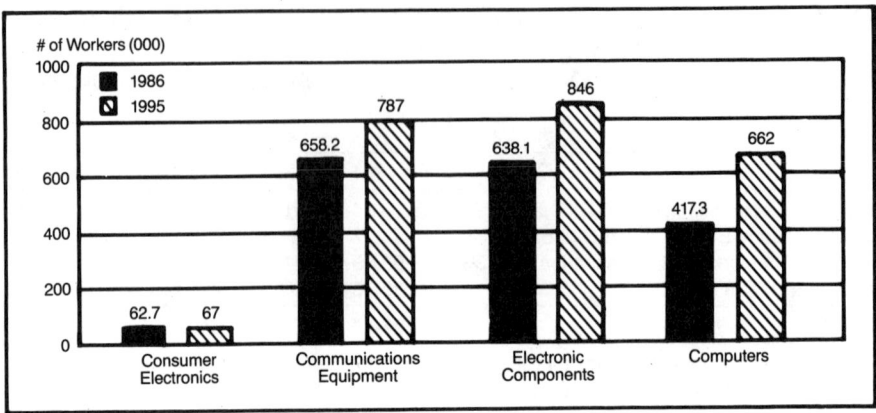

Figure 4. *Employment in electronics industries (thousands of workers), United States, 1986 vs. 1995. Source: U.S. Bureau of Labor Statistics.*

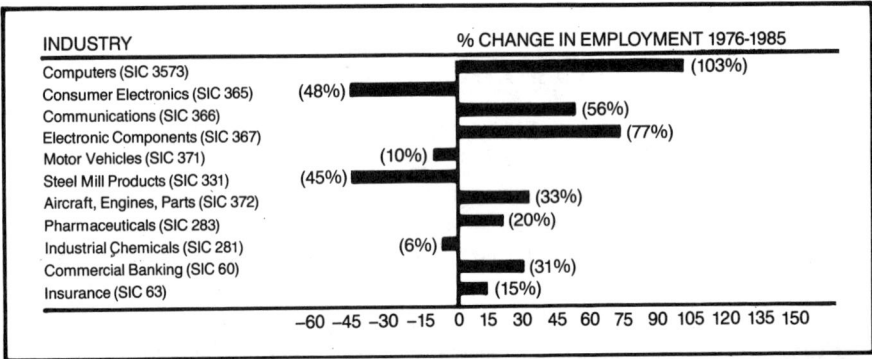

Figure 5. *Comparative employment by industry, United States 1976-1985. Source: U.S. Department of Labor, Bureau of Labor Statistics.*

machines based on the packaging of the component. The board with inserted components is soldered using a soldering technique (for example, wave solder). The PCB when tested (in-circuit and/or functional), is ready for mounting in a unit or a system.

The various key processes that are involved in realizing electronic products are shown in Figure 7. These key product realization processes (as shown in the figure) are so closely tied to each other that they cannot be examined

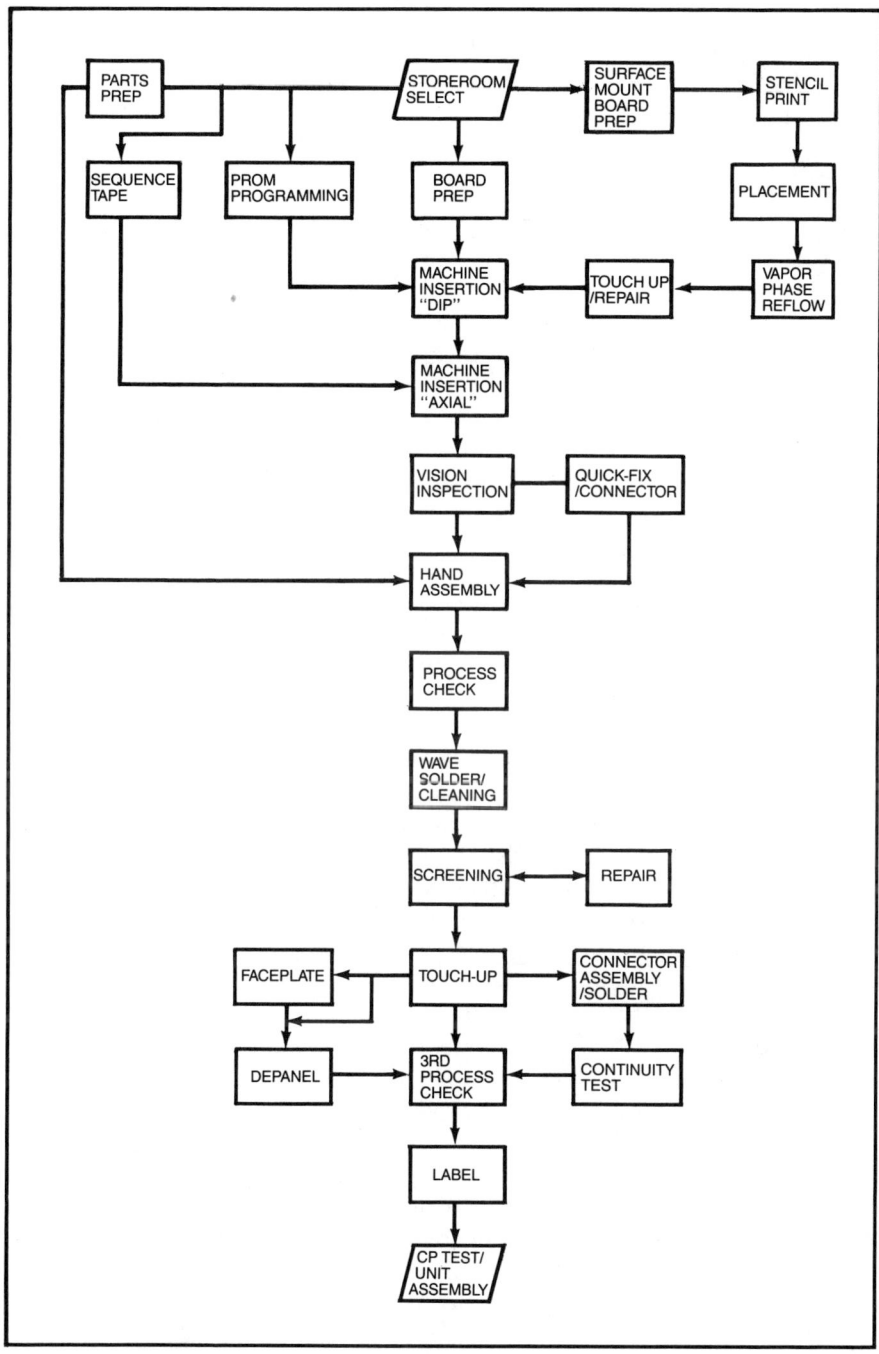

Figure 6. *Printed circuit board assembly process.*

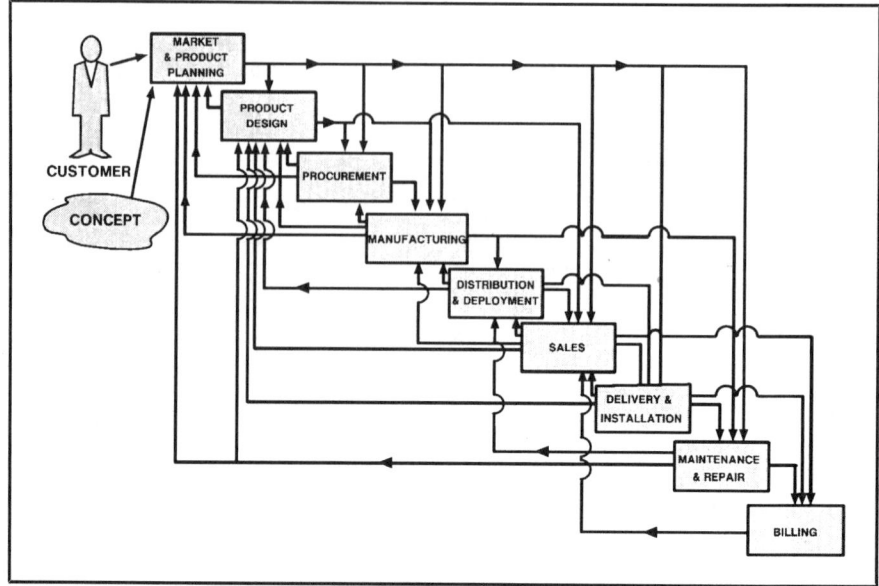

Figure 7. *Key product realization processes.*
Source: Ballakur and Pratt (1987).

separately. This section describes some of the interactions of these product-realization processes.

A successful electronics product is driven not only by technological superiority but also by perceived market needs of potential customers. These requirements are then converted into products (hardware, software) in the product design process. During this phase, early manufacturing involvement, early procurement involvement, and early deployment involvement must be carried out to ensure a smooth transition from design to manufacture to distribution.

To physically build the products in the required quantity, volume manufacturing operations are necessary. Procurement processes must ensure that adequate supply and timely delivery of raw materials, components, subassemblies, etc., is guaranteed to meet the manufacturing requirements. Knowing what to build or what components to order requires good material management and customer service processes. These processes work together to promise the customer the delivery of the product in the required quantity and at an established time. Adequate distribution and deployment operations are essential. For delivering the product, to the customer. The sales force must be knowledgeable about the product, and ensure that customers are satisfied with the service. When the product is sold and the payment is collected, the product realization process does not end. Adequate customer service (such as maintenance and repair) need to be provided throughout the product life cycle to ensure future sales.

DEFINITION OF PRODUCTIVITY AND QUALITY MANAGEMENT

Productivity and quality management is an integrated process involving both management and employees with the ultimate goal of managing the design, development, production, transfer, and use of the various types of products or services in both the work environment and the market place. The process requires total involvement of everyone in planning and analysis, measurement, evaluation and control, and improvement and monitoring of productivity, and quality at the source of production or the service center.

The components of the comprehensive productivity and quality management model (CPQMM) that has been tested in several industries, including the electronics industry, are presented in Figure 8. A major requirement of CPQMM is to ensure that there is a balance between productivity and quality results. This means that both productivity and quality goals must compliment each other and be managed in one total pool to achieve the desired results. The premise CPQMM is that productivity and quality are connected and inseparable. The actions geared toward productivity improvement should also be used for quality improvement and vice-versa. The following implementation steps are recommended for the CPQMM:

Step One. Classify the work processes into cell units and understand product mix, procedures, systems requirements, and technology interaction and provide personnel training.

Step Two. Understand the input and output components of each cell, workflow patterns, and interdependencies among service and production variables.

Step Three. Develop and implement measurement methods for productivity, quality, and system variables. Adopt a total measurement approach that considers all input and output elements.

Step Four. Develop and implement planning and analysis tools for short-term and long-term impact. Set up data base for information from all work process cells.

Step Five. Collect data periodically, synthesize data, and compute indices and values needed to understand the performance in productivity and quality of all work process cells.

Step Six. Perform root-causes analysis for service, production, and systems-related problems. Develop and implement improvement actions and monitor each work cell on an on-going basis.

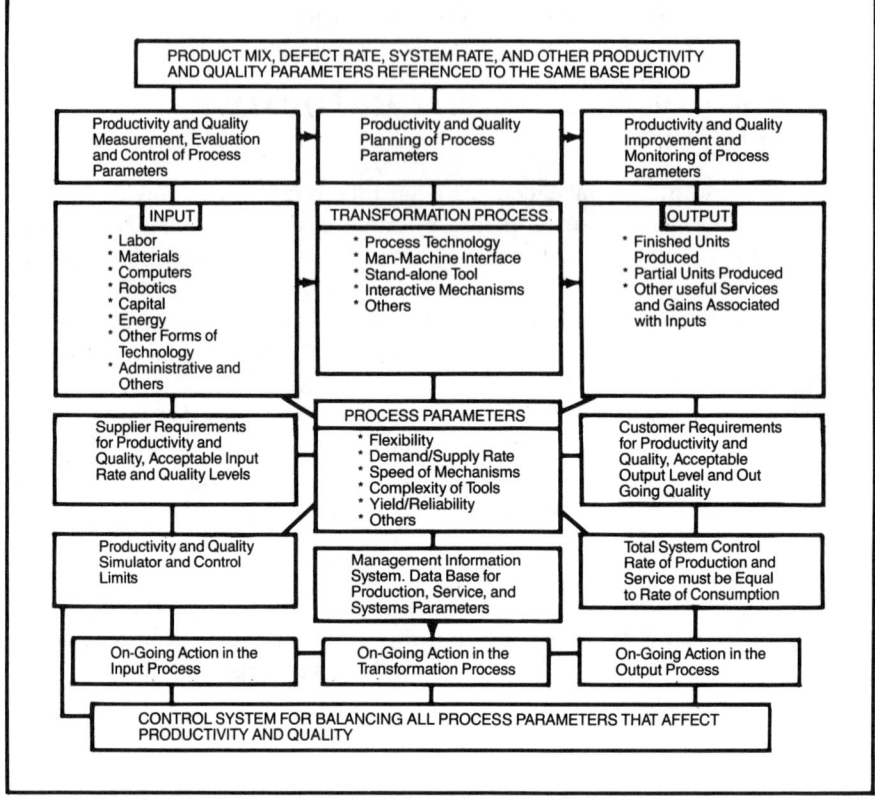

Figure 8. *Components of CPQMM.*
Source: Edosomwan (1987a, 1986a).

The effective system management of productivity and quality in the work environment also requires an on-going effort directed toward the following:

- Effective and efficient allocation of all resources to produce output and problem solving.
- Dedication to removing barriers and bottlenecks of good ideas that reduce cost.
- Commitment to providing adequate employee and management development.
- Simplification of work processes and procedures.
- Use of realistic measures for productivity and quality.

- Total commitment to excellence and prevention techniques as a way of quality control.
- Total commitment to continuous productivity and quality improvement efforts. Recognize that productivity and quality improvements do not come from a one-time effort.

Successful gains in productivity can also be obtained through teamwork that focuses on problem identification and resolution, process monitoring and control, and implementation of connective actions.

DEFINITION OF PRODUCTIVITY

Productivity can be defined as a unique relationship between output produced by an organizational operational unit and the input used by the operational unit to produce the output. The following are three types of productivity definitions that are commonly used:

Total productivity: A ratio of all measurable output to all measurable input at the task or organization level.

Total productivity of electronics assembly task i, in site j, in period t. $=$ $\dfrac{\text{Total measurable output of electronic assembly task } i, \text{ performed in site } j, \text{ in period } t.}{\text{Total measurable input of electronic assembly task } i, \text{ performed in site } j, \text{ in period } t.}$

Where: $(i = 1,2,\ldots n, j = 1,2,\ldots m, t = 1,2,\ldots k)$

Total factor productivity: The ratio of all measurable output of a task or operational unit to the sum of the associated labor and capital input factors.

Partial productivity: The ratio of all measurable output of a task or operational unit to one class of input.

The total productivity measure is recommended because it takes into account the total output and input. The total factor productivity and the partial productivity should not be used alone because they could provide misleading performance trends. They should be used in conjunction with the total productivity measure to pinpoint specific input responsible for increases or decreases in total productivity. Productivity management and performance management compliment each other. Performance management may be described as a unique process that is concerned with the measurement, planning, implementation, and evaluation of improvement actions. Productivity management is concerned with the overall relationship between all inputs and outputs used by an organization's operational unit, as well as the on-going management of the measurement, evaluation, planning, analysis, improvement, and monitoring processes.

DEFINITION OF QUALITY

Quality can be defined as a measure of how well the final output of individual, task, or operational unit is fit for use. The important parameters involved in fitness for use are quality of design, quality of conformance, availability, and field service. The quality parameters are shown schematically in Figure 9.

QUALITY COSTS

Quality costs can be classified into four categories:

Category One: Internal failure costs associated with defective products, components, and materials that fail to meet quality requirements and cause manufacturing losses in both direct and indirect efforts.

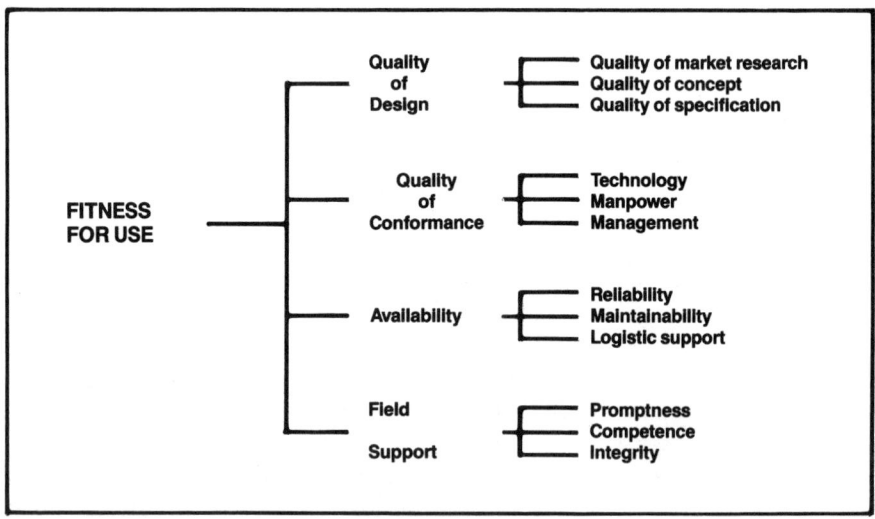

Figure 9. *Interrelationships between fitness for use parameters. Source: Juran (1980), pp. 2-9.*

Category Two: External failure costs generated by defective products shipped to customers.

Category Three: Appraisal costs associated with measuring, evaluating, or auditing products, components, and purchased materials to ensure conformance to quality requirements and specifications.

Category Four: Prevention costs associated with designing, implementing, and maintaining a quality system capable of anticipating and preventing quality problems before they generate avoidable costs.

PROCESS CONTROL IN ELECTRONICS ASSEMBLY

To control the various processes in electronic assembly, it is important to understand the common and special causes of variation that could come from process design, changes to process, and natural uncontrollable random and nonrandom sources. Statistical process control (SPC) is recommended as a technique for use as a feedback mechanism and to provide information about the following:

- process characteristics and variables;
- process performance;
- action on process inputs;
- transformation process; and
- action on process output.

The definition and concepts of SPC is presented in some of the papers included in Part 1. A specific methodology for implementing SPC in an electronics assembly environment is presented by Edosomwan in Chapter 1.

BENEFITS OF PRODUCTIVITY AND QUALITY MANAGEMENT IN ELECTRONICS ASSEMBLY ENVIRONMENT

Electronics assembly processes are constantly changing and relying heavily on the use of sensitive parts and sophisticated tools and equipment. In addition, companies designing, manufacturing, and marketing electronic products are operating in a highly competitive environment. Further, the rate of technological obsolescence is much greater in the electronics industry than in other industries. On-going productivity and quality management efforts are required for survival in this high technology industry. Electronic firms that institute an effective productivty and quality management program can derive the following benefits:

1. Lower prices to consumers for goods and services, because the cost of production is reduced through less rework and gains in productivity.
2. Effective use of resources so that more goods and services are produced for reasonable amounts of expended resources, higher real earnings for employees, and a reduction in increases of total wages without significantly offsetting gains in total productivity.
3. Enhancement of organizational strength for dealing with internal operating weaknesses and external competition.
4. Greater profitability, because quality improvement leads to less rework, reduced scrap, better use of tools and equipment, and reduced work-in-process inventory—that, in turn, leads to higher productivity. Minimizing the cost of manufacturing improves the profit margin for goods and services. The production of high quality products also improves sales, because the satisfied consumer will buy more and will recommend the product to other consumers.
5. More social benefits through increased public revenues from organizations.

REQUIREMENTS TO SUPPORT PRODUCTIVITY AND QUALITY IMPROVEMENT AT THE SOURCE IN ELECTRONIC ASSEMBLY

The task boundaries involved in on-going productivity and quality management are presented in Figure 10. Productivity and quality improvement at the source requires an effective management system of the production or service input, process, and output at the individual, task, operational unit, and organizational levels. In addition, both management and workers need to be aided with tools and techniques to perform the following:

1. On-going problems analysis and resolution at the source.
2. On-going assessment and measurement of task parameters, inputs, processes, and outputs.
3. On-going real time feedback on process performance.
4. Real-time support from team members and management for problem resolution.
5. On-going training on new methods and techniques to support the complex work environment. The training requirements include both basic job skills and cross-training for multiple skills.
6. Support system that encourages decision making and problem solving at the production source.
7. Clear definition of roles and responsibilities at the task operational unit and organization levels.

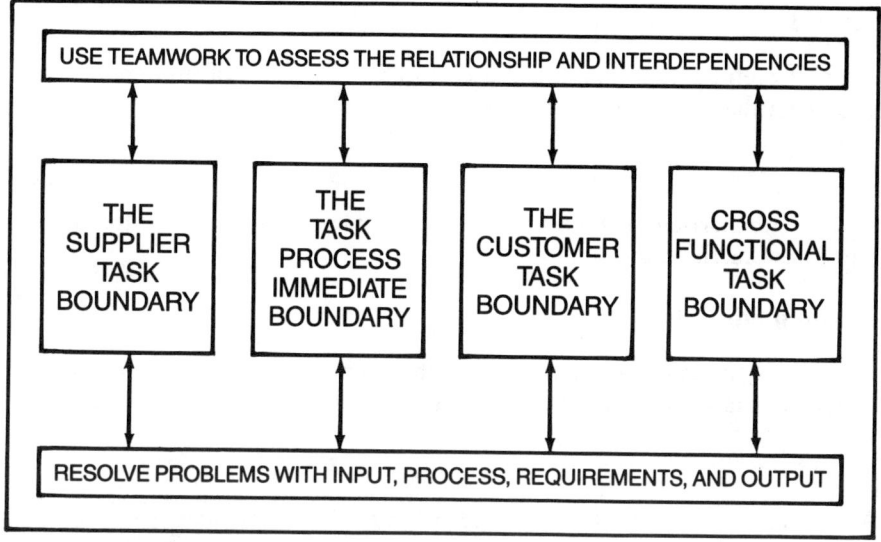

Figure 10. *Productivity and quality management task boundaries.. Source: Edosomwan (1987a,b).*

8. Teamwork within the work cell and across boundaries.
9. On-going enhancement of the communication channels at the task, operational unit, and organizational levels.
10. Techniques for productivity and quality measurement, evaluation, control, planning, analysis, improvement, and monitoring.

Assessment Tools and Techniques for Productivity and Quality Management in Electronics Assembly

The following techniques and tools are recommended for measuring, controlling, and evaluating productivity issues at the source.

Control Chart

A method of monitoring the output of a process or system, through sample measuring of a selected characteristic and analyzing its performance over time.

Fishbone Chart

A cause-and-effect diagram for analyzing problems, issues, and the factors that contribute to them.

Equal Productivity Curve

A visual method of comparing the cost of two methods of production, each producing the same output.

Cost Curves

A visual comparison of the amount of material and other supplies that can be purchased for the same sum.

Productivity Ratios

Productivity measures provide the basis to compare total output obtained from a particular task with the total partial input used.

Histogram

A bar graph displaying a frequency distribution. This is usually used to categorize product defect levels or distribution.

Scatter Diagram

A graph displaying the correlation of two characteristics. The illustrations for the various tools and techniques are presented in Table 4.

PASIT Technique

The production and service improvement technique (PASIT) requires that the production or service rate must be equal to the consumption rate at the task level.

Let:

P_{ijt} = Production rate of task i, in site j, in period t
W_{ijt} = Work in process of task i, in site j, in period t
C_{ijt} = Consumption rate of task i, in site j, in period t
S_{ijt} = Supplier input to task i, in site j, in period t
T_{ijt} = Total process requirements of task i, in site j, in period t

An optimally balanced production level can be obtained to satisfy the PASIT requirement when:

$$\text{Pijt} = \text{Cijt}$$
$$\text{Wijt} < \text{S, Wijt} = 0 \text{ (for ideal state)}$$

and $\text{Sijt} = \text{Tijt} = \text{Cijt}$

where: (i   = 1,2,...n, j = 1,2,...m, t = 1,2,3,...k)
        (s   = optimal steady-state, work-in-process level)

The PASIT technique is recommended for production worker use at the task level to ensure that only what is needed is produced.

TABLE 4 – Examples of Assessment Tools and Techniques for Productivity and Quality Management at the Source

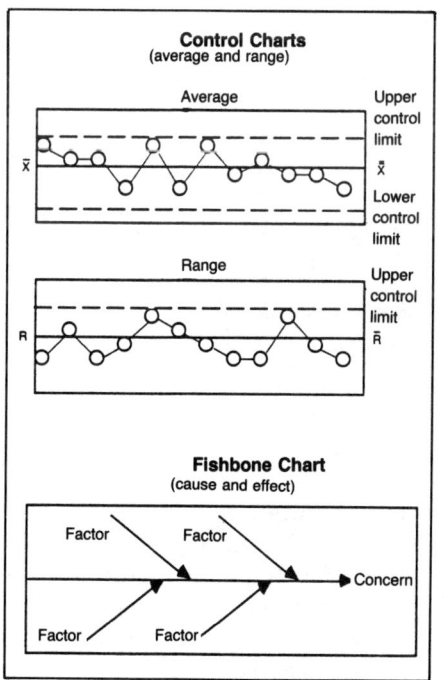

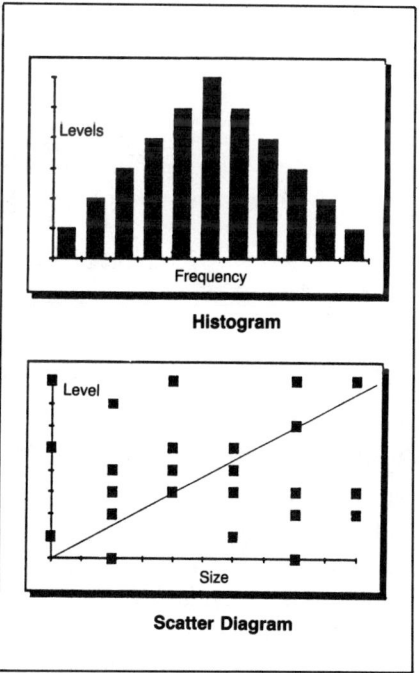

TABLE 4 (contd.) – Examples of Assessment Tools and Techniques for Productivity and Quality Management at the Source

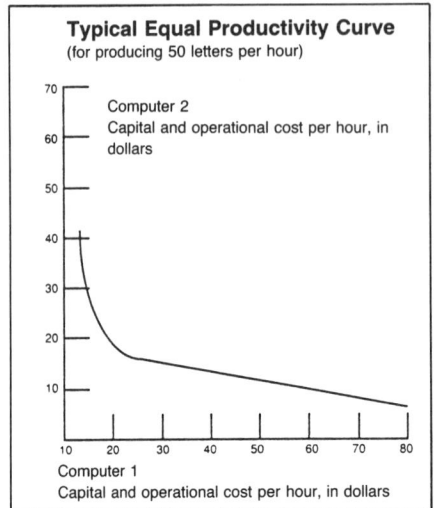

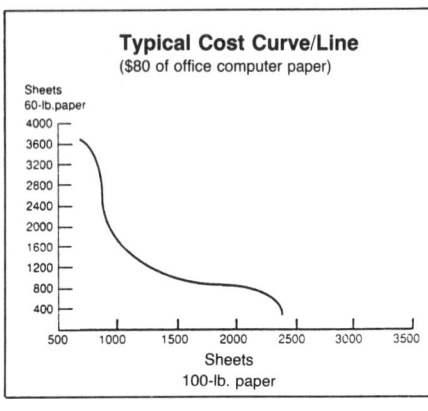

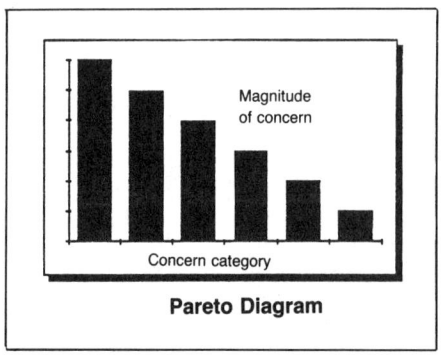

Pareto Diagram

JIT DEFINED

JIT is "a disciplined approach to improving overall productivity and quality through: *respect for people* and *elimination of waste* (AIAG 1984)." In this context, *respect for people* implies employee involvement, education and training, teamwork, and recognition of employee contributions by management. According to Fujio Cho of Toyota Motor Corporation, *waste* refers to "anything other than the *minimum* amount of equipment, materials, parts, space, and worker's time, which are *absolutely essential* to *add value* to the product."

NEED FOR JIT IN ELECTRONICS MANUFACTURING

Electronics industries, in spite of their rapid growth, are facing stiff foreign competition, notably from Japan and Korea. The ability of a company to offer the desired product at a competitive cost and quality level is paramount to the company's success. JIT offers a means for the company to beat the competition and attain world-class status.

Factors that establish the need for JIT in electronics industry are as follows:

- Short product life cycles are the way of business for many electronics manufacturers. Therefore, lead times need to be shortened and inventory must be low.
- A major proportion (about 70% to 80%) of the cost of goods sold in electronics industries is materials cost. In turn, high materials cost can lead to excessive costs of inventory and quality. Therefore, electronics manufacturers must implement JIT to achieve high quality and lower inventory to decrease total costs.
- A result of the combined effect of short product life cycles and high materials cost is the high level of material obsolescence. This requires that companies continuously reduce their lead times and inventory and effectively manage new product introductions.
- Finally, technological advances occur in this industry at rapid and unpredictable rates, which cause wide fluctuations in the marketplace. Therefore, electronics manufacturers must be highly responsive to technological changes. They must be able to produce several products which are at different stages of their product life cycles and ensure that the products reach the market quickly.

JIT offers several potential benefits to electronic manufacturing, including the following:

- significant reductions in manufacturing and procurement lead times;
- productivity increases with respect to direct and indirect labor;
- reduced cost of quality (including scrap/rework reduction);
- increased inventory turnover ratio due to reductions in raw materials, work-in-process, and finished goods inventory investment;
- reductions in floor space;
- increased manufacturing flexibility; and
- improved ability to service customer needs.

JIT applications in Japan have attracted a lot of attention from U.S. industry. Several U.S. electronics manufacturers have applied JIT and have also experienced significant improvements. A few examples of improvement are listed in

Table 5. A summary of overall benefits realized through JIT applications in various industries is presented in Table 6.

TABLE 5 – JIT Applications in U.S. Electronics Manufacturing.
Source: Sepehri (1986), Sandras (1987)

COMPANY	Productivity Improvements	Set-up Time Reduction	Inventory Reduction	Quality Improvements	Lead Times
Apple Computer (Macintosh)	Labor force: - inspection: 25 to 9 - Assembly: 22 to 4	30% estimate	90% in WIP and raw materials	scrap/rework: 10% incoming materials: 20%	Components: Less than 2 weeks.
General Electric (Switch gear)	Direct labor prod'y: 7% Indirect labor	56 min. to 1.5 min. in die change	WIP: 10 wks to 1.5 wks	—	Weekly lots to daily lots
Hewlett-Packard (Computer Systems Division)	Std. Hrs: 87 to 39	30%-45% in manual set-ups	PC Assembly: 30 days to 1.5 days	TQC implemented	PC Assembly: 15 days to 1.5 days
Motorola (Industrial Electronics)	6%-30%		5.6 wks to 2.6 wks	15% increase in on-time delivery	Significant reduction
Tellabs (PCB Assembly)			Savings: $3.25 million	scrap/rework reduction: 90%	96% reduction

TABLE 6 – Potential Benefits of Just-in-Time Applications.
Source: Rath and Strong (1986).

PAYBACKS	Automotive Supplier	Flexible Packaging	Mechanical Equipment	Electric Cons. Goods
Manufacturing Lead Time Reductions	89%	86%	83%	85%
Productivity Increase				
- Direct Labor	19%	50%	5%	n/a
- Indirect/Salary	60%	50%	21%	38%
Scrap/Rework Reduction	50%	63%	33%	25%
Inventory Reduction				
- Raw Material	35%	70%	73%	50%
- Work-In-Process	89%	82%	70%	85%
- Finished Goods	61%	71%	0%	90%
Space Reduction	53%	n/a	n/a	80%
Additional Capacity	n/a	36%	n/a	na/
Set-up Reduction	75%	75%	75%	94%
Purchased Material Price Reduction	n/a	7%	6%	n/a

ELEMENTS OF JIT

This section presents the elements of JIT that should be considered when implementing a JIT program. Figure 11 presents a framework that illustrates these elements. This section briefly discusses each of the elements.

SUPPLY-CHAIN MANAGEMENT	SUCCESSFULLY ORGANIZING JIT	HOUSEKEEPING
JIT PRODUCTION EXECUTION	JIT MANUFACTURING Visibility Simplicity Predictability Continuous Improvements	CELLULAR MANUFACTURING
JIT PRODUCTION PLANNING	TOTAL QUALITY CONTROL REDUCTION	SET-UP REDUCTION

Figure 11. *Elements of JIT.*

CONTINUOUS IMPROVEMENTS

The concept of continuous improvements in products and processes is central to all successful JIT implementations. JIT is not just a pull technique, but a management process continually seeking higher levels of performance through excellence in operations execution. As a part of this improvement effort, all material and information flow processes should be simplified. Techniques such as design for manufacturability and cellular manufacturing are meant to attain this simplification in products and processes. Also, all processes should be visible and predictable. This implies that variances should be minimized in the inputs and outputs of all operations. Lack of predictability causes production disruptions such as material shortages, unplanned machine breakdowns, unmanaged design changes, and changes in customer order configurations. The cumulative cost of these disruptions is high and must be eliminated. This can only come about through an in-depth understanding of cause- and -effect relationships in all design and manufacturing processes, as well as in the supporting business processing such as sales, order processes, cost accounting, and distribution.

HOUSEKEEPING

A JIT program begins with good housekeeping. A clean environment stimulates productivity. Further, good housekeeping provides visibility of all problems on the shop floor. When problems are exposed, eliminating the root causes of these problems can be targeted. Good housekeeping also means that the layout of the workplace is well organized. This implies that there is a place for everything and that everything is in its place. Additionally, policies and procedures for attendance, tardiness, and cleanliness must be implemented to encourage a healthy environment for all workers.

CELLULAR MANUFACTURING

The flow of materials within the factory largely influences the complexity introduced in managing it. JIT emphasizes simplicity. Cellular manufacturing is an approach for streamlining the material flows and creating "focused factories" or "cells."

In the United States, many electronic factories are typical job-shops. In these shops, each shop order of parts is routed through a series of operations in different functional departments with intermediate storage between operations. This involves complex and ineffective production control procedures, high levels of in-process inventory, long manufacturing lead times, and excessive material handling. The net result is a manufacturing process with high overhead costs and inherent inefficiencies that can adversely affect productivity and customer service.

Cellular manufacturing offers a meaningful alternative for manufacturing job-shops. Because cells process assemblies with similar processing requirements ("part families"), the tooling and fixture requirements of equipment in each cell can be standardized. This leads to reduced set-up times, and, hence, provides an opportunity to reduce lot sizes, trim work-in-process inventories, and shorten manufacturing lead times. Cellular manufacturing, therefore, provides a basis for JIT production (Hall (1983), Schonberger (1983).

SET-UP REDUCTION

Set-up reduction is a key ingredient in implementing a JIT program. Set-up time should be considered as waste since it is not directly adding any value to the product. Further, production lot size is a function of the set-up time. Smaller lot sizes are economically feasible if set-up times are small. Several benefits result from reduced set-up times and smaller lot sizes, including lower inventory investment, shorter lead times, and increased flexibility to manufacture according to customer demands.

Set-up time reduction can be achieved through careful analysis of the set-up process. Shigeo Shingo, a consultant at Toyota, uses an approach called single minute exchange of dies (SMED) through which set-up times can be reduced to less than a minute. Important concepts in this approach are: 1) separation of internal set-ups (i.e., actions that require the stoppage of the equipment) from external (off-line) set ups; 2) conversion of internal set-up elements to external set-up elements; and 3) elimination of adjustments using quick-clamping devices. A detailed discussion of these issues is available in a book by Shingo (1981).

TOTAL QUALITY CONTROL

Total quality control (TQC) is an integral part of JIT. It provides the means for controlling and improving quality through the use of statistical techniques. Quality management methods have progressed through several phases: 1) quality inspection; 2) statistical process control; 3) quality improvement; and 4) product and process designs for quality (Fuchs (1986), Schonberger 1986).

Quality inspection involves testing the product at the end of the line. The customer, on receiving the product, must inspect the product before accepting it. Inspection techniques such as acceptance sampling and sampling inspection plans were developed to measure and report quality.

Statistical process control (SPC) is a more improved technique than quality inspection. For SPC, the process parameters are measured and periodically checked to ensure that the process is in control or that the lot produced is good. Further, in traditional quality inspection techniques, attention is given to production falling between engineering limits (even if the mean was close to the upper or lower limit). In contrast, quality control via SPC requires the use of statistical limits for which the standard deviation is the key parameter.

Quality improvement techniques go beyond quality inspection or SPC because they emphasize continuous improvements in the product and process quality. For example, fishbone charts (also known as "Ishikawa diagrams") help to identify the root causes of problems and improve the process by eliminating the root causes.

The objective of designing products and processes for quality is to adjust the product and process variables to ensure the desired quality of the output. For example, Dr. Taguchi's robust design technique significantly reduces variations in the output of a process. It uses statistical experimental design to identify the process variables that affect the mean and variance of the output. These variables are then adjusted to obtain the desired mean and variance.

JIT PRODUCTION PLANNING

Production planning refers to all the functions involved in developing feasible materials and capacity plans for the shop floor and for vendors. The nature of these functions can vary significantly from one JIT environment to another. For example, the planning functions for a repetitive process making a few products can be different from those for a low volume-high product-mix environment. In spite of those differences, a number of common planning concepts emerge under JIT. As a result of reductions in lead times and inventory levels and simplifications in product structures, the materials planning functions are simplified under JIT. In particular, process simplifications under JIT allow the planning systems to move away from large lot/weekly time buckets to small lot/daily planned rates.

A framework for production planning in a JIT environment is presented in Figure 12. To start, production plans for end-item families are developed by apportioning gross requirements among families of items through the planning bills of materials. The planning bills are derived from family-level forecasts and the backlog of actual customer orders. The family production plans are disaggregated into daily planned rates for the end items. These rates are different from the traditional MRP-II processes in that end items are not scheduled in lumpy intermittent batches. Instead, the firm planned order line of a rate-based master production schedule (MPS) displays a nearly constant build plan for each item in a given time interval. Resource planning is performed to develop leveled plans that allow balanced flow of materials across critical manufacturing cells.

The final assembly schedule (FAS) establishes the actual build program for the final assembly shop in the short-term. Planning periods within FAS are short, covering at most one day's requirements. FAS can be stated as daily build rates for the end-item configurations. In some cases, FAS may take the form of mixed-model sequences of end items. In the medium to long term, where the MPS is driving the plan, the length of the planning horizon can vary from one day to one week. Time fences, regulating either the MPS or the FAS, are used to "freeze" the production program over a brief time. Ideally, the "frozen period" can be as short as one day. Frozen schedules are easiest to construct for repetitive manufacturing processes.

The daily planned production rates for end items are used to determine the planned rates for the subassembly shops that supply materials to the final assembly shop. They are also used to determine the planned delivery rates for vendors. An explosion of the bill of materials may be performed to determine the requirements for the vendors and the shop floor. However, compression of the bill due to product simplification should reduce the computational burden. In addition, the need for netting will diminsh as the on-hand inventory levels approach zero.

In complex subassembly shops that produce a large variety of parts in low volumes, the daily rates can be further disaggregated into production sequences for the cells. These sequences can specify how much of each product to make.

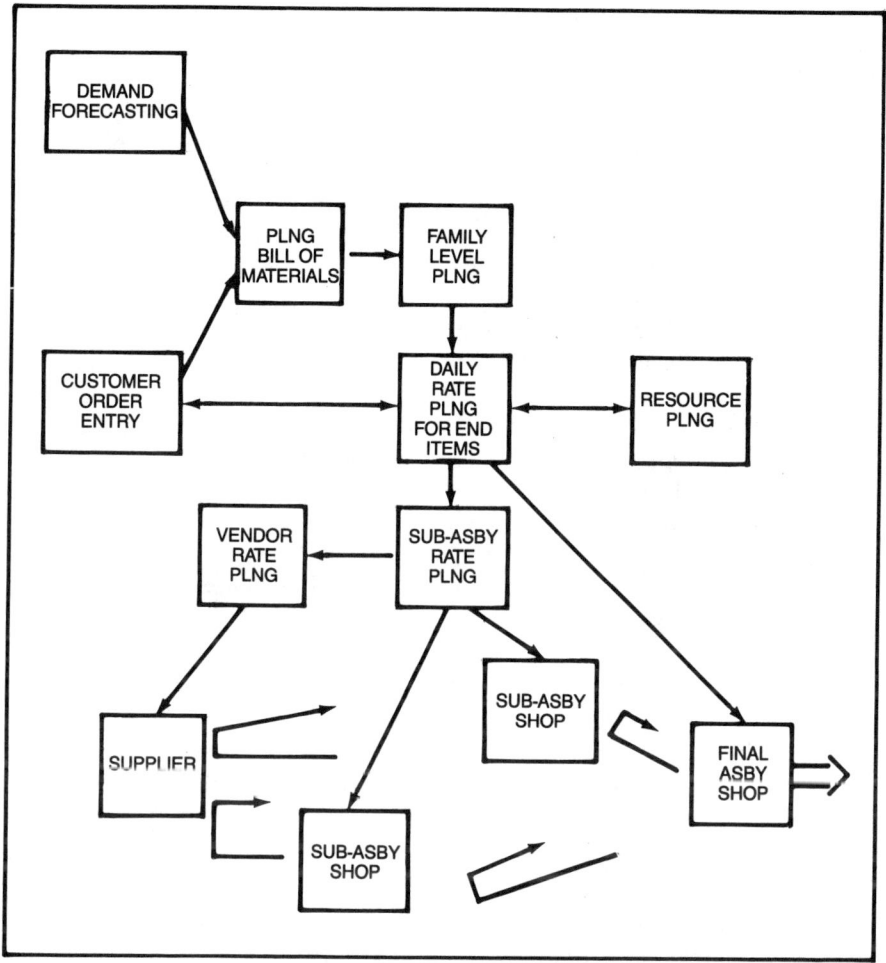

Figure 12. *A JIT production planning framework.*

The pull signals from downstream operations/cells specify when to begin production.

The planned rates for materials procurement should be sent to the JIT vendors. These rates are blanket orders for materials against which pull signals will be generated to deliver components at desired delivery points. The medium- to long-term material and capacity plans should be shared with the vendors to promote vendor partnership programs.

JIT PRODUCTION EXECUTION

An important element of JIT is its "kanban" system of production execution. Essentially, the kanban system is a simple information system that regulates production flow throughout the manufacturing network, consisting of the factory and its suppliers. Kanbans are the signals that trigger production.

JIT is often associated with "pull" production control instead of the traditional schedule-driven "push" system. "Pull" means that all subsequent operations withdraw the required quantity of parts from their preceding operations at the appropriate time. The preceding operation then produces the parts withdrawn by the subsequent operation. Often, there are two types of kanbans: 1) production kanbans that specify the type and quantity of items which the preceding operation must produce; and 2) withdrawal kanbans that specify the type and quantity of items which the subsequent operation should withdraw from the preceding operation. Additional details on various types of kanbans can be found in Monden (1986).

A pure pull production across the entire factory is easiest to implement for a repetitive manufacturing process. However, all electronic manufacturing processes are not repetitive. Subassembly shops in electronic assembly factories often produce a large variety of parts in low volumes. Moreover, integrated circuit manufacturing operations tend to be complex and manual with long lead times. In these situations, it may be meaningful for some of the nonrepetitive subassembly shops to continue to operate in a push mode, while the final assembly shop operates in a pull mode. Outputs from the push-type subassembly shops can be sent to a buffer that is subsequently consumed by the final assembly operations through pull signals.

SUPPLY-CHAIN MANAGEMENT

Supply-related activities constitute a large portion of the total operating cost of a company. The cost of poor quality of incoming parts or late deliveries can be devastating to the downstream manufacturing processes. The major concepts in supply management are mentioned as follows.

Vendor Selection

Selection of a few excellent vendors leads to improved management control from the customer's standpoint. This enables purchasing to make better procurement decisions based on quality, delivery, and price. Reducing the vendor base helps develop long-term vendor partnerships with suppliers and also reduces administrative costs in the purchasing departments. A data base of preferred suppliers and all relevant information on components should be maintained to help product designers and purchasing staff make better decisions.

Vendor Partnership Program

Interaction between a vendor and its customer must exist at all management levels. At the operating level, vendors must provide the customer with data on their product and process quality. The customers must share their projected material and capacity plans with the vendors. The customer should provide precise and clear specifications of what parts they need from the vendors and must release the procurement orders on time. The upper level management of the two companies must also maintain communication to facilitate the relationship.

Vendor Evaluation

There is a need for a comprehensive plan for vendor evaluation that is based on several performance criteria, not price alone. These criteria are on-time vendor delivery performance, quality of incoming parts, length of procurement intervals, lateness and earliness of deliveries, and vendor service. A vendor certification program must consider all of these factors. As the vendor-customer partnership continues to evolve and the vendor performance improves, the nature of conducting vendor evaluations will change and will be based on greater mutual trust. For example, daily multiple deliveries of parts will directly feed the production line without any incoming inspection and storage.

SUCCESSFULLY ORGANIZING A JIT PROGRAM

There are several ways that companies can begin to implement JIT. In this section, one successful method for organizing a JIT program is presented. The sequence of activities is divided into three phases discussed as follows:

Phase 1: Opportunities Identification

The first phase involves initial planning to identify and prioritize broad areas for improvements. As a start, management must examine the competitive strategies of the business unit and then decide what improvements should be made in the existing products and processes to implement these strategies. This will require a broad evaluation of market requirements as well as fixed and variable cost structures of the factories. The products and processes must be analyzed to quickly identify major areas of improvement. A small working group of operating managers, engineers, and staff professionals should be formed for this analysis. An initial project plan should be created that includes the overall project goals, major steps and milestones, resource requirements, and projected cost-benefits. Specifically, it is important to identify a set of financial and operating metrics (e.g., operating costs, inventory turns, lead times, and quality levels) that will be used by management to track improvements. The target values of these metrics and corresponding due dates should be identified.

Phase 2: Buy-In and Detailed Planning

Perhaps the most important element in a JIT program is to obtain a publicly visible commitment from top management and to have a "champion" at every management level. Personnel education and training should begin at all levels. This will include executive training seminars on JIT for management, as well as detailed training at appropriate levels for all others. The small working group from Phase 1 should be expanded into an implementation team divided into several subgroups. Detailed implementation plans with deliverables and milestones should be developed by each subgroup. It is desirable to start developing the implementation plans for a small pilot line. An upper level factory manager should be appointed as the full-time project manager. It should be noted that buy-in by management is not enough. Buy-in by the operators is crucial to the success of the project.

Phase 3: Implementation

It is important to start soon with a pilot project on a small production line. The pilot project would typically consider materials flow, materials planning and scheduling, as well as product design issues, and associated information flows. Pull productions should begin at the final assembly shop. It is desirable to include within the pilot the materials flow across the receiving area, storeroom, shop floor, and shipping area. To ensure success of the pilot project, operators who are volunteers need to be on the project. They should be cross-trained for running different operations. Responsibility should be given to operators to stop an operation if they find a defect, and solve the problem with other members of the team. At this stage, emphasis should be placed on operational improvements such as reductions in lead times, number of defects, and other waste, instead of immediately rushing into implementation of excessive automation. The improvements resulting from these efforts should be tracked regularly by employing a set of operational and cost metrics. The lessons learned from the pilot effort should be carefully documented so that they can be used as the JIT effort expands within the factory. Ultimately, the project scope should include both the supply and the distribution operations.

JIT AND BUSINESS OBJECTIVES

The ultimate purpose of JIT is to improve a company's competitive position. Several steps are necessary to achieve this goal. To start, the competitive strategy of the business should be clearly articulated. For example, does the business wish to compete as a least-cost producer or as a provider of highly differentiated products with unique features and best quality? It is misleading to specify that a company wants to simultaneously minimize costs, minimize lead times, and maximize the number of unique product features because of the trade-offs among those objectives. Therefore, precise quantitative targets should be set as far as possible to avoid any confusion.

JIT should be viewed as a broad operational strategy with a goal of realizing the business objectives. A number of corporations have successfully embraced JIT as a companywide means to improve all aspects of their businesses. In this context, JIT is much broader than the mechanics of pull production or set-up reduction. JIT encompasses broad corporate decisions in facility location, capacity management, vertical integration, human resources, multiplant production planning, scheduling, and quality. The interfaces between design and manufacturing, between sales and manufacturing, and between manufacturing and distribution must also be well managed as a part of JIT.

Finally, in any JIT program, the competitive and financial objectives of a business must be clearly linked with the operational objectives such as inventory turns, lead time, and number of defects. This will be especially helpful in convincing the top level management about the benefits realized through JIT. In this way, the JIT program will not be just another productivity improvement project in a remote corner of a factory. Each and every individual and operation will function in a synchronized fashion, so that the total effect is a significant improvement in the competitive position of the overall business.

EDOSOMWAN'S GUIDELINES FOR PRODUCTIVITY AND QUALITY MANAGEMENT

The following 10 guidelines are recommended for managing productivity and quality in a competitive business environment.

1. Encourage both management's and employees' commitment to productivity and quality management at the production or service source.
2. Implement measurement, control, and evaluation techniques for productivity and quality. Control charts, fishbone diagrams, histogram, productivity ratios, equal productivity curves, cost curves, and scatter diagrams are examples of some of the techniques that can be used.
3. Implement planning and analysis for productivity and quality. The techniques selected for this purpose must be used to monitor short-term and long-term productivity and quality planning.

The editors thank Arunabha Chatterjee of AT&T Bell Laboratories for his assistance in preparing a portion of the JIT materials.

4. Provide a focal point for coordinating productivity and quality projects and programs. Implement improvement and monitoring techniques for productivity and quality at all levels of the organization.
5. Provide management and employee training for directing productivity and quality improvement efforts.
6. Institute a productivity and quality team that includes all levels of the organization. Provide adequate communication channels for productivity and quality projects.
7. Allow for productivity and quality goals that are measurable, attainable, and open to opportunity. Provide adequate focus on cost avoidance to avoid cost overrun in productivity and quality projects implementation.
8. Assign priority to goals and problems. Implement measures to monitor the contribution of each unit of the organization.
9. Provide feedback mechanism for information among people, processes, and procedures. Facilitate cooperation among organization units and individuals.
10. Provide real-time recognition for a job well done and share the gains from productivity and quality improvement. Motivate people by providing them with meaningful job content, autonomy, and training required to perform their job. Reward all accomplishments in a timely manner.

EDOSOMWAN'S PRINCIPLES FOR PRODUCTIVITY AND QUALITY MANAGEMENT *

The following 12 process-oriented improvement principles are recommended when designing, implementing, and managing productivity and quality improvement projects.

1. The approach to improving the process requires that management take a leadership role in creating the right awareness and prioritizing resources to the various improvement projects.
2. The theme that should be communicated to everyone in the organization is: Continuous improvement is needed in the process.
3. The scope of the improvement effort should include all activities and tasks performed in the organization.
4. The scale of the improvement activities should specify that everyone is responsible.
5. The training provided should include, but not be limited to, problem-solving techniques, decision-making techniques at the source of production or service, statistical process control, productivity and quality measurement, evaluation and improvement techniques, customer requirements management, planning tools, and project management.

6. The style should focus on a prevention system for defects and process bottlenecks. Inspection and nonvalue add operations used for detecting defects should be discouraged.
7. The awareness for everyone should focus on a vision of the improved process, excellent quality product, and service.
8. The method(s) employed should be based on demonstrated results in providing adequate process yield and output.
9. The standard set should require that everyone should do the job right the first time at the source.
10. The measures used for productivity and quality should include, but not be limited to, the following: partial and total factor and total productivities; percent defective; customer satisfaction indexes; cost of quality; error rate indexes; process failure rate; and line control indexes.
11. The rewards and recognition for improving productivity and quality should be given to both management and employees.
12. Teamwork approach—comprising of all functional areas—and participative management style should be encouraged. Focus on providing effective cultural support systems, consensus building, and adequate mechanisms for recognizing team performance.

* Source: Edosomwan, J. A. (1988). *International Industrial Engineering Conference Proceedings.* Reprinted with permission.

Suggested Readings

Automotive Industry Action Group. 1984. *JIT Definition and Requirements/ Delivery Systems.* Automotive Industry Action Group. Southfield, MI.

Ames, R. 1983. Zero inventories crusade. *APICS 26th Annual Conference Proceedings.* November. American Production and Inventory Control Society. Falls Church, VA.

American Production and Inventory Control Society. 1984. *Readings in Zero Inventory.* American Production and Inventory Control Society. Falls Church, VA.

American Production and Inventory Control Society. 1983. *Twenty-sixth Annual International Conference Proceedings. New Orleans, LA.* American Production and Inventory Control Society. Falls Church, VA.

Ashton, F. A. 1985. Master production scheduling for orginal equipment automotive suppliers. *Production and Inventory Management.* 26(4): 71-82.

Ballakur, A. 1986. Streamlining material flows for just-in-time production. *International Industrial Engineering Conference Proceedings. Boston.* Institute of Industrial Engineers. Norcross, GA.

Ballakur, A. and M. K. Pratt. 1987. Integration of product and process design in high technology equipment production. *Third IEEE/CHMT International Electronic Manufacturing Technology Symposium Proceedings. Anaheim, California.* October.

Ballakur, A. and H. J. Steudel. 1987. A within-cell utilization based heuristic for designing cellular manufacturing systems. *International Journal of Production Research.* 25(5): 639.

Ballakur, A. and H. J. Steudel. 1987. A dynamic programming based heuristic for machine grouping in manufacturing cell formation. *Computers and Industrial Engineering.* 12(3): 215.

Bhote, K. R. 1987. *Supply Management—How to Make U.S. Suppliers Competitive.* American Management Association. New York.

Brooks, R. B. 1987. How to integrate JIT and MRP-II. *American Production and Inventory Control Society 30th Annual Conference Proceedings.*

Burgam, P. M. 1984. JIT: on the move and out of the aisles. *Manufacturing Engineering,* June: 65-71.

Chatterjee, A., M. A. Cohen, and W. L. Maxwell. 1984. Manufacturing flexibility: models and measurements. *Proceedings of First ORSA/TIMS Conference on Flexible Manufacturing Systems.* Ann Arbor, MI.

———. 1984. JIT software capitalizes on in-house experiences. *CIM Technology.* Winter: 15-16.

———. 1985. Automaker preps for just-in-time. *CIM Technology.* Spring: 13.

Cohen, M. A. 1985. Manufacturing strategy—a competitive weapon for American industry. *Wharton Journal.* 38-43.

Cook, J. 1984. Kanban American-style. *Forbes.* October.

Crobsy, L. B. 1984. The just-in-time manufacturing process: control of quality and quantity. *Production and Inventory Management.* 25(4): 21-33.

Deming, W. E. 1982. *Quality, Productivity, and Competitive Position.* Massachusetts Institute of Technology. Cambridge.

Deming, W. E. 1986. *Out of the Crisis.* MIT Press. Cambridge.

Edosomwan, J. A. 1987a. *Integrating Productivity and Quality Management.* Marcell Dekker Inc. New York.

Edosomwan, J. A. 1987b. Managing productivity and quality in a production environment. *Proceedings for the IBM Symposium on Reliability and Quality.* IBM Technical Institute. Thornwood, New York.

Edosomwan, J. A. 1987c. A program for managing productivity and quality. *Industrial Engineering.* 19(1): 64-68.

Edosomwan, J. A. 1987d. The challenge for industrial managers: productivity and quality in the workplace. *Industrial Management.* 29(5): 25-27.

Edosomwan, J. A. 1987e. Understanding the connection between productivity and quality in a competitive business environment. *Proceedings from the IFS Conference on Statistical Process Control. Birmingham, England.* November.

Edosomwan, J. A. 1986a. Productivity and quality management—a challenge in the year 2000. *Fall Industrial Engineering Conference Proceedings. Boston.* December.

Edosomwan, J. A. 1986b. Statistical process control in electronics assembly. *Fall Industrial Engineering Conference Proceedings. Boston.* December.

Edosomwan, J. A. 1986c. Statistical process control in group technology production environments. *SYNERGY Proceedings.* June 17-19, 1986. Society of Manufacturing Engineers (SME). Universal City, CA.

Edwards, J. N. 1987. Integrating MRP-II with JIT: an update. *American Production and Inventory Control Society 30th Annual Conference Proceedings.*

Electronics Industries Association. 1987. *Electronic Market Data Book.* Electronics Industries Association. Washington, D.C.

Electronics Industries Association. 1986. *Electronic Market Data Book.* Electronics Industries Association. Washington, D.C.

Esrock, Y. D. 1985. The impact of reduced set-up time. *Production and Inventory Management.* 26(4): 94-101.

Fuchs, E. 1986. Quality: theory and practice. *AT&T Technical Journal.* 65(2): 4.

Gitlow, H. S. and P. T. Hertz. 1983. Product defects and productivity. *Harvard Business Review.* 61(5): 132-141.

Grant, E. L. and R. S. Leavenworth. 1974. *Statistical Quality Control.* McGraw-Hill, Inc. New York.

Greico, P. L. 1984. JIT - does it work? *CASA/SME AutoFact Proceedings.* 23.13-23.17.

Hall, R. W. 1983. *Zero Inventories.* Dow Jones-Irwin. Chicago, IL.

Hall, R. W. 1987. *Attaining Manufacturing Excellence.* Dow Jones-Irwin. Chicago, IL.

Harrison, J. M. and M. Finley. 1984. Hewlett-Packard personal office computer division. *Case # S-DS-81.* Graduate School of Business, Stanford University. Stanford.

Hayes, R. H. 1981. Why Japanese factories work. *Harvard Business Review.* 59(4): 57-66.

———. 1985. *HP JIT: Just-In-Time Manufacturing.* (Brochure). Hewlett-Packard Manufacturing Systems.

Huang, P. Y. and B. L. W. Houck. 1985. Cellular manufacturing: an overview and bibliography. *Production and Inventory Management.* 26(4): 83-93.

Juran, J. M. 1964. *Managerial Breakthrough.* McGraw-Hill. New York.

Juran, J. M. and F. M. Gryna, Jr. 1980. *Quality Planning and Analysis.* McGraw-Hill Book Company. New York.

Kackar, R. N. and A. C Shoemaker. 1986. Robust design: a cost-effective method for improving manufacturing processes. *AT&T Technical Journal.* 65(2): 39.

Kear, F. W. 1987. *Printed Circuit Assembly Manufacturing*. Marcel Dekker, Inc. New York.

Kelleher, J. P. 1984. Reducing lead times: assumptions, techniques, benefits. *Annual International Industrial Engineering Conference Proceedings*. Institute of Industrial Engineers, Norcross, GA.

Kendrick, J. W. and the American Productivity Center. 1984 *Improving Company Productivity: Handbook with Case Studies*. The John Hopkins University Press. Baltimore, MD.

Kenfield, J. E. 1985. A nine step approach to JIT implementation. *Autofact 7 Conference Proceedings*. CASA/SME. Dearborn, MI.

Kim, T. M. 1985. Just-in-time manufacturing system: a periodic pull system. *International Journal Of Production Research*. 23(3):553-562.

Leahy, J. A. 1984. Toyota production system-just in time, not just in Japan. *CIM Review*, Fall: 59-66.

Monden, Y. 1986. *Applying Just In Time: The American/Japanese Experience*. Industrial Engineering and Management Press. Norcross, GA.

Monden, Y. 1981a. What makes the Toyota production system really tick? *Industrial Engineering*. 13(1): 36-46.

Monden, Y. 1981b. Adaptable kanban system helps Toyota maintain just-in-time production. *Industrial Engineering*. 13(5): 29-46.

Monden, Y. 1981c. Smoothed production lets Toyota adapt to demand changes and reduce inventory. *Industrial Engineering*. 13(8): 42-51.

Monden, Y. 1981d. How Toyota shortened supply lot production time, waiting time, and conveyance time. *Industrial Engineering*. 13(9): 22-30.

Monden, Y. 1983. *Toyota Production System*. Industrial Engineering and Management Press. Norcross, GA.

Nakane, J. and R. W. Hall. 1983. Management specs for stockless production. *Harvard Business Review*. 61(3):84-91.

Nellemann, D. O. 1982. 'Just-in-time' vs. 'just-in-case' production/inventory systems concepts borrowed back from Japan. *Production and Inventory Management*. 23(2): 12-20.

Phadke, M. S. 1986. Design optimization case studies. *AT&T Technical Journal*. 65(2): 51.

Plenert, G. 1985. Are Japanese production methods applicable in the United States? *Production and Inventory Management*. 26: 121-129.

Rice, J. W. 1982. A comparison of kanban and MRP concepts for the control of repetitive manufacturing systems. *Production and Inventory Management*. 23(1): 1-14.

Sandras, W. A. Jr. 1987. Accelerated JIT/TQC implementation case studies. *American Production and Inventory Control Society Conference Proceedings*.

Sandras, W. A. Jr. 1984. Linking MRP and JIT: the best of the Occident and the Orient. *Readings in Zero Inventory*. American Production and Inventory Control Society. Falls Church, VA.

Sandras, W. A. Jr. 1983. Continuous flow customized production. *American Production and Inventory Control Society 30th Annual Conference Proceedings*. American Production and Inventory Control Society. Falls Church, VA.

Schonberger, R. J. 1987. *World Class Manufacturing Casebook—Implementing JIT and TQC*. The Free Press. New York.

Schonberger, R. J. 1986. The quality concept: still evolving. *National Productivity Review*. Winter: 81.

Schonberger, R. J. 1983. Plant layout becomes product-oriented with cellular, just-in-time production concepts. *Industrial Engineering*. 15(11): 66-71.

Schonberger, R. J. 1983. Selecting the right manufacturing inventory system: western and Japanese approaches. *Production and Inventory Management*. 24(2):33-44.

Schonberger, R. J. 1982. *Japanese Manufacturing Techniques—Nine Hidden Lessons im Simplicity*. The Free Press. New York.

Schorr, J. E. and T. F. Wallace. 1986. *High Performance Purchasing*. Oliver Wight Limited Publications.

Schroer, B. J., J. T. Black, and S. X. Zhang. 1984. Microcomputer anaylses 2-card kanban system for just-in-time small batch production. *Industrial Engineering*. 16(5):54-65.

Sepehri, M. 1987. *Quest for Quality*. Industrial Engineering and Management Press. Norcross, GA.

Sepehri, M. 1986. *Just-In-Time, Not Just-in-Japan - Case Studies of American Pioneers in JIT Implementation*. American Production and Inventory Control Society. Falls Church, VA.

Sepehri, M. 1985a. Stockless production supported by flexible manufacturing in job shops. *Annual International Industrial Engineering Conference Proceedings*. Institute of Industrial Engineers. Norcross, GA.

Sepehri, M. 1985b. How kanban system is used in an American Toyota motor manufacturing. *Industrial Engineering*. 17(2): 51-56.

Sepehri, M., J. Carlson, and E. Manrique. 1985. Material management at Toyota Japan and U. S. *1985 Annual International Industrial Engineering Conference Proceedings*. Institute of Industrial Engineers. Norcross, GA.

Shingo, S. 1981. *Study of 'Toyota' Production System from Industrial Engineering Viewpoint*. Japan Management Association.

Sloma, R. S. 1980. *How to Measure Managerial Performance*. MacMillan. New York.

Sullivan, L. P. 1984. Reducing variability: a new approach to quality. *Quality Progress*. 17(7): 15-21.

Swann, D. M. 1985. Planning and inventory control in low volume environments. *International Industrial Engineering Conference Proceedings.* Institute of Industrial Engineers. Norcross, GA.

Sugimori, Y., F. C. Kusunoki, and S. Uchikawa. 1977. Toyota production system and kanban system: materialization of just-in-time and respect-for-human system. *International Journal of Production Research.* 15: 553.

Taguchi, G. and Y. Wu. 1980. *Introduction to Off-Line Quality Control.* Central Japan Quality Control Association. Nogoya, Japan.

Waller, L. 1985. Just-in-time manufacturing starts paying off in U.S. *Electronics.* August: 26.

Waters, C. R. 1984. Why everybody's talking about 'just-in-time'. *Inc.* 6(3): 77-90.

Wheelright, S. C. 1984. Manufacturing strategy: defining the missing link. *Strategic Management Journal.* 5: 77-91.

Wilson, G. T. 1985. Kanban scheduling: boon or bane? *Production and Inventory Management.* 26(3): 134-142.

Winfield, G. 1985. Just-in-time manufacturing: a case study. *CIMCOM '85 Proceedings.* CASA/SME MS85-339. Dearborn, MI.

Wortmann, J. C. and W. Monhemuis. 1984. Kanban—its use as a final assembly scheduling tool within MRP II. *Operating Research '84.* Elsevier Science Publishers. Amsterdam.

———. 1986. *Documentation on JIT Manufacturing Survival.* California. Xerox Computer Services.

Part 1

Productivity and Quality Management in Electronics Assembly

This section presents selected chapters that offer methodologies, frameworks, and techniques for improving productivity and quality in electronics assembly. The type of program designed for improving productivity and quality is an essential ingredient for ensuring success. An example of a step-by-step methodology that is applicable in designing a productivity and quality program in a printed circuit board assembly environment is provided in the article written by Edosomwan (1987). Work and task complexity are known impediments affecting productivity and quality in the work place. The more complex the task to be performed, the greater the difficulty of ensuring optimal gains and improvements in productivity and quality. Fuller (1985) discusses techniques for eliminating complexity from work. The author presents a complexty model and describes how it can be used to enhance quality and productivity.

Wilhelm (1986) presents an approach for devising an effective quality control plan for small-lot assembly of printed circuit boards. Wilhelm's approach has potential application in developing a quality control plan for evaluating alternative strategies for testing at component, in-circuit, function, and system levels. A conceptual framework for productivity management that is applicable at the organization level is presented by Sink (1983), who further discusses a system for managing performance in an organization. Managers and decision makers in general are often faced with the task of balancing resources to ensure gains in productivity, which is no easy task. Sumanth and Genie (1985) describe the effect of productivity in management decision making. They further present a total productivity management model that has potential use in electronics assembly. Edosomwan (1985) presents a task-oriented total productivity measurement model for electronics assembly. A case study is also presented to highlight how the model can be implemented in the real world.

Hauck and Ross (1985) discuss several methods for improving productivity. The methods suggested by the authors include, among other methods: total people involvement in programs that improves productivity and quality, open-door policy, quality circles, rewards, wage incentives, productivity gainsharing, and competent management team effort. Schonberger (1982) outlines the techniques behind the Japanese success in improving total quality. He emphasizes giving production workers responsibility in the Japanese industry. A case study on how quality cost analysis can be used for management improvement is presented by Blank and Solorzano (1978). Bellefeuille (1987) offers a process (PICA cycle) for total quality control and improvement based on experience from electronics assembly processes. He emphasizes the need for planning, acting, measuring, and implementing actions that improve quality. The PICA cycle is presented as an aid to decision makers and practitioners who are involved in quality improvement at the organization level. Phadke (1987) discusses how the application of the Taguchi methods can significantly shorten product development cycle, improve quality, and reduce costs at the company level. A methodology for designing experiments for quality improvement is presented by Box and Bisgaard (1987). Part 1 Matrix summarizes the chapters presented in this section.

Part 1 Matrix

Chapter Number	Title	CONTENT CLASSIFICATION						ELECTRONICS ASSEMBLY PROCESSES						
		Model	Quantitative Technique	Strategy/Methodology	Case Study	General Theory	Design	Receiving & Store	Inspection	Kitting	CP Assembly Mfg.	Equip. Assembly	Packing	Mat'l Handling
1	A Program for Managing Productivity and Quality			X	X	X			X		X			
2	Eliminating Complexity from Work: Improving Productivity by Enhancing Quality		X	X		X			X		X	X		
3	An Approach for Devising an Effective Quality Control Plan for Small Lot Assembly of Printed Circuit Boards			X					X					
4	Much Ado About Productivity: Where Do We Go From Here?	X		X		X	X				X	X		
5	Total Productivity Management for Competitive Decision-Making in the Electronics Industry	X		X		X	X				X	X		
6	Pathways to Productivity Improvement			X		X					X			
7	Production Workers Bear Major Quality Responsibility in Japanese Industry			X			X				X	X		
8	Using Quality Cost Analysis for Management Improvement		X	X		X		X	X		X	X		
9	A Management Process for Implementing Total Quality Control and Improvement				X				X					
10	Design Optimization Case Studies				X				X					
11	The Scientific Context of Quality Improvement: A Look at the Use of Scientific Method in Quality Improvement	X				X					X			
12	How to Measure Productivity in an Electronic Printed Circuit Board Assembly	X	X	X	X		X				X	X		

1

A Program for Managing Productivity and Quality

Johnson Aimie Edosomwan

In recent years, the management of productivity and quality has emerged as a major new business strategy in many organizations. This is driven by a number of factors, including:

- Global intensive competition in all industrial sectors. There is increasing demand for a better quality product or service at the existing price or at a lower price.
- Increasing cost of production and services. New ways to improve the productivity of labor, materials, energy, capital and technology are being sought.
- Increasing consumer education and awareness of quality. Acceptance of product produced or services rendered relies on conformance to requirements or specifications, which means all output produced is expected to be defect-free.

As shown in Figures 1 and 2, poor quality goods and low rates of productivity growth can have significant impact on the survival of any economic unit.

The management of productivity and quality in organizations can provide the following benefits:

1. Lower prices to consumers for goods and services because the cost of production is reduced through less rework and gains in productivity.

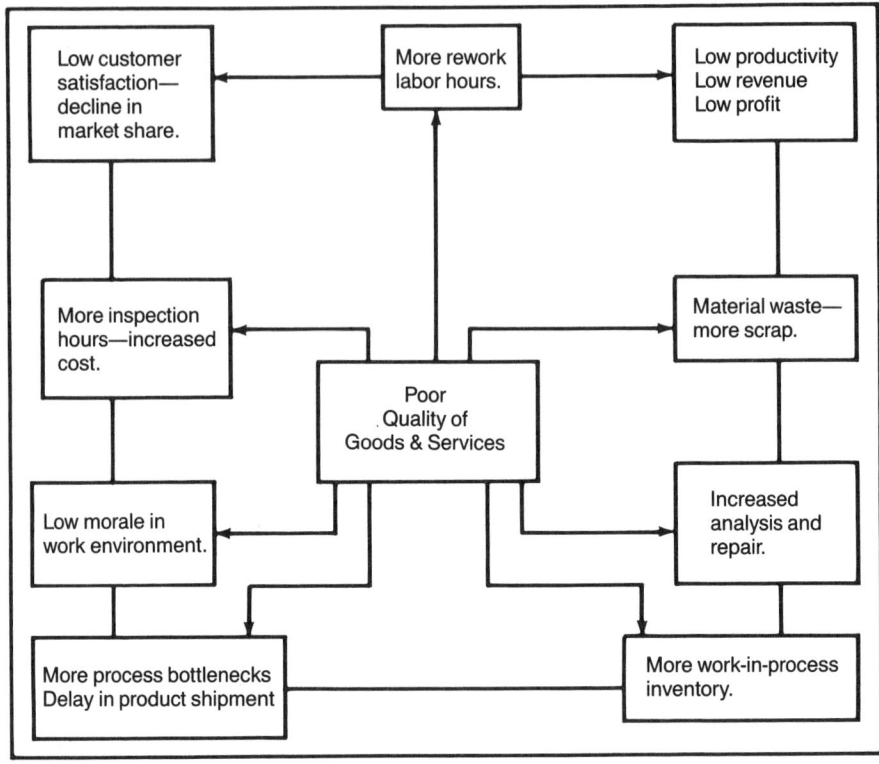

Figure 1. *Impact of poor quality in an economic unit.*

2. Effective utilization of resources, so that more goods and services are produced for reasonable amounts of expended resources.
3. Higher real earnings for employees. The reduction in the cost of production of goods and services will allow increases in wages without significantly offsetting gains in total productivity.
4. Enhancement of organizational strength for dealing with internal operating weaknesses and external competition.
5. Greater profitability, because quality improvement leads to less rework, reduction in scrap, better utilization of tools and equipment and reduced work-in-process inventory—which in turn leads to higher productivity. Minimizing the cost of manufacturing improves the profit margin for goods and services. The production of high quality products also improves sales, because the satisfied consumer will buy more and will recommend the product to other consumers.
6. More social benefits through increased public revenues from organizations.

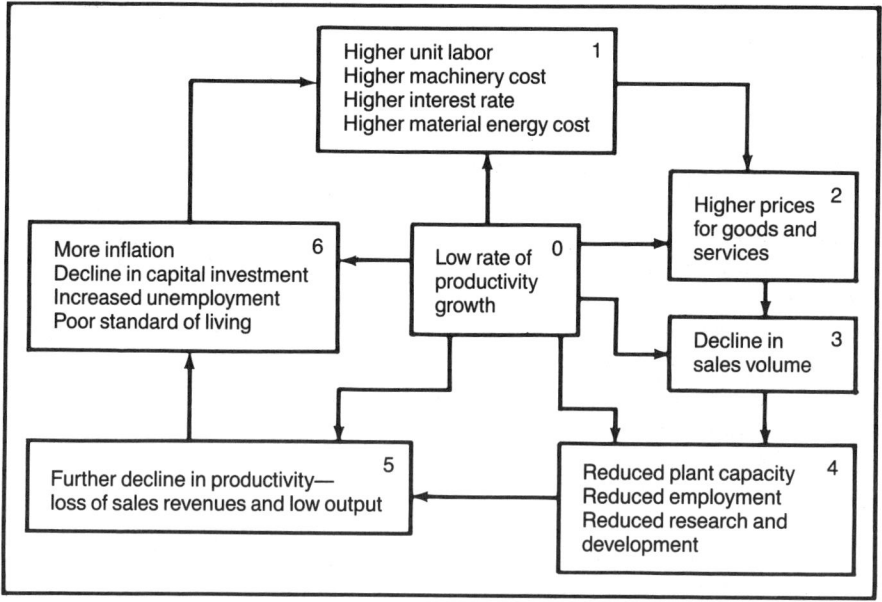

Figure 2. *Cyclic effects of low productivity growth.*

BASIC DEFINITIONS

Productivity

Productivity is generally viewed as a measure of output relative to either one input, two inputs or total input. The "measure" also pertains to how well resources are utilized.

Three basic forms of productivity have been accepted by most researchers and practitioners Kendrick and Creamer (1965), Sumanth (1979), and Edosomwan (1983; 1985; 1986). These forms can be defined as follows:

1. *Total productivity:* the ratio of total output to all input factors.
2. *Total factor productivity:* the ratio of total output to the sum of associated labor and capital (factor) input.
3. *Partial productivity:* the ratio of total output to one class of input.

Quality

A commonly accepted definition of quality is "fitness for use." This term has valuable meaning to both producer and customer. The producer views fitness for use in terms of the ability to process and produce with less rework, less scrap, minimal down time, high productivity, etc. From the customer viewpoint, fitness for use includes product durability, availability of spare parts, identity, comfort, etc.

There are two aspects of quality: the quality of design and the quality of conformance. Quality of design is quality obtained through changes in or manipulation of design parameters. Differences in quality result from differences in items such as size, materials used, tolerances in manufacturing, reliability, equipment utilized and temperature.

Quality of conformance is a measure of how well the product conforms to the specifications and tolerances required by the design. Factors such as training, production process, motivation levels, procedures and quality assurance systems can affect the quality of conformance.

Quality characteristics are elements of fitness for use that typify the variety of uses of a given product. They may be of several types, including:

- Time-oriented (serviceability, reliability, maintainability).
- Sensory (color, taste, beauty, appearance, etc.).
- Structural (frequency, weight, length, viscosity, etc.).
- Commercial (warranty).
- Behavioral or ethical (fairness, honesty, courtesy, etc.).

Improvements in quality are made by examining the design and conformance phases and their associated characteristics and applying techniques and methods such as changes in design, timing, inspection procedures and process control procedures.

PRODUCTIVITY AND QUALITY MANAGEMENT

Productivity and quality management is an integrated process involving both management and employees with the ultimate goal of managing the design, development, production, transfer and use of the various types of products or services in both the work environment and marketplace. The process requires total involvement of everyone in the planning and analysis, measurement, evaluation and control, and improvement and monitoring of productivity and quality at the source of production or service center.

The productivity and quality management triangle (PQMT) shown in Figure 3 encompasses an information system that provides information relevant to the planning and analysis process, the performance and measurement process, and the implementation of corrective actions and techniques to improve productivity and quality.

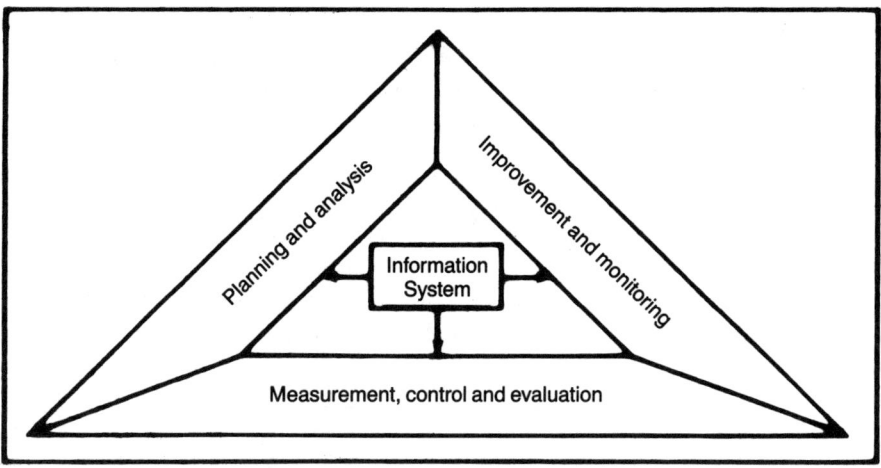

Figure 3. *Productivity and quality management triangle.*

Measurement, control and evaluation make up the first stage in the PQMT triangle. Once a set of measures has been developed, plan targets and strategies are formulated. Based on these strategies, short-range objectives are formulated, and operational improvement techniques are then used to implement short- and long-range objectives.

In order to assess the extent to which improvements have been successfully implemented, the measurement, control and evaluation step is performed again. This triangular relationship thus continues for as long as the comprehensive productivity and quality program exists in the organization.

CASE STUDY

A case study will now be presented to illustrate the relationship between productivity and quality.

A manufacturing company that produces printed circuit boards (PCBs) to customer order was interested in improving the quality of the PCBs. The parts and components used for the assembly of the boards were from different suppliers. They were assembled in a group machining process. The PCB production process operated at a yield of 63%; i.e., with 37% of the output nonconforming.

To solve this problem, the firm instituted eight inspection and rework processes. The direct and indirect manufacturing cost due to inspection and rework was approximately $288,000 per week. The added inspection and rework stations created additional problems: increased product cycle time, more work-in-process inventory, a decline in labor productivity, increased handling, and additional defects created by the inspection and rework process itself.

These problems also affected the overall utilization of resources such as manpower and energy. The profit margin of the firm was affected significantly.

METHODOLOGY AND RESULTS

A flow diagram of the PCB manufacturing process was drawn as shown in Figure 4. A careful analysis revealed that there was no control in place to monitor the production process. Control charts were developed and implemented in all operations. Edosomwan (1985, 1986, 1987).

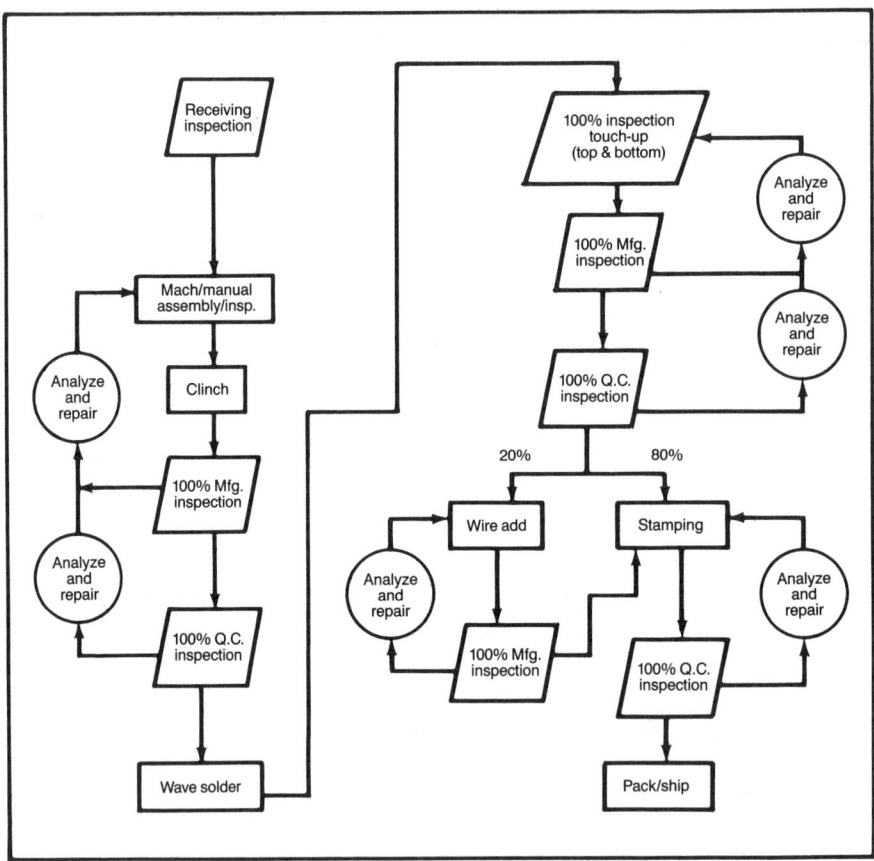

Figure 4. *Process flow diagram of PCB assembly before SPC implementation.*

The statistical process control mechanisms implemented enabled process performance to be monitored and properly. Corrective actions were taken to ensure good quality at the source of production. Data were gathered and analyzed in real time. The implementation of statistical process control (see Figure 5) led to significant improvements in both productivity and quality.

It was also observed that tremendous amounts of paperwork at each work station were impeding labor productivity. Work simplification techniques were used to reduce 24 forms to five. The bottleneck problems between operations were resolved by analyzing production time, queueing time, arrival time and waiting time for jobs. This enabled production line balancing.

Employee involvement in and commitment to quality improvement at the source of production enabled production errors to be determined and resolved

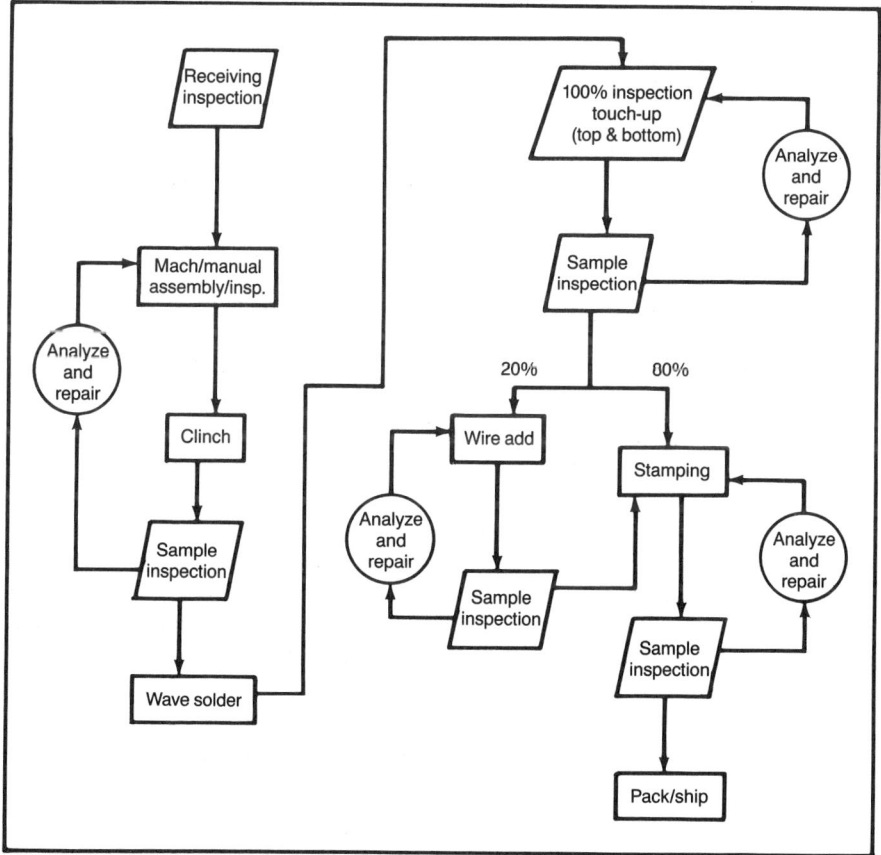

Figure 5. *Improved process flow diagram of PCB assembly.*

in real time. The hours needed for, and cost of, inspection were significantly improved. Inspection costs per week was reduced from $288,000 before process control was implemented to $72,000 after. The first pass yield improved by 35% from 63% to 98%. Labor productivity improved from 68 units per hour to 120 units per hour, and the total productivity index climbed from 0.723 to 1.478.

This case study indicates that productivity and quality are connected. Techniques implemented to improve quality also improved productivity, and vice versa.

A TEN-STEP APPROACH

The ten-step productivity and quality improvement program outlined below is a practical system that can be implemented in any organization. Each step will now be described.

- **Step 1:** *Management and employees' commitment to productivity and quality management at the source of production or service*—The total commitment of both management and employees to quality and productivity improvement through a defect and error prevention program is essential. No levels of defect are acceptable. The concept of "inspecting quality into the product" should be avoided. The commitment from everyone should be "Do it right the first time at the source of production or service."
- **Step 2:** *Implement measurement, control and evaluation techniques for productivity and quality*—A total productivity measurement system should be implemented at both the task and the firm level. Such a system enables an organization to relate partial productivity components to total productivity and pinpoint specific areas for improvement. Edosomwan, Kendrick and Creamer, and Sumanth have recommended such a measurement system (see "Suggested readings").

  Institute statistical process control techniques where possible. Develop measurements for percent defective, rework, engineering change and impact on quality, and measure the variability of the process on an ongoing basis. The process control charts should be understandable by everyone as a measure of process effectiveness. The author has implemented such techniques in a group technology production environment (see "Suggested readings"). The cost of quality and productivity improvement should be calculated periodically. Actions should be implemented to correct high failure rates and and errors at the source of production.

- **Step 3:** *Implement planning and analysis for productivity and quality*—Based on the measures obtained in Step 2, short term and long term planning for zero defect and productivity gains should be developed. Juran and Gryna and the author have developed a formalized approach for quality and productivity planning (see "Suggested readings").

- **Step 4:** *Implement improvement and monitoring techniques for productivity and quality*—Many types of improvement techniques are available. Error removal at the source of production or service seemed to work well. This technique requires a clear definition and identification of a specific error affecting quality and productivity. The technique recommends resolution of such an error before commencing on any further work activity. Enhancement of processes through technological innovation, task analysis, employee involvement, and work simplification should be encouraged at all levels.
- **Step 5:** *Provide management and employees' training for directing productivity and quality management*—Everyone in the organization should be made to understand the meaning of quality and productivity in the survival contest for profit. A typical training package should include techniques for productivity and quality measurement, control and evaluation; planning and analysis; improvement and monitoring; and root cause analysis. Experiment design and construction and interpretation of controls are other areas for useful understanding.

One key benefit of the training program is that it creates awareness by everyone of the issue of quality and productivity. Understanding of the goals and objectives facilitates getting commitment from everyone.

- **Step 6:** *Institute a productivity and quality team that includes all levels of the organization*—A productivity and quality improvement team should be formed that includes members from each sector of the production and service processes. The team should meet regularly to resolve potential problems. Brainstorming techniques should be encouraged at the team level.
- **Step 7:** *Allow for productivity and quality goals that are measurable, attainable and open to opportunity*—A participatory management approach should be used in goal setting. Employees and teams should be involved in setting quality and productivity improvement goals for the organization. Checkpoints should be put in place to measure any potential deviations from the original goals.
- **Step 8:** *Prioritize goals and problems*—Quality and productivity issues can be very complicated and deep-rooted. In such situations the Pareto 80%:20% principle should be used to classify goals and problems. Each goal and problem should be matched with implemented action and expected benefits on a real-time basis.
- **Step 9:** *Provide feedback mechanism for information among people, processes and procedures*—A management information system should be implemented. This system should incorporate a data base on productivity and

quality trends, issues and key accomplishments. The data should be available to analysts and team members working on the improvement of productivity and quality.

- **Step 10:** *Provide real-time recognition for a job well done, and share the gains from productivity and quality improvement*—Money is not the only source of motivation. However, compensation and promotion should be related to good performance. Ideas implemented for productivity and quality improvement should be rewarded by management through such means as awards, recognition among peers, bonus checks, additional technical challenge and promotion. A simple thank-you for a job well done can go a long way toward reinforcing the commitment of an employee to the improvement of productivity and quality. A suggestion program that rewards employees for their ideas should be encouraged.

CONCLUSIONS

Productivity and quality management is not a one-time action. It is an ongoing process that uses organized programs, sound techniques and common sense for the improvement of resources. For a productivity and quality management program (PQMP) to be successful, it must be designed to allow universal participation; provide leadership by example; give information, facts and ownership at all levels; and provide communication channels, feedback for performance, and continuous orientation and training of the work force.

Suggested Readings

Deming, W. E. 1982. *Quality, Productivity and Competitive Position*, MIT Press.

Edosomwan, J. A. 1985. Computer-Aided Manufacturing Impact On Productivity, Production Quality, Job Satisfaction and Psychological Stress. *Proceedings: 1985 Fall Industrial Engineering Conference*. IIE. Chicago. December 8-11. p. 469.

Edosomwan, J. A. 1986. A Conceptual Framework For Productivity Planning. *Industrial Engineering*. January.

Edosomwan, J. A. 1987. *Integrating Productivity and Quality Management*. Marcel Dekker Inc. January.

Edosomwan, J. A. 1985. A Methodology For Assessing The Impact of Computer Technology on Productivity, Production Quality, Job Satisfaction, and Psychological Stress In A Specific Assembly Task. Doctoral Dissertation. Department of Engineering Administration. George Washington University. January. (sponsored by Social Science Research Council, U.S. Department of Labor and IBM Corp., under grant numbers SS-36-83-21 and IBM 2J2-2K5-722271-83-85).

Edosomwan, J. A. 1985. Quality at the Source Of Productivity, IBM technical working paper. December.

Edosomwan, J. A. 1986. Statistical Process Control In Group Technology Production Environment. *SYNERGY '86 Proceedings*, SME and APICS.

Edosomwan, J. A. 1985. A Task-Oriented Total Productivity Measurement Model For Electronic Printed Circuit Board Assembly. International Conference on Electronics Assembly Proceedings. Santa Clara, CA. October 7-9.

Edosomwan, J. A. 1985. A Ten-Step Approach For Productivity And Quality Planning. IBM technical working paper. November.

Edosomwan, J. A. 1985. "Total Productivity and Quality Improvement at the Source of Production." working paper. ASEM Conference.

Juran, J. M., and F. M. Gryna, Jr. 1980. *Quality Planning and Analysis*. McGraw-Hill Book Co. New York.

Kendrick, J. W., in collaboration with the American Productivity Center. *Improving Company Productivity Handbook with Case Studies*. The Johns Hopkins University Press. Baltimore, MD.

Kendrick, J. W., and D. Creamer. 1965. Measuring Company Productivity. Handbook with Case Studies (Studies in Business Economics, No. 89). National Industrial Conference Board. New York.

OEEC: Terminology of Productivity Para. 2,2 rue Andre-Pascal, Paris-16, 1950.

Sumanth, D. J. 1979. Productivity Measurement and Evaluation Models for Manufacturing Companies. Doctoral Dissertation. IIT. Chicago. August. (University microfilms, Ann Arbor, MO, No. 80-03, 665).

Reprinted from the January 1987 issue of *Industrial Engineering*.

2

Eliminating Complexity From Work: Improving Productivity by Enhancing Quality

F. Timothy Fuller

With the emergence of Japan as the worldwide quality and productivity leader in a number of industries, many U.S. manufacturers are embarking on companywide programs of quality and productivity improvement. Many of these programs have been sparked by the teachings of W. Edwards Deming, the man who has been given credit for teaching the Japanese his powerful philosophy of making decisions based upon statistical principles. Hewlett-Packard is one of those companies studying and attempting to implement Deming's philosophy of managing better.

The Computer Systems Division of HP began in 1981 to use Deming's methods in the manufacturing of its line of HP3000 general purpose business computers. A consultant, familiar with Deming's work, had helped guide a number of successful projects in the department that assembled and tested printed circuit boards. The results of these projects were the virtual elimination of solder joint defects and associated rework, reduction in component insertion defects, improvement in manufacturing cycle time, and reductions of inventory and space requirements.

After studying the results of the initial projects, managers in the division were beginning to realize that every time defects were reduced, productivity rose measurably. This increase in productivity could often be attributed to reduction in rework that followed the reduction of defects.

In late 1983 management was looking for more ways to improve productivity of the circuit board assembly process. The problems associated with late delivery of materials seemed to be a likely candidate for study and improvement. This study led to some manufacturing changes that produced startling improvements in productivity.

The success of these efforts convinced the author that tremendous productivity gains could be achieved by reducing the unnecessary work, or complexity, introduced by defects in the quality of materials, tools, equipment, and other process variables. This article is concerned with finding, measuring, and eliminating complexity in the work place.

COMPLEXITY IN MANUFACTURING: THE BACK ORDER PROBLEM

The assembly process for printed circuit boards in the Computer Systems Division consisted of a number of steps, beginning with gathering a kit of parts, continuing with auto inserting, hand loading, wave soldering, and back loading, and ending with testing. Boards were built with lot sizes of 20 to 200 and were controlled through work orders issued through the material requirements planning (MRP) system. Work orders were started as close as possible to the time that kits were issued, even if some of the parts were missing.

When the assembly process was started before all parts were in hand, the process would generally proceed as far as the wave solder operation. If the missing parts had not arrived by this step, the partially completed boards were pulled off the line and stored on shelves. When the missing parts arrived, the partially completed assemblies were brought back to the line and the assembly process continued.

The logic behind the process of building incomplete kits was related to two beliefs held by management:

1. It is important to keep people busy working, even if the overall task cannot be completed. If production workers are idled by missing parts, labor hours are wasted.
2. In order to meet production schedules, the assembly should proceed as far as possible so that when the missing part arrives, it can be quickly inserted and the lot of boards can be expedited through the remainder of the process.

Data which had been collected on the parts back order problem showed that, on average, about 98 percent of the kit parts were in the stores area when the work orders were pulled. Materials management felt that due to the number of vendor problems and the variation in the production schedule, this performance was acceptable. However, from the point of view of the production department, it was less than desirable. Since the majority of the kits required as many as 100 different parts, numerous kits had back-ordered parts when they were pulled to the production floor. The data showed that about 75 percent of all the kits pulled had one or more back orders when delivered and that, on the average, each kit had from one to three missing parts.

Production management had been working with the materials group for some time to improve the availability of parts but was unable to achieve higher than the 98 percent in-stock level.

Awareness of the Problem

Although management was aware that back orders were a problem, it was not considered serious enough to make improvement a high priority objective. Most similar assembly operations in the company were experiencing the same degree of difficulty in procuring parts. One of the reasons for this was high demand in the chip market, which was causing a number of suppliers to miss promised shipping dates or to allocate scarce parts among their customers instead of sending complete orders.

One particular event raised the back order problem to a higher priority within the management group. The assembly department of the Computer Systems Division had been experiencing a higher than normal demand for completed boards and was having difficulty meeting the production schedules. A neighboring division was faced with less than expected demand for its products and had a number of surplus production workers. An agreement was worked out to borrow some of these workers to help the assembly operation.

After a week or two in their temporary assignment, the loaned workers approached management with complaints about the working conditions in the assembly department. Their comments included statements like, "We don't like working here. Things are too disorganized," and "Every time we start working on something, we run out of parts and have to find something else to work on. We didn't have these problems at our other job."

The assembly management was quite concerned over these comments, especially because it had thought that the department was quite well organized and that morale was relatively high. When asked about the differences between the two departments, the loaned employees stated that in their own department no work orders were started until the kits were complete.

The New Process

The department manager thought about these comments and decided to try an experiment to eliminate the problems associated with back orders. He proceeded to modify the parts-pulling process as follows:

1. Stores would continue to pull kits of parts according to the MRP schedule.
2. Complete kits were to be delivered to the assembly area as usual.
3. Incomplete kits were to be placed on shelves in the hallway with a note indicating which parts were missing.
4. When the back orders were filled, the completed kits would be delivered to the assembly floor.

The production control supervisor confronted the assembly management with the prediction that "If you let work orders sit around and don't start working on them, you will never meet your production schedules," but the assembly manager held firm and the experiment was begun on a Monday. Immediately, material began to build up on the shelves in the hallway. The work load in the assembly area began to slow noticeably. When work-in-process began to flow out of the area, the supervisors showed some nervousness as they saw their people idle more and more often. For the first week very little new material flowed into the department.

One day the department manager found a supervisor rummaging through the incomplete kits in the hallway trying to combine two partial kits to make one full kit in order to give some work to his people. The manager asked him not to do that. "Instead," he said, "why not do some training? Hold your staff meeting. If you've nothing else to do, take your crew to a movie. Just don't be concerned if your people aren't busy. I'm not measuring supervisors on how busy their people are any more." He also requested that they not expedite late parts. He suggested that they wait until the kits were complete and then do their best to build them as quickly as possible.

Soon the material in the hallway became noticeable to higher division managers. "You can't have a million dollars worth of expensive RAM's sitting in the hallway like this. There's no control," they said. Moreover, the division managers were not convinced that the experiment would work. But they supported its continuation. A compromise was reached whereby the incomplete kits would be stored in a special area that could be more tightly controlled.

In addition to the lack of work in the assembly department, other changes became noticeable. The work-in-process shelves gradually emptied as more of the old back orders were filled. The department manager decided that it might be possible to eliminate some of the shelves. Some of the idle production workers were given the task of dismantling the shelves and getting rid of them.

Significant pockets of vacant space opened up in various parts of the department, and it began to take on a cleaner look. Three weeks after the experiment began, almost all of the work-in-process shelves had been emptied.

The Remarkable Results

As the experiment went into its fourth week, the manager noticed that the production workers were still often idle, even though work had begun to flow through the process again. A quick check of the production output showed that weekly production had climbed back to the level maintained before the experiment began. He also noted that a number of the production workers had been loaned to work in other departments.

Concerned, he asked for a review of the actual hours of work being recorded to build a set of boards and compared this to the current labor standards. Incredibly, the amount of labor to assemble a kit of boards had been cut nearly in half by this single process change.

It appeared that as much as half the activities of about sixty people had been to set up and take down jobs, expedite, move material, count material, and do other tasks that were unnecessary in the new process.

As the department began to adjust to the new procedure, other problems began to surface. As the work-in-process queues disappeared, an extreme variability in work load became visible. At times the workers were almost idle; at others they were inundated with work. Previously, the work-in-process queues had hidden the variation. Production control was called in to study the problem; as a result of the study, lot sizes were reduced significantly. Smaller lots of each board type would be delivered several times each week. High-volume assemblies would be delivered daily. It also became apparent that the new process could not tolerate significant downtime of critical equipment, such as the automatic insertion and automatic test equipment.

The data now showed that a significant reduction in cycle time had been achieved. With the reduction in lot sizes not yet in effect, the cycle time appeared to average five and one-half days, down from sixteen and one-half days before the experiment. Sorting the cycle time data by lot size revealed that further improvements would be achieved as smaller lots reached the assembly area.

The manager decided to collect more data to see how the new process affected his ability to expedite critical boards that were late because of missing parts. Data from before the experiment showed that if a lot of boards was partially assembled up to the wave solder step, it would take approximately two days to expedite them through the process when the missing parts arrived. Data now showed that a lot of boards could be expedited through the entire process, from start to finish, in less than twelve hours. Production control's prediction had been proven wrong.

Clearly, the experiment was a success. Significant improvements in every measure of productivity had been achieved by improving the quality of the incoming kits. It should also be noted that no additional work was required of anyone outside the production department. The data showed a tight link between quality and productivity. Improved quality had eliminated the need for many complex process steps. Less complexity meant less work required to produce a given output.

Let's now look at a model of this process change and describe in detail how this improved quality leads to a reduction in complexity and increased productivity.

THE COMPLEXITY MODEL

Figure 1 is a process flow diagram of a simple assembly process. The process is designed to have three steps: get the kit, assemble it, and move the material. If one asked supervisors to draw a flow diagram of such a process, most of the diagrams would look like this one. However, if one actually followed the flow of material through the shop, one would probably find many more steps in the process than are shown here. The extra steps would in most cases be related to unexpected problems such as late parts, defective parts, and poor procedures.

Why would most supervisors leave out these critical steps? One reason may be that figure 1 represents the most common path through the process. Another reason may be that the process was designed this way by the supervisor and, due to lack of knowledge of the process, he or she thinks it operates this way, or at least wishes it would operate this way. In any case, we know that in the real world problems do come up and they have to be dealt with.

Let's now add a quality problem to the perfect process shown in figure 1. Suppose that when a worker goes to pick up a kit, one of the three parts is missing. Also assume that our standard operating rule is to try to keep busy and work around problems as best we can. How could we redraw the figure 1 diagram to show the additional steps needed to handle the problem of missing parts? Figure 2 shows how this new process might look.

Across the top of the diagram one additional step has been added to the process, an inspection step. The person who picks up the kit of parts now is

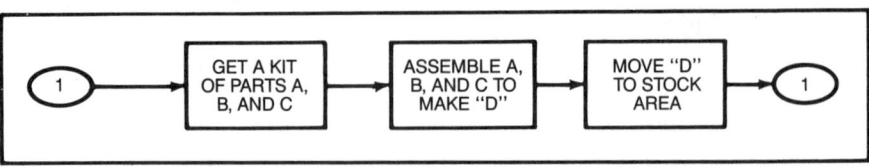

Figure 1. *The perfect assembly process.*

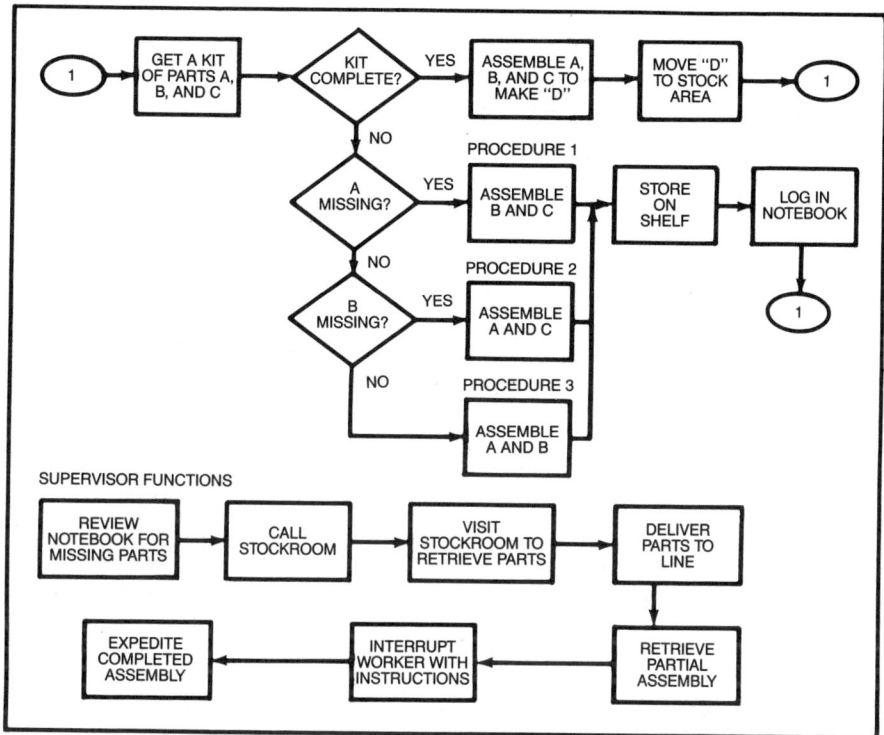

Figure 2. *An assembly process with errors complexity.*

required to make a decision: if all the parts are in the kit, the standard process applies; if one of the three parts is missing, there are some different steps to follow. Let's suppose that inspection finds that one of the parts is missing. Now we must know which one is missing, because in order to partially build the assembly, we need to know which of three special procedures to follow.

The next step is to assemble the parts on hand and then find a place to store the material until the missing parts show up. In order to keep track of all the work in process, a special log or computer entry may be required to describe the location of the WIP.

The job has thus become more challenging for employees. More training is required because there are more than twice as many process steps for them to perform. More space is required to store the WIP. A cabinet may be needed to store the procedures. In some cases an employee with a higher level of skills may be required to perform the work.

Also, another process has been added for the supervisor. In the perfect process shown in Figure 1, the supervisor could spend all available time hiring, training, and otherwise helping his or her employees develop good work skills. In the first process the supervisor played no part in the actual accomplishment of the tasks. Once he or she was trained, the worker had complete control of all the steps of the process.

In our new process the supervisor has many new jobs to do. The supervisor may have a "hot list" of all the critical parts he or she is waiting for and another list of customers with the most urgent needs for late assemblies. The supervisor now becomes an expediter in order to attempt to satisfy his or her customers. The supervisor will also likely be the one who goes to get the critical parts the minute they arrive in the stockroom, the normal delivery procedure being much too slow.

Once the parts are in hand, the supervisor must find someone to install them. He or she must decide who should be interrupted and must ask the employees who are selected to put away their current work and set up and perform the critical work. The supervisor may help by finding and retrieving the partially completed assembly. When the job is complete, the supervisor may be the one to deliver it to the customer, as the delivery system may again be too slow.

Enumerating The Extra Work

Now we have two processes, and the second one just described clearly involves more work than the process described in Figure 1. In the second process, each time a kit is received an extra inspection step is required; so even the error-free process takes a little longer. In addition, when a missing part is discovered, many extra steps are required. The second process will always take more time than the first and will require several times more work than the first if every kit has one part missing.

However, we have only begun to enumerate the extra work associated with the second process. Let's assume that at every step of the process, errors can be made. In the simple process, errors could be made in three places. The worker could get the wrong kit of parts; a mistake could be made in the assembly step; and the completed assembly could be delivered to the wrong place. The frequency of errors will depend on a number of things, but the quality of the initial training and the amount of practice the worker has had will certainly be the main contributors.

With relatively few different types of possible errors, the recovery process for each error can be described and practiced. Therefore, we can assume that the simple process will probably have relatively few errors, each of which can be quickly corrected.

But in our second process we have a different situation. We can expect a few more errors in the standard procedure (the top line of steps in Figure 2) because there is an extra step in each repetition of the process. What can we expect in the special steps required to handle missing parts?

Since some of these process steps are performed infrequently, the worker may have little chance to practice them. In addition, the initial training may not cover all the possible steps in the process. This implies that the error rate may be substantially higher in these nonstandard steps. Now consider what happens when the worker tries to recover from a second-level error.

Suppose that part "B" is missing from the kit and an attempt to put "A" and "C" together is made. Let's also suppose that a mistake is made when the entry is made in the log that records the location of the partially completed assembly. Now when the parts arrive and the partially completed assemblies can't be found, what process should be followed to find the assemblies and get things straightened out?

It is likely that a new procedure will be invented on the spot to handle what has now become a crisis. It is at this point when things really begin to go wrong. Tempers get short, one person blames another for the problem, and so on.

Let's now define process complexity as being extra process steps that are required to deal with external errors ahead of the process or extra process steps to recover from error in the process, or internal errors. Reducing external and internal errors improves productivity through the following sequence:

1. Error reduction permits elimination of some process steps, such as disposition of faulty material, and reduction of the number of times that some process steps, such as rework, need be repeated.
2. Now that less rework steps are being performed, there is less chance of internal errors. This reduces some lower level rework steps. Fewer rework steps at this lower level lessens the chance of internal errors at that level and therefore reduces the number of rework steps at a still lower level, and so on.

In sum, reducing errors can lead to elimination of work at multiple levels of the process and therefore highly leverages productivity improvement.

Experience has shown that eliminating errors will produce extensive gains in productivity that far exceed potential gains achieved by trying to improve the efficiency of an error-ridden process. Automation of an assembly process that is full of errors will likely force everyone into a crisis situation. Implementation of Just-in-Time manufacturing techniques (JIT or Kanban) without first reducing quality problems will likely have similar results.

Our model has suggested that the addition of one external quality problem to a perfect process can introduce a significant amount of complexity that substantially reduces productivity. In the majority of departments, whether manufacturing or administrative, most standard processes have far more steps than the simple one in our model. In addition, many types of errors can flow into the process and many other types can be introduced into the process itself. Every error requires extra process steps to deal with it. If the error is not discovered in the process, the customer will likely find it and will be required to deal with it.

This implies that in most processes in our offices and factories where no long-term process improvement efforts have been in place, most of the activities undertaken by people are part of the complexity and few activities represent the "real work" that people would like to be doing. As William Conway, former president of Nashua Company, put it, "There's just not much work in anything."

HOW TO FIND THE COMPLEXITY

We have shown that in a typical operation or department, much of the work being done might be complexity that has been introduced by errors. Unfortunately, much of this complexity is usually not apparent to the manager of the department. We have been doing these unnecessary tasks for so long that we see them as part of the standard process.

Some people have jobs that are largely the result of errors which have been introduced into the system. Consider these examples:

1. A person who opens and restocks customer returns;
2. A customer service representative who follows up on customer complaints;
3. A collector who calls customers who are late in paying for merchandise;
4. An expediter of late parts or products; and
5. An inspector who looks for defects.

All people who are engaged in performing a standard process spend some portion of their time solving problems. All people make mistakes and must correct them. However these activities are seen as normal parts of their jobs, and no special notice is taken of them. If a copy machine breaks down occasionally and sometimes produces poor copies, working around the inconveniences is considered the mark of a good, resourceful employee. Each employee builds into this job some informal procedures to overcome the little problems faced each day.

Only when several copy machines break down at the same time and there are loud complaints does management grasp that there is a problem that needs to be solved. Now something will be done, even if it is only a temporary solution to get the work moving again. Let's explore some techniques we can use to begin to find and measure the complexity in an operation. Then we can discuss some techniques for removing it.

The "Real Work " Model

Figure 3 depicts that part of a typical employee's time during which no work is possible. The circle represents the total eight-hour work day. The shaded area is an estimate of the time that is lost due to sanctioned benefits, company-sponsored activities, and unsanctioned business that take the employee out of the work place. Some examples of sanctioned benefits are vacation time, coffee breaks, and sick leave. Company-sponsored activities include staff meetings, training, United Fund meetings, and fire drills. Unsanctioned personal business includes unscheduled rest breaks, personal phone calls, late arrival, and early departure.

It has been the author's experience that this unavailable time is as high as 25 percent of the total time in large organizations with a full range of employee benefits. This leaves approximately 75 percent of the eight-hour day that potentially could be used for doing work.

Complexity Caused by External Errors

Now let's make some estimate of the activities that are going on during the remaining hours. From the back order case discussed earlier, we can estimate that, on average, people spend up to half their working time fixing problems caused by errors introduced into their process from other sources. The activities comprising this time are designated in Figure 4 as complexity due to external errors. Added to time unavailable for work, it further reduces the

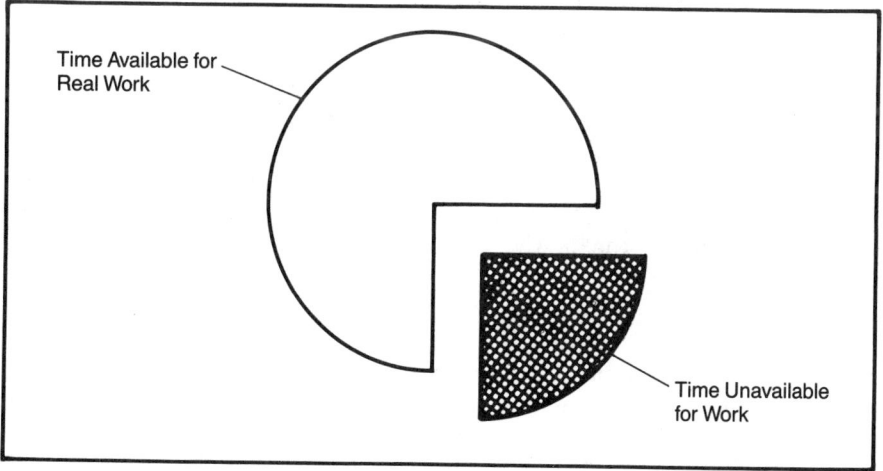

Figure 3. *Amount of time not available for work—complexity model.*

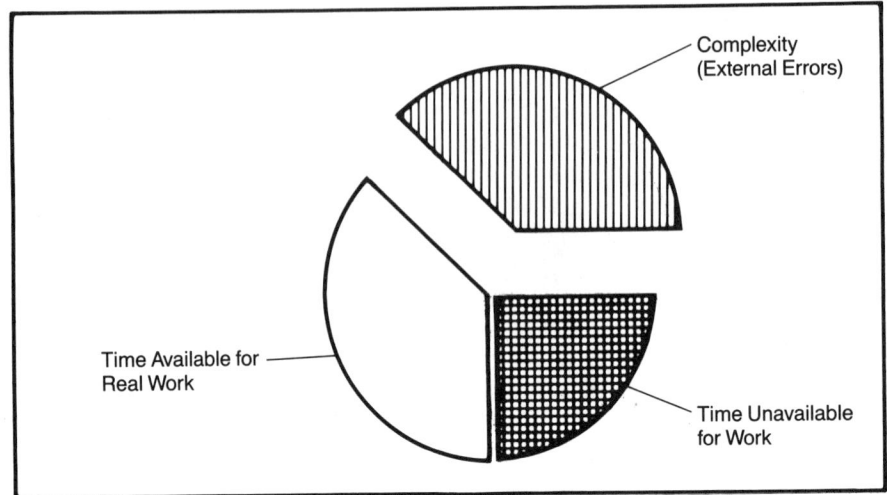

Figure 4. *Amount of time lost due to external errors—complexity model.*

amount of time available for "real work," which can be defined as activities that an organization is in business to carry out and that, in the absence of errors, would still be performed. Now only 35-40 percent of people's time is available for real work. Table 1 lists examples for five departments of activities representing real work and complexity resulting from external errors.

Complexity Caused by Internal Errors

Referring back to the complexity model presented earlier, we recall that even a process using high quality materials will still have internal process errors. So we can expect a number of errors while people are doing real work.

Some of these errors are mistakes in carrying out the steps in the process, while others are problems with tools, supplies, equipment, and other items associated with the process. We might say that the real work activities are made up of "subactivities," some of which are complexity caused by errors within the process. By breaking up activities into very small parts and adding up those that are rework for internal errors, we might estimate that as much as 75 percent of the real work is complexity. In Figure 5 the shaded area of the circle has again been increased, this time to account for the subactivities devoted to fixing internal problems. Now less than 10 percent of people's time is available for real work.

TABLE 1 – Examples of Real Work and Externally Induced Complexity in Five Departments

Department	Complexity	Real Work
Accounting	Collecting overdue accounts resulting from carrier delivery problems	Mailing invoices
Production	Rework of faulty incoming materials	Assembly
Marketing	Handling customer complaints about poor quality materials	Helping customers but
R&D	Redesign due to market research error	Asking customers about their needs
Personnel	Handling a lawsuit from employee who was mistreated in manufacturing.	Training new managers

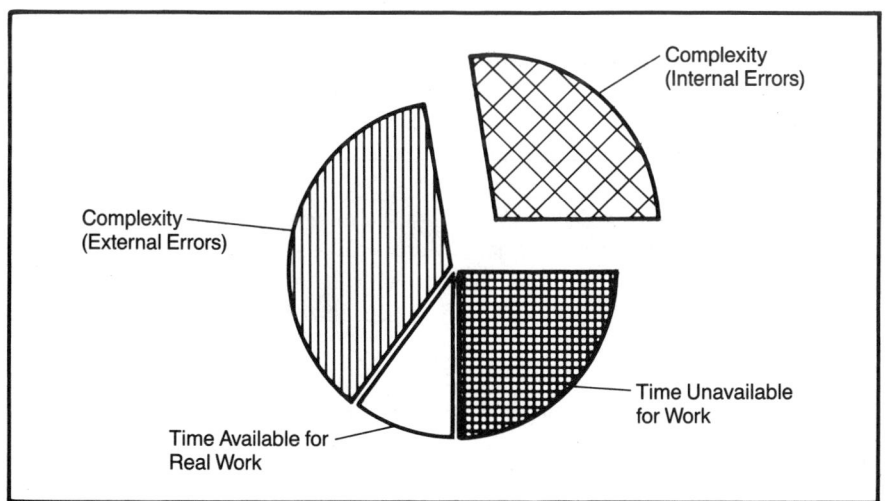

Figure 5. Amount of time lost due to internal errors—complexity model.

Following are some subactivities categorized as complexity or real work:

Activity: Training new managers (Personnel).
Subactivities: Reserving a conference room (real work).
　　　　　　　Resolving a meeting room conflict caused by a mistake in scheduling (complexity).
　　　　　　　Leading a group discussion (real work).
　　　　　　　Having a discussion about why the handouts can't be read because of copy machine problems (complexity).
　　　　　　　Watching a videotaped lecture (real work).
　　　　　　　Waiting for all the participants to arrive so the meeting can start (complexity).

Again, by visualizing the perfect process, one can sort these subactivities into two categories: real work (subactivities that would be required even if everything were to run perfectly) and complexity (subactivities that could be eliminated if the process were to run perfectly every time). It might follow that if one could break the subactivities into even smaller pieces, one could repeat the analysis and could categorize more activities as complexity at each iteration. Again, it appears that there truly may not be very much work in anything.

HOW TO MEASURE COMPLEXITY

To decide where to start the quality and productivity improvement process, it helps to have some idea of the relative amounts of complexity in various parts of one's organization. The more complexity in an area, the more quickly and easily significant improvements can be made.

Often, one can make a rough estimate of the level of complexity in an operation by just walking around the work area and making visual observations of certain conditions. Seven conditions indicating a high level of complexity, and seven corresponding conditions indicating a low level, are listed below:

Indicators of high complexity:
1. Lots of work-in-process materials. Many shelves in the work area to hold material.
2. Many people walking from place to place, standing in a line waiting for something, standing idle.
3. Work areas that are in disarray. Dusty boxes on the floors, bookcases full of dusty binders, and desks and walls covered with little scraps of paper serving as reminder notes.
4. People who can give only a brief, vague explanation of what they are working on and why it is important.

5. Humorous signs taped to the walls that say things like, "You want it when? Ha! Ha!" or "A clean desk is a sign of a sick mind."
6. In office areas, piles of processed and unprocessed documents stored in the work area.
7. Supervisors and managers pacing around the area trying to find out what's going on, ascertain who made a critical mistake, and expedite late orders.

Indicators of low complexity:
1. A small amount of work-in-process material. Few shelves in the work area to hold material.
2. Few people walking around carrying materials. Most people working at a steady, relaxed pace. No one waiting in line at copy machines, office supplies, stores.
3. Work areas that are neat. Everything in a department has a place and a use. People using time management systems instead of scraps of paper. Desk tops containing only what the person is working on at the time.
4. People on the production floor or in an office area who can give complete descriptions of what they do, why they do it, who their customers are, and what's important to those customers.
5. The most common item displayed on department walls are monthly performance graphs, daily control charts of defects, Pareto charts of defects and problems.
6. In office areas all documents are received, processed, and filed. In baskets are clean.
7. Supervisors and managers who are relaxed, walking around the area talking with employees, asking them what they are working on, and looking for ways to make their employees' jobs easier and more satisfying.

After looking at these items, a manager should have an idea of the overall level of complexity. However, it may be more difficult to accurately categorize activities as real work or complexity and measure them. Simple work sampling can be used to make a good estimate of activities that are being performed because of errors introduced from outside the process and subactivities that are part of internal process complexity.

How to Perform Work Sampling

The advent of cheap, multifunction electronic watches has made work sampling simple and easy for almost anyone to do. The basic idea is to look periodically at what a person is doing so that a list of activities can be developed and the relative frequency of each measured. If we ignore non-work-related activities, the list can be sorted into the two categories of real work and complexity due to *external* problems.

Then subactivities can be similarly grouped into real work and complexity related to *internal* process errors. When these data have been prepared, management can pull together interdepartmental task groups to eliminate the external errors. Work group improvement teams led by a supervisor can address the internal problems by solving the ones over which they have control and collecting data on the others so that management can take the proper action.

Work Sampling Process

Step I. Select the process to be studied. This may be determined from the data gathered previously by the department walk-through.

Step II. Procure a watch that has a "repeating countdown" function. This function allows setting a countdown timer to a particular number of minutes and seconds. When the countdown feature is turned on, the watch counts down to zero, beeps one or more times, resets itself automatically, and begins the countdown again.

Step III. Determine the sampling procedure and the sample period. In some cases one may wish to look only at the activities of a single person. If so, the person will wear the watch, start the watch each morning, and turn it off at the end of the day. Each time the employee hears the beep, he or she is to immediately stop working and make several entries in his or her log or check sheet. The employee should record the time, place, activity, and subactivity.

This procedure will be most successful if a list of the major activities is determined in advance so that sorting will be easy. Determine the number of observations needed. In general, the more activities that a person might be doing, the more observations are required to obtain a true picture of what the person is working on. In most cases 100 should be enough for one person. A larger department may require several hundred observations spread over many people during an interval of such length that weekly and monthly activities can be recorded.

It is important that the beep of the watch be a surprise to the work sampling subject. If the employee anticipates the beep, he or she is likely to modify his or her behavior in some way that will distort the data. Ideally, the turning on and turning off times of the watch should be random. But since few watches have the capability to generate random beeps, the countdown timer should be set at an interval long enough and odd enough so the individual will be surprised when it beeps. Good results have been obtained with settings of twenty-three, forty-one, and forty-seven minutes but not with sixty minutes.

No matter what the setting, subjects are bound to change their behavior to some degree because of the study. However, this is potentially beneficial if the person is permanently imbued with an interest in studying the activities being performed.

Step IV. Train the worker. The work sampling process can be quite threatening to a person who does not understand how the data are to be used. The following points should be made clear to the person at the outset of the project:

1. The data should be used by the person doing the work to make improvements in his or her own process where he or she has control of it. For instance, the worker can control his or her personal business. The worker may also be able to improve the way work is done, within limits. He or she will be rewarded for helping in the project, especially if improvement suggestions are made.
2. The data will be used by management to look for system-type errors, either internal or external, and to eliminate them from the system. This will make the worker's job easier so that more of his or her time can be allocated to more productive activities.

Step V. Start taking observations. Visit with the worker after a few hours to make sure that he or she understands how to set and control the watch and that the watch is functioning properly. Check to see that the data are being recorded in the proper format.

Step VI. Analyze the data. After the required number of observations have been recorded, summarize the data. Some of the activities will fall into the unsanctioned personal time category and should be grouped separately. The issue of unsanctioned personal time is a highly sensitive area for the employee and should be handled carefully—management must take care not to criticize the person's use of time in order to encourage accurate data reporting. If the worker has any control, seeing the sorted data is usually good motivation to make changes for improvements in the use of time.

Now go through the list of activities. Decide for each activity whether it belongs in the category of real work or complexity due to external problems. Put together a Pareto chart for the top ten activities with an annotation on each bar showing the category.

Sort the sub-activities within the real work category, determining for each subactivity whether it is real work or complexity due to internal errors.

The manager should now have an excellent understanding of the amount of complexity in the department and the potential for improvement. He or she should also have an excellent understanding of the types and effects of errors from inside and outside the process. This exercise will usually motivate the manager to make a number of obvious improvements shortly after seeing the data. More data collection and tracking of process variables can be started to begin removing the causes of the more subtle errors.

TWO CASE EXAMPLES

The complexity model can be used to detect quality problems in clerical-related as well as manufacturing-related processes. Below, case examples are provided of the application of the model in a sales office and an order processing function. The data collection techniques differ in some ways from those proposed above, but in fundamental respects they follow our methodology.

Marketing Associates in a Sales Office

Approximately thirty clerical and professional people worked in a Hewlett-Packard office taking orders for the company's products over the telephone. Management felt that a large amount of the work being performed was related to resolving problems caused by mistakes in processing and shipping the orders. It was decided that a study of the people's activities should take place so that management could have a better idea where the major problems were. Then, action could be taken to reduce them.

The work sampling plan was set up as follows:

1. The supervisor would wear the sampling watch.
2. When the watch beeped once every forty-two minutes, the supervisor would walk around a group of about ten people and ask each one what activity he or she was currently performing. Out of area or nonwork activities would be excluded from the study. If an employee was away from his or her desk when the supervisor came to collect data, no entry would be made for that person.
3. The study would cover a three-day period.

After three days of collecting data, the supervisor had a notebook containing 130 observations of the activities of 10 people. The date, time, and activity were recorded. Subactivities were not recorded.

The activities were then grouped by major category and counted. No attempt was made to determine whether the cause of any problem-related activities was internal to the group or external. However, the data suggested that most problems were caused by activities of people outside the department.

The data were then grouped and sorted by frequency. The supervisor was asked the following question about each activity: "If there were no errors in the process and everything were running perfectly, would you be working on this activity?" If the answer was "no," that activity was categorized as complexity. If the answer was "yes," that activity was categorized as real work.

The real work and complexity activities were then counted and compared. Figure 6 shows the relative size of the two categories of work according to the

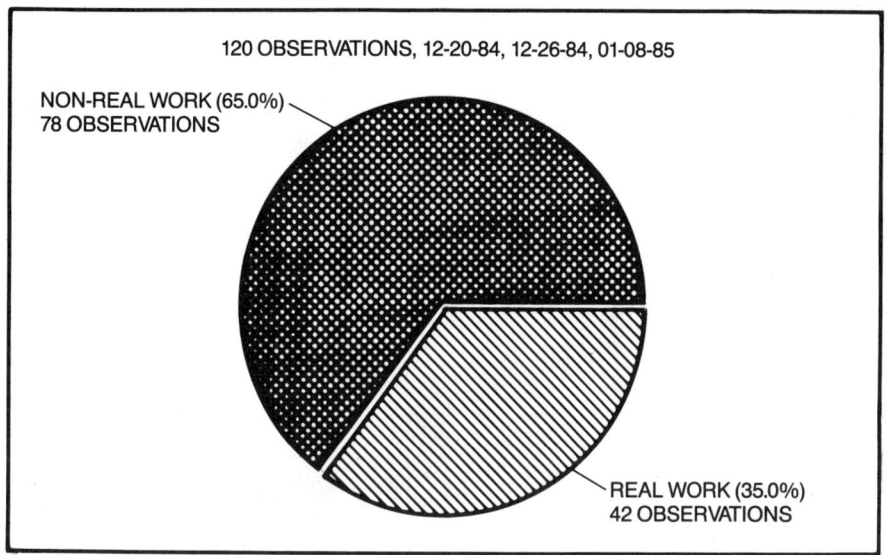

120 OBSERVATIONS, 12-20-84, 12-26-84, 01-08-85

NON-REAL WORK (65.0%)
78 OBSERVATIONS

REAL WORK (35.0%)
42 OBSERVATIONS

Figure 6. *Marketing associates: how time available for real work was spent.*

number of activities in each category. The data showed that the supervisor classified 42 of 120 observations as real work and 78 as complexity.

Figure 7 shows the relative frequency of the seven most likely activities that were being performed by the marketing associates. The seven activities, in descending order of frequency, were:

1. Processing customer returns (complexity).
2. Entering orders into the computer system (real work).
3. Converting orders to fix a problem (complexity).
4. Making changes to orders (real work).
5. Expediting shipments (complexity).
6. Answering questions from customers about the status of orders (complexity).
7. Taking orders over the telephone (real work).

Three of the seven most frequent activities were judged to be part of the standard process of taking orders and therefore were classified as real work. The most frequent activity was processing merchandise that was being returned by customers. The reasons given by customers included wrong product, duplicate shipment, and wrong quantity. This activity was categorized as complexity.

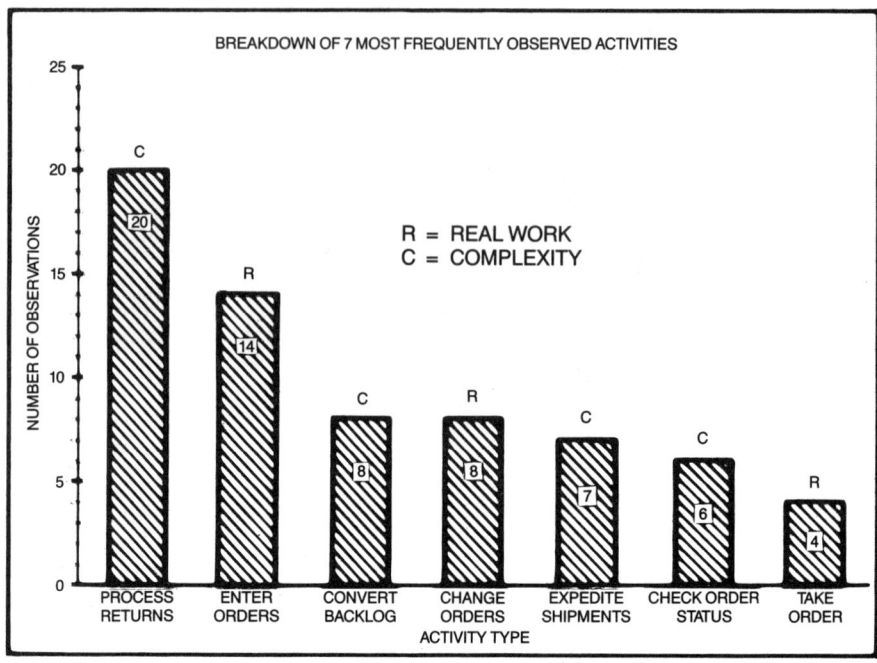

Figure 7. *Marketing associate work sampling results.*

Upon seeing the data, the supervisor had several reactions. One was that "15 percent of the time my people are processing customer returns. This is equivalent to six people. This is far too many, and we need to first streamline the way we process returns and then see what we can do to eliminate them." Immediately, the supervisor made changes in the work procedures to improve the processing of returns. The supervisor felt that seeing the data sorted in the form of a Pareto chart helped motivate her to make the change. At the same time, a task force was formed to reduce the number of products returned. One person from each department that could ameliorate the problem joined this team.

Employees Processing Orders in a Factory

The second case example concerns a group of clerical and professional people working in a Hewlett-Packard factory processing orders received from the sales office. Some people entered orders into the computer system, some matched orders with available products, and others shipped and invoiced the orders.

Management believed that a great deal of the work time was being spent fixing problems. It was felt that, as a result, employees were working a substantial amount of overtime and that morale was going down because people could see no end to the heavy work load. Management decided to study the activities of the people to see if the situation could be improved.

A work sampling study was set up with the following rules:

1. The supervisor would wear the watch, which would be set to beep every forty-one minutes. If the supervisor was to be out of the area, some other member of the department would wear the watch.
2. At the beep the person with the watch would roll a twenty-sided die to select three workers to be observed.
3. The person with the watch would ask each of the three people selected what they were working on at the moment. If any of the selected employees was out of the area, a note would be made of this fact. Upon returning, the employee was to be asked where he or she had been and what activity he or she had been engaged in. If the person was not working that day, no data were to be recorded.

Work sampling was carried out over a period of six days. During that time, 265 observations were made and recorded. The activities were grouped, counted, and classified in the same manner as in the previous case example. Figure 8 shows the division of the activities into real work and complexity. Of the 265 activities, 113 were classified as real work and 152 as complexity.

Figure 9 shows the nine activities that were observed most frequently. The activities, in descending order of frequency, were as follows:

1. Acknowledging order ship dates to customers (real work).
2. Sending messages through electronic mail (complexity).
3. Making computer entries to ship products (real work).
4. Processing customer returns (complexity).
5. Resolving billing mismatches (complexity).
6. Working on miscellaneous problems (complexity).
7. Working on quality improvement projects such as preparing graphs or training (real work).
8. Processing credits for goods returned or to correct other problems (complexity).
9. Out of area, miscellaneous problems (complexity).

"Sending messages" was classified as complexity because the purpose of most messages was to explain problems or order status and because most messages

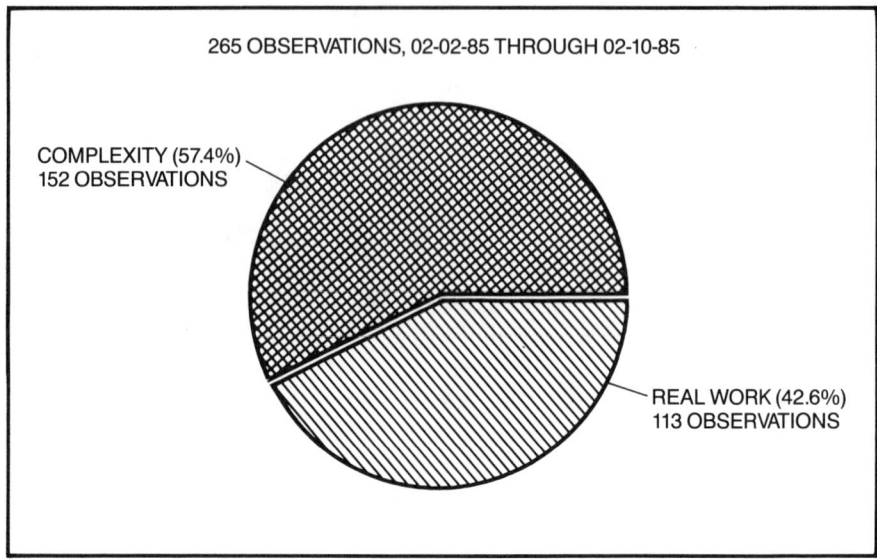

Figure 8. *Factory order processing: how time available for real work was spent.*

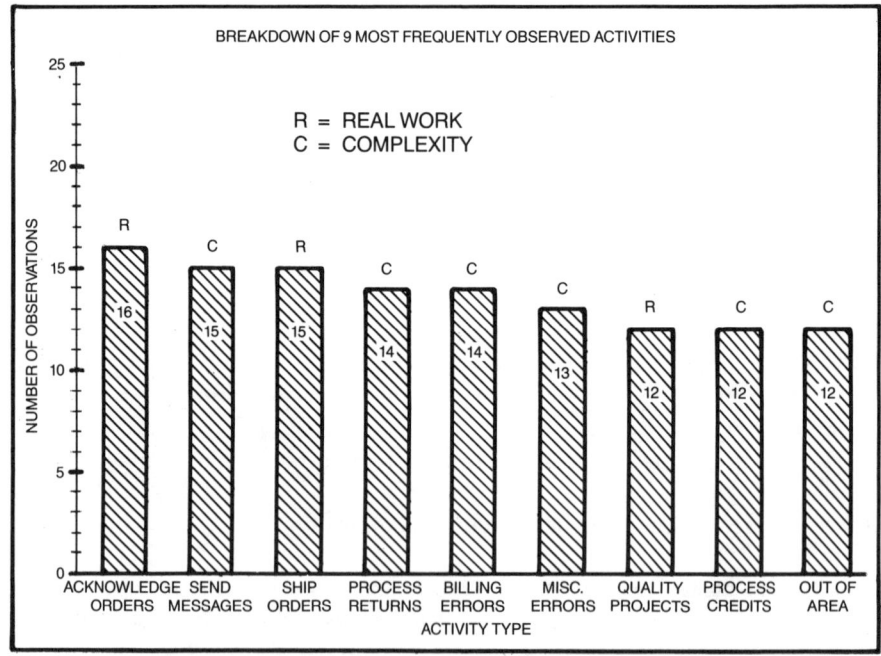

Figure 9. *Factory order processing time study results.*

were the result of errors in the process. Six of the nine most frequent activities were classed as complexity.

The data generated from this study confirmed management's belief that the people in the order processing department were spending a large portion of their time recovering from problems. A study was done to find the reasons for the large number of messages that were sent each day, and it was found that most were requests for status on orders. A project was begun to reduce order turnaround to improve customer satisfaction and to reduce the need to send messages asking for order status. Other projects were started to study the causes of the large number of customer returns and billing errors.

CONCLUSION

With experience in productivity improvement efforts comes increasing awareness that the bulk of the work we do in most large organizations is devoted to fixing problems. The data presented here suggest that far more than half of an employee's day may be spent either away from the work place or in the work place performing tasks that would be unnecessary if the quality of materials, tools, equipment, and other process variables were improved.

Elimination of errors in factory and clerical processes can have a dramatic impact on raising worker productivity, often at little cost, as was shown in the example of back-order kit parts.

The complexity model may be applicable to other business processes, as indicated by the case examples. One need only look carefully at the activities that are performed to see that many of the tasks we carry out can be eliminated if we improve the quality of the processes of which they are a part.

Only management can make the process changes that can reduce the complexity in its organizations. Collecting and showing management work sampling data based on the complexity model can help motivate them to take needed action.

Reprinted with permission from *National Productivity Review*, Vol. 4 No. 4, pp. 327-344. Autumn 1985. Executive Enterprises, Inc., New York, NY.

3

An Approach for Devising an Effective Quality Control Plan for Small-Lot Assembly of Printed Circuit Boards

Wilbert E. Wilhelm

Printed circuit boards (PCBs) are fundamental subassemblies in most industrial and military electronic systems. PCB producers face particularly vigorous international competition relative to price, quality, delivery lead time, and design sophistication. An effective quality control (QC) plan (Strumar 1985) can contribute to each of these capabilities and, therefore, is important to the success of any business.

Compared with the features of other products, PCBs have a unique set of quality characteristics. First, PCBs are typically populated with a large number of electronic components—perhaps several hundred. Each component may be subject to a variety of physical defects and, perhaps, numerous defects that affect electronic function. In addition, "good" components may interact through circuitry to cause defects. There may thus be a "logic" of defects; for example, one defect may mask another so that the second can be detected only after the first is repaired. Furthermore, it is well-known that electronic components have a high rate of infant mortality so that a component which tests "good" may become defective after a short period of service.

PCBs are typically produced in a dynamic environment in which new designs are introduced frequently as component technology and product applications change. State-of-the-art designs may be subject to frequent engineering changes before reaching maturity.

The relatively small size and high degree of PCB complexity challenges the workmanship capabilities of manufacturing operations. Few standards exist so that each PCB may require unique components and assembly procedures. Defects can be introduced one at a time (e.g., the wrong component can be inserted) or in vast numbers (e.g., as the PCB is flow-soldered).

This chapter's objective is to describe a systematic means of analyzing QC plans used for producing PCBs. Mathematical models developed by Wilhelm (1986) are described, and implications for managers of PCB assembly are reviewed.

Systems that produce small-lot quantities are the focus of this discussion, because they pose unique problems to manufacturing managers and because industrial and military markets require production predominantly in small-lot quantities. Such systems typically produce several thousand different product designs each year. Each design may pose unique quality problems, because of the number and types of components used, as well as circuit complexity.

A systems approach is advocated to ensure total QC. The role that each element plays in quality must, therefore, be identified and incorporated in the plan. PCB design, component quality (as supplied by the vendor), assembly processes, material handling operations, storing procedures, soldering processes, test, inspection, and rework all contribute to quality. In addition, total quality also encompasses system-level test and warrantee support for the customer.

Alternative strategies may be used for testing, which could be implemented in four levels: component, in-circuit, function, and system. Components may be tested on receipt from vendors, or defective ones could be identified by tests after assembly. Similarly, an in-circuit test could be performed or replaced by a comprehensive function-level test. Alternately, all testing could be done at the system level. This testing might require lengthy diagnostic time to identify which components are defective, but would eliminate the time required to test good components at earlier levels. According to *Assembly Engineering* (1985), trends are to incorporate component-level testing and to do comprehensive testing using automated equipment.

Testing does not produce quality, thus the system's perspective must trade-off costs of testing with those of assembly processes. Processes used for preassembly, assembly, soldering, and rework operations must also be evaluated as part of the QC plan. In addition, the level of process control must be planned along with selecting vendors, processes, and testing strategy.

The effectiveness of a QC plan may be measured in several dimensions: average outgoing quality level (AOQ) of PCBs, product yield (alternatively, the number of rework operations/PCB), and total cost including fixed (investment) cost, variable cost/PCB, inventory carrying cost, and warrantee support. The plan should maximize AOQ subject to business-imposed goals for cost.

The systematic description of quality presented here is incorporated in a mathematical model (Wilhelm 1986) that might be used to evaluate the impact of various design alternatives on system performance. Relative impacts of various system elements and of the system structure are emphasized.

PRIOR RESEARCH

Previous literature has rarely dealt specifically with the QC of PCB assembly, although the literature available does provide insightful views of certain problems.

Buzacott and Cheng (1982) developed both analytical and simulation models to study the effects of various factors on yield and AOQ, including tests designed to detect some—but not necessarily all—existing defects, imperfect tests, and rework of rejected boards before retesting. They advocated a system perspective to achieve the best possible performance, ensuring the quality of components supplied by vendors, the capabilities of assembly processes, and the effectiveness of test and rework operations.

Buzacott and Cheng's models are based on the philosophy that assembly processes can only create defects and that tests identify defects and remove them. Consequently, their models are composed of three basic types of logical structures: assembly operations, test operations, and merge blocks (which are required to accumulate several defect streams into one).

Dooley (1983) developed a comprehensive model called Prediction of Assembly, Rework and Test Yields (PARTY). The viewpoint of his model is that "defects" flow through assembly, not PCBs as such. His model incorporates a number of important, practical features, such as rework with accompanying recycling of PCBs, imperfect inspection, and multiple types of defects. Dooley demonstrates his model using a hypothetical example—model outputs describe test results, types of rework performed, and types of defects identified.

PARTY is an APL-based model that maintains an array to descibe the flow of defects in three registers:

1) DNF, which tallies defects not yet discovered by test/inspect including those introduced by rework;
2) DFD, which tallies defects identified but not yet corrected; and
3) DFX, which records corrected defects.

By inputting the probabilities that particular defects occur, tests identify them correctly and rework operations make necessary correction. PARTY apparently calculates the expected number of PCBs that will cycle through test and rework operations and the number of defects which may be permitted to escape detection/correction.

Dooley provides an insightful discussion of the types of defects that can occur and certain relationships among different types of defects. For example, he points out that defects can result from rework operations and that one defect can "mask" another.

Richards (1984) applied Jacksonian queuing models (to the extent possible) and simulation to study the configuration of rework operations. He tested two configurations—one incorporated a separate rework station near test and the other required defective items to be returned to the assembly operator for rework. In principle, the latter design would provide feedback on job performance and could motivate the assembler to improve performance. Such intangible factors were not included, however, because there is, apparently, little data that describe these phenomona. Richards' study related to electronic products in general, but could be considered relative to PCBs.

Trends in testing philosophy and in the capabilities offered by automatic test equipments are reviewed in *Assembly Engineering* (1985). In addition, this article categorizes a variety of catastrophic defects that may affect PCBs and provides further insight into the capabilities of current and future equipment.

Recently, Buzacott and LeBlanc (1986) have studied the management of test/rework operations. They applied priority queuing models to study the effects of different sequencing rules at test stations. For example, should tested, reworked PCBs be sequenced before others that have not yet been tested?

ASSEMBLY SYSTEM DESCRIPTION

Generally, the flow of materials in PCB assembly systems starts with the receipt of components from vendors. Components may be tested, prepared for assembly (e.g., by tinning leads), and then inventoried until needed. Subsequently, a set of components may be withdrawn from inventory and accumulated for used in producing a particular customer order. Preassembly operations then prepare the board and components for assembly, perhaps by bending component leads or by affixing them in a particular sequence on a tape to allow automatic insertion.

A variety of assembly methods may then be used, depending primarily on production volume and the type of components involved. High-volume assembly can justify "hard" automation equipment to insert components. Mid-volume production systems may employ semi-automatic insertion machines or "flexible" automation (e.g., robotics). Low-volume production may be accomplished manually. In fact, even though overall production level is relatively high, a PCB may require insertion of only a few "odd form" components by exceptional (e.g., manual or, perhaps, robotic) methods.

Insertion equipment is generally specialized to a particular type of component, of which there is a wide variety, including DIP, SIP, axial, radial, and cans. In addition, the new surface-mount technology (SMT) requires that components be "onserted" rather than inserted. Most systems use the flow- (or wave-) soldering process, although different processes may be used, depending on the type of component (e.g., SMT components may be soldered by a vapor phase process) or production volume (e.g., manual methods may be used in low-volume situations). Some components may be heat-sensitive and must be inserted and soldered by hand after flow soldering.

After soldering, boards are cleaned and sent to in-circuit, then functional, testing. Defective boards are reworked, then retested. Finally, completed boards may be coated before shipment to final assembly. Throughout the assembly process, components and partially assembled PCBs are transported by material handling equipment and stored, perhaps in potentially degrading environments.

The primary purpose of this discussion is to emphasize the myriad of details that may influence quality. There may be no single, best way of producing PCBs—system design varies from facility to facility. Even within one facility a variety of methods may be used, depending upon production volume, equipment available, and product design. Nevertheless, a general approach may be developed to model the quality features that are prevalent in most systems, as well as others which may assume specific forms in each particular production system. Such an approach will be described.

MODELING APPROACH

The philosophy of modeling described in this chapter relies on making suitable abstract representations of component defects, test capabilities, and rework procedures. Abstraction is necessary to cope with the myriad aspects of system composition and operation as described previously.

All types of defects that might occur must be identified and categorized. *Assembly Engineering* (1985), for example, lists seven types of catastrophic faults: out of tolerance, dead, or partially dead device, wrong component, backward component, or missing component, and shorts and opens. Groups separated by semi-colons may likely be caused by the component (or, perhaps, damage during soldering), the assembly process, or the soldering process, respectively. This list is correct, but not rich enough to categorize all types of defects, because some which should be corrected may not be catastrophic and many— particularly those involving component and circuit function—may have logical relationships among each other. For example, the capability of performing one function may not indicate the ability to perform another (e.g., the integrity of one memory location may be independent of another), one defect may "mask" another, or one defect may preclude another (e.g., a missing component cannot also be inserted backward). Dooley (1983) describes other types of logical relationships that are possible.

The types of components and PCB circuit complexity also influence the number and type of defects that may occur, as well as the difficulty involved in testing for them ("test" is used as a comprehensive term, including both inspection and testing). Each specific process selected for preassembly, assembly, and soldering has a characteristic proclivity of introducing defects of certain types, as does equipment used for material handling and storage. Significant changes in defect occurrence may be made by upgrading processes. Important, but perhaps less comprehesive, changes can be made in a given process, for example, by training operators, computerizing quality checking, or introducing some other type of process control. Note that process control could be viewed as adding a testing station to screen for defective products. However, process control is philosophically different from testing, and the distinction will be made throughout this chapter.

A complete categorization would identify a set of defects that may be introduced by each type of process, a set of defects to which each type of component is vulnerable, and a mapping from the former to the latter. PCB quality must include the quality of all the individual components that populate it.

Even though similar types of defects may occur in all PCB assembly shops, each produces a different set of products and may, therefore, require a unique QC plan. Considering the difficulty of collecting shop-floor data and the sometimes imprecise identification of defects, accurately estimating the rates at which defects occur can be difficult. To be useful in devising quality plans, quantitative models must, therefore, permit sensitivity analysis to evaluate the impact that individual defect rates have on the structure of the QC plan.

Testing may be unreliable failing to detect existing defects, as well as identifying defects which, in fact, do not exist. Furthermore, each test must be designed to detect certain types of defects, and each test station must be assigned a specific set of tests to perform. Partial function tests may be designed purposefully, because a limited number of functions may be good indicators of other capabilities. However, if the untested defect exists, much time may be required at a downstream test (e.g., at the system level) to diagnose the problem and pinpoint the responsible component.

Defective components are usually scrapped, but PCBs are typically reworked—they are not scrapped routinely. Rework operations are generally done manually and require a high level of skill. They may entail repair, replacement, or other activity. A rework operation may fail to correct the identified defect and, worse yet, it may introduce one or more defects. For example, additional defects may be introduced by handling, by poor control of the heat from hand-soldering, or by inserting the wrong component.

Virtually any type of defect may be introduced at rework. This error potential creates a real problem, because reworked boards are typically retested at the current location, but they may not be recycled upstream to be completely retested. Some of these defects may cause problems, others will be passed along to system-leveling testing, and/or the customer. Because of these intricacies, rework is a crucial process that must be considered carefully in the QC plan.

Specific assumptions that are invoked to structure models are as follows:

1. Multiple types of defects may be incurred.
2. Defects are added independently at a station and among stations (i.e., defects added at one station do not affect the number added at another station).
3. The presence of one type of defect will not affect the probability that another type of defect will be present (unless there is a logical relationship between the two types of defects).
4. Tests are not perfect.
5. Rework is not perfect.
6. All rework is done at specified rework stations apart from assembly stations.
7. Components are available for replacement of defectives at rework stations.

MODEL DESCRIPTION

The model is described using the following indices:

i = 1,2,....,I board type
j = 1,2,....,J component type
k = 1,2,....,K defect type
l = 1,2,....,L (t) index for test/rework cycles at test t
m = 1,2,....,M index for processes
t = 1,2,....,T index for test operations,

sets

$D(kt)$	=	the set of defects that can be identified by test t
$N(k)$	=	set of components vulnerable to defect type k
$O(km)$	=	the set of defects that may be introduced by process m
$U(jk)$	=	the set of defect types to which a component of type j is vulnerable
$V(jm)$	=	the set of vendors that supply component type j

and other notations, in order of presentation,

$r(jkm)$	=	Pr (component of type j does not incur a defect of type k at process m)
$v(jm)$	=	portion of component j requirements supplied by vendor m
$P(jk0)$	=	Pr (component of type j is good with respect to defect type k from vendors)

Q(jkt)   =   Pr (component of type j passes test *t* with respect to defect type k)

e(jkt)   =   Pr (test calls component good | the component is good)

f(jkt)   =   Pr (test calls component good | the component is bad)

S(jk*t*)   =   Pr (component of type j passes test t* with respect to defect types k*)

I(j)   =   Pr (component of type j will be scrapped at incoming test)

g(jk)   =   Pr (component type j is good with respect to defected type k)

h(j)   =   Pr (accepted component of type j has no defects)

P(jkm)   =   Pr)Component j is good with respect to defect type k after process m)

g'(jklt)   =   Pr (component of type j is good with respect to defect type k incoming to cycle 1 at test t)

a(jklt)   =   Pr (component of type j is accepted with respect to defect type k on cycle 1, test t)

b(lt)   =   Pr (accept board in cycle 1 of test t | 1-1 cycles occurred)

The model is based on the particular view of defects, components, boards, assembly and rework processes, and test operations described previously. Defect types must be defined to be mutually exclusive and collectively exhaustive. Each type of PCB consists of a set N' of components that must be analyzed to identify subsets of components which are identical with respect to the defect classification, as well as processing, test, and rework characteristics. Each of these mutually exclusive sets is composed of components vulnerable to some defect type, k, [N (k)]. A particular type of component j can be associated with different types of defects, because it can fulfill different logical functions. The quality control plan can be evaluated only by modeling the quality of all component types in the set N' for each type of board (an i subscript to indicate board type is omitted throughout for clarity of presentation).

Each process, m, can only add defects—perhaps several types of them [O(km)]. For example, an assembly operation may insert the wrong component or damage another already in place; soldering may result in voids as well as heat-damaged components. Assume that the incidence of different defects is independent at each process and among processes and that r(jkm) describes the quality achieved by process m. Processes include component vendors, preassembly, assembly, soldering, cleaning, coating, burn-in, handling, and rework.

Each test station, t, is capable of detecting a certain set of defects (D(kt). This set may be composed of some defects that are the primary focus of a particular test, as well as others which may be identified as a by-product. A test may be fallible, specifying a defect when in fact it does not exist or failing to specify an existing defect.

If a defective component is identified at incoming test, it is not allowed to enter the assembly process. However, defects identified at other test stations may be corrected by rework operations, which repair and/or replace the component in question. Rework is a process that may or may not yield an acceptable result—it may also lead to additional defects through handling or poor workmanship.

After rework, the PCB is usually retested. PCB quality, the probability that the board is good with respect to defect type k, may change at each test/rework cycle. It is assumed that the PCB is scrapped if a total of $L(t)$ test/rework cycles does not produce an acceptable board. $L(t)$ can be determined on the basis of either economic or technical considerations.

The quality history of a component can readily be traced throughout assembly. Each vendor that supplies components of type j is a process, m, which supplies portion $v(jm)$ of requirements. Each component supplied has probability $r(jkm)$ of being "good" with respect to defect type k. If components from the set of vendors are pooled for assembly, a measure of the quality of component j from the pool is

$$P(jk0) = \Sigma \ r(jkm) \ v(jm)$$

in which the summation is over $V(jm)$.

Incoming components may be subject to a series of tests designed to eliminate defectives. A test operation will label a component "good" if the component is, in fact, good and is identified as such, or if it is bad and not correctly labeled. The effect of test t can be measured by

$$O(jkt) = P(jk0) \ e(jkt) + \overline{P}(jk0) \ f(jkt) \tag{1}$$

the overbar is used throughout and indicates the complement of a probability — i.e.,

$$\overline{P}(jk0) = 1 - P(kj0).$$

Test effectiveness is indicated by the conditional probabilities e and f. The probability $(1-e)$ is the chance of incorrectly rejecting a good component, a type I error. The chance of a type II error, f, indicates the probability of accepting a component that is actually defective.

The probability that a component will pass a series of independent tests in which each checks for different types of defects is

$$S(jk*t*) = \Pi\Pi \ O(jkt)$$

in which the first product is over all tests in the set t*, and the second is over the set of defects, D(kt), that may be identified by test t. The probability that a component will be scrapped at incoming test is

$$I(j) = 1 - S(jk*t*).$$

Even though the component passes the series of tests, it may not be good because tests are fallible. The probability that an accepted component of type j is "good" relative to defect k after screening at incoming test is

$$g(jk) = P(jk0) \, \Pi \, e(jkt)$$

in which the product is over the tests which check for defect type k. Considering all types of defects to which component j is vulnerable, the probability that it has no defects after the series of tests is

$$h(j) = \Pi \, g(jk)$$

in which the product is over the set U(jk*). The conditional probability

Pr [the component is good with respect to all k* defect types | it is accepted by tests t*] $= h(j)/S(jk*t*)$.

Accepted components are stored in inventory and subsequently withdrawn to form a kit of components required to assemble PCB type i. Handling, storing, and accumulation are all processes that may introduce defects. In addition, all preassembly, assembly, soldering, and cleaning operations are processes that have the same structural effect on product quality. Presumably, existing defects have no effect on processes and are not detected at processes.

In general, the effect of any process, m, is to alter the probability that a component of type j is "good" relative to defect type k,

$$P(jkm) = Pr[\text{component } j \text{ was good before process } m] \, r(jkm).$$

Expanding this, the quality of the board is described by

$$A(km) = \Pi\Pi \, P(jkm)$$

in which the first product is over the set that is the intersection of U(jk) and O(km) and the second is over all affected components, N(k).

After a series of assembly processes, m*, the quality of a component of type j becomes

$$g(jk) \leftarrow g(jk) \, \Pi \, r(jkm) \tag{2}$$

in which the product is over the m* processes.

After assembly, PCBs are subjected to in-circuit and function tests and subsequently to a systems-level test. Each of these tests, t, is designed to detect certain types of defects (although others may be identified as by-products). If test t rejects a board for any reason, it must be reworked and subsequently retested. A total of L(t) test/rework cycles are permitted at test t; presumably a board rejected L(t) times will be scrapped.

Rework is a process that requires a high level of skill. It may involve repairing and/or replacing components and may introduce errors that did not exist or fail to repair identified defects. Wilhelm (1986) shows that these complex processes can be modeled by expressions of the following form.

Component quality incoming to the first cycle at test t, which is assumed to check for defect k, is

$$g'(jklt) = g(jk)$$

in which g(jk) is described by Equation 2. At the 1th test/rework cycle [1 = 2,3,..., L(t)] the board has been rejected 1-1 times, the probability of accepting a component is

$$
\begin{aligned}
a(jklt) &= g'(jklt)e(jkt) \\
&+ \overline{g}'(jklt) \, f(jkt)
\end{aligned}
\tag{3}
$$

and the probability of accepting the PCB is

$$b(lt) = \Pi\Pi\Pi \, a(jklt)$$

in which the products are (respectively) over the sets D(kt), U(jk) and N(jk). The set of probabilities $\{b(lt)| \ 1 = 1,2,...\}$ can be used (Wilhelm 1986) to define a proability distribution, $\{B(lt)| \ 1 = 1,2,...\}$, which gives the probability that the PCB will be accepted on the 1th cycle.

Consider a rework operation that repairs an identified defect such as a solder bridge, a missing component, or a damaged board. Because the test may be fallible, the component may be either good or bad. Due to workmanship, a good component may or may not result. In addition, the rework operator may introduce new defects as a result of handling and/or workmanship.

A component may complete repair in a "good" state if it were bad, identified bad, and repaired correctly; if it were good, misidentified, and repaired "correctly"; or if it were good, identified as such, and subsequently not damaged during repair (of other components). Assuming independence of defects among components, events on the lth rework cycle determine

g'(jk,1+l,t) = Pr[component of type j is good with respect to defect type k on entering test t for the (1+l)st cycle]

which is used in Equation 3 to model the (1+l)st test (with 1 ⟵ 1+l).

The probability that the PCB will be accepted on test cycle 1 and that the component will be good is

$$g(jk) \leftarrow g'(jklt)\ e(jkt)\ b(lt)/a(jklt). \tag{4}$$

Test t, along with accompanying rework, may, therefore, be viewed as a device that transforms incoming g values to those given in Equation 4. Wilhelm (1986) describes changes in g' and B values that are required to model rework operations which involve replacement, either repair or replacement, and the correction of defects that may preclude or be associated with others. Furthermore, system-level tests may be modeled in a similar manner. In-warrantee service may also be viewed as another "test" operation in which failures are recycled through in-plant test/rework operations. In all cases, the structural form of Equation 4 can be used to model test/rework operations.

This description has shown that straightforward probability expressions may be used to model a broad variety of processes, tests, and rework cycles found in PCB assembly systems. The end result is a means of measuring the quality of outgoing PCBs as a function of each element in the system. Selecting a particular set of vendors, processes, process controls, tests, and rework operators will lead to an AOQ that can easily be calculated.

Sensitivity analyses may then be conducted to evaluate alternative OC Plans that may lead to an improved AOQ. In particular, the QC engineer is typically interested (Wilhelm 1986) in maximizing AOQ subject to a constraint on cost (e.g., investment or perhaps cost/PCB) (Rauof, et. al. 1983).

CONCLUSIONS

This chapter has described a unique model that incorporates the following set of features required to measure the quality of PCB assemblies: multiple defect types, multiple components/product, fallible tests, rework workmanship, and component/board quality relationships. In particular, a variety of rework operations may be modeled, including repair and/or replacement and those that deal with logical relationships among defect types.

The models might be applied as an aid in developing a QC plan by evaluating alternative strategies for testing at component, in-circuit, function, and system levels. The impact of test efficiency, as well as that of the skill of rework operators, may be evaluated using models. Perhaps most importantly, the model will evaluate trade-offs to achieve quality objectives balancing testing strategy, process capabilities, process controls, vendor selection, and workmanship.

Suggested Readings

Buzacott, J. A. and D. H. W. Cheng. 1982. Quality models of assembly systems. *Proceedings of the Winter Simulation Conference.* pp. 361-371.

Buzacott, J. and B. LeBlanc. 1986. Quality models of printed circuit board assembly. Presented at the TIMS/ORSA Joint National Meeting, Los Angeles. April 14-16.

Dooley, B. J. 1983. A model for the prediction of assembly, rework and test yields. *IBM Journal Res. Devel.* 27(1): 59-67.

Raouf, A., J. K. Jain and P. T. Sathe. 1983. A cost-minimization model for multi-characteristic component inspection. *IIE Transactions.* 15(3): 187-194.

Richards, J. 1984. Integrated assembly, testing and rework in automated electronics manufacturing. *1984 Annual Industrial Engineering Conference Proceedings*, pp. 495-501.

Schwartz, W. 1985. Soldering, cleaning, test and rework in electronics assembly. *Assembly Engineering.* 28(5): 66-74.

Strumar, A. J. 1985. Flexible manufacturing systems for electronics. *International Electronics Assembly Conference Proceedings.* Institute of Industrial Engineers. Norcross. October.

Wilhelm, W. E. 1986. Optimal quality plans for small-lot assembly of printed circuit boards. Working Paper. Department of Industrial and Systems Engineering, The Ohio State Univ. January.

This paper is based on work supported by the National Science Foundation under grant number DMC-8500898.
Reprinted from 1986 International Electronics Assembly Conference Proceedings

Wilbert E. Wilhelm is an Associate Professor in the Department of Industrial and Systems Engineering at The Ohio State University.

4

Much Ado About Productivity: Where Do We Go From Here?

D. Scott Sink

Is the recent, often-frenzied search for the secret of productivity and how to increase it really "much ado about nothing?" Consider, if you will, the following statements:

- Productivity is a concept American managers are charged up about because they have lacked the discipline, fortitude and skills in the past 20 years to realistically define, measure and manage performance.
- American managers have become complacent and are searching for easy, simple answers for complex problems.
- Productivity is looked to as a panacea for all our various performance problems.
- Productivity, the Fountain of Youth, the Golden Fleece, Shangri-La, the Emerald City of Oz, the pot of gold at the end of the rainbow and, of course, Japan's secret to "success" are all illusions that American managers seem determined to continue to search for.
- Productivity ≠ performance.
- Performance is a broad concept comprised of at least seven criteria: effectiveness, efficiency, quality, productivity, quality of work life, innovation and profitability.

- No two managers or organizations will or necessarily should weigh equally all these criteria.
- Moreover, performance will probably be operationalized uniquely in different organizations; it will mean something different, it will be measured differently and it will be managed differently.
- It is not an easy job to define performance, or to manage it. It is easy to lose sight of what performance means; it is an elusive goal.
- Many organizations and managers are "in search of excellence;" few in America have achieved it consistently in the past 20 years.
- Performance in America in the 1960s and '70s, in many firms, came to mean short-term profits. Performance was reduced to simply a matter of raising prices faster than costs (i.e., putting it to your customers faster than your suppliers put it to you). The price recovery process became a major management tool in the '60s and '70s.
- American managers speak with an imprecision about performance and productivity. This imprecision has caused clouded, misguided, inconsistent, poorly thought through operationalizations of performance and productivity measurement, evaluation, control and improvement at critical lower levels in the organization.
- Most managers give only lip service to productivity measurement.
- We do not lack creativity and invention in American organizations either in product or process. We do lack innovation (applied, implemented creativity).
- "What we need to learn from the Japanese is not what to do, but to do it." (Peter Drucker)
- You do not have to measure productivity to improve productivity.

FROM AWARENESS TO CHANGE

I hope these opening statements arouse emotional reactions. Self awareness is the first step toward positive change. We need to realize that "we have met the productivity enemy and he is us."

In the past eight years I have been researching, consulting, writing and lecturing on the topic of productivity management. I have worked with managers from organizations of all types and sizes who have come to my seminar entitled "The Essentials of Productivity Management: Planning, Measurement, Evaluation, Control, and Improvement" in search of answers and solutions. There is nothing wrong with that; however, a large number of those managers have been looking for answers to the wrong questions.

This chapter will focus on productivity basics. I will present a very disciplined view of productivity and focus on how productivity relates to the broader concept of performance. The chapter will perhaps raise more questions than it provides answers for. That is not all bad. In my opinion many managers have lost sight of the forest for the trees. I hope this chapter will help clarify the issues.

Productivity ≠ performance. Conceptually, at least, I don't suspect there would be much argument with this statement. Yet my experience suggests that most managers operate on the assumption that productivity does equal performance. The basic problem with this is that the term performance is almost always defined in a very broad, multi-attribute fashion; for example, "the performance of an individual is comprised of a relationship between ability, effort and attitude."

Productivity, on the other hand, is a rather well defined and delimited term representing a relationship between the outputs produced by an organizational system over some period of time and the inputs required to produce those outputs.

I am convinced that many of today's managers' problems arise from a lack of clarity and discipline in relation to what they actually want the systems they are managing to do. This chapter will examine the relationship among these various components of organizational performance and discuss the role productivity measurement might play in the design of an organization's performance control systems.

BASICS OF PRODUCTIVITY

Productivity is an extremely abused and misused term. This is because there has been no disciplined attempt to stand up and say, "that is what it is and that's all it is." The "half-truth" rhetoric floating around about productivity is absolutely amazing. It has become such a significant buzz-word that almost every discipline and profession imaginable has grabbed onto it and begun to use it in an attempt to further market and promote its own often myopic "solutions." The need for synthesis, clarification, disciplined definitions and a generic conceptual framework is quite evident.

In order to provide adequate conceptual development, the general management process, organizational systems performance measurement/control basics, and productivity basics will be briefly examined. The objective is to assist you in more clearly viewing the relationship between productivity, performance and control systems design and development.

Whether you are a first line supervisor or a chief executive officer, your job entails all the components of the general management process depicted in Figure 1. Organizational systems performance measurement, evaluation and control represents a critical component in that general management process. There are obviously many ways to classify control systems. They can be classified in terms of the resource they are intended to manage. Financial control systems (accounting, comptroller, budgets, etc.), production control systems

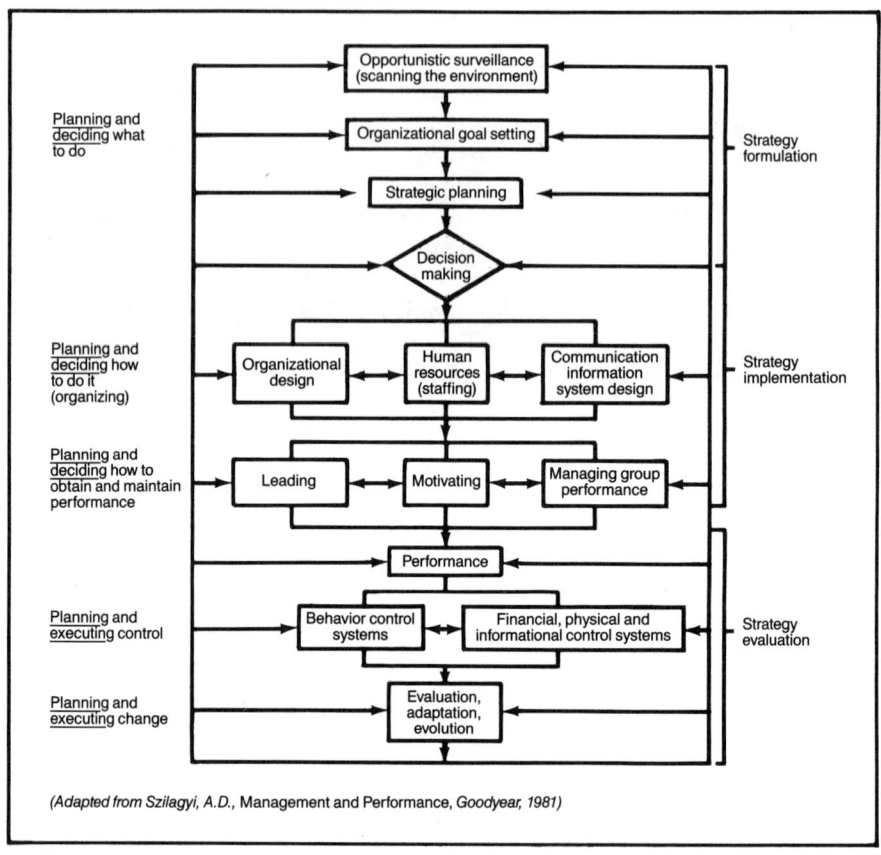

Figure 1. The general management process.

and behavioral control systems (performance appraisal, policies and procedures, etc.) are examples of this type. We also classify control systems with respect to the type of organizational system performance they are attempting to control or manage.

Seven Performance Measures

In general, there are at least seven distinct, although not necessarily mutually exclusive, measures of organizational systems performance. They are:

- Effectiveness
- Efficiency

- Quality
- Productivity
- Quality of work life
- Profitability
- Innovation

Every organization, every manager in one form or another monitors, evaluates and controls one or more of these seven measures.

Note that productivity is only one measure of performance for a given system. It is *not* clear that it is the most important, or even necessarily a critical, measure of performance for all systems.

The problem of designing a control system is a multi-attribute or multi-criterion one. No two organizational systems or managers are likely to weigh these measures equally. Moreover, the characteristics of the actual development process of a control system will vary significantly from organization to organization and from manager to manager. Some will be very explicit, rational, systematic and pragmatic while others may be very implicit and subjective. Some will focus on only one measure of performance, while others will truly be multi-attribute in character. Some systems will be the result of an explicit, systematic, consciously thought through design process while others will be characterized more by a "random walk" process comprised of add-ons to an inherited system.

One important job of a manager is to determine what the appropriate priorities or relative weights are for each performance measure; how to measure, operationally, each measure; and how to link the measurement system to improvement (in other words, how to most effectively use the control system to bring about appropriate changes or improvements).

The priorities or weightings for each of these performance criteria will vary according to several factors:

- Size of the system;
- Function or objectives of the system;
- Type of system;
- Maturity of the system in terms of management, employees, technology, organizational structure and processes;
- The environment (political, economic and social characteristics).

Managers at all levels and in all organizations are striving to achieve results through people. People are at the heart of successful accomplishment of objectives and of effective and efficient use of resources. C.I. Barnard (see Suggested Readings) states that "essential to the survival of an organization are the willingness to cooperate, the ability to communicate, and the existence and acceptance of purpose." There is an appealing simplicity to his perceptions.

In 1953, Peter Drucker identified seven key result areas. He pointed out that in order to be successful in the long run, an organization would have to manage these seven key result areas.

In 1982, Peters and Waterman, in their book entitled *In Search of Excellence*, identify several attributes that "excellent" firms in the U.S. exhibit and manage carefully. The accompanying sidebar summarizes those eight attributes.

One might conclude that the seven organizational system performance criteria presented at the outset of this section are at the heart of both Drucker's and Peters' and Waterman's observations. Figure 2 shows the relations among these three conceptualizations of organizational system performance. For any size or type of organizational system, these seven performance criteria represent the basic areas that managers, supervisors, presidents, vice presidents, directors, etc., should be focusing their management efforts on.

THE MANAGERIAL TASK

The basic management functions of planning, organizing, leading, controlling and adapting exist at all levels of management and supervision. Every manager and supervisor has the task of controlling the systems performance he or she is

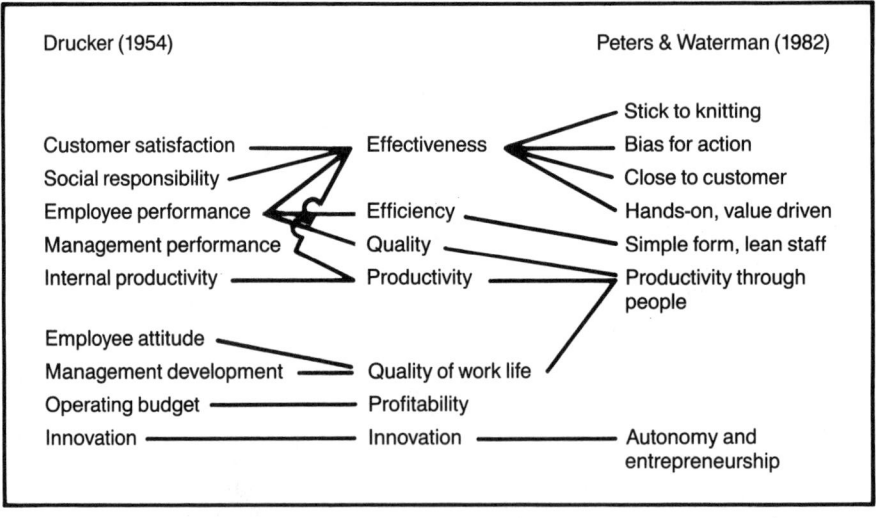

Figure 2. *Organizational systems performance criteria.*

responsible for, and should be held accountable for performing this management function in a professional, effective and efficient fashion. The control systems managers design, develop, implement and maintain should not only monitor organizational system performance, but also indicate when, where and perhaps why performance is or isn't what it should be.

When we speak of control systems, we are not referring to quality control systems, cost/budget control systems and the like that have been designed to control specific resources or results of production and to react to situations rather than manage them in a proactive sense. When we speak of control system design, do not allow yourself to develop a stereotyped image of that system. Let's not fall into the same trap we did with quality control. Note that Japanese quality management control systems are designed to proactively build it right the first time.

By control systems design, we mean a combination of monitoring subsystems integrated with proactive, in-process control and feedback systems. Such systems would involve all levels of management, supervision and employees in a way that would increase the probability that behaviors would change for the better.

It is clear that these concepts are new and different, and therefore will encounter resistance. The solution is not simple; however, these concepts can and will work if developed correctly.

Most managers and supervisors inherit a "control system" and simply make it work, or they "band-aid" one together. A serious effort to rethink the purpose and logic of control systems needs to take place. Furthermore, the impact of these control systems on behaviors surely needs examination in American organizations. We have a habit of "rewarding A while hoping for B." A simple analogy will help to clarify these concepts.

Imagine yourself as the designer for the control/instrument panel of a Boeing 747 jetliner. Think through some of the issues you would consider in terms of deciding:

- What needs to be on the panel?
- Where does each instrument, dial, knob, indicator need to be in relation to the pilot and copilot?
- How big does each instrument need to be?
- How can we ensure the proper relationship between system performance indicators on instruments and physical control by the pilot with knobs, dials, control stick, throttle, etc.?
- How can we design a control system that does not overload the pilot (manager)?
- What control aspects can we automate so as to relieve the pilot (manager) from the routine decisions?

An airplane has categories of performance measures or criteria—navigational, communication, engine performance, aircraft attitude and, of course, control response. Your organizational systems also have categories of performance criteria—effectiveness, efficiency, quality, productivity, quality of work life, innovation and profitability.

THE PRODUCTIVITY RELATIONSHIP

Productivity is a relationship (usually a ratio or an index) between output (goods and/or services) produced by a given organizational system and quantities of inputs (resources) utilized by the system to produce that output. As you can see, it is a very simple concept. You take what a given organizational system produces or creates, quantify it and put it in the numerator of an equation. You then take the specific resources (labor, capital, materials and/or energy) utilized to create those outputs and put them in the denominator of the equation. You end up with an operational measure of the concept of productivity.

It seems so simple and easy, and often it is. For instance, if an organizational system has clearly measureable outputs and identifiable and measurable inputs that can be matched temporally to the production of outputs (this would imply, for example, reasonably short cycle times), productivity measurement is quite routine. There are models and programs available that will even crunch out all the ratios for you.

However, if outputs are a little hard to measure, input resource utilization is hard to match up with outputs for a given period of time, input and output mix or type is constantly changing, data are difficult to obtain or not even available, etc., as is often the case, productivity measurement can be difficult and frustrating.

As a simplifying mechanism for making the presentation of productivity basics more efficient, Figure 3 has been developed. This illustration depicts the basic productivity management process, incorporating the definitions and concepts presented to this point.

Starting at the top of the figure, you will notice the planning and decision process denoted by the diamond-shaped symbol. Directly underneath that is a basic systems flow model for an organizational system. Input variables flow into the system and are transformed into new states represented by goods and/or services which are outputs, and these outputs are then delivered to "customers."

Note that input variables are procured by some "procurement function" and carry with them quality attributes, quantity attributes and financial attributes. That is, the input variables or resources come into the organizational system with specific "price tags," with certain quality characteristics and in specific volumes or quantities. For the time being, we will treat the transformation process(es) as a "black box," keeping in mind that—particularly for productivity

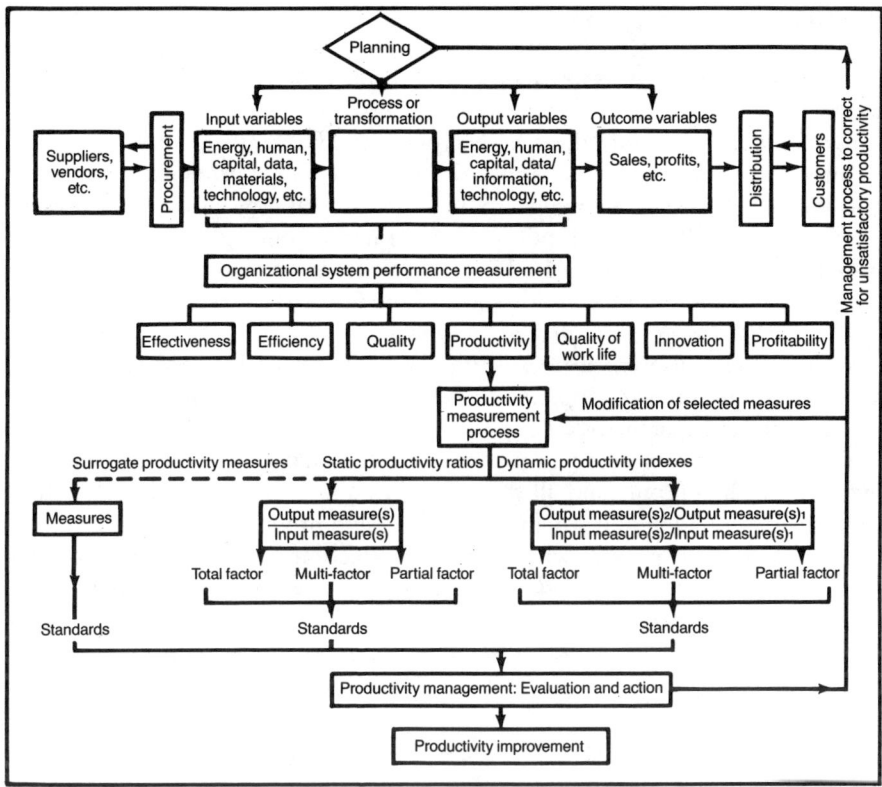

Figure 3. *Productivity measurement process.*

improvement—we will need to make the "black box" a "glass box." That is, we will have to understand and analyze specific transformations in order to improve productivity.

Output variables, or transformed input variables, also carry with them quality attributes, quantity attributes and financial attributes. That is, the output(s) (goods and/or services) have associated levels of quality, have "price tags'" and come out in specific volumes and quantities.

Outcome variables reflect the results achieved after the output has been successfully distributed to "customers" (sales, profits, customer satisfaction, etc.). It is particularly important not to confuse outcomes with outputs. Persons in the organizational system have control over the attributes of the output; however, they often lack control over outcomes. Keep definitions clearly in focus.

RATIOS AND INDEXES

The productivity measurement process is broken down in the lower part of Figure 3.

Conceptually, there are three basic types of productivity measures. First two major categories: *static productivity ratios* and *dynamic productivity indexes.* You recall that we are defining productivity as a ratio of output measures to input measures. If all the output and all the input from a given organizational system get into the equation, we have a *total factor static productivity ratio.* If some or all of the output and only some of the input get into the equations, we have a *partial factor static productivity ratio.*

We call these ratios static because they are like a snapshot. For instance, we take a snapshot of what happened in July 1982 for a given plant and construct the ratio(s) from that snapshot.

The other basic measure of productivity is called a *dynamic productivity index.* If all the outputs and all the inputs from a given organizational system get into the equation and we also compare static productivity ratios from one period (say a base period) with current productivity ratios, we have a *total factor dynamic productivity index.* If, however, we manage (for whatever reason) to capture all or some of the outputs and only some of the inputs in the two ratios, we have a *partial factor dynamic productivity index.* An example of how a dynamic productivity index would be structured is as follows:

$$\frac{\text{Outputs 1982/Inputs 1982}}{\text{Outputs 1978/Inputs 1978}}$$

A third measure of productivity, which does not adhere strictly to the common definition of the term, is called a *"surrogate" productivity measure.* A surrogate productivity measure represents factors that are not included in the concept of productivity, but are highly correlated with productivity (customer satisfaction, profits, effectiveness, quality, efficiency, etc.). Most managers today operationalize productivity in this manner.

After the productivity measurement process is completed, data can be collected and management can begin to get a feel for the range and variability of the numbers. Standards can be established using norms (industry or internal), engineered methods, historical data, etc.

Evaluating the Data

The productivity improvement process in its simplest form is one of evaluation of productivity data (although in most cases today we are simply evaluating performance data, not necessarily productivity data) and planning managed interventions that you suspect will improve productivity and, eventually, other performance measures. Many managers today muddy the waters with respect to productivity because they do not carefully delineate the differences between productivity definitions and performance concepts, productivity measurement and performance measurement, productivity improvement and performance improvement, etc.

Interestingly, in a typical audience of 100 managers, if you ask the question "how many of your organizations measure *productivity*?" approximately 95 will raise their hands. If you then ask them to show you or tell you exactly *how* they measure productivity you find that fewer than five of them actually are measuring it as we have defined it. Close examination reveals that most of their measures are efficiency measures, quality measures, profitability measures, etc., not productivity measures.

Is there anything wrong with this? Not necessarily. They may well know what performance means and what it takes to compete and survive. They may well be measuring, evaluating and controlling those factors it takes to compete and survive.

However, the era of management just being an art is over. Management is truly an art *and* a science today. In order for it to progress as a science, and for management to succeed in the '80s and '90s, there will have to be greater precision of terms, concepts, techniques and approaches.

I am personally disheartened when I hear a Ford manager tell me that quality and productivity are synonymous (or, as they put it, "Quality is Job 1"). That is nice rhetoric and good PR for the American public, and it may even be appropriate given the tarnished image American cars developed during the 1960s and '70s. But in a group of managers talking about performance and productivity, we need to be more precise with our terms.

Just to set the record straight, I am not just picking on a Ford manager. I could easily insert a GM manager and employee involvement, a Motorola manager and participation, etc.

Productivity in a very, very loose sense has become the call-to-arms for whatever we have performed most poorly at during the past decade or so. Think about the quality image at Ford, Lordstown for GM, the Quasar plant for Motorola, etc.

We sorely need to get back to basics. We can start by asking ourselves these questions:

- What is performance in the organizational system I am responsible for?
- How do we now operationalize (measure, evaluate and control) performance?
- Do we indicate what we really want by what we measure, evaluate, control and reward for?

- Are our performance measurement, evaluation, control and improvement systems doing what we want them to? Are they effective and efficient?
- Is productivity an element that is missing in our performance management systems?
- Would productivity measurement give us new, valuable insights into whether we are effective, efficient, meeting quality specs, innovative?
- Would productivity measurement improve managers' diagnostic capabilities and, therefore, their improvement effectiveness?

Is recent interest in productivity simply much ado about nothing? The answer is both yes and no. Yes in the sense that most managers are really concerned about a broader issue than productivity—a more complex performance concern— but hope productivity is the key that will unlock simple solutions and answers. Many managers are interested in productivity because they are looking for an easy solution to problems with effectiveness, efficiency, quality, quality of work life, innovation and, of course, profitability. This is, of course, naive.

Interestingly enough, many managers come to productivity short courses and seminars with these overly simplistic expectations and return disenchanted and unsatisfied when the instructors' views on the subject are not as imprecisely formulated as their own ideas are.

In many organizational systems, productivity is a relatively lower priority criterion. For instance, I am convinced that in the case of white collar, professional type employees the term "white collar productivity" is a perfect example of a confusion of purpose. Productivity, as strictly defined, is important in white collar areas; however, operationally speaking, performance can be more straightforwardly improved by measuring, evaluating and controlling effectiveness, quality, innovation, quality of work life and efficiency—in that order. If you manage those five criteria of performance well in white collar areas, productivity almost becomes a moot issue. "White collar productivity" is a buzzword developed to sell books, obtain consulting contracts, get papers accepted at conferences and in journals, etc.

On the other side of the coin, the answer to the question raised in this chapter is no. Productivity at a conceptual level is like baseball, hot dogs, apple pie and Chevrolet (Honda?). Who can argue against it? It is important. However, there has to exist some pragmatic, operational reality. What is it really? How do we define it? How do we measure it? Can we become better managers if we measure, evaluate, control and improve it directly?

For some organizational systems, productivity represents a significant gap between past and current organizational system measurement, evaluation and control systems. I am convinced that many (although not all) organizations would benefit from developing productivity measurement systems and integrating them into the total performance management system.

We clearly need less rhetoric about productivity and more disciplined thought and development of productivity measurement, evaluation, control and improvement systems. American managers need to do a better job in the area of strategic planning for performance management—defining what it is, setting two-to-five-year goals and developing effective management systems to achieve those goals. In the context of such a process, and with full understanding of what its role really is—and is not—productivity most certainly is a worthwhile concern.

Suggested Readings

Barnard, C. I. 1938. *The Functions of the Executive.* Harvard University Press. Cambridge, MA.

Drucker, P. F. 1980. *Managing in Turbulent Times.* Harper & Row. NY.

Drucker, P. F. 1954. *The Practice of Management.* Harper & Row. NY.

Morris, W. T. 1979. *Implementation Strategies for Industrial Engineers.* Grid Publishing Co. Columbus, OH.

Peters, T. J. and R. H. Waterman. 1982. *In Search of Excellence.* Harper and Row. NY.

Sink, D. S. 1982. Building a program for productivity management: a strategy for IEs. *Industrial Engineering.* October.

Sink, D. S. 1984. *Productivity Management: Planning, Measurement, Evaluation, Control and Improvement.* John Wiley and Sons. NY.

Reprinted from the October 1983 issue of *Industrial Engineering.*

5

Total Productivity Management for Competitive Decision-Making in the Electronics Industry

David J. Sumanth
Idsa Genie

Today, virtually every industry in the world is affected by the microelectronic revolution that started about 40 years ago. The micro-processor, integrated-circuit technology is evident all around us—it is found in microwave ovens, automobiles, TV sets, cameras, wristwatches, air travel, pacemakers, sports, entertainment, defense, government operations, education, religion, communications, stockmarkets, war, and peace—just to name a few! Yet, except for space electronics and possibly a few others, we have lost a major portion of our technical prowess in consumer electronics. One by one, we seem to be losing the marketshare in the electronics industry. Even in the microchip industry, we cannot claim leadership any more. What is more troublesome is that the infrastructure of these basic consumer electronic products is a foundation for a host of other industrial products and services. For example, robotic capabilities are only as powerful as their microelectronics. The Japanese Fifth Generation Supercomputer Project is not only a threat to the U.S. computer industry, but also to our industrial infrastructure.

In the years ahead, with an already-eroded manufacturing base, the one potent weapon with which we can still make the difference is a refreshing, realistic, and reliable management philosophy. If only all the management

107

concepts taught and practiced during the last 40 years were relevant for the 1970s and 1980s, we would not be so concerned today with competitiveness, standard of living, federal and trade deficits, and the stockmarket crash of October 1987. One such management philosophy has been in the open literature since 1979 (Sumanth 1979), and since then more than 500 students have been exposed to this philosophy at the University of Miami and thousands of others in the practicing world have been taught this philosophy during the past nine years. Students are already having some visible impact in each of their own ways. The time for productivity management is never more urgent than for the electronics industry in America—today!

Productivity management has received unprecedent attention for the last 13 years, but particularly, in the last seven. In 1981, the need for an association of productivity managers was felt so strongly that the American Productivity Management Association was formed in Skokie, Illinois, early in 1982. Several corporations now have productivity directors, productivity coordinators, and even vice-presidents of productivity.

The role of managers in managing productivity has been emphasized by many, including Peter Drucker, who pointed out that "making resources productive is the specific job of Management as distinct from the other jobs of the 'manager': Entrepreneurship and Administration" (Drucker 1980).

Y. K. Shetty (1982) also emphasized that "productivity is the ultimate responsibility of the manager". Yet the task is not simple. There must be a conscious effort by academia and practitioners to develop the necessary philosophy and strategy that would integrate the "total productivity factor" into the existing decision-making processes. This chapter attempts to present an approach toward such an effort.

Before presenting the proposed approach to management decision making, the major techniques that have been used to date, the diversity of organizational goals often pursued by managers, and the importance of management's role in increasingly productivity will be reviewed.

TRADITIONAL MANAGERIAL TECHNIQUES IN DECISION MAKING

Analytical techniques used so far by managers can be categorized into three groups (Massie and Douglas 1973):

1. managerial economics;
2. managerial accounting; and
3. management science.

Managerial economics is management's application of economic principles in the decision-making process. The most common analytical aids in this category include break-even analysis and demand analysis. On the other hand, the analytical aids used in managerial accounting include financial ratio analysis, funds analysis, and tailored-cost analysis.

The management science approach provides managers with conceptual frameworks and analytical techniques based on quantitative concepts.

In implementing these analytical aids, a manager usually has a choice of the following:

1. deterministic models;
2. probabilistic models; and
3. simulation.

Organizational Goals for Managerial Decision-Making

Objectives provide practical direction for an organization and are also the basis for its evaluation of performance. Therefore, they must be set carefully before decisions are made using any of the aforementioned analytical aids. Management needs to have its company's objectives stated and understood clearly to achieve them. As Oxenfeldt, et. al. (1978) pointed out, "Objectives are—or ought to be—integral to the organization's entire process".

Through the years, several authors have identified a set of objectives that a company should strive for. Peter Drucker (1954) considers the following eight objectives important goals:

1. market share;
2. innovation;
3. productivity;
4. physical and financial resources;
5. profitability;
6. manager performance and development;
7. worker performance and attitude; and
8. public responsibility.

Ericson (1969), on the other hand, distinguishes the goals of a business organization as follows:

1. high productivity;
2. industrial leadership;
3. organizational stability;
4. profit maximization;
5. organization efficiency; and
6. organizational growth.

England (1967) identifies four levels of organizational goals based on behavioral importance and content:

1. maximization criteria—organizational efficiency, high productivity, and profit maximization;
2. associate status goals—organizational growth, industry leadership, and organizational stability;
3. intended goals—employee welfare; and
4. low-relevance goals—social welfare.

Even though productivity has been stated as a desirable goal to pursue, Shetty by conducting a survey, has found (1979), that the three most desirable goals recognized by the respondents were profit, growth, and market share.

The survey was conducted among 193 companies from Business Week's list of the largest industrial and nonindustrial firms. The companies were divided into four basic industrial groups: chemicals and drugs, packaging materials (containers and papers), electrical and electronics, and food processing. The results of the survey are based on 82 respondent companies and show that efficiency as a goal appears only in the container and paper industry when the five most frequently cited goals of corporations are listed by industrial groups.

The results compiled by company size indicate that the productivity goal does not appear in the most frequently cited five goals of corporations for which sales exceeded one billion dollars.

Individual corporations may emphasize productivity as an objective, but that it did not appear in the group as a whole is an important observation, considering that "total productivity" is a powerful measure of competitiveness.

Importance of Management's Role in Increasing Total Productivity

According to Drucker (1980), during the last 10 to 15 years, the productivities of all resources have actually decreased in all major developed countries. Yet productivity is one of management's major responsibilities. As previously noted, the number one corporate goal in most companies is profitability, followed by growth. These goals have led to development strategies that stimulate growth and profit without considering how the productivity of the firm is being affected (Shetty 1982).

Shetty (1982) has also pointed out two important aspects of management's role that have been responsible for little growth in productivity. The first is the reward system that encourages short-term performance and penalizes long-term investment. The second is that executives, who are mostly finance-oriented and law-oriented, tolerate and even encourage production systems which are designed to meet growing market demands and financial goals rather than the requirements for highest possible quality and productivity improvements.

English and Marchione (1983) identified the causal variables that have a direct impact on an organization's performance and for which management has control. These causal variables are found in company structure, processes, and leadership. However, management has often failed to recognize their control over these variables.

Need for Total Productivity Emphasis in the Electronics Industry

For almost 110 years, productivity of direct labor has been overemphasized. The scientific management of Frederick Taylor took this emphasis to even greater heights, because of the higher labor costs in his era. Today, most companies still spend millions of dollars setting up time standards for direct labor to "control" that cost. Today, in many cases, direct labor accounts for only about 10% to 15% of the total cost. The costs of human resources—clerical staff, professionals, and managers—have not been "attacked" as vigorously for many reasons, including the inability to define and measure such people's work accurately. What companies often miss seeing is the "big picture." For electronic products for which material and component costs can be 45% to 50%, why don't managers bring together all employees and vendors as a team and reduce such costs, rather than wasting hundreds and thousands of man-hours justifying the time standards? There are very important places for time standards, in manpower planning, equipment requirements planning, assembly line balancing, etc. But using time standards to control labor performance, in our opinion, has been one of the greatest blunders of American management. Overemphasis on such standards has distracted management and even industrial engineers from concentrating their energies, time, and efforts on improving product and process quality, reliability, effectiveness, and innovation. The principal author of this chapter, after 20 years of industrial engineering experience, has seen how soon the time standards become obsolete and how they prevent workers from participating in a creative, innovative process of what he calls "improving the improved." The incentive plans established for direct labor several years ago, in some companies, are still a source of litigation and arbitration. The final output is generated not just by direct labor, but by the joint impact of other human, material, energy resources, and capital. Unfortunately, the labor productivity measure and such other partial productivity indicators have ended up giving incentives to the wrong groups, blaming the wrong people, increasing the layers of middle management, and unnecessarily making the operation complex and wasteful.

Proposed Approach to Management Decision Making

In view of the importance of management's role in increasing "total productivity," there is need for a type of management style that considers total productivity a primary goal. "Total productivity management" is suggested here as a supplement, or even an alternative, to current management decision-making approaches.

Total productivity management (TPMGT) is a "formal management process," involving all levels of management and employees with the ultimate objective of reducing the cost of manufacturing, distributing, and selling of a product or service through an integration of the four phases of the Productivity Cycle, namely productivity measurement, productivity evaluation, productivity planning and productivity improvement (Sumanth 1980). This formal productivity management concept is based on the total productivity model developed by Sumanth (1979, 1985). The model is summarized in Table 1. It emphasizes total productivity as opposed to the traditional, exclusive emphasis on partial productivity measures such as labor productivity (output per man-hour). The definition of total productivity, as used in this model, is the ratio of sum of total tangible output elements (in physical or value terms) to the sum of total tangible inputs (in cost terms). Figures 1 and 2 show the elements of tangible output and tangible input elements, respectively. The model also defines a break-even

TABLE 1 – The Total Productivity Model
Sources: Sumanth (1979; 1984; 1985)

$$TPFt = \frac{OF_t}{IF_t} = \frac{O_{it}}{I_{it}}$$

TP_{BEP} = Break-even point of total productivity

$$TP_{it} = \frac{O_{it}}{I_{it}}$$

PP_{ijt} = Partial productivity of product (operational unit) i in period t with respect to input type j

$$PP_{ijt} = \frac{O_{it}}{I_{ijt}}$$

j = Input type: H, M, C, E, X

H = Human input

$$TP_{BEP} = 1 - \frac{\text{working capital}}{\text{total input}}$$

M = Material input

WHERE,

C = Capital input

OF = Total tangible output of the firm

E = Energy input

IF = Total tangible input of the firm

X = Other expenses input

O_i = Tangible output corresponding to product (operational unit) i

t = Given period

I_{ij} = Tangible input of type j corresponding to product (operational unit) i

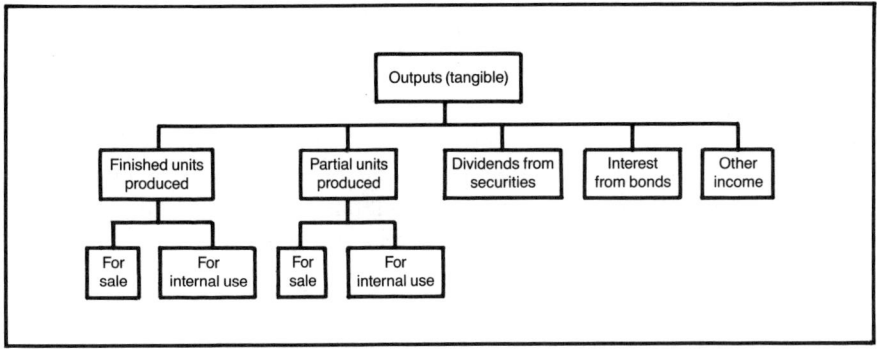

Figure 1. Components of tangible output in the total productivity model.

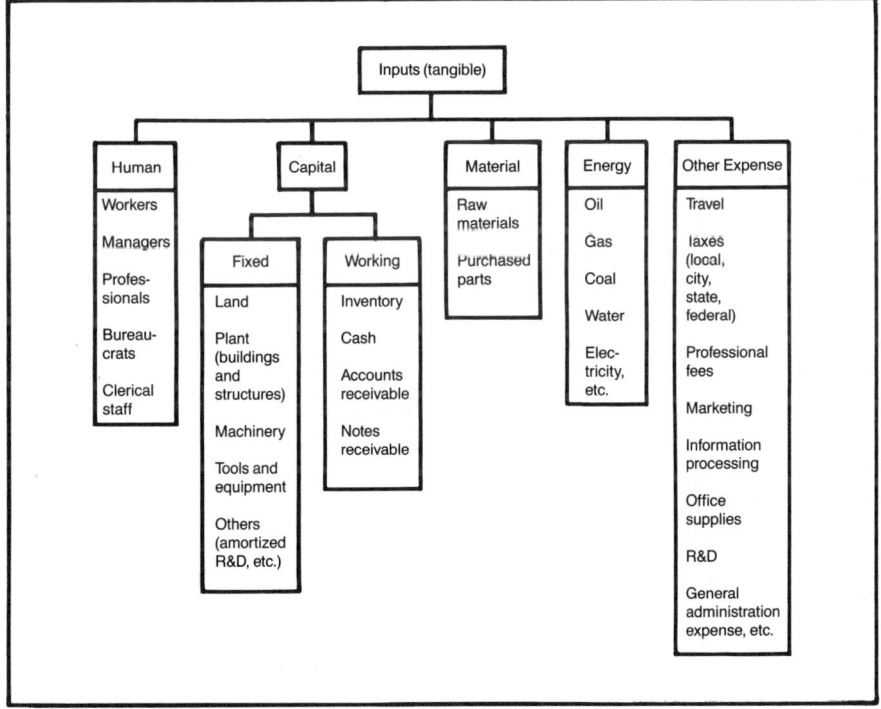

Figure 2. Components of tangible input in the total productivity model.
Sources: Sumanth (1979; 1984; 1985; Ch. 8)

point for total productivity, assuming linearity. Figure 3 shows the relationship between productivity-oriented profit and total productivity. The practical implication of this relationship is that a company can determine the level of total productivity needed to obtain certain a level of profit and vice-versa. The management of the firm can objectively plan its profits based on the ability to reach certain preestablished targets of total productivity. The management strategy to use TPM is shown in Figure 4. Some of the practical principles of TPMGT are synthesized in Sumanth's (1986) previous work.

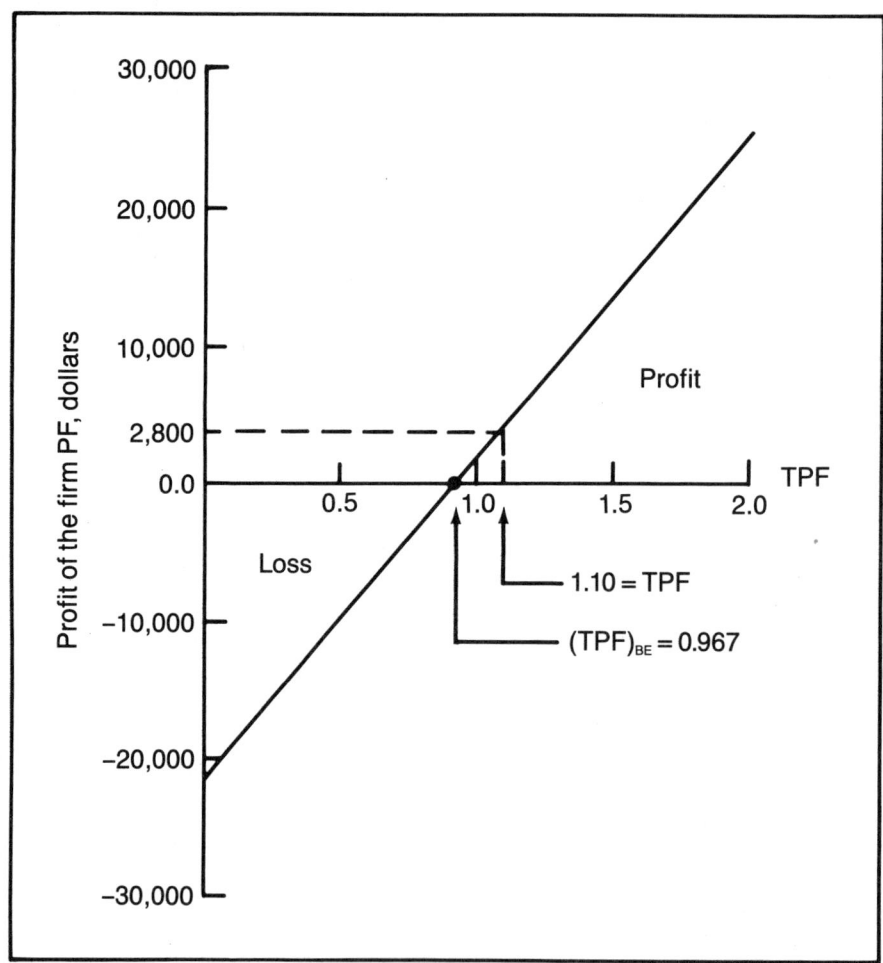

Figure 3. *Total profit of a firm as a function of its total productivity. (PF = Firm's profit; TPF = Firm's total productivity) Sources: Sumanth (1979, 1984, 1985; Ch. 8)*

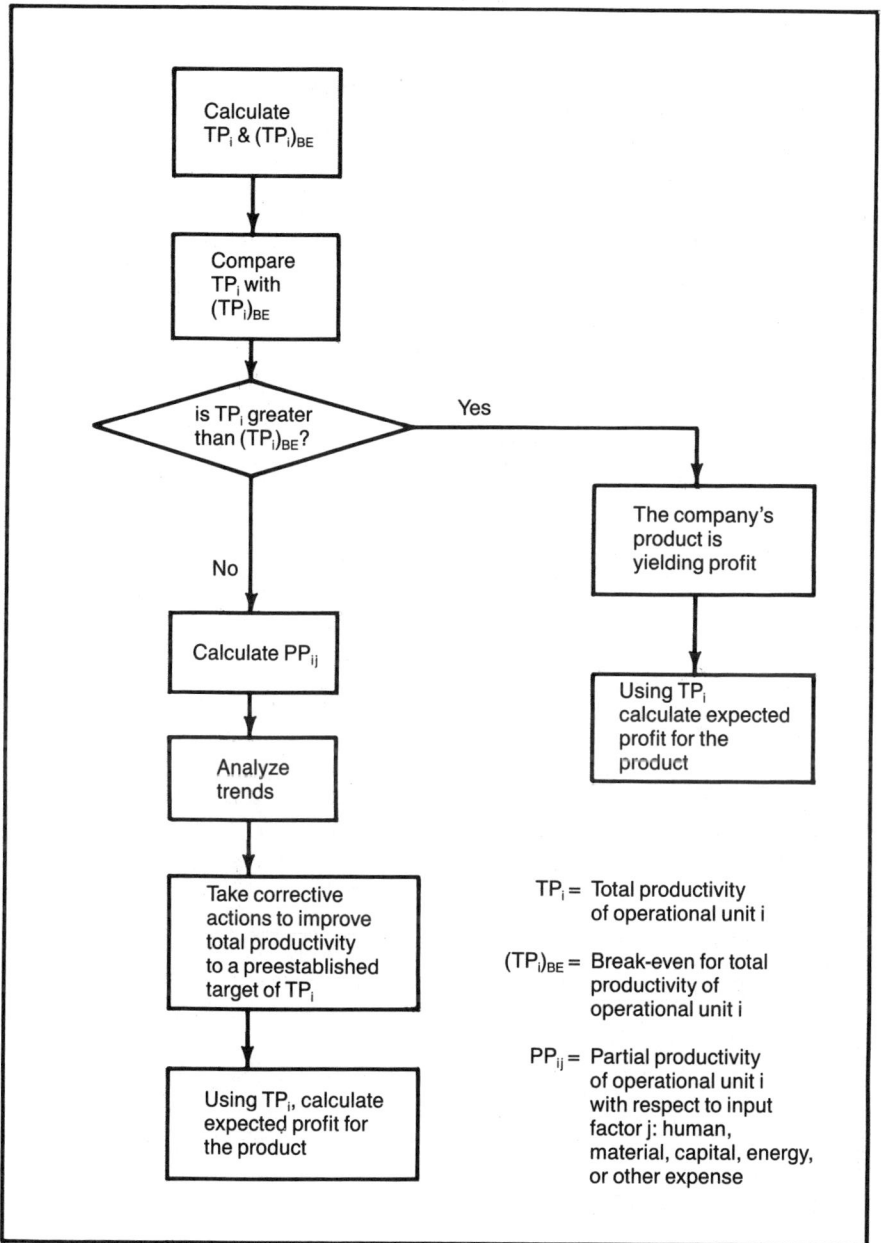

Figure 4. *Flow chart of total productivity approach.*

Relationship Between Total Productivity and Other Management Goals

Figure 5 shows how traditional management goals can be affected by controlling the single factor: total productivity. An increase in the total productivity of a firm means an improvement in the product quality and service and a decrease in the production costs, which in turn will yield greater marketshare and profits.

A greater marketshare usually translates into a sales growth and even leads to multinational operations. As the profit margin becomes larger, the money available for research and development is greater, and this in turn will help improve production systems and procedures and encourage the development of new technologies and new products. Creating and producing new products diversifies the company. Also, an increase in the profit margin leads to greater financial stability. This greater financial stability will benefit the welfare of the

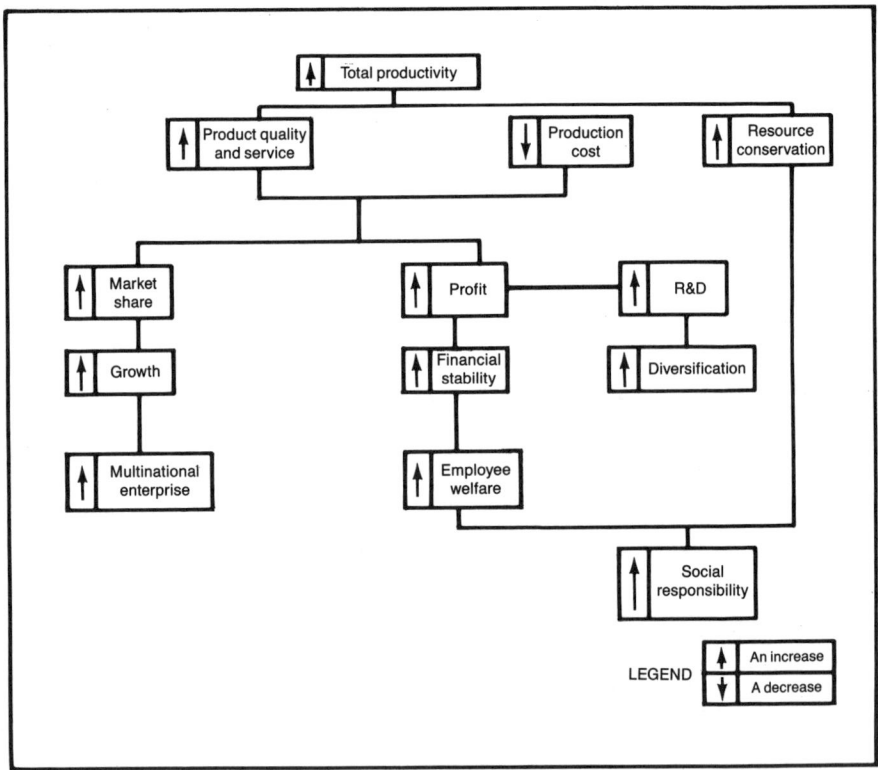

Figure 5. *Relationship between total productivity and other organizational goals.*

employees, because jobs will not only be more stable but also wages and salaries can be increased as a result of greater profits. Better use and conservation of resources is another result of an improvement in total productivity, and this, along with an increase in employee welfare, will help the company in meeting its social responsibility.

Focus on Total Productivity, not Labor Productivity

After reviewing the implications of an increase in total productivity, management should include improvement of total productivity as one of the firm's primary goals, because by attaining this goal, several others are automatically achieved. Total productivity should be the concern of every one in a company, from top management down to the lowest operator. For a productivity process to be successful, maximum and consistent participation of all employees must be encouraged. However, the overall strategy of the firm should be agreed on at the top echelons of management.

Management should take a total productivity approach to decision making in addition to the traditional managerial economics, managerial accounting, and managerial science. The concepts used in total productivity approach are total productivity, break-even analysis for total productivity, and linear relation between profit and total productivity. A flow chart of this thinking is shown in Figure 6. This total productivity approach, if implemented correctly, will provide a common direction to enable the achievement of multiple goals such as

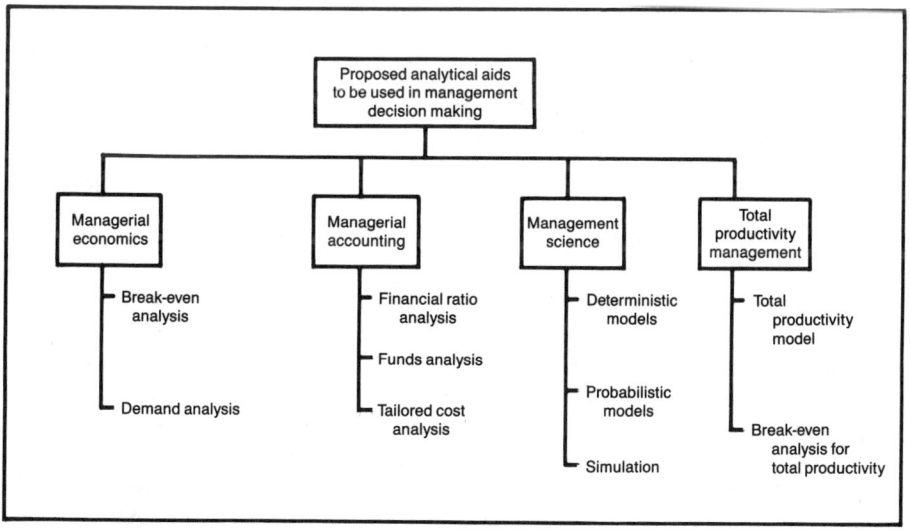

Figure 6. *Proposed analytical aids to be used in management decision making.*

marketshare, innovation, profitability, efficiency, effectiveness, growth, stability, and social welfare. Corporate and academic interest in a total productivity concept is becoming more evident in the United States now than three years ago and a number of case studies are coming to light (Sumanth 1987). Clearly, this is a good sign, but many Fortune 500 companies are still unaware of this approach.

CONCLUSION

Productivity (in general, and total productivity in particular), for the most part, is not included as a primary goal in a firm's objectives, even though it has been identified as a desirable goal to pursue. The crucial role of management in increasing productivity is often overlooked, and, therefore, its importance is often unrecognized. In this chapter, a total productivity approach has been proposed to superimpose current decision-making processes.

The benefits derived from using the TPMGT approach are as follows:

1. Multiple goals can be achieved simultaneously by increasing the total productivity of an enterprise (Yavuz and Sumanth 1987).
2. The management of a firm can objectively plan its profit level based on its ability to reach a given total productivity level by using the relationship between profit and total productivity.
3. The management of a firm can determine if its total productivity level is in the region that yields profits by using the concept of break-even point for a firm's total productivity.
4. Management is forced into a long-term emphasis on quality, profitability, marketshare, growth, innovation, and such other factors.
5. Technology planning can be integrated much more effectively and rationally with corporate strategy, business strategy, manufacturing strategy, and marketing strategy.

Incorporating total productivity as an important corporate criterion of success will, in our opinion, help companies and organizations to better compete in today's complex and competitive market environment. We believe that the total productivity approach to managerial decision making is a viable alternative to the traditional decision making tools. If only these traditional techniques had worked in the American electronics industry we wouldn't be having such a problem competing with the Japanese and German electronics industries. Is it time to try the TPMGT approach to managerial decision making? We believe that this approach is more than an analytical tool—it is a new philosophy of management that makes an organization work in a total-systems perspective rather than in a sub-optimal, segmented, self-defeating manner. TPMGT provides both a quantitative and philosophical base to manage enterprises. It persuades management, employees, vendors, and customers to operate as a team, without giving away individual creativity—this approach is beyond the barriers of culture. It can be tailor made to fit each corporate culture. If there is one management tool that the Japanese have not yet tried, it is TPMGT. Thus, if the U.S. electronics industry and others want to regain their competitive posture in world markets, perhaps they should seriously consider trying TPMGT and a practical way to start is with the measurement phase.

Suggested Readings

Drucker, P. F. 1980. *Managing in Turbulent Times.* Harper & Row. New York.

Drucker, P. F. 1954. *The Practice of Management.* Harper & Row. New York.

England, G. W. 1967. Organization goals and expected behavior of American Managers. *Academy of Management Journal.* 10:107-117.

Ericson, R. F. 1969. The impact of cybernetic information technology on management value systems. *Management Science.* 16:B-48.

English, J. and A. R. Marchione. 1983. Productivity: a new perspective. *California Management Review.* 25(2):57-66.

Massie, J. and J. Douglas. 1973. *Managing—A Contemporary Introduction.* Prentice Hall. Englewood Cliffs, N.J.

Oxenfeldt, A. R., D. W. Miller, and R. A. Dickinson. 1978. *A Basic Approach to Executive Decision Making.* AMACOM. New York.

Shetty, Y. K. 1982. Management's role in declining productivity. *California Management Review.* 25(1):33-47.

Shetty, Y. K. 1979. New look at corporate goals. *California Management Review.* 22(2):71-79.

Sumanth, D. J. 1979. Productivity measurement and evaluation models for manufacturing companies. Doctoral Dissertation. Illinois Institute of Technology, Chicago.

Sumanth, D. J. 1980. Productivity management—a challenge for the 80's. *ASEM Proceedings, First Annual Conference, 1980.* St. Louis, MO. American Society for Engineering Management. Rolla, MO.

Sumanth, D. J. 1984. *Productivity Engineering and Management.* McGraw-Hill, New York, and McGraw-Hill, Singapore. 1985.

Sumanth, D. J. 1986. Some principles of total productivity management. *Proceedings of XXVII National Convention. IIIE,* pp. 161-172.

Sumanth, D. J. (ed.) 1987. *Productivity Management Frontiers—I* (Book of Refereed Papers for the First International Conference on Productivity Research). Elsevier Science Publishers. The Netherlands.

Sumanth, D. J., and J. A. Edosomwan. (In press.) *Productivity Measurement Guide for Companies and Organizations.* Unipub. New York.

Yavuz, F. P. and D. J. Sumanth. 1987.Company-level investigation of performance measures as related to total productivity. In: D. J. Sumanth (ed.) *Productivity Management Frontiers—I.* Elsevier Science Publishers. The Netherlands. pp. 191-202.

This chapter is adapted from the authors' previous work titled: "The Productivity Factor in Management Decison-making," published by the Institute of Industrial Engineers in the *1985 Annual International Industrial Engineering Conference Proceedings.* Reprinted with the permission of the Institute of Industrial Engineers.

David J. Sumanth is Professor and Founding Director of the Productivity Research Group in the Department of Industrial Engineering at the University of Miami.

6

Pathways to Productivity Improvement

Warren C. Hauck
Timothy L. Ross

Productivity in the United States made impressive gains during the mid 1980s following a decade in which it grew at a dismal average, annual rate of less than 0.1%. Many factors contributed to this turnaround in productivity growth. Among those noted by The Conference Board, an industry-supported, non-profit economic research institute, were the following:

1. The incentives to invest in new plants and equipment, which were created by the Economic Recovery Tax Act of 1981.
2. Widespread deregulation — especially in transportation and communications.
3. A significant drop in the rate of inflation (and in the associated interest rates).
4. New information and communications technologies.
5. Improved labor-management relations, which were frequently reflected in concessions and/or cooperative cost and quality improvements designed to enhance the competitive position of a company.

Most of these factors are present in the general economic environment. The response to most of these factors by managers is generally reactive — it seems that a proactive posture which builds on these factors and uses our industrial engineering skills would be more effective. Accordingly, this will be the focus of this chapter.

There are many pathways available for increasing an enterprise's productivity. Among those methods discussed in my earlier book, *Motivating People to Work: The Key to Improving Productivity* (Hauck 1984), are suggestion programs, value analysis/value engineering, labor-management participation teams, quality circles, quality of worklife programs, work measurement, individual wage incentives, multiple-factor incentive plans, group incentive plans, and various forms of productivity gainsharing. Each of these methods may benefit your company depending upon site-specific factors.

CATEGORIES OF METHODS

Essentially, methods for improving productivity can be divided into three categories: involvement, reward, and combination. The initial category does not encompass and direct financial payment, but seeks to involve employees in some of the decisions that affect their overall working environment. This can be as simple as the classic "open door policy" or as structured as a quality circle program. The second category seeks to motivate workers by the payment of result-related bonuses — that is, suggestion awards based on a percentage of annual savings and/or standard hour/wage incentive plans. The final category seeks to combine participation with financial gain — the most common form of this is productivity gainsharing.

Involvement

The Theory X and Theory Y managerial styles that were originally identified by Douglas McGregor are good explanations for autocratic behavior of supervisors and managers. Generally, employees working in a Theory X type of organization will fulfill managers' expectations by actually becoming lazy. If employees aren't recognized for their contributions to the enterprise, there is little or no incentive for them to make any suggestions and/or to work harder. We are all familiar with the change in attitude and output that frequently occurs when a new manager is appointed. In one insurance company's regional office, replacing an office manager who had truly been a Marine Corps sergeant with a more reasonable person had many beneficial results. Turnover of employees dropped dramatically, casual absenteeism nearly vanished, and morale increased as soon as the new up-from-the-ranks manager made her friendly attitude known. The employees gave more attention to customer service, and there was more cooperation between departments. Although it is difficult to measure the change with any degree of precision, it was obvious to all concerned that productivity has increased.

Open-Door Policy

In most offices and in some manufacturing operations, employees can get involved if various levels of supervision are readily available to them. This can take the form of a stated policy that the manager is generally available without appointment for discussion on any topic of the employee's choosing. Although the possibility exists that this option might be abused, effective managers can prevent this from happening by courteously discouraging such employees. One plant manager notes that his policy of hearing from any employee led to many productivity improvements, including the identification of a lightly supervised, third-shift clean-up crew that was sleeping during half of their scheduled work period. His prompt and confidential handling of that situation did much to convince all of the employees that "a fair day's work" was required.

Quality Circles

One definition of a quality circle is a semistructured method that allows the worker to participate in improving the job environment and product quality, instead of merely performing the job (Enrick et al. 1983). This is done by allowing employees to volunteer to meet in a group of 6 to 10 area-associated persons to jointly pinpoint, analyze, and solve problems. Such employees receive training to ensure that their joint efforts are effective. Our survey indicates that the most popular types of training are problem solving (96%), group dynamics (73%), and statistics (56%) (Wolfe et al. 1984). Obviously there is an element of "recognition" present in all of the quality circles' activities—this is generally found to be sufficient motivation for continuing contributions from the involved employees. However, the survey indicated that 13% of the plans did make monetary awards to members of the quality circles.

Reward

Although much has been written about the motivational impact of nonmonetary labor-management activities, generally, well-designed financial award plans do more to enhance the productivity gains obtained through employee efforts. We can see this in many of our daily activities: the cab driver who will make an extra effort to get us to the airport in time to catch our plane and in performance-based executive bonus plans and in MBO-related awards for middle managers. Even in academia there has been a growing effort to reward merit with pay increases based on achievements in teaching, research, and service. A basic human characteristic is that extra compensation generates increased output — and vice versa.

Suggestion Systems

A program designed to elicit ideas from the company's employees nearly always has a *quid pro quo* — cost reductions and value improvements are rewarded with money (or something of value to the employee). The individuals involved include the suggestor, the evaluator, and the implementer: for the plan to succeed, each of these parties must be motivated. We have found that delays in evaluation and nonimplemented approved suggestions will tend to "dry up" the input of new ideas. This has been the focus of many revitalization programs, such as the one at United States Steel Corporation (Alexander 1982). Research at Indiana University indicates that quality circles may generate greater savings than suggestion systems; however, both should be used, because they provide alternative methods for accomplishing the same objective. Their concurrent use presents opportunities for both individual and group efforts, but the lack of financial awards for quality circles' proposals can lead to some motivational problems. Obviously, the design of all types of employee-based productivity improvement programs must be carefully integrated to ensure that they are compatible.

Wage Incentives

Payment by results is the term used in England for wage incentives. It captures the essence of this type of motivational technique. Though specific designs may vary among companies, the most general approach is to establish a standard of output per unit of time that reflects the use of a "best method" of operation. The resulting output is measured in standard hours, which are compensated at the base hourly rate. A recently published survey indicates that wage incentives are used for operations activities on a "continuing and routine" basis in 41.8% of the companies — this is down slightly from the 45.2% reported a decade earlier (Starr 1986). Motivation under wage incentives is not "automatic," and employees sometimes resist installations and/or revisions in standards following changes in product, equipment, or operating conditions. When this occurs, good management must find supplemental ways for achieving the required output. One general foreman returned from vacation to find that spooling operators were not reaching incentive-performance levels on a new line of sound-Twin 8 movie film. A review of the standards showed that they were consistent with the existing silent-Twin 8 and optical sound 16-mm films. The new standards incorporated an adjustment for the magnetic stripe that was on the new product. Despite this equitable adjustment, the department forelady and the assigned industrial engineer had not been able to get the more than 20 operators involved to even meet the base standard of output. The resulting reduction in product transferred to finished goods was causing the marketing forces to become very upset. The solution implemented by the general foreman was simplicity itself: he announced that a two-pound box of candy would be given to the first employee to exceed the standard for a full day. It was sufficient marginal motivation — especially when he amended the offer (at the employees' request) to award a box of candy to everyone who exceeded the standard on the following day. Sometimes using wage incentives is not this simple, but represents more of a strategic approach.

The federal sales department of the company managed to win the annual contract for sheet x-ray film from the Department of Defense, this volume was just about equal to the available capacity of the firm's visual inspection unit. Through some mix-up in forecasting the competition's pricing strategy, the sales group also received a parallel contract from the Veterans Administration. The combined annual quantities greatly exceeded the current capacity. Various short-term means of expanding the capacity were explored in great depth, and all but one solution were rejected for valid reasons. The remaining solution involved placing the visual inspection operations on incentives tailored to the two-person group assigned to each conveyorized inspection line. There was great initial resistance from the quality assurance professionals and only limited support from the sales group who had "caused" the problem. The wage incentive system was based on MTM-derived standards for "sheets accepted" and the "sheets rejected" (the latter involves additional paperwork to ensure corrective actions are pinpointed for prior operations). Provisions were made for a supervisory recheck of all "rejects," and known rejects were introduced into some batches. Any discrepancies were the basis for disciplinary action starting with the loss of incentive earnings for one day. The results achieved were a near doubling of the output from this "bottleneck" operation and the on-time delivery of the monthly requirements on both contracts. Obviously, where creative managerial efforts can bring about an installation, wage incentives can be most effective in improving productivity.

COMBINATION: PRODUCTIVITY GAINSHARING

Productivity gainsharing merges the "best features" of the preceeding methods. In one "Productivity Opinion Survey" sponsored by the Institute of Industrial Engineers, nearly 30% of the formal incentive programs were of the gainsharing category (Hauck 1986). Apparently, there is a significant rise in this category as the standard wage-incentive plans become less viable. From the name, "productivity gainsharing," we obtain the following simple definition of these programs: a productivity-based plan that measures the change in productivity from an appropriate base period and shares the resulting gains between the company and all participants in the plan.

Use of Plans

The Institute's survey shows that only 30% of the plans in use are the commonly known ones — IMPROSHARE[R], Rucker[R], and Scanlon. This parallels our own experience which shows that the remaining 70% are especially designed plans tailored to site-specific factors. However, all plans of this final type are worth the extra efforts required to develop them. Respondents to the Institute's survey gave them a higher rating than for wage-incentive plans for a question asking whether the expense was justified (88.9% vs. 80.5%). The special gainsharing plans also outscored the average for three commonly known gainsharing plans by a much wider margin (92.9% vs. 81.0%) (Hauck 1986). Because the commonly known gain-sharing plans have been covered in the literature of the past decade, we will focus on the features of the "other" specially designed plans that we have observed. As an outline for such a review Frost's Four Principles can be used: identification, involvement, equity, and competent management.

Identification

In this first phase of gainsharing, a spirit of cooperation and mutual trust should be developed. Given the adversarial relationship that exists between many management personnel and the union which represents their employees, this essential pre-condition for a truly successful gainsharing plan is sometimes difficult to obtain. In one branch plant of a medium-sized conglomerate, the union's executive committee bluntly informed us that they had been lied to for so long that they would not be willing to enter into a gainsharing plan. Accordingly, we informed the rather forthright and autocratic general manager that no further efforts would be fruitful unless he was transferred back to their corporate headquarters or until he retired as scheduled about two years in the future. In other cases, the "normal resistance" can be overcome by convincing all concerned that productivity improvement is truly a "survival objective." One independent union in a large midwest city reached this conclusion after watching their membership drop from 1,500 to 325 over a five-year period. The union enlisted our help in analyzing a company-developed proposal that was shown to be one-sided. Fortunately, the company could recognize the need for employee involvement and accepted most of the proposed changes.

Involvement

This principle extends well beyond the initial design and "sale" of the plan to all participants. It generally includes multilevel departmental committees tied to a plant-wide steering committee. These committees make decisions about implementing productivity improvement ideas and serve as a two-way communications network on related matters. (By design, the committees' actions exclude any which are covered by the collective bargaining agreement.) In a machined products plant in northern England, the steering committee included the managing director, a few key members of his staff, union officials, and representatives of their technical departments (Hauck and Ross 1984). The meeting was conducted in a imperious manner by the managing director. The lack of true involvement and resultant participant resentment made it easy to predict that the plan would fail, as it did about one year later. In many other cases, we have seen good involvement efforts "pay off " in many ways, even when the nature of the gainsharing calculation did not yield a reward to the participants over many financial periods. This observation is supported by the recent Institute survey which shows that 58.2% of the respondents chose "personal recognition" as the most effective way of encouraging employee involvement (Starr 1988). This is double the 28.9% who selected a "money reward"; the comparable figures for 1983 were much closer, 63.3% and 57.6% respectively. Apparently we are moving toward greater acceptance of this principle of involvement. Despite this, we feel that the long-term success of a productivity gainsharing plan must provide for periodic financial awards to all participants in the plan.

Equity

A rational distribution of the gains acquired through the plan must be given to all participants in the plan. This equitable decision depends in part on the nature of the productivity measurement being used. Generally, the broader the measure, the lower the percentage of the gains allocated to the participants. In the earliest form of the plan, the "labor-only" Scanlon plan, 100% of the savings went to the employees. Such decisions are obviously intertwined with not only the measurement, but also with the ability of the participants to influence the changes. One example of the complexity that can evolve is a silver mine in Canada. The mine's outputs from a variable ore body include large tonnages of copper, significant amounts of silver, and a fair amount of gold. All three of the products have shown a wide variability in market prices, and a bimodal demand had evolved for leached and unleached slurry. Inputs required per 1,000 tons show a great deal of seasonal variability, reflecting the climatic impact on these mainly outdoor operations. Proprietary constraints do not permit the discussion of the "solution" that evolved, but you can begin to appreciate that this principle of equity is closely bound to the prior two principles: identification and involvement. Likewise, competent management is also a related requirement.

Competent Management

Implementing a productivity gainsharing plan involves many decisions requiring keen judgment. Once a plan is in place, it must be periodically reviewed to ensure that changes in product mix, made-or-buy decisions, additional capital investments, and other factors are suitably integrated into the productivity gainsharing plan to maintain the desired equity in distributing gains to employees and to the company. This periodic review supports the need for "competent management" in achieving success. This need may be illustrated by the actions taken by Volvo's management in their automotive assembly operations throughout Sweden (Hauck and Ross 1985). Starting with their new plant in Klamar, they evolved and installed facilities that permitted small work groups to individually design jobs and control their own performance. At a later date, the company installed a corporatewide profit-sharing plan. More recently, they have supplemented this with plant-level, multifactor gainsharing plans which vary somewhat between plants (Hauck and Ross 1987). At Kalmar (passenger cars) and at Tuve (cab-over-engine trucks), the factors are as follows:

1. Reducing man-hours per vehicle (including "white-collar" workers).
2. Improving the weekly quality index (computed from a small sample of completed vehicles).
3. Lowering overhead costs.
4. Increasing inventory turnover.

The first two factors are also used at Boros (bus chassis) along with "deliveries made on schedule" and "lack of required adjustments following normal assembly." That Volvo management did not follow the lead of much smaller Danish companies which also include a material bonus in their plans probably reflects their "competent" decision that Volvo's extensive vertical integration precluded the need for such an adjustment. Less than competent management would be likely to make "off-the-cuff" decisions that would adversely affect the ongoing operation of a well-designed productivity gainsharing plan.

CONCLUSION

The pathways to improved productivity are many and varied. Some good results can be achieved from a reactive approach to changes in the economic environment. The better results are likely to come about from a proactive mind-set in a group of competent managers. Many of these methods are covered in the literature and can be considered readily available to effective and interested industrial engineers. Rather recently, there has been increased interest in productivity gainsharing plans that typically involve several functions within the organization. Recent survey information suggests that gain-sharing plans which are tailored to meet site-specific factors are twice as effective as the more commonly known, semi-proprietary plans. Obviously, to develop such plans involves interdisciplinary efforts coupled with "cradle-to-grave" participants involvement. For industrial engineers to provide effective leadership in this area, it seems essential that they carefully study the available publications in the field.

Suggested Readings

Alexander, W. 1982. Suggestions are alive and well at U.S. Steel. *Annual International Industrial Engineering Conference Proceedings.* Institute of Industrial Engineers. Norcross, GA.

Enrick, N. L., R. H. Lester, and H. E. Mottley. 1983. Quality circles: motivation through participation. *Industrial Management.* 25(2): 1-5.

Hauck, W. C. 1984. *Motivating People To Work: The Key To Improving Productivity.* Industrial Engineering and Management Press. Norcross, GA.

Hauck, W. C. 1986. Factors influencing the choice of financial incentives. *Northeast Decision Sciences Institute Proceedings.* Williamsburg, VA.

Hauck, W. C., and T. L. Ross. 1987. Sweden's experiments in productivity gainsharing: a second look. *Personnel.* 64(1): 61-67.

Hauck, W. C., and T. L. Ross. 1985. Volvo's new solution at Kalmar: multifactor productivity gainsharing. *American Institute of Decision Sciences (AIDS) Proceedings.* Las Vegas, NV.

Hauck, W. C., and T. L. Ross. 1984. England's approach to increased productivity: added value schemes. *Industrial Management.* 26(2): 15-21.

Starr, S. 1988. Seventh annual survey of IIE members' views on productivity contrasts with JIIE survey. *Industrial Engineering.* 20(4): 60-64.

Starr, S. 1986. Productivity survey results. *Industrial Engineering.* 18(1): 91-95.

Wolfe, D., W. C. Hauck and G. Varney. 1984. Quality circles: the U.S. experience. *Midwest Academy of Management Proceedings.* South Bend, IN.

Reprinted from *1985 Fall Industrial Engineering Conference Proceedings.*

Warren C. Hauck is Professor of Management at Bowling Green State University in Ohio.

7

Production Workers Bear Major Quality Responsibility in Japanese Industry

Richard J. Schonberger

In 1945 Japan's self image was at an all-time low. The Japanese had lost the war; had bombed-out cities, ports and factories to rebuild; had United States' occupation forces dictating vital elements of the recovery; and had a worldwide pre-war reputation for exporting low-quality "junk."

As demoralizing as these circumstances surely were, they presented some salutary opportunities, one of which was to rebuild in a way that enabled Japan to overcome its image as an exporter of shoddy goods.

In that regard, General MacArthur's occupation forces provided welcome aid. The United States military advisers who assisted in the rebuilding of Japanese industry told the Japanese industrialists about statistical quality control and lot-acceptance sampling, which had come into wide use among American companies that contracted with the United States military establishment for the production of war materials.

The Japanese were eager assimilators of any information about quality control that they could lay their hands on, and the information supplied by the occupation forces whetted Japanese appetites for more. In 1949, the Japanese Union of Scientists and Engineers invited W. Edwards Deming, a United States authority on quality control, to Japan to "tell us more." Deming has visited Japan many times since, as has Joseph M. Juran, an equally well known United States authority.

The rest is spectacular history—the ascendency of Japan to world leadership in quality, productivity and export success. The Japanese were not content merely to put United States' quality concepts into practice. They molded their own by modifying the American concepts through trial and error on the shop floor.

This article explains the quality concepts fashioned by the Japanese and offers commentary on the prospects for their adoption in Western industry.

QUALITY RESPONSIBILITY

The term *total quality control* (TQC) is an apt descriptor of today's quality practices in Japan. The Japanese may have borrowed the term from an American book by A.V. Feigenbaum entitled *Total Quality Control*. A thematic sentence from Feigenbaum's book is: "The burden of quality proof rests . . . with the makers of the part."

A small change in wording translates this statement into a basic precept of TQC as practiced in Japan today: The *responsibility for* quality rests with the markers of the part. "Burden of proof " has a defensive ring to it. "Responsibility for" alters the basis of quality control by making quality a fundamental manufacturing objective, carrying with it offensive policies, concepts and procedures.

For most Western manufacturers, the business of production is output. Performance reports and salaries depend on it. Quality questions are referred to the quality control department.

It is a woeful state of affairs when workers need not answer for or become involved with quality to much of an extent—a cop-out that is bound to encourage undesirable worker behaviors, which no amount of humanistic leadership is likely to be able to offset.

If Western manufacturers are to close the quality gap with the Japanese, there is no better way to begin than by transferring primary responsibility for quality from the QC department to production. Not only is this relatively easy to do, but it also brings quality responsibility back to its natural home.

There is no reason to believe that making quality a primary responsibility of manufacturing is workable in Japan but not here. On the contrary, there are recorded cases of dramatic quality improvements in North American plants where quality responsibility has been so placed.

The Sanyo TV plant in Forrest City, Arkansas, which is the subject of a Harvard case study (see "For further reading") is one of the most noteworthy.

Sanyo bought the old Warwick TV facility in 1977, when the plant's quality was so poor that its sales had dried up and nearly all production had ceased. Sanyo employed the same workers and proceeded to effect a change from abominable to very high quality output.

A critical first step was making the former chief of quality control the new chief of manufacturing. No other act could have signaled so swiftly and surely management's resolve to place first priority on quality.

The key to unlocking a firm's quality potential is assigning primary responsibility to production. But that responsibility is by no means absolute. Total quality control calls for *total involvement* in quality improvement, from top management to the janitor and including both line and staff personnel.

Some Japanese companies prefer the term *company-wide quality control* (CWQC) over total quality control, because CWQC conveys the idea of total involvement. But TQC is a broader term that can represent involvement not only of people, but also of functions, equipment, processes and so forth.

It is difficult for Americans to get used to the idea that a responsibility can be shared. Akio Morita, the chief executive officer of Sony Corporation, cites the difficulty of "persuading our American engineers and managers to go onto the production floor and mingle with foremen and workers."

We Americans have become so specialized that we are reluctant to set foot on someone else's turf. In TQC and CWQC, quality is everybody's turf. But quality is closer to production than to other functions, and therefore production must have primary responsibility.

SETTING A TARGET

The mere act of assigning QC responsibility to production is likely to trigger a flurry of immediate activity to create quality standards, measures, reports and controls and to bring about substantial long-term quality improvement. To accelerate the rate of quality improvement, there are a number of TQC concepts and techniques that can be implemented.

First of all, a long-term objective or target and an operational plan for getting there need to be established. In the early years of the Japanese quality control crusade, there was no particular QC target. Rather, the Western notion of an acceptable quality level (AQL) was a standard basis for quality planning.

Today the term AQL is likely to draw looks of wry amusement from knowledgeable Japanese. The AQL—a key input into the U.S. military standard sampling tables that excited Japanese industrialists in the 1950s—has long since fallen out of favor. In its place is a true long-term target, which is, simply, *no defectives*.

The American term *zero defects* (ZD) sounds as if it might express the same idea, and the Japanese are fond of the term. However, the sloganism that has attended zero defects as used in the United States has severely watered down the impact of the term among Westerners. "No defectives" enjoys wholesale acceptance in Japanese industry, not as a term or slogan, but as a target that does not seem beyond the bounds of reason.

In Japanese electronics, current quality talk is in terms of defect levels in parts per million (ppm). This is a fraction of a percent, and it runs off the usual AQL-oriented sampling tables.

The operational goal for moving toward the target of no defectives may be gleaned from the writings of Juran, who said: "Over the years the accumulated experience [of the Japanese] has developed its own imperative—the precious *habit of improvement.*" By contrast, our Western focus is on static standards and on quality control by means of minimizing variances from standard.

How did the Japanese break away from the control mode and adopt the habit of improvement and the target of no defectives? Part of the answer lies in the obsession of the Japanese with overhauling attitudes about quality from top to bottom.

In the 1950s, upper Japanese managers throughout industry received extensive quality training, and in the next decade the message was spread to foremen and workers. A milestone year was 1962 when a new quality control journal for foremen, called *Gemba To QC* (Quality Control for Foremen), began publication. Foremen all over Japan received the journal, which was written in plain shop talk; the foremen read about and proceeded to try out quality control techniques on the shop floor—and they are still doing so.

The habit of improvement began and continues because shop people were made the central part of the quality team. Total quality control evolved on the shop floor through trial and error. The TQC concepts that remain to be explained emerged from that environment.

PREVENTION, NOT DETECTION

A phrase often heard among Western quality people is: "You can't inspect quality into the product." But judging by the number of QC department inspectors on the shop floor drawing statistical samples in Western plants, we try to do just that. We try to detect the defectives through inspection, and rectify the quality problems of the lots by removing the defectives.

By contrast, the Japanese are firmly committed to prevention in the first place. The related Western slogan, "quality at the source," has become an article of faith in Japanese TQC. The Japanese implement this article of faith by employing the principles discussed below.

Process Control

Process control, a well known concept in classical Western QC, means controlling the production process by checking the product quality or the process during production—and stopping the process if it goes out of control.

Our Western QC books give advice on how to select which processes are to be so controlled; the assumption is that only a limited number of processes may be selected because of the cost of performing the quality checks—whether they are performed by QC department inspectors or the worker.

In Japanese TQC every work station is a control point, and workers, not QC department inspectors, are responsible for doing the checking if automatic checking devices are not available. A company cannot afford to assign a QC department inspector to every station, but workers are at every station, and it is natural for them to inspect their work.

Visible, Measurable Quality

Deming and Juran preached the gospel of "measurable standards of quality" to the Japanese, who not only adopted it wholeheartedly, but also appreciably extended the concept. The Japanese want quality to be measurable and *visible*— even to the untrained eye.

Consequently, quality records and ongoing quality improvement activities are expressed in easy-to-understand charts and diagrams and displayed on chalk boards, pasteups, electric signs and the like throughout the Japanese factory. The displays give recognition to the workers' achievements, keep other plant employees and managers informed and impress outside visitors.

Some of the outside visitors are customers, and in the Japanese system customers do not just tour. They audit. And they insist on easy-to-understand visual signs of quality so that the plant's quality commitment is obvious and discussable.

Takashige Yamane, manager of quality control at Mitsuboshi Belting Company of Kobe, Japan, has told (in a personal interview) of the great apprehension that impending customer audits provoke among Mitsuboshi employees.

A customer—usually an auto company—will send a team of perhaps ten buyers, engineers and quality control experts. They investigate everything from defect rates to recreation programs for employees, and they generally leave a list of 200 or so demerits. Mitsuboshi values the customer's business, so the company strives to correct the demerits right away.

Insist on Compliance

The customers, of course, are insisting on compliance with their quality-oriented recommendations. In the Japanese system of purchasing, buyers generally have the clout to be insistent, because suppliers often sell to just one or a few buyer companies and cannot afford the loss of a customer.

Insistence on compliance also applies internally; i.e., management insists that quality goods be produced. The Japanese word *hadome* is often used to refer to the minimum quality level that *must* be complied with.

We have quality goals (e.g., the AQL) in American companies, too, but in many companies the manufacturing operation learns that its rewards and penalties are based on achievement of output goals, not compliance with quality standards.

Recently, however, many American companies have made a quality commitment, and some are installing measures of quality that are as tough as our usual measures for cost variance, labor efficiency and output.

Line-Stop Authority

If management is to insist on compliance with quality standards, workers must have the authority to stop the production line in order to comply. In Western firms line-stop authority may be given to workers in certain food and drug industry plants, but rarely elsewhere. Without such authority, workers press on, allowing errors—missing bolts, thin paint jobs, etc.—to creep in.

In Japan, line-stop authority at the worker level is common, and the concept is being introduced among North American workers employed by some Japanese subsidiary plants.

For example, at the Kawasaki motorcycle plant in Lincoln, Nebraska, final assembly lines are strung with yellow "help" lights and red "stop" lights. Workers may push a button to turn on one of the lights and, in the case of the red light, automatically stop the line until the problem can be solved.

In more capital intensive processes in Japanese plants, equipment is typically outfitted with automatic devices to detect bad quality output or such problems as excessive tool wear that are likely to result in bad quality.

Self-Correction of Errors

In Western plants, rework lines and field rework facilities are common. Feigenbaum refers to these facilities as the "hidden plant" and estimates that they consume from 15% to as much as 40% of productive capacity (see "For further reading").

In Japanese factories, rework is the responsibility of those who made the bad items. Rework calls for backtrack handling and is often too disruptive to be fitted in right away. Therefore, the worker or work group responsible may have to work late if there is rework to be done. They usually make sure that none is required.

Expose Problems

A defective part, in TQC, brings forth smiles as well as frowns. The smiles arise because the defect exposes one more problem to investigate and possibly prevent from ever happening again.

Exposing problems is valued to the degree that Japanese managers will deliberately remove buffer inventories or workers from a production line so that small delays—for quality or other reasons—begin to cause parts shortages.

The *causes* of the parts shortages are recorded, analyzed and eliminated. Workers generally are happy to contribute to the solution of the problems, because their resolution is rewarding and the failure to resolve them often means unplanned overtime to catch up and meet daily schedules.

100% Check

Sampling from completed lots and sampling on the production line are still fundamental in Western quality control. But in Japanese TQC the emphasis is on 100% quality checking, especially for finished goods, and where feasible for component parts. Emphasis on 100% checking plus process control at every process is a potent combination for prevention of defectives.

Project-by-Project Improvement

The quality-oriented visual displays that are everywhere in Japanese plants include extensive summary data on past and present improvement projects.

The displays list the projects, name project team members, indicate progress or results, give recognition and publicity for past achievements and provide other information. Many employees are involved in one improvement project after another for their entire working careers.

Top plant managers generally review proposals for new projects, assign project team members and a leader, frequently review project progress and sometimes join project teams. They also shower the project team with praise as each milestone is reached.

SUPPORTING CONCEPTS

The eight "accelerators" of the pace of quality improvement described above need good support for maximum effectiveness. Five TQC concepts that provide excellent support are discussed below.

Quality Control Department's Role

In assigning primary QC responsibility to production, the company does not disband the QC department. Instead, QC assumes a secondary role and shrinks in size. The smaller department monitors production processes to see that standard procedures are followed, assists in the removal of error causes, accompanies purchasing personnel on audits of supplier plants and so forth.

In some Japanese plants there is still QC department involvement in quality inspections after final assembly, especially where torture testing or technically difficult inspections are necessary. But quality inspections in other stages of manufacture are left up to the production people.

Also, the Japanese try to eliminate receiving inspections and rely instead upon self monitoring and extensive process control in the suppliers' plants. Kawasaki, Lincoln, has QC department inspectors assigned to the receiving dock for items bought in the United States, but not on the separate receiving dock for items imported from Japan. The latter items are presumed to be good, and are, because the producing companies in Japan employ TQC, including rigorous process control.

Small Lot Sizes

Small production and purchase lot sizes are critical in assuring that defectives are caught early.

For example, if a lot of just ten widgets is made and immediately delivered to the next stage of manufacture, any defectives will be found—and the causes can be investigated—by the time the ten pieces are used up. But if one thousand widgets are made and delivered, it is necessary to go through (use up) 100 times as many pieces to find out if there are defectives.

Small lot sizes and quick deliveries are normally thought of as inventory control factors. Indeed, the Japanese just-in-time inventory and production control techniques are based on small lot sizes. But the quality benefits are at least as important as the inventory control benefits, and Japanese quality people embrace small lot sizing as fervently as inventory and production people do.

Housekeeping

Japanese streets and subways are generally trash-free, and so are Japanese factories. But the fetish for plant cleanliness and tidiness is not just a matter of inbred habit. It is a basic component of total quality control.

At the aforementioned Sanyo plant in Arkansas, quality was made the first priority, and one of the first actions taken by the new management, according to the Harvard study, was to "clear out the plant over a weekend, clean it and polyurethane the floors. Not only did this make the whole plant look cleaner and brighter, but it also reduced the dust in the air which sometimes caused equipment to gum up or interfered with the connections of electronic parts." Sanyo believes that cleaning the plant was instrumental in the rapid recovery of the plant's capability to produce quality TVs.

Many American plants are clean, too, but many others are as dirty and messy as some of our city streets are. While street cleanliness is mainly an esthetic concern, plant cleanliness is an important contributor to quality—and to worker safety and equipment "up time" as well.

Total Preventive Maintenance

Good plant housekeeping is closely related to (and is actually a form of) preventive maintenance (PM). Regarding PM as applied to equipment, the Japanese zealously follow the recommendations of our Western authorities: Keep careful records on breakdowns and machine usage and establish procedures for regular maintenance based on those records.

But Japanese preventive maintenance goes farther. Some Japanese companies, especially in the auto industry, employ what is sometimes called "two-shifting," whereby two eight-hour production shifts have three 3 1/2- to 4-hour maintenance shifts fitted in before, in between and after them.

Such a commitment to preventive maintenance keeps machines up and functioning properly, which prevents defects and improves process yields considerably. Two-shifting plans are in the works or already in operation at some plants of Ford Motor Company, 3M Corporation and Control Data—based on those companies' studies showing that process yield improvements gain back nearly all of the production lost in reducing from three to two production shifts.

Another element in Japanese "total preventive maintenance" is airplane-pilot style machine checking: Workers commonly follow a checklist in fully checking out their machines each morning before beginning production.

Finally, in the Japanese system, machine operators tend to perform a good deal of the routine PM, including janitorial work around their own work areas, and also some repair maintenance.

Less-Than-Full-Capacity Scheduling

Any impression that the Japanese sacrifice the meeting of production quotas in their quest for high quality is erroneous. To assure that both quality and production goals are achievable, the Japanese set daily schedules at less than full capacity.

This allows workers to slow or stop production for quality problems, provides time for rework and discussion of problems after the day's schedule has been met and prevents problems that might otherwise arise from haste. Problem prevention, in turn, reduces the need to slow or stop production.

TOOLS AND AIDS

The Japanese quality crusade began over 30 years ago with an early focus on American statistical and other tools and aids to quality control. The Japanese have not forgotten or neglected the basics. But the concepts discussed above seem to have become the dominant factors in Japan's current excellence in quality.

Some tools and aids, less dominant but still important, are discussed below.

Bakayoke

Bakayoke is a Japanese word that means, roughly, "foolproof." As applied to quality control, *bakayoke* may refer to devices attached to machines to automatically check for abnormals or for malfunctions such as tool or bearing wear. Some *bakayoke* automatically stop the machine when a problem is detected. These devices are vital in high-volume operations in which 100% checking by the worker is infeasible.

N = 2

The concept of statistical random sampling inspections is ingrained in Western QC thought. But practical-minded workers practicing QC on the shop floor in Japanese factories have cast off the concept. Where inspecting 100% is technically difficult or not affordable, the Japanese approach is to inspect just the first and the last piece; therefore, the sample size, N, equals 2.

The reasoning is that you want to detect a process change; if the first and last pieces are good, the process probably did not change and the intervening pieces probably are good.

A typical Western sample size of eight randomly selected pieces will indicate the quality between the first and eighth pieces, but that span usually will not provide a check on the process for early and late pieces. N = 2 is cheaper and provides more comprehensive process control.

Statistics and Other Analytical Aids

Japanese foremen and workers use statistical control charts, histograms and other such aids with ease—as if they were paint sprayers, welding torches and drill presses. It has taken many years of training for statistical techniques to have found a place in workers' tool kits; that it could have happened at all seems remarkable to Western observers.

In addition to the traditional Western QC aids, the Japanese also employ the Ishikawa diagram (see "For further reading"), commonly called a fishbone chart because of its shape. A quality problem is selected for analysis; the quality improvement project group determines the major, secondary and tertiary factors causing the problem; and the factors are entered on the major, secondary and tertiary "bones of the fish;" i.e., lines branching off from a center spine bone.

The chart may then be hung out in the work area to remind the workers of problem areas, trigger the collection of problem frequency data and get them thinking of possible solutions.

Quality Control Circles

Quality control circles began in some Japanese companies in the early 1960s as study groups in which foremen shared what they learned from the magazine Gemba To QC with their workers. Others formed later for both training and problem analysis purposes.

The sheer numbers of QC circles and their visibility in Japanese factories have given many Western visitors the impression that QC circles are a key to quality and perhaps worker commitment as well. That view seems highly exaggerated. The preceding TQC concepts appear to be more basic; the true role of QC circles in Japan is more to generate ideas for wringing the last defects out of the processes.

QC circles provide a convenient formal organizational mechanism to house the improvement projects mentioned above under the topic "project-by-project improvement;" however, an ad hoc project team formed to work on a specific project can be just as effective.

TQC IN THE WEST

It took a full decade, the 1950s, for Japanese management to be trained in quality control. It took the next decade to train Japanese foremen and workers. Juran does not think the West will be able to catch up with Japan's level of quality before the end of the decade, because we must go through similar massive training.

Juran may be right. On the other hand, QC training may take less time in the United States. For one thing, we can learn from the Japanese example, which is convincing, and can adopt Japanese TQC concepts and techniques that took years to develop to their present state. Also, the fact that Japanese subsidiaries in this country have been adopting and "proving out" TQC techniques with American workers and managers gives us a running start.

In addition, America has a vast network of professional societies that are effective in retraining their constituencies.

Furthermore, U.S. business and engineering colleges work more closely with industry than is the case in any other country—especially Japan, where academicians are more confined to "ivory towers."

As Philip Crosby's book title proclaims, *Quality Is Free*, and American industry is beginning to realize it. William Conway, chief executive officer of Nashua Corporation, puts it this way (as quoted by Ringle):

> As quality goes up, so does productivity. Consider the impact on overall levels of productivity if everyone and every machine in your company

performed properly the first time, every time. The same number of employees would be handling much larger volumes of work. The high cost of inspection would be directed into productive activities. Rework, downgrading, scrapping would be eliminated. Administrative efficiency would be much higher.

American industry is generally aware of the quality problem, consumers are quality conscious, and TQC training and implementation is well under way in some American companies. The American and Western quality crusade has begun.

Suggested Readings

Crosby, P. B. 1979. *Quality Is Free: The Art of Making Quality Certain*, McGraw-Hill. New York.

Feigenbaum, A. V. 1961. *Total Quality Control: Engineering and Management:* McGraw-Hill. New York.

Harvard Business School. 1981. Sanyo Manufacturing Company—Forrest City, Arkansas. Case number 2-682-045.

Ishikawa, K. 1972. *Guide to Quality Control*. Asian Productivity Organization. Tokyo. pp. 19-30.

Juran, J. M. 1981. Product quality—a prescription for the west; Part I: training and improvement programs. *Management Review*. 6 June. Vol. 70, No. 6. pp. 8-14.

Nellemann, D. O., and L. F. Smith. 1982. 'Just-in-time' vs. just-in-case production/inventory systems: concepts borrowed back from Japan. *Production and Inventory Management*. Second Quarter. pp. 12-21.

Ringle, W. M. 1981. The American who remade 'made in Japan.' *Nation's Business*. February. pp. 67-70.

Sasser, E. 1981. Quality: a presentation to the A.I.D.S. group. unpublished summarization of materials presented at the national conference of the American Institute for Decision Sciences. Boston. November 18-20.

Schonberger, R. J. 1982. *Japanese Manufacturing Techniques: Nine Hidden Lessons in Simplicity*. The Free Press. New York.

Tsurumi, Y. 1981. Productivity: the Japanese approach. *Pacific Basin Quarterly*. (Summer No. 6.). pp. 7-9.

8

Using Quality Cost Analysis for Management Improvement

Lee Blank
Jorge Solorzano

Most of the decisions concerning management personnel, organization structure, and promotion recommendations do not take product quality into account. Once the people have been selected to fill the positions in management and quality control, it is generally assumed that the quality assurance program will operate efficiently and economically. In fact, even if the company has a long-standing high service record in product quality, the cost to the company and customer for this level of quality is not of great concern to management or users.

QUALITY COST STUDY

A quality cost system has been defined as the business of assigning a monetary value to things that go wrong.[1] That is, a quality cost analysis is done to determine how much it costs to maintain a certain level of quality in the manufactured items or services of the company. The quality costs are collected and categorized so that the specific purposes for the costs can be determined and analyzed. Juran and Gryna[2] and the American Society for Quality Control[3] have presented data collection approaches.

The categories used for quality cost analysis are defined in terms of quality assurance functions, rather than with specific regard to departments in the organization. Therefore, there will undoubtedly be costs which are classifiable as quality costs that originate in departments other than quality. The two main quality cost categories are discretionary costs and consequential costs.

Discretionary Costs

These are all sums of money which must be spent to assure that a desired level of quality is present in the service or finished product. There are two types of quality costs which may be incurred at the discretion of the management. They are:

1. Prevention costs: Management must decide how much to spend to prevent quality problems which degrade product quality below a minimum acceptable level.
2. Appraisal costs: These costs are incurred to determine the actual quality level of the finished product prior to its placement on the market.

Consequential Costs

These are sums of money which must be spent to *correct* the service or product because the quality level is too low. These costs often occur because management did not avoid them by incurring a discretionary cost. The categories are:

1. Internal failure costs: These are costs which must be spent to improve the quality level of the product *prior to delivery* to the customer.
2. External failure costs: These costs must be incurred to correct all the product found to be below some minimum quality level *after delivery* to a user.

Figure 1 is a simplified sketch of a production and inspection system. The main sources of each category of quality costs are shown. Only prevention costs originate from all components of the production system, since it is everyone's responsibility to prevent quality problems, but only some people's responsibility to discover quality faults.

Of the costs defined above, prevention is one of the most important because it can be used to reduce total quality cost. However, due to its less specific definition, prevention expenditures are harder to quantify than other quality costs. Therefore, the actual data collection may be done in the following order:

1. Appraisal costs.
2. Internal failure costs.
3. External failure costs.
4. Prevention costs.

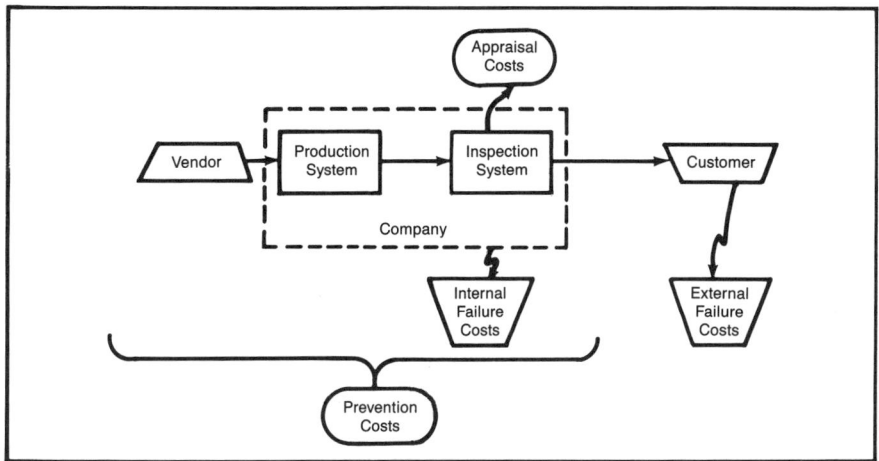

Figure 1. *Sources of quality costs by category.*

This order is suggested because with it the engineer can use the more easily collected appraisal (inspection) costs as a learning mechanism to classify data, how to read company accounting records, and who to go to in the company for quality cost information.

ASSESSING QUALITY COSTS

A more detailed discussion of what types of costs fit into each of these four categories follows. You may, in practice, want to refer to References [2] and [3] for further definition and collection techniques of these costs.

Appraisal Costs

These are usually the largest quality expenditure for a company because management commonly feels (consciously or unconsciously) that more inspection and testing will undoubtedly improve quality. Appraisal costs do absolutely nothing to prevent or correct defective product. They are spent only to discover that it does exist. Characteristically, a quality cost study will indicate that as much as 50%, and, if you have a quality-minded management, as little as 10% of quality costs are appraisal.[2,4]

Data collection here is fairly simple. If you determine the following you will have most of the quality appraisal costs:

- Pay for all full-time and spot check inspectors, supervisors, testers, and clerical workers in incoming material and in-process inspection.

- Costs to own or lease and maintain all materials and equipment used for inspection and testing of product.
- Cost of processing and evaluating all appraisal data.
- Cost of all quality audit efforts on in-process or finished goods.
- Costs to perform field tests.

Internal Failure Costs

The expense to correct defective product prior to its release is usually about one-half of the total failure cost (internal plus external). You can usually anticipate that a discovered and corrected internal failure will be less expensive than an external failure, because the customer is not involved at all with the internal failure. Since the cost of product liability has risen so dramatically in the last few years, it is more important than ever to prevent or find a major portion of defectives prior to release from company control.

Since appraisal costs are incurred only by the discovery of production-generated failures, internal failure costs are also incurred in the production system to remove the mistake in the most economic way. If the following costs are determined, most of the internal failure costs will be accounted for:

- Cost of all items scrapped because they miss quality specifications.
- Total cost of reworking (completing some missed production operations) and repair (correction to salvage the product) for defects found prior to or during appraisal.
- Expenditures incurred by failure analyses on company and vendor materials and processes.
- Costs to re-appraise product which has been reworked or repaired.
- Loss in revenue due to reduced price for product of poor quality.

External Failure Costs

This cost may be large for two rather unrelated reasons. First, the company must bear the expense of correcting to the customer's satisfaction and the company's ability, the failures discovered after release of the product from company control. Second, lost revenues because customers will not purchase the product due to their, or another customer's, past bad experiences. The first cost can be estimated from records by thorough data collecting, but the second falls in the area of loss-of-goodwill and is quite difficult to quantify. The total of internal and external failure costs is usually 50% to 90% of total quality costs.[2]

External failure costs are more difficult to obtain than internal failure costs. However, the following components include most expenses:

- Costs of processing, repairing, and replacing product returned for quality reasons.

- Costs incurred for customer contact and service to handle quality complaints.
- Costs of replacement during warranty period.
- Total cost of testing, legal services, settlement charges, etc., for product liability problems.

Prevention Costs

The actual dollar values spent in prevention of quality problems are quite difficult to obtain. Estimates must be made, especially when quality costs are being collected for the first few times. Prevention costs occur in several departments because many different types of personnel assist in the avoidance of future defective material. The first quality cost study will probably be a relatively poor estimate of actual prevention costs. However, do not be surprised if costs in the category seem very low when compared to other costs, especially appraisal. Historically, it has been found that prevention is only 0.5 to 5% of total quality system expenditures.[2,4] The percentage is low, not because prevention costs are so hard to collect that much of them are missed, but because such a large proportion of the quality dollar is spent trying to find and fix failures, instead of preventing them. The majority of prevention costs will be included if the following are determined:

- Costs and salaries to plan better quality by engineering, process control, training, etc.
- Costs of designing and developing equipment, processes, and systems to measure and control quality.
- Expenses incurred in all types of quality training.
- Costs to improve quality by special studies, vendor relations, data analysis, etc.
- Costs incurred by departments other than quality assurance to improve quality.

INCREASED PREVENTION COSTS?

Prevention costs are usually the smallest category and the most neglected, because management finds it difficult to evaluate the profit results of planning, trouble shooting, designing, communicating, etc. It is much easier to *count heads*, as it were, and thereby increase total quality costs in the categories of appraisal and failures. The expenses incurred in the name of prevention offer a large amount of quality-leverage, because prevention is so much more powerful and long lasting for the quality image of the product. For example, if prevention costs are incurred to find and correct a quality-morale problem among assembly personnel, the quality of all future product is improved. On the other hand, if appraisal costs which are also discretionary, are incurred to correct the problem, internal failure costs, which are consequential, will usually increase. But, and this is important, morale will be further decreased because:

1. The defective product must be repaired.
2. The employee's and management's dissatisfaction with each other will increase.
3. The entire quality image of the company will suffer.

This spiral effect can only be broken with time, effort, and discretionary expenditures to *prevent* further quality and morale reductions. The generally observed relation between discretionary, consequential, and total quality costs is shown graphically in Figure 2.

Management has a definite responsibility to determine how much the company will spend to improve quality. This decision should be made consciously using the advice of quality assurance personnel. The spending level should be reviewed periodically by responsible management. The final decision on future prevention spending should be made by management personnel who are organizationally above quality control, because they will be able to make the decision from a more informed position. Besides, this review of the scope of the quality commitment will make management more conscious of quality assurance's role in the company. The management should conclude that, even though prevention costs may be harder to justify and measure, these costs will do much to reduce the entire cost of quality.

COLLECTION OF QUALITY COSTS

Once you have determined what data you need to perform a quality cost analysis, you have to determine how to get it. Accounting records will help you initially. However, the object of a well organized quality cost system is to set up and maintain a set of quality cost centers which will gather and report the desired quality cost data in the future. Typically, the presently maintained records of a company are not arranged to readily reflect quality costs on such items as scrap cost, all types of inspection costs, quality engineering costs, vendor relations costs, etc. It will take a while to determine which records and reports should be kept so that an ongoing quality cost analysis can be maintained.

• Be prepared to be disappointed with the initial results, because of the monumental tasks that must be performed at this plant.

The quality situation presented in this case study makes it relatively easy to determine the quality costs for each category and analyze the problems in the plant. However, it is not the ease or difficulty of quality cost analysis that is

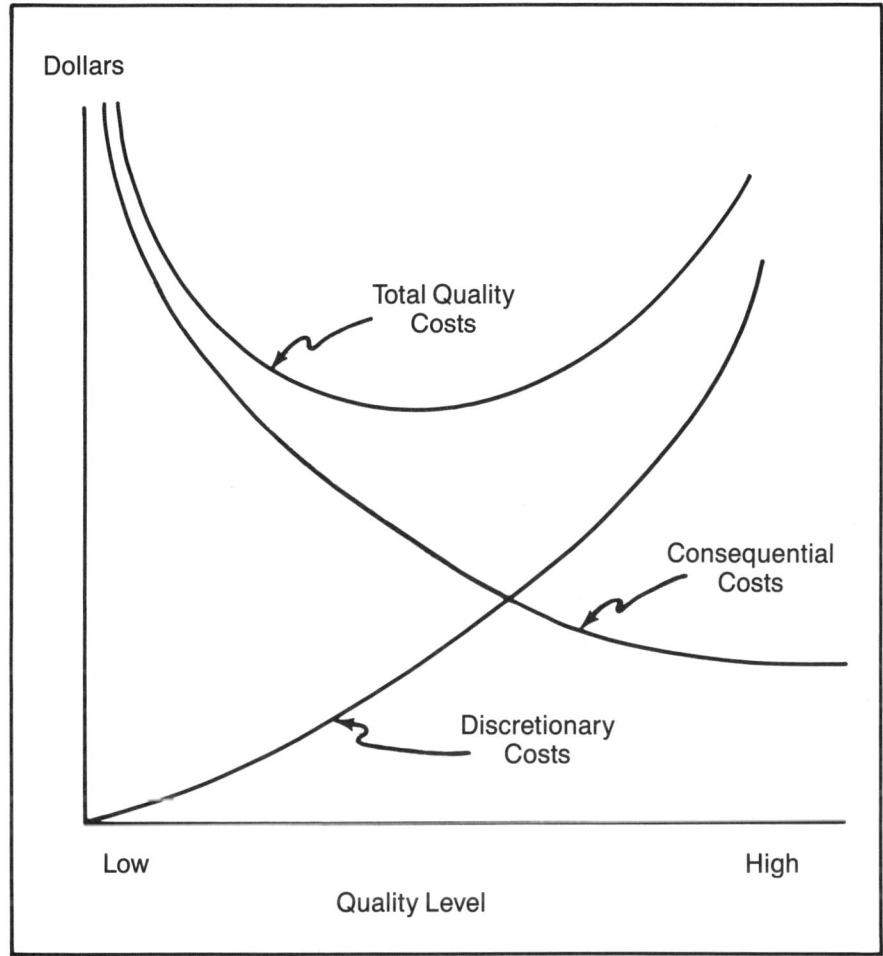

Figure 2. *Effect of discretionary and consequential spending on total quality cost.*

important. It is how the results are used to better the quality image of the product.

Industrial engineers working in quality control must strive to make this function strong. Quality control should make management—top and middle—conscious of:

- The present quality situation in the plant.
- What quality control can do to improve the situation in terms of defect presentation. This may mean the initiation of a basic quality cost study.

- How important management is to the quality improvement trends of the plant because of the power over spending. This definitely means that quality control should suggest ways for top management to allow quality control to improve quality.
- The fact that increased defect prevention spending is an effective way to reduce total quality costs.

CASE STUDY

Since many industrial engineers and, in fact, most managers are new on the quality cost scene, a basic cost collection and analysis case study is presented. Assume that you are a new plant manager in an electronics product manufacturing plant located outside the United States, but you report organizationally to the main United States office. You know that there has been quality deterioration at the plant, that there have been virtually no quality control personnel changes, and that previous management had not been able to correct the situation. What would you do to improve quality? This case study is an example of what should happen if you had a quality cost study performed.

Except for plant management, most of the employees are nationals. The organization chart for quality control is given in Figure 3. Of the two supervisors of product acceptance one is in charge of process control inspection and the other supervises all final product inspection and testing. The supervisor of salvage operations manages all personnel who determine the cause of detected internal failures. He is also responsible for rework of these products, because these operations have been separated from the main production line due to

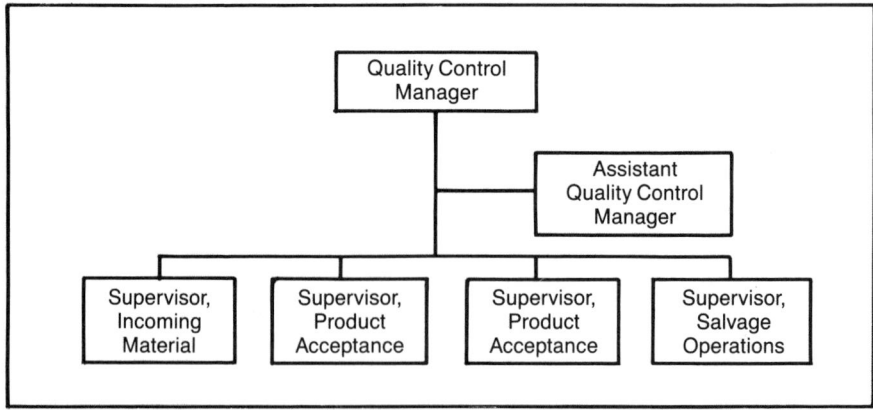

Figure 3. *Organization chart for quality control.*

high scrap rates. The economics of this arrangement are not studied here, but the arrangement makes quality cost data collection simpler.

Since this is a foreign operation, once product is shipped stateside it would not be returned to the plant in the event of an external failure. Repair is done at a U.S. location for all manufacturing locations, and charges are not maintained for each plant. Therefore, there are no external failure costs that can be used specifically in this analysis.

Data Collection

To collect the pertinent quality cost data the following things were done.[5]

Objective of the Analysis

The objective of the analysis was determined as follows: To observe trends in cost data (for each category and total) and to relate these results to management turnover and style. This objective was selected because turnover in management positions had been larger than expected.

Time Allocations for All Quality Personnel

The percentage of time spent by each person was categorized as appraisal, internal failure, or prevention. Table 1 presents this breakdown for all supervisory and management personnel. The breakdown is very important since it is used to proportion quality costs correctly among the three categories. The individuals were asked to give estimates of these percentages for each category.

Itemizing Quality Cost Data

Quality cost data was collected and categorized for each month from May 1975 to May 1976. Appraisal data was determined first, as suggested earlier. Table 2 details this data for the month of October 1975 only. The time percentage values of Table 1 were used for all 13 months of data since monthly fluctuations were not considered extreme. Tables 3 and 4 present the October data for internal failures and prevention. Note that the purpose for each cost is the same as listed under the description of each category in the beginning of this paper.

Percentage Breakdown of Costs

The percentage of total costs expended for each category was determined for each month. For October 1975 the results are shown in Table 5. The percentage values are characteristic of most organizations as discussed previously. Failures account for about 67% and appraisal about 25% of total costs, while

TABLE 1 – Time Percentage Breakdown for Quality Cost Purposes

Quality Cost Category	Costing Purpose	QC Manager	Assistant QC Manager	Supervisors, Product Acceptance	Supervisor, Incoming Material	Supervisor, Salvage Operations
				Percent of Time for Each Purpose		
Appraisal	Incoming	35%	10%		90%	
	In-Process/Product			40%		
	Audits		10	10		
	Data Analysis	5	10	10	10	
Internal Failure	Rework/Repair	5				60%
	Scrap					40
Prevention	Planning	30	40			
	Process QC			30		
	Training	10	10	10		
	Miscellaneous	15	20			
Total		100%	100%	100%	100%	100%

TABLE 2 – Summary of Appraisal Costs for October 1975

Purpose	Function	Percent of time	Monthly cost	Monthly total
Incoming	Manager	35	$ 560	
Material	Asst. Manager	10	80	
Testing and	Supervisor	90	495	
Inspection	Technician	100	310	
	6 Inspectors	100	1,320	$ 2,765
In-Process	QC Supervisors	40	$ 440	
Testing and	3 Technicians	100	930	
Inspection	20 Inspectors	100	4,400	5,770
Set-up for	Eng. Manager	40	$ 160	
Testing and	3 Engineers	40	920	
Inspection	Technician	40	124	1,204
Quality	Asst. Manager	10	$ 80	
Audits	QC Supervisor	10	110	
	Technician	100	310	500
Equipment	2 Engineers	10	$ 175	
Calibration	5 Tech. (testing)	10	155	
	2 Tech. (production)	30	186	516
Inspection	Manager	5	$ 80	
Data	Asst. Manager	10	80	
Analysis	2 Supervisors	10	100	
	2 Analysts	100	500	770
Materials			$ 200	200
Total				$11,725

TABLE 3 – Summary of Internal Failure Costs for October 1975

Purpose	Function	Percent of time	Monthly cost	Monthly total
Scrap	Supervision	40	$ 220	
	2 Operators	100	400	
	Reports	—	14,100	$14,720
Repair	Supervision	60	$ 330	
and Rework	3 Operators	100	8,600	
	Utilities	—	275	
	Overhead	—	275	9,480
Failure Analysis	32 Technicians	100	$ 9,920	9,920
Total				$34,120

TABLE 4 – Summary of Prevention Costs for October 1975

Purpose	Function	Poroont of time	Monthly cost	Monthly total
Quality	Manager	30	$480	
Planning	Asst. Manager	40	320	$ 800
Process	2 QC			
QC	Supervisors	35	$385	385
Equipment	Eng. Manager	20	$320	
Design	Test Engineer	60	330	650
Training	Manager	10	$160	
	Asst. Manager	10	80	
	2 QC Supervisors	10	110	350
Other	Design Engr. Mgr.	30	$480	
Departments	2 Design Engrs.	30	330	810
Misc.	Manager	15	$240	
Prevention	Asst. Manager	20	160	
Costs	Secretary	30	90	
	Analysts	30	440	930
Total				$3,925

TABLE 5 – Breakdown of Quality Costs for October 1975

Category	Cost	Percentage	Cumulative
Prevention	$ 3,925	7.9%	7.9%
Appraisal	11,725	23.6	31.5
Internal Failure	34,120	68.5	100.0
	$49,770	100.0%	

prevention is predictably low at 8%. Figure 4 is a plot of cumulative percentage expenditures by category for each month. Also shown are the total quality expenditures for each month. It can be seen that, even though total spending in March went up, the appraisal portion decreased slightly.

Analysis of Data/Management

The data collection phase has generated some good elementary quality cost figures to be used and manipulated in the detailed analysis. There are several bases that can be used to analyze quality costs. Most of these are covered in publications already referenced, especially [3]. However, since the relation to quality costs and management was the study objective, quality costs were first normalized to production by using the standard cost of production system already established at the plant. This also neutralizes the inflation effect. For each category an index of the quality cost per $1000 of standard cost of production was computed using the formula

$$\text{Category index} = \frac{\text{Quality cost in category}}{\text{Standard cost}/\$1000}$$

For example, in October standard cost of production was $2,250,000. The prevention index is

$$\text{Index} = \frac{3925}{2,250,000/1000} = 1.7$$

The usual quality cost analysis would proceed with a study of graphs and figures similar to Figures 4 and 5. We can see that prevention spending has

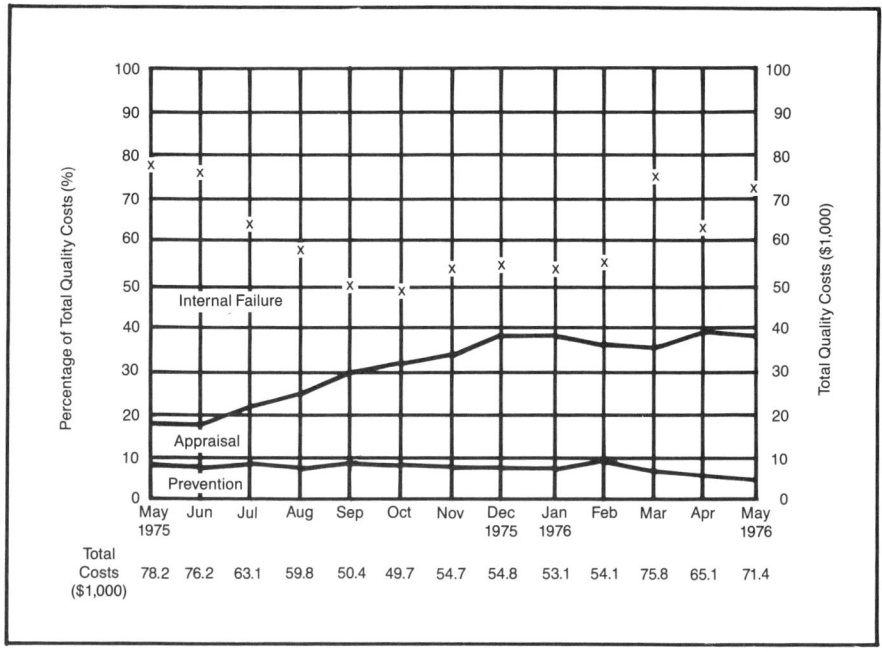

Figure 4. *Percentage of quality costs expended in each category.*

continued at about the same level and appraisal has grown steadily. Figure 5 indicates that the failure index decreased dramatically and then began a steady ascent in 1976. This means that the total cost of scrap and rework went down rapidly and then started to climb again. Computation of separate index values for scrap and rework shows that the value of the scrap index (in terms of standard cost of production) went from 39.0 in May 1975 to 4.9 in December while the rework index was relatively constant. In terms of actual quality dollars expended on the scrap function (see Table 3) the plant went from $49,950 in May 1975 to $12,200 in December, a 76% decrease! Note however, that the appraisal index steadily increased during this period.

In order to accomplish the objective, it was decided to introduce management personnel changes into the analysis. The times at which pertinent position turnovers or abolition took place are shown below.

January 1975	Plant manager
March 1975	Manager of materials procurement
April 1975	Supervisor, product acceptance
June 1975	Manager, manufacturing

July 1975	Design engineer
August 1975	Manager, manufacturing engineering
September 1975	Design engineer
February 1976	Production superintendent
May 1976	Plant manager

There has been only one quality control position turnover since January 1975—one supervisor of product acceptance. Of course, the large number of personnel changes indicate an organization in constant transition.

At the beginning of this case study you were asked to assume that you are a new manager of a plant which had a history of quality problems and virtually no quality personnel turnover. You are the new manager assuming the position in May 1976 and have requested that a quality cost study be performed.

Analysis of past quality improvement programs shows that the previous plant manager asked quality control to do an analysis to decrease the scrap. The results are shown in the reductions in the failure percentage of total costs, Figure 4, and index, Figure 5. The preventive cost category decreases throughout the study period. This is due to a cost reduction program which resulted in the loss of personnel in design engineering and the removal of the position of manager, manufacturing engineering (August 1975). This required reorganization of production and process control with increased workloads and decreased communication in several areas, including quality assurance.

In the last few months of 1975 a trade of prevention for appraisal costs was made, Figure 5. This management decision came when the effects of the scrap analysis of 9 months earlier were diminishing and product modifications were just introduced on the production line. Rather than troubleshooting the production problems by preventive analysis, the number of inspections was increased so most defects could be found and repaired. Once again the internal failure index started to climb rapidly. This trend continues through the remainder of the period. Note that none of these trends changed, even with new personnel in design (September) and production (February).

There are other indicators that can be used to analyze the situation; some good in one way, but poor in others. For example, total quality costs decreased through October, Figure 4, and remained fairly steady into February 1976. However, total production was fluctuating from September to March, so the indices of Figure 5 give a better idea of quality cost trends since they are production normalized. Generally, Figure 5 shows a fading preventive program, increased appraisal to try to overcome the diminishing effect of past effective preventive-type analysis. These results are characteristic of a weak quality assurance program, and a plant management attitude which believes that tighter control means more measuring to detect and repair defects. This generic type

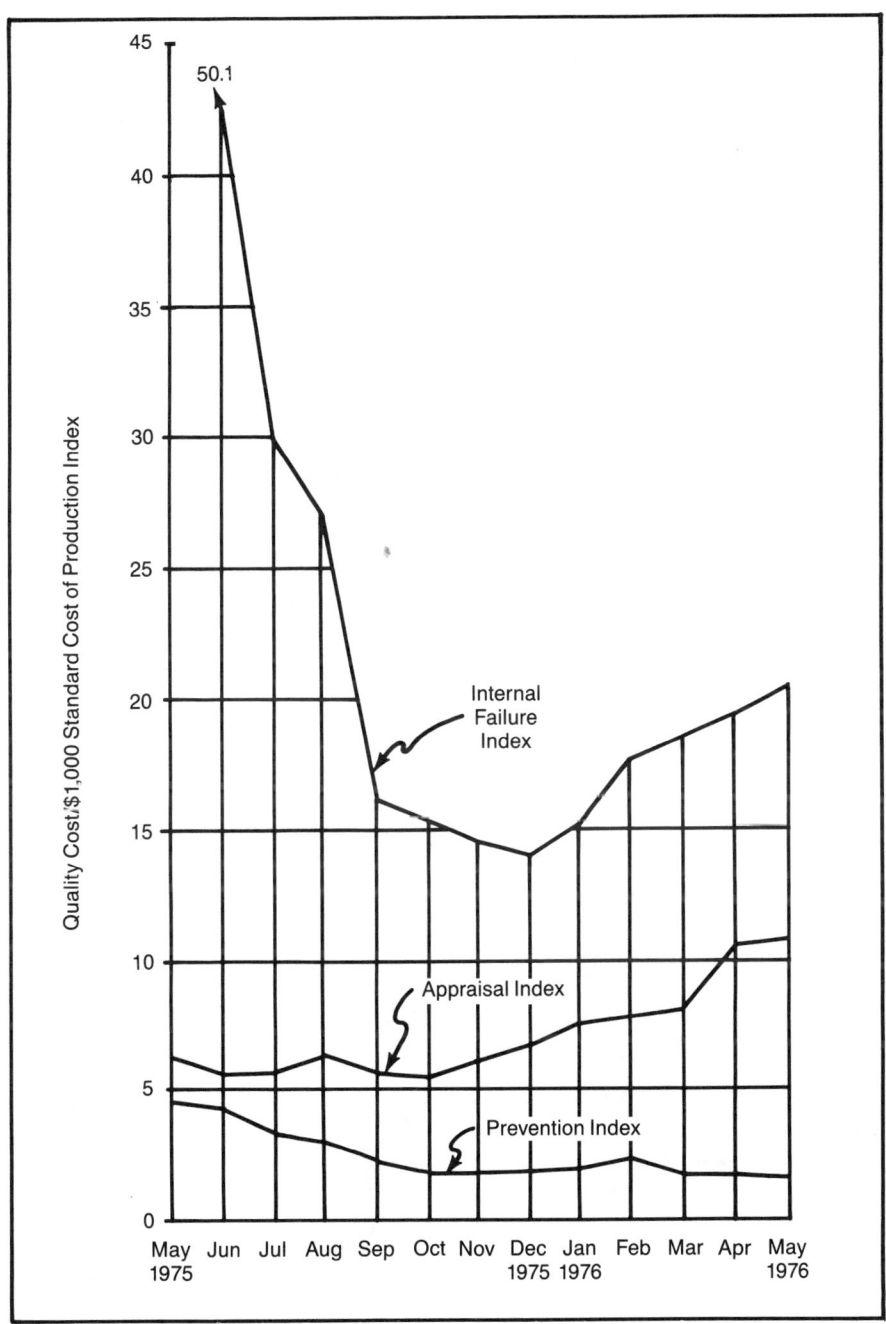

Figure 5. *Index value by category of quality cost per $1000 of standard cost of production.*

of quality philosophy has helped in causing the present large staff of 32 technicians in failure analysis, Table III, and a separate function for salvage operations, Figure 3, which has put quality control in the repair business. This can only increase the tension between production and quality control.

Now the new plant manager (you) must face a very bleak situation—weak quality analysis on a newly modified product, production and quality working against each other, low morale, and a relatively green staff in several key management positions.

Suggested Future Actions

Naturally, the new plant manager has many things to do—cost reduction, staff development, improvement of plant reputation, quality improvement, morale building, etc. Quality must be considered no less important than any of these other factors. In fact, a success in quality improvement may help some of the other situations. Some of the first actions the new manager should take are listed.

- Recognize that there is experience present in the existing quality control staff. Ask for support, help, and guidance from these people.
- Initiate a special quality study to determine defect sources, training needs, etc. This will initially increase prevention costs.
- Return the salvage operation function to production where it belongs.
- Make special efforts to improve communication between design, production, and quality control; at first through you, then expect continued efforts to improve quality.
- Request that quality control continue the collection, analysis, and reporting of quality cost information.
- Establish and maintain quality objectives for the entire plant and specific appropriate people.
- Sincerely realize that top management guides the quality image of the plant.

Suggested Readings

Special Report. 1975. Money (mun′e) n., 1: a measure of value. *Quality Progress.* September, 1975, pp. 24-25.

Juran, J. M. and F. M., Gryna, Jr. 1970. *Quality Planning and Analysis.* McGraw-Hill Book Company. New York. pp. 37-69.

ASQC. 1971. *Quality Costs—What and How.* American Society for Quality Control. Milwaukee, WI, Second edition.

Feigenbaum, A.U. 1961. *Total Quality Control.* McGraw-Hill Book Company. New York.

Solorzano, J. 1976. Investigation of Quality Costs for Improved Management. (Unpublished Masters Thesis). University of Texas at El Paso.

Reprinted from February 1978 issue of *Industrial Engineering.*

9

A Management Process for Total Quality Control and Improvement

Joseph H. Bellefeuille

Electronic goods have become world market commodities. Their demand has spread rapidly around the globe. Electronics manufacturers can no longer rely solely on introducing new product features to hold their customer base. Customers have new "benchmarks" by which to judge goods and services and will no longer accept old performance standards. They look to the supplier that has demonstrated faster installation and "turn on" time and better on-line system reliability. Hence, to maintain their existing customer base and to capture new customers, manufacturers need continuous quality improvement of all operational processes. In this environment, as long as better is possible, good is never good enough.

This mention of customer is not casual. Those organizations that fully understand "world class" competition never take customers for granted. They develop policies that hold customers at the focal point of everything they do. Their strategy is to align their human resources so that each employee fully understands his or her role and how it fits into the "big picture."

To accomplish this understanding, businesses employ total quality control (TQC). Total Quality Control is an interlocking arrangement of procedures and practices which ensures that all employees in every department are adequately trained and directed to continuously implement congruent improvements in quality, service, and total cost. TQC enables an organization to meet or exceed its customers' expectations and environmental requirements consistently while also maximizing resource effectiveness.

This chapter presents a management process for implementing TQC. The management process spells out how policies can be deployed, how capabilities of individuals and organizations can be assessed and improved, and how a continuous cycle of quality improvement can be implemented.

THE TQC MANAGEMENT PROCESS - UNDERSTANDING AND DEPLOYING TQC POLICIES

One of the first steps for implementing TQC is to understand the need for successful deployment of TQC policies. Companies are not pursuing excellence until they involve and challenge their people. To tap into the human resource, management has to provide clear instructions that direct, inspire, and motivate all employees. These instructions take the form of policy statements.

A "policy" must have a customer focus and must embrace all employees. It must be concisely written. At the highest level of a company, a policy statement cannot be specific enough to be considered a goal. Each organization within a corporation must interpret the policy, making it more specific and meaningful to their employees. As the policy is deployed down and throughout the entire corporation, it should take on more and more specific meaning, until all employees can relate to it and understand the roles that they must play for the corporation to gain and maintain a competitive edge.

After an organization has grasped the policy, the entire organization has to be paced, tracked, and coordinated. This is accomplished by recording individual actions that must be taken, along with the persons responsible for those actions, and the target date for completion. By maintaining a series of interlocking "action records" the entire organization's improvement can be coordinated. This ensures congruent improvements.

For a person or organization to be accountable for the necessary actions, they must exhibit sound problem-solving competencies. An understanding of these competencies is essential for successfully executing the TQC management process. The following presents an approach for assessing and improving problem-solving capabilities required for TQC.

ASSESSING AND IMPROVING PROBLEM-SOLVING CAPABILITIES

To improve an individual's or organization's capabilities, various levels of competence and their relationships with team and task skills must be understood. Only then can results be identified and improved. Individuals and isolated departments will possess some of the required skills. As individuals and organizations go through a transition from lower skill levels toward "world-class" skills, they will go through distinct and identifiable competency levels. The various levels of competency can be analyzed in numerous ways. The following discussion introduces a conceptual framework for understanding levels of competence. Competence can be classified into four different levels based on consciousness—individuals' realization of their TQC problem-solving abilities—and competence—ability to understand, solve, and improve quality problems. The different levels and their main characteristics are detailed in Table 1.

TABLE 1 – Levels of Competence

Level of Competence	Main Characteristics
I. *Unconscious* Incompetence	• Tolerates problems. • Does not see problems as opportunities for improvement.
II. *Conscious* Incompetence	• Recognizes problems as barriers to success. • Problem-solving skills limited to reactions only ("fire fighter").
III. Conscious *Competence*	• Addresses and solves problems. • Focuses only on "good news" data. • Ignores "bad news" data. • Leaves opportunities undetected.
IV. *Unconscious* Competence	• Uses all information to recognize and expose problems as opportunities for improvement. • Able to solve problems quickly, tapping the appropriate "expert" resources.

The classifications in Table 1 help to place competence of individuals and organizations into a structured framework. This framework is useful for understanding and assessing which persons or organizations lack a particular skill.

The next section relates the four levels of competence to "team" and "task" skills so that skills can be improved for a successful TQC implementation.

RELATING COMPETENCE WITH SKILLS

Unconscious competence is synonymous with "world class" performance. To achieve this status, companies have to undergo a cultural transformation. This requires skill development along two seemingly orthogonal axes. Table 2 is a matrix with rows representing team skills and columns representing task skills. These skills are marked "low" and "high." Each element of the matrix in Table 2 has a letter (A through J) that represents the cell number and a Roman numeral which illustrates the level of competence as discussed previously. To analyze isolated tasks or processes, traditional techniques use statistical methods, flow charting, and other similar task skills. These techniques are usually used in the privacy of one's office or desk. As these skills are developed through training one can imagine progress from cell A to cell B. This imagery will enable the individual to bring about moderate process improvements. However, this technique alone will not raise the person to a Level II - conscious incompetence. Without team building skills, efforts to deal with problems that cross organizational boundaries will fail. As another example, attempts to bring people together to identify problems will have little impact without adequate task skills. This might be the situation shown in cells A & D. Once the team reaches cell

TABLE 2 – Team vs. Task Skills

Team Skills	HIGH	II G	III H	IV J
		I D	II E	III F
	LOW	I A	I B	II C
		LOW		HIGH

Task Skills

G, it is able to focus on a specific problem using brainstorming, nominal group technique, etc. A lack of task skills limits the team's ability to pinpoint the causes of identified problems.

Existence in Level II cells (C, E, & G)—conscious incompetence—is a painful struggle. Problems are seen as barriers to success. They are often considered the "other guy's" fault. Level III organizations are characterized by their exclusive focus on "good news" data. "Bad news" data are ignored or withheld from wide publication. This causes many opportunities for improvement to go undetected.

In the lower three levels (I - III) employees generally react to situations by expending the minimum acceptable effort. In this bureaucratic environment, finding an excuse that is preventing success is equally acceptable as a successful result. Often excuses are sought more enthusiastically than solutions. The fourth level (IV) is characterized by a "can-do attitude." We refer to this as an entrepreneurial environment. Entrepreneurs see problems as precious gems that mark the path to success.

Organizations are not monolithic. Hence, some individuals and departments will exhibit levels of competence higher or lower than most in the organization. A comparison of the behavior of individuals with the tables allows appropriate action to be taken to improve the situation. Perhaps some team skill training, then some task skill training, and then some more team skill training will eventually transform the organization to Level IV. One youthful sailor referred to this as being much like tacking a craft into the wind to gain forward momentum—as one's company "sails toward world-class waters."

EDUCATION, EDUCATION, EDUCATION, EDUCATION

All employees must be exposed to varying depths of understanding of broad-based statistical problem-solving and team-building skills. The objective is to achieve a working knowledge of the various techniques. The necessary level of understanding depends on the job function and vertical position within the organization. As an example, a vice-president of engineering may well be required to have a working knowledge of simple statistical methods, a familiarity with design of experiment techniques and an awareness of orthogonal array analysis. Engineers on the vice-president's staff might be called on to have a good working knowledge of all and enough theoretical understanding of design of experiment to be able to significantly alter its use. The deep understanding of orthogonal array analysis would be left to the technical specialists within or even outside of the corporation. Employees in less technical environments would require correspondingly less understanding of the techniques at all levels.

The educational effort required for this transformation is similar to the story about the three most important attributes to consider in buying real estate.

Location, location, and location are the three most important considerations in buying a new home for family or business. To bring about corporate transformation, the first four considerations are education, education, education, and education. The organizations that are vying for world-class status realize this and take steps to cope with it by establishing education councils, school boards, etc. One company asked their middle managers to become deans of education for their organization. They formed an internal "quality college." Education must not be confined to formal meetings. Only about one fourth to one third of an adult's education comes from the formal classroom. The supervisor must continue educating subordinates on a one-on-one basic through actual work.

Implementing TQC policies provides direction for the organization. Understanding competency and its relationship to team and task skills provides a means of assessing and improving an organization's competence. Thus, the requirements for control—target, detection, and correction—are fulfilled. How do we implement TQC? TQC goes beyond control—it is characterized by the "continuous implementation of congruent improvements throughout the entire company." What, then, is the essence of the "management process?"

IMPLEMENTING TQC THROUGH A CONTINUOUS PROCESS— PICA CYCLE

Those organizations that have a bias toward doing things—getting on with it—seem to have a great tendency to succeed. They move quickly toward excellence. This bias toward doing has been characterized by the sequence "ready-fire-aim." In this context, the idea is not to over analyze a situation, but rather size it up, do something, see how it works, and try again based on what was learned.

In this chapter, the approach is a structured, cyclical-phased method of accomplishing improvements. This cycle has been referred to by Ishikawa (1985) as the Deming Cycle and by Deming (1982) as the Shewhart Cycle. We will refer to it here as the plan-implement-check-act (PICA) cycle. Improvement is a continuous process of planning for improvement, implementing the plans, checking to see what the outcome is, and taking appropriate action to either

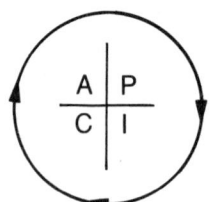

Figure 1. PICA Cycle

correct toward the target or continue on with the original plans. The next few sections explore the PICA cycle in detail.

Plan

Plan, when defined as a noun, means "an orderly arrangement of parts of an overall design or objective," and when defined as a verb it means "to arrange the parts of a design or objective." In the context of this chapter, a result of the planning phase will be considered as the noun definition and the process of planning as the verb definition.

A planning process is a series of decisions that reduce or eliminate the gap between what is happening—or may happen—and what we would like to see happen.

Arnold (1978) suggests this gap manifests itself as the following:

1. a situation that is wrong and needs to be corrected;
2. a situation that is threatening and needs to be prevented;
3. a situation that is inviting and needs to be accepted; and
4. a situation that is missing and needs to be provided.

The planning and decision-making process begins when this gap is perceived and ends with an orderly arrangement of actions that will narrow or eliminate it. The process can be carried out in any number of ways.

A recommended method for action is the following:

1. Determine the issues: Why is a decision necessary?
2. State the purpose: What needs to be decided?
3. Set criteria: What needs to be achieved, preserved, or avoided?
4. Establish priorities: What are the absolutely necessary criteria? What other criteria should be met?
5. Search for solutions: How can criteria be met?
6. Test the alternatives: How does each solution "stack-up" against the criteria?
7. Trouble-shoot the decision: What could go wrong? How could the decision be improved?

These steps employ distinct skills and techniques and require skill-training and practice.

Implement

There are two facets to an implementation process—one is education and training and the other is the implementation of work. Education and training improve skill levels and increase awareness of the need for change. Work can be implemented in parallel with education. The implementation and checking phases provide on-the-job training that is a vital reenforcement of the classroom training.

If the planning assumptions used during the previous phase were completely valid and invariant, the implementation phase would be painless and error-free. Changes in the environment are continuous. Therefore, plans cannot keep pace with implementation. The best strategy is to seek consensus by involving the implementors in the planning process. Thus, the intent of the plan can be carried out. A point of caution—problems arising from the implementation phase can appear throughout the improvement process and afterward in the control phase. Control is a stable state that is desirable for a process between periods of improvement. It is important to follow the intent of the plan and carefully observe the results.

Check

Once plans have been made and implemented and the results are forthcoming, we must check to see that the changes are bringing about the desired results. It is nearly always necessary to take some sort of action in one part of the system or another whenever implementing changes. This is the subject of the next phase of the PICA cycle and cannot be carried out properly without careful analysis and understanding of the results. The checking phase calls for deploying a full range of skills from simple observation to sophisticated statistical analysis. This may require the use of more sophisticated measuring techniques or instruments than would be used for normal operations. For instance, a simple go-no-go gauge may be employed for inspecting printed wiring boards during fabrication, whereas, during check out of a planned change, a dial indicator and a micrometer may be used to establish what the process capability is. The object is to compare actual results with planned, expected, or desired results.

Act

Once the analysis is complete and understood, action must be taken—therefore, this is called the act phase of the PICA cycle. The appropriate action may range from taking steps to put the changed process under control to abandoning the plans and starting over. A point of caution—often when the exact, expected results are achieved, an undesirable event will occur somewhere else in the system. This, too, must be coped with and should not necessarily be cause for abandonment.

Control is not a static state. A process under control will tend to drift into higher and higher levels of disorder as sources of variation creep into the process. Therefore, it is often appropriate to commence a new PICA cycle as soon as the process is put under control. A good strategy is to apply successful

PICA cycles to other parts of a process or organization. In this way a company can exponentially spiral to higher and higher achievement levels.

In the context of this chapter, the PICA cycle may take several weeks to a few months to progress through. The key is to see each phase as it fits into the continuous improvement sequence. Do not get "bogged down" by considering a phase as an end in itself. Anticipate events in the next phase or even the next cycle. This anticipation gives a good perspective on appropriate level of detail and care being taken in the present work.

The PICA cycle can be employed at all levels in the corporate structure. In fact, when employed and tracked with "action records," the cycle can tie detailed operations improvement work into corporate strategy and direction. This provides a systematic approach to continuous improvement from the bottom to the top of the corporation. When applied to detailed operations improvement, the cycle duration may be measured in hours or even smaller increments of time. Again, this is part of a larger improvement process. Avoid the temptation to skip steps—keep going.

SUMMARY AND CONCLUSION

We commenced our examination of the management process by establishing a need to change. The change has to focus on the customer and embrace all employees. Those who prosper get on with the business of improvement. As companies progress from a general bureaucratic state to an entrepreneurial environment, organizations and individuals progress from a state of relative incompetence to a degree of excellence characterized by extreme competence. The very definition of excellence suggests continually expanding one's competence. This chapter has examined four levels of competence ranging from unconscious incompetence to unconscious competence. It is helpful to understand these four phases of competence through which individuals and organizations progress as they transform to exellence—continuous improvement. Improvement herein is described as a process that continues through the PICA cycle over and over—literally forever—on the path to entrepreneurial excellence.

Excellence is a transformation that calls for a paradigm shift in the minds of all employees, a complete change in how we think about our fellow employees and our bosses. We all work in a process, and we are all experts about the details and problems in that process. It is our obligation to work with those responsible for improving the process in which we work. Those responsible for designing the process—the supervisor, the manager, the vice-president, the owner, the board of directors—are obliged to view their daily problems in a completely revolutionary way to continuously improve it.

This new idea of cooperation supplants the old idea of adversarial management and starts at the top and progresses down and throughout the organization.

This direction is imparted to the organization through policy. Understanding the management process is key to bringing this much-desired change to the electronics industry.

Suggested Readings

Arnold, J. D. 1978. *The Art of Decision Making.* AMACOM. New York.
Deming, W. E. 1984. *Quality, Productivity, and Competitive Position.* MIT, Center for Advanced Engineering Study. Cambridge, MA.
Ishikawa, K. 1985. *What Is Total Quality Control?* Prentice-Hall, Inc. Englewood Cliffs, N.J.
Juran, J. M. 1964. *Managerial Breakthrough.* McGraw Hill. New York.

Joseph Bellefeuille is an Engineering Manager at AT&T Technologies.

10

Design Optimization Case Studies

Madhav S. Phadke

Designing high quality products at low cost is an economic and technological challenge to the engineer. A systematic and efficient way to meet this challenge is a new method of design optimization for performance, quality, and cost. The method, called "robust design," has been found effective in many areas of engineering design. In this paper, the basic concepts of robust design will be discussed and two applications will be described in detail. The first application illustrates how, with a very small number of experiments, highly valuable information can be obtained about a large number of variables for improving the life of router bits used for cutting printed wiring boards from panels. The second application shows the optimization of a differential operational amplifier circuit to minimize the dc offset voltage by moving the center point of the design, which does not add to the cost of making the circuit.

AN ECONOMIC AND TECHNOLOGICAL CHALLENGE

A goal of product or process design is to provide good products under normal manufacturing conditions and under all working conditions throughout intended life. Further, the cost of making the product (including development and manufacturing) must be low, and the development must be speedy to meet the market needs. Achieving these goals is an economic and technological challenge to the engineer. A systematic and efficient way to meet this challenge is the method of design optimization for performance, quality, and cost, developed by Genichi Taguchi of Japan. Called "robust design," the method has been found effective in many areas of engineering design in AT&T, Ford, Xerox, ITT, and other American companies. In this paper we will describe the basic concepts of robust design and describe the following two applications in detail:

- A router bit life improvement study
- Optimization of a differential operational amplifier circuit

The first application illustrates how, with a very small number of experiments, highly valuable information can be learned about a large number of variables for improving the life of router bits used for cutting printed wiring boards from panels. The study also illustrates how product life improvement projects should be organized for efficiency. This case study involved conducting hardware experiments, whereas in the second case study a computer simulation model was used. The second application shows the optimization of a differential op-amp circuit to minimize the dc offset voltage. This is accomplished primarily by moving the center point of the design, which does not add to the cost of making the circuit. Reducing the tolerance could have achieved similar improvement, but at a higher manufacturing cost!

PRINCIPLES OF ROBUST DESIGN

Genichi Taguchi views the quality of a product in terms of the total loss incurred by a society from the time the product is shipped to the customer. The loss may result from undesirable side effects arising from the use of the product and from the deviation of the product's function from the target function. What is novel about this view is that it explicitly includes the cost to the customers and the notion that even products that meet the "specification limits" can impart loss due to nonoptimum performance. For example, the amplification level of a public telephone set may differ from cold winter to hot summer; it may differ from one set to another; also it may deteriorate over a period of time. A consequence of this variation is that a user of the phone may not hear the conversation well and that an expensive compensation circuit may have to be provided. The quadratic loss function can estimate with reasonable accuracy the loss due to functional variation in most cases. For a broad description of the principles of robust design see References 2 to 5.

Sources of Variation

Robust design is aimed at reducing the loss due to variation of performance from the target. In general, a product's performance is influenced by factors that are called noise factors. There are three types of noise factors:

1. *External*—factors outside the product, such as load conditions, temperature, humidity, dust, supply voltage, vibrations from nearby machinery, human errors in operating the product, and so forth.
2. *Manufacturing imperfections*—the variation in the product parameters from unit to unit, inevitable in a manufacturing process. For example, the value of a particular resistor in a unit may be specified as 100 kilohms, but in a particular unit it turns out to be 101 kilohms.
3. *Deterioration*—when the product is sold, all its performance characteristics may be right on target, but as years pass by, the values or characteristics of individual components may change, leading to product performance deterioration.

One approach to reducing a product's functional variation is to control the noise factors. For the telephone set example, it would mean to reduce allowable temperature range, to demand tighter manufacturing tolerance or to specify low-drift parameters. These are costly ways to reduce the public telephone set amplification variation. What then is a less costly way? It is to center the design parameters in such a way as to minimize sensitivity to all noise factors. This involves exploiting the nonlinearity of the relationship between the control factors, the noise factors, and the response variables. Here, control factor means a factor or parameter over which the designer has direct control and whose level or value is specified by the designer.

Note that during product design one can make the product robust against all three types of noise factors described above, whereas during manufacturing process design and actual manufacturing one can reduce variation due to manufacturing imperfection, but can have only minor impact on variation due to the other noise factors. Once a product is in the customer's hand, warranty service is the only way to address quality problems. Thus, a major portion of the responsibility for the quality and cost of a product lies with the product designers and not with the manufacturing organization.

Steps in Product/Process Design

Product and process design are complex activities involving many steps. Three major steps in designing a product or a manufacturing process are system design, parameter design or design optimization, and tolerance design. System design consists of arriving at a workable circuit diagram or manufacturing process layout. The role of parameter design or design optimization is to specify the levels of control factors that minimize sensitivity to all noise factors. During this step, tolerances are assumed to be wide so that manufacturing cost is low. If parameter design fails to produce adequately low functional variation of the product, then during tolerance design, tolerances are selectively reduced on the basis of cost effectiveness.

The Design Optimization Problem

A product or a process can be represented by a block diagram (figure 1) proposed by Taguchi and Phadke.[3] The diagram can also be used to represent a manufacturing process or even a business system. The response is represented by y. The factors that influence the response can be classified into four groups as follows:

1. Signal factors (M): These are the factors that are set by the user/operator to attain the target performance or to express the intended output. For example, the steering angle is a signal factor for the steering mechanism of an automobile. The speed control setting on a fan and the bits 0 and 1 transmitted in communication systems are also examples of signal factors. The signal factors are selected by the engineer on the basis of engineering knowledge. Sometimes two or more signal factors are used in combination; for example, one signal factor may be used for coarse tuning and one for fine tuning. In some situations, signal factors take on a fixed value, as in the two applications described in this paper.
2. Control factors (z): These are the product design parameters whose values are the responsibility of the designer. Each of the control factors can take more than one value; these multiple values will be referred to as levels or settings. It is the objective of the design activity to determine the best levels of these factors. A number of criteria may be used in defining the best levels; for example, we would want to maximize the stability and robustness of the

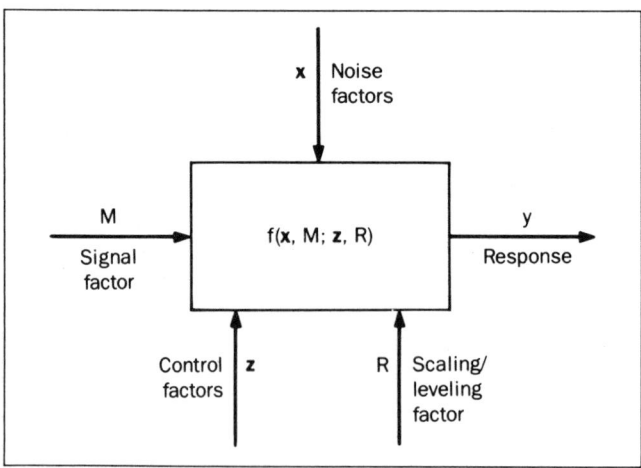

Figure 1. *Block diagram of a product or process.*

design while keeping the cost to a minimum. Robustness is the insensitivity to noise factors.

3. Scaling/leveling factors (R): These are special cases of control factors that can be easily adjusted to achieve a desired functional relationship between the signal factor and the response y. For example, the gearing ratio in the steering mechanism can be easily adjusted during the product design phase to achieve the desired sensitivity of the turning radius to a change in the steering angle. The threshold voltage in digital communication can be easily adjusted to alter the relative errors of transmitting 0's and 1's. Scaling/leveling factors are also known as adjustment design parameters.

4. Noise factors (x): Noise factors, described earlier, are the uncontrollable factors and the factors that we do not wish to control. They influence the output y and their levels change from one unit of the product to another, from one environment to another, and from time to time. Only the statistical characteristics of the noise can be known or specified, not their actual values.

Let the dependence of the response y on the signal, control, scaling/leveling, and noise factors be denoted by

$$y = f(x, M; z, R)$$

Conceptually, the function f consists of two parts:
$g(M; z, R)$, which is the predictable and desirable functional relationship between y and M, and $e(x, M; z, R)$, which is the unpredictable and less desirable part. Thus,

$$y = g(M; z, R) + e(x, M; z, R)$$

In the case where we desire a linear relationship between y and M, g must be a linear function of M. All nonlinear terms will be included in e. Also, the effect of all noise variables is contained in e.

The design optimization can usually be carried out in two steps:

- Find the settings of control factors to maximize the predictable part while simultaneously minimizing the unpredictable part. This can be accomplished through an optimization criterion called signal-to-noise (S/N) ratio. In this step the variability of the functional characteristic is minimized.
- Bring the predictable part, $g(M; z, R$, on target by adjusting the scaling/leveling factors.

Design problems come in a large variety. For a classification of design problems and the selection of S/N ratios see Reference 3.

ROUTER BIT LIFE IMPROVEMENT

The Routing Process

Typically, AT&T printed wiring boards are made in panels of 18 × 24 inch size. Appropriately sized boards, say 8 × 4 inches, are cut from the panels by stamping or by routing. A benefit of routing is that it gives good dimensional control and smooth edges, thus reducing friction and abrasion during circuit pack insertion. However, when the router bit gets dull, it produces excessive dust, which cakes on the edges and makes them rough. In such cases, a costly cleaning operation is necessary to smooth the edges. But changing the router bits frequently is also expensive.

The routing machine has four spindles, all synchronized in rotational speed, horizontal feed (x-y feed) and vertical feed (in-feed). Each spindle does the routing operation on a separate stack of panels. Two to four panels are usually stacked for cutting by a spindle. The cutting process consists of lowering the spindle to an edge of a board, cutting the board all around using the x-y feed of the spindle, and then lifting the spindle. This is repeated for each board on the panel.

Our objective in this experiment was to increase the life of the router bits, primarily in regard to the onset of excessive dust formation. The dimensions of the board were well in control and were not an issue.

Selection of Control Factors and Their Levels

Selecting appropriate control factors and their alternate settings is an important aspect of optimization. Prior knowledge and experience about the process is used in this selection. The alternate settings are called levels. It is a good practice to choose these levels wide apart so that a broad design space is studied in one set of experiments and there is a potential for major improvement. For the routing process, the eight control factors listed in Table 1 were chosen.

Suction is used around the router bit to remove the dust as it is generated. Obviously, higher suction could reduce the amount of dust retained on the boards. The starting suction was 2 inches of mercury—the maximum available for the pump. We chose 1 inch of mercury as the alternate level, with the plan that if a significant difference in the dust was noticed, we would invest in a more powerful pump. Related to the suction are suction foot and the depth of backup slot. The suction foot determines how the suction is localized near the cutting point. Two types of suction foot were chosen: solid ring and bristle brush. Underneath the panels being routed is a backup board. Slots are precut in the backup board to provide air passage and a place for dust to temporarily accumulate. The depth of the slots was a control factor in this study.

TABLE 1 – Control Factors and Levels for the Routing Process

Factor	Level			
	1	2	3	4
A. Suction (in of Hg)	1	2 *		
B. x-y feed (in/min)	60*	80		
C. In-feed (in/min)	10*	50		
D. Type of bit	1	2	3	4*
E. Spindle position‡	1	2	3	4
F. Suction foot	SR	BB*		
G. Stacking height (in)	3/16	1/4*		
H. Depth of slot (mils)	60*	100		
I. Speed (rpm)	30,000	40,000*		

† Spindle position is not a control factor. In the interest of productivity, all four spindle positions must be used.
* Denotes starting condition for the factors.

Stack height and x-y feed are control factors related to the productivity of the process; that is, they determine how many boards are cut per hour. The 3/16-inch stack height means three panels were stacked together, while 1/4-inch stack height means four panels were stacked together. The in-feed determines the impact force during the lowering of the spindle for starting to cut a new board. It could influence the life of the bit by causing breakage or damage to the point. Four different types of router bits made by different manufacturers were used. The router bits varied in cutting geometry in terms of the helix angle, the number of flutes, and the point. Spindle position was not a control factor. All spindle positions must be used in production, otherwise productivity would suffer. It was included in the study so that we could find best settings of the control factors to work well with all four spindles.

In addition to the variation from spindle to spindle, the noise factors for the routing process are the bit-to-bit variation, the variation in material properties within a panel and from panel to panel, the variation in the speed of the drive motor, and similar factors.

The Orthogonal Array Experiment

The full factorial experiment to explore all possible factor-level combinations would require $4^2 \times 2^7 = 2048$ experiments. Considering the cost of material, time, and availability of facilities, the full factorial experiment is prohibitively large. However, it is unnecessary to perform the full factorial experiment because processes can usually be characterized by relatively few parameters. An orthogonal array design with 32 experiments was created from the L_{16} array and the linear graphs given in Reference 1. The array appears in Table II. This design allowed us to obtain uncorrelated estimates of the main effect of each control factor as well as of the spindle position, and the interactions between x-y feed and speed, in-feed and speed, stack height and speed, and x-y feed and stack height. This information about the effect of control factors was used to decide the best setting for each factor. The 32 experiments were arranged in groups of four so that for each group there was a common speed, x-y feed, and in-feed, and the four experiments in each group corresponded to four different spindles. Thus each group constituted a machine run using all four spindles, and the entire experiment could be completed in eight runs of the routing machine.

The study was conducted with one bit per experiment; thus a total of only 32 bits was used. During each machine run, the machine was stopped after every 100 inches of cut (100 inches of router bit movement in the x-y plane) so that the amount of dust could be inspected. If the dust was beyond a predetermined level, the bit was recorded as failed. Also, if a bit broke, it was obviously considered to have failed. Otherwise, it was considered as having survived.

Before the experiment was started, the average bit life was estimated at around 850 inches. Therefore, to save time, each experiment was stopped at 1700 inches of cut, which is twice the estimated original average life, and the survival or failure of each bit was recorded.

Table 2 gives the experimental data in hundreds of inches. A reading of 3.5 means that the bit failed between 300 and 400 inches. Other readings have similar interpretations, except the reading of 17.5, which means survival beyond 1700 inches, the point where the test was terminated. There are 14 readings of 0.5, indicating extremely unfavorable conditions. There are eight cases of life equal to 17.5, indicating very favorable conditions. During experimentation, it is important to take a broad range for each control factor so that roughly equal numbers of favorable and unfavorable conditions are created. In this way, much can be learned about the optimum settings of control factors.

Analysis of the Life Data and Results

Two simple and separate analyses of the life data were performed to determine the best level for each control factor. This first analysis was to determine the effect of each control factor on the failure time. The second analysis was performed to determine how changing the level of each factor changes the survival probability curve (life curve).

The first analysis was performed by the standard procedure for fractional factorial experiments given in Suggested Readings (Hicks and Box). In this

Table 2 – Experiment Design and Observed Life for the Routing Process

Experi-ment No.	Suction A	x-y feed B	In-feed C	Bit D	Spindle E	Suction foot F	Stack height G	Depth H	Speed I	Observed life*
1	1	1	1	1	1	1	1	1	1	3.5
2	1	1	1	2	2	2	2	1	1	0.5
3	1	1	1	3	4	1	2	2	1	0.5
4	1	1	1	4	3	2	1	2	1	17.5
5	1	2	2	3	1	2	2	1	1	0.5
6	1	2	2	4	2	1	1	1	1	2.5
7	1	2	2	1	4	2	1	2	1	0.5
8	1	2	2	2	3	1	2	2	1	0.5
9	2	1	2	4	1	1	2	2	1	17.5
10	2	1	2	3	2	2	1	2	1	2.5
11	2	1	2	2	4	1	1	1	1	0.5
12	2	1	2	1	3	2	2	1	1	3.5
13	2	2	1	2	1	2	1	2	1	0.5
14	2	2	1	1	2	1	2	2	1	2.5
15	2	2	1	4	4	2	2	1	1	0.5
16	2	2	1	3	3	1	1	1	1	3.5
17	1	1	1	1	1	1	1	1	2	17.5
18	1	1	1	2	2	2	2	1	2	0.5
19	1	1	1	3	4	1	2	2	2	0.5
20	1	1	1	4	3	2	1	2	2	17.5
21	1	2	2	3	1	2	2	1	2	0.5
22	1	2	2	4	2	1	1	1	2	17.5
23	1	2	2	1	4	2	1	2	2	14.5
24	1	2	2	2	3	1	2	2	2	0.5
25	2	1	2	4	1	1	2	2	2	17.5
26	2	1	2	3	2	2	1	2	2	3.5
27	2	1	2	2	4	1	1	1	2	17.5
28	2	1	2	1	3	2	2	1	2	3.5
29	2	2	1	2	1	2	1	2	2	0.5
30	2	2	1	1	2	1	2	2	2	3.5
31	2	2	1	4	4	2	2	1	2	0.5
32	2	2	1	3	3	1	1	1	2	17.5

* Life was measured in hundreds of inches of movement in x-y plane. Tests were terminated at 1700 inches.

analysis of variance, the effect of censoring was ignored. The results are plotted in Figure 2. The following conclusions are apparent from the plots:

- 1-inch suction is as good as 2-inch suction.
- Slower x-y feed gives longer life.
- The effect of in-feed is small.
- The starting bit is the best of the four bit types.
- The differences among the spindle positions are small.
- A solid ring suction foot is better than the bristle brush type.
- Lowering the stack height makes a large improvement.

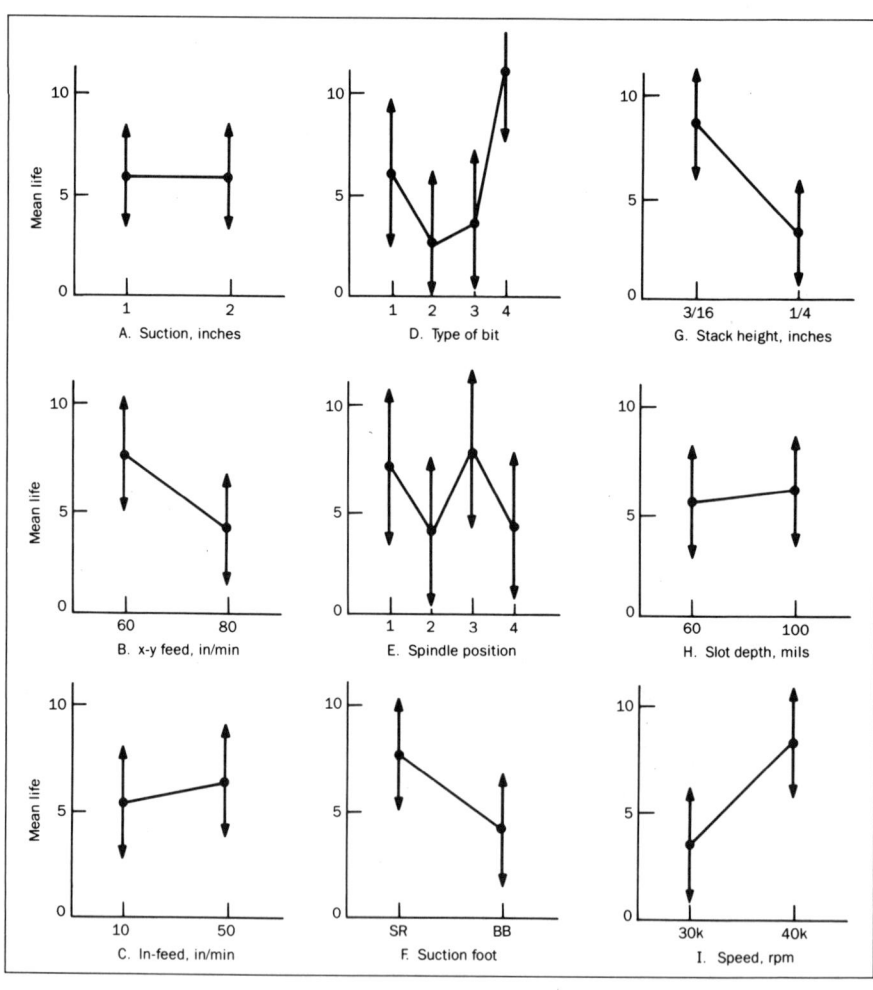

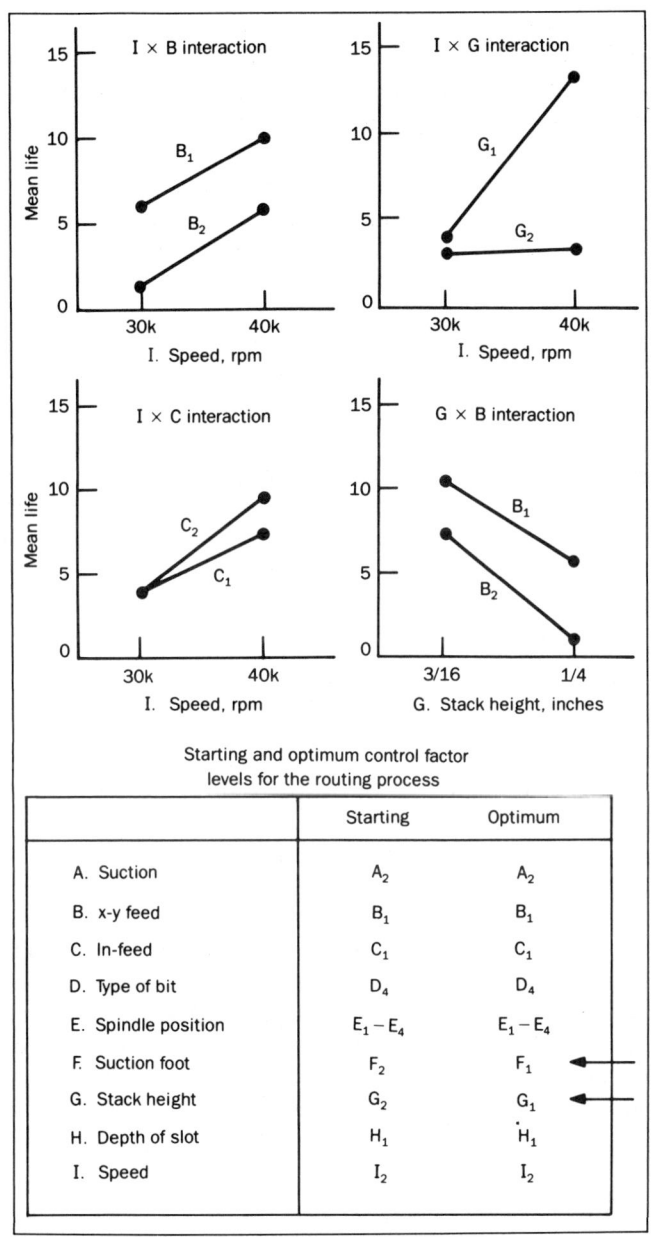

Figure 2. *Average factorial effects are shown in first table. Selected two-factor interactions are given in second table. Mean life is given in hundreds of inches. The 2σ limits are also shown. The table indicates the starting and optimum control factor levels for the routing process.*

- This change, however, raises machine productivity issues.
- The depth of slot in the back-up material has negligible effect.
- Higher rotational speed gives improved life. If the machine stability permits, even higher speeds should be tried in the next cycle of experiments.

The only two-factor interaction that is large is the stack height versus speed interaction. However, the optimum settings of these factors suggested by the main effect are consistent with those suggested by the interaction. The best factor level combination suggested by the above results and the starting factor level combination are tabulated in Figure 2.

Using the linear model of Reference 1 and taking into consideration only the terms for which the variance ratio is large, that is the factors B, D, F, G, I and interaction I × G, we can predict the router bit life under starting, optimum, or any other combination of factor settings. The predicted life under the starting conditions is 860 inches and under optimum conditions is 2200 inches. Because of the censoring at 1700 inches, these predictions are obviously likely to be on the low side. The prediction for optimum conditions especially is likely to be much less than the realized value. From the machine logs, the router bit life under starting conditions was found to be 900 inches while the confirmatory experiment under optimum conditions yielded an average life in excess of 4150 inches.

In selecting the best operating conditions for the routing process, one must consider the overall cost, which includes not only the cost of router bits but also the cost of machine productivity, the cost of cleaning the boards if needed, and so forth. Under the optimum conditions listed in Figure 2, the stack height is 3/16 inch as opposed to 1/4 inch under the starting conditions. This means three panels are cut simultaneously instead of four panels. The lost machine productivity due to this change can however be made up by increasing the x-y feed. If the x-y feed is increased to 80 in/min, the productivity of the machine would get back approximately to the starting level. The predicted router bit life under these alternative optimum conditions is 1700 inches, which is twice the predicted life for starting conditions. Thus a 50 percent reduction in router bit cost can be achieved while still maintaining machine productivity. An auxiliary experiment would be needed to precisely estimate the effect of x-y feed under the new settings of all other factors. This would enable us to make an accurate economic analysis.

Survival Probability Curve

The life data can also be analyzed in a different way, by the minute analysis method described in Reference 1, to construct the survival probabability curves for the levels of each factor. To do so, we look at every 100 inches of cut and note which router bits have failed and which have survived. Treating this as 0-1 data, we can determine factorial effects by the standard analysis method. Thus for suction levels A1 and A2, the survival probabilities at 100 inches of cut were estimated to be 0.44 and 0.69. Likewise the probabilities are estimated for each factor and also for each time period: 100 inches, 200 inches, etc. These data, plotted in Figure 3, graphically display the effects of factor level changes on the entire life curve. The conclusions from these plots are consistent with the conclusions from the analysis described earlier.

Plots like Figure 3 can be used to determine the entire survival probability curve under a new set of factor level combinations such as the optimum combination. See Reference 1 for the method of calculation.

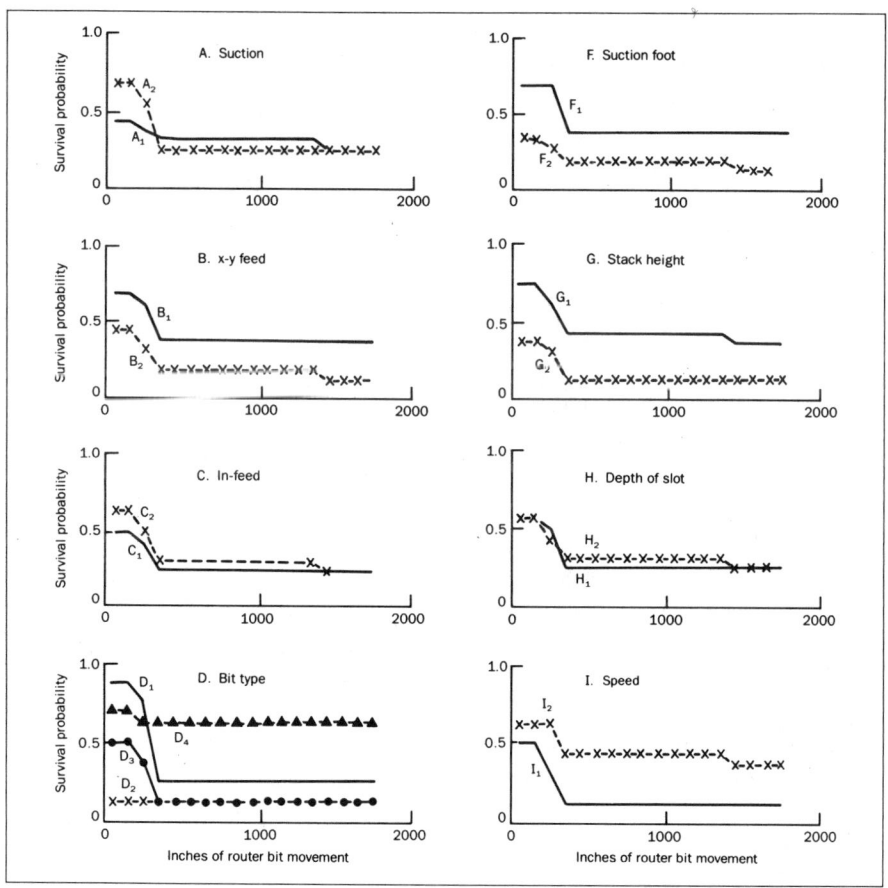

Figure 3. *Survival probability curves for router bits.*

Notice that in this method of determining life curves, no assumption was made regarding the shape of the curve—whether it follows a Weibull or a lognormal distribution, for example. Also, the total amount of data needed to come up with the life curves is small. In this example it took only 32 samples to determine the effects of eight control factors. For a single good fit of a Weibull distribution one typically needs several tens of observations. So the approach used here can be very beneficial for reliability improvement projects.

There are, of course, some caveats. First, as in any fractional factorial experiment, one needs to guard against the interactions among the various control factors. But this difficulty can be overcome through the confirmatory experiment. Second, the method for determining the statistical significance of the differences between the life curves for different factor levels needs more research.

CIRCUIT DESIGN OPTIMIZATION

The differential operational amplifier circuit is commonly used in telecommunications. An example is as a preamplifier in coin telephones, where it is expected to function over a wide temperature range. An important characteristic of this circuit is its offset voltage. (If the offset voltage is large, then the circuit cannot be used over long loops between the central office and the telephone). So the optimization objective was to minimize the offset voltage. The balancing property of the circuit makes the offset voltage small under nominal conditions. What needs to be minimized is the effect of tolerances and temperature variation on the offset voltage.

In the circuit diagram (Figure 4), there are two current sources, five transistors, and eight resistors. This differential operational amplifier circuit is made as part of a larger integrated circuit.

Control and Noise Factors

The balancing property of the circuit dictates the following relationship among the nominal values of the various circuit parameters: RFP = RFM, RPEP = RPEM, RNEP = RNEM, AFPP = AFPM, AFNP = AFNM, SIEPP = SIEPM, and SIENP = SIENM. The circuit parameter names begining with AF refer to the alpha parameter of the transistors and those beginning with SIE refer to the saturation currents of the transistors. Further, the gain requirements of the circuit dictate the following ratios of resistance values RIM = RFM/3.55 and RIP = RFM/3.55. These relationships among the circuit parameters are called tracking relationships.

There are only five control factors for this circuit: RFM, RPEM, RNEM, CPCS, and OCS. The transistor parameters could not be specified for this design because the manufacturing technology was preselected, and it dictated

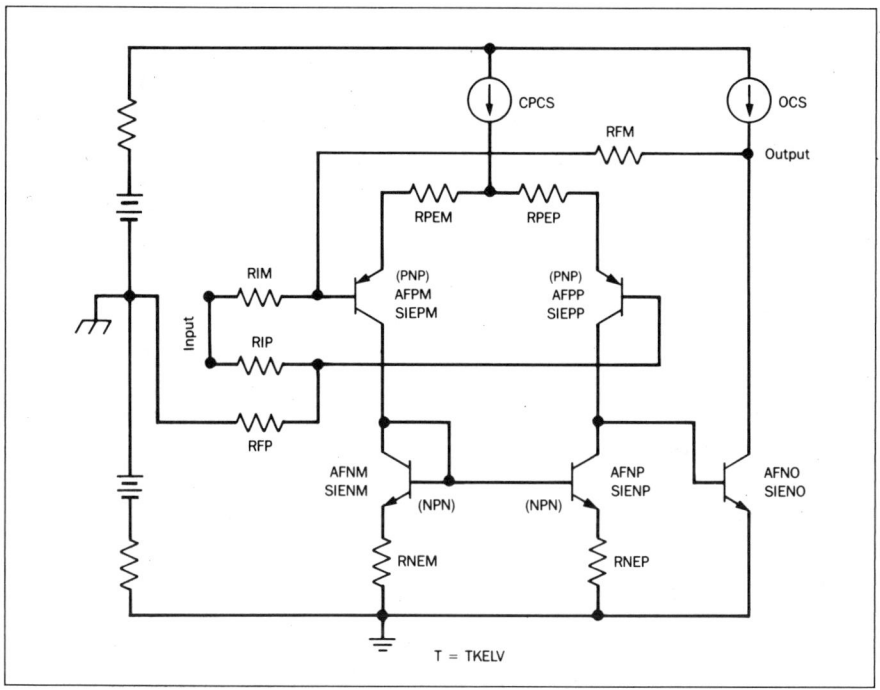

Figure 4. *Circuit diagram for a differential operational amplifier.*

the nominal values of all transistor parameters. Also, the tracking relationships determine the nominal values of the remaining resistors.

The number of noise parameters is 21. They are the tolerances on the eight resistors, 10 transistor parameters corresponding to the five transistors, two current sources and the temperature. The mean values and tolerances for these noise factors are given in Table III. The mean values of only the first five parameters are determined by the circuit design.

The tolerances listed are the 3σ limits. Thus for RPEM the σ is 21/3 = 7 percent of its nominal value. Further, the interdependency or the correlation among the noise factors is also expressed through tracking. To see this point, let us look at the relationship between RPEP and RPEM. The 21 percent tolerance for RPEM represents the specimen-to-specimen variation within a lot and the lot-to-lot variation. But on a given specimen, the two resistors RPEM and RPEP are located physically close together. So there is less variation between the two resistances. This is the origin of the correlation between the two resistances. Suppose in a particular specimen RPEM = 15 kilohms. Then for that specimen RPEP will vary around 15 kilohms with 3σ limits equal to 2

TABLE 3—Noise Factors for the Differential Op-Amp Circuit

Name	Mean	Tolerance	Levels (multiply by mean) 1	2	3
1. RFM	71 kilohms	1%	0.9967	1.0033	
2. RPEM	15 kilohms	21%	0.93	1.07	
3. RNEM	2.5 kilohms	21%	0.93	1.07	
4. CPCS	20 μA	6%	0.98	1.02	
5. OCS	20 μA	6%	0.98	1.02	
6. RFP	RFM	2%	0.9933	1.0067	
7. RIM	RFM/3.55	2%	0.9933	1.0067	
8. RIP	RFM/3.55	2%	0.9933	1.0067	
9. RPEP	RPEM	2%	0.9933	1.0067	
10. RNEP	RNEM	2%	0.9933	1.0067	
11. AFPM	0.9817	2.5%	0.99	1	1.01
12. AFPP	AFPM	1/2%	0.998	1	1.002
13. AFNM	0.971	2.5%	0.99	1	1.01
14. AFNP	AFNM	1/2%	0.998	1	1.002
15. AFNO	0.975	1%	0.99	1	1.01
16. SIEPM	3.OE-13 A	Factor of 7	0.45	1	2.21
17. SIEPP	SIEPM	Factor of 1.214	0.92	1	1.08
18. SIENM	6.OE-13 A	Factor of 7	0.45	1	2.21
19. SIENP	SIENM	Factor of 1.214	0.92	1	1.08
20. SIENO	6.OE-13 A	Factor of 2.64	0.67	1	1.49
21. TKELV	298 K	15%	0.94	1	1.06

percent of 15 kilohms. If for another specimen RPEM is 16.5 kilohms (10 percent more than 15 kilohms), then RPEP will vary around 16.5 kilohms with 3σ limits equal to 2 percent of 16.5 kilohms.

The saturation currents are known to have long-tailed distributions. So the tolerances are expressed as multiplicative instead of additive. In other words, these tolerances are taken to be additive in the log domain.

Evaluation of Mean Squared Offset Voltage

For a particular design, that is, for a particular selection of the control factor values, the mean squared offset voltage can be evaluated in many ways. Two common methods are:

- Monte Carlo simulation. Random number generators are used to determine a large number of combinations of noise factor values. The offset voltage is evaluated for each combination and then the mean squared offset voltage is calculated. For obtaining accurate mean squared values, the Monte Carlo method usually needs a large number of evaluations of the offset voltage, which can be expensive.

- Taylor series expansion. In this method, one finds the first derivative of the offset voltage with respect to each noise factor at the nominal design point. Let $x_1, \ldots, x_k$ be the noise factors, with variances $\sigma_1^2, \ldots, \sigma_k^2$, respectively. Let v be the offset voltage. Then the estimated mean square offset voltage is

$$ r = \sum_{i=1}^{k} \frac{\partial v}{\partial x_i} \; \sigma_i^2 $$

Second-order Taylor series expansion is sometimes taken if curvatures and correlations are important. When the tolerances are large so that the nonlinearities of v are important, the Taylor series approach does not give very accurate results.

In this application, however, we used the approach suggested by Taguchi. The orthogonal array, L_{36}, taken from Reference 1, was used to estimate the mean squared offset voltage as a standardized measure to be optimized. Simulation studies reported by Taguchi during his trips to AT&T Bell Laboratories have shown that the orthogonal array method gives more precise estimates of variances and means when compared to the Taylor series expansion method.

For the resistance and current source tolerances two levels were chosen, situated one standard deviation on either side of the mean. These noise factors were assigned to columns 1 through 10 of the L_{26} matrix (Table 4). For the 10 transistor parameters and the temperature, three levels were chosen, situated at the mean and at $\sqrt{3/2}$ times the standard deviation on either side of the mean. These noise factors were assigned to columns 12 through 22 of the matrix L_{26}. The submatrix of L_{36} formed by columns 1 through 10 and 12 through 22 is denoted by $\{J_{jl}\}$ and is referred to as the noise orthogonal array.

Each row of the noise array represents one specimen of differential op-amplifier with different values for the circuit parameters in accordance with the tolerances. Let v_j be the offset voltage corresponding to row j of the noise orthogonal array. Then the mean square offset voltage is estimated by

$$ r = \frac{1}{36} \sum_{j=1}^{36} v_j^2 $$

TABLE 4 – L_{36} Orthogonal Array

No.	1	2	3	4	5	6	7	8	9	10	11	12	13	14	15	16	17	18	19	20	21	22	23
1	1	1	1	1	1	1	1	1	1	1	1	1	1	1	1	1	1	1	1	1	1	1	1
2	1	1	1	1	1	1	1	1	1	1	1	2	2	2	2	2	2	2	2	2	2	2	2
3	1	1	1	1	1	1	1	1	1	1	1	3	3	3	3	3	3	3	3	3	3	3	3
4	1	1	1	1	1	2	2	2	2	2	2	1	1	1	1	2	2	2	2	3	3	3	3
5	1	1	1	1	1	2	2	2	2	2	2	2	2	2	2	3	3	3	3	1	1	1	1
6	1	1	1	1	1	2	2	2	2	2	2	3	3	3	3	1	1	1	1	2	2	2	2
7	1	1	2	2	2	1	1	1	2	2	2	1	1	2	3	1	2	3	3	1	2	2	3
8	1	1	2	2	2	1	1	1	2	2	2	2	2	3	1	2	3	1	1	2	3	3	1
9	1	1	2	2	2	1	1	1	2	2	2	3	3	1	2	3	1	2	2	3	1	1	2
10	1	2	1	2	2	1	2	2	1	1	2	1	1	3	2	1	3	2	3	2	1	3	2
11	1	2	1	2	2	1	2	2	1	1	2	2	2	1	3	2	1	3	1	3	2	1	3
12	1	2	1	2	2	1	2	2	1	1	2	3	3	2	1	3	2	1	2	1	3	2	1
13	1	2	2	1	2	2	1	2	1	2	1	1	2	3	1	3	2	1	3	3	2	1	2
14	1	2	2	1	2	2	1	2	1	2	1	2	3	1	2	1	3	2	1	1	3	2	3
15	1	2	2	1	2	2	1	2	1	2	1	3	1	2	3	2	1	3	2	2	1	3	1
16	1	2	2	2	1	2	2	1	2	1	1	1	2	3	2	1	1	3	2	3	3	2	1
17	1	2	2	2	1	2	2	1	2	1	1	2	3	1	3	2	2	1	3	1	1	3	2
18	1	2	2	2	1	2	2	1	2	1	1	3	1	2	1	3	3	2	1	2	2	1	3
19	2	1	2	2	1	1	2	2	1	2	1	1	2	1	3	3	3	1	2	2	1	2	3
20	2	1	2	2	1	1	2	2	1	2	1	2	3	2	1	1	1	2	3	3	2	3	1
21	2	1	2	2	1	1	2	2	1	2	1	3	1	3	2	2	2	3	1	1	3	1	2
22	2	1	2	1	2	2	2	1	1	1	2	1	2	2	3	3	1	2	1	1	3	3	2
23	2	1	2	1	2	2	2	1	1	1	2	2	3	3	1	1	2	3	2	2	1	1	3
24	2	1	2	1	2	2	2	1	1	1	2	3	1	1	2	2	3	1	3	3	2	2	1
25	2	1	1	2	2	2	1	2	2	1	1	1	3	2	1	2	3	3	1	3	1	2	2
26	2	1	1	2	2	2	1	2	2	1	1	2	1	3	2	3	1	1	2	1	2	3	3
27	2	1	1	2	2	2	1	2	2	1	1	3	2	1	3	1	2	2	3	2	3	1	1
28	2	2	2	1	1	1	1	2	2	1	2	1	3	2	2	1	1	3	2	3	1	2	3
29	2	2	2	1	1	1	1	2	2	1	2	2	1	3	3	3	2	2	1	3	1	2	1
30	2	2	2	1	1	1	1	2	2	1	2	3	2	1	1	1	3	3	2	1	2	3	2
31	2	2	1	2	1	2	1	1	1	2	2	1	3	3	3	2	3	2	2	1	2	1	1
32	2	2	1	2	1	2	1	1	1	2	2	2	1	1	1	3	1	3	3	2	3	2	2
33	2	2	1	2	1	2	1	1	1	2	2	3	2	2	2	1	2	1	1	3	1	3	3
34	2	2	1	1	2	1	2	1	2	2	1	1	3	1	2	3	2	3	1	2	2	3	1
35	2	2	1	1	2	1	2	1	2	2	1	2	1	2	3	1	3	1	2	3	3	1	2
36	2	2	1	1	2	1	2	1	2	2	1	3	2	3	1	2	1	2	3	1	1	2	3

Since the most desired value of the offset voltage is zero, the appropriate S/N ratio to be maximized for optimizing this circuit is

$$\eta = -10 \log_{10} r$$

Optimization of the Design

Orthogonal array experimentation is also an efficient way to maximize a nonlinear function—in this case the maximization of η with respect to the control factors. The control factors and their alternate levels are listed in Table 5. The L_{36} array was used to simultaneously study the five control factors. (In this case, the array L_{18} would have been sufficient.) The factors RFM, RPEM, RNEM, CPCS, and OCS were assigned to columns 12, 13, 14, 15, and 16, respectively. The submatrix of L_{26} formed by columns 12 through 16 is denoted by $\{ I_{ik} \}$ and is referred to as the control orthogonal array.

Each row of the control orthogonal array represents a different design. For each design the S/N ratio was evaluated using the procedure described under "Evaluation of Mean Squared Offset Voltage." The simulation algorithm is graphically displayed in Figure 5. Standard analysis of variance was performed on the η values to generate Table 6. The effect of each factor on η is displayed in Figure 6. From Table 6 it is apparent that only RPEM, CPCS, and OCS have an effect on η that is much bigger than the error variance. The effect of RPEM is the largest, and there is indication that reduction in its value below 7.5 kilohms could give even more improvement in offset voltage. For both current sources, 10 μA to 20 μA seems to be the flat region, indicating that we are very

TABLE 5 – Control Factors for the Differential Op-Amp Circuit

Label	Name	Description	Levels 1	2	3
A	RFM	Feedback resistance, minus terminal (kilohms)	35.5	71	142
B	RPEM	Emitter resistance, PNP, minus terminal (kilohms)	7.5	15	30
C	RNEM	Emitter resistance, NPN, minus terminal (kilohms)	1.25	2.5	5
D	CPCS	Complementary pair current source (μA)	10	20	40
E	OCS	Output current source (μA)	10	20	40
				Starting design	

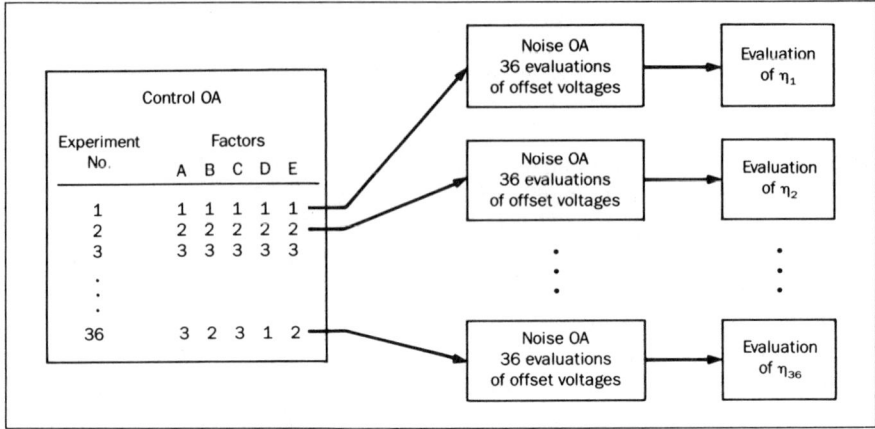

Figure 5. *Simulation algorithm.*

near the best values for these parameters. Also, the potential improvement by changing these current sources from 10 μA to 20 μA seems small. Thus we chose the following two designs as potential optimum points:

- Optimum 1: Only change RPEM from 15 kilohms to 7.5 kilohms. By the procedure in "Evaluation of Mean Squared Offset Voltage," the value of η for this design was found to be 33.70 dB compared to 29.39 dB for the starting design. In terms of the rms offset voltage this represents an improvement from 33.9 mV to 20.7 mV.
- Optimum 2: Change RPEM to 7.5 kilohms. Also change both CPCS and OCS to 10 μA. The η for this design was computed to be 35.82 dB, and rms offset voltage was seen to be 16.2 mV.

In the discussion so far, we have paid attention to only the dc offset voltage. Stability under ac operation is also an important consideration. For a more elaborate study of this characteristic, one must generate more data like those in Table 6 and Figure 6. The optimum control factor setting should then be obtained by jointly considering the effects on both the dc and ac characteristics. If conficts occur, appropriate trade-offs can be made using the quantitative knowledge of the effects. In our study, we simply checked for ac stability at the two optimum conditions. For sufficient safety margin with respect to ac stability, we selected optimum 1 as the best design and called it simply the optimum design.

TABLE 6 – Analysis of Variance for $\eta = -10 \log_{10}$ **(mean squared offset voltage)**

Control factor	Level means			Sum of squares	Degrees of freedom	Mean square	F
	1	2	3				
A. RFM	26.5	26.4	25.3	9.9	2	4.95	0.5
B. RPEM	30.3	26.4	21.5	463.7	2	231.85	25.0
C. RNEM	25.1	25.8	27.3	29.9	2	14.95	1.6
D. CPCS	27.5	27.1	23.6	111.1	2	55.55	6.0
E. OCS	27.3	27.0	23.8	87.5	2	43.75	4.7
Error				231.6	25	9.26	

Overall mean = 26.05

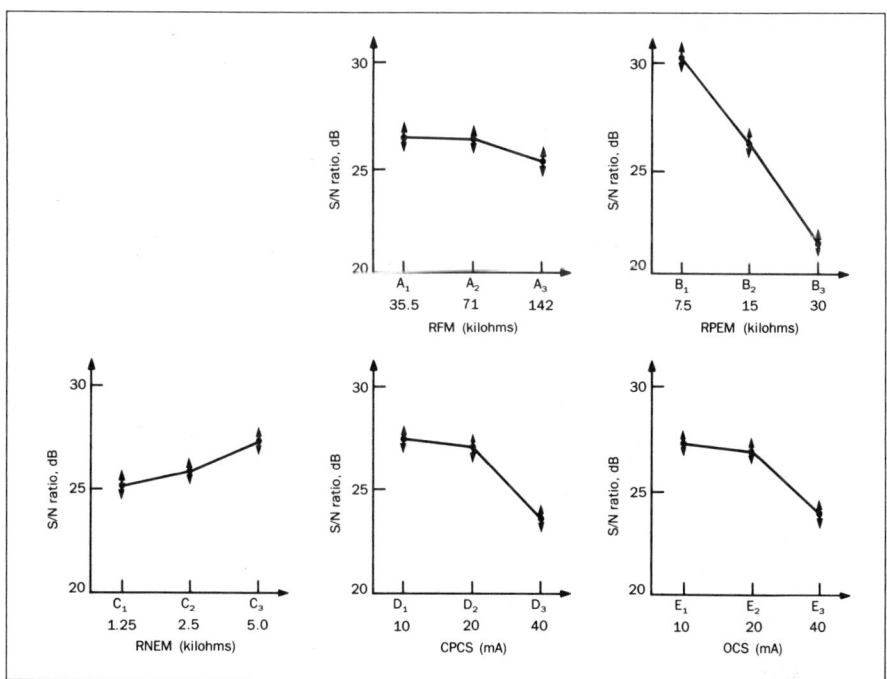

Figure 6. *Average effects of control factors on* $\eta = -10 \log_{10}$ *(mean squared offset voltage). The* 2σ *limits are indicated by the arrows.*

Tolerance Design

With the sensitivity to noise minimized for the dc offset voltage, the next step is to examine the contribution to the mean squared offset voltage by each noise source. By performing analysis of variance of the 36 offset voltages corresponding to the noise orthogonal array for the optimum control factor settings, we obtained Table 7, which gives the breakdown of the contribution of each tolerance to the total mean squared offset voltage. The table also gives a similar breakdown of the mean squared offset voltage for the starting design. The reduction in the sensitivity of the design to various noise sources is apparent.

The table exhibits the typical Pareto principle. That is, a small number of tolerances account for most of the variation in the response. In particular, SIENP has the largest contribution to the mean squared offset voltage. Hence it is a prime candidate for reducing the tolerance, should we wish to further reduce the mean squared offset voltage. AFNO, AFNM, and SIEPP have moderate contribution to the mean squared offset voltage. The rest of the noise factors contribute negligibly to the mean squared offset voltage. So if it will yield further manufacturing economies, relaxing these tolerances should be considered carefully.

TABLE 7 – Breakdown of Mean Square Offset Voltage by Noise Sources

Noise factor*	Contribution to mean square offset voltage (10^{-4} V^2)	
	Starting Design	Change RPEM to 7.5 kilohms
SIENP	6.7	2.3
AFNO	2.2	0.8
AFNM	1.6	0.4
SIEPP	0.5	0.5
RPEP	0.2	0.1
AFPP	0.1	0.1
Remainder	0.2	0.1
Total	11.5	4.3

* The six largest contributors to the mean squared offset voltage are listed here.

Reducing Computational Effort

The optimization of the differential op-amp discussed earlier needed $36 \times 36 = 1296$ evaluations of the circuit response—namely the offset voltage. Although in this case the computer time needed was not an issue, in some cases making an evaluation of each response can be expensive. In such cases, significant reduction of the computational effort can be accomplished by using qualitative engineering knowledge about the effects of the noise factors. For example, we form a *composite noise* factor. The high level of the composite noise factor corresponds to the combination of the individual noise factor levels that give high response. The *low* and *nominal* levels of the composite noise factor can be defined similarly. Thus the size of the noise orthogonal array can be drastically reduced, leading to much smaller computational effort during design optimization. Further reduction in computational effort can be obtained by taking only two levels for each noise factor; however, generally we recommend three levels. For tolerance design, we need to identify the sensitivity of each noise factor. Therefore a composite noise factor would have to be dissolved into its various components.

CONCLUSION

The robust design method has now been used in many areas of engineering throughout the United States. For example, robust design has lead to improvement of several processes in *very large scale integration (VLSI) fabrication:* window photolithography, etching of aluminum lines, reactive ion etching processes, furnace operations for deposition of various materials, and so forth. These processes are used for manufacturing 1-megabit and 256-kilobit memory chips, 32-bit processor chips, and other products. The *window photolithography* application documented in Reference 8 was the first application in the United States that demonstrated the power of Taguchi's approach to quality and cost improvement through robust process design. In particular, the benefits of the application were:

- Fourfold reduction in process variance.
- Threefold reduction in fatal defects.
- Twofold reduction in processing time. This resulted from the fact that the process became stable so that time-consuming inspection could be dropped.
- Easy transition of design from research to manufacturing.
- Easy adaptation of the process for finer line technology, which is usually a very difficult problem.

The *aluminum etching* application was an interesting one in that it originated from a belief that poor photoresist print quality leads to line width loss and to undercutting. By making the process insensitive to photoresist profile variation and other sources of variation, the visual defects were reduced from 80 percent to 15 percent. Moreover, the etching step could then tolerate the variation in photoresist profile.

In *reactive ion etching* of tantalum silicide, the process gave highly nonuniform etch quality, so only 12 out of 18 possible wafer positions could be used for production. After optimization, 17 wafer positions became usable—a hefty 40 percent increase in machine capacity. Also, the efficiency of the orthogonal array experimentation allowed this project to be completed in the 20-day deadline. Thus in this case $1.2 million was saved in equipment replacement costs not counting expense of disruption on the factory floor.[9]

The *router bit life improvement* project described in this article led to a two- to fourfold increase in the life of router bits used in cutting printed wiring boards. The project illustrates how reliability or life improvement projects can be organized to find best settings of control factors with a very small number of samples. The number of samples needed in this approach is very small yet it can give valuable information about how each factor changes the survival probability curve.

In the *differential op-amp circuit optimization* example described in this article, a 40-percent reduction in the rms offset voltage was realized by simply finding a new design center. This was done by reducing sensitivity to all tolerances and temperature, rather than reducing tolerances, which could have increased manufacturing cost.

Here the noise orthogonal array was used in a novel way—to efficiently simulate the effect of many noise factors. This approach can be beneficially used for evaluating designs and for system or software testing. Further, the approach can be automated and made to work with various computer-aided design tools to make design optimization a routine practice.

This approach was also used to find optimum proportions of ingredients for making *water-soluble flux*.[10] By simultaneous study of the parameters for the wave soldering process and the flux composition, the defect rate was reduced by 30 to 40 percent.

Orthogonal array experiments can be used to tune hardware/software systems.[11] By simultaneous study of three hardware and six software parameters, the response time of the UNIX® operating system was reduced 60 percent for a particular set of load conditions experienced by the machine.

Under the leadership of American Supplier Institute and Ford Motor Company, a number of automotive suppliers have achieved quality and cost improvement through robust design. Many of these applications are documented in Reference 12.

These examples show that robust design is a collection of tools and comprehensive procedures for simultaneously improving product quality, performance, and cost, and also engineering productivity, Its widespread use in industry is bound to have a far-reaching economic impact.

Acknowledgments

The two applications described in this chapter would not have been possible without the collaboration and diligent efforts of many others. The router bit life improvement study was done in collaboration with Dave Chrisman of AT&T Technologies, Richmond Works. The optimization of the differential op-amp was done jointly with Gary Blaine, a former member of the technical staff at AT&T Bell Laboratories, and Joe Leanza of AT&T Information Systems. The survival probability curves were obtained with the software package for analysis of variance developed by Chris Sherrerd of Bell Laboratories. Rajiv Keny and Paul Sherry of Bell Laboratories provided helpful comments on this paper.

Suggested Readings

Box, G.E.P., W. G. Hunter, and I. S. Hunter. 1978. *Statistics for Experimenters—An Introduction to Design, Data Analysis and Model Building*. John Wiley & Sons. Inc. New York.

Hicks, C. R. 1973. *Fundamental Concepts in the Design of Experiments*. Holt, Rinehart and Winston. New York.

Kackar, R. N. 1985. Off-Line Quality Control, Parameter Design, and the Taguchi Method. Journal of Quality Technology. October. pp. 176-188.

Katz, L. E. and M. S. Phadke. 1985. Macro Quality at Micro Cost. AT&T *Bell Laboratories Record*. November.

Lin, K. M. and R. N. Kackar. Wave Soldering Process Optimization by Orthogonal Array Design Method. *Electronic Packaging and Production* [to appear].

Pao, T. W., M. S. Phadke, and C. S. Sherrerd. 1985. Computer Response Time Using Orthogonal Array Experiments. *Conference Record*. Vol 2. IEEE International Communications Conference. Chicago. June 23-26.

Phadke, M. S. 1982. Quality Engineering Using Design Experiments. *Proceedings of the American Statistical Association*. Section on Statistical Education. Cincinnati. August. pp. 11-20.

Phadke, M. S., R. N. Kackar, D. V. Speeney, and M. J. Grieco. 1983. Off-line Quality Control in Integrated Circuit Fabrication Using Experimental Design. *The Bell System Technical Journal*. Vol. 1, No. 5. May-June. pp. 1273-1309.

Proceedings of Supplier Symposia on Taguchi Methods. April 1984, November 1984, and October 1985. American Supplier Institute, Inc. Romulus, MI.

Taguchi, G. 1978. Off-Line and On-Line Quality Control Systems. International Conference on Quality Control. Tokyo.

Taguchi, G. and M. S. Phadke. 1984. Quality Engineering through Design Optimization. Conference Record. GLOBECOM 84 Meeting. IEEE Communications Society. Atlanta. November. pp. 1106-1113.

Taguchi, G. and Yu-In Wu. 1979. *Introduction to Off-Line Quality Control,* Central Japan Quality Control Association. Meieki Nakamura-Ku, Nagaya, Japan, (Available in English through American Supplier Institute, Inc.)

Reprinted with permission from the *AT&T Technical Journal.* Copyright 1986 AT&T. March/April 1986. Vol-65, No. 2.

11

The Scientific Context of Quality Improvement: A Look at the Use of Scientific Method In Quality Improvement

George E. P. Box
Soren Bisgaard

The United States is facing a serious economic problem. Its role as the world's most powerful industrial nation is being threatened. Automobiles, cameras, color televisions, computer chips, and machine tools are just some of the product categories in which the United States has lost at least 50% of its market to Japan over the last ten years.[1] It's clear that the Japanese are doing a number of things right, but a major reason for their competitive edge in high-quality, low-cost products is that they use statistical methods. Quite simply, the Japanese have a secret ingredient: they do it and we don't.

That simple statement, "they do it and we don't," is the crux of this issue. W. Edwards Deming's campaign is directed toward alerting management to the vital need to produce an environment in which statistical tools can be used to produce this quality revolution. Without such an environment, detailed discussion about techniques is fruitless. However, the Japanese are using modern approaches that are different from those that have been practiced in this country. The philosophy used to be that control charts and inspection at the end of the production line were enough. Now the emphasis is on moving upstream, building good quality into the products and processes, instead of trying to inspect bad quality out. That is a profound difference.

Quality is a full-time job for everyone in the company. It requires teamwork and the use of scientific methods. This article will focus mainly on the scientific context of quality improvement and give a nontechnical overview of some important new ideas of modern quality improvement.

SCIENTIFIC METHOD

One might think that "scientific method" is something that physicists do—complicated equations and intricate theories. But scientific method is used anytime anyone tries to solve a problem in a systematic way. Scientific method is just an accelerated version of the ordinary process whereby human beings find things out.

Producing good quality and increasing productivity at low cost is achieved by learning about processes. This was clearly Shewhart's intention when he said "The three steps [specification, production, and judgement of quality] constitute a dynamic scientific process of acquiring knowledge ... mass production viewed in this way constitutes a continuing and self-corrective method for making the most efficient use of raw and fabricated materials."[2] He never intended quality control to be just passive inspection.

What are the essential elements in discovering something? Two things are needed: first, a *critical event*, one that contains significant information. But no one is going to see or exploit it unless there is a *perceptive observer* present. These two things, a critical event and a perceptive observer, are essential.

Most events are not critical and do not offer much from which to learn. They are just part of the ordinary things happening all the time. But every now and then something happens that can be learned from. These events are comparatively rare—as are perceptive observers. Ideally, they should possess not only natural curiosity, but also training in the relevant area of expertise. Hence, we are dealing with the coming together of two rare occurrences. The probability of these two coming together purely by chance is very small. That's why, in looking back at the 12th, 13th, and 14th centuries, we guess that technological change occurred very slowly.

Technological change started to occur much more rapidly around 300 years ago. One of the main reasons was the use of scientific method, which increased the probability of the critical event and the perceptive observer coming together. There are two ways to achieve this. One is to make sure that naturally occurring informative events are brought to the attention of the perceptive observer (informed observation), and the other is to increase the chance of an informative event actually occurring (directed experimentation).

Informed Observation

As an example of informed observation, suppose there was going to be a solar eclipse and that the best location to see it was in southern Australia. To increase the probability of learning from the eclipse, there should be people knowledgeable about astronomy and solar eclipses in southern Australia at the right time, equipped with the necessary instruments to make meaningful observations. That's what's meant by increasing the probability that a naturally occurring event would be observed and appreciated by a perceptive observer.

A quality control chart serves the same purpose. People knowledgeable about the process look at it and ask, "What happened at point x?" In the words of Walter Shewhart, they look for an assignable cause. With a quality control chart, data from the process are not buried in a notebook, a file drawer, or a computer file. No, the charts are put up on the wall where the people who know about the process can observe, ask what happened, and take action: "If that's where Joe spat in the batch, then let's change the system so Joe cannot spit in the batch anymore." Thus, the people working with a system can slowly eliminate the bugs and create a process of continuous, never-ending improvement.

Every process generates information that can be used to improve it. This is perhaps simple to see when the process is a machine. But the philosophy applies equally to a hospital ward, to customer billing, to a typing pool, and to a maintenance garage. Every job has a process in it, and every process generates information that can be used to improve it.

Tools for Observation

One can rant and rave about why it takes a typing pool so long to type documents, but Deming tells us that it is useless just to grumble at people. Management must give employes tools that will allow them to do their jobs better and encouragement to use these tools. In particular, they must collect data. How long do typing pool documents sit around? How long does it take to type them? How are they being routed? How much delay is caused by avoidable corrections? How often are letters unnecessarily labeled "urgent"?

Kaoru Ishikawa created an excellent set of tools that helps people get at the information being generated in processes. He calls them "The Seven Tools": check sheets, the Pareto chart, the cause-and-effect diagram, histograms, stratification, scatter plots, and graphs (in which he includes control charts). All are simple and everybody can use them.

Consider some of these tools and how they might be used to examine a problem. Suppose that in one week's production of springs, 75 are defective. That's informative, but not of much help for improving the process. Suppose all the defective springs are sorted into those having defects because of cracks, because of scratches, because the dimensions were not right, and so on, and listed on a check sheet. The results can then be displayed using the Pareto chart in Figure 1a. In this instance, 42 out of the 75 springs were discarded because of cracks. Clearly cracks are an important problem. Pareto charts focus attention on the most important things to work on. They separate the vital few from the trivial many.[3]

The next step is to figure out what could be causing the cracks. The people who make the springs should convene around a blackboard and make a cause-and-effect diagram listing possible causes, and causes of causes. Somebody may believe that the inspection process itself produces cracks. Another questions whether the gages for measuring cracks are set right. Someone doubts whether inspectors agree on what a crack is. It's suggested that the source of cracks could be in the process of assembly. It is pointed out that two types of springs, A and B, are made; do cracks occur equally in both? The foreman believes that the reason for cracks may be that the hardening temperature is sometimes not set correctly. He also thinks that the temperature of quenching oil is important.

The final cause-and-effect diagram resembles the skeleton of a fish, and is often called a "fishbone" chart. Its main function is to facilitate focused

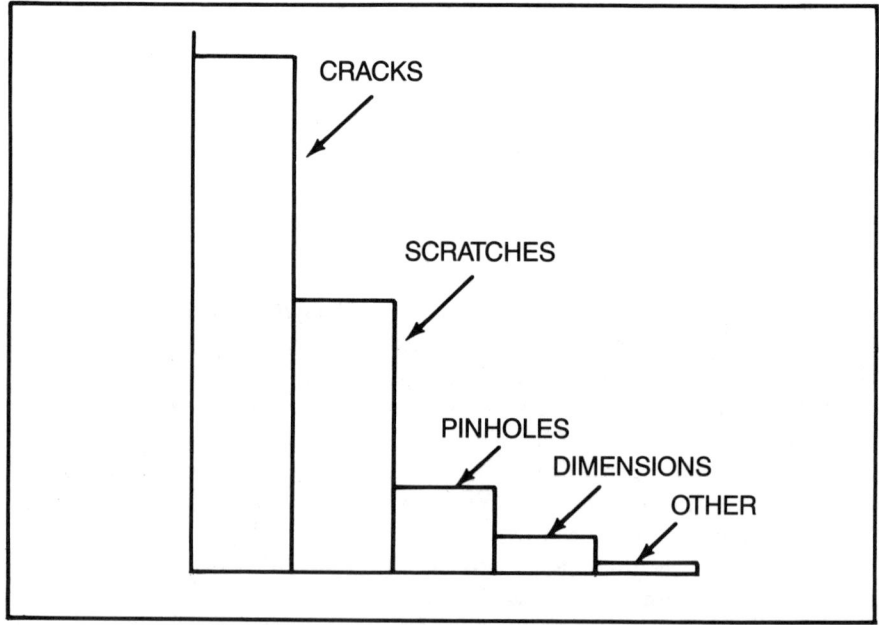

Figure 1a. *Pareto diagram of defective springs from one week's production.*

brainstorming, argument, and discussion among people about what might cause the cracks. Often some of the possibilities can be eliminated right away by simply talking them over. Usually there will be a number of things left and it will be unknown which items cause which problems. Using data coming from the process can often help sort out the different possibilities with the seven tools.

For example, suppose the cracks have been sorted by size. The histogram, shown in Figure 1b, can then be used to show how many cracks of different sizes occurred. This is of great value for conveniently summarizing the crack size problem. But much more can be learned by splitting, or stratifying, the histogram in various ways. For example, the fact that there are two types of springs provides a chance for stratification. In Figure 1c, the original histogram has been split into two, one for springs of type A and one for those in type B. It is clear not only that most of the cracks are in springs of type A, but also that the size of the type A cracks is on the average larger and that the spread is also larger. Now the question is, "What's special about type A springs?" If we know who inspected what, the data could also be stratified by inspector.

Scatter diagrams can also be helpful. If hardening temperature varies quite a bit, the crack size might be plotted against the hardening temperature. A control chart recording the oven temperature can tell how stable the temperature is over time and whether a higher frequency of cracks is associated with any unusual patterns in the temperature recording.

Clearly. these seven tools are very common-sensical, yet very powerful. By helping people to see how often things happen, when they happen, where they happen, and when they are different, the work force can be trained to be quality detectives producing never-ending improvements.

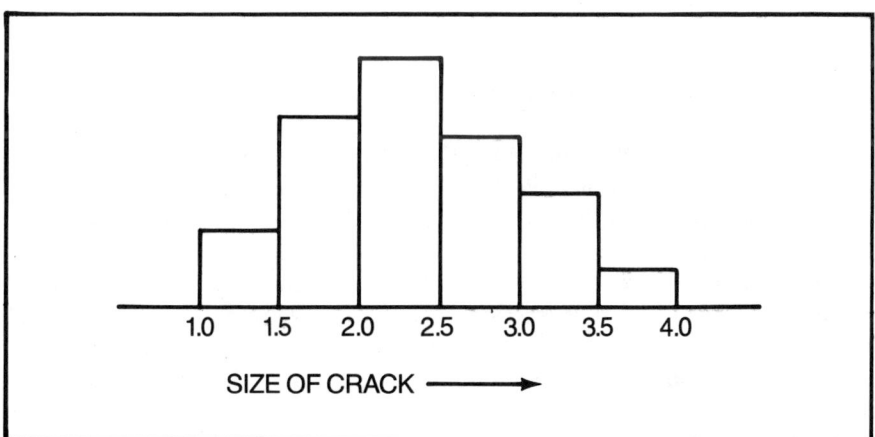

Figure 1b. *Histogram of crack size from production of springs.*

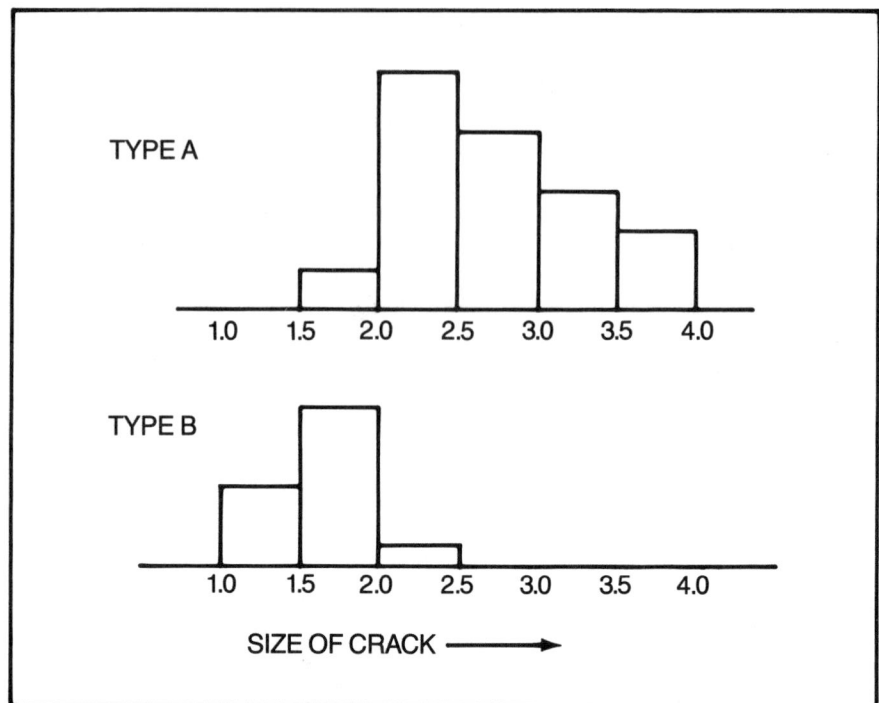

Figure 1c. *Stratification of crack size in springs. Stratified according to springs of type A and B. This stratification shows that there is a clear difference in the number and size of cracks in the two types of springs. This provides evidence for continued detective work.*

DOES IT WORK IN THE UNITED STATES?

People often ask, "Will this work in the United States?" One interesting example is Matsushita, a Japanese manufacturer of television sets and other electronic devices, which some years ago bought an American television plant just outside Chicago. When Japanese management first took over, the reject rate was 146%. That may sound extraordinary; it means that most television sets were pulled off the line to be repaired at least once—and some were pulled off twice. By using tools like the ones described above, the reject rate was brought down to 2%. The change did not happen overnight. It took four or five years of meticulously weeding out the causes of defects. However, when a study group from the University of Wisconsin, led by the late Professor William G. Hunter, visited the plant in the spring of 1985, there was not a Japanese in sight. Only Americans were employed, and there seemed to be no question that they preferred the new system. The managers said that the plant was a madhouse before. They could not find any time to manage because they were always putting out fires. The changes had dramatically improved product quality, productivity, and morale—and the managers got time to manage.

Directed Experimentation

Informed observation is not enough in itself, because no amount of fine-tuning can overcome fundamental flaws in products and processes due to poor design. Directed experimentation, the other component of the scientific method, is needed to ensure that we have excellently designed products and processes to begin with. Only by designing quality into the products and processes can the levels of quality and productivity achieved in Japan be achieved in the United States.

In the past, quality control was thought to concern only the downstream side of the process, with an emphasis on control charts and inspection schemes. Now the emphasis is moving upstream, focusing attention on getting a product and a process that are well designed so that seldom is anything of unsatisfactory quality produced. This reduces the need for inspection and provides economic gains. This push for moving upstream, strongly recommended 50 years ago by Egon Pearson,[4] is embodied in the Japanese approach. Directed experimentation is needed to reach this goal.

On a recent study mission to Japan, researchers from the Center for Quality and Productivity Improvement at the University of Wisconsin and Bell Labs were shown a new copying machine. This machine can produce copies of almost any size at an enormous rate. It had taken the company three years to design it. The chief design engineer reported that 49 designed experiments, many of which had involved more than 100 experimental runs, had been conducted to get the machine the way it is. It took a lot of effort to get the engineering design right, but now they have developed a copier that will very seldom fail or give trouble—and that is its own advertisement.

Statistical experimental design was invented in the early 1920s by R.A. Fisher in England. In her biography. Joan Fisher Box describes how in 1919 R.A. Fisher went to work at a small, at that time not very well known, agricultural research station near London called Rothamsted.[5] The workers at Rothamsted were interested in finding out the best way to grow wheat, potatoes, barley, and other crops. They could have conducted their experiments in a greenhouse, carefully controlling the temperature and humidity, making artificial soil that was uniform, and keeping out the birds. By doing so they would have produced results that applied to plants grown under these very artificial conditions.

However, the results would most likely have been entirely useless in deciding what would happen on a farmer's field.

The question, then, was how to run experiments in the "noisy" and imperfectly controlled conditions of the real world. Fisher showed how to do it, and his ideas were quickly adopted worldwide, particularly in the United States. This is perhaps one reason why at least one American industry—namely agriculture—still leads the world.

It slowly became clear that Fisher's work on experimental method constituted a major step forward in human progress and that his results were by no means confined to agriculture. He showed for the first time how experimentation could be moved out of the laboratory. His methods quickly produced relevant, practical findings in many important fields including medicine, education, and biology. In particular, his ideas were suitably developed for the industrial setting,[6,7,8,9] and over the years. Many statistically designed experiments have been run in industry in the United States, Great Britain, and many other countries. Statistical design is a potent tool: it is almost invariably successful. Unfortunately, application has been very patchy in the West, usually because of a lack of management understanding and support. In Japan, on the other hand, management requires that such methods be used to develop high-quality products and processes that will rarely go wrong.

Statistically Designed Experiments

Consider a specific example of designed experiments. Suppose we want to redesign the springs mentioned earlier to eliminate the problem with cracks. The temperature of the steel before quenching, the carbon content of the steel, and the temperature of the quenching oil are important factors. Thus, the effect on the number of cracks of varying the three factors will be studied: temperature of the steel before quenching (whether 1,450°F or 1,600°F is better), content of carbon (whether 0.50% or 0.70% is better), and the temperature of quenching oil (whether 70°F or 120°F is better).

In the old days it was believed that the "scientifically correct" way to conduct an experiment was to vary just one factor at a time, holding everything else fixed. Suppose we do that in this situation, starting with steel temperature. Since we want reasonably reliable estimates of the effects of the factors, we run four experiments at 1.470°F steel and four experiments at 1,600°F steel, fixing the carbon content to 0.50% and using 70°F quenching oil. Suppose that using an accelerated life test the percentage of springs without cracks is as follows:

	1,450°F	1,600°F	% Difference
Percent springs	72	78	6
without cracks	70	77	7
	75	78	3
	77	81	4
		Average difference	5

The average difference in the number of cracks between 1,450°F and 1,600°F is 5%. From this it might be concluded that 1,600°F is the better temperature. But is it? Remember, we fixed the carbon content at 0.50% and used 70°F quenching oil. After these eight trials all that can really be said is that it appears to be better to use the high steel temperature if the carbon content is 0.50% and the oil is 70°F. If somebody asks whether changing the steel temperature would produce the same reduction in cracks with 0.70% carbon steel or with 120°F oil, the honest answer is that we don't know. To study the effect of changing the carbon content in the same way will require an additional eight runs. After that all we could say is that for the particular fixed choice of steel temperature and oil temperature, there is a certain change in response when the carbon content is changed. The same difficulty would apply to a third set of eight runs meant to determine the effect of changing the quenching oil temperature. Thus, after 24 experiments, all we'd know would be the effect of each variable at one particular combination of settings of the other two. Additional experiments would be needed to find out more.

Factorial Designs

It was Fisher's idea that it was much better to vary all the factors simultaneously in what he called a factorial design. Using such a design, we would run just one set of eight experimental trials to test all three variables, as shown in Table 1. Notice the pattern in the columns of this table headed T, C, O: the steel temperature is switched every run; carbon in pairs; the oil temperature in sets of four. One way to interpret the results of such an experiment is to display the data at the corners of a cube, as in Figure 2.

The trials with low steel temperature are shown on the left side of the cube, and the trials with the high steel temperature on the right. Low carbon trials are on the bottom, and high carbon trials on the top. Trials with 70°F oil are on the front and those run with 120°F oil on the back. For example, for the third trial in Table 1. the steel temperature was low (1,450°F), carbon was high

TABLE 1 – The experimental design for testing three factors in eight runs using Fisher's idea is simultaneously varying all factors according to a two-level factorial scheme. Each line in the table corresponds to one single run. In the columns labeled T, C, and O it is indicated how the three factors should be adjusted.

Run	T Steel Temp.	C Carbon	O Oil Temp.	Springs without cracks	Day Run
1	1,450°F	0.50%	70°F	67%	1
2	1,600°F	0.50%	70°F	79%	2
3	1,450°F	0.70%	70°F	61%	2
4	1,600°F	0.70%	70°F	75%	1
5	1,450°F	0.50%	120°F	59%	2
6	1,600°F	0.50%	120°F	90%	1
7	1,450°F	0.70%	120°F	52%	1
8	1,600°F	0.70%	120°F	87%	2

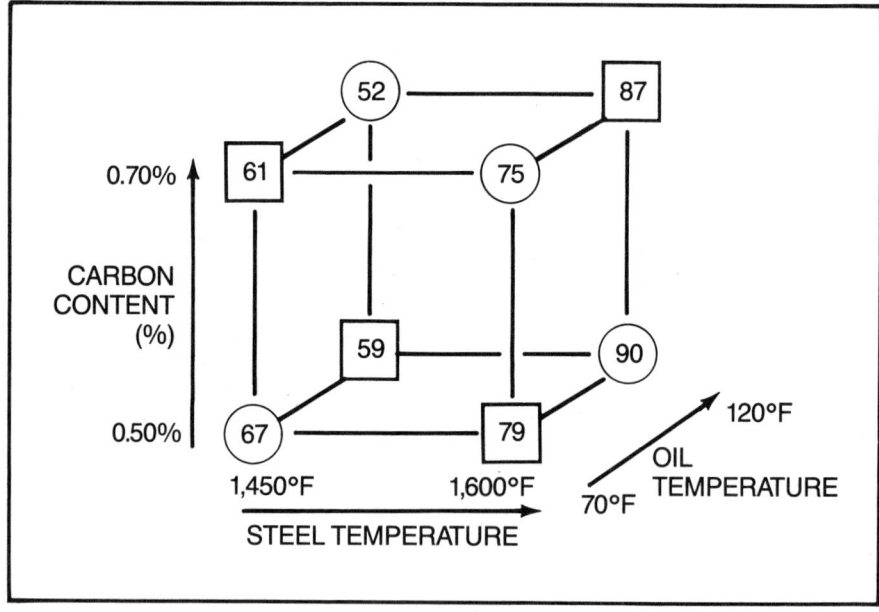

Figure 2. Two-level, three-factor designed experiment.

(0.70%). and the oil temperature was low (70°F). That takes us to the left top front corner of the cube, where the proportion of springs without cracks is 61%.

What can be gleaned from all this? Suppose we want to find the effect of steel temperature. We assess the effect of changing steel temperature by examining the pairs of readings in Figure 2 (67,79), (59,90), (61,75), and (52,87). In each pair, one of the readings reflects an experiment run at high steel temperature and the other experiment at low steel temperature—but within each pair the other factors remain constant. Therefore, as with the one-at-a-time experiment, four comparisons allow us to calculate the effect of changing only steel temperature:

Steel Temperature Effect:		
79 - 67	=	12
75 - 61	=	14
90 - 59	=	31
87 - 52	=	35
Average	=	+23

Comparing the four corresponding pairs of numbers along the top-bottom axis and four pairs on the back-front axis permits the same calculations for carbon content and oil temperature. This gives an average carbon effect of −5.0 and an average oil temperature effect of +1.5. Another way to calculate the same quantities would be to take the average of the four numbers on the left face and subtract that from the average on the right face. Likewise, to calculate the effect of carbon, subtract the average of the four numbers on the bottom face from the average of the four numbers on the top. The difference between them is the effect of carbon. Thus, using a factorial design in only eight runs allows us to test each of the three factors with the same precision as in a one-factor-at-a-time arrangement that has three times the number of runs.

Interactions

The quantities discussed so far are the averages, or main effects, of the single factors. If they are denoted by the letters T, C, and O, we find from the data:

	T (Steel temperature)	+23.0
Main Effect	C (Carbon)	−5.0
	O (Oil temperature)	+1.5

Even more information can be extracted. For example, look at the individual steel temperature effect differences again: 12%, 14%, 31%, and 35%. The cube in Figure 2 shows that the first two numbers represent contrasts on the front of the cube where 70°F oil was used, while the last two came from the pairs on the back of the cube where 120°F oil was used. This indicates that the effect of steel temperature is different depending on whether 70°F or 120°F quenching oil temperature is used. Therefore, steel temperature (T) and oil quenching temperature (O) interact. This T × O interaction is calculated as follows:

$$\underline{\begin{array}{l} \text{Steel temp. effect with } 120°\text{F oil} = (31 + 25)/2 = 33 \\ \text{Steel temp. effect with } \ 70°\text{F oil} = (12 + 14)/2 = 13 \end{array}}$$

$$\begin{array}{ll} \text{Difference} & 20 \end{array}$$

$$\text{T × O interaction effect} = 20/2 = 10$$

Similar calcuations can be done to produce the T × C and O × C interactions: T = 23.0; C = −5.0; O = 1.5; TC = 1.5; TO = 1.0; and OC = 10.0. This single eight-run design has not only helped estimate all the main effects with maximum precision, it has also helped determine the three interactions between pairs of factors. This would have been impossible with the one-factor-at-a-time design.

Blocking and Randomization

The practical engineer might, with reason, have reservations about process experimentation of any kind. Processes do not always remain stable or in statistical control. For example, the oil used for quenching deteriorates with use. Steel, while having the same carbon content, may have other properties that vary from batch to batch. He might therefore conclude that his process is too complex for experimentation to be useful.

However, what could be more complicated and noisy than the agricultural environment in which Fisher conducted his experiments? The soil was different from one corner of the field to the other, some parts of the field were in shadow while others were better exposed to the sun, and some areas were wetter than others. Sometimes birds would eat the seeds in one part of the field. All such influences could bias the experiment results. Fisher used blocking to eliminate the effect of inhomogeneity in the experimental material and randomization to avoid confounding with unknown factors.

To understand blocking, suppose with the above experiment that only four experimental runs could be made in one day. The experiments would then have to be run on two different days; any day-to-day differences could bias the results. But suppose we ran the eight runs in the manner shown in Figure 2 (see also Table 1) where the data shown in circles were obtained on day one and those in squares obtained on day two.

Notice that there are two runs on the left side and two runs on the right side of the cube that were made in the first day. Similarly, there are two runs on the left side of the cube and two runs on the right side of the cube that were made on the second day. With this balanced arrangement, any systematic difference between days will cancel out when the temperature effect is calculated. A similar balance occurs for the other two main effects and even for the interactions.

A skeptical reader could add five to all the numbers marked as squares on the cube, pretending that the process level changed from one day to the next by that amount. He will find that recalculation of all the main effects and interactions produces the same results as before. As one can see, the beauty of Fisher's method of blocking is that inhomogeneities such as day-to-day, machine-to-machine, batch-to-batch, and shift-to-shift differences can be balanced out and eliminated. Effects are determined with much greater precision. Without blocking, important effects could be missed or it would take many more experiments to find them.

It might seem that blocking permits experiments to be run in a nonstationary environment at no cost. However, this is not quite true. From an eight-run experiment one can calculate at most eight quantities: the mean of all the runs, three main effects, three two-factor interactions, and a three-factor interaction. Using this arrangement we have associated the block difference with the contrast corresponding to the three-factor interaction. Thus, these two effects are deliberately mixed up, or in statistical language, the block effect and the three-factor interaction are confounded. In many examples, however, the three-factor interaction will be unimportant.

What about unsuspected trends and patterns that happen within a block of experimental runs? Fisher's revolutionary idea was to assign the individual treatment combinations in random order within the blocks. In our example, the four runs made on day one would be run in random order, and the four runs made on day two would also be run in a random order. In this way the experimenter can avoid the possibility of unsuspected patterns biasing the results.

Fractional Designs

If we wanted to look at several factors, a full factorial design might require a prohibitively large number of runs. For instance, to test eight factors each at two levels, a full factorial design would require $2^8 = 256$ runs. It slowly became clear to the early researchers that important conclusions could be drawn from experimental designs that used only a carefully chosen piece (or fraction) of the full factorial design.

As early as 1934, L.H.C. Tippett, who had worked closely with R.A. Fisher, used a fractional factorial design to solve a problem in textile manufacturing. His design was a 1/125th fraction using only 25 runs instead of the full $5^5 = 3,125$ required by the full factorial. The first systematic account of how to choose appropriate fractions appeared in 1945 by D.J. Finney, who was working at

Rothamsted Experimental Station with Frank Yates. Fisher's successor, Fractional factorials are particular examples of what C.R. Rao, in his extension of these ideas, called orthogonal arrays.[10] Another important contribution to this field came out of operations research work in Britain in World War II by Burman and Plackett (Figure 3).[11]

For illustration of fractionation, look again at the cube in Figure 2. Instead of the full eight-run design, suppose we ran only the four runs marked with circles. This arrangement, which is shown in Figure 3, has a very special property. Imagine a bright light shone along the steel temperature axis. A complete square of four points would be projected for the factors carbon and oil temperature. As Figure 4 shows, the similar phenomenon occurs if the light is shone down the "carbon" axis or the "oil-temperature" axis. Suppose now that we wanted to test three factors, a Pareto effect was expected: it was fairly certain that, at most, two factors would be important enough to worry about. If just the four runs shown in Figure 3 are run, we would have a complete 2^2 design in whichever two factors turned out to be important.

That example is not particularly interesting because it is so small, but the idea can be extended to a much larger number of factors. Table 2 is a design that can be used to test seven factors A, B, C,..., G in only eight runs. The high level of a factor is indicated by $(+)$ and the low level by $(-)$. It is simple to check that if any two columns are chosen, there will be a complete set of factor combinations $(-)$, $(+ -)$, $(- +)$, $(+ +)$ repeated twice in those factors. Thus, if the engineer believes that there will be a strong Pareto effect, so that no more than two factors of the seven tested will be important, he can use the design to, assure that whichever two turn out to be influential, he will have them covered with a complete duplicated factorial design.[8]

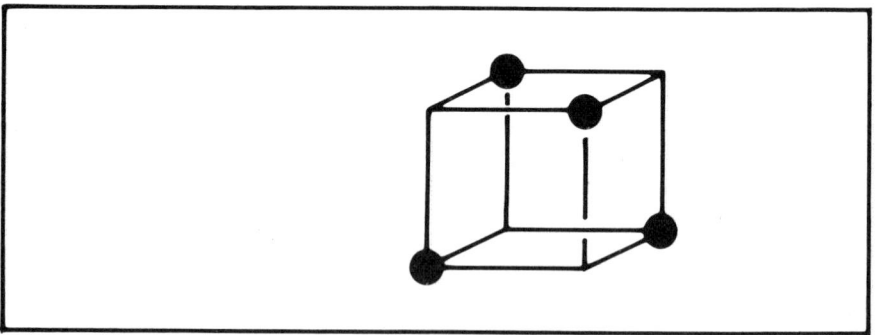

Figure 3. *Half-fraction of an eight-run factorial design.*

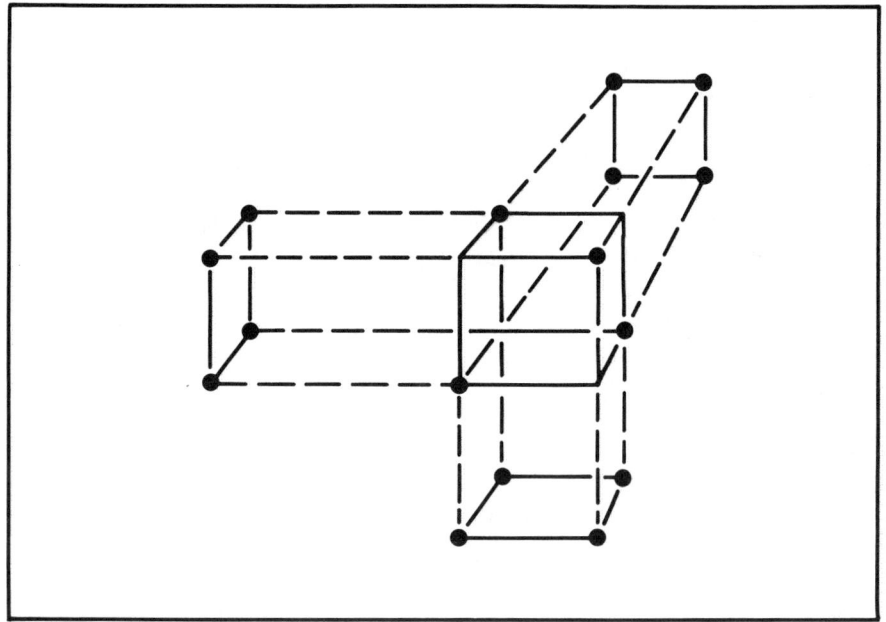

Figure 4. *The projections of a fractional factorial in two dimensions.*

TABLE 2 – An eight-run two-level fractional factorial experiment for varying seven factors. One line in the table corresponds to a single run. A minus sign in a column means that the factor listed above the column should be adjusted at its low level. A plus sign indicates that the factor should be adjusted at its high level in that particular run.

Run	Factors						
	A	B	C	D	E	F	G
1	−	−	+	−	+	+	−
2	−	−	+	+	−	−	+
3	−	+	−	−	+	−	+
4	−	+	−	+	−	+	−
5	+	−	−	−	−	+	+
6	+	−	−	+	+	−	−
7	+	+	+	−	−	−	−
8	+	+	+	+	+	+	+

Larger Fractions

The ideas discussed above can be carried even further. Table 3 shows an experimental arrangement involving 16 runs and 15 factors—a fully saturated fractional factorial. If the experimenter thinks that not more than two factors will be important, these 16 runs can be used to screen as many as 15 factors each at two levels, denoted by + and −, using all 15 columns to accommodate factors. For illustration, consider an experiment by Quinlan concerning quality improvement of speedometer cables.[12] The problem was that speedometer cables can sometimes be noisy. One cause is shrinkage in the plastic casing material. Hence, an experiment was done to find out what caused shrinkage. The engineers started with 15 different factors—liner material, liner line speed, wire braid type, braid tension, wire diameter, etc.—labeled A through O. They suspected that only a few of the factors listed would be important, so they decided to test 15 factors in 16 runs. For each run, 3,000 feet of plastic casing was produced and four sample specimens were cut out and the percentage shrinkage measured. The average shrinkage is shown on the right of Table 3.

TABLE 3 – Quinlan's experiment on speedometer cable shrinkage. The columns of the original experiment have been recoded and reorganized to conform with standard Yates order. The response variable y is the average of four observations.

Run	H	D	-L	B	-J	-F	N	A	-I	-E	M	-C	K	G	O	y
1	−	−	+	−	+	+	−	−	+	+	−	+	−	−	+	0.4850
2	+	−	−	−	−	+	+	−	−	+	+	+	+	−	−	0.5750
3	−	+	−	−	+	−	+	−	+	−	+	+	−	+	−	0.0875
4	+	+	+	−	−	−	−	−	−	−	−	+	+	+	+	0.1750
5	−	−	+	+	−	−	+	−	+	+	−	−	+	+	−	0.1950
6	+	−	−	+	+	−	−	−	−	+	+	−	−	+	+	0.1450
7	−	+	−	+	−	+	−	−	+	−	+	−	+	−	+	0.2250
8	+	+	+	+	+	+	+	−	−	−	−	−	−	−	−	0.1750
9	−	−	+	−	+	+	−	+	−	−	+	−	+	+	−	0.1250
10	+	−	−	−	−	+	+	+	+	−	−	−	−	+	+	0.1200
11	−	+	−	−	+	−	+	+	−	+	−	−	+	−	+	0.4550
12	+	+	+	−	−	−	−	+	+	+	+	−	−	−	−	0.5350
13	−	−	+	+	−	−	+	+	−	−	+	+	−	−	+	0.1700
14	+	−	−	+	+	−	−	+	+	−	−	+	+	−	−	0.2750
15	−	+	−	+	−	+	−	+	−	+	−	+	−	+	−	0.3425
16	+	+	+	+	+	+	+	+	+	+	+	+	+	+	+	0.5835

Fifteen effects can be computed from these 16 numbers, just as was done for the simpler three-factor experiment. A simple way to find out which of the 15 effects really are active is to plot the effects in order of magnitude along the horizontal scale as an ordinary dot diagram. The points are then referred to their respective order of magnitude along with vertical scale of a normal probability plot. To interpret this plot, all one has to know is that effects due to random noise plot on this paper as a straight line, so that points falling off the line are likely to be due to real effects. In this case only factors E and G, the type of wire braid, and the wire diameter, turn out to be important. This is a typical example of Juran's vital few (E and G) and trivial many.

People often worry about the risk of using highly fractionated designs. The problem is that if main effects and higher-order effects exist, it will be impossible to tell from a highly fractionated experiment whether an observed effect is due to a main effect or an interaction. The effects are mixed up or confounded with each other. Nevertheless, fractional factorials often work very well. There are at least three different rationales for this fact. One is that the experimenter can guess in advance which interaction effects are likely to be active and which are likely to be inert. However, this rationale seems to require rather curious logic; it says: while we cannot guess which main (or first order) effects are important, we can guess which interactions (or second order) effects are important. A second rationale is that by using transformations, interaction effects can be eliminated. That is occasionally true, but many interactions cannot be transformed away. For example, no transformation can eliminate the interaction effect that occurs when two metals are combined to form an alloy. The third rationale is based on the Pareto ideas and the projective properties of fractionals. The most appropriate use of fractional designs is for purposes of screening when it is expected that only a few factors will be active. It may be necessary to combine them with follow-up experiments to resolve any ambiguities. Highly fractionated designs should not be used at later stages of experimentation when the important factors are known.[8]

Response Surface Methods

Much experimentation ultimately concerns the attainment of optimum conditions—for example, achieving a high mean, low variance, or both, of some quality characteristic. One method for solving optimum condition problems involves two phases.[6] In phase one the experiment moves from relatively poor conditions to better ones, guided by two-level fractional factorial designs. Phase two begins when near-optimal conditions have been found, and involves a more detailed study of the optimum and its immediate surroundings. The study in phase two often reveals ridge systems, much like the ridge of a mountain that can be exploited, for example, to achieve high quality at a lower cost. In phase two curvature and interactions become important. To estimate these quantities, response surface designs were developed that are often assembled sequentially using two-level factorial or fractional factorial designs plus "star points."[6] These "composite" designs are not of the standard factorial type, and for quantitative variables, have advantages over three-level factorials and fractions.

When the initial conditions for the experiments are rather poor, great progress can be made in phase one with simple two-level fractional factorial designs. The idea of sequential assembly of designs—developing the design as the need is shown—can also be used with fractions. Thus, if after running a fractional design the results are ambiguous, a second fraction can be run selected to resolve those ambiguities.[8]

TAGUCHI'S CONTRIBUTIONS TO QUALITY ENGINEERING

The Japanese have been very successful in building quality into their products and processes upstream using fractional factorial designs and other orthogonal arrays. In particular, Genichi Taguchi emphasizes the importance of using such experimental design in:

- minimizing variation with mean or target.
- making products robust to environmental conditions.
- making products insensitive to component variation.
- life testing.

The first three categories are examples of what Taguchi calls parameter design. He also promotes novel statistical methods for analyzing the data from such experiments using signal-to-noise ratios, accumulation analysis, minute analysis, and unusual application of the analysis of variance (ANOVA). However, some of these methods of analysis are unnecessarily complicated and inefficient. American industry can benefit by adding Taguchi's good quality ideas to those developed in the West and combining them with more efficient statistical methods.

Managing Experimentation

Planning and running experiments and collecting data are hard to do well, and are often traumatic. The process involves talking to many people and getting them to work together, to understand what is required of them, and so forth. One of the key reasons why Rothamsted Experimental Station was so successful was that they had a great deal of experience with running good agricultural experiments. Companies should focus on acquiring similar expertise in running good programs of experimentation and should regard it a subject in its own right. This activity must have management support and understanding.

Managing experimentation and managing the construction of a house, for example, are fundamentally different, because each stage of experimentation can lead down new avenues. One involves serendipity; sometimes even the objective of an investigation can change. For example, in one investigation whose goal was to increase the yield of a process, none of the factors appeared to have any effect. However, one of the factors was residence time. Since this factor had no effect, it meant that the process could be greatly speeded up without any loss. The objective, therefore, changed from increasing yield to decreasing residence time. The message here is that flexibility is a very important element of managing an experimental program.

Fringe Benefits

What can a company gain from all this effort? Better quality products and improved productivity are the first things that come to mind, but that's not the whole story. Improved morale is equally important. When employees can help improve quality, they become much more interested in their jobs—enormous sources of latent energy are released. Give employees the education and tools they require to improve processes. With everyone in the organization pulling together. United States industry will be able to regain its competitive position. The necessary conditions are enthusiastic support from an educated management, companywide education from the chief executive officer to the janitor, and constancy of purpose.

Acknowledgment

We are grateful to Sue Reynard, Stephen Jones, and Judy Pagel for their help in preparing this paper.

References

1. Wheelwright, S. C. 1984. Strategic Management of Manufacturing. in *Advances in Applies Business Strategy*. JAI Press, Inc.
2. Shewhart, W. A. 1939. *Statistical Methods from the Viewpoint of Quality Control*. W. E. Deming, ed. (Washington, D.C.: Graduate School, Department of Agriculture).
3. Juran, J. M. 1964. *Managerial Breakthrough*. McGraw-Hill. New York.
4. Pearson, E. S. 1935. Discussion of Tippett's paper. *Journal of the Royal Statistical Society Supplement*. No. 2.
5. Box, J. F., R. A. Fisher. 1978. *The Life of a Scientist*. John Wiley & Sons. New York.
6. Box, G. E. P. and K. B. Wilson. 1951. On the experimental attainment of optimum conditions. *Journal of the Royal Statistical Society Supplement*. No. 23.

7. Box, G. E. P., L. R. Connor, W. R. Cousins, O. L. Davies, F. R. Himsworth, and G. P. Sillito 1954. *Design and Analysis of Industrial Experiments*. O. L. Davies, ed. Oliver and Boyd. London and Edinburgh.
8. Box, G. E. P., W. G. Hunter, and J. S. Hunter. 1978. *Statistics for Experimenters*. John Wiley & Sons. New York.
9. Box, G. E. P. and N. R. Draper. 1978. *Empirical Model-Building and Response Surfaces*. John Wiley & Sons. New York.
10. Rao, C. R. 1947. Factorial experiment derivable from combinational arrangements of arrays. *Journal of the Royal Statistical Society Supplement*. No. 9.
11. Burman, J. P. and R. L. Placket. 1946. The design of optimum multifactorial experiments. *Biometrika*. No. 33.
12. Quinlan, J. 1985. Product Improvement by Application of Taguchi Methods. Paper presented at Third Supplier Symposium on Taguchi Methods. American Supplier Institute, Inc. Dearborn, MI.

Suggested Readings

Box, G. E. P. 1986. Studies in Quality Improvement: Signal to Noise Ratios, Performance Criteria, and Statistical Analysis: Part I. Report No. 11, Center for Quality and Productivity Improvement. University of Wisconsin-Madison.

Box, G. E. P. and N. R. Draper. 1969. *Evolutionary Operation—A Statistical Method for Process Improvement*. John Wiley & Sons. New York.

Box, G. E. P. and C. A Fung. 1983. Some Considerations in Estimating Data Transformations. Technical Summary Report #2609. Mathematics Research Center. University of Wisconsin-Madison.

Box, G. E. P. and C.A. Fung. 1986. Studies in Quality Improvement: Minimizing Transmitted Variation by Parameter Design. Report No. 8. Center for Quality and Productivity Improvement. University of Wisconsin-Madison.

Box, G. E. P. and S. Jones. 1986. Discussion of Testing in industrial emperiments with ordered categorical data. by V. N. Nair. *Technometrics*. Nov. Vol. 28. No. 4.

Box, G. E. P. and J. G. Ramirez. 1986. Studies in Quality Improvement: Signal to Noise Ratios, Performance Criteria, and Statistical Analysis: Part II. Report No. 12. Center for Quality and Productivity Improvement. University of Wisconsin-Madison.

Box, G. E. P. and R. D. Meyer. 1986. An analysis for unreplicated fractional factorials. *Technometrics*. February. Vol. 28. No. 1.

Cochran, W. G. 1950. The comparison of percentages in matched samples. *Biometrika*. No. 37.

Daniel, C. 1959. Use of half-normal plot in interpreting factorial two-level experiments. *Technometrics*. November. Vol. 1. No. 4.

Daniel, C. 1962. Sequences of fractional replicates in the 2^{p-1} Series. *Journal of the American Statistical Association*. No. 57.

Davies, O. L. 1947. *Statistical Methods in Research and Introduction*. Oliver and Boyd. London and Edinburgh.

Deming, W. E. 1986. *Out of the Crisis*. MIT Press. Cambridge, MA.

Hamada, M. and C.F.J. Wu. 1986. Discussion of testing in industrial experiments with ordered categorical data by V. N. Nair. *Technometrics*. November. Vol. 28.

Hunter, W. G., J. O'Neill, and C. Wallen. 1986. Doing More with Less in the Public Sector: A Progress Report from Madison, Wisconsin. Report No. 13. Center for Quality and Productivity Improvement. University of Wisconsin-Madison.

Ishikawa, Kaoru. 1976. *Guide to Quality Control*. Asian Productivity Organization. Tokyo.

Joiner, B. L. and P. R. Sholtes. 1986. The quality manager's new job. *Quality Progress*. October.

Nair, V. N. 1986. Testing in Industrial Experiments with Ordered Categorical Data. *Technometrics*. November. Vol. 28.

Snedecor, G. W. and W. G. Cochran. 1980. *Statistical Methods*. Seventh Edition. The Iowa State University Press. Ames, IA.

Taguchi, G. and Y. Wu. 1980. *Introduction to Off-Line Quality Control*. Central Japan Quality Control Association. Nagoya, Japan.

Tippett, L. H. C. 1934. Application of statistical methods to the control of quality in industrial production. *Journal of Manchester Statistical Society*.

Youden, W. J. 1969. Experimental Design and ASTM Committees. *Material Research and Standards* (1961). Reprinted in *Precision Measurements and Calibration*. No. 1. United States Department of Commerce. National Bureau of Standards.

Zelen, M. 1959. Factorial experiments in life-testing. *Technometrics*. August. Vol. 1. No. 3.

12

How to Measure Productivity in an Electronics Printed Circuit Board Assembly Environment

Johnson A. Edosomwan

This chapter presents a task-oriented, total productivity model that incorporates a new method of inputs and outputs in a printed circuit board production environment. The inputs are labor, materials, energy, capital, computer expenses, robotics expenses, and other administrative expenses. The outputs are finished and partial units produced. A new method of allocating overhead expenses to the input components is also developed. The model was tested in a printed circuit board production environment. Two groups of five subjects were studied. The two groups were trained on the manual and computer-aided methods of assembling the printed circuit boards. Data on total and partial productivities were collected daily for five weeks during a posttraining performance period. The analysis of variance technique was used to determine whether the productivities of the two methods differed. The results of the study showed that when the manual method of assembling printed circuit boards was changed to the computer-aided method, there were increases in total and labor productivities. However, there was a corresponding decrease in the productivities related to energy, capital, computer operating expenses, and administrative expenses. The results also showed that there was no significant impact on material productivity. The practical problems encountered while implementing the total productivity model are discussed, as well as ways to correct these problems.

INTRODUCTION

For companies to effectively compete in the world market and contribute to the national growth rate of productivity by both short-term and long-term contributions, they must institute a formal productivity measurement system. Such a system can have the following important benefits:

1. Productivity measurement provides an important motivation for better performance, because it helps to identify the basis to measure individual task, project, product, or customer output. It provides the basis for planning the profit level in a company.
2. Productivity measurement highlights, through indices, areas within the company that have potential improvement possibilities. Productivity values and indices also provide a way of detecting deviations from established standards on a timely basis to correct such deviations.
3. Productivity measurement creates a basis for effective supervision of necessary actions and improves decision making through better understanding of the effect of actions already taken to address a given problem.
4. Productivity measurement can be used to compare the performance levels of individual, work groups, tasks, projects, departments, and the firm as a whole.
5. Productivity measurement facilitates better short-term and long-term resource planning. It also simplifies communication by providing common measures, language, and concepts with which to think, talk, and evaluate the business in quantitative terms.

BRIEF DESCRIPTION OF PRODUCTIVITY MEASUREMENT APPROACHES IN COMPANIES

Economic studies reported in the literature have used three types of productivity measures at the company level.

1. *Partial productivity* is the ratio of total output to one class of input. Output per man-hour (labor productivity) is the best example of a partial productivity and is commonly used. Most productivity indices published by the U.S. Department of Labor (1980) are partial measures. Melman (1956), Mundel (1976), and Turner (1980) have also used such measures. Craig and Harris (1973), Siegel (1976), Sumanth (1979), Edosomwan (1980), and Kendrick (1984) agreed that partial measures of productivity, such as labor productivity, could not be interpreted as an overall productivity measure because it did not take into account all input costs.
2. *Total factor productivity* is the ratio of net output to the sum of associated labor and capital (factor) inputs. Mali (1978), Kendrick and Creamer (1965),

and Taylor and Davis (1977) have recommended and used this measure. One great disadvantage of this measure is that it omits the cost of materials, which is one of the vital inputs in business.

3. *Total productivity* is the ratio of total output to all input factors. Sumanth (1979), Kendrick and Creamer (1965), Craig and Harris (1973), and American Productivity Center (1978) have proposed such measures.

Other measures of productivity proposed in the literature include utility index (1978) and array approach (1976). The available theory of productivity has not been concerned with productivity measurement at basic functional levels, such as by task or project. Methodologically, aggregation techniques have been used to measure the overall productivity of the firm. At the firm level, such aggregation techniques provide different results, depending on the type of allocation criteria used, individual resources considered, overhead cost, the definition of fixed and working capital, and floor output elements, as well as identifying the contribution to the output of a given input. Some of the total productivity measures such as that of Kendrick and Creamer (1965), Sumanth (1979), and Craig and Harris (1973) provided the basis for quantifying some of the tangible input and output elements considered in the task-oriented total productivity model. However, the new technological version in which computer and robotic operating expenses are important input components are omitted in these models. Also, most of the authors neglected intangible elements because of the difficulty that arises in measuring them. Evident from the discussion presented so far, there is need for an appropriate productivity measurement model to be developed at the basic levels, such as by task or project. Aggregation techniques would then be used to extend such a measure to the customer, product, department, and company levels.

THE TASK-ORIENTED TOTAL PRODUCTIVITY MODEL (TOTPM)

An incremental analysis is somewhat implicit in the task-oriented productivity model developed by Edosomwan (1985). The measures derived from this model are in the form of an index that intuitively has the following properties and advantages:

1. Indices derived use the broadest possible inputs (labor, materials, energy, robotics, computers, capital, and other administrative expenses) and output (finished and partial units produced).
2. Productivity indices derived vary with changes in task parameters, resources used, and output obtained from transformation of resources.

3. Productivity indices derived are comparable over time and can objectively be used to measure the productivity of tasks, customers, products, projects, work groups, departments, divisions, and companies.
4. Productivity indices derived provide a means of focusing attention on key problem areas for productivity improvements.

Definitions

The key definitions associated with TOTPM are as follows.

Task

At the basis level, a task is a unit of work accomplished primarily at a single location (site), by a single agent, during a single time period, producing useful output from some resources available.

Total Productivity

Total productivity is the ratio of total measurable output (total finished and partial units produced) to the sum of all the measurable inputs (labor, materials, capital, energy, robotic expenses, customer expenses, and other administrative expenses) used for producing the printed circuit boards.

Partial Productivity

Partial productivity is the ratio of total measurable outputs (total finished and partial units of printed circuit boards produced) to one class of measurable input (example: man-hours used for producing printed boards).

The input and output components of the task-oriented total productivity model are shown schematically in Figures 1 and 2, respectively.

Derivation of Productivity Values and Indexes Notations

LET

i   = Printed circuit board (PCB) assembly task (manual or computer-aided method)
j   = XYZ manufacturing plant location (j = 1,2,3,...m) (model can be applied simultaneously in more than one site)
t = Study period t = 1,2,3,...,n (n = 5 weeks)
Oijt   = Total quantity of PCB produced (total output) by task i, in site j, in period t
Fijt   = Finished task (finished units of PCB produced by task i, in site j, in period t)

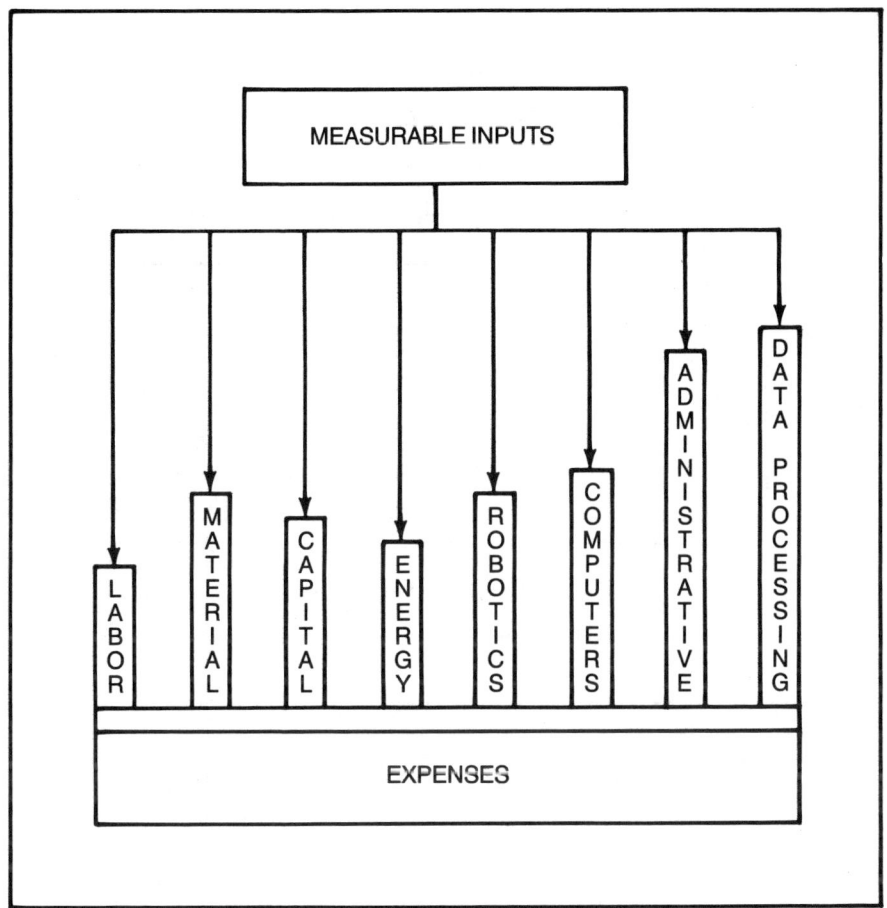

Figure 1. *Input components considered in the total productivity model.*

PCijt = Percent completion of partial units of PCB produced by task i, in site j, in period *t*

Pijt   = Partially completed task (partial units of PCB produced) by task i, in site j, in period *t*

SPijt = Base period selling price per unit for a PCB produced by task i, in site j, in period *t*

CRijt = Variable computer-related expense input in dollars used by task i, in site j, in period *t*

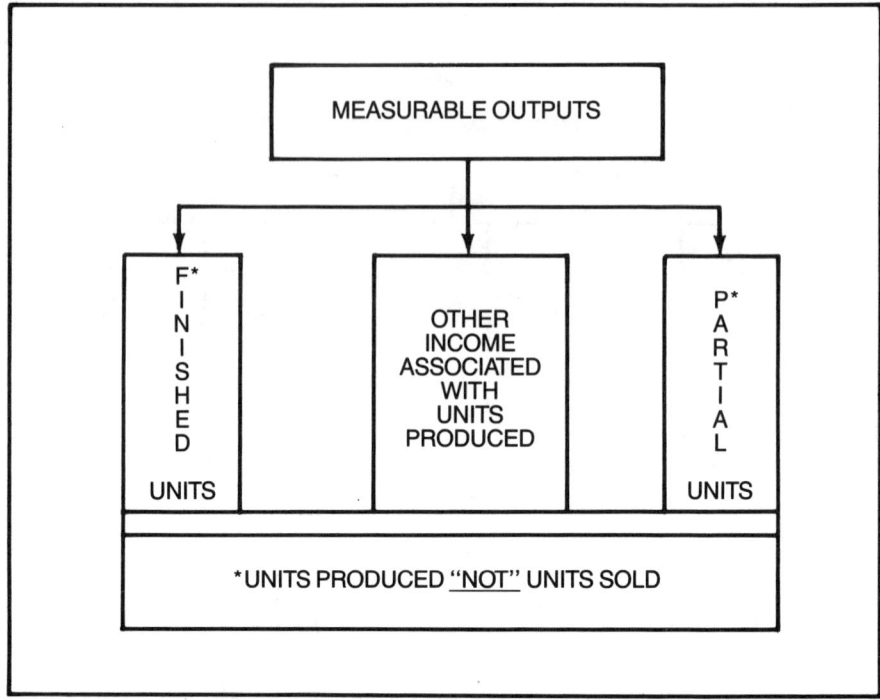

Figure 2. *Output components considered in the total productivity model.*

RRijt = Variable robotics-related expense input in dollars used by task i, in site j, in period t

Lijt   = Labor hours input used by task i, in site j, in period t

LRijt = Labor rate per hour used by task i, in site j, in period t

Mijt   = Material and purchase parts expense input in dollars used by task i, in site j, in period t

Cijt   = Capital-related expense (includes fixed and working capital such as cash, accounts receivable, tools, plants, buildings, amortized, R&D, etc.) input in dollars used by task i, in site j, in period t (capital is computed using the lease value concept)

Eijt   = Energy-related expense (includes electricity, solar energy, water, coal, gas, etc.) input in dollars used by task i, in site j, in period t

OEijt = Other administrative-related expense input in dollars used by task i, in site j, in period t (other expenses include travel, taxes, professional fees, marketing, R&D, general administration, others)

Yijt = Total productivity of task i, in site j, in period t
TFijt = Total factor productivity of task i, in site j, in period t
Pijt = Partial productivity of task i with respect to one input in site j, in period t
Iijt = Total input of task i, in site j, in period t
Tb   = Base period time. Base period time is a reference period in which the output and input in monetary terms is reduced. Thus, the total productivity of a task expressed as $ output/$ input in constant dollars

The total and partial productivities for printed circuit board assembly using the computer-aided method are expressed as follows:

Yijt = Total productivity of task i (computer-aided method of assembling PCB) performed in site j, in period t

$$Yijt = \frac{\text{Total measurable output of task i, performed in site j, in period } t}{\text{Total measurable input of task i, performed in site j, in period } t} \quad (A1)$$

$$Yijt = \frac{Oijt}{Iijt} \quad (A2)$$

$$Yijt = \frac{(Fijt)(SPijt) + (Pijt)(SPijt)(PCijt)}{(CRijt) + (Iijt)(LRijt) + (Mijt) + (Eijt) + (OEijt) + (Cijt)} \quad (A3)$$

Pijt = Partial productivity of task i, with respect to labor input in site j, in period t

$$Pijt = \frac{\text{Total measurable output of task i, performed in site j, in period } t}{\text{Measurable labor input of task i, performed in site j, in period } t} \quad (A4)$$

$$Pijt = \frac{(Fijt) SPijt) + (Pijt)(SPijt)(PCijt)}{(Iijt)(LRijt)} \quad (A5)$$

IMPLEMENTATION METHODOLOGY

The task-oriented productivity model was used in XYZ plant to measure and compare partial and total productivities of the manual and computer-aided method of assembling printed circuit boards. The following steps were taken during implementation of the model:

Step 1: Familiarization Sessions

In this step, a formal study of the XYZ printed circuit board manufacturing process, procedures, and cost-accounting system, types of boards produced, types of operations, tasks performed, and the J-2000 computer operation was conducted. The key personnel in the various departments were introduced to and taught the purpose of the productivity measurement system.

Step 2: Task Significance, Input/Output Analysis

The significance of each task performed, the extent of the manual method of assembly, computer usage, and the associated input/output relationship of both tasks were determined.

Step 3: Development of Allocation Criteria for Overhead Expenses

Although several overhead allocation criteria exist, the most generally used traditional cost-accounting and industrial engineering approach is the proportional contribution of direct hours to allocate overhead expense to product or task. This approach required daily, direct labor recording and the derivation of burden rate from net expense and direct hours. Therefore, product cost included labor and burden. Allocating overhead is, therefore, done using direct hours and one burden rate applied to all products produced by the manufacturing plant unit or organization. In direct-hours allocation criteria some product costs could be misallocated, especially in situations where the manufacturing plant produced more than one product. Edosomwan (1985) and Kendrick (1984) pointed out that allocation criteria used for output and input elements could vary, depending on the type of production environment, cost elements, accounting information, and managerial preferences. For the purpose of our study at XYZ, proportional contribution to total number of printed circuit board insertions was used as an allocation criteria for overhead expenses. The allocation criteria was preferred to other allocation schemes because energy usage, machine utilization, and component preparation time varied with the number of insertions. For example, printed circuit board Mark I type required 218 insertions, as opposed to printed circuit board Mark II type with 120 insertions. Mark I, therefore, used more energy than Mark II—similarly, Mark I used more machine time and more labor time for precomponents inserted. The various printed circuit board types and the number of insertions in each printed circuit board are presented in Table 1.

TABLE 1 – Summary of Printed Circuit Board Insertion Rate at XYZ Plant

Printed Circuit Board Type	Number of Insertions in Each Printed Circuit Board Type
*Mark I	218
Mark II	120
Mark III	262
Mark IV	98

* Mark I was the printed circuit board type considered in our study.

Step 4: Base Period and Deflators Selection

Generally, the input and output elements of the task-oriented total productivity model are deflated to remove the effect of price changes. If needed, the deflators are obtainable from the "Monthly Labor Review," published by the Bureau of Labor Statistics, Washington, D.C. The deflators are chosen based on their correlation with the various input and output elements. In this study, the deflators were not required because the study period was only for five weeks, not long enough to have significant variation in price changes. When deflators are not available for real time, they could be forecasted. Edosomwan (1980) presented a methodology for forecasting deflators and the criteria for selecting base period.

Step 5: Data Collection Design

In this step, various forms and instruments were designed to capture input and output elements needed for the task-oriented productivity model. In designing these forms, various factors, such as how often data was collected, information required, computation steps, scale of measurement, and clarity, were considered.

Step 6: Personnel Training and Testing of Data Collection Instruments

In this step, the subjects were trained for one week on how to perform both the manual and computer tasks, how to complete the productivity measurement forms, safety regulations, and XYZ operating procedures. The data collection instruments designed in Step 3 was pretested with five workers and two managers and revised for better clarity. The operators were selected at random to perform the manual and computer-aided printed circuit board assembly tasks.

Step 7: Data Collection, Synthesis, and Computations

Using the various forms designed in Step 3, input and output components needed for the productivity model were collected periodically for each individual worker, was well as for each group that performed the manual and computer-aided tasks. The synthesized data were used to compute the values and indices for partial and total productivities.

Step 8: Trend Analysis and Interpretation of Findings

The partial and total productivities for each task obtained by period were analyzed periodically to ascertain why they increased or decreased. The analysis of variance technique was also used to interpret data collected.

RESULTS

The input and output components, total and partial productivities by period both the manual and computer-aided tasks, are shown in Tables 2 and 3, respectively. A summary of the research findings is also presented in Table 4. At $\alpha = 0.05$ level of significance, there was a significant difference in total and labor productivities between the two tasks. The computer-aided method increased total and labor productivities more than the manual method of assembling the printed circuit boards. The programming sequence of the computer-aided method enabled operators to work at a faster pace, and with the aid of the rotating bin and special handling fixtures on the machine, operators were able to assemble two printed circuit boards simultaneously. On the other hand, the manual method required operators to work at their own pace and also required more hand motion because of the way the work stations were arranged.

The computer-aided method was also found to have decreased energy, capital, computer operating expense, and other expense productivities. In addition to the energy used for the lighting fixtures, the computer-aided method also used direct electricity to run the rotary bin and the movement of the assembly fixtures. The manual method only used energy for the lighting fixtures. Under the lease-concept approach for capital, the computer-aided method used more capital than the manual method to support tools, maintenance, programming, safety, and indirect effort supervision costs. This affected both computer-operating expenses and other expense productivities. No significant difference in material productivity was observed. Although the results have been used to compare the computer-aided and manual methods of assembly printed circuit boards, the trends for each task by period are also essential, especially for investigating areas which need productivity improvement. The total productivity trend for the manual and computer-aided tasks are presented in Figure 3. In this situation, the decision maker is faced with either using the computer-aided

TABLE 2 – Output and Input Components for the Manual and Computer-Aided Tasks

Items	Period 1 Manual Task	Period 1 Computer Task	Period 2 Manual Task	Period 2 Computer Task	Period 3 Manual Task	Period 3 Computer Task	Period 4 Manual Task	Period 4 Computer Task	Period 5 Manual Task	Period 5 Computer Task
Quantity of finished PCB produced	40	55	38	53	35	54	39	53	40	54
Percent completion of partial units of PCB										
Quantity of partial PCB produced										
Average selling price per unit of PCB ($)	1,000	1,000	1,000	1,000	1,000	1,000	1,000	1,000	1,000	1,000
Total output ($)	40,000	55,000	38,000	53,000	35,000	54,000	39,000	53,000	40,000	54,000
Labor expense input ($)	85	85	84	84	84	84	84	84	84	84
Material expense input ($)	4,147	5,775	3,990	5,565	3,675	5,617	4,147	5,513	4,200	5,517
Capital Expense Input ($)	63	73	56	67	60	71	61	70	62	73
Energy expense input ($)	86	96	86	96	86	96	86	96	86	96
Robotics expense input ($)										
Computer expense input ($)		98		101		97		100		104
Other admin expense input ($)	4,417	5,082	4,417	5,111	4,417	5,093	4,417	5,088	4,417	5,107
Total input ($)	8,798	11,111	8,633	10,923	8,322	10,961	8,795	10,851	8,849	10,877

Note: No partial units of printed circuit boards (PCB) were produced during the study period. No robotic expense was incurred.

method to accept gains in total and labor productivities, but must investigate and continue to improve the other partial productivities (energy, capital, computer operating expense, and other expenses), or use the manual method to assemble the printed circuit boards in an environment where labor is inexpensive. Other factors that could influence the method of production selected are impact on employment, job satisfaction, psychological stress, and ergonomics. These factors are discussed by Edosomwan (1985).

Problems Encountered During Implementation

The problems presented here are mainly intended to form a guide to companies setting up a productivity measurement program for the first time. Noted,

TABLE 3 – Total and Partial Productivities for the Manual and Computer-Aided PCB Tasks

Items		Period 1 Manual Task	Period 1 Computer Task	Period 2 Manual Task	Period 2 Computer Task	Period 3 Manual Task	Period 3 Computer Task	Period 4 Manual Task	Period 4 Computer Task	Period 5 Manual Task	Period 5 Computer Task
Total productivity	Value	4.55	4.95	4.40	4.85	4.21	4.93	4.43	4.88	4.52	4.96
	Index	1.00	1.00	0.97	0.98	0.93	0.99	0.97	0.99	0.99	1.00
Labor expense productivity	Value	470.59	647.06	452.38	630.95	416.67	642.86	464.29	630.95	476.19	642.86
	Index	1.00	1.00	0.96	0.98	0.89	0.99	0.99	0.98	1.01	0.99
Material expense productivity	Value	9.65	9.52	9.52	9.52	9.52	9.61	9.40	9.61	9.52	9.79
	Index	1.00	1.00	0.97	1.00	0.97	1.01	0.97	1.01	0.97	1.03
Capital expense productivity	Value	634.92	753.42	678.57	791.04	583.33	760.56	639.34	757.14	645.16	739.73
	Index	1.00	1.00	1.07	1.05	0.919	1.01	1.01	1.00	1.02	0.98
Energy expense productivity	Value	465.11	572.92	441.86	552.08	406.98	562.50	453.48	552.08	465.12	562.50
	Index	1.00	1.00	0.95	0.96	0.88	0.98	0.97	0.96	1.00	0.98
Robotics expense productivity	Value										
	Index										
Computer expense	Value		561.22		524.75		556.70		530.00		519.23
	Index		1.00		0.94		0.99		0.94		0.93
Other administrative expense productivity	Value	9.06	10.82	8.60	10.37	7.92	10.60	8.83	10.42	9.06	10.57
	Index	1.00	1.00	0.95	0.96	0.87	0.98	0.97	0.96	1.00	0.98

however, that the problems associated with starting a productivity measurement program could vary depending on the type of production environment. Based on the implementation of the task-oriented total productivity model at XYZ plant, we found that the existing data collection system was not able to provide the input and output components needed for the task-oriented total productivity method. A computer program was developed to enable us to collect input and output by task at the time of production. A session was held with the cost-accounting department, and the cost-accounting system was slightly modified. Similar to what occurs with other studies, employees initially believe that a study brings layoffs, changes in procedures, etc. We experienced such

TABLE 4 – Summary of Research Findings

Variable	Difference in Mean Productivity Values Between Manual & Computer Task	Impact on Total and Partial Productivities		Significant at α = 0.05
Total productivity	0.49	C+	M−	YES
Labor expense productivity	182.92	C+	M−	YES
Material expense productivity	0.09	C+	M+	NO
Capital expense productivity	124.12	C−	M+	YES
Energy expense productivity	113.91	C−	M+	YES
Computer operating expense productivity	538.38	C−	M+	YES
Other administrative expense productivuty	1.87	C−	M+	YES

C+ = Computer-aided task had positive impact on productivity.
C− = Computer-aided task had negative impact on productivty.
M+ = Manual task had positive impact on productivity.
M− = Manual task had negative impact on productivity.
* = No significant difference in productivity was observed

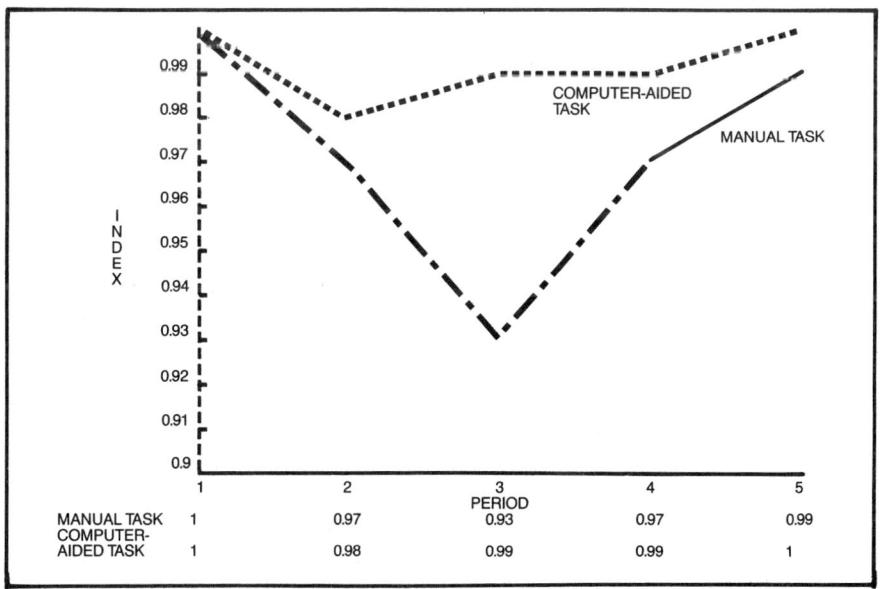

	1	2	3	4	5
MANUAL TASK	1	0.97	0.93	0.97	0.99
COMPUTER-AIDED TASK	1	0.98	0.99	0.99	1

Figure 3. *Total productivity trend for the manual and computer-aided tasks.*

feelings and resistance to change at the beginning of our study, but they faded eventually as our study objectives were communicated to and supported by both management and employees. Initially, data collection was very difficult, because we had to rely on persons who had to help with data collection in addition to their normal duties. The teamwork approach adopted to implement the model eliminated the problem of "it's not my program" syndrome. Specific tasks were assigned to specific people, and meetings were held regularly to discuss and resolve potential problems.

Conclusions

The concept of a formal approach to productivity measurement for printed circuit board production is still in its infant stages of conceptualization and rationalization. This chapter has presented a total productivity measurement model that takes into account all the measurable factors of input and output. Although the model was sensitive in the one test case of the printed circuit board assembly task, the model potentially could be applied in other types of tasks performed in both manufacturing and service organizations. However, minor modifications are needed in the model notations and data collection instruments. A computer program is also available to ease the computational burden.

To maintain the productivity measurement program once it has already been established using the task-oriented total productivity model, the following are needed:

1. Periodic collection. This will depend on whether the productivity indexes are established on a monthly, quarterly, or yearly basis.
2. Computations of productivity indices for each task and for the firm, and trend analysis for specific areas to be investigated for improvement.
3. Comparison of tasks for the same base period.
4. Development of historical data on new tasks to permit sales/cost/profit analysis before the tasks are introduced to the productivity measurement program.
5. Comparison of productivity trend patterns with the actual profit figures by period.
6. Training sessions for key personnel in the various departments that will be involved in maintaining the productivity measurement program.

Suggested Readings

American Productivity Center. 1978. Productivity and the industrial engineer. Region VIII, American Institute of Industrial Engineers Conference. Chicago, October 27. (Seminar Notes)

Craig, C. E. and C. R. Harris. 1973. Total productivity measurement at the firm level. *Sloan Management Review*. 14(3): 13-29.

Dewitt, F. 1976. Productivity and the industrial engineer. *Industrial Engineering*. 8(1): 20-27.

Edosomwan, J. A. 1985. A methodology for assessing the impact of computer technology on productivity, production quality, job satisfaction, and psychological stress in a specific assembly task. Doctoral Dissertation. The George Washington University.

Edosomwan, J. A. 1985. Computer-aided manufacturing impact on productivity, production quality, job satisfaction, and psychological stress. *1985 Fall Industrial Engineering Conference Proceedings*.

Edosomwan, J. A. 1980. Implementing of a total productivity model in a manufacturing company. Masters Thesis. University of Miami.

Edosomwan, J. A. and D. J. Sumanth. 1988 (in press). A practical guide for productivity measurement in organizations working manual. Unipub. New York.

Kendrick, J. W. 1984. Improving company productivity handbook with case studies. The Johns Hopkins University Press. Baltimore.

Kendrick, J. W., and D. Creamer. 1965. Measuring company productivity: handbook with case studies. *Studies in Business Economics*. National Industrial Conference Board. New York: No. 89.

Mali, P. 1978. Improving total productivity: MBO strategies for business, government, and non-profit organizations. John Wiley and Sons. New York.

Melman, S. 1956. Dynamic factors in industrial productivity. John Wiley and Sons. New York.

Mundel, M. E. 1976. Measures of productivity. *Industrial Engineering*. 8(5): 32-36.

Siegel, I. H. 1976. Measurement of company productivity in improving productivity through industry and company measurement. National Center for Productivity and Quality of Working Life, Washington, D.C.: U.S. Government Printing Office. Series 2: 15-25.

Stewart, W. T. 1978. A yardstick for measuring productivity. *Industrial Engineering*. 10(2): 34-37.

Sumanth, D. J. 1979. Productivity measurement and evaluation models for manufacturing companies. Doctoral Dissertation. I.I.T.

Taylor, B. W. III and R. K. Davis. 1977. Corporate productivity — getting it all together. *Industrial Engineering*. 9(3): 32-36.

Turner, J. A. 1980. Computers in bank clerical functions: Implication for productivity and the quality of working life. Doctoral Dissertation. Columbia University.

U.S. Department of Labor. 1980. Bureau of Labor Statistics. *Monthly Labor Review.* 103(1): 40-43.

The author is grateful to the Social Science Research Council (SSRC) (U.S. Department of Labor) and International Business Machines Corporation (IBM) for supporting this work under their grant numbers SS-36-83-21 and IBM-2J2-2K5-722271-83-85, respectively. Any opinions, findings, conclusions, or recommendations expressed in this publication are those of the author and do not necessarily reflect the views of the SSRC and IBM. Many thanks to Dr. John W. Kendrick, Dr. Robert C. Waters, Dr. James N. Mosel, Pat Lopez, and Gerald Talen for their continued encouragement and help.

Part 2

Just-In-Time Applications in Electronics Assembly

This section presents case studies and implementation experiences from a number of electronics assembly environments. Even though the implementations differ in exact steps taken to become a successful practitioner of just-in-time (JIT), there are several lessons that can be learned. Kenfield (1985) describes a nine-step JIT implementation cycle at Hewlett-Packard (HP). He describes the various steps of the JIT cycle and relates it with HP's own experiences. A detailed implementation case study at HP's Computer System Division is presented in the next chapter by Sepheri and Walleigh (1986). Using examples from a wave solder project and an automatic insertion project, they demonstrate the relationship of total quality control to reduction of inventories.

Hedden and Richardsen (1987) present a JIT success story at MICROTEL Ltd., Canada. Using many of the well-known JIT principles, MICROTEL reduced their power-printed circuit card assembly from 23 weeks to 2 weeks and improved their first pass test yields by 30 percent. JIT manufacturing has also been implemented at a major surface mount technology (SMT) facility. McCLelland (1987) presents a case study at Philips Radio Communication Systems, Cambridge, England where a variety of products from mobile radios to pagers are assembled. JIT application reduced their PCB leadtimes by 96 percent and work-in-process by 80 percent, while requiring only 70 percent of the original floor space. Application of JIT is not restricted to simple products produced in high volumes on progressive assembly lines. Sandras (1986), in an article on low volume/high mix JIT, describes a case study of how JIT was successfully applied to a complex low volume product that had disconnected job-shop type process flow.

A number of other leading corporations, including Digital Equipment, Motorola, General Electric, Texas Instruments, Apple, and Honeywell, claim significant productivity and quality improvements through JIT implementation. Cortes-Comerer (1987) presents an excellent summary of the major JIT impact areas and describes some of the successes reported by some of the previously mentioned corporations.

Additionally, this chapter presents issues and methods involved in modeling JIT [Trevino and McGinnis (1986)], JIT purchasing [Parmelee (1986)], JIT cost accounting [McIlhattan and Anderson (1987)] and managing JIT programs [Leahy (1985)]. Part 2 summarizes the papers presented in this section.

Part 2 Matrix

Chapter Number	Title	CONTENT CLASSIFICATION					ELECTRONICS ASSEMBLY PROCESSES							
		Model	Quantitative Technique	Strategy/Methodology	Case Study	General Theory	Design	Receiving & Store	Inspection	Kitting	CP Assembly Mfg.	Equip. Assembly	Packing	Mat'l Handling
13	A Nine-Step Approach to JIT Implementation			X		X								
14	Quality and Inventory Control Go Hand in Hand at Hewlett-Packard's Computer Systems Division													
15	JIT Manufacturing at Microtel Limited				X	X	X	X	X		X			
16	Philips Gets SMT Just-In-Time			X			X				X			
17	Low Volume/High Mix JIT		X			X	X							
18	JIT is Made to Order				X									
19	A Mathematical Procedure to Support the Design of "Pull" Production Strategies in Electronics Assembly	X		X			X							
20	The New Ballgame: JIT Purchasing							X			X	X		
21	Cost Management Impact of JIT			X		X								
22	Management Issues In Just-in-Time and Optimized Production Technology			X		X					X			

13

A Nine-Step Approach
To
JIT Implementation

John E. Kenfield

Worldwide competitive pressures are demanding improvements of productivity and quality in all manufacturing companies. In response, many companies have implemented techniques popularized by the Japanese to reduce inventories, reduce cost, shorten manufacturing times, and improve quality. Most of these techniques are bundled under the heading of just in time (JIT).

This chapter offers a definition of JIT and explores the elements of JIT by looking at a nine-step approach to its implementation. Some of the accomplishments within Hewlett-Packard will also be discussed.

JUST-IN-TIME-MANUFACTURING

What is JIT?

Generally, JIT is a manufacturing philosophy and strategy for survival, growth, and excellence through improved productivity and quality. Whether a company's motivation is survival, growth, or excellence depends on where they are relative to the competition.

In its narrowest sense, JIT is a set of material movement techniques that result in only the right material being at the necessary place at the necessary time (Hall 1983). Many of the techniques that lead to lower inventories and increased velocity of material through the manufacturing process will be discussed later. In a nutshell, JIT is only common sense attention to detail and an overwhelming desire to excel.

The major objectives of JIT are the following: 1) streamline the process to shorten cycle times; 2) improve quality by eliminating the reasons for rejects; 3) strive for lot sizes of one through short set-ups; 4) implement "demand pull" material movement with a Kanban signaling technique; 5) eliminate waste everywhere; and 6) make JIT a continuous improvement effort.

"Waste" in the manufacturing environment is ". . . anything other than the minimum amount of equipment, material, parts, space, and workers time which are absolutely essential to add value to the product. . ." (American Production and Inventory Control Society 1984). The key concept is that the nonvalue added elements are waste.

What are the Benefits?

The paybacks from a JIT approach to productivity and quality are reduced inventories, savings in space, manufacturing cycle time reductions, and permanent solutions to problems. These all mean lower manufacturing costs that yield more profit and a competitive advantage. The gains aren't small, incremental improvements. As we'll see, the percentage improvement numbers are big.

THE NINE STEP IMPLEMENTATION

The circle diagram (Figure 1) is an attempt to condense almost all of the concepts of JIT. The arrows are meant to indicate progression through the nine steps, but note that the order of the steps is not critical. A second important concept to note is that no step is ever done. The mindset of continuous improvement requires that each step and the entire process is ongoing.

Step 1: Education and Awareness

As previously mentioned, the order of the steps is not critical. However, there is one exception! All levels of the organization must understand JIT concepts, potential benefits, and the activities because, if they aren't actively participating, they will be ultimately affected. There must be total commitment to the program and a top management "champion," as these are consistent characteristics of the successful cases.

Another characteristic of success is to start small. Start a pilot project in an area that can be isolated from material flows in the rest of the plant but has similar manufacturing processes. This will be the demonstration project to convince the skeptical. Don't analyze the effort to death, though. Start somewhere and start now.

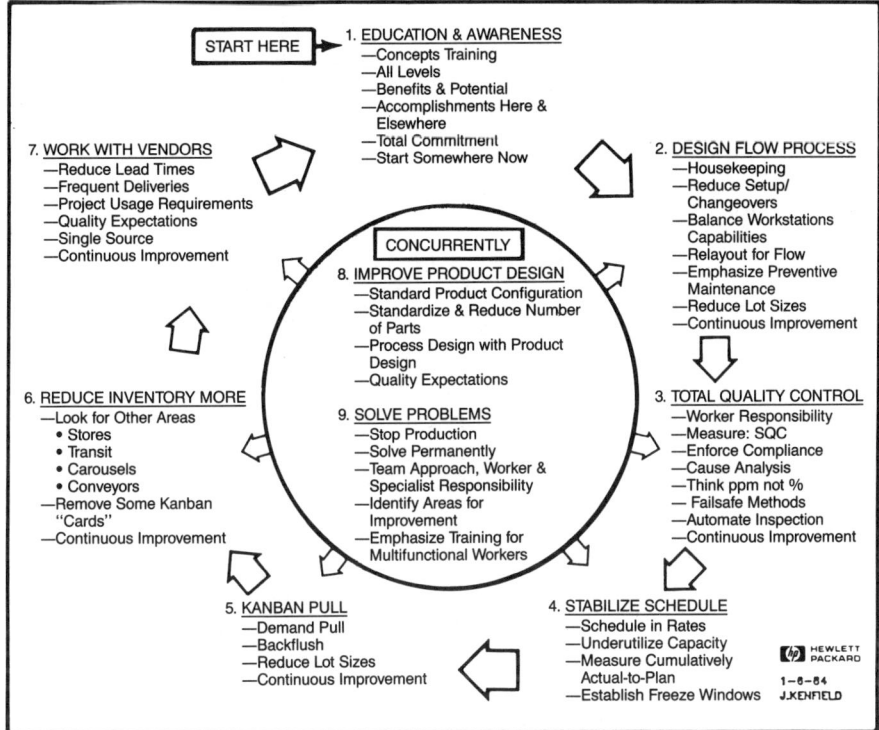

Figure 1. *How to accomplish repetitive manufacturing—Nine steps to improve quality and productivity.*

Step 2: Design the Flow Process

Usually, the next step involves improving or setting up a material and manufacturing flow process. Set-up or changeover times are attacked without mercy—this is the secret to reducing lot sizes. Work station capacities are balanced and designed for smooth flow.

Much can be done with existing equipment and machinery to shorten set-up time. By careful analysis of the existing procedure, many improvements can be identified. Do as much preparation and put-away as possible while the machine is running. Eliminate adjustments, and modify equipment and tools for fast exchange. Using these techniques, and not much money, set-up times can be reduced 90%.

Another often used approach is to eliminate stockrooms and locate material at its point of use. Extra counting, sorting, paperwork, and material movement does not add value. Linking related processes also reduces material handling.

A smoothly running process cannot tolerate unexpected downtime. Consequently, equipment preventive maintenance must have attention. Ideally, the operators should have preventive maintenance responsibility as part of their daily routine. The objective is to make equipment more capable of its intended function when it is old rather than when it was new.

Step 3: Total Quality Control

If quality is built into every step of the production process, from design through preproduction preparation to shop floor execution, a quality product automatically follows. Process variance is reduced by attention to equipment stability, tools, set-up reproducibility, material, and standardized procedures.

Process and products quality are monitored with statistical quality control (SQC) techniques such as control charts and Pareto analysis. Casual analysis is used to identify the sources of problems. The production worker is given the tools and responsibility for quality, and the measurements are used for improvement, not to place blame.

Step 4: Stabilize the Schedule

JIT production is scheduled by rates, not work order lots with due dates. The rates may be units per hour, shift, or day. Mixed model production is scheduled with a consistent repetition of the mix to create a consistent flow of the material used. A stable schedule means that it is frozen for some period and there is ± zero deviation from the schedule when measured cumulatively. To accomplish zero deviation, some capacity must be reserved to recover from problems and production down times.

Step 5: Kanban Pull

"Demand pull" production control replaces "schedule push". This means that material is not issued and production is not started based on a schedule, but instead is controlled by demand from the next operation. Kanban is the signaling technique used to pull material through the manufacturing process. The signal may be a card requesting replenishment material, an empty square, or any technique that indicates demand. The using work centers always send for material from producing work centers. Parts are made only when authorized. There may be extra "cards" in the system to allow a small buffer against minor disruptions and problems. Standard-sized containers are used for material movement to control lot sizes and reduce the need for counting and recounting.

Another JIT concept relating to material movement is the backflush or postdeduct method of accounting for material usage. Periodically, the count of units completed is used to explode the bill of material to provide detail of components used. Unusual material usage is captured by exception reporting.

Step 6: Reduce Inventory More

Once the process has improved, quality has improved, the schedule is stabilized somewhat, and demand pull is in operation, other reasons for inventory waste can be targeted. Storerooms, material handling methods, such as conveyors, carousel storage, and remotely located processes, create and hold inventory. Once a process is stable, the extra Kanban signals that maintain buffers can be removed so that the next generation of problems can be uncovered. This step is reemphasis of continuous improvement.

Step 7: Work with Vendors

The suppliers' role in JIT includes frequent delivery of small lots, shorter lead times, and 100% reject-free material or parts. This is generally addressed with close, partership-like relationships with fewer vendors. Single sourcing, sharing forecasts of requirements routinely, and gathering and sharing quality data at the source are also elements of supplier relationships.

More frequent deliveries of smaller lots is just the extension of in-plant JIT demand pull to the suppliers. A couple techniques that have been used include scheduled truck routes to pick up mixed loads from regular suppliers and the use of consolidation points to improve the logistics efficiency.

A couple of cautions are in order when addressing the supplier linkage. The manufacturers should have inside processes operating with effective JIT methods before extending the challenge to the suppliers. That is why this step of the implementation is placed later on the chart. Secondly, a supplier education and awareness program will often be needed to ensure their complete understanding of what is changing and why. JIT supplier involvement does not mean pushing the inventory back onto the supplier, but often that is the initial response and action.

Step 8: Improve Product Design

Steps 8 and 9 are placed in the center of the chart because these elements of JIT are implemented and operated concurrently with the other elements and affect other steps. Step 8 is meant to imply that of attention to a products' manufacturability during its design will contribute greatly to the productivity and quality goals.

Designing standard products (no built-in options) from preferred standard parts supports the strategy of fewer vendors, lower inventory, and process flow manufacturing. Designing for automation permits that extension, but will also greatly improve the manufacturing process even if never automated. One of the problems in the JIT environment is implementing engineering changes. In general, engineering changes that can't be avoided are grouped and applied at a time where there is a major schedule change.

Step 9: Solve Problems

This element of the JIT philosophy deals with people: their role, responsibility, and training. Permanently solving problems means stopping production when they are encountered, working as a team to recognize and solve problems, and having the skills necessary to offer flexibility. Multifunctional workers offer that flexibility because they can operate many different machines, participate in or do set-ups, do preventive maintenance, deal with problems, and even have material control responsibility.

JIT IS NOT

There are a number of invalid reasons one might mistakenly use to pursue a JIT program.

JIT is not an inventory program, a program just for suppliers, a new fad, a material management project, a cultural phenomenon, a program that displaces MRP, or a panacea for poor management. JIT is a philosophy and strategy for improved productivity and quality.

HEWLETT-PACKARD EXPERIENCES

One of the first divisions of Hewlett-Packard to implement the JIT elements on a broad scale was the Vancouver (Washington) Division, which produces work station printers. During a period where shipments grew 20%, the following improvements were achieved when compared with traditional batch manufacturing. Work-in-process inventory was down 82%, floor space required went down 40%, scap and rework went down 30%, and labor efficiency improved 50%.

At the Roseville (California) Networks Division where printed circuit assemblies for computers are produced, similar improvements were achieved during a shipment growth of 40%. Work-in-process inventory went down 70%, manufacturing cycle time from beginning to end of the process was cut to one fifth of the previous year's time, and productivity was up 23%.

The Loveland Instrument Division in Colorado installed a demand-pull technique along with other JIT elements and realized the following gains. Workmanship defects are down 35%, work-in-process inventory was cut in half, 44 hours of data entry time has been eliminated through system simplifications, and start-to-finish assembly time was reduced from 40 days to 8 days.

These improvements are characteristic of the gains realized in one company through JIT techniques. JIT is an important element for success in today's competitive business environment.

Suggested Readings

Hall, R. W. 1983. *Zero Inventories* Dow Jones - Irwin. Homewood, Illinois.

American Production and Inventory Control Society (APICS). 1984. *Zero Inventories Crusade, An Introduction.* APICS. Falls Church, VA.

Reprinted from 1985 *International Electronics Assembly Conference Proceedings.*

John E. Kenfield is Corporate Program Manager for Manaufacturing Quality Systems, Hewlett-Packard Company, Palo Alto, California.

14

Quality and Inventory Control Go Hand In Hand at Hewlett-Packard's Computer Systems Division

Mehran Sepehri
Rick Walleigh

Hewlett-Packard, a worldwide manufacturer of electronics devices and business computers, is known as a pioneer of just-in-time and total quality control techniques in the U.S. Several divisions of Hewlett-Packard have been very innovative in adapting these techniques to their operations.

The Computer Systems Division, located in northern California, has achieved impressive improvements in quality and inventory levels through a combination of just-in-time and total quality control. The product, the series 68 HP 3000 business computer, is a high performance machine with low volume production. It is the largest business computer Hewlett-Packard currently markets. A tremendous number of components go into the system, and the units go through extensive testing and a complex manufacturing process.

The Computer Systems Division is organized into the following departments:

- *Printed circuit board assembly*—Blank printed circuit (PC) boards and components are assembled by means of automated and manual component loading methods, wave and manual soldering, and various other operations, including masking, forming, back loading and assembly of fabricated parts.

- *PC test*—Completed PC assemblies are turned on and tested using both automated and manual test equipment. Defective and misloaded components, solder defects and internal board problems are identified and repaired in this area.
- *Cable assembly and test*—Cables and harnesses, which connect PC boards inside the computer, are assembled through the use of both manual and semiautomated techniques and equipment. Assemblies are also tested here and repaired if necessary.
- *Final assembly and test*—Subassemblies and fabricated parts are assembled here to make completed systems, which then undergo automated testing. Some defective subassemblies are repaired in this department; others are returned to manufacturing areas to be reworked.
- *Support departments*—Departments that provide support to the production areas include other processing, production control, purchasing, incoming inspection, stores, information systems, process engineering and production engineering.

Prior to June 1982, when the productivity improvement activities started, the Computer Systems Division was a typical manufacturer operation with a large number of back orders and inflated inventory stocks. A series of programs were implemented, including statistical quality control (SQC)—which later became total quality control (TQC)—machinery cycle time reduction using SQC, simplification of the process, a policy of not starting incomplete kits and a just-in-time (JIT) production process.

When the total quality program had been in place for more than a year, most process and quality problems had been eliminated and the manufacturing environment was ready for just-in-time implementation. The production manager had read about Japanese manufacturing concepts and just-in-time systems, and attempted to interest the rest of the group in implementing such concepts.

Initially, most people were skeptical about just-in-time, and several argued against it. There were about 200 late back orders at any one time, and implementing such a system seemed impossible.

After attending a number of seminars about just-in-time and visiting other plants that had implemented it, a few managers decided to investigate the applicability of the concepts in the plant. They sketched an overall work plan for reducing inventory and implementing a "demand pull" system. Implementation was slow but steady, and the early results were promising.

PROBLEM AWARENESS

Before just-in-time was implemented, the assembly process for printed circuit boards in the Computer Systems Division consisted of a number of steps, including gathering a kit of parts, auto inserting, hand loading, wave soldering, back loading and final testing. Boards were built in lot sizes from 20 to 200 and were controlled by work orders issued from the material requirements planning (MRP) system. The biggest problems were material storage and product backlog. Work was issued even if some of the parts were missing.

The MRP system, used for material scheduling, was based on batch quantities and average lead times. The average production cycle time was three weeks, with actual time usually a few days longer. Production and materials groups were continuously expediting parts.

When the assembly process was started before all parts were in hand, the process would generally proceed as far as the wave solder operation. If the missing parts had not arrived by this step, the partially completed boards were pulled off the line and stored on shelves. When the missing parts arrived, the partially completed assemblies were brought back to the line and the assembly process continued.

Data collected on parts back order problems showed that on the average, 98% of the parts were in the stores area when work orders were pulled. Materials management felt that because of vendor problems and variations in the production schedule, this performance was acceptable. However, from the point of view of the production department, it was the greatest obstacle to improving productivity.

Since most kits required as many as 100 different parts, numerous kits had back ordered parts when they arrived on the production floor. Data showed that about 75% of all the kits pulled had one or more back ordered parts when delivered and that on average, each kit had one to three missing parts.

Production management had been working with the materials group for some time on improving parts availability, but had been unable to raise the in-stock level above 98%. Although management was aware of the back order problem, it was not considered serious enough to make improvement a high-priority objective. Most similar assembly operations in the company were experiencing the same amount of difficulty in procuring parts. One reason for this was high demand in the chip market, which was causing a number of suppliers to miss promised shipping dates or to ration scarce parts to their customers.

After observing many problems with back orders and incomplete kits, the assembly manager and his group decided to implement a new process whereby no work order with missing parts would be started. They modified the parts-pulling process as follows:

1. Stores would continue to pull kits of parts according to the MRP schedule.
2. Complete kits were to be delivered to the assembly area as usual.
3. Incomplete kits were to be placed on shelves in the hallway with notes indicating which parts were missing.
4. When the back orders were filled, the completed kits would be delivered to the assembly floor.

The experiment was started on a Monday. Immediately, material began to build up on the shelves in the hallway. The work load in the assembly area began to slow noticeably. As work-in-process began to flow out of the area, supervisors showed some nervousness as their people became more and more idle. For the first week, very little new material flowed into the department.

Soon the material in the hallway became noticeable to higher division management. A compromise was reached under which the incomplete kits would be stored in a special area that could be more tightly controlled.

The work-in-process shelves gradually emptied as more of the old back orders were filled. The department manager decided that it might be possible to eliminate some of them. Significant pockets of vacant space opened up in various parts of the department, and it began to take on a cleaner look.

As the experiment went into the fourth week, the workers were still often idle even though work had begun to flow through the process again. A quick check of the production output showed that weekly production had climbed back to the level that was maintained before the experiment began. Incredibly, however, the amount of labor required to assemble a kit of boards had been cut nearly in half by this single process change.

As the department began to adjust to the new procedure, other problems began to surface. As work-in-process queues disappeared, an extreme variability in workload became visible. Sometimes the workers were almost idle; at other times they were inundated with work. Previously, the work-in-process queues had hidden the variation.

Production control was called in to study the problem. As a result of the study, lot sizes were reduced significantly. Smaller lots of each board type would be delivered several times each week. High volume assemblies would be delivered daily.

Additional attention was given to maintaining equipment. The reliability of critical equipment, and therefore its use, increased due to increased awareness of equipment maintenance.

Data from before the experiment showed that if many boards were partially assembled up to the wave solder step, when the missing parts arrived it would take approximately two days to expedite them through the process. New data showed that a lot of boards could be expedited through the entire process, from start to finish, in less than twelve hours.

The data showed a tight link between quality and productivity. Improved quality had eliminated the need for many complex processes. Less complexity meant that less work was required to produce a given output.

TOTAL QUALITY CONTROL

Total quality control covers statistical quality control (SQC) and process enhancement activities. Key to total quality control are providing feedback and understanding both the failures and the work that is completed. At the Computer Systems Division, information is returned immediately from the customer (i.e., the next station) on the types of failures, their locations on the board and their probable causes.

In PC assembly, failure is traced to final assembly, auto insertion or hand load stations. Failure information is entered into a Lotus program, and the parts per million (ppm) defect rate is plotted over time for different processes. Workers have access to microcomputers to retrieve information or to generate graphs to help them understand what conditions exist within a process.

The goal of the total quality control program is zero defects. In December 1984, wave solder production reached the zero defect level for the first time. Although this was not repeated in the next six months, it proved that a zero defect level was achievable.

The Computer Systems Division began implementing SQC in its manufacturing area in July 1982. The purpose of SQC was to improve product and process and, in turn, increase customer satisfaction at minimum cost.

Thirteen months after implementation of SQC began, a number of significant improvements had been achieved, including a reduction in defects caused by defective solder joints, elimination of a large number of defects associated with the automatic insertion of components, and improvements in PC assembly cycle time. For example, solder joint failure rate was reduced from 5,000 ppm to less than 10 ppm a year later.

Automatic insertion-related problems were slashed from an astronomical 30,000 ppm to a respectable 5,600 ppm. Similar dramatic improvements were achieved in cycle times.

The SQC process begins with the identification of a specific quality control problem. Data are then recorded and analyzed using various charts and diagrams. Based on these data, process changes are implemented, and the results of these changes are analyzed to determine their effectiveness.

THE WAVE SOLDER PROJECT

The first project undertaken was an attempt to measure the quality of the joints being produced by the wave solder process. Many defective solder joints were being detected in later stages of manufacturing, even though all soldered boards were inspected and touched up immediately after the wave solder process.

The initial step was presenting a training class to the operators on the wave solder team. They were instructed on how to collect data and plot points on X-bar and R control charts. The training emphasized that this was to be a team problem-solving effort and that management was firmly committed to the belief that most defects were caused by problems inherent in the process itself, and were not the fault of the operators. It was also emphasized that the operators'

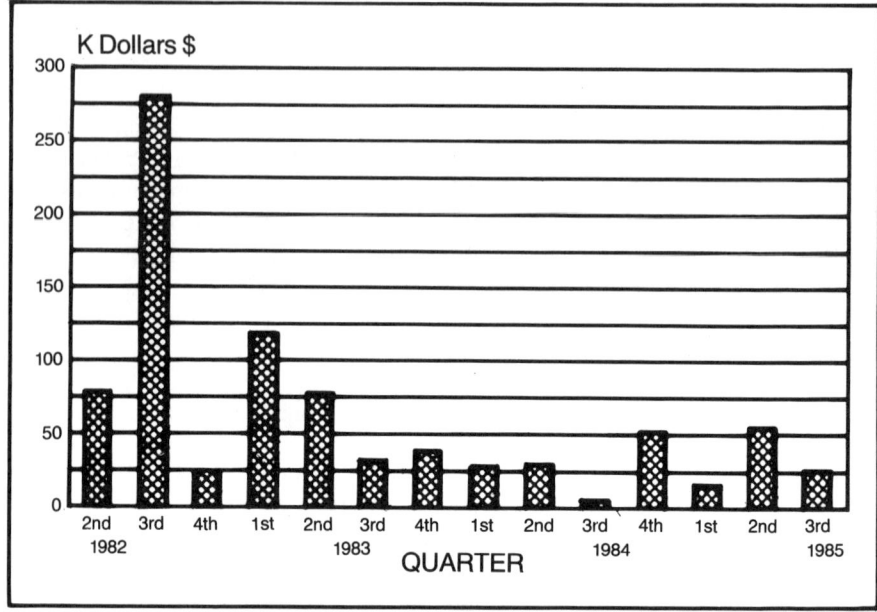

Figure 1. *CSY manufacturing scrap material.*

responsibility was to help identify problems and that corrective action would be the job of management.

After a few days it became clear that the process was severely out of control and that approximately 1.8% (18,000 ppm) of the solder joints were defective. A process engineer was called in to study current operating procedures.

A matrix was developed to show the values of the critical variables, such as conveyor speed and solder temperature, that could be adjusted for each type of assembly. New operating procedures were then written and displayed. An immediate reduction in defects was noted as the new procedures were put into effect by the operators.

Pareto analysis (the 80/20 rule) of the defects showed that a large number were blow holes and that high defect rates seemed to be associated with certain lots of raw boards. A raw board manufacturing problem was suspected, and a meeting was called with the supplier to discuss it.

The wave solder group decided to begin incoming solderability testing of a sample board from each lot; if the sample showed poor solderability, the supplier was instructed not to ship the lot until the problem could be found and corrected. This procedure eliminated blow holes as a major category of defects.

Two months of work and some simple process changes reduced the defect rate to approximately 2,000 ppm.

Further analysis of the data showed that insufficient solder in the holes was now the major cause of defects. Insufficient board temperature was the most likely cause, so an experiment was performed. After it was determined that modifying the equipment to increase preheat would be difficult, procedures were changed to run boards through the preheater twice. Although this practice was quite out of the ordinary, the fact that it was developed by the operators themselves helped ensure that it was followed faithfully.

Data on the revised process showed that other categories of defects seemed to have been reduced also. At the same time, the operators observed that the conveyor occasionally halted briefly and that boards that were on the conveyor during these halts tended to have more defects. Maintenance was called in and the problem was corrected. Control charts now indicated that the defect rate was lower than 100 ppm.

TOWARD ZERO DEFECTS

As the wave solder team became more expert at collecting and analyzing defect data, a feeling was spreading that a defect-free solder process was possible. During the next few months a number of small projects were carried out. Solder rack dimensions were checked and a number of fixtures were either repaired or discarded.

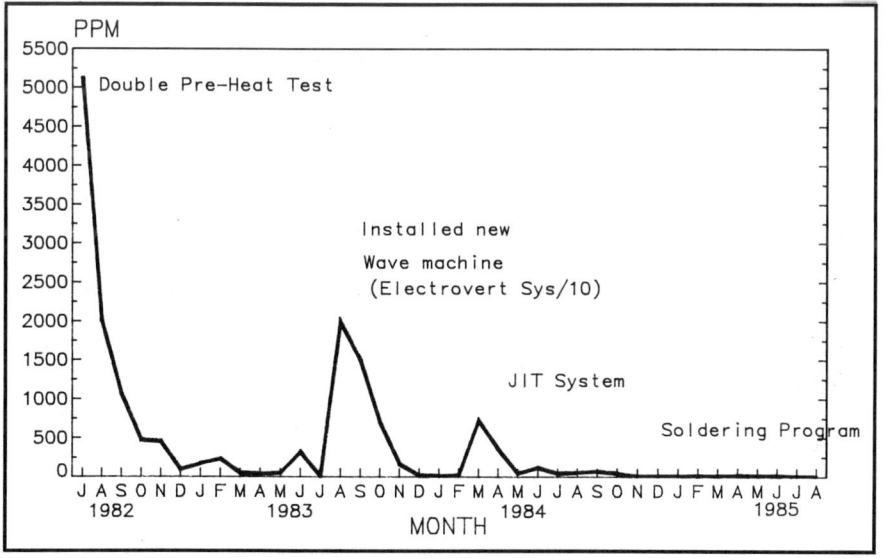

Figure 2. *In-process wave solder defects.*

It was determined that partially loaded assemblies that were sometimes placed on shelves for several weeks awaiting the arrival of back ordered parts often had higher levels of defects. A process change was made to reduce lead time by delaying board loading until all components were on hand. During the month of July 1983, the defect rate was calculated to be 6 ppm.

With reduction of the defect rate and elimination of the touch-up operation, solder-related defects found in the turn-on area declined from 0.05 defects per board to 0.01 defects per board. It should be noted that fewer defects were discovered in the test area than were counted by the solder operators. The criteria used by the latter were very tight, and many of the defects they counted were extremely minor.

AUTOMATIC INSERTION PROJECT

Another SQC project involved the process of automatic insertion (AI) of integrated circuits into boards. Kits of parts were delivered to the machine, and approximately 10 to 50 boards were loaded in a batch. After the components were loaded onto the boards, an inspection was performed.

When mistakes were found, the inspector replaced the incorrect parts. The machine often malfunctioned and caused parts to be misinserted or damaged. The machine operator spent a significant amount of time inserting parts and clearing jams from the machine.

The equipment was often out of service while the maintenance department worked to correct problems. Breakdowns lasting several days were not uncommon while replacement parts were air-shipped from the manufacturer's factory. Often large queues of work-in-process were formed, and weekend overtime was required to catch up with the workload. The process had been operating in this way for approximately two and one-half years.

In September 1982, the supervisor of the AI department began introducing SQC techniques into the process in hopes of improving the situation. A check sheet was developed so that the operators could begin collecting data. It listed 12 of the most common problems they had been experiencing. For this project only machine-related problems were studied.

The operators were given some basic instruction in SQC methods, and data collection procedures were established. It was found that defects were occurring at the rate of approximately 3% (30,000 ppm) and that there was significant variability in the process. Pareto analysis showed that two classes of defects related to integrated circuit insertion accounted for the majority of the defects.

Based on these data, two changes were made in the process. First, maintenance was asked to add magazine cleaning and rotator adjustment to its regular semimonthly preventive maintenance procedures. Second, springs were checked, and worn or damaged springs were replaced. Further data collection indicated that these two classes of defects had been reduced.

Data collected for the month of November 1982 indicated that the major problem was "phantom stoppage" of the machine. Poor design of the hole sensing unit in the machine was found to be the cause of that problem. An improved rotator had been designed by the factory and was ordered.

After several months of work, the mean defect rate was running at 1.5% (15,000 ppm). Up-time on the equipment was now better than 90%, up from less than 50% at the beginning of the project.

While watching the insertion operation, the supervisor noticed that the boards tended to vibrate rapidly as the insertion head completed each insertion cycle. With assistance from the tooling department, support tabs were added to the tooling plate to improve board stability.

During this time a concern was raised that the hole diameters on raw boards might be smaller than specified and could be causing misinsertions. A study of hole diameters, carried out by the operators, ruled out undersized holes as a defect cause. The rotator-to-performer alignment was improved, and the rotation chamber was readjusted. These actions helped lower the defect rate to approximately 0.75% (7,500 ppm) for the month of May 1983.

In June it was noticed that the defect rate was higher than normal when plastic and ceramic parts were mixed in the same lot. Adjusting the interposer and the height of the preform chute allowed the machine to insert both types of parts.

Turnaround time for batches of boards processed through the auto insertion department was now less than five hours, down from an average of 20 hours earlier in the project. The defect rate recorded for June 1983 was 0.56% (5,600 ppm).

During the first nine months of the auto insertion project approximately 80% of misinsertion defects were eliminated from the process. The changes made to reduce defects produced equally dramatic improvements in turnaround and equipment uptime. The working environment was improved for both operators and maintenance people by management's emphasis on the collection and study of data, not on the blaming of operators and technicians for misinsertions.

SQC was also used by the production staff in the assembly area to find a way of reducing kitting errors. Before SQC was implemented, employees often spent much of their time filling out requisitions for missing parts, returning excess parts to stock and exchanging wrong parts before starting work on a batch of assemblies.

To reduce kitting errors, a team was formed of supervisors from both the assembly and stores areas. This group collected data using control sheets similar to those used in the AI project and then compiled a list of possible causes for the material discrepancies. As a result of the team's activities, the material discrepancy rate dropped from 1.3 errors per kit to fewer than 0.2 errors per kit. In addition, assembly productivity improved, because less time was spent correcting material discrepancies.

SQC also improved the subcontracting process. The subcontractor had been allocated 180 sq ft near the stores area for inspecting kits before they were sent out for assembly; however, as a result of improvements in the kitting process, the subcontractor was able to eliminate the inspection procedure.

Two important conclusions can be drawn from the results of these projects. First, significant process improvements can be realized in a short period of time through the use of simple, accurate methods of collecting and analyzing data. Second, a team approach helps departments understand one another's working methods and make improvements that will benefit all.

Acknowledgements

The authors wish to acknowledge the contributions of Bob Tellez, production section manager in the computer systems division of Hewlett-Packard, and Tim Fuller, quality assurance manager in the direct marketing division of H-P.

Reprinted from February 1986 issue of *Industrial Engineering*.

15

JIT Manufacturing at Microtel Limited

Robert P. Hedden
Malcolm Richardsen

CASE STUDY

This chapter will review the production of GTD-5 EAX digital telephone central office systems in the Brockville, Ontario, Microtel facility, with special emphasis on the production of printed circuit cards and line frames. Just-in-time (JIT) manufacturing techniques were used initially for a "power" printed wiring card (PWC). This pilot project was so successful that the entire card assembly facility was restructured for JIT assembly. Some of the benefits which resulted in card assembly included the following:

1. space savings of 5,000 square feet;
2. manufacturing lead time reduction of 13 weeks to one week on the power card;
3. power card composite assembly yield improved from virtually zero to 70%;
4. first-time yields in PWC electrical test improved from 70% to 90%;
5. productivity as measured by the ratio of standard hours to actual hours improved from 0.40 in 1984 to 0.60 in 1987, or up 50%; and will reach 70% in 1988;
6. a $3 million reduction in work-in-process inventory.

To understand the project implementation some background on Microtel and an introduction to terminology is necessary.

BACKGROUND

Microtel Limited, Canada, as a wholly owned subsidiary of British Columbia Telephone, is part of the GTE Corporation. With a history that spans more than 80 years, Microtel today develops, designs, and manufactures state-of-the-art telecommunications systems that are used extensively throughout Canada, the United States, and internationally. Microtel's services to communications carriers, government, utilities, and industry range from system consulting, design, and manufacturing to turnkey installation of complete systems.

For example, as a leader in satellite communications systems, Microtel has entered into a joint venture with CANAC to participate in an extraordinary turnkey operation in fulfilling the communications requirements involved in the design, manufacture, deployment, and maintenance of the new North Warning Communications System. This system replaces the distant early warning line and involves Microtel's Spacetel and System 51 product lines.

Microtel's Brockville, Ontario, Network Product business unit is forging a leadership position in the supply of central office, digital, PABX, and network products equipment. The GTD-5 EAX central office switch is now servicing every operating telephone company west of the Ontario border, plus Quebec Telephone, which has the largest number of remote switches of any system in the world.

Other Microtel divisions are as follows:

1. Manutronics, which provides a vertically integrated custom electronic manufacturing service to high grade commerical customers.
2. Lentronics, which manufactures analog multiplex single- and multi-channel systems.
3. Microtel Learning Services, which conducts standard and customer training courses.

PRODUCT

System

The GTD-5 EAX is the industry's only totally modular family of digital telecommunications switching systems. Because all GTD-5 EAX systems were designed at the same time they feature similar hardware and software. A switch can grow easily by increments to handle 150,000 central office subscriber lines and is capable of functioning as a host for a number of remote line units and multiplexer pair-gain systems, which are intended primarily for use in sparsely populated areas.

Hardware

The GTD-5 system hardware consists primarily of a set of printed circuit cards inserted into a back panel with connectors as part of a nine file, eight-foot high sheet metal frame. The primary growth frame is what Microtel refers to as the analog line unit frame (ALUF). An ALUF is part of the peripheral equipment that performs the analog to digital conversion (or vice-versa) between the analog telephone instrument and the digital network section of the switch. This encoding and decoding function is done by what is called a CODEC circuit, at one per telephone line. Each printed circuit card is equipped with eight line circuits. The nine-file frame holds up to 96 line cards, thus providing a maximum of 768 lines per frame (8 x 96 = 768). One power card is inserted into the frame for each two files.

Because the capacity of Microtel's Brockville facility is geared to produce a finite number of GTD-5 EAX lines, the line frame, line card, and power card were prime candidates for JIT manufacture using the work-cell concept for assembly and test. Because other departments manufacture components and subassemblies used in these parts, we will continue to push the JIT style "pull" system back as far as the first level of fabrication.

JIT

Because definitions vary for this term, we will clarify the context in which we use it here. JIT in the narrow sense means to have only the necessary part, at the necessary place, at the necessary time, and no material should be in the plant unless value is being added to it. The objective is to eliminate waste of material, labor, machine time, and money. Most of the views of what constitutes a JIT improvement program can be condensed to three key elements: people, production techniques, and total quality control. There are many concepts and programs embodied in each of the three elements. Those which we used will be identified as we progress.

INTRODUCTION OF JIT MANUFACTURING

Late in 1983, Microtel was actively researching the possibility of introducing JIT manufacturing as part of a philosophy which stressed the continued pursuit of excellence. This led to a contract with a consultant (Kenneth J. McQuire, Inc.) to assist with the JIT project introduction into the Brockville factory in early 1984. The Japanese ideogram TAKUMI is applied to this continuing search for excellence. TAKUMI symbolizes an all-encompassing drive for everyone to do the very best they can in everything they do, however menial the task. From a systems perspective, TAKUMI includes implementing quality improvement programs (QIPs), manufacturing excellence teams, computer-integrated manufacturing (CIM), MRP II systems, statistical process control (SPC), Complete Operation By "Required Date" Assigned (COBRA) projects, and other programs in addition to JIT.

With the consultant's help, we put together a technical action committee consisting of representatives from eight different departments within the manufacturing operation. The purpose of this team was to develop a hands-on learning experience with JIT manufacturing. This learning experience involved producing and procurring the smallest number of units or minimum material required to meet the shortest manufacturing cycle possible at all production stages. The objectives included the following:

1. reducing inventory costs;
2. solving recurring problems to provide quality improvements;
3. shortening manufacturing lead times to the customer; and
4. involving factory personnel in a work-cell team approach to improve the day-to-day production operations.

The action committee led to a pilot project definition guide for the GTD-5 EAX power-printed circuit card. There were many reasons why this particular card was selected.

1. Each of the three levels of card assembly had the dubious distinction of being in the top 10 worst assembly yields at final electrical test. Consequently, the composite yield (yield of each of the three cards multiplied) was extremely low, and any quality improvement would show up quickly.
2. This multiple-level assembly had a manufacturing lead time of 13 weeks. Reduced lead time would result in substantial inventory-carrying cost savings.
3. The power card is used in the line frame; therefore, the requirements were predictable week by week and work load could be scheduled within ±10% each week.
4. Other areas within the plant were involved in a direct manufacturing relationship. Some of these were thick-film hybrid assembly, magnetics, and sheet metal. This meant that JIT concepts could be extended to these areas and further reduce manufacturing lead time and other work-in-progress (WIP) inventory costs.
5. No mechanization was required—that is, no substantial capital investment—so implementation would be relatively simple.

In retrospect, the power card was an excellent choice, as reductions in manufacturing lead times were soon achieved and a dramatic improvement in quality resulted.

Before introducing the work cell, assembly operations in the PWC assembly area had consisted of the following:

1. Kitting of small resistors, diodes, capacitors, and other components for manual assembly.
2. Mechanically inserting small axial lead resistors, diodes, capacitors, and integrated circuits (dual in-line packages called DIPs).
3. Assembling manually other components for which operators worked on a conveyor line that moved each card through wave solder, clean, solder touch-up, reclean, and an unscrambling and unloading station.
4. Conducting additional post-wave solder assembly, visual inspection, and in-circuit testing.

Functional electrical testing and repair were then performed in a separate department. Each manual assembly operator did their job independently of other operators, thus having no sense of ownership in the quality of the finished product. Feedback from the test department was usually too little and far too late for assembly to respond. Therefore, when the same assembly was built several weeks or months later, the same or similar assembly errors would occur. The work-cell concept, which combined the test department with the assembly department, cross-trained people in each other's job, and used a U-shaped arrangement where possible to improve corrective feedback communication, addressed these deficiencies. However, achieving all of our objectives was not simple. An early critique by the consultant pointed out that improved housekeeping was required, that the pull system was not evident, that performance measurements were not satisfactory, and that quality issues reporting was not the same as acting on on-line quality problems. He also suggested that productivity improvements be measured in units per time period, because benefits reporting would lag by three months in the financial reporting system or may not be included in the "accounting" methodology.

Over the next several months the technical action committee worked with the people on the shop floor to meet these objectives:

1. reduce cycle time;
2. reduce inventory;
3. improve equipment availability;
4. develop performance measurements;
5. increase cost-effectiveness;
6. reduce the cost of quality—that is, rework and repair;
7. improve scheduling and work-load leveling; and
8. increase employee participation.

The pilot project on the power card was so successful that by late 1984 Microtel completely restructured its in-line card assembly facility into a work-cell format. This consisted of five U-shaped assembly and test work cells manufacturing:

1. power cards;
2. line cards;
3. analog/digital cards;
4. digital memory cards; and
5. miscellaneous electrical/mechanical cards.

To minimize start-up costs, products were dedicated to the facility rather than facilities dedicated to the products. Because electrical test sets are very expensive, to dedicate facilities to products (the desirable long-range JIT goal) would have been prohibitive from a capital investment standpoint.

The changeover to the five U-shaped assembly cells provides the following benefits:

1. common nontechnical labor grades for operators by cross-training;
2. reduced supervision;
3. space reduction of 5,000 square feet; and
4. manufacturing lead time reduction of 13 weeks to one week on the power card, three to one week on the line card, and four to two weeks off other cards.

Lead time reduction for the power and line cards were quite spectacular because we were able to use a KANBAN-type "pull" system for in-house manufactured components as required directly in the area without having to go through the main inventory store. These components included magnetic devices, hybrid assemblies, and sheet metal items. In a pull system, shop orders are not used to authorize production. Fabrication is linked to the final assembly schedule, and parts are pulled from the producing work centers. Some form of visual signal, such as a card or empty tote pan, is used to indicate when a standard quantity of parts is to be produced. In the case of the line and power cards, this visual signal was the appearance of eight-foot high nine-file frame.

The line card work cell was set up to include final line and trunk frame stuffing (card insertion) and functional system testing. This has resulted in what could be considered Microtel's ultimate work cell. A small group of people are working as a team to produce a very high quality finished product of considerable value. Consolidating the frame test operation in this work cell has eliminated the so called "ping-pong" card. The line cards would previously pass the card test, but fail at the frame test, resulting in back and forth traveling between the card assembly department and the function system test department—hence, the name "ping-pong".

PEOPLE INVOLVEMENT

From the beginning, the technical action committee encouraged the personnel working in the work cell groups to be totally involved. For the work cell concept to work, it was recognized that all nontechnical operators would need to be cross-trained to provide complete interchangeability. This meant operators in assembly, solder touch-up, postwave assembly, and visual inspection would learn each other's job. As expected, some felt comfortable with this concept and some did not. Gradually, as the months went by, the work-cell personnel made it work for the group. In addition to on-the-job cross-training, operators were trained off-line to improve job knowledge. To do this, the production department built a fully equipped conference/training room with a VCR.

Courses were given on JIT concepts, soldering, PWC modifications, SPC techniques, and other subjects to involve the operators in the company's dedicaton to excellence on the job. The conference room is also used for bi-weekly meetings with the line personnel to discuss and find solutions to problems encountered in the work cells during the previous two or three days.

To incorporate some of the principles of total quality control, the third major element of JIT, the operators were encouraged to take full responsibility for quality as well as production. This meant that they would be responsible for finding and correcting errors, stopping production when quality problems arise, clearing component shortages, finding permanent solutions to reoccurring problems, doing their own material handling, and being responsible for wave soldering and cleaning. Technicians testing the cards would involve the nontechnical operators immediately when an assembly error was found. This instant feedback was primarily responsible for the very substantial improvements of first-time yields through the electrical test sets.

QUALITY IMPROVEMENTS

Quality improvements in the power card cell were spectacular because of the multiple level nature of the assembly. Composite yields improved from virtually zero to 70% for the three-level assembly, with individual printed wiring card first-time yields close to 90%. Overall, first-time yields through PWC electrical test sets improved from 70% to 90%. This reflects the work cell operator participation in almost eliminating assembly errors.

PRODUCTIVITY IMPROVEMENTS

Productivity improvements as measured by the ratio standard hours per actual hours, where actual includes both indirect and direct hours, have improved as follows:

	1984	1985	1986	1987	1988
Productivity Factors	0.40	0.43	0.50	0.60	0.70 (target)

The trend is favorable, and additional improvement will require full cooperation and total involvement of the work cell personnel.

INVENTORY REDUCTION

We do not have a system of accurately measuring inventory reduction plantwide, which could specifically attribute benefits as resulting from reduced manufacturing lead time. Overall inventory reduction figures tempered by a knowledge of product lead times and dollar value indicate that WIP inventory is down 15% to 25%. Main stores inventory reductions from lead time reductions are more difficult to estimate, but a 5% to 10% reduction seems reasonable at this time.

ADOPTING JIT IN OTHER AREAS

Acknowledging JIT implementation in card assembly, other departments studied the possibility of adopting cellular-type manufacturing. This resulted in new factory layout for metal fabrication, printed circuit board fabrication, back panel numerical control wiring and frame assembly, and functional system test. Because of the nature of our business, *the dedicated facilities to product concept* has not been achieved for individual work cells. What has been done on a factorywide basis has been to provide the framework for a verbal KANBAN system, which is used to pull in-house manufactured components, purchased components, and subassemblies to the final assembly and test of switch frames. This is done at biweekly meetings where the functional system test supervisor indicates what the frame requirements are for final test. Stores, frame assembly, and printed wiring personnel set their priorities so that jobs issued meet the shipping schedule.

RESULTS SPEAK FOR THEMSELVES

JIT manufacturing is still relatively new at Microtel, but its benefits for day-to-day operation are numerous and are summarized as follows.

1. Power cards manufacturing time is down from 13 weeks to five days.
2. Quality levels are at 95%.

3. Production efficiency levels are at 85% and rising.
4. Decision making and problem solving is taking place at the shop floor level with minimum involvement from management.
5. In certain areas of the GTD-5 line, the manufacturing interval has been cut in half from 18 weeks to nine weeks. Early 1988 targets are five weeks.
6. WIP inventory has been reduced $3 million.
7. WIP inventory is 100% controlled—parts are not ordered until needed.
8. On-line mass card assembly errors have been eliminated because part faults are spotted immediately.
9. Return and repair cost have been reduced significantly.
10. Communication has improved dramatically.
11. JIT cell members meet regularly, solve most problems among themselves, and assess their own improvements.
12. JIT cell members are now able to do a variety of jobs, whereas previously they were assigned one task.
13. Reorganization of the Brockville floor layout opened up 10,000-square feet of space.

FUTURE IMPROVEMENTS

In keeping with the continued pursuit of excellence (TAKUMI), we set up a manufacturing excellence team to focus on improving the overall productivity of the manufacturing operation. One of the team's mandates is to move JIT back to the supplier's facilities. A pilot project will start with one large-use, in-house manufactured component, the loop sensor magnetic device. These are assembled eight per line card. This part is assembled along an in-line work cell in equal quantities each week, which makes it an ideal candidate.

Initially, 12 different part numbers will be purchased from six outside vendors. Job lots will be manufactured and delivered weekly with no provision for inventory. Parts received will go directly from incoming inspection to the loop sensor assembly line. To measure savings generated, some means of assessing the reduced cost of quality and true inventory savings needs to be developed. These savings will offset increases in the "accounting" standard cost caused by small-lot manufacture and weekly shipments.

Other goals include a gradual move away from the large facility work cell to a small flexible work cell set-up to make a specific family of products and incentives for work cell operators to participate in productivity improvements. These improvements include the following:

1. improving job methods;
2. implementing more SPC;
3. introducing more quality improvement programs

4. expanding CIM;
5. improving KANBAN-type pull system;
6. reducing material handling costs;
7. implementing a suggestion program;
8. instituting work load leveling; and
9. providing additional on-going training for department and factorywide interchanging of people.

Our goals also include continued reductions in all inventory and expansion of the KANBAN pull system to the whole factory. The introduction of MRP II systems will assist in meeting this long-term goal.

Suggested Readings

Hall, R. W. 1983. *Zero Inventories* Dow Jones-Irwin. Homewood, Illinois.
Monden, Y. 1982. *Toyota Production System*. Institute of Industrial Engineers. Norcross, GA.
Schonberger, R. J. 1982. *Japanese Manufacturing Techniques: Nine Hidden Lessons in Simplicity*. The Free Press. New York.
Schonberger, R. J. 1986. *World Class Manufacturing: The Lessons of Simplicity Applied*. The Free Press. New York.

Robert P. Hedden is a CIM Marketing Specialist at IBM Corporation in Endicott, New York.

J. Malcolm Richardsen is a senior project coordinator in manufacturing engineering (advanced technology) for Microtel Ltd. in Brockville, Ontario, Canada.

16

Philips Gets SMT Just-In-Time

Stephen McClelland

High volume telecommunications equipment is expected to be one of the biggest users of surface mount technology over the next decade in Europe, so it was no surprise to find Philips Radio Communication Systems Ltd (PRCS), Cambridge, England headquarters is one of the largest SMT facilities. The size advantages offered by SMT are probably most clearly demonstrated by one of the products Philips makes here: pocket radio pagers. Together with modern custom analogue integrated circuits, SMT has enabled a pager to take up little more than the volume of a cigarette lighter. Just a few years ago, a fully transistorised version would have been the size of a hip flask and needed a substantially larger pocket. Nevertheless, the production of pocket pagers is not a simple application of such technology since the products themselves have special features. To begin with, they are necessarily customised according to the segment of the frequency band they work in, and consequently must be calibrated and tested as they are assembled. Philips has also adopted a full just-in-time philosophy for component and board make-up (the first fully integrated one this writer has seen), and according to them some 28 different PCB-sized products can be accommodated in the system from week to week.

These represent a variety of product types from mobile radios like the M294E (about 200 per week) to the pagers, the small PG32 and the larger numeric and alphanumeric pagers, the PG32N and PG32A, respectively. Some 100,000 pagers were produced in 1986, the first year of issue. The production line centres around the placement machines: Philips now has four machines: two Dynapert MPS 500 (manual type) and two MPS 500 (automatic operation).

These are sufficient to maintain moderate production volumes, but more important, to offer the flexibility that a constantly changing product mix requires. Very high speed machines, like those that parent company Philips itself makes, are uneconomic for any but the highest production rates and require considerable set-up time to function at all.

Even so, surface mount device consumption in the Cambridge line is relatively large—a total of 10 million chips will be placed this year, rising to a predicted 62 million in 1990 as new products come on stream, and older ones are upgraded using SMT. Whilst the four placement machines handle most of the placement, an IBM 7545 robot is retained to locate very high pin count (64- and 80-lead) packages.

The company maintains solder paste application facilities by means of a DEK 245 printer; this is able to achieve a registration accuracy of ±0.004in. Reflow is by vapour phase technology, due to its relative maturity three years ago when the SMT production line was planned, but the company anticipates switching over to the much improved infra-red reflow ovens now available. Although the reflow alternatives are comparable in performance, Philips engineers take the view that IR offers no consumables cost (a very high overhead in the case of vapour phase) and is probably more amenable to in-line processing which will eventually be adopted by the company.

The line has a small rework area: two operators use a hot nitrogen jet for desoldering, and the area is fully anti-static. In fact, an overriding concern with quality control has enabled a very low defect rate to be achieved; a total component-related defect count probably comes to about 100 ppm, but if solder-related defects only are considered, the figure drops to about 35 ppm. No special techniques are used to achieve this: an ultrasonic cleaning bath is available but is only used on rare occasions when certain boards are used with gold plated connectors. All board material is standard FR-4 epoxy glass fibre except for one product about to be manufactured which has—for reasons of circuit operation at high frequency—a Teflon glass substrate. In fact, the products are part of an integrated production system with each individual pager being bar coded; during calibration, each pager must be calibrated so that the calling frequency bands and pager addresses are unique. The pagers operate between frequency limits of 138 and 174 MHz.

The surface mount technology line works on a just-in-time basis for component supply and batch of one production is not only planned, it is actually utilised, because of the particular nature of the radio systems which frequently involve specific customised variations.

PRCS has implemented this throughout its manufacturing facilities at Cambridge and the approach covers not only surface mount assemblies but also that of conventional lead through boards and even the full mechanical assembly required to complete the finished product. So extensive was the company's plan that a team of management consultants, Handley Walker, was called in to draw up a detailed design for the new production methodology. The need for flexible product assembly had been clear for some time in the case of, for example, the PFX portable radio, of which there are already 18,000 variants in the field. A full analysis of the JIT possibilities for this was done and the results reported. Although the product is a six-board assembly, for example, it became clear that at least 75% of the assembly was common across the full range.

The first step was to implement PCB population up to the 75% commonality level and to use JIT methods for both the PCB and main assembly activities. In taking the batch size down from 50 to one, leadtime was reduced by 96% and work-in-progress time by 80%, and because scattered inventories were eliminated, only 70% of floorspace was required by the new lines. The consultants also report that the job satisfaction of the line workers had also increased significantly with the introduction of JIT methods.

This first stage was the start of a strategy to be dubbed the Flexible Assembly Systems (FAS) which also was to incorporate the latest in production line technology at a cost of about £2.5 million, although the consultants point out that substantial improvements in manufacturing cost can be achieved for less. In the case of these portable radios, three assembly lines were eventually implemented: the first line assembles and tests all of the PCBs of the PFX or PF85 product, and the other two lines are responsible for the final assembly and test operations. The latter lines can potentially build all the radio variants. As in the pager case, bar coding is used extensively; the bare boards are tested and given a bar code identifier. The code gives the precise customisation information for each order. At each stage of the board's progress down the line, the code is read and a display informs the operator at that station if a particular job needs to be done on that particular board.

Both conventional (lead-through) components and SMDs are used for the radios. A two-bin system is in general operation for the low volume (unique) components and high volume common components, with each bin also bar-coded. When the first bin is emptied the operator enters the bar code into the stores computer, which then activates the stores to deliver a replacement bin by means of an automatic guided vehicle system. This computer control system has initially already determined that the components are fully available to compete a product through a 'CAN BUILD' routine. It is also responsible for monitoring the quality of the production process and giving real time information about pass rate, so that corrective action can be taken immediately and locally on the line. The computer system is built up from a LAN, based on IBM equipment, and the software consists of standard IBM protocol access to e.g. material planning and customer ordering packages.

Suggested Readings

Wilson, I. B. Achieving flexibility in assembly through technology and people. Handley Walker Management Consultants. UK.

First published in *Assembly Automation*, Vol. 7, No. 1, pp. 32-34, 1987. Reprinted Courtesy of IFS (Publications) Ltd., Bedford, UK.

17

Low Volume/High Mix JIT

William A. Sandras, Jr.

One of the most common misconceptions regarding JIT is that it applies only to simple products produced in high volumes on progressive assembly lines. It is true that JIT is applicable to these products. It is also true that progressive assembly lines provide certain production advantages over other assembly layouts, if your product meets the conditions required for this approach to manufacturing (see Figures 1a, 1b, and 1c).

FUNDAMENTAL PRODUCTION LINE CHARACTERISTICS		
CONSIDERATIONS	CONNECTED (Repetitive Assembly Line)	DISCONNECTED (Job Shop)
LINE BALANCE	Equal Time	Different times
ROUTINGS	Fixed	Variable
WORKSTATIONS	Dedicated, connected	Generic, disconnected
PROCESS	Product dependent	Product independent
OPERATIONS	Linked	Decoupled
EXPANSION	Replicate, re-engineer	Add workstation
FLOW	Simple, rapid	Complex, unclear

Figure 1a. Fundamental production line characteristics.

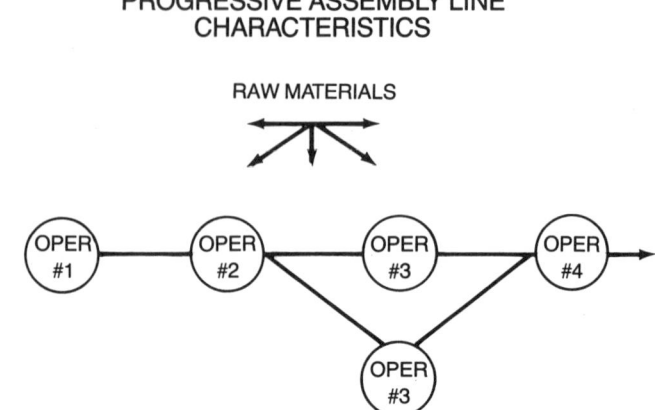

PROGRESSIVE ASSEMBLY LINE
CHARACTERISTICS

RAW MATERIALS

OPER #1 — OPER #2 — OPER #3 — OPER #4

OPER #3

- OPERATION LINES MEASURED IN SECONDS
- FLOW IS SIMPLE TO UNDERSTAND AND OBSERVE
- OPERATIONS CONNECTED; WHEN ONE PERSON STOPS, ALL STOP
- TRAINING OF NEW PEOPLE MINIMIZED
- # OF PEOPLE => # OF OPERATIONS, DEDICATED WORKSTATION
- LOT SIZE = 1, WORK-IN-PROCESS MINIMAL
- PROCESS VULNERABLE TO MATERIAL AND LABOR SHORTAGES
- LINE BALANCING IS CRITICAL
- MORALE AND JOB ENRICHMENT REQUIRE CAREFUL ATTENTION
- CAPACITY INCREASED BY DUPLICATING LINE OR RE-ENGINEERING
  LINE BALANCE FOR MORE PEOPLE
- PROCESS ACCOUNTING USED FOR MATERIAL AND LABOR TRACKING
- PRODUCTION PROCESS DEDICATED TO ONE PRODUCT
  (OR VERY SIMILAR PRODUCTS)
- FIXED ROUTINGS
- RAPID FEEDBACK ON PROBLEMS, RE-WORK MINIMIZED
- SET-UP PROBLEMS MINIMIZED WITH DEDICATED STATIONS
- RATE SCHEDULING

Figure 1b. *Progressive assembly line characteristics.*

If your product does not lend itself to progressive assembly techniques, does that mean JIT is not possible? Certainly not! The applicability of JIT is not dependent upon the selection of a narrow choice of production layout and assembly techniques. One approach may be better suited to specific product characteristics than another, but JIT can apply to a wide variety of production

BATCH ASSEMBLY LINE

RAW MATERIALS/SUB-ASSEMBLY/FGI

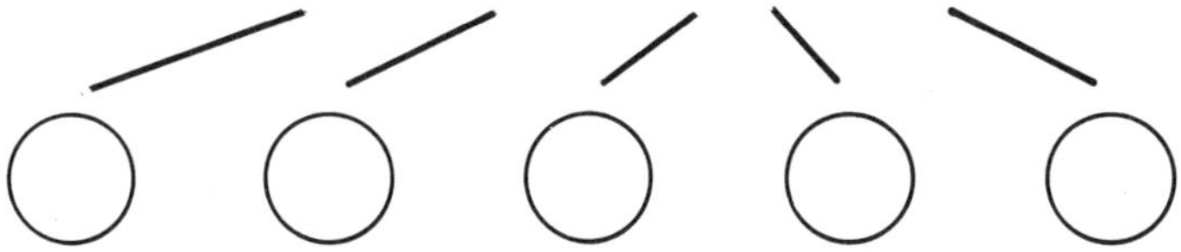

- OPERATION TIMES MEASURED IN HOURS—DAYS
- FLOW IS DISCONNECTED, NOT OBVIOUS
- OPERATIONS DISCONNECTED, WHEN ONE PERSON STOPS, OTHERS CONTINUE
- PEOPLE USUALLY CROSS-TRAINED ON SEVERAL OPERATIONS
- # OF PEOPLE <= # OF OPERATIONS, GENERIC WORKSTATIONS
- LOT SIZE > 1, SET BY EOQ, PERIOD'S NEED
- FLEXIBILITY AND INVENTORY PROTECT AGAINST MATERIAL AND LABOR SHORTAGES
- LINE BALANCE NOT A FACTOR
- MORALE AND JOB ENRICHMENT MAY BE HIGH UNLESS BATCHES TOO LONG
- CAPACITY INCREASED BY ADDING PEOPLE IN INCREMENTS OF 1
- WORKORDER BATCH ACCOUNTING USED TO TRACK MATERIAL AND LABOR
- PRODUCTION PROCESS IS FLEXIBLE, CAN BUILD VARIOUS PRODUCTS
- VARIABLE ROUTINGS
- DELAYS IN QUALITY FEEDBACK DUE TO INVENTORY LEVELS
- SET-UP AMORTIZED OVER LOT SIZE
- BATCH SCHEDULING

Figure 1c. *Batch assembly line.*

flow choices (see Figure 2). It is true however, that as flows are simplified, as processes become more standardized and as manufacturability improves, that JIT will draw production flows closer and closer to the progressive assembly line approach. Whether or not the assembly line approach is likely to ever be adopted is irrelevant regarding the practicality of beginning the JIT journey.

This chapter will address an actual implementation of JIT in a job shop, work bench environment. Also covered will be the role of Total Quality Control and Manufacturing Resource Planning. To better illustrate JIT in a low volume/high mix environment, a case study of the HP9000 Series 500 computer produced at Hewlett-Packard's Fort Collins Systems Division will be used. Today, similar

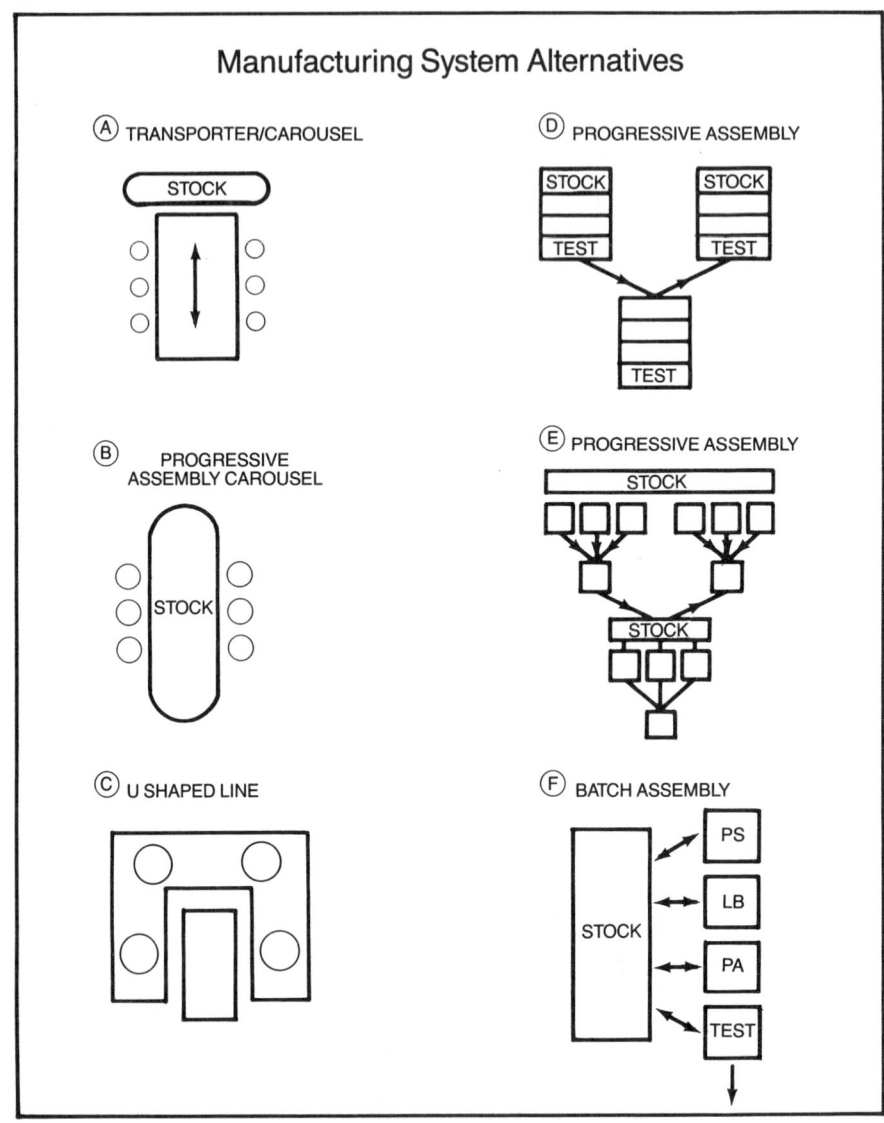

Figure 2. *Manufacturing system alternatives.*

approaches to low volume/high mix JIT can be found at Tektronix, Harris, NCR, McDonnell Douglas and in other companies.

INTRODUCTION TO HP/FSD

Hewlett-Packard has chartered the Fort Collins Systems Division (HP/FSD), located in Fort Collins, Colorado, USA, to design and manufacture engineering workstation computers. The division contains the full spectrum of business functions including manufacturing, design, quality, marketing and finance. HP/FSD first implemented JIT in a job shop environment on the emerging HP9000 product line. HP/FSD also played a strategic role in YHP's Deming prize winning efforts and therefore had an early opportunity to learn TQC. In short order, HP/FSD became HP's most advanced division in job shop JIT, and the most advanced TQC division in the United States. Later development expanded JIT to HP/FSD's repetitive assembly lines.

Today, virtually all HP/FSD products and subassemblies are planned using MRPII and executed using JIT principles. TQC is used by all functions in the facility. All software products, final assemblies and subassemblies are manufactured using a workorderless pull system (with the exception of circuit board loading where a pull system is used to initiate production but material is still issued from a workorder generated picklist). Lower volume, complex products are manufactured using JIT principles in a job shop, workbench, disconnected flow approach. Higher volume, less complex products are produced using JIT principles in a progressive assembly line, connected flow approach. This study will concentrate on the low volume, high mix production processes.

HISTORICAL PERSPECTIVE OF MRPII, JIT AND TQC AT HP

Hewlett-Packard implemented its first Manufacturing Resource Planning (MRPII) system at its Colorado Springs division in 1972. Activity in Just-in-Time (JIT) and Total-Quality-Control (TQC) began in 1980. The first two JIT projects began at newly formed divisions located in Vancouver, Washington (VCD) and Greeley, Colorado (GLD). These two divisions used repetitive manufacturing assembly lines for their relatively low mix, higher volume products. In 1981, a third startup division, located in Fort Collins, Colorado, also began a JIT project. This time JIT was to be applied to a complex, state of the art, highly featured low volume product line. Today, more than half of HP's 54 divisions throughout the world have active JIT and TQC projects underway and the rest have begun serious studies prompted at least partially by the competitive results achieved by their sister divisions. The emphasis for TQC within HP began with a tops down emphasis. The spread of JIT within HP began as a grass roots effort but now has earned top management recognition and support. Today, new divisions install MRPII systems as routinely as they do telephone systems. MRPII is viewed as a powerful and necessary tool. JIT and TQC are still relatively new concepts but will most likely be viewed in the routine same manner as MRPII in a few years.

THE JIT/TQC CONCEPT AT HP/FSD

Manufacturers process material, inspect it, move it and put it in a queue or into stock. We are in business to add value, not to add cost with no value added. According to JIT philosophy, all inspection is 100% waste, all movement is 100% waste, and all queues and stock are also 100% waste; and the portion of the process that generates scrap and rework is also waste. But there is a difference between inventory (or inspection, or movement) being wasteful and being necessary—given today's processes. But the process is not a given! Processes can and should change to eliminate the need for the remaining waste.

A fundamental JIT technique called kanban is one of HP/FSD's primary methods used to drive the process changes. The process of reducing the number of kanbans in the system:

• exposes the next problem (waste);
• prioritizes process change efforts;
• works between every customer/vendor relationship;
• provides micro controls;
• provides feedback to individuals; and
• is consistent with bottom line improvement.

As problems are exposed by reducing the number of kanbans in the system, we have a desire to re-add inventory to the system to minimize the inconvenience caused by the problem. With the adoption of JIT the option of adding inventory to insulate us from the problem is voluntarily relinquished. With JIT, we must learn to live with the problem or eliminate it. Because JIT does not allow us to forget about the problem, eliminating the waste becomes a high priority. Eliminating the problem is easier than ignoring the consequences. Necessity is the mother of invention, and reducing the number of kanbans creates the necessity to invent process changes to eliminate the waste. JIT forces problem solving.

Does JIT = zero inventory? Does JIT = one at a time? Perhaps, but the concept of JIT may be best described by understanding that JIT = one less at a time! Kanban is used to drain inventory and expose the problems in priority sequence. TQC is used to trace to the root cause of the problem and permanently eliminate the waste. Then kanbans are reduced again. By reducing the number of kanbans in the system, we will drive change resulting in:

• lower setup times;
• smaller lot sizes;

- increased delivery frequency;
- less material handling;
- lower transportation costs and distances;
- higher quality of product and processes;
- reduced paperwork;
- improved preventative maintenance;
- increased flexibility of processes and personnel;
- fewer departments and specialists;
- improved communications and teamwork;
- improved ability to forecast;
- smoother schedules;
- improved manufacturing ability to respond to change;
- fewer suppliers; and
- fewer part numbers.

At HP/FSD, JIT and TQC are two sides of the same coin. JIT exposes and motivates us to eliminate the problems. Rigorous multi-level problem solving in the form of TQC is used to eliminate the problem.

Phase I

The HP9000 Series 500 desktop mainframe is a very powerful 32x32 bit computer. Each computer is about the size of a large typewriter with a display screen on top and has 6 million logical configurations (a choice of 6 keyboards x 3 displays x 8 base power options x . . . x but only 1 color combination fortunately!) HP9000/500 sales are relatively low compared to traditional "repetitive" JIT products.

The initial manufacturing objectives were to maximize product and process quality, maximize asset utilization, maximize manufacturing flexibility and to minimize product cost. It was decided that the best strategy was to apply MRPII and JIT to the HP9000 product line—despite the fact that it was a complex low volume product. In the summer of 1981, Phase I began with preparations to manufacture the first HP9000—and the first unit was actually produced using JIT principles in October of that same year.

After studying the fundamental characteristics of repetitive assembly lines (see Figure 1a), it was decided that connected flow assembly line production would not be appropriate for the HP9000. Batch production was also unacceptable. Therefore, Phase I of the HP9000 JIT/TQC implementation consisted of a closed-loop transporter and a carousel for stockroom material located on the production line (see Figures 3a and 3b). This seemed like the best way to build a complex product, maintain flexibility and achieve one at a time production with kanban control.

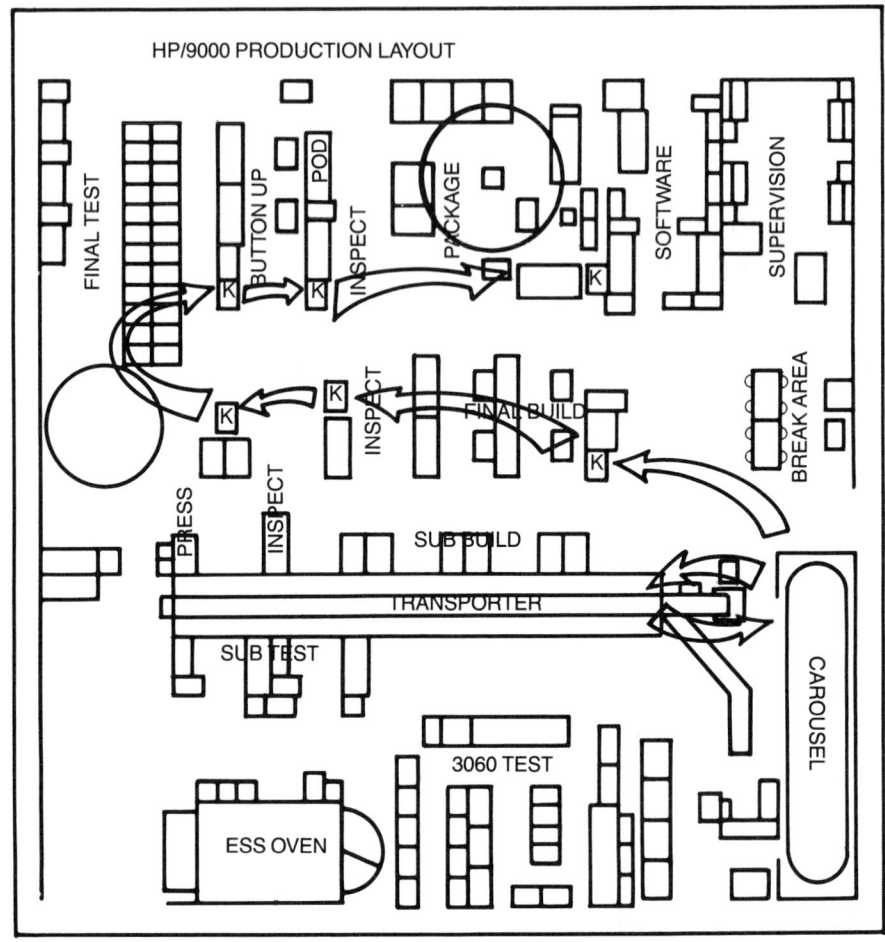

Figure 3a. *HP9000 production layout.*

The transporter/carousel approach lasted for 21 months. During that time work in process decreased 50% (from four initial kanbans per assembly down to two, on the average; build lot sizes continued to be one). MRPII and JIT were successfully linked and extra features included to facilitate operation in a job shop environment.

Phase II

Phase II began in April 1984 when it becomes necessary for the HP9000/500 manufacturing line to relocate. Moving provided the opportunity to reexamine

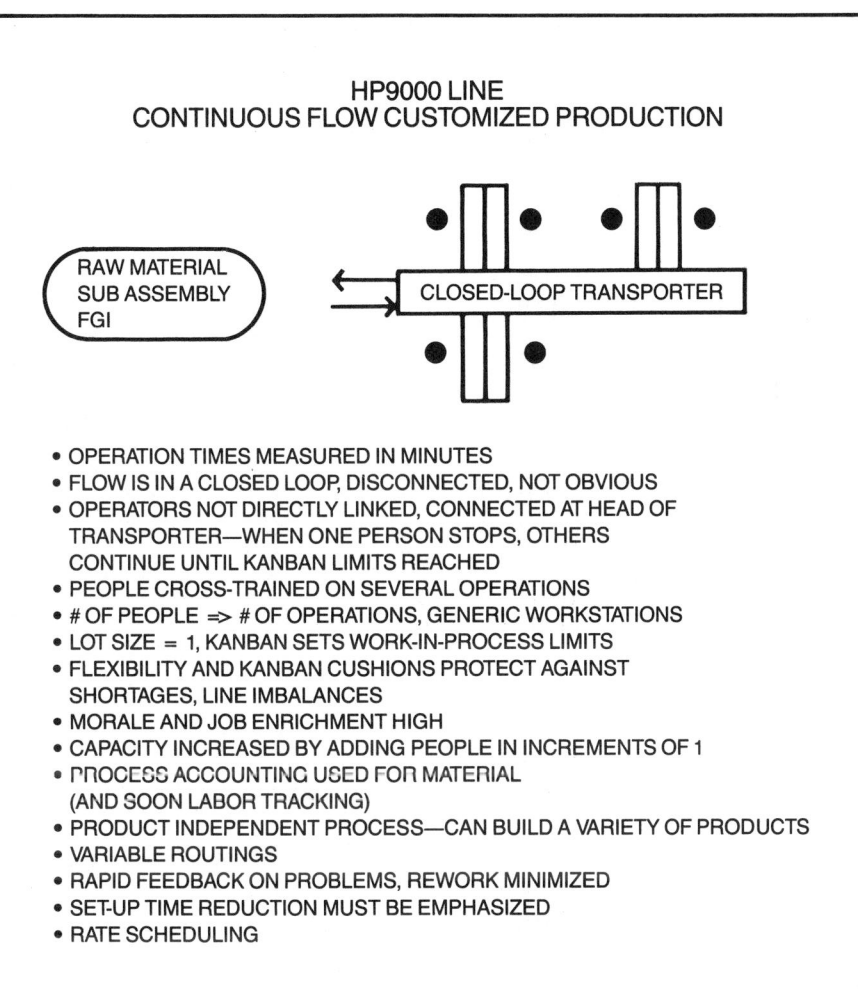

Figure 3b. *HP9000 line—Continuous flow customized production.*

the JIT knowledge gained over the last 30 months. FSD's objectives were simplified, but essentially the same. They were to maximize product quality, maximize product delivery and minimize product cost. The new Phase II layout is shown in Figures 4a, 4b, and 4c. While at first glance the flow may appear to be a connected flow repetitive approach, one should realize that the major assembly and final assembly areas contain groups of people each working at generic workbenches. Today, HP/FSD builds six totally different products in the same production area as the HP9000/500. For Phase II the transporter and

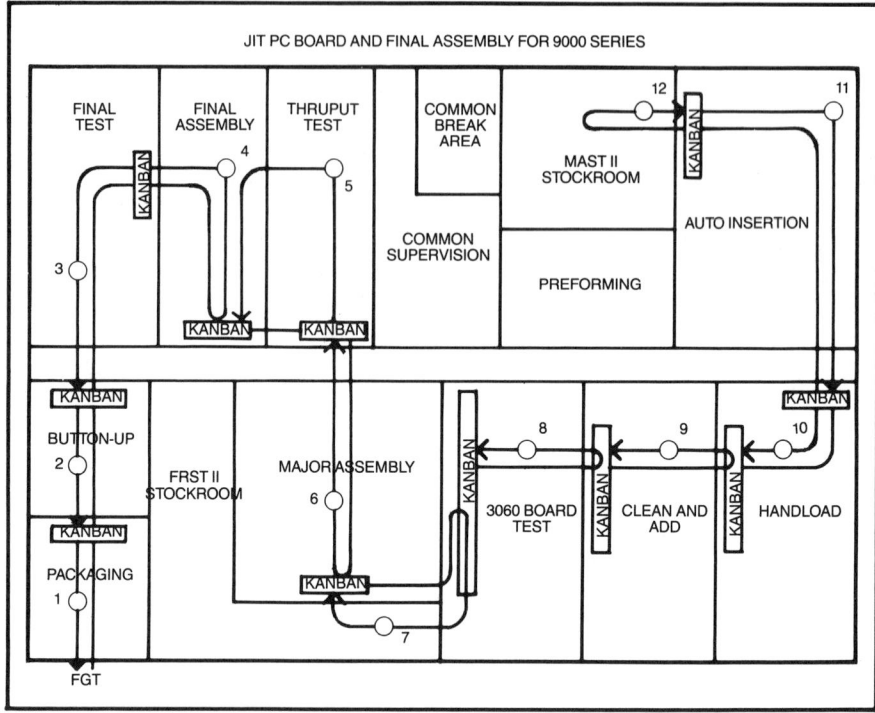

Figure 4a. *JIT PC board and final assembly for 9000 series.*

carousel were removed resulting in improved flexibility and a better return on a lower asset base due to less machinery and equipment and space. (In the Phase I layout, the transporter and carousel operated in the major assembly area.)

During Phase II, kanban quantities were further lowered, to a single kanban for most mechanical and electro-mechanical assemblies, and even to zero in some cases. A zero kanban quantity was achieved for some assemblies by using a generic kanban card in conjunction with a broadcast system similar to that used in the automotive industry. Build quantities continued to be one at a time. For circuit board assemblies, the number of kanbans varied, but in most instances production was initiated when four kanbans were accumulated. Generic kanbans have proved to be a key ingredient for practical application of JIT in environments where many products are manufactured in relatively low volume.

Also during Phase II, production output increased approximately 30%, while efforts continued to be successful in eliminating the causes of waste and poor quality. As a result, work in process decreased 75%, thruput times shortened

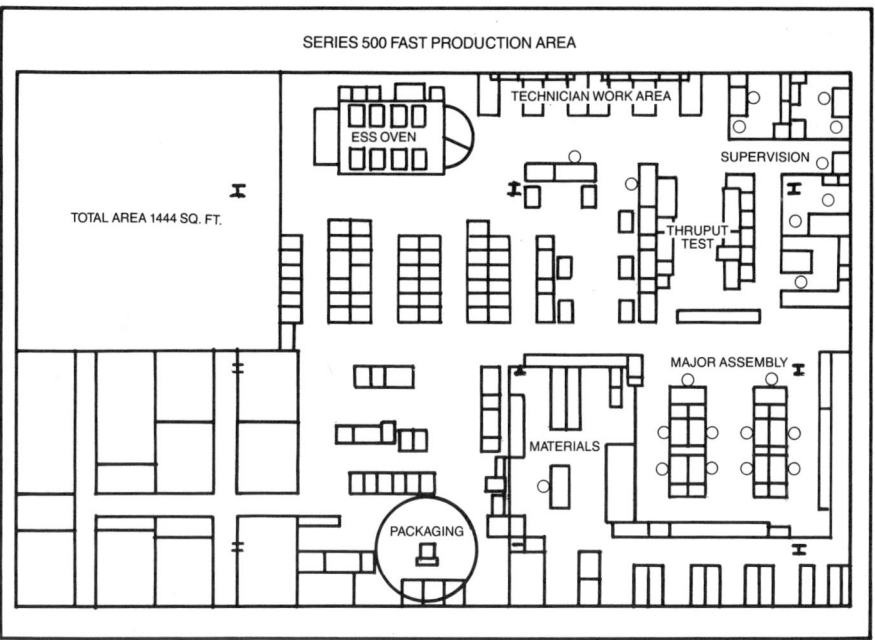

Figure 4b. *Series 500 fast production area*

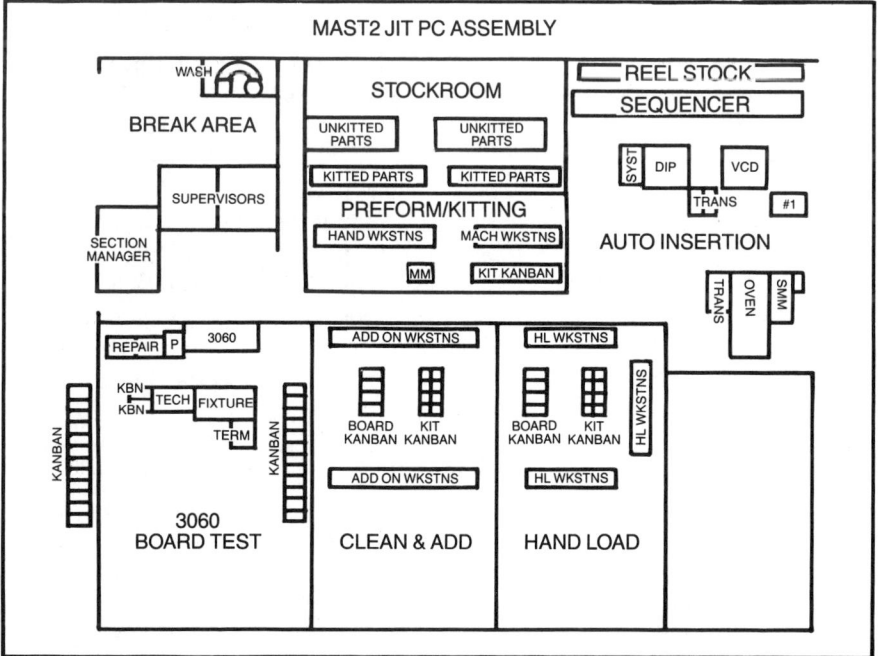

Figure 4c. *MAST2 JIT PC assembly.*

accordingly and scrap and rework fell over 50%. Phase II results provided management with the proof that JIT and TQC were powerful tools, and when properly applied to the manufactured process, these tools could positively and significantly impact key manufacturing objectives. The Phase II implementation provided the insight on how a "pull system" could work from the actual customer order all the way back to the stock room. (In Phase II all purchased parts were being planned using MRPII and most released for delivery using a purchase order, but preparations were beginning to link the JIT pull process into the supplier base.) Significant results were achieved during Phase I and the substantial benefits were continuing during Phase II of the JIT/TQC journey.

Phase III

Phase III of the JIT/TQC implementation began in the fall of 1984. The objective of Phase III was to convert all areas of the factory to JIT and TQC—not just manufacturing but support functions as well. Realizing the impact of this significant change, standardizing processes and interfaces between key areas of manufacturing was a prime consideration. The design phase took approximately five months to complete with the majority of the effort spent on the material flow process and on a new layout. Implementation of the new designs took three months while production activities continued (during the period two new major products were also introduced).

For Phase III, the production of a large number of printed circuit boards (PCB's) would need to be converted to a "pull system". In Phase II, 40 PCB's associated with the HP9000/500 production area were converted to JIT; however in Phase III, 240 PCB's affecting all products would be converted to JIT. The entire PCB production area would be relayed out with improved communication and material flows.

The new layout for the combined PCB area is shown in Figure 5 (the HP9000/500 final assembly area remained the same except for the fact that new products were added into the process; other JIT final assembly production areas were established during Phase II for higher volume lower mix products but discussion of those areas is beyond the scope of this paper). The Phase III PCB layout emerged with a "U" shape design that allows the input and output points of the PCB process to be within ten (10) feet of each other. In addition, all of the work cells in the process used a basic "U" shape work flow to enhance material and information flow. The design of the material flow process is considerably different from that chosen in Phase II. In Phase II, each process operation was situated between input and output kanban racks of partially completed PCB's (see Figure 4c); each PCB resided in a dedicated pigeon hole in the rack. When a downstream operation "pulled" a PCB out of a dedicated pigeon hole in its input kanban rack, the empty hole served as the authorization or trigger for the supplier workstation to replace the consumed part. The supplier

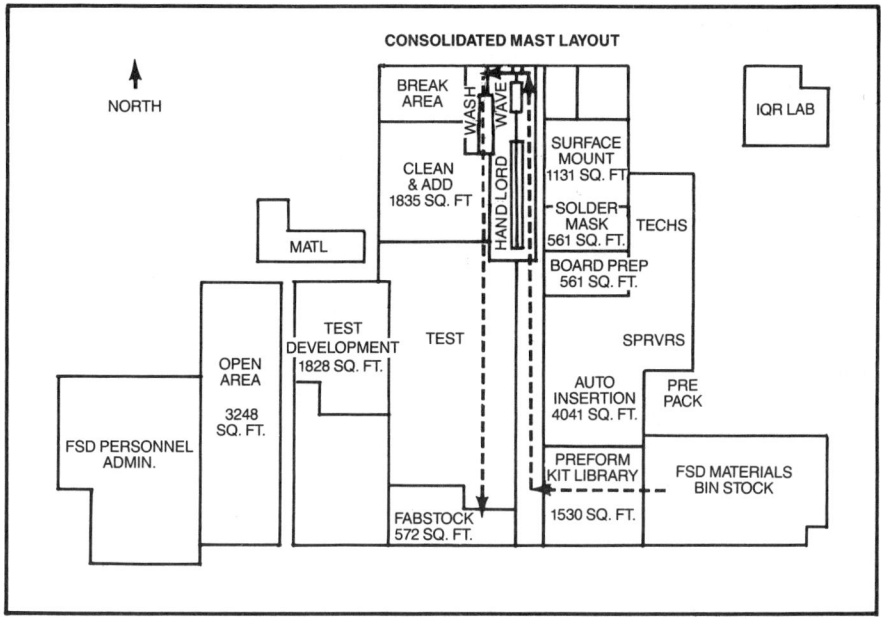

Figure 5. *Consolidated mast layout.*

operation would then withdraw a partially completed assembly of the same type from a specific pigeon hole in their own input kanban rack, complete the appropriate operation and then place their finished PCB in the empty pigeon hole of their output kanban rack (the input rack for the downstream operation). This process would continue all the way back to the material stock areas.

With only 40 PGB's in the process for Phase II, this method of material flow was very effective. It also served as a clear example of how the "pull" process functioned. However, in Phase III a process simulation indicated that these kanban racks would take up considerable space and hold significant amounts of inventory since 240 PCB's were now involved. Improvements for Phase III resulted in just one input (materials) and one output kanban rack (Fabstock; PCB FGI) with dedicated (labeled) pigeon holes. All intermediate kanban locations were generic (non-labeled). The entire process would be treated as if it was one large workcell. The Phase II method is referred to as the "short pull" (using dedicated kanbans) and the Phase III approach as the "long pull" (using generic kanbans for intermediate operations).

The actual "long pull system" works as follows:

1. Final Assembly issues a kanban authorization to move or deliver more PCB's from "Fabstock" to a specific final assembly area. PCB's are delivered on two hour intervals.

2. When a lot size of a given type of PCB's is consumed by final assembly, a kanban card is issued to materials to start a lot of PCB's into the beginning of the PCB process.
3. Fabstock levels for each PCB are set based on demand rates and the assembly process cycle times.

In order for this "long pull" process to be successful, it is essential for the assembly process cycle time (thruput time) to be minimized otherwise, shortages to final assembly might be created due to increased exposure to problems and changing needs, eventually resulting in more inventory in the process.

Phase III resulted in inventory being reduced by a factor of 3.5 times, cycle time went from 16 days to 2 days (two shifts), workmanship quality improved by a factor of 2.5 times, responsiveness to our customer needs improved, and morale which is difficult to quantify improved significantly.

What is Phase IV?

It is still too soon to tell, but it will likely apply JIT principles in the office areas (TQC is already being applied in all functional areas). And of course, increased emphasis in linking more and more external suppliers into the JIT network.

CONCLUSION

Can JIT be applied to high volume products? To low volume/high mix products? Can JIT be used in assembly line environments? In disconnected flow, job shop work bench environments? The answers to all of these questions is yes. Is material handled the same in each instance? Are the product flows the same? No, but the JIT/TQC principles are identical. JIT is not limited to manufacturers that produce nearly identical products using fixed path assembly lines. What is important is that production processes repeat, regardless of the product they are making or the patch of the material flow. JIT does apply to low volume/high mix operations, and continually improving manufacturing competitiveness is the benefit of successful application of JIT to these complex environments.

Reprinted with permission, America Production and Inventory Control Society, Inc., "Low Volume/High MIX JIT" by William Sandras, *Twenty-ninth Annual International Conference Proceedings*, October 1986, pp. 295-299.

18

JIT is Made to Order

Nhora Cortes-Comerer

At Hewlett-Packard Company's plant in Cupertino, California, home of the HP 3000 Series 68 minicomputer, the time it takes to assemble a set of 31 printed-circuit boards has been slashed from 15 days in 1982 to 11.3 hours in 1986. During the same period, the inventory of circuit board work in process was cut from $670,000 to $20,000, and the number of back orders was reduced from an average of 200 to two.

The secret to this success story? "Just-in-time manufacturing," or JIT, a technique that in simple terms calls for the reduction of inventory by having materials ready at each point in the manufacturing process, just in time to be used. Viewed back-to-front—in the fashion in which JIT is typically implemented—this translates into production and delivery of finished goods just in time to be sold, subassemblies just in time to be assembled into finished goods, fabricated parts just in time to go into subassemblies, and purchased materials just in time to be transformed into fabricated parts, according to Richard J. Schonberger, president of Schonberger & Associates, a management consulting firm in Seattle, Washington, that specializes in manufacturing.

Implementing JIT, however, is far from simple. JIT challenges traditional manufacturing practices in the United States, with particular impact on four areas:

- *Inventory management.* In Western industry the attitude prevails that a certain percentage of inventory should be kept between assembly stations to serve as a buffer, or security against unforeseen events, and to keep per-part costs low. JIT views this accumulation as hiding problems. The flow of materials from one workstation to another is also handled differently. Instead of "pushing through" parts into final assembly, JIT "pulls" them into the system.

- *Configuration of the work center.* JIT calls for a shift from manufacturing in large lots to the small-batch, or small-lot, approach. This often means reducing the amount of time it takes to set up equipment and reconfiguring the production plant from one that focuses on job-shop fabrication-with equipment arranged by function-to one in which each production line produces a different part or group of parts in sequence. In these aspects JIT integrates with current production trends, such as flexible manufacturing.

- *Customer-supplier relations.* Companies that adopt JIT gradually thin out the number of suppliers they use, eventually arriving at a sole source for each part or product. In another deviation from standard practice, JIT eliminates the incoming inspection of goods from suppliers.

- *Management-labor relations.* As opposed to more conventional methods, JIT advocates full shutdown of the production line as often as necessary to expose quality problems, and places this responsibility squarely on the shoulders of the line operators. A specialized workforce becomes multifunctional as workers are retrained to operate several machines.

Finally, JIT complements efforts to integrate and automate every aspect of manufacturing, from design to customer service. Important elements of these trends are designing for manufacturability, computer-integrated manufacturing, and quality control. In fact, JIT experts advise against implementation of JIT unless a company has made a full commitment to quality.

The overhaul is well worth the effort, according to the increasing number of companies that have adopted JIT techniques. Schonberger has compiled an honor roll listing 84 departments or divisions of companies that have achieved 5-, 10-, or 20-fold reductions in inventory, using JIT. The notables include the Digital Equipment Corporation's plant in Colorado Springs, Colorado, which assembles Winchester disk drives and which registered a drop in work-in-process inventory from $5 million to $900,000. Inventory reductions of 60 percent were reported at an IBM plant in Owego, New York, which makes computer products for the United States Government. A Motorola plant in Seguin, Texas, which produces electronic controllers for the automobile and appliance industries reported a 75 percent reduction.

Other JIT pioneers like Hewlett-Packard, Texas Instruments, Tektronix, Honeywell, General Electric, and Apple Computer claim improved product quality and higher worker productivity.

FIRST USED TO BUILD TIN LIZZIES

Although considered a Japanese phenomenon, just-in-time manufacturing had its roots in the automotive industry in the United States in the early 1900s. Henry Ford first used JIT principles in the manufacture of Tin Lizzles at the company's plants in Highland Park, Michigan. Here parts were mass-produced just in time for assembly, and the assembly lines pulled work to the next assembly stations just in time for the next step.

In Japan, Toyota is considered the pioneer in developing JIT production, starting in the 1960s. Today JIT "is a way of life in most manufacturing companies in Japan," according to Mehran Sepehri, a visiting associate professor in the Institute of Safety and Systems Management at the University of Southern California in Los Angeles. In 1981 JIT principles returned to the United States as a result of a management information exchange program between Toyota and the General Electric Company.

According to Robert E. Sessions, JIT project manager for General Electric's Corporate Production Resources Consulting Operation in Bridgeport, Connecticut, what he and his colleagues witnessed in Japan was not just a method of inventory control but a philosophy that looked at inventory as a way of hiding problems. Eliminate inventory, the reasoning went, expose the problems for correction, and you have greater productivity. Thus, Toyota was not only talking about reducing inventory by 75 percent, but also about increasingly weekly output by 30-40 percent and reducing defects in parts by 90 percent.

The first JIT projects in the United States were initiated in five General Electric businesses' upon Sessions's return in 1981. That same year, Kawasaki in Lincoln, Nebraska, and Toyota in Long Beach, California, changed their standard manufacturing operations to JIT.

REGAINING THE COMPETITIVE EDGE

JIT techniques have been credited with impressive results in the automotive industry, as well as in the aircraft and machine tools industry, according to M. Eugene Merchant, Director of Advanced Manufacturing research at Metcut Research Associates Incorporated, a consulting company in Cincinnati, Ohio. Even bigger payoffs have been realized in the low-volume production of mechanical parts than in high-volume industries, he noted. More recently, the electronics industry in the United States has been making greater use of JIT techniques, in particular since Japan's competitive advantage in that industry has been attributed in part to its use of just-in-time manufacturing.

"Our survival in world markets is at stake," Sessions said. Other JIT proponents, like Sepehri, believe that with a predominance of U.S. products being manufactured offshore, JIT may be the catalyst that will help the United States regain its competitive edge.

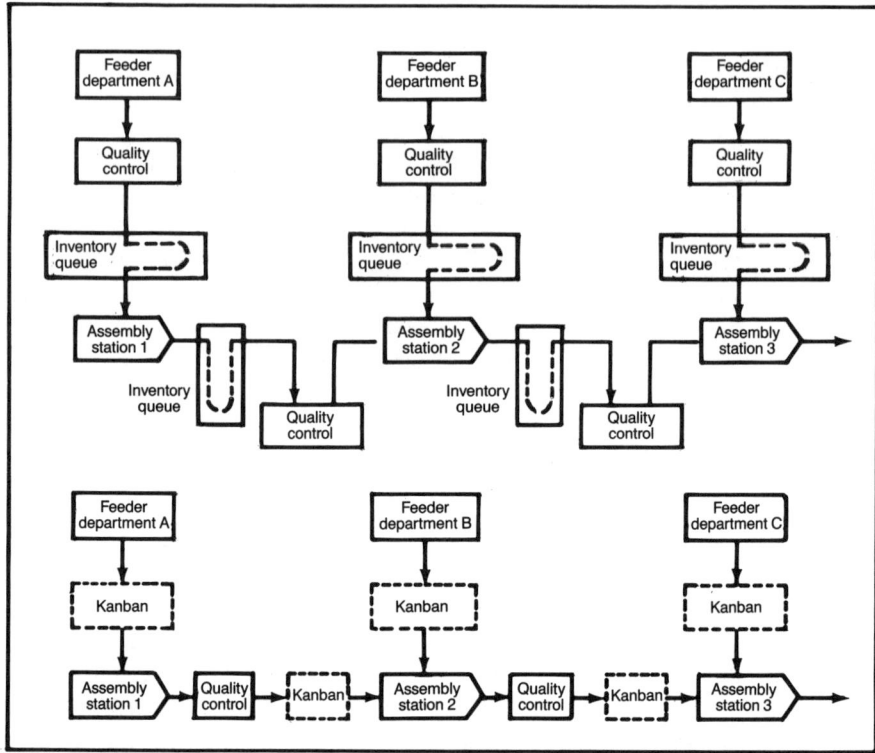

Under traditional manufacturing methods (top), parts are produced in large quantities at one time to reduce the cost of setting up equipment. Parts leaving feeder departments from the fabrication side of manufacturing pass through quality control before "queueing" up for assembly. Inventory in the queue contains a certain percentage of buffer stock as protection against product defects and other unforeseen problems. As work on a part ends at one station, the worker "pushes" the part through to the queue before the next assembly station, and so on to completion. JIT manufacturing (bottom), on the other hand, is based on the production of goods in small lots, with minimal or zero inventory. Testing and inspection are no longer relegated to a quality control department but become the responsibility of the individual line operator. Some physical signal a kanban, must exist to tell the operator to "pull" a part or a lot into production or to deliver more parts. A finished part, or a small lot of parts, remains at an assembly station until the next worker is ready to "pull" it into production.

JIT's emphasis on reducing the lead time for a product, Sepehri noted, is particularly important in the electronics industry. "Product life cycles are so short," he elaborated, "that if a part is stored longer than one year, the chances are 50 percent that it will become obsolete."

Also, since 80 to 90 percent of the cost of making electronic products is attributed to materials, a system that promises to reduce inventory is attractive.

Sepehri also indicated that the wide fluctuations in the electronics market-place make it hard to predict what products will be in demand. The flexibility of JIT allows for fast production of custom products, quick response to customer requests. "This," Sepehri emphasized, "is more valuable than inventory reduction."

CHIPPING AWAY AT INVENTORY

Inventory control, nevertheless, is the cornerstone of JIT, and reducing inventory by switching from the production of large job orders to small batches and then chipping away steadily at any buffer inventory is often a first step in implementing JIT. In manufacturing complex microelectronic circuits at the Tektronix hybrid circuits organization in Beaverton, Oregon, for example, clusters of similar equipment that typically performed one function on a batch of 4000 parts were redistributed among several separate production lines that assemble a hybrid circuit in sequence, from queues, or lots, of as few as 10 parts.

The next important step, according to Robert Erb, the organization's director of quality, was to gradually reduce the number of parts in the queues, to discover hidden problems and operate with minimal inventory. "We would start production on a line with a queue of 40 or 50 parts," he explained, in a simplified example. When a problem with a part or within the process occurred, the line would be stopped instead of discarding the part or ignoring the problem. "When you've managed that bottleneck," he said, "take more and more parts out of the queue, and then go on to the next bottleneck." The process gets more difficult and the line stops more often as the queue gets smaller.

Managing bottlenecks, he elaborated, means constantly looking at everything—engineering, equipment, and the process—and asking, How can we do this faster or better? A good example would be a wire bonder operating at five parts per hour. The line rate, however, might be 10 parts per hour. "You ask yourself, what do I have to do: Reduce set-up time? Speed up the machine? Put in new equipment?" Erb said.

A year after initiating JIT manufacturing, Tektronix reported a 98 percent yield on one of its components, a complex time-delay hybrid circuit used in its 1240 logic analyzer, at annual savings of $175,400. A Tektronix report puts the industry yield for a similar product at 85 percent. The company also reported reducing the time to manufacture a typical product from 40 days to five, and cutting the work-in-process inventory by 80 percent.

THE ONE-MINUTE SETUP

JIT advocates small-lot production on the basis that it allows production of a daily mix of products that more closely matches demand. Until the introduction of flexible manufacturing systems, however, small-batch production generally was frowned upon, on the basis that it was too expensive to set up conventional, nonautomated equipment for a short production run. Today, however, the driving force behind flexible manufacturing is to make it economically justifiable—and technically possible—to produce even a single unit.

Production need no longer be limited to dedicated, high-volume production lines making just one product. The same equipment can be set up to produce one part or sets of parts in the just-in-time mode for assembly, according to research director Merchant. Flexible automation is a prime tool for companies that implement JIT, he said, and represents an untapped potential for others.

For example, the Tektronix hybrid group produces about 4,000 components daily. But each line produces only 100 to 1,000 components. Honeywell's process control division in Fort Washington, Pennsylvania, is another mid- to low-volume JIT operation that produces temperature and humidity recorders and pilot valves in quantities of 10,000 to 20,000 per year.

Setup changes need not be expensive or time-consuming, JIT practitioners have discovered. Setups that take hours in a traditional approach take minutes in General Electric's JIT operations, reported Sessions. "It is a matter of methods, tooling, and training," he continued. General Electric has reduced setup times by as much as 97 percent in its JIT operations.

At Texas Instruments' Schottky department in Sherman, Texas, setup costs were reduced by using software rather than a manual process in the wire bonding of ICs when switching from one type of IC to another. The changes were made off line, explained department manager David Van Winkle, while production was still running on the line. This department is a high-volume operation providing standardized families of integrated circuits to customers like Hewlett-Packard, Wang, Digital Equipment Corporation, and IBM.

In Japan, where a campaign to reduce setup times has been waged since the early days of JIT, the emphasis is on achieving "single-digit" setup times—that is, under 10 minutes each. This drive has gained so much momentum, observed Sepehri, that a popular book now is *Single Minute Exchange of Dyes*, by Shiego Shingo, a Japanese manufacturing expert.

FROM THE CLUSTER TO THE CELL

By breaking up machine clusters into cellular production lines, JIT makes optimal use of yet another manufacturing concept: group technology, or cellular manufacturing. In a more conventional setting, for example, a company that makes integrated circuits would group together all component mounting equipment in one part of the plant. Curing ovens would be in another, and so on. A cell configuration would call for breaking up the clusters and creating several cells, each containing one piece of each type of equipment, in the sequence used.

Restructuring into cells in a plant that makes dishwashers was one of the basic measures taken by General Electric when the company first implemented JIT. General Electric is also using this approach in the manufacture of circuit boards in a military electronics plant in Daytona Beach, Florida. An added benefit of cellular manufacturing, Sessions said, is that it gives the operator more control over the manufacturing process. In one area where General Electric has applied cellular technology, Sessions said, the company has achieved a 46 percent reduction in direct labor, a 67 percent reduction in work-in-process inventory, and zero defects.

There are cases in cellular technology, Sessions noted, where one operator supervises a group of machines, sometimes up to 50. In making a shaft, for example, he said, an operator may be walking a part through three or four different operations—turning, milling, and drilling—each one on a different machine until the part is complete. As the operator takes one part out of a machine, he puts another one in. At any-time, the operator may be making three parts or even operating three different cells. This operator is also responsible for inspecting each part following each operation.

As in some other applications of cellular technology, the parts Sessions referred to are small, therefore portable. In addition, he said, JIT eliminates conveyors, which consume space and encourage the holding of inventory. Since the JIT concept calls for keeping a part moving and not laying it down once production has started, the operator is constantly on his or her feet rather than sitting. Because of this, manufacturing cells are more efficient if they are U shaped rather than being laid out in a straight line. A U-shape saves space and makes it more convenient for the operator to move from machine to machine.

Restructuring the work center into cells also reduces lead time. "If, in theory, it takes five weeks to produce a product," Sepehri commented, "that time can be cut to one week because the work center is right on the floor and you are not moving parts around 20 times, starting with storage 20 miles away."

Manufacturing of the Apple Macintosh computer in Fremont, California, he pointed out, is a particularly good example of more fundamental restructuring. The Macintosh production line uses many JIT principles. Originally, the company's manufacturing facilities were scattered around the world, with final assembly taking place at a single plant. For the Macintosh, however, all the manufacturing was moved to one highly automated factory located close to product design and engineering. Because of the reduction in lead time, it takes only 27 seconds to assemble each unit.

WHEN PUSH COMES TO PULL

Before JIT was implemented at Texas Instruments' Schottky department, said department manager Van Winkle, "TI typically used push-through production," based on "an educated guess at what customers wanted. Now," he continued, "nothing is put on the production line unless something [sic] says we need it."

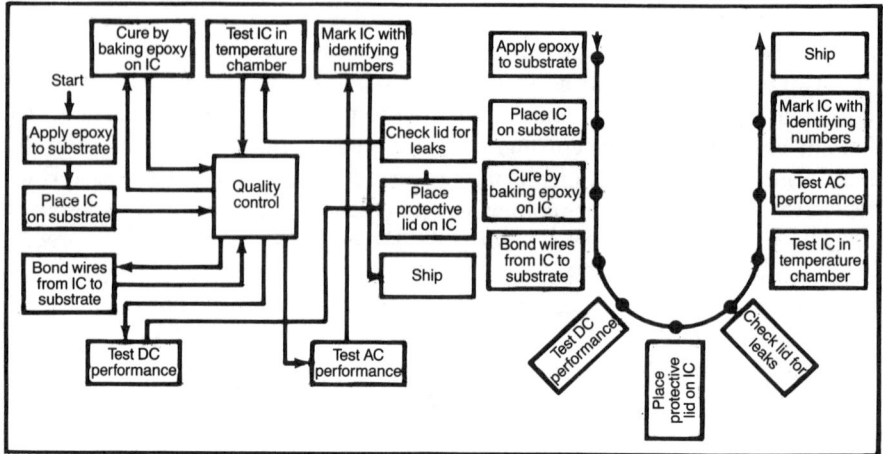

JIT makes use of the cell concept to arrange equipment progressively in U-shaped production lines that can produce one part or family of parts. At Tektronix in Beaverton, Oregon, equipment needed to manufacture a hybrid integrated circuit was formerly clustered by function, with each function scattered among various plant locations (left). All wire bonders, for example, would be grouped in one place and could handle as many as 4000 parts. A part would pass through quality control several times on its way to another process. Manufacturing in small batches meant breaking up the clusters into several U-shaped cells that typically contain one of each type of machine (right). Operators handle parts in queues of 10, and each part is handled one step at a time. Workers learn to operate all the equipment, and can hop over an operation while waiting for work to be completed at a station.

Used in conventional manufacturing, the push-through system is based on a computer-generated master schedule that details what materials, parts, or products should be bought or produced on a long-term, interim, or short-term basis. Because production estimates are based on sales forecasts and backlog orders, rather than actual orders, JIT users criticize push-through production as a way of keeping buffer inventory "just in case" it is needed.

In conventional production, a part is pushed through from one workstation to another and on to final assembly, where it becomes part of the finished goods inventory. Push-through production, said Sessions, is a way of keeping people or equipment productive, whether a request has been received for a product or not.

In "pull" production, by contrast, "nothing is produced until parts are withdrawn for final assembly," Sessions explained. A finished part or a lot of parts remains at an assembly station until the next worker is ready to "pull" it into production.

The signal that tells an operator to pull is the *kanban*—a Japanese word that means visible record or visible plate, according to Schonberger. It can be a card, which includes information about a certain part, or a physical *kanban*, like a container that holds only a certain number of parts, or space on the floor-for example, squares-that when empty will be a signal to "pull"—produce or deliver more parts.

In Japan the latest evolution of kanban relies on sensors to read bar codes and on vision systems to identify parts by computer, according to research director Merchant. Parts are delivered by automated guided vehicles or conveyers.

GREEN LIGHT ON QUALITY

By flipping a switch that turns on a red light, a worker assembling a Hewlett-Packard Winchester disk drive at the company's plant in Greeley, Colorado, can shut down a production line. The action may signal a serious problem, such as a part that does not insert properly or material that is missing. More significantly, it is a signal for labor and management to converge to analyze and solve the problem.

Similar systems exist at most plants using JIT. A yellow light signals a line slowdown and a green light means all systems are go. At General Electric's dishwasher plant in Louisville, Kentucky, said Sessions, workers on the assembly line can pull one of two handles: a green handle to release the dishwasher to the next assembly station or a red handle to shunt the washer aside and stop the assembly until the problem is corrected. One result has been a 60 percent improvement in product quality, according to Sessions.

The signal system is important, said Sessions, because it visually informs management and labor of the status of production. Beyond that, other JIT users claimed, it provides instant feedback and requires immediate correction of a problem; it means less scrap and fewer parts to rework; and it involves workers in decision making and troubleshooting.

Although stopping production is costly, it is cheaper than completing large lots with defects. Mark Oman, a production manager at Hewlett-Packard's Greeley division, illustrated this point. Before JIT, he related, a test of one queue of 800 circuit board assemblies revealed that an IC had been inserted incorrectly. Thus parts had to be removed from all 800 boards. Under JIT, a maximum of three boards are on the queue. In the event of a similar problem, there is immediate feedback and only three boards need to be reworked.

Fully stopping a production line also occurs under traditional manufacturing, but usually the stoppage is not immediate, according to Rick Hoole, a consultant with Pittiglio, Rabin, Todd & McGrath, a management consulting firm with offices in Wellesley, Massachusetts, and Mountain View, California "JIT forces a faster reaction and the problem is more visible," Hoole noted. JIT also aims at 100 percent inspection of every part by the operator, according to Sepehri. In contrast, conventional techniques call for statistical sampling or end-of-line inspection, which he claimed is not as effective.

In Japan, according to manufacturing consultant Schonberger, management is usually pleased when many yellow lights are aglow. If they are not, management assumes the line is moving too slowly or that there are too many workers. Some of the workers may then be assigned elsewhere until the yellow lights start coming on again. If all of them are lit, it is a signal to add workers or slow down the line.

Another practice used in Japan and being adopted in the United States is to build time for line stops into the daily production schedule. The goal is to prevent errors by overpressured workers or overtaxed equipment, according to Schonberger. If too many line stops have occurred during the day, however, workers may be asked to stay after hours to meet the day's quota; they may or may not be paid overtime.

THE SUPPLIER SCENARIO

At Honeywell's process control division in Fort Washington, Pennsylvania, 95 percent of the production lines using JIT carry no inside inventory and have switched to receiving delivery of small lots from suppliers on a daily and weekly basis, according to Gene Goldberg, the division's principal quality engineer. Suppliers must provide defect-free products, since there is no inspection of incoming materials. There were some initial problems in getting suppliers to meet these demands, Goldberg observed, but they have been ironed out by patiently going back to the supplier and reinforcing the agreement.

Coordinating with suppliers is usually the last thing JIT implementors attempt, and with good reason. Suppliers generally resist agreeing to just-in-time delivery, for doing so means they must convert their own operations to JIT mode or face holding buffer inventory at their own facilities and expense. Questions also arise about whether the JIT practice of sole-sourcing—using only one supplier— will lead to noncompetitive pricing and stifle competition to improve parts quality and performance. Related concerns include what to do in the event a sole supplier goes out of business or if there is a strike.

The supplier-customer relationship has the potential of becoming adversarial, admits Sepehri. But that need not happen. JIT practice, he maintained, advocates sharing of resources and forging a relationship of trust. Furthermore, he believes, JIT will change the character of competition. "It will be the supplier who can deliver quality, rather than price, who will survive," he said. For companies genuinely concerned about the perils of sole-sourcing, he advised selecting the best supplier for each product, but using several suppliers overall.

Indeed in Japan the ties between customer and supplier are so close that suppliers customarily set up shop near the plants they serve. The practice is so pervasive that a July 14, 1986, *Business Week* article reported that a proposed Toyota facility in Georgetown, Kentucky, will bring with it a contingent of 10 to 12 affiliated suppliers.

In the United States, manufacturers are also trying to develop close relationships with suppliers. Harley-Davidson Motor Company, in York, Pennsylvania, for example, held sessions with suppliers to explain JIT objectives. Some suppliers later called on Harley-Davidson for assistance in implementing their own JIT programs. At Hewlett-Packard's Greeley, Colorado division, suppliers were included in team-building sessions as part of JIT implementation, according to company spokesman Jim Hasl.

As JIT supplier and manufacturer, Texas Instruments sees the issues from both sides of the coin. To meet JIT customers' demands for reduced lead time, Texas Instrument's Schottky department began implementing JIT two years ago. According to department manager Van Winkle, the amount of time that elapses from receipt of an order to delivery has been reduced from 20 weeks to six or eight. The window for shipping also has shrunk, from two weeks to one day.

Van Winkle summed up the customer-supplier issue in this way: "When you have to depend on a product to be there on a given day, and there is limited opportunity for rework, you just can't afford to deal with many suppliers."

CULTURAL ADJUSTMENTS

JIT will spotlight a slow worker as easily as it exposes other problems in a production line. When that happens, it is not unusual for the worker—accustomed to working at a different pace, without yellow and red lights flashing—to report feeling like he or she is in a pressure cooker, said Hewlett-Packard's Oman. On the other hand, the thought of sitting idle because the system calls for it also terrifies people, and the tendency among employees is to "push" ahead.

To overcome such hurdles, Hewlett Packard's Greeley, Colorado division implemented team-building sessions headed by an industrial psychologist. The key, said Oman, is to get the worker to become a participant in solving the problems, and to put blame on the process, not the people.

At the Tektronix hybrid circuits organization, said director of quality Erb, about 10 percent of the workers cannot adjust and are transferred to other jobs. When new people are hired, two or three of the operators on the line are selected to interview candidates. In the case of internal transfers, he said, there is a trial period of three or four days on the line.

Even when little worker resistance is encountered, he said, it is hard to get people to follow the rules without exception. "Old habits creep back, " he said. "It can be difficult to get operators to shut down a line promptly, to correct problems immediately, or to get immediate response from engineering."

Tektronix feels strongly enough about the advantages of JIT, he commented, "to have established a set of rules. How well these rules are followed becomes part of a manager's and a worker's performance appraisal."

Workers are not the only ones required to adjust, commented Sepehri. Unions also must change their thinking about how workers operate under JIT, he said. On the one hand, JIT tactics encourage workers to become multifunctional; on the other, there are times when a production line may have to be idle. Unions and management, he remarked, need to realize that it is no longer possible for incentive pay to be based on the number of units that a worker produces.

Old habits die hard for management also. Implementing JIT, Sepehri commented, may not involve an ethnic cultural change, but it does challenge an organization to alter its company culture and allow the worker to become a participation in decision making. This involves putting "trust and responsibility in the hands of the worker. This type of commitment," he stressed, "does not happen overnight."

DEFINING TERMS

Flexible manufacturing. The use of computers to automate manufacturing operations, such as changing the type and quantity of manufactured products, through minimal changes in hardware and/or software.

Pull production. A method of handling the flow of materials from one stage of manufacturing to another by having operations in the downstream stage of production draw work from the previous stage; removal of a piece for final assembly typically initiates the back-to-front chain reaction.

Push-through production. A method of handling the flow of materials from one stage of manufacturing to another by moving work along to the next stage upon completion of an operation.

Setup time. The time elapsed between production of the last piece of one model and the first piece of the next model, due to modifications in manufacturing equipment, such as the changing of tools and dyes.

Work-in-process inventory. The total amount of goods in all phases of manufacturing between release as raw materials and logging as finished stock.

Suggested Readings

The themes of just-in time manufacturing, total quality control, and world-class manufacturing are discussed in two books by Richard J. Schonberger: *Japanese Manufacturing Techniques*, 1982, and *World Class Manufacturing*, 1986, both published by the Free Press, a division of Macmillan Inc., New York.

A series of case histories that deals specifically with the JIT experience in the United States is presented in *Just-in-Time, Not Just in Japan*. Most of the book was written by Mehran Sepehri, and it includes studies of JIT implementation at several U.S. electronics companies. The book was published by, and is available from the American Production and Inventory Control Society, 500 W. Annandale Rd., Falls Church, VA 22046-4274. It forms part of the proceedings of the Apics Zero Inventory/Just in Time Seminar, which was held July 21-23. *Winning the Productivity Race*, by Bernard N. Slade and Raj Mohindra (Lexington Books, Lexington, MA. 1985) discusses the methods that allow Japan to move technology from the lab to the marketplace two to five times faster than any other country.

19

A Mathematical Procedure to Support the Design of "Pull" Production Strategies in Electronics Assembly

Jaime Trevino
Leon F. McGinnis

Most just-in-time (JIT) success stories reported in the literature are the result of improvements in production, purchasing, and management practices. Interestingly, in most cases, companies still "push" products from process to process, ignoring the further benefits that could be achieved if the "pulling" procedure which characterizes the Toyota Production System that was implemented (McGinnis et al. 1985; McGinnis and Trevino 1986).

On the other hand, the few companies using "pull" production strategies have addressed the fundamental design decisions using some of the recommended rules of thumb, or trial and error. As an example, a common practice is to use unit load sizes less than or equal to 1/10 of the daily production requirements and to compute the number of KANBANS using equations that are a function of the "pulling" rate and the time to a KANBAN request (Hall 1985; McGinnis et al. 1985; Monden 1983; Schonberger 1982; Schonberger and Schniederjans 1984; Sugimori et al. 1977; Wantuck 1981).

Additionally, the literature has suggested that lots of size one can be purchased or produced if the set-up and delivery times are reduced. However, lots of size one could be economically justified only for very simple production systems, with low throughput requirements and very small set-up times, or for production systems with excess capacity and dedicated production lines.

Consequently, a scientific procedure to address the fundamental design decisions of JIT production systems is needed, and the purpose of this chapter is to explain a design procedure for pull production strategies. This design procedure is based on mathematical modeling and supported by simulation.

We will present the characteristics, requirements, and some applications of the JIT production system. Additionally, the different JIT production strategies being used by American companies are explained. We will summarize the fundamental design issues of pull production strategies and include a hypothetical example of an assembly scenario that is used to illustrate the design procedure we present. We will explain the different variations of the assembly scenario and the possible modeling extensions. Finally, we present some conclusions about the design procedure.

BACKGROUND

Traditionally, American manufacturing systems have been based on a production schedule, generated from customer orders or forecasted demands, and a specific procedure (e.g., material requirements planning) to compute the purchasing and production lot sizes. Following the production schedule, economic order and production quantities are pushed through the production processes.

Interestingly, as most of the lot sizing procedures used are based on inventory carrying and set-up (ordering) costs, and, furthermore, as set-up and ordering-delivery times have been usually long, large lot sizes lead to large inventories (high inventory carrying costs), large waiting times (high inventory carrying costs), high storage and space costs, long lead times, and small inventory turnovers. Additionally, large lot sizes can lead to large scrap or rework costs (McGinnis et al. 1985; McGinnis and Trevino 1985).

In contrast, the JIT production system is based on reducing inventory queues to minimize investment and storage and space cost, to shorten production lead times, to provide faster response to demand changes, and to uncover any product quality or machine reliability problems. This is achieved through the following (McGinnis et al. 1985):

1. a frozen production schedule for the finished products;
2. JIT purchasing, JIT production, and JIT delivery of small lots with zero defects; and
3. a "pulling" procedure using cards or KANBANS.

To achieve JIT manufacturing, the traditional purchasing, production, and management practices need to be changed. Specifically, a JIT production program requires adopting the following (McGinnis et al. 1985)

1. short set-up times and quick changeovers;
2. zero defectives;
3. efficient preventive maintenance programs;
4. efficient material handling and transportation systems;
5. location of suppliers and plants for JIT purchasing and JIT delivery;
6. plant layout for JIT delivery;
7. flexible workforce; and
8. improved vendor relationships.

Applications

Since 1977, American companies have been studying and implementing the JIT production system. Today, nearly all of the top 1,000 industrial firms in the United States are involved in JIT programs and for most of them it seems that the investments required are more than offset by the benefits accomplished.

Examples of the benefits reported by some U.S. companies that have implemented "pull" production strategies are as follows:

1. The Hewlett-Packard (HP), Vancouver, Washington, facility that produces workstation printers reduced work in process (WIP) by 82%, increased shipments by 20%, reduced space by 40%, increased labor efficiency by 50%, and reduced scrap and rework by 30%. The Roseville (California) Networks Division that produces computer-printed circuit assemblies reduced WIP by 70%, reduced lead time to one fifth of the previous year's time, and increased productivity by 23%. Other H-P JIT applications have also been reported (Kenfield 1985; McGinnis et al. 1985; Walleigh and Sepehri 1986).
2. The Westinghouse Electric Corporation, Asheville, North Carolina, facility since 1984 has reduced inventories by 45%, improved productivity by 30%, reduced product warranty costs by 35%, and saved 40% in manufacturing space (McGinnis et al. 1985; *P&IM Review* (1986).
3. The Apple Computer, Fremont, California, facility that produces the Macintosh personal computer reduced quality rejects by 27% in a month, achieved inventory turns of 20 to 30, reduced space by 35%, and reduced inventories by 65% (Sepehri 1985; *P&IM Review* (1986).

Pull Production Strategies

Even when all the JIT requirements are not fully satisfied, some elements of the JIT approach can be implemented. Three distinct strategies for JIT production can be identified.

There are companies in which the basic JIT requirements have been satisfied, but no KANBANS are used. Consequently, although products are produced, purchased, and delivered in small lots, they are pushed from process to process following the traditional American manufacturing approach. This strategy can be referred to as a "modified-push" production strategy (McGinnis et al. 1985; Trevino and McGinnis 1986).

In other companies, the basic JIT requirements have been satisfied, and withdrawal KANBANS are used to retrieve parts from supplying processes, but products are purchased and produced according to a daily production schedule. Furthermore, the daily production schedule is followed using production rules that consider the production capacity of the supplying process(es) and the withdrawal pattern of the consuming process(es). This strategy can been referred to as a "push-pull" production strategy (McGinnis and Trevino 1984; McGinnis and Trevino 1985; Schonberger 1982; Trevino 1986; Trevino and McGinnis 1986).

Finally, there are some companies in which all the basic JIT requirements have been satisfied, and withdrawal and production KANBANS are being used. In the plants using this strategy, hourly or daily supplier deliveries are commonly used, and, in many instances, the warehousing, receiving, and inspection functions have been eliminated. Because consuming processes pull the parts from the output buffer of supplying processes (inside or outside the plant), it can be referred to as a "pull" production strategy (Schonberger 1982; Trevino 1986; Trevino and McGinnis 1986).

DESIGN ISSUES

Once the basic requirements of the JIT production system have been satisfied and the company is willing to freeze the production schedule of the finished products and implement the KANBAN control system, the following fundamental design issues need to be addressed (Trevino 1986).

1. How many KANBANS should be used for each part between supplying and consuming processes (if physical KANBANS are used) or which production strategy should be followed (if electronic KANBANS are used)?
2. How many parts should be produced per set-up (lot size) and how many parts per container or pallet (unit load size) should be transported between supplying and consuming processes?

3. How many inventory buffers should be used between processes and how many storage positions should be assigned to each buffer?
4. How many containers or pallets should be used for each part?
5. What handling and production capacity should be provided for each process?

Note that the answers to 2-5 will have an impact on investment decisions.

HYPOTHETICAL EXAMPLE

The manufacturer of high-end engineering workstations produces a small number of basic machine types, each with a variety of models. In the final assembly process, the purchased components—chassis, case, fasteners, power supply, and wiring harnesses—and the manufactured subassemblies—printed circuit cards—are delivered to manual workstations where final assembly and initial hot testing are performed.

The printed circuit cards are assembled in one of four card assembly lines, depending on the type of card. The four card assembly lines are main cpu or motherboards, memory cards, video display cards, and peripheral controller cards. These subassembly operations, in turn, receive the components and bare boards from outside vendors or off-site, company-owned fabrication operations.

In each card assembly cell, cards are produced in batches following a pull production strategy. In other words, every time a full container of cards is required in one of the workstations of the final assembly line, the full container of cards is pulled from the output buffer of the cell, and the production of another batch of identical cards is authorized. Cards are moved in containers and after testing are stored in the output buffer.

Furthermore, because a nonsynchronous assembly procedure is being used in the final assembly line, and each workstation is capable of assembling the different subassemblies, components, and parts that form an engineering workstation, each card is pulled one by one from the containers stored at each workstation of the assembly line during the day.

The company has achieved all the prerequisites for JIT, and now is concerned with the details of implementation on the shop floor. If the pull procedure is going to be implemented between the workstations of the final assembly line and the card assembly cells, the following questions need to be resolved:

1. What should be the basic lot size (LS) for each type of card?
2. What should be the unit load size (ULS) for each type of card?
3. How many production KANBANS (NPK) should be used between each assembly line workstation and each card assembly cell?
4. How much buffer space should be provided for each type of card at the output buffer of each assembly cell?

5. What should be the throughput capacity of each assembly cell?
6. What should be the assembly capacity of each workstation at the final assembly line?

DESIGN PROCEDURE

Trevino presents a methodology for addressing the six design questions listed previously (1986). This methodology employs as a key element the probability of stockout at the assembly workstation, which is constrained to a specific maximum value. Design alternatives (i.e., assembly capacities, lot sizes, unit load sizes, and number of KANBANS) that satisfy this constraint are identified through stochastic analyses and then evaluated with regard to total cost to select the preferred alternative. Due to space limitations, the methodology will not be presented here, but further details can be found in the literature (Trevino 1986; Trevino and McGinnis 1986).

Assembly Scenario to Model

To illustrate how the design procedure works, assume that the memory cards (MCs) are pulled from the output buffer of the MC assembly cell (P2) by one of the workstations of the final assembly line (P1) to be assembled with the purchased components and the other cards and form the engineering workstation (EW). The other printed circuit cards also are pulled from their assembly cells, so the procedure is applicable for all the printed circuit cards and assembly cells.

An inventory buffer of MCs is maintained between the two processes to support the pulling procedure that works as follows.

1. Every time a MC is needed at P1, it is pulled from the MC container (see Figure 1). If the container is empty, a full container is retrieved from the buffer, provided one is available. If the buffer is empty, a stockout condition is reached, and the operator of P1 waits for the arrival of more MCs from P2. (Every time a full container of MCs is retrieved from the buffer, the empty container and the production KANBAN that was attached to the full container are sent to P2.)
2. The arrival of KANBANS (and empty containers) authorize P2 to process another full container of MCs. A full container in this case is equivalent to the number of parts to produce per set-up (LS) and to carry in a container (ULS). Thus, in this case, the LS is equal to the ULS.
3. Once a full container of MCs is produced, it is placed in the buffer, if the system is not in stockout, or it is used immediately by the operator of P1, if he is waiting for MCs. In the latter case, the empty container that was waiting at P1 and the arriving KANBAN are sent immediately to P2.

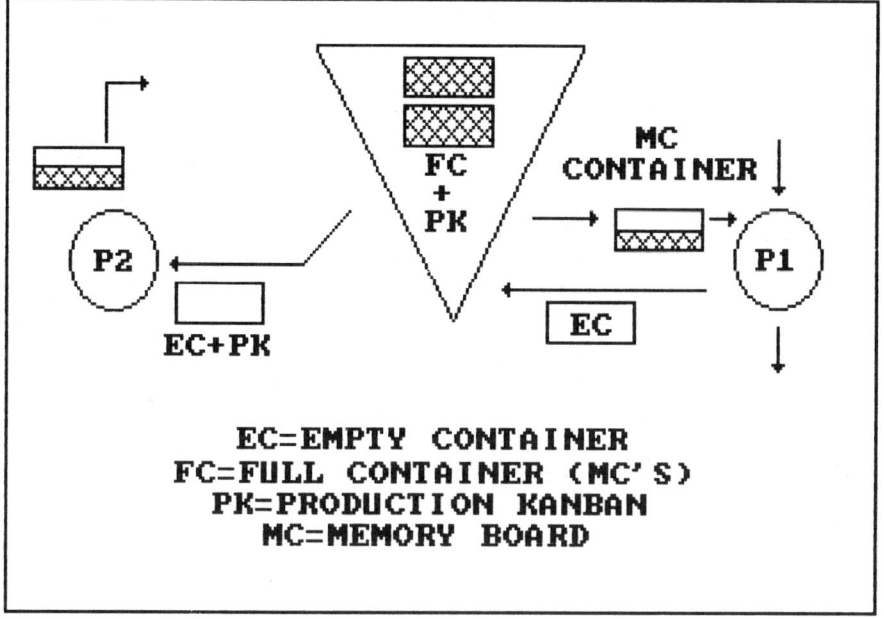

Figure 1. *Assembly scenario to model.*

The fundamental design parameters of this assembly scenario are as follows:

1. the capacity of P1 (μ_1);
2. the capacity of P2 (μ_2);
3. the ULS and LS for MC's; and
4. the NPK for MCs to use between P1 and P2.

Assumptions

1. MCs are pulled from the MC container by P1 one by one.
2. The arrival rate of "kits" of purchased components (pulling rate of MCs) is Poisson with parameter λ.
3. MCs are being produced at P2 according to a pull production strategy with continuous review and n = 1 (production of a container of MCs every time a production KANBAN and an empty container arrive to P2).
4. The assembly time per container at P2 is exponentially distributed with parameter μ_2/ULS.
5. The assembly time at P1 is exponentially distributed with parameter μ_1.
6. Production KANBANS are served in order of release (FCFS queue discipline).

7. No handling system is used between the two processes (both processes are close to each other).
8. P1 is dedicated to the assembly of engineering workstations and P2 to the assembly of MCs.

The Procedure

A detailed description of the design procedure is available (Trevino 1986) and will not be repeated here. The procedure's objective is to find the design parameters that eliminate assembly interruptions due to stockouts and minimize the annual total cost of the system.

The annual total cost of the system includes the following: 1) the annual inventory carrying cost of MCs at (cost of maintaining inventory of MCs at buffer and container); 2) the annual storage cost (storage equipment and space cost required by each container of MCs); and 3) the annual assembly equipment cost (assembly capacity cost at P2).

Mathematically, the design procedure can be stated as follows:

Minimize TC
Subject to
(1) $\qquad$ PS < PSC
(2) $\qquad$ ULS ≥ 1
$\qquad$ ULS < CCMC
(3) $\qquad$ $\mu_2 > \lambda$
(4) $\qquad$ $\mu_1 > \lambda$
(5) $\qquad$ NPK > 0

where,

$$
\begin{array}{rcl}
TC &=& \text{Annual total cost.} \\
PS &=& \text{Probability of stockout of MCs.} \\
PSC &=& \text{Probability of stockout criterion.} \\
CCMC &=& \text{Container capacity in MCs.} \\
\lambda &=& \text{MCs' pulling rate.}
\end{array}
$$

TC and PS are nonlinear functions of ULS, μ_2, NPK, and λ; constraint one is used to eliminate assembly interruptions due to stockouts, and constraints three and four are used to achieve the daily assembly requirements.

Illustration of Procedure

Assume that 100 engineering workstations (EWs) are going to be assembled per day at each assembly workstation. Also assume that the pulling rate is Poisson.

If at least one MC is available at the MC container or buffer, P1 begins the assembly. The assembly time is exponentially distributed with $\mu_1 = 110$ EWs/day. Every time a MC is required and no MCs are available at the MC container, a full container of MCs is pulled from the buffer (if available) and a production KANBAN is released for processing at P2. If no MCs are available at the MC container and buffer, the kit of purchased components, the handling vehicle transporting the kit, and the server of P1 wait for the arrival of more MCs from P2. Assume that the assembly time at P2 per container is exponentially distributed with

$$\mu_2 = 120/\text{ULS} \quad \text{containers/day.}$$

1. Determine the ULS and NPK that minimize the annual total cost of the system and maintain a PSC <0.01.
2. If μ_2 is unknown, determine the ULS, NPK, and μ_2 that minimizes the annual total cost of the system and maintain a PSC <0.01.
3. For the optimum design of 2), find the utilization at P2, the storage requirements, and the average WIP of MCs.
4. Show how the optimum design can be changed to increase the utilization at P2.
5. Show the effect of changes on the pulling rate of EWs on the optimum design combination.
6. For a PSC <0.01, show the effect that changes on the assembly capacities and ULSs will have on the minimum annual total cost. Show also how the design parameters can be modified to increase the utilization at P2 or to reduce the storage requirements.

Example Solution

A computer program was developed to implement the design procedure. Results of a run with the data provided previously are presented in Table 1.

1. As seen in Table 1, if $\mu_2 = 120$ MCs/day, the minimum NPK for a PSC <0.01 would be 25. This minimum cost design combination (ULS = 1, $\mu_2 = 120$, and NPK = 25) would cost $1,039.16 per year (assembly and storage equipment cost plus inventory carrying cost) and would ensure the daily assembly requirements are achieved. Additionally, if this design combination is used, one MC stockout will occur for every 100 kit arrivals (on the average).
2. If μ_2 is unknown, Table 1 can also be used to determine the optimum design combination. As seen in Table 1 and Figure 2, the annual total cost decreases as μ_2 increases until $\mu_2 = 180$ and then it begins increasing. Then, the design combination that minimizes the annual total cost would be ULS = 1, $\mu_2 = 180$, and NPK = 7, with a cost of $585.

TABLE 1 – Design Procedure Results for Assembly
Scenario Modeled (PSC< = 1% and ULS = 1)

μ_2	UTILP2	AWIP	NPK	SC	EC	ICC	TC
120	0.833333	20.00001	25	500	239.16	300.0002	1039.16
140	0.714285	10.5	13	260	279.02	157.5	696.52
160	0.625	7.333334	9	180	318.88	110	608.88
180	0.555555	5.75	7	140	358.74	86.25001	584.99
200	0.500000	5	6	120	398.6	75.00001	593.6
220	0.454545	4.166667	5	100	438.46	62.5	600.96
240	0.416666	4.285714	5	100	478.32	64.28571	642.6057
260	0.384615	3.375	4	80	518.18	50.62501	648.805
280	0.357142	3.444445	4	80	558.0401	51.66667	689.7068
300	0.333333	3.5	4	80	597.9	52.50001	730.4
320	0.3125	2.545455	3	60	637.76	38.18182	735.9418
340	0.294117	2.583334	3	60	677.62	38.75001	776.37
360	0.277777	2.615385	3	60	717.48	39.23077	816.7108
380	0.263157	2.642857	3	60	757.34	39.64286	856.9829
400	0.25	2.666667	3	60	797.2	40.00001	897.2

3. Although the optimum design of 2) minimizes the annual total cost and uses a reasonable NPK, the utilization at P2 would be equal to 55% (see Table 1). In addition, seven storage locations will be required at the buffer and six MCs would be carried on inventory on the average (see Table 1).

4. The utilization of the operator of P2 can be increased to 83% if more production KANBANS are used (see Table 1) to justify using a smaller assembly capacity for P2. Specifically, if the design combination ULS = 1, μ_2 = 120, and NPK = 25 is used (assembly capacity reduced in 60 MCs/day, but NPK increased by 18), the utilization of the operator of P2 can be increased to 83%. Obviously, more production KANBANS are going to require more containers and storage locations and the average number of MCs on inventory will also increase (see Figure 3). Therefore, the annual total cost would increase to $1,039.16.

5. As seen in Table 2 and Figure 4, if the pulling demand changes by a small amount, the system would handle the changes using more or fewer production KANBANS, depending on the nature of the pulling demand change. However, if drastic pulling demand increases occur, more assembly capacity and NPK will be required. Likewise, if drastic pulling demand decreases occur, fewer production KANBANS and the same assembly capacity will suffice to achieve the system goals (with a low utilization at P2), or fewer

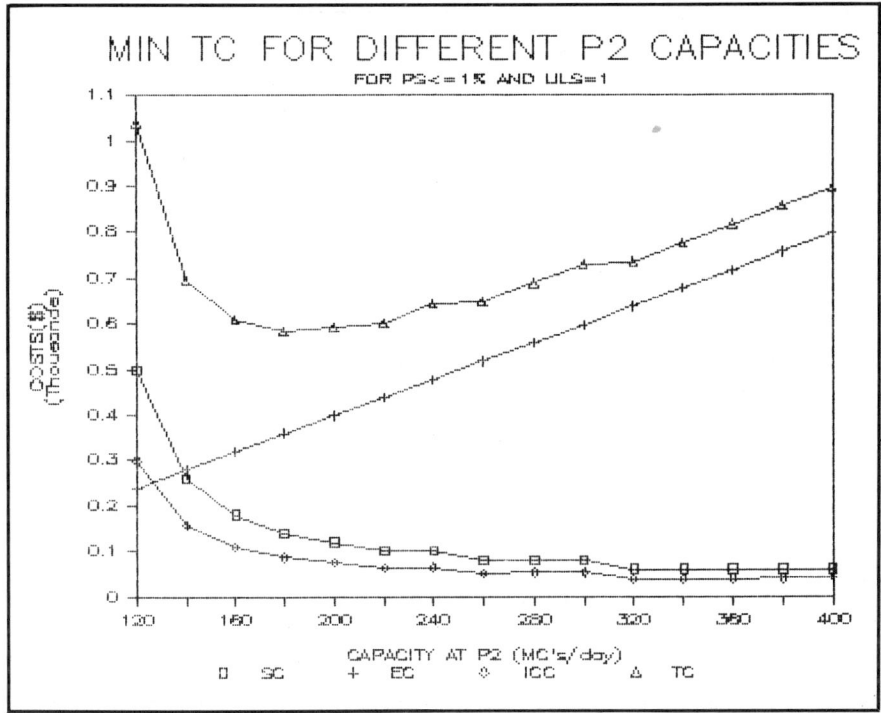

Figure 2. *Minimum annual total cost for different P2 assembly rates.*

production KANBANS and a lower assembly capacity could be used if a higher use at P2 was desired.

6. The effect of the ULS on the minimum annual total cost for the given PSC criterion is seen in Figure 5. As a larger ULS is used, the storage cost decreases and the inventory carrying cost and annual total cost increase. Then, the ULS that minimizes the annual total cost and satisfies the system objectives is the "minimum possible" ULS. In this case, as the set-up time is not relevant (P2 is dedicated to the assembly of MCs), the "minimum possible" ULS can be as small as one.

EXTENSIONS

The model presented previously can be extended to analyze the following design variations:

1. handling considerations between assembly workstations and card assembly cells with KANBANS or electronic signals (Trevino 1986);

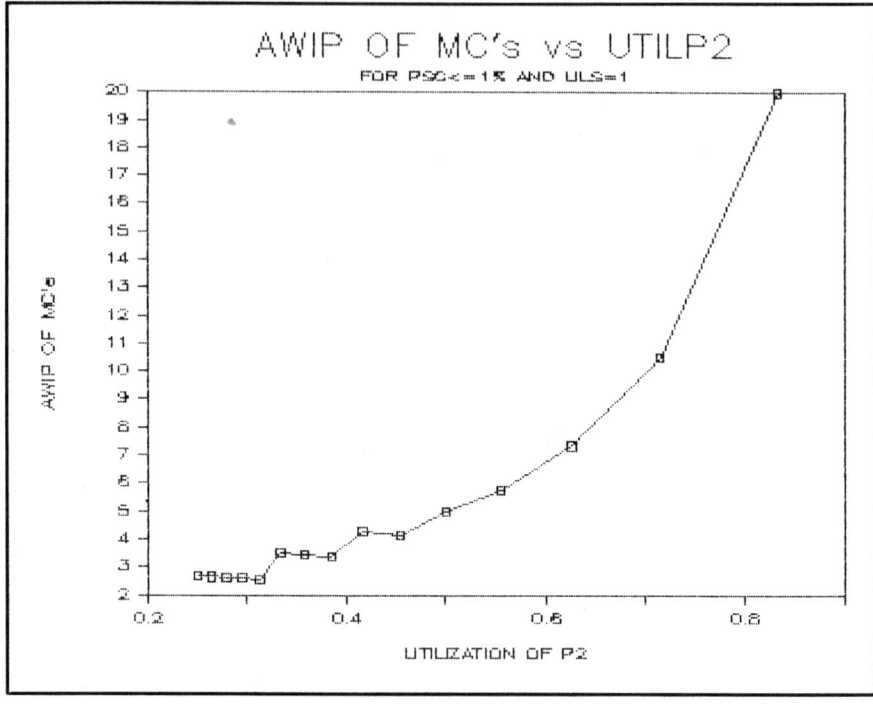

Figure 3. *Average WIP of MCs vs. utilization at P2.*

**TABLE 2 – Effect of "Pulling" Demand
Changes on Optimum Design
(PSC< = 1% and ULS = 1)**

μ_2	P2C	NPK	ULS	MINTC
50	100	6	1	394.3
100	180	7	1	584.99
150	240	9	1	768.32

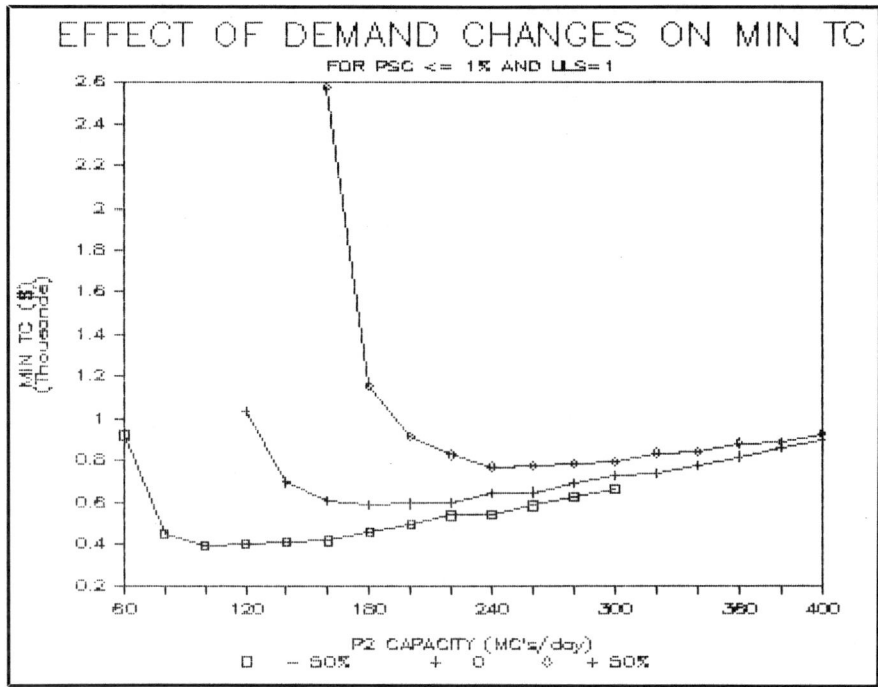

Figure 4. Effect on demand changes on the minimum annual total cost.

2. assembly cells producing more than one card (set-up considerations) (Trevino 1986);
3. other continuous review policies (Trevino 1986);
4. periodic review policies (Trevino 1986);
5. push-pull production strategies (McGinnis and Trevino 1984; McGinnis and Trevino 1985; Trevino 1986; Trevino and McGinnis 1986);
6. kits of purchased components arriving in batches (Trevino 1986);
7. other nonsynchronous assembly strategies (Trevino 1986); and
8. quality considerations (Trevino 1986).

CONCLUSION

The JIT approach to manufacturing can potentially significantly improve quality, reduce lead times and inventories, and improve responsiveness. However, adapting the KANBAN control system, one of the most important characteristics of the JIT approach, requires specifying some fundamental system design decisions for which only rough guidelines have existed in the past.

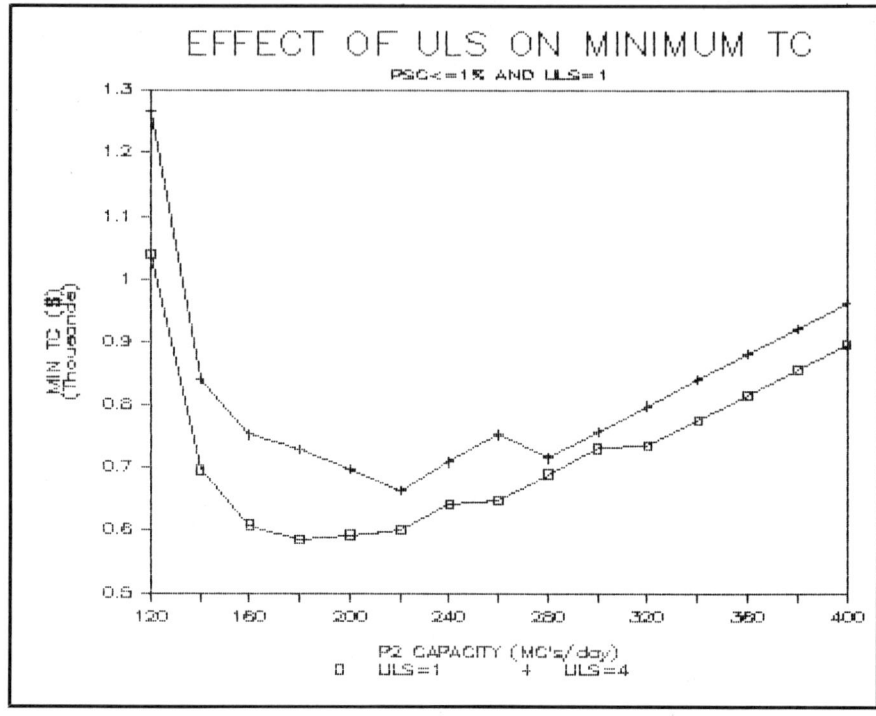

Figure 5. *Effect of the ULS on the minimum annual total cost.*

In this chapter, we have identified the variations of the KANBAN control system, have discussed the pull design problem in the context of printed circuit card assembly, have presented a mathematical procedure to solving the design problem, and have identified some extensions of the procedure.

Suggested Readings

Hall, R. W. 1983. *Zero Inventories.* Dow-Jones-Irwin. Homewood, Illinois.
Kenfield, J. E. 1985. A nine step approach to JIT implementation. *International Electronics Assembly Conference.* October 7-9. Institute of Industrial Engineers. Norcross, GA.
McGinnis, L. F., J. Trevino, and J. A. White. 1983. A bibliography on work-in-process inventory management. *TR-83-05, Technical Paper.* Material Handling Research Center. Georgia Institute of Technology. Atlanta, GA.

McGinnis, L. F., and J. Trevino. 1984. Analysis of a lot sizing procedure for a single-card kanban system. *TR-84-02, Technical Paper.* Material Handling Research Center. Georgia Institute of Technology. Atlanta, GA.

McGinnis, L. F. and J. Trevino. 1985. Electronic spreadsheet implementation of a lot sizing procedure for a single-card kanban system. *TR-85-14, Technical Paper.* Material Handling Research Center. Georgia Institute of Technology. Atlanta, GA.

McGinnis, L. F., Z. Toro-Ramos, and J. Trevino. 1985. A review of the Toyota production system. *RS-85-01, Technical Paper.* Material Handling Research Center. Georgia Institute of Technology. Atlanta, GA.

McGinnis, L. F., and J. Trevino. 1985. Integrated analysis of nonsynchronous assembly systems used in the electronics industry. *International Electronics Assembly Conference.* October 7-9. Institute of Industrial Engineers. Norcross, GA.

McGinnis, L. F., and J. Trevino. 1985. Lot sizing and unit load sizing for just-in-time production systems. *TIMS/ORSA Joint National Meeting.* November 4-6. Atlanta, GA.

McGinnis, L. F., and J. Trevino. 1986. Just-in-time systems. *The 36th Annual Material Handling Short Course.* March 6. Atlanta, GA.

Monden, Y. 1983. *Toyota Production System.* Industrial Engineering and Management Press, Institute of Industrial Engineers. Norcross, GA.

P&IM Review. 1986. JIT urgency revealed during APICS summit. *P&IM Review.* 6(8): 26-30.

Schonberger, R. J. 1982. *Japanese Manufacturing Techniques.* The Free Press. New York.

Schonberger, R. J., and M. J. Schniederjans. 1984. Reinventing inventory control. *Interfaces.* 14(3): 76-83.

Sepehri, M. 1985. A machine builds machines at Apple computer's highly automated Macintosh manufacturing facility. *Industrial Engineering.* 17(4): 60-67.

Sugimori, Y., K. Kusunoki, F. Cho, and S. Uchikawa. 1977. Toyota production system and kanban system—materialization of just-in-time and respect-for-human system. *International Journal of Production Research.* 15(6): 553-564.

Trevino, J. 1986. Design procedures for JIT production systems. Unpublished Doctoral Dissertation. Georgia Institute of Technology. Atlanta, GA.

Trevino, J., and L. F. McGinnis. 1986. Electronic spreadsheet implementation of a lot sizing procedure for a single-card kanban system. *8th Annual Conference on Computers and Industrial Engineering.* March 19-21. Orlando, FL.

Trevino, J., and L. F. McGinnis. 1986. Analysis of pull production system. *TIMS/ORSA Joint National Meeting.* October 27-29. Miami, FL.

Walleigh, R., and M. Sepehri. 1986. H-P division programs reduce cycle times, set stage for ongoing process improvements. *Industrial Engineering.* 18(3): 74-81.

Wantuck, K. A. 1981. The ABCs of Japanese productivity. *Production and Inventory Management Review and APICS News.* September 1981: 22-28.

This work was supported by the National Science Foundation, Grant no. ISI-8300965, the Georgia Institute of Technology, and the member companies of the Material Handling Research Center.

Reprinted from the *1986 International Electronics Assembly Conference Proceedings.*

Jaime Trevino is Assistant Professor of Industrial Engineering and Research Associate of the Integrated Manufacturing Systems Engineering Institute and JIT Revitalization Forum at North Carolina State University.

Leon F. McGinnis is Professor of Industrial Engineering and Director of the Computer Integrated Manufacturing Systems Program at the Georgia Institute of Technology.

20

The New Ballgame: JIT Purchasing

Bob Parmelee

Human nature causes us to deal very skeptically with new, unfamiliar concepts, and, as many of us can testify, JIT manufacturing and JIT purchasing have not been immune from skepticism and resistance. But fortunately, JIT is very different from some of the other programs which have come and gone—and embodies unique techniques which promote success. The purpose of this chapter is to discuss some high impact JIT techniques and review some of the reasons why just-in-time has changed the rules of the purchasing game. The chapter is based largely on existing programs at one of Digital Equipment Corporation's (DEC's) largest plants, a manufacturer of computer disk drives and controllers.

WHAT'S SO DIFFERENT ABOUT JIT?

Most of us are used to dealing with strategies that deal with "results" or measurements rather than tools. For example, MRPII Programs might declare "Class A" status if plant achieves 95% performance in a long list of metrics such as vendor delivery performance. But, "How to do it" advice is usually absent. JIT doesn't work this way.

JIT strategies deal with a list of things that you "do" and methods or techniques that you implement. The emphasis is not on achieving "95% performance", but on a longer view of continuous improvement towards an unstated goal of eventual 100% performance. JIT precepts recognize that 95% isn't good enough any more. Also recognized is the concept that it is useless to set out arbitrary numerical goals without providing the workforce the tools to do the job.

From this viewpoint, JIT is entirely consistent with W. Edwards Deming's 10th Law, which states: "Eliminate numerical goals, posters and slogans for the work force, asking for new levels of productivity without providing methods."

How many of us have heard or read these words without pausing to think about what they mean?

How many of us have actually changed our behavior and what we ask our people to do, based on the concept of enshrining methods as goals, not numbers?

As Deming further points out, management's obligation is not to incite and exhort but to "provide a road map to improvement."

Just-in-time manufacturing and JIT purchasing techniques provide just such a road map. By following the JIT path you can undoubtedly move your firm toward a stronger competitive position. Working together in well planned, vendor-customer teams, we can all work together in reestablishing the competitiveness of American manufacturing.

It is clearly beyond the scope of this chapter to provide a complete analysis of all of the various aspects of JIT purchasing. However, there are many excellent references. (For example, "JIT Purchasing" by Larry Giunipero, in the NAPM *Guide to Purchasing.*)

Instead, I will concentrate on several high-impact aspects of JIT purchasing that capture the essence of the concept. The first area I will address is the central core area of forecast and release purchasing.

JIT FORECAST AND RELEASE PURCHASING

A comparison of the conventional purchasing process (Figure 1) and the forecast and release process (Figure 2) reveals the simplicity and cost effectiveness of JIT. Because of our tendency to specialize and compartmentalize, we often lose sight of the total process involving the acquisition of material and its provisioning to WIP. Within the conventional process there are often delays for approval of internal purchase requisitions (IPR's) and later, for approvals and data entry of purchase orders or change orders (PO/CO's).

Because of the delays inherent in the process, it is a common occurrence to have several versions of the open order file: the planner's version, the buyer's version and the supplier's version. An additional defect of the conventional process is that the purchase orders may be severely limited as forecasting tools because of a customer's normal tendency to limit the time span covered by purchase orders to the vendor lead time. After all, applying the conventional wisdom, if a vendor quotes an eight week lead time, why bother to provide 50 weeks of demand visibility? When employing the traditional methods, purchasing is often relegated to a clerical role of trying to keep up with the paperwork.

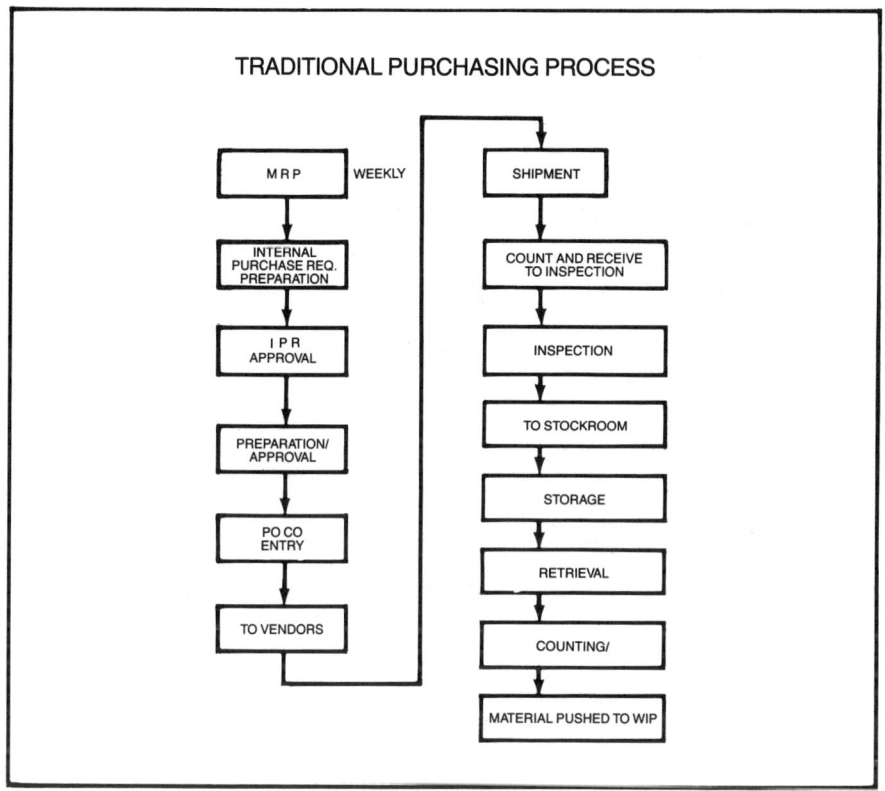

Figure 1. *Traditional purchasing process.*

There never seems to be enough time to focus on vendor relations and broader procurement activities.

Most companies have attempted to streamline this awkward flow by creating planner/buyers with both inventory and purchasing responsibilities, but this does not alter the basic clumsiness of the overall process.

Within the JIT flow, there are two major processes, forecasting and releasing, which operate virtually independently. Forecasting is driven by MRP (or at least a bill processor) and consists of part number requirements sorted by vendor. These scheduled requirements may be transmitted by normal means or electronically. The customer should have little reluctance in passing the MRP output directly to a vendor if the planning system has integrity, such as good BOM's, excellent record accuracy, etc. A schedule developed in this manner is much more accurate and timely than purchase orders developed in

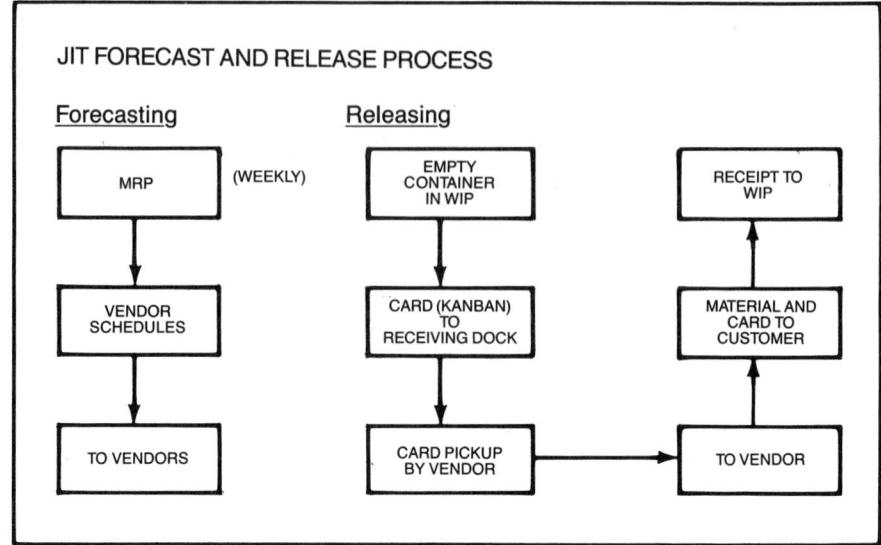

Figure 2. *JIT forecast and release process.*

the conventional manner and also provides demand data out to the planning horizon limits of the particular MRP system.

The release system fine tunes the process. The employment of pull systems to release only what is actually needed couple the vendor tightly to the customer production line. But, for the system to operate properly, the customer must carry only minimal inventory. Low inventories protect the supplier from abnormal demand fluctuations caused by inventory policy game-playing, and more significantly, low inventories enforce problem solving and enforce superb levels of quality and delivery from the vendor.

The best known example of the release process is the Kanban card system popularized by Toyota. The card process is accurate, simple, fast and dependable. It also is very cost effective with both the materials planner and the buyer removed from the ordering loop. Generally, card releasing is performed by material handlers using the simple, mechanical activity of delivering the card to a receiving dock and, in at most a few hours, the vendor truck driver. Another advantage of the card is that it is easily bar codeable since the quantity, part number, etc., do not change. Bar coding eliminates terminal data entry and promotes greater accuracy. Figure 3 illustrates an internal card design. A single bar code ward updates the inventory, purchasing and accounts payable systems. Moreover, the terminal transaction time of two seconds replaces a terminal data entry time of approximately 150 seconds.

RELEASE CARD IN USE AT DEC'S COLORADO
SPRINGS PLANT

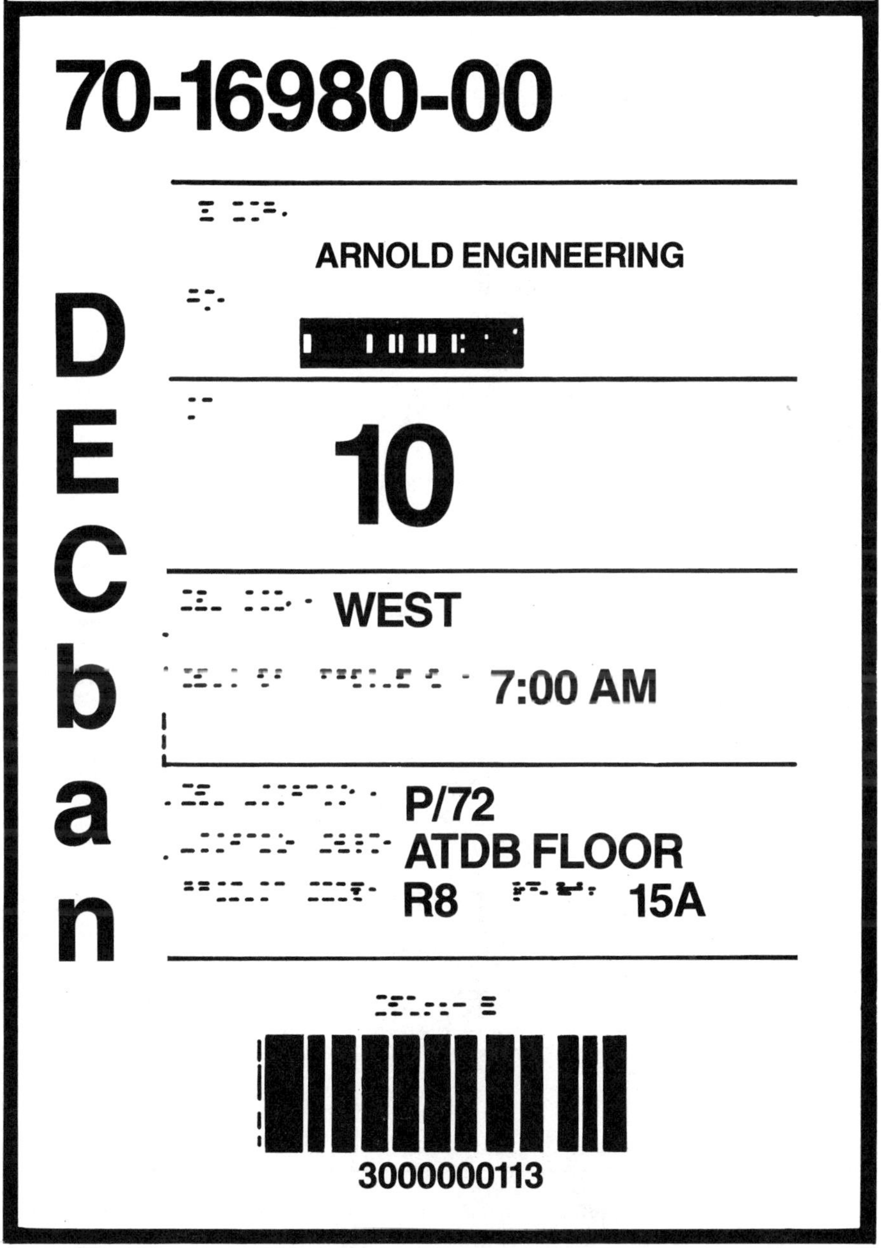

70-16980-00

D
E
C
b
a
n

ARNOLD ENGINEERING

10

WEST

7:00 AM

P/72
ATDB FLOOR
R8   **15A**

3000000113

Figure 3. *Release card in use at DEC's Colorado Springs plant.*

Unfortunately, the Kanban process begins to break down when remote vendors are involved. Difficulties exist in reliably transmitting the cards to the vendors, employing recyclable containers, using common freight carriers, etc. Figure 4 illustrates a pilot release system which was attempted at DEC and discarded because of complexity. When using remote vendors, the best approach is probably to authorize shipment of the first time bucket(s) in the MRP vendor schedule. This is clearly a second best alternative but with good discipline, MRP is self-adjusting and can be a rough substitute for a pure pull system.

A much more serious consequence of a remote vendor base is the necessity to carry stockroom inventory to compensate for the uncertainty of long distance distribution cycle time. Of course the vendor can carry the inventory in a local warehouse, but the associated costs will be passed on to the customer.

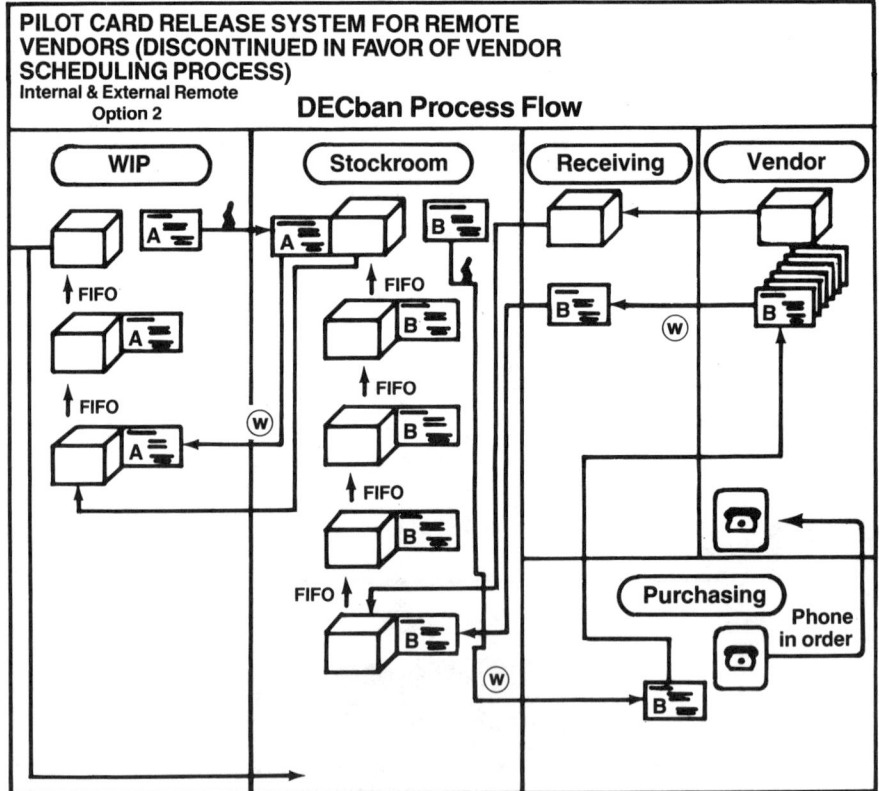

Figure 4. *Pilot card release system for remote vendors (discontinued in favor of vendor scheduling process).*

CUSTOMER ASSISTANCE TEAMS

The old conventional wisdom held that the customer didn't really care how the vendor operated his shop, just as long as he shipped "reasonably" on time with "acceptable" quality. The trouble with this concept was that it didn't work. Delivery surprises and quality fiascos were inevitable because many vendors employed push manufacturing systems and failed to control their process. In today's zero inventory, JIT environment, the customer can no longer tolerate the output of unreliable manufacturing processes and must be willing to help train vendor personnel in statistical process control (SPC) and JIT manufacturing. There is ample evidence that this team approach can have handsome payoffs both for the customer and the vendor.

For example, one of DEC's largest plants sent an assistance team to help teach such concepts as SPC, JIT, workcells, etc. to a major sub-assembly vendor. After the departure of the team, the vendor continued vigorous implementation efforts. After a period of six months, the vendor released the results depicted in Table 1.

TABLE 1 – Before and After (6 Months): Results of JIT/TQC Implementation at Vendor 1

PRODUCT: ELECTRONIC SUB-ASSEMBLIES			
	BEFORE	AFTER	Δ
Mfg. Cycle Time	13 wks	10 days	Reduced 89%
Assy. Labor Hrs.	22 hrs	18 hrs	Reduced 18%
Queue Hours (Per assy)	2200 hrs	240 hrs	Reduced 89%
BOM Levels	7	4	Greatly Simplified
Rework as % of Total Inventory	15-20%	Zero	Rework Eliminated
Materials IL Personnel	11	5	Reduced 55%
Inventory	$1.1M	$360K	Reduced 67%
Cost	NA	NA	Improved 17%

IMPACT OF JIT ON SUPPLIERS

The previous success story illustrates that JIT is not just a scheme to get the supplier to hold inventory, but instead, offers much tangible, internal advantage to the supplier. Many vendors recognize this and convert to JIT without any customer persuasion. The results of just such a conversion are indicated in Tables 2 and 3. The reduction in indirect labor is particularly noteworthy. Cost improvements resulting from these improvements were, in large part passed onto the individual customers as noted in Table 4.

TABLE 2 – JIT Impact on Vendor 2

	BEFORE (1983)	AFTER (1985)
Facilities (sq. ft.)	180K	120K
Direct Labor	530	375
Overhead	35%	10%
Overtime	12%	.5%
Packaging ($/Month)	3500	500
Shipping Volume $M	12	20

TABLE 3 – JIT Impact on Indirect Headcount at Vendor 2

	BEFORE	AFTER
Purchasing	5	3
Engineering	27	11
Quality	60	15
Order Control	17	7
Documentation	6	2
Personnel	6	2
Payroll	2	1/2

TABLE 4 – JIT Cost Reductions Passed on
to Customers (Vendor 2)

CUSTOMERS	ANNUALIZED COST REDUCTIONS ($K)
A	385
B	93
C	112
D	209
E	607
F	172

THE ROLE OF INVENTORY REDUCTION TO ENFORCED PROBLEM SOLVING

There is often a great distrust of the "enforced problem solving" doctrine of JIT manufacturing. To the uninitiated, it sounds suicidal to eliminate inventories in order to force the vendor to perform. It is often argued that, "we will convert to JIT as soon as the vendor achieves 100% delivery and 100% quality". The problem is that this will never happen unless we first eliminate inventories!

This paradox is understandable when we consider the behavioral aspect of the problem. If a buyer has six weeks of inventory on the shelf, does he really care if the vendor is a week or so late? If he decides to follow up, the vendor may very well pose the age-old question, "When do you really need it?" The sad reality is that, in using conventional practices, we actually "train" our vendors to be late and train our buyers to accept sloppy performance.

A vendor, operating in a conventional, high-inventory mode, will never attain 100% performance in quality and delivery because, in the final analysis, the customer doesn't really require high levels of performance. Figure 5 illustrates the JIT approach to quality, delivery and inventory. Improvements in inventory promote quality and delivery improvements, and vice versa.

JIT AND REDUCED LEAD TIMES

As has been shown above (Table 1), JIT manufacturing techniques can have very large effects on vendor lead time. When overall acquisition lead time is analyzed, the chief causes are not distribution lead time, order placement time,

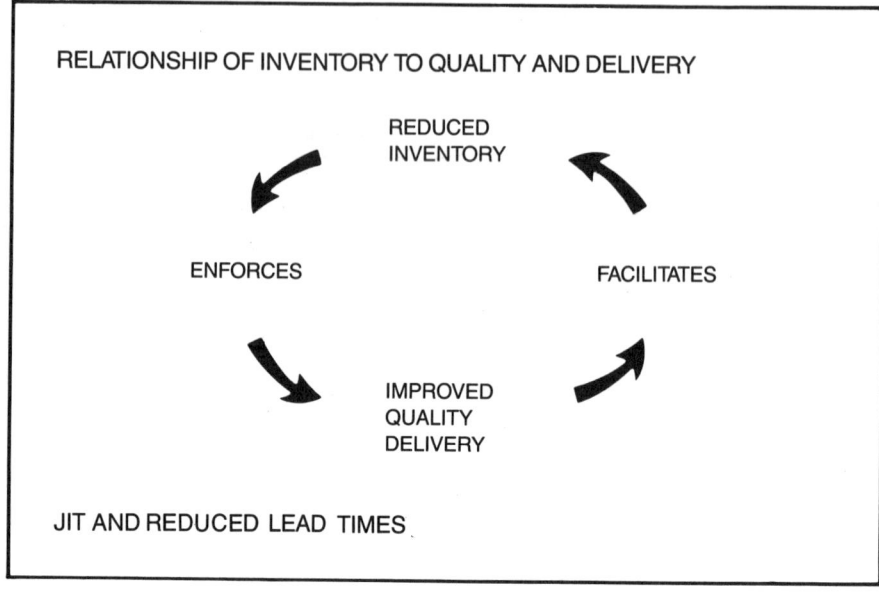

Figure 5. *Relationship of inventory to quality and delivery.*

receive/inspect time, etc. (although each is important) but predominately, vendor manufacturing time. If the analysis is taken one step further to analyze what causes manufacturing time, the principal problem is queue time. As has been often documented, queue time normally accounts for 90% or more of actual floor manufacturing time (see Figure 6). JIT is a powerful lead time reduction tool because it directly attacks queue time, the primary component of lead time.

There are a number of JIT techniques which reduce queue time. Pull system mechanics can directly reduce queues particularly when allied with intensive set up reduction programs and workcells or group technology programs.

There are an entire set of simple and effective JIT programs which directly impact most, if not all, of the causal elements of lead time, or cycle time. A partial listing of these is provided in Table 5. Any serious effort to reduce lead times, internally or externally, should include JIT as the central core strategy.

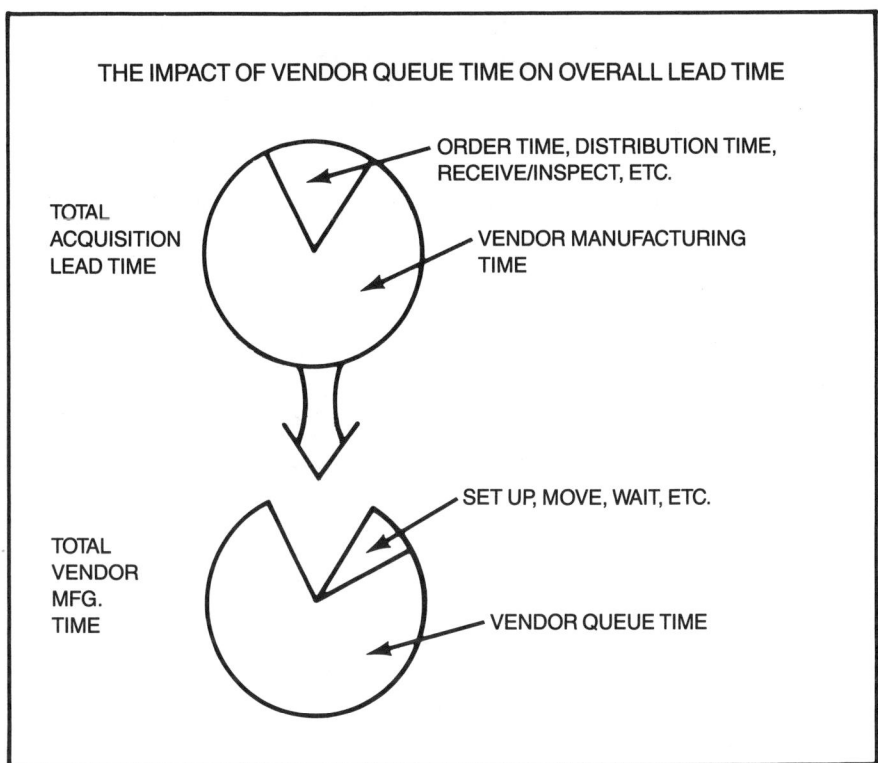

THE IMPACT OF VENDOR QUEUE TIME ON OVERALL LEAD TIME

Figure 6. *The impact of vendor queue time on overall lead time.*

CUSTOMER - VENDOR RELATIONS

It has often been observed that JIT purchasing seems to be associated with a closer customer-vendor relationship, often euphemistically termed a partnership. Some companies recognize the potential benefits of this close relationship and attempt to "decree" closer ties into existence. But simply announcing good relations doesn't make them happen; they must be based on broad mutually supporting day-to-day business practices such as those embodied in JIT purchasing.

One of the keys to establishing excellent vendor relations is eliminating annual bidding wars and awarding business to a sole source supplier for the duration of the business or product. Companies including Digital Equipment Corporation and Harley-Davidson have tried to cement these awards with long term contracts, with mixed results. The biggest problem is that long term

TABLE 5 – The Impact of Various JIT Techniques on Acquisition Lead Time.

Lead Time Causal Element	JIT Technique to Address
Order Placement Time	Forecast and Release Purchasing.
Vendor Material Issue Time	Eliminate job order kitting. Replace with pull techniques and backflush
Vendor Setup Time	Setup reduction programs, workcells, group technology
Vendor Queue Time	Pull systems, workcells, group technology, setup reduction
Vendor Production Cycle Time	Setup reduction programs, workcells, pull systems, U shaped lines, SPC
Vendor Move/Wait Time	Workcells, group technology, repetitive manufacturing processes
Vendor Rework, Inspection Time	SPC. Elimination of queues
Vendor Packaging	Recyclable Standard Containers
Distribution Vendor to Customer	Local sourcing
Customer Receiving counting and transaction time	Standard containers, bar code
Customer Inspection	Eliminated. Rely on Vendor SPC programs
Customer Storage/Retrieval/ Kitting/Counting	Eliminate workorder systems and kitting. Issue from Receiving directly to WIP

contracts are very difficult to write. If you water down the contract with all sorts of escape clauses to provide for contingencies, e.g., moving the end product production to Asia, then you are severely limiting the applicability of the contract. More often than not, long term contracts die on lawyer's desks.

At DEC, the current thinking is to incorporate long term business philosophies concerning the process of awarding long term business, JIT and SPC into non-contractual business agreements.

By the way, awarding long term sole source business does not mean "giving away the store" to the suppliers or failing to push for cost reductions as aggressively as in the past. To the contrary, in exchange for long term business, vendors are going to have to be willing to commit to such practices as volume-based experience curves, open books and customer assistance programs to find ways to mutually reduce costs. A valuable approach which Toyota employs is to partially base the award for future business on the vendor's past record of submitted cost reductions. This is a particularly valuable approach in that a strong incentive is provided to the vendor to constantly identify and pass on cost reductions.

Beyond the longer term rituals involving contracts or business agreements there are aspects of JIT purchasing which stimulate close relationships on a daily basis. Simply stated, the "buffer" inventories which we used to buffer ourselves from the vendor's process are gone; we have no choice but to get involved and stay involved. Consider the dilemma faced by a buyer in a JIT environment who decides to drop a vendor because a competitor's product is a few pennies cheaper. In the past, we could pull the tooling, shift it to the competitor and live off our inventory while the new vendor got up and running. But how is this accomplished in a zero inventory environment? The answer is that it can't be, at least not without a great deal of pain and effort. In a JIT environment, the buyer has to get out and solve the real problem.

THE ROLE OF THE BUYER

This leads us into the broad area of the buyer's new JIT role. As pointed out above, the absence of inventory forces the buyer to be on his toes and aware of what is happening at the vendor. He needs to make many more visits, even becoming as "familiar as the furniture" in the suppliers offices. The buyer will undoubtedly be more involved in vendor surveys and periodic reviews of vendor capacity and vendor JIT or SPC Programs.

Many companies are now looking for buyers to do much more than simply working "business" issues while expensive engineers work the process control reviews and new proposed engineering changes. For example, at Harley-Davidson buyers must also be engineers and must work all of the technical and quality problems as well as more traditional issues.

But, if the buyer is spending all his time managing the vendor and working vendor relations, then how is he going to find time to process all those change orders and all those debit memos? The answer is that these activities are gone! MRP-driven forecasting and material-handler managed releasing relieves the buyer from his old order placing chores. And quality at the source provided by vendor process control programs relieve the buyer of having to deal with rejected material and the material review board.

OVERCOMING INERTIA - IMPLEMENTING JIT AND JIT PURCHASING

A senior manufacturing executive was once overheard to say "Well, if JIT is so great, why aren't we already doing it?" This is the same attitude that greeted the pioneers in automobiles and aviation. Even though today we live in a technological age, and most of us readily accept new advancements in cartridge disk players, video cassette recorders, etc., JIT is an entirely different category, much more difficult to understand and with many more profound implications. The JIT true believer has to be patient, persistent and understanding particularly if his firm is not in an obvious crisis and not desperately searching for new ideas.

In J. M. Juran's book, *Managerial Breakthrough*, the author argues that there really are three classes of people which fit into a bell curve (Figure 7).

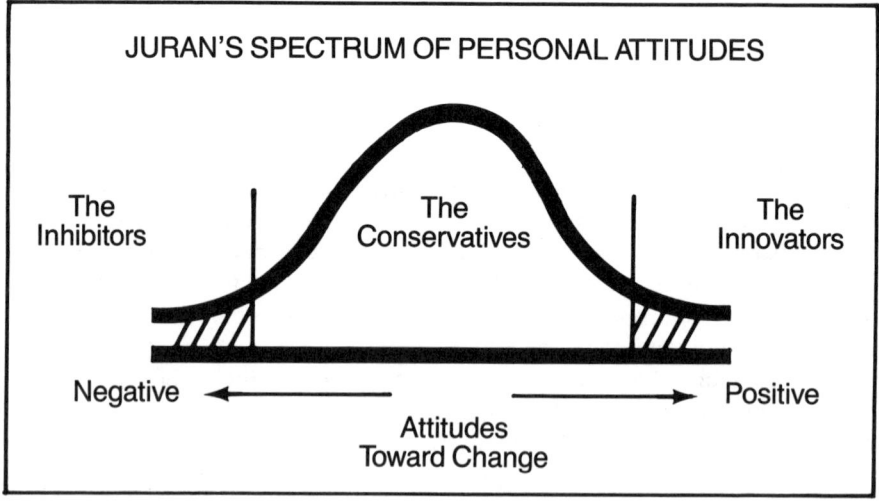

Figure 7. *Juran's spectrum of personal attitudes.*

1. A little band of Innovators. They exhibit an itch for change.
2. A great center of Conservatives. They are neither for nor against change. They are just for results.
3. A little band of Inhibitors. They are negative on ideas for change.

Inhibitors may resist any logical argument to implement JIT, possibly because they have publicly staked out contrary positions and would stand to "lose face" or perhaps their views have become part of a personal religion. The inhibitor may be totally immune from persuasion through logical reasoning.

Conservatives really have a "show me" attitude. They are seldom the first to try a new concept but will adopt new ideas if the results are clearly promising. Pilot JIT Programs should be particularly effective in convincing Conservatives.

Innovators frequently consist of those who have been part of successful efforts in the past to improve on the status quo. They have seen innovation succeed; they know it can succeed again.

If this analysis of managerial attitudes is somewhat correct, then a reasonable approach for implementing a JIT program would be to:

1. Identilfy and form alliances with innovators.
2. Educate and inform the conservatives.
3. Circumvent the inhibitors.

The key approach to introducing JIT purchasing, knowing that conservatives are the most numerous category, is to focus on education and awareness.

CONCLUSION

I think we can see that the advent of JIT zero inventory operating practices have created a whole new ballgame in purchasing. JIT isn't just a rehash of old ideas in a new envelope, but represents a fundamentally revolutionary way of doing business - a roadmap for continuous improvement.

Key to the success of JIT is the emphasis on specific methods as goals rather than numbers. The methods are based on an understanding of human nature which make them appealing and natural.

In addition, I think we've seen that JIT purchasing is a truly comprehensive strategy which is bringing about a long overdue strengthening of the bonds between customer and supplier. JIT purchasing clearly addresses quality and delivery in a very impressive way, and these improvements in quality, inventory and productivity lead inevitably to major improvements in cost and leadtime. Just-in-time purchasing is a key program that can strengthen the competitive position of your firm and American manufacturing.

Reprinted with permission, American Production and Inventory Control Society, Inc., "The New Ballgame: JIT Purchasing" by Bob Parmelee, *Twenty-ninth Annual International Conference Proceedings*, October 1986, pp. 498-503.

21

Cost Management Impact of JIT

Robert McIlhattan
Alan Anderson

In recent years, American manufacturers have been responding to increased competitive pressures by adopting new approaches to their manufacturing processes. One of the more celebrated approaches to improving manufacturing that has been applied by many organizations has been just-in-time (JIT). The underlying philosophies of JIT are focused on making the manufacturing process more efficient and productive. JIT is reshaping the physical nature of the production environment and changing both the behavioral patterns of production costs and the means with which financial executives must measure and control these costs. The purpose of this chapter is to identify some of the changes that are being applied by financial executives to traditional cost management systems to make them better management tools, consistent with the JIT philosophy.

Each of the changes referenced have been implemented by organizations that have had success with applying the JIT philosophies to their business and have realized that changes were necessary to their cost management systems in order to enable them to keep pace with the process changes resultant through the adoption of JIT.

Some of the areas of cost management that the JIT philosophies will significantly impact include the following:

- The identification of cost drivers.
- The number of product cost elements.

- The application of product costs.
- The nature of performance measures.

Each of these areas of impact will be outlined more thoroughly throughout this article. Cost management changes and solutions referenced are examples of how specific JIT organizations enhanced their cost management systems to complement JIT methods. While specific solutions varied from one organization to another, common elements of how change was initiated existed in each of the organizations that successfully aligned their cost management systems and manufacturing processes. These elements are referenced in the conclusion of this chapter as a means of helping other financial executives initiate similar changes within their organizations.

COST ACCOUNTING VERSUS COST MANAGEMENT

Before outlining some of the more significant impacts of JIT on cost management, a brief statement is necessary regarding the difference between cost accounting and cost management.

Cost accounting focuses on the process and procedures necessary to summarize the costs of an activity with the purpose of reporting and/or booking the results to the financial statements. Typically, the primary focus of a cost accounting system deals with inventory valuation, journal vouchering, and in the case of standard cost systems, variance reporting.

Cost management, on the other hand, deals with the "management" of cost, whether or not the cost has direct impact on inventory or the financial statements. As such, cost management has a broader definition than cost accounting and, as I will attempt to illustrate throughout this chapter, is consistent with the JIT philosophy of a total company-wide commitment to the continual reduction of cost, regardless of whether or not that cost can be measured in purely financial terms. This is also supportive of the JIT philosophy of cost avoidance versus cost control or reporting.

It has been this concept of total cost management that has helped lead financial executives toward many of the changes outlined in greater depth below.

THE IDENTIFICATION OF COST DRIVERS

Perhaps the greatest impact of JIT on an organization and the cost management system is that it focuses management's attention on *non-value added process*. A non-value added process is defined as any activity or procedure that is performed within a company that does not add value to a product. An example may help clarify this definition:

Assume that the lead time associated with manufacturing a saleable product is comprised of the following general steps:

- Process time—the amount of time that a product is actually being worked on.
- Inspection time—the amount of time spent either assuring that the product is of high quality or actually spent reworking the product to an acceptable quality level.
- Move time—the time spent moving the product from one location to another.
- Queue time—the amount of time the product waits before being processed, or moved, or inspected, or whatever.
- Stock time—the amount of time a product spends in stock before further processing or shipment.

Of these five steps, only process time actually adds value to the product. All other activities—Inspection Time, Move Time, Queue Time, and Stock Time— add cost but no value to the product and therefore are deemed as "non-value added" processes within the JIT philosophy.

In many organizations, the amount of process time is much less than 10% of the total manufacturing lead time and cost (costs associated with longer lead times include shortages, obsolescence, and expediting) associated with manufacturing a saleable item. Therefore, over 90% of the manufacturing lead time associated with a product adds cost, but no value to the product. It is this promise that leads to the JIT philosophy that reducing lead time will reduce total cost.

In order to assist in this process, financial executives in JIT environments have begun to identify the causes for the time and cost associated with the non-value added elements of manufacturing a process.

The key impact on traditional cost accounting is that cost management systems now need to identify the cause of costs, i.e., the "cost drivers," in addition to capturing the resultant costs.

Some of the possible cost drivers analyzed are listed in Table 1 below:

TABLE 1 – Potential Cost Drivers

Number of labor transactions	Number of accessories
Number of material moves	Number of vendors
Number of total part numbers	Number of units scrapped
Number of parts received in a month	Number of engineering change
Number of part numbers in an	notices
average product	Number of process changes
Number of products	Number of units reworked
Average number of options	Number of direct labor employees
Number of schedule changes	Number of new parts introduced

In all cases, the organizations determined that there was a direct correlation between the number of transactions and the cost of production. Additionally, Hewlett-Packard determined that many of their costs were a direct function of: the number of vendors used, the number of engineering changes to their product, and the total number of part numbers they maintained.

In few cases is there a direct correlation between direct labor head count and total production costs.

By refocusing the cost management system to identify the true driving force behind non-value added activities, the financial executives were able to assist the manufacturing managers in eliminating the product design and manufacturing process inefficiencies that were at the root of the product cost issues.

Product designs were simplified reducing engineering changes and part numbers, which correspondingly reduced financial problems associated with excess stock, obsolescence, storages, rework, and other associated costs. Vendors were reduced, improving quality and delivery schedules, and reporting transactions were either eliminated completely (in the case of direct labor) or reduced to an obsolete minimum, thereby eliminating support costs for clerical activities associated with transaction process, error correction, waiting time, and moving time.

Through the identification of the costs associated with non-value added activities, financial executives in each of these organizations have been able to help identify the true "drivers" of these activities and costs leading to their reduction or elimination.

THE NUMBER OF PRODUCT COST ELEMENTS

One of the other impacts that JIT is having on cost management systems is the reduction in the number of cost elements for a product. The philosophies of

JIT, while applicable to virtually any industry or process, have had their greatest successes in the industries that, because of the nature of their products and processes, have adopted standard cost systems.

Most traditional standard cost systems maintain standard cost elements for material, direct labor, and manufacturing overhead. More sophisticated standard cost systems many times maintained more than these elements. As manager's needs for better cost information increased, overhead costs were typically broken into more finite elements in order to better control production costs. Standard product cost elements associated with variable overhead, fixed overhead, set-up, material acquisition, energy, direct labor overhead, and others were added to cost systems in an effort to obtain better visibility and control over product related costs.

However, one of the primary philosophies of JIT is to identify the cost drivers associated with production costs. Once identified, the concept of striving for continual improvement in the reduction of product cost through design and process improvements on a daily basis eliminates the need to define multiple cost elements.

For these organizations, the JIT philosophies have helped them recognize that the issue is the elimination and prevention of costs, not simply the reporting of cost elements. The organizational acceptance and awareness that any cost, regardless of its nature, should be reduced, has focused attention that the design and process improvements necessary to successfully implement JIT will reduce cost through the enhancements themselves. The reduction of cost elements help people focus on total product cost as opposed to individual elements. Additionally, the reduction of cost elements further reduces the support costs associated with their reporting, calculation, maintenance and control.

It should be noted that while the number of product cost elements defined within the cost systems for JIT organizations declined, they all retained their standard cost systems. In fact, IBM is in the process of converting some facilities from a weighted average actual cost system to a standard cost system. The application of the standard cost system changed in that it was no longer used as widely to measure performance (as we will see later in this article). However, standard cost systems are still important as a tool for valuing inventory and cost of sales and as a tool to estimate potential future costs associated with design and/or process changes. Therefore, an additional impact of JIT is that standards are used to a much greater extent as a tool to prevent costs before they arise as opposed to report against once they have been incurred. Again, fewer cost elements will suffice in order to meet their purpose.

THE APPLICATION OF PRODUCT COST

As stated, one of the key characteristics of JIT is the adoption of manufacturing cells dedicated to the production of single or similar products or major components. In addition to the primary objective of the reduction of manufacturing lead time, manufacturing cells also change the nature of product costs and introduce alternative methods of applying production costs to specific products flowing through each cell.

The vast majority of traditional cost accounting systems in place today apply indirect manufacturing costs to products based upon either the direct labor hours or dollars charged to a specific product. A JIT environment challenges this practice in two significant areas:

1. The vehicle used to charge and collect labor hours (or dollars) to a specific product in most traditional environments is the factory work order. As individuals work on specific jobs, they change their time to the factory work order which is associated with the item being manufactured. Costs are therefore accumulated as the factory work order travels through the product process. Within a JIT environment, there may be no factory work orders. Daily production schedules are provided for each cell and typically only finished items are reported by the cell over the course of the day. No detail reporting is performed (which again is consistent with the philosophy of reduction of transactions and lead time). Therefore, the total of all related costs are applied to the day's production, not individual jobs and tasks.

2. In a JIT environment, direct labor may not have a correlation to other manufacturing costs and, as previously stated, is usually included in the total conversion costs. Within a JIT environment, alternative methods of applying cost to a product may be more appropriate.

    For example, many JIT users apply total conversion cost based on velocity through a manufacturing cell. Velocity is based on the theoretical number of units that can be produced within a cell over a given period. Theoretical capacity is used because it is consistent with the JIT philosophy of continual improvement towards perfection with no allowance for inefficiency or downtime. Based on velocity, a cost per hour is computed for a given cell. Therefore, a day's production is costed simply by multiplying the number of units produced by the cost associated with the hours required to produce that day's production. It does not matter whether the hours are direct labor, set-up, queue, or machine hours, the concept is that "time is money" and that the longer it takes to produce something, the more it will cost.

It should be noted that the other application methods do exist depending on the nature of the manufacturing cell process. These may include material usage, equipment costs, or more imaginative items identified as true cost drivers such as a number of transactions, quality, or number of engineering change orders.

A second major impact of JIT on the area of applying product costs is the increase in the amount of production cost that can be directly applied to a product. This phenomenon is a function of the adoption of manufacturing cells and the dedication of those cells to singular or similar products. Table 2 helps illustrate this point.

Within a JIT environment, a fundamental goal is the reduction of total product cost. In order for the cost management system to measure success in this area, allocations must be eliminated to the greatest extent possible. As most financial executives are well aware, the greater the degree of allocations, the less reliable (or accepted) the information is for decision making purposes.

As illustrated in Table 2, JIT helps eliminate allocations through the implementation of manufacturing cells dedicated to singular or similar product production. However, many JIT organizations including Caterpillar, Harley Davidson, Omark, and IBM have begun to adopt new cost management methods of associating total production costs, including support function costs, directly to products in an effort to reduce allocations and increase cost information reliability and responsibility, i.e., for "ownership" of product costs.

The benefit of enhancing cost management systems to be more compatible with the philosophies of JIT and resultant process changes can be seen through these reactions by controllers who have adopted direct charging concepts:

- "Improves product manager's ownership of expense."
- "Improves accuracy of product cost."
- "Improves expense information to manage product cost: increases visibility and awareness of expense items."
- "Allows for flexibility in changing environment, improved sourcing decisions, and competitive analysis."
- "Allows for cost reductions and improves competitiveness."

TABLE 2 – Direct Versus Indirect Costs

	Traditional Environment	J-I-T Environment
Direct labor	Direct	Direct
Material handling	Indirect	Direct
Repairs and maintenance	Direct	Direct
Energy	Indirect	Direct
Operating supplies	Indirect	Direct
Supervision	Indirect	Direct
Production support services	Indirect	Largely direct
Building occupancy	Indirect	Indirect
Insurances and taxes	Indirect	Indirect
Depreciation	Indirect	Direct

THE NATURE OF PERFORMANCE MEASURES

As organizations begin to adopt a company-wide commitment to total cost management, the performance measurements used to monitor improvement and motivate personnel begin to change.

Traditional measures that are commonplace in many cost accounting systems are not appropriate within the JIT philosophy of cost management. In fact, in some cases they may encourage actions that are contrary to the spirit of JIT. Four such examples are:

- Direct labor efficiency
- Direct labor utilization
- Direct labor productivity
- Machine utilization

These measurements are inappropriate for the following reasons:

1. They all promote building inventory beyond what is needed in the immediate time frame.
2. Emphasizing performance to standards gives priority to output, at the expense of quality. Relatively few companies even adjust results to reflect bad parts. Utilizing standards for performance measurements can be somewhat limiting relative to continuous improvement. Once standards are attained, people usually feel that they have "arrived."
3. Direct labor in the majority of manufacturers accounts for only between 5%-15% of total product cost. Traditional cost managers have run with very tight direct labor control and relatively loose overhead control. Frequently, direct labor head count reductions have been more than offset by overhead increases.
4. Using machine utilization is similarly inappropriate because it encourages results in building inventory ahead of needs. Focusing on this measurement has frequently resulted in utilizing expensive equipment and sometimes entire plants around the clock thinking this would maximize ROI. The fact is that under this scenario, virtually no time is allowed for preventative maintenance; equipment is run flat out until it breaks down. When it does break down, there is considerable disruption that ripples throughout manufacturing. This results in unnecessary costs and, in fact, reduction in ROI instead of its maximization.

Table 3 below highlights some performance measures that may be appropriate for a cost management system that is consistent with the JIT philosophies.

TABLE 3[1] – Performance Measures Traditional Versus JIT

Traditional	JIT
• Direct Labor	• Total Head Count Productivity
• Efficiency	• Output—Total Head count
• Utilization	(direct, indirect,
• Productivity	administrative personnel)
• Machine Utilization	• Return on Net Assets
• Inventory Turnover or Months on Hand	• Days of Inventory
• Cost Variances	• Product Cost, especially relative to competitors' costs
• Individual Incentives	• Group Incentives
• Performance to Schedule	• Customer Service
• Promotion based on *seniority*	• Promotion based upon increased knowledge and capability
	• Ideas generated
	• Ideas implemented
	• Lead time by product/product family
	• Setup reduction
	• Number of customer complaints
	• Response time to customer feedback
	• Machine availability
	• Cost of Quality

(Left margin letters: C H A N G E D   N E W)

[1] "The Spirit of Manufacturing Excellence," September, 1986, Ernest C. Huge, Alan D. Anderson.

Table 3 lists new JIT performance measures that may be appropriate for inclusion in a cost management system within a JIT environment. Specific performance measures are dependent on the unique business environment and process being managed. For example, Harley Davidson has adopted the following ten measurements to assess their manufacturing effectiveness:

1. Schedule Attainment
2. Manning Requirements

3. Conversion Costs
4. Overtime Requirements*
5. Inventory Levels
6. Material Cost Variance
7. Scrap/Rework
8. Manufacturing Cycle Time*
9. Quality Level
10. Productivity Improvements

* Measures of Flexibility

Conversely, a different Fortune 100 company has adopted the seven measures and goals listed below for measuring their effectiveness in an integrated circuit facility:

Measure	Goal
1. Unit Cost—cell $/hr—theoretical units/hr	$1.00
2. Cycle Time—Through cell with no downtime	3 days
3. On time delivery	100%
4. Quality	0 defects
5. Linearity—Ability to meet daily schedule	0% deviation
6. Inventory turns	75
7. Scrap	0

While the measures are different for these two organizations there are similarities. In both cases, non-financial indicators were used to measure performance as part of the cost management system. This is consistent with the identification of true cost drivers outlined earlier in this article and with the JIT focuses on quality and lead times.

Both measurement systems were proposed and maintained by the financial executive in these organizations. In each case, the financial executives took the initiative to propose more effective ways of monitoring performance and reducing overall cost and worked closely with manufacturing to refine these proposals and establish a "team" approach to performance measurement.

Both performance measurement systems were simplified from their predecessor traditional systems. Simple, easy to understand measures were implemented so that everyone on the organization could understand their intent and interrupt their results. Additionally, measurement results were posted in the factory so that everyone in the organization could be more aware and in tune to company improvements in these areas.

FOCUS ON SIMPLIFICATION

In each of the major areas of cost management outlined above, specific company solutions differ. The impact of true cost drivers within organizations will differ from one company to another; the number of product cost elements necessary to properly report and control costs may differ between companies; the ability to directly apply costs to product and the types of cost being applied to products will change based on organizational differences and products being manufactured; and, as we have seen, performance measurements differ between companies.

However, without exception, simplification was one of the primary objectives of every financial executive that enhanced his cost management system for JIT. The focus on the two or three true cost drivers within an organization, the reduction in standard product cost elements, the reduction of allocations and the establishment of fewer, easy to understand, key performance measures are all examples of the simplification in cost management system.

JIT strives for design and process simplification, because the proponents of JIT recognize that with simplification comes better management. Better management allows for better quality, better service, and less cost. The same principles are true for cost management systems. Traditional cost accounting systems have a tendency to be very complex, with many transactions and reporting of data. Simplification of this process enables the cost accounting system to be used by all individuals within an organization, transforming the cost "accounting" system into a cost "management" system that can be used to support and drive the company toward the manufacturing excellence promoted by JIT.

HOW TO INITIATE CHANGE

The management accountant can learn to think "Just-In-Time accounting" to keep in step with desirable changes in management style. It is not possible to redo a whole accounting system and/or philosophy in one quantum leap. However, the management accountant must begin to adjust to changing technology, new management philosophies, the changing demands of information. Nothing is more discouraging than to hear financial people say "we can't accommodate that change because of our accounting system." Accountants would do well to become familiar with the latest management philosophies and advances in technology so that they can help their firms stay in step and "Just-In-Time" as well.[2]

Each of the organizations cited in this article were able to heed the above advice, but their approaches were different and the time frames over which change was initiated were different. Each organization had its own series of

[2] "Just-In-Time"—The Accounting Implications Ragnor Seglund and Santiago Ibarreche *Management Accounting* August 1984.

successes and failures before focusing on the cost management direction necessary to support JIT in its environment.

However, in listening to each of the financial executives relate their story regarding the cost management changes they made, a set of common steps emerged as to how change occurred. The major steps taken are outlined below:

1. The perception of the accounting function needed to be changed. In each company, the accounting function was perceived as a control function, a function that reported when things were bad. Each company had to change this perception (or reality) from one of control of operations to cooperation with operations to reduce costs. The financial executives took the time and effort necessary to understand what JIT was and joined the operating professionals in an effort to implement the philosophies in their companies.

2. The financial executives became an integral part of the manufacturing and engineering project teams. In each organization, one of the first steps was for a cost manager to sit in on all product design meetings, manufacturing engineering meetings, and production control and planning meetings on a regular basis. The step is critical in that it accomplished these essential objectives: a) it educated the accounting personnel as to the key elements of the engineering and manufacturing process; b) it raised the awareness of the operating personnel as to the implications of their actions on total cost, raising the total cost awareness throughout the company; and c) perhaps most importantly, it served as the basis for eliminating the communication barriers between accounting and operations. Each of these objectives helped pave the way for obtaining the interfunctional cooperation necessary for the JIT philosophy to timely work in a company.

3. True cost drivers were identified. Once accounting became more aware of operations and once operating personnel became more aware of the cost implications of their actions, those processes or engineering issues that truly determined cost could be identified, seregated, and attacked by all management.

4. Each company reanalyzed its application of costs to products and implemented a higher level of direct charging. This elimination of allocations gave a clearer picture of true product costs and raised the responsibility of costs to managers. Again, as with other steps, this one is predicted on accounting's understanding of the engineering and production process.

5. Performance measures were altered to help motivate the entire operating group toward positive results. Individual performance measures were reduced to encourage a team concept and further team cooperation. Personnel were trained and informed as to the meaning of performance measures and were rewarded for improvements in operating results.

6. All systems were simplified. Traditional accounting systems were analyzed and redesigned to be more reflective of the simplification principles of JIT. Information flows and reports were simplified to focus on the critical processes and measures.

The simplification increased awareness and, as stated, allowed all of management to focus on only a few issues which greatly increased the benefits associated with their actions. Additionally, the simplification of the system helped provide flexibility. As process changes took place as a result of JIT advances, cost drivers changed, and performance measurements changed. Simplification of the cost management system enabled the financial executive to change reports and information flows to reflect the new manufacturing process quickly and inexpensively, maintaining their synergy between the accounting system and the manufacturing process.

Each of these steps outlining how financial executives can enhance their cost systems to be more supportive of the JIT philosophy can be initiated by virtually any organization. However, the first step is for the financial executive himself to adopt the primary JIT principle of continual improvement within the organization and cost management process.

Reprinted with permission, American Production and Inventory Control Society, Inc. "Cost Management Impact of JIT" by Robert McIlhattan and Alan Anderson, *Thirtieth Annual International Conference Proceedings*, 1987, pp. 614-617.

22

Management Issues Just-In-Time and Optimized Production Technology

James A. Leahy

This chapter will document some of the techniques and analyze the operational impact of reduced inventory systems. In addition, it will frame the managerial requirements for success in implementing such a system.

Essentially, a reduced inventory manufacturing environment represents a significant change in the traditional style of managing a manufacturing organization. Examples of such systems are "just-in-time (JIT) production and optimized production technology (OPT). I will present the important issues that management must address for any reduced inventory implementation to succeed in delivering productivity benefits.

JIT: TOYOTA'S PRODUCTION SYSTEM

Toyota Motor Company developed and pioneered the JIT approach to manufacturing management. Today, this production system is the model for nearly all major Japanese manufacturing operations. Toyota's philosophy is that there is no inconsistency between "highest quality" and "lowest cost" products. The operational goal is simple: reduce cost by eliminating waste. The system achieves this goal by using the absolute minimum amount of equipment, material, parts, and labor required for production. Any resources beyond that represents, in effect, increased cost.

Two rules are basic to the Toyota approach: 1) allowing problems in the manufacturing process to surface; and 2) conforming to requirements at each operation.

Toyota's Operating Techniques

The four operating techniques of the Toyota production system are Jikoda, JIT, Production Leveling, and Hyo-Jyn-Sagyo. These interconnected techniques represent a systematic approach to achieving the system's end benefits of productivity and quality improvements.

Jidoka

This technique consists of four practices: 1) concentrating on basics; 2) using self-stop (quality at the source); 3) managing by sight; and 4) emphasizing human relations. In its simplest form, Jikoda is the essence of increasing efficiency and use at each operation. By following these practices, there is a marked increase in worker job satisfaction and product quality.

JIT

The production and inventory control system at Toyota allows parts and processes to come together "just-in-time" to produce the specific products required in the exact quantities demanded. This system disregards conventional production philosophy concerning lot production and conveyance. To implement JIT, Toyota uses three production principles: 1) withdrawal by subsequent processes; 2) model leveling, and 3) a Kanban system. In fact, the Toyota system is often mistakenly labeled KANBAN, as if this were the key element of their production system. In reality, however, KANBANS can be thought of as "factory money," the medium of exchange required to buy parts. KANBAN is a communication mechanism specifying the necessary units, quantities, and timing required to connect parts and processes throughout the system.

In theory, a JIT system using KANBANS can only be achieved if there is an uninterrupted flow of defect-free parts throughout the system. Thus, quality is the backbone of the system—from the purchase of parts right through to the finished product. This emphasis on quality indicates the importance of the Jikoda practice of "self-stop" (quality control at the source) in implementing effective JIT production.

Production Leveling

This technique attempts to reach the goal of today's production, which is yesterday's sales. In such a way, finished goods inventory build-up is minimized.

Production leveling uses frequent small inventory rebalances to follow the marketplace demand curve. This Toyota technique contrasts sharply with the typical United States approach of supplying products to the marketplace. From Toyota's view, the true strength of a company is how close production can match sales.

Hyo-Jyun-Sagyo

The fourth operating technique used in the Toyota production system identifies opportunities for productivity improvements. For an additional operation, the objective is to increase the value added portion of the job by eliminating waste from an individual worker's activities. The key to this process is operator participation, as each worker is the expert at his or her job and, thus, the best source for improvement. In theroy, Hyo-Jyun-Sagyo is similar to time and motion studies. The key distinction is that Hyo-Jyun-Sagyo is operator-driven. Getting employees involved in identifying improvement opportunities clearly increases job satisfaction.

Implementing Toyota's Operating Techniques

What does it take to implement the operating techniques that Toyota has used to improve quality and lower cost? First, managers should not emphasize Hyo-Jyun-Sagyo. This technique requires ideal labor-management cooperation. Often, this condition has to be developed over time. Secondly, managers should also remember that production leveling is a strategy which any business can initiate. On the other hand, the Jidoka and JIT techniques require changes at the operational level and can be more difficult to implement.

Initially, to implement Toyota's techniques, a company must focus on the basic Jidoka principles, because JIT cannot be implemented without quality parts. As shown in Figure 1, surfacing problems and conforming to requirements, inherent in Toyota's system, result in two main ages of impact: 1) a productivity dip; and 2) the need to locate the "root cause" of a problem. Each impact area, in turn, leads to other issues, eventually leading to a common subject that will be addressed.

Productivity Dip

The first impact of surfacing problems lies in a dip in production. Through the Jidoka principle of self-stop, workers stop the line when necessary to ensure product quality. These stops will have an unfavorable impact on productivity, because the number of units produced will be less. Consequently, a productivity dip results from the problems surfacing in the production process. The underlying idea is that it is worthwhile to accept the pains of this dip in

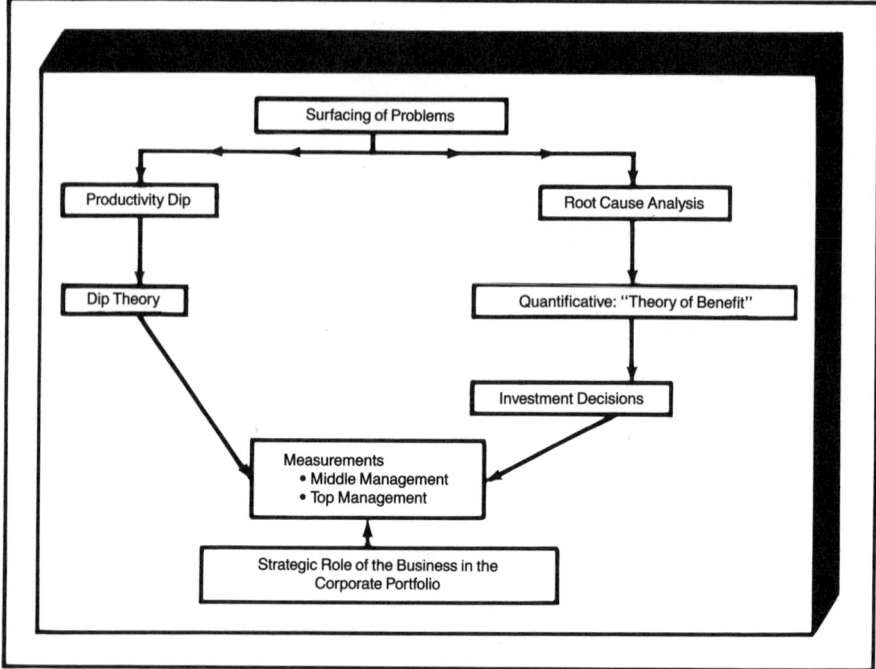

Figure 1.

productivity. To avoid line stops is less efficient in the long haul than immediately stopping the line and making improvements. Management must realize that the inevitable result of self-stop is lost output and a negative effect on the measurements of individual foremen, operating managers, and plant managers.

Measurement: Middle Management

Production volume is the main priority in most manufacturing operations. "Cost" is the second priority, while quality is a junior partner to the two. It's not that manufacturing management does not want good quality and better quality—they certainly do. It's simply that when problems arise there's no question what the decision will be—measurements dictate behavior.

Decisions must be made during the productivity dip in the initial stages of implementing Toyota's operating techniques. "Today's production . . .", "hours per equivalent unit . . .", "%UDL to ADL . . ." are the measures that dominate the daily lives of manufacturing personnel. Measurements that bias decisions more toward output and cost reductions than toward quality attainment and improvements are inconsistent with the goals of the Toyota Production System. If manufacturing produces a better quality product, the value of that product will be greater in the marketplace. Therefore, performance measurements should be based on quality.

Another measurement issue is related to the budgeting process. Some performance measures motivate supervisors to budget for as many operators as possible even though the effect is a build-in operating inefficiency. In the Toyota system, measurements are geared toward quality. Moreover, action is deemed more important than measurement. In Many United States companies, there is nearly obsessive emphasis on quantified (often unnecessary) detail, leaving managers at all levels spending an inordinate amount of time reviewing insignificant performance measurements. At Toyota, information—similar to any other resource—is provided JIT on the shop chalkboards, not on backroom reams of white paper. For their part, manufacturing management spends their time on the shop floor helping workers solve problems, rather than staying awake during a steady stream of endless meetings.

Root Cause Analysis

The second main impact (see Figure 1) of the surfacing of problems is the root-cause analysis required to solve the problem. Toyota calls its scientific approach "5W1H." 5W is five "whys." Repeating why five times reveals the true cause of a problem and, thus, how to solve it (1H = How). Finding this root cause is essential, otherwise effective action cannot be taken.

Many causes may be at the root of a surfaced problem. Two possible sources are current work methods and the condition of equipment. To simplify the point, I will focus on the equipment root cause. Suppose a problem can be resolved by repairing or replacing the presses, so investment is required. Two factors will determine whether the problem can be resolved; the quantification process and the basis for investment decisions.

Quantification - A Theory Of Benefit

Quantifying the cost of an investment—replacing a press, for example—is relatively easy. Quantifying the value of quality improvements that result from the investment is more difficult. All benefits, not just the tangible ones, should be considered. An example of a less tangible benefit would be to assess the effect of an investment on brand image and end-user commitment—how the business will benefit from investing in any item which reduces in-warranty service calls.

Another intangible benefit is the effect of the change on employees. Investment demonstrates to all employees (both labor and management) that the business intends to act forcibly to improve product quality. This investment commitment by the company, reasonably speaking, leads to a more positive employee attitude which, in turn, should also lead to better product quality.

Going back to the example of replacing a press (the root cause), assume that all the benefits of such an investment are identified (see Figure 2 for a summary of these benefit factors). Now, a new press has to be purchased. This situation leads to the next two issues that are best viewed in tandem; the investment decision and measurements of top management.

Investment Decisions and Measurements of Top Management

Investment decisions are intricately related to the measurement of top management performance. Similar to foreman, top management has "numbers to make." The difference is that their measurement is "net income" rather than "units of production." Moreover, because these executives *have* reached high levels in the company, one thing is certain: they make their numbers. "Profit-oriented" on executive personnel forms translates to "I make the assigned numbers."

Yet, some decisions clearly have extended periods over which the benefit is accumulated. The issue is the incentive (or lack thereof) for a manager to pay (in terms of reduced current period net income) for such benefit. Any dip in productivity threatens to reduce the profit of the moment, which, in turn, jeopardizes the general manager's evaluation, incentive compensation, and career. In such a circumstance, the measurement promotes behavior inconsistent with the goals of the Toyota production system.

	Internal	External
More Tangible	• Eliminates maintenance of excess production capacity (people and equipment) • Reduces queue time • Reduces raw material and work-in-process inventory levels • Improves quality measurements • Reduces scrap and rework • Demonstrates management commitment to quality • Improves space utilization • Provides an operational discipline	• Decreases service call rates (in warranty) • Improves ability to serve • Increases end-user brand commitment • Increases price elasticity
Less Tangible	• Provides a more stable operation • Improves employee commitment • Increases employee job satisfaction	• Provides faster and better feedback from end-users • Decreases service call rates (out of warranty) • Improves brand image

Figure 2.

A larger problem than this "short-term orientation" is the resulting effect of employees morale. How can each individual become emotionally involved with his job (Toyota calls it creating an environment in which people can "work for a will") when the business leader makes decisions, not for long-term business benefit, but to attain numerical measurements on which that leader is evaluated? Such situations cause cynical reactions from both labor and management and contribute to a breakdown of trust and a general lack of team play.

Implications for Corporate Strategic Planning

Net income targets, on which top management performance is measured, are based on the corporate business strategy. Often, the analytic "portfolio management" approach is used to siphon resources from the perceived mature or declining businesses and to invest those resources in more growing areas. At first glance, this strategy appears to be basic, sound manipulation of resources— "financial synergy."

What is lacking is a strategy to generate "organizational synergy," to stimulate the ability of a business to retain its vitality and high morale, to sustain its entrepreneurial flavor. To do so requires allowing divisional businesses to reinvest most earnings in their own operations. Thus, change in corporate strategy may be required if the techniques of the Toyota production system are to be implemented.

Labor-Management Relations

Certainly, any business can implement some of the Jidoka principles and improve quality and productivity, to a certain extent. Full benefits from the system, however, can be achieved only through a positive labor-management relationship. Such a relationship can begin with an action plan that suggests some new approaches to traditional problem areas. As in many issues, creative thinking may simply mean that there's no particular reason in doing things the way they have always been done. Top management must take the lead in initiating this dialogue with labor on items that surface when the Jidoka principle is implemented. Properly handled, these exchanges could represent the start of a positive working relationship.

OPT: OPTIMIZED PRODUCTION TECHNOLOGY

Developed by Dr. Eli Goldratt, the OPT philosophy is that "the sum of local optimums is not equal to the global optimum." Similar to a basketball team, if each player tries to maximize his own statistics (measurements), overall team play will suffer. The same holds true for the diverse parts of a manufacturing organization.

> **Scheduling Observations**
>
> • Balance the flow-not capacity
>
> • Constraints determine non-bottleneck utilization
>
> • Activation is not equal to utilization
>
> • A lost bottleneck hour is a system hour
>
> • A saved non-bottleneck hour is an opportunity hour
>
> • Bottlenecks govern throughput and inventory
>
> • Transfer batch should not always equal process batch
>
> • Process batch should be variable, not fixed
>
> • Set the schedule by examining all constraints simultaneously

Figure 3.

Originally, OPT was merely a software scheduling system consisting of a series of nine general observations on scheduling (Figure 3) and three operational measures—throughput, inventory, and operating expense.

A key consideration is that one cannot balance a plant's capacity. From this observation comes two concepts: 1) the objective should be to balance flow of products through the facility; and 2) there are two types of resources in any facility: constraints in the flow—bottlenecks—and "nonbottlenecks." Therefore, operational goals are to simultaneously increase throughput while decreasing inventory and operating expense.

OPT's Operating Techniques

The methodology of OPT is, simplistically stated, a five-step procedure: 1) modeling the manufacturing product-process flow; 2) identifying bottlenecks; 3) verifying data on parts processed by bottlenecks; 4) scheduling these resources forward in time; and 5) back-scheduling the remaining resources based on the bottleneck schedule.

Implementing The OPT Management System

To implement the schedules based on the OPT concepts, performance measurements throughout the plant must be modified. Thus, the following example:

Capacity	200 hours	200 hours
Demand	200 hours	150 hours
Utilization	100%	75%
	X	Y

In this example, there are 200 hours of capacity available in the month, and the demand on each resource is 200 and 150 hours, respectively. (Obviously, resource X is a bottleneck.) Because the demand on resource Y (a nonbottleneck) is less than its capacity, the use of Y is 75%. However, if Y had material to process and was motivated to do so by the measurements imposed, the resource could be activated to 100%. The only effect on the total system, however, would be an increase in inventory in front of the bottleneck.

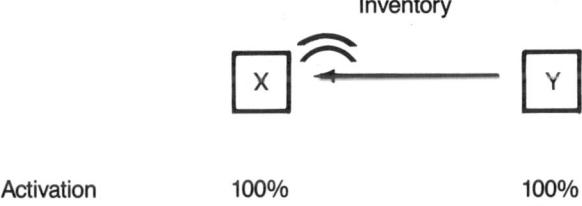

Activation	100%	100%

From a conceptual viewpoint, OPT distinguishes between use and activation. Instead of local efficiencies being the basis of measurement, adherence to the schedule is key. In this way, the Y resources (the majority in most plants) do not simply activate and build unneeded and costly inventory.

Implementing an OPT management system approach in a business requires an understanding of some key interfaces with three other areas: marketing, engineering, and financial control. I will address each area, in turn, from a conceptual view.

Marketing Interface

If a reliable schedule can be generated, a schedule based on the nine observations of Figure 3, then one result is a more predictable delivery schedule.

Estimating when the product will arrive at the end-user's doorstep is no longer needed. But the benefit goes further. What's a reliable due date worth in the marketplace? Clearly reliable delivery enhances brand image.

Moreover, the "sorting out" of resources into bottlenecks and nonbottlenecks soon unclutters the production area. With this usually comes reduced lead times. Furthermore, many businesses carry high levels of expensive finished goods inventory. Not only are predictable delivery and shorter lead times beneficial competitively, a business may well reassess its finished goods inventory stocking policy.

A final interface with marketing is pricing. If, for example, a company had excess resource capacity at every operation that produced a particular product, the cost of producing an incremental unit would be only the material costs and perhaps some energy costs, as well. Locating such products and pricing with this concept in mind can lead to a significant increase in the bottom line.

Engineering Interface

Distinguishing between bottleneck and nonbottleneck resources is important in prioritizing process improvement activity. Because bottlenecks control the throughput of the plant, improving the process there is of key importance. If it formerly took 10 minutes to process a part on a bottleneck and an improvement changed that to nine-minutes, it would be significantly increasing the production of that bottleneck resource. Consequently, the entire system benefits.

Another approach is to "offload" capacity from the bottleneck. By changing routings, certain jobs could be processed on Y resources, which, by definition, have idle capacity. By doing so, the throughput of the entire system, increases by the amount of the product rerouted while the level of inventory decreases by that same amount.

A third area requiring interfacing with engineering lies in the area of capital investment. If investment is being made for quality reasons, that is justifiable. Many investments, however, are being made on the basis of "cost reductions." But is that sound reasoning? What if this savings was calculated on a Y resource, one that has inherent idle time? If so, the investment will be adding capacity on a resource that already has capacity available. Consequently, the "savings" could be elusive and, despite the sharp penciled justification, will not make it to the bottom line.

Financial Control Interface

From the examples cited in the interfaces with marketing and engineering, clearly pricing and investment should be considered carefully. Thus, standards and set-up must be implemented.

Setting standards on the shop floor is usually guided by the fastest machine available. Again, assume that the machine in question is a bottleneck (X) with a time standard of 10 minutes/part. We have an older, reliable machine (a nonbottleneck) that can do the same job in 15 minutes/part. Would it be better to take material waiting in front of the X machine and process it on the Y machine? If so, would the foreman of that area be motivated to do so?

A second issue is set-up. Most companies desire to save set-ups everywhere. Isn't there much more benefit to having long runs on bottlenecks than on resources with extra capacity? In fact, on these nonbottlenecks, perhaps the batch size could be cut, thereby increasing the number of set-ups and be of benefit in balancing the flow? But how are we motivating and managing our people?

SUMMARY

JIT is similar to OPT in many respects. Although, JIT is a method of scheduling, it's really a set of diverse techniques to improve operating practices.

The OPT approach begins with some observations on scheduling. Based on these concepts, reliable schedules are generated. Yet, the impact is more greatly felt in functional areas distant from the shop floor. Decisions in the marketing, engineering, and financial control areas can affect "the best laid plans."

Both systems require a firm understanding of potential "productivity dips" and the issue of measurements seen both from the foreman's daily motivation to the rising-star general manager's career aspiration. Secondly, both systems surface unavoidable issues on pricing, capital expenditures, and financial control. It is no coincidence that Dr. Ohno, founder of the Toyota system, "does not allow" cost-accounting ideas on his shop floor.

Both systems take leadership, hard work, and significant reworking of how you run your business. The full potential from either system requires a steady belief in its eventual benefits. There's no panacea, yet by using good techniques and innovation tools, a forward-thinking manager can spark the resurgence of America's traditional national strength—its manufacturing base.

Reprinted from *1985 Fall Industrial Engineering Conference Proceedings.* Originally appeared in *Zero Inventory Philosophy and Practices Seminar Proceedings,* October 1984, pp. 334-335. Reprinted with permission, American Production and Inventory Control Society, Inc.

James A. Leahy

Part 3

Electronics Assembly Support Systems

Electronic assembly production environments have a number of support systems to make them function smoothly. For example, material handling systems provide the backbone of a network of production equipment and ensure that the right kind of material is presented in the right fashion at the right time. Likewise, a materials management system synchronizes the production activities at different stages of the factory floor and also triggers the coding of the required materials from vendors.

This section presents a collection of chapters on various aspects of electronics assembly. Hales (1985) presents a number of long-range facilities planning strategies and tactics for electronics industries. He stresses the importance of integrating the product and process engineering functions through careful facilities design. McKenna (1985) describes the use of a conceptual model to evaluate the capabilities of different PC assembly lines at Hewlett-Packard. This application resulted in a manufacturing system that processes more than 100 different types of printed circuit assemblies in an area of 12,000-square feet.

A 10-step planning procedure for developing and implementing material handling systems is presented by Cullinane (1984). The steps are based on an "Understand-Design-Build-Install" system development life cycle. Various material handling equipment alternatives for progressive-build electronics assembly are examined in the chapter by Orr, et al. (1985). In their discussion, they use various criteria that are critical in evaluating alternative solutions. In the next chapter, Daebler (1987) discusses in depth the flexibility offered by wire-guided automated guided vehicles (AGVs) and explains their applications in auto-insertion, soldering, and cleaning facilities.

Electronics assembly lines constantly face changes in production volumes due to the nature of the business. Chow (1986) of IBM presents a simple method to resolve such problems using a direct-access material handling system.

To support production, a successful kitting approach is highly essential. Conrad and Pukanic (1986) outline various kitting issues and identify different ways to meet kitting needs in electronics assembly.

Product testing is usually a major component of electronics assembly capital expense and is also the most time-consuming process. Garcia-Rubio (1987) of AT&T Bell Laboratories examines the major product testing problems in the industry and presents an overview of the short-term and long-term solutions.

Wilson (1985), in a case study on receiving inspection at Rockwell International Corporation, presents a comprehensive program involving supplies for simplifying and/or eliminating incoming inspection.

Hixon (1985) describes a 10-step improvement program for solving packaging problems. Following these steps may ensure product quality as perceived by the customers.

Warren (1986) addresses the implementation of MRP-II, a widely used materials management system. He discusses some common implementation problems and presents a case study of a company using MRP-II. Part 3 matrix summarizes the chapters presented in this section.

Part 3 Matrix

Chapter Number	Title	CONTENT CLASSIFICATION					ELECTRONICS ASSEMBLY PROCESSES							
		Model	Quanti-tative Tech-nique	Strategy/ Method-ology	Case Study	General Theory	Design	Receiv-ing & Store	Inspec-tion	Kit-ting	CP Assem-bly Mfg.	Equip. Assem-bly	Packing	Mat'l Hand-ling
23	Time has Come for Long-Range Planning of Facilities Strategies in Electronics Industries			X	X		X	X	X	X	X	X	X	X
24	Conceptual Model Aids in Development of Manufacturing System for PC Assembly Shop	X	X								X	X		
25	Developing a Material Handling Plan for Electronics Assembly			X			X							X
26	Material Handling Equipment Alternatives Examined for Progressive Build in Light Assembly Operations					X	X				X	X		X
27	Automatic Guided Vehicles Move PCBs Between Assembly Stations					X					X			
28	Design for Line Flexibility		X				X	X			X	X		X
29	Process Approach to Planning a Successful Kitting System is Outlined		X							X				
30	Key Problems and Solutions in Electronics Testing			X			X		X		X	X		
31	Electronics Parts Handling in Receiving Inspection		X		X			X						
32	Ten-Step Guide Provides an Orderly Procedure for Solving Packaging Problems			X									X	
33	How to Implement MRP II in the Electronics Industry					X	X	X	X	X	X	X	X	X

23

Time Has Come for Long-Range Planning of Facilities Strategies in Electronics Industries

H. Lee Hales

A few years ago, it was fashionable to say, "It can't be done," when discussing the need for long range facilities planning. Young electronics firms riding successive waves of microprocessor technology were particularly prone to the "What, me worry?" attitude. And many did experience profitable and rapid growth.

Except for the recession in 1981-82, all segments of the electronics industry have shown more than ten years of steady growth—in sharp contrast to the rest of our economy. As a result, electronics has emerged as one of our largest manufacturing industries—on a par with automotive, primary metals, aerospace and petrochemicals.

But lately, conditions have changed. The talk is of shake-outs, bankruptcies and accelerated moves to off-shore production. Historically healthy firms are showing losses. A number of recent start-ups have closed or been acquired and folded into larger firms.

Slowly, inexorably, the electronics industry is maturing. Yesterday's touted new technologies have become today's price-shopped commodities. Diverse product offerings have coalesced around various design standards. The resulting uniformity has shifted competitive attention to price and service, pulling the Japanese and other foreign producers into our domestic markets.

Electronics firms must now pay unprecedented attention to manufacturing. For big companies, production issues have become at least as important as product design and performance, or marketing savvy. Hand work and lab-bench approaches are rapidly giving way to refined and often automated techniques.

To remain competitive, some firms have radically altered their work flows and job designs. Others have simply abandoned much of their own production, relying instead on lower-cost sources in Mexico and the Far East.

With this new attention to cost and productivity comes a new agenda for facilities planning. In the early days the issue was simple—how fast could we find some more space? "Making do" was the order of the day. Office parks, spec warehouses, even vacant supermarkets were common facilities solutions. Some firms did not know how many leases or how much space they had, let alone how well it was being used.

Today the issues are consolidation, economies of scale and automation. Resolving them in a profitable manner requires strategic thinking and sound long-range plans.

CORPORATE STRATEGIES

Good strategies rest on a series of hierarchical decisions about goals, directions and resources. Corporate strategy decides which businesses will be pursued. The attractiveness of various markets is weighed against the competitive strengths and weaknesses of the firm.

Each current business is reviewed. Is it a keeper or a loser? Should it receive funds for reinvestment in facilities and equipment? Or should it be "harvested" or sold to provide cash for other, more attractive opportunities?

Individual business strategies focus on which products and markets to pursue, and how best to compete: will our growth come from acquisitions, or from internal product development? How should products and markets be assigned among divisions and units? What will be our emphasis—on price? On product performance? On delivery and service? Should engineering be centralized or decentralized? What about manufacturing and distribution?

The business profile shown in Figure 1 is a useful way to summarize both corporate and business strategies. Each of the firm's businesses, divisions or product lines is classified according to its competitive position and the maturity of the markets it serves. When this is done, certain natural business and investment strategies emerge. These are:

- *Grow and defend*: for well positioned businesses in new and mature markets.
- *Specialize*: for units that are not influential in the market as a whole.
- *Up or out*: for marginal operations.
- *Get out*: for nonviable, typically older businesses in older markets.

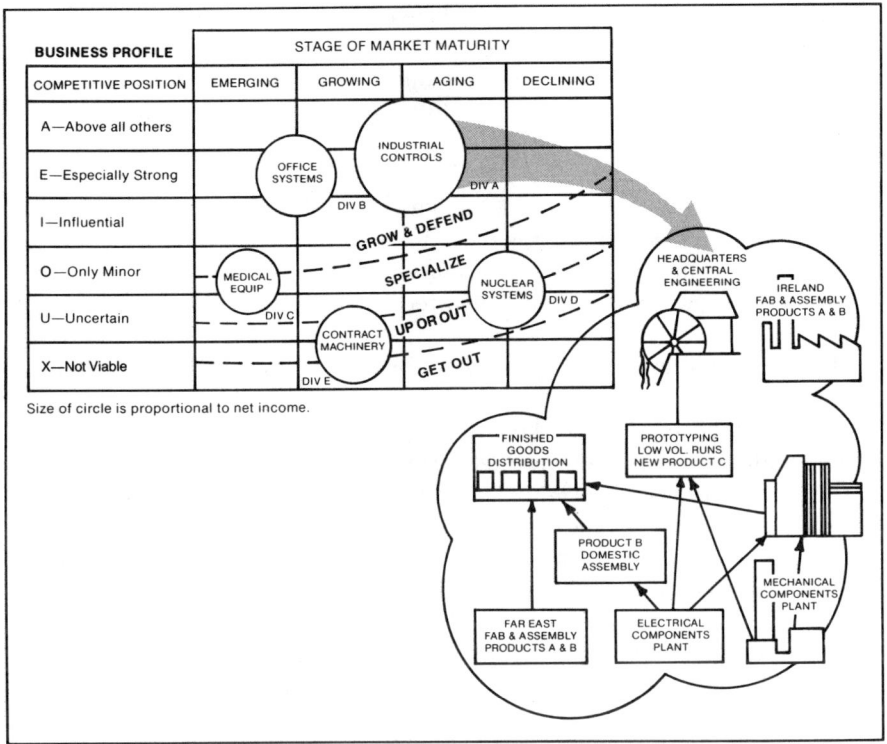

Figure 1. Portfolio of businesses for diversified firm.

In Figure 1, we see a highly diversified firm. As its industrial controls business matures, we would expect it to generate cash for reinvestment in medical equipment and office systems. Meanwhile, nuclear systems and contract machinery will be phased out if they cannot find profitable niches.

The business profile is constructed around a firm's "strategic business units." These can be defined in several ways.

For a company like Xerox, focused largely on the single market of office automation, the profile might show product lines such as large copiers, printers, electronic typewriters and computer systems. For a company like Apple Computer, focused largely on microcomputers, the profile might reflect customer markets such as office, school and home. For a semi-conductor firm like Intel, the profile might be technology-based: ECL, TTL, gate array, standard cell, etc.

The business profile and its natural investment strategies can be directly extended to provide *corporate-level facilities strategies*. Each of a firm's sites and plants can be classified along the same dimensions as the businesses they house. The result is a set of natural facilities strategies that are consistent with the business strategies above:

- *Expand and upgrade*: for facilities assigned to well positioned business units.
- *Selective improvement*: to strengthen units in a specialization mode.
- *Make do*: for "up or out" business units.
- *Sell or reassign*: for units being divested or phased out.

Ideally, as businesses age and decline, or lose their competitive advantage, their facilities will be reassigned to stronger units in emerging or growth modes. This assumes, of course, that the facilities are physically suitable and properly located with respect to their new occupants.

This kind of long-range *facilities management* is possible only with a corporate-level facilities and real estate function—one that is plugged into the ongoing strategic planning of the firm. In an industry where "real estate" often means the space on a circuit board, this approach will not arrive overnight.

Some of the largest and best managed electronics firms have long recognized the value of long-range facilities plans and investments. Over the past five years, IBM alone has spent over $13 billion on land, buildings and equipment.

Hewlett-Packard, Motorola, Texas Instruments and Digital Equipment all have strong facilities programs that receive top management's attention. But in many firms, especially younger ones, this attention stops with the terms of a lease, and major decisions are often delegated to divisional levels. As a result, many beautiful facades on offices and plants are hiding chaos within.

MANUFACTURING MATTERS MOST

Business-level strategies drive functional strategies in marketing, in finance and, of course, in manufacturing. As a rule, manufacturing strategy is the single most important influence on long-range facilities plans. Much could be said about strategies for manufacturing, but for our purposes here we will focus on the key facilities issues that must be resolved by the strategic planning effort.

We should note in passing that many manufacturing strategies are weak when it comes to facilities. Too often, they concentrate on capacity and process technology and ignore or skirt issues of site location and development.

When this happens, major facilities decisions are often made "on-the-fly" in the middle of crash projects. The results of such expediency are compromise and occasionally costly mistakes.

Table 1 shows some of the issues that are addressed by manufacturing strategy. In this example, compiled from various sources, we see how IBM supports its business goal of high volume at low cost.

Note the facility implications. Businesses with a different goal—low volume market nichers, for example—have a different strategy and different implications. From the standpoint of facilities planning, a complete manufacturing strategy will give explicit consideration to the following issues:

- *Size of plant*—and by implication, the number of sites required. Some managements favor small plants of a few hundred employees. Others operate effectively with 5,000 or more at a single site. There is no industry-wide answer, and specific companies change their philosophies from time to time.
- *Geographic centralization*—a matter of top management philosophy, tempered by cost patterns and logistics. At one time, much of Apple Computer's production came from its Dallas facility. With the arrival of a new president, John Sculley, and the introduction of the Macintosh computer, Apple chose to re-concentrate in Silicon Valley.

Clearly, our long-range planning must be alert to such changes and their implications for specific sites. Closely related is the desire for physical centralization on a single site or under a single roof. Many electronics firms are now moving from scattered, often leased, locations into central locations. Such moves, of course, require a clear policy on plant size.

- *Offshore/on shore*—another matter of top management philosophy, subject again to the realities of costs and logistics. Motorola has made a move with its U.S. 1 facility in Chandler, AZ, to keep leading-edge semi-conductor production on-shore. Priam, a maker of large capacity disk drives, has made a similar management decision, choosing to automate extensively and stay in Silicon Valley.

But the prevailing philosophy among makers of peripheral products—terminals, printers, disk drives—has been to move off shore. Control Data, Harris, ITT-Qume, Centronics, Seagate, Tandon and others have opted in the past year for the lower wage and materials costs of Taiwan, Singapore and other Far East locations.

- *Vertical integration*—deciding what will be made and what will be bought; also the number and location of suppliers, and the frequency of their deliveries. A few years ago Tektronix was vertically integrated, making many components that similar firms bought out. Now the company entertains more outside purchases. This makes for significant changes in the utilization, size and mission of the firm's components facilities.

TABLE 1 – IBM: Manufacturing Strategies for the '80s

Goal: High volume/low cost
Strategies:

- Close suppliers could reduce the need for purchased parts storage.

- In time this will lead to fewer parts in production and less need for stockroom storage.

- Early manufacturing involvement:
  - —Design influence
  - —Process verification
  - —Sourcing decisions
- Design for automation:
  - —Minimize number of parts
  - —Eliminate fasteners
  - —Self-alignment; no adjustments
  - —Symmetrical where possible
  - —Avoid parts that interfere
  - —Rigid, stiff parts
  - —Close tolerance
  - —One-sided assembly
- Limited models/features:
  - —Stable product design
  - —Group engineering changes for "model year"
  - —Customize in distribution centers
- Build to plan:
  - —Finished goods owned by sales
  - —Continuous flow manufacturing
  - —Supplier integration
  - —Zero defects
  - —Reduce work-in-process
  - —Management by sight
  - —Multi-skilled team
  - —Focused team
- Defect free at shipment

- Need close physical proximity between design and production engineering. Ideally these will be in the same building. Avoid separate sites.
- Will achieve simplification and greater throughput, even in manual mode.
- Primary goal is robotic assembly.
- Should yield greater output per square foot of plant space.

- Greater emphasis on dedicated lines and cellular layouts.

- Choice between progressive build and stationary build.
- More reason than ever to segregate low-volume products.

- Potentially reduced emphasis on traditional AS/RS. Less need for storage and staging.

- *Internal organization*—how will facilities be assigned? By product line? By process? By geographic zone? Will they belong to a central manufacturing group, or be assigned to product divisions?
- *Product and process technology*—these change more predictably perhaps than the policy matters above, but no matter how gradual the changes, new technology poses some of the thorniest problems for the facilities planner. It seems that in every aspect the electronics industry is currently beset by new technology. Upstream, in semiconductor and circuit board operations, we have the moves toward 1 micron design, 6-inch wafers, CMOS semicustom circuitry, and surface mounted devices.

Downstream, in assembly and distribution, we have the moves toward robotics, automated handling, and the new "management technology" of just-in-time production. Some of these developments will increase the capacity of existing facilities. Others will make existing facilities obsolete. Planners must clearly consider both implications.

LONG-RANGE FACILITIES PLANS

The long-range facilities plan begins where manufacturing strategy leaves off. Armed with current policy decisions on the issues just discussed, the facilities planner can now develop a long-range plan for each site or major structure. The purpose of this plan is to provide direction and infrastructure, so that day-to-day and short-range decisions can be made more quickly and effectively.

In most industries, "long-range" implies a horizon of five to seven years or more. But in electronics, the pace of technological change and the rate of market growth have been so rapid that two or three years is often "long-range." We should probably be more concerned with a state of mind, and a way of thinking, than with a specific calendar period.

The most effective approach relates long-range decisions to capacity and aggregate demand. How do we support 50% growth? Or 100% growth? What if growth flattens? The answers to these questions should be sound without too much concern over the year or quarter in which such events come to pass.

Long-range plans consist primarily of words, not drawings. In their simplest form they contain three elements:

1. The facility mission statement.
2. The basic growth plan.
3. The master site concept.

The long-range plan is not a simply short-range plan with extra years tacked on. Short-range plans seek to provide specific capacity for known products and processes. We project specific space needs, based on product sales forecasts. The finished plan is a set of drawings and layouts, ready for implementation.

In contrast, a good long-range plan is general and conceptual—and ready for subsequent short-range detail. It should be versatile enough to accommodate unknown products and processes not yet developed. The long-range plan is driven, at best, by projections of historical activity instead of current sales forecasts.

This type of planning was difficult a few years ago when many firms lacked the history required for useful space projections. This is no longer a problem for firms that have survived and grown over the past five or 10 years.

FACILITY MISSION STATEMENTS

The mission statement is perhaps the most important element of the long-range facilities plan. It is derived directly from manufacturing strategy, and sets the ground rules for project decisions on plant layout, building expansion, material handling and process improvement.

The mission statement is always less than a page in length, and may be a single paragraph. It should be part of the published manufacturing strategy, and may be included as a supporting document in business and corporate-level plans. Three examples appear in Figure 2. Each represents a situation common to electronics firms.

In the first example—Plant #1—we see a case in which one facility (often the company's first large plant) evolves into a proving ground for new products and processes. Back in the 1970s, Digital Equipment's Westfield plant served this role for DEC's peripheral products business. As the business grew, new products were moved out to dedicated locations such as Phoenix and Colorado Springs.

The second example of the Texas site epitomized the mission of a highly focused single-product plant. Tandem Computer's Austin terminals plant or Apple Computer's Macintosh plant might have a similar mission statement, although neither employs or requires "field customization."

This particular twist—shipping a "neuter" device that is configured later at a distribution center—is appropriate for some types of electronic systems and products. It appears in the IBM strategy described in Table 1.

The third mission statement is appropriate for companies with large, shared sites. Examples would include Motorola's headquarters in Schaumburg, IL; Martin-Marietta's sites in Orlando; Tektronix's headquarters in Beaverton, OR; and Texas Instruments in Dallas, to name but a few.

In such cases, the mission statement should clarify the role of major parcels and structures within the site, so that each occupying unit can move ahead with a measure of independence. Such prior clarification can save months on some project decisions.

BASIC GROWTH PLANS

Armed with a current mission statement, our next task is to select a basic growth plan. We should avoid the tendency at this point to get lost in precise growth projections.

The most practical approach is to work with simple multiples of current capacity, again without a great deal of regard for the specific time at which such multiples will come onstream. What is our growth plan to double current output? To triple it? To increase it five-fold over the next three to five years?

MISSION STATEMENT FOR PLANT #1

The primary purpose of Plant #1 will be the initial, pilot production of all new products. It will act as an incubator and a proving ground for new manufacturing techniques. Because of its proximity to Corporate Research and Development, Plant #1 will also serve as a prototype shop and test facility.

Existing production activities will be moved from Plant #1 to other sites over the next 18 months. As future products reach moderate volumes (40 units per day), they will be transferred to other manufacturing sites.

MISSION STATEMENT FOR LOS ANGELES FACILITIES

The Los Angeles complex will provide facilities for three purposes: 1. corporate headquarters; 2. all production of military electronics; 3. final assembly and distribution of industrial control systems.

Corporate headquarters will include the executive offices, administration, engineering, R&D, and the corporate training center.

The North Campus will serve as a central facility for all military electronics activities, including the testing labs.

Industrial control facilities will be limited to final assembly and finished goods warehousing. Sub-assemblies and components will be made in other locations or purchased from outside suppliers.

MISSION STATEMENT FOR TEXAS SITE

The Texas site will be dedicated exclusively to the production of standard Model 5 computers, including sub-assemblies and major components. For the next three years, all Model 5 production will be at this site. Future versions of Model 5 will be made at the Texas Site unless their design involves radical departures from current versions.

All customization of Model 5 computers will take place at field distribution centers, using kits supplied from the plant. No custom or special units will be produced on site.

Figure 2. *Variety of plant missions.*

What is the impact of a change in process technology? These are the kinds of questions that should be answered by the growth plan.

Experience shows that there are six ways to provide for future growth. These are pictured in Figure 3. Briefly, they are:

1. Grow into a reserved area on an existing plant site.
2. Acquire a neighboring site to house future growth.
3. Relocate completely to a larger site and grow there.
4. Relocate partially to another site, continuing to grow in both locations.
5. Decentralize—relocating activities and products to several separate sites.
6. Consolidate—a variation of plan number 3, in which several dispersed operations relocate and grow at a single large site.

The consolidation plan is currently popular in the electronics industry. Xerox Corporation's Diablo subsidiary used this plan when constructing its 500,000 square foot plant in Fremont, CA.

This new site allowed Diablo to vacate 13 scattered facilities throughout the Silicon Valley. Consolidation also permitted extensive automation in material handling. The resulting cost savings have made it possible, so far, to keep production of daisy wheel printers and electronic typewriters in the U.S.

All growth plans require some form of "land banking" in which land is acquired well in advance of need. Often, major utilities and roads will be installed at the outset to lessen delays and compress the time it takes to complete specific projects.

Burr Brown Corporation a rapidly growing instrument maker in Tucson, provides a good example. The company now owns a 225,000-square foot main facility on a 20-acre site, which gives it a 4:1 land-to-building ratio. Burr Brown leases another 100,000 square feet at various Tucson sites.

If the latter were consolidated onto the existing site, the ratio would drop to 3:1—much too low for a rapidly growing firm. Instead, the company has purchased a 113-acre site to cover expansion needs. Together with the existing facilities, this puts the land-to-building ratio at a far-sighted 18:1.

The opportunity to expand into a neighbor's facility is often overlooked or underrated. In long-range planning it is OK to "covet thy neighbor's plant space." In fact, it is good practice to figure out early on how we might use such space, and how much we would be willing to pay for it. Sometimes we can move our neighbor for less than we can move ourselves.

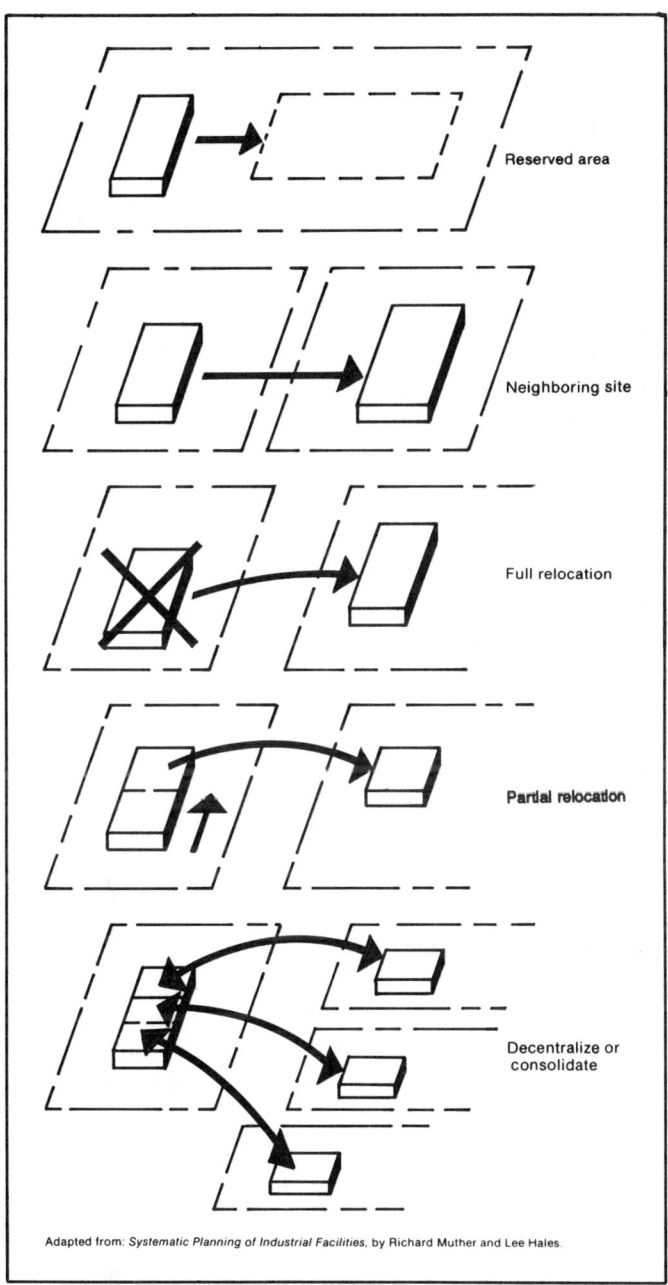

Reserved area

Neighboring site

Full relocation

Partial relocation

Decentralize or consolidate

Adapted from: *Systematic Planning of Industrial Facilities*, by Richard Muther and Lee Hales.

Figure 3. Basic growth plans.

Once a growth plan is established, it should be communicated throughout the organization and become a factor in all decisions that relate to capacity expansion and capital investment.

MASTER SITE CONCEPTS

The final component of a long-range plan is the master site concept. Its simplest expression requires but a single word: zone, block or duplicate. All site plans are ultimately variations of these simple themes.

With the zone concept, each major activity—fabrication, assembly, storage, office, support—is given an independent zone of expansion, outward from a central or common core. The spine concept used at Texas Instruments' Lubbock facility is a good example. Building modules extend in comb-like fashion from a shared service corridor or spine. Modules can be added or expanded independently of one another, and with no disruption to the spine. (See Figure 4.)

The block concept is appropriate when activities are placed "campus-style" in separate structures. Blocks of site space are then reserved around each building to allow for independent expansion.

The duplicate concept is used for the special case of modular production processes. When capacity is fully used, a second module is constructed with the appropriate space for each activity—fabrication, assembly, storage, office, support.

Each concept should be applied to the full acreage of the site, no matter how distant such full occupancy might be. Then, working back to the present, the planner can define meaningful phases and increments of the site's development.

This effort requires very little detail. The final plan can typically be shown clearly on an 8½ × 11-inch sheet of paper. A nice rendering, however, in the general manager's office or the main lobby, is a good way to build commitment and support.

PLANS FOR MULTIPLE FUTURES

A final point is in order. When making long-range plans, we must acknowledge that our chances of accurately predicting the future are slim, at best. So it makes good sense to develop a plan that will work for a number of possible futures.

Procedurally, the best way to do this is by starting with the "official future"—the one described as most likely in the current business plan. Two or three long-range facilities plans should be defined to support this base case. Then, by varying the assumptions made in the business plan, or by adding factors that were not included, we can construct three or four additional futures.

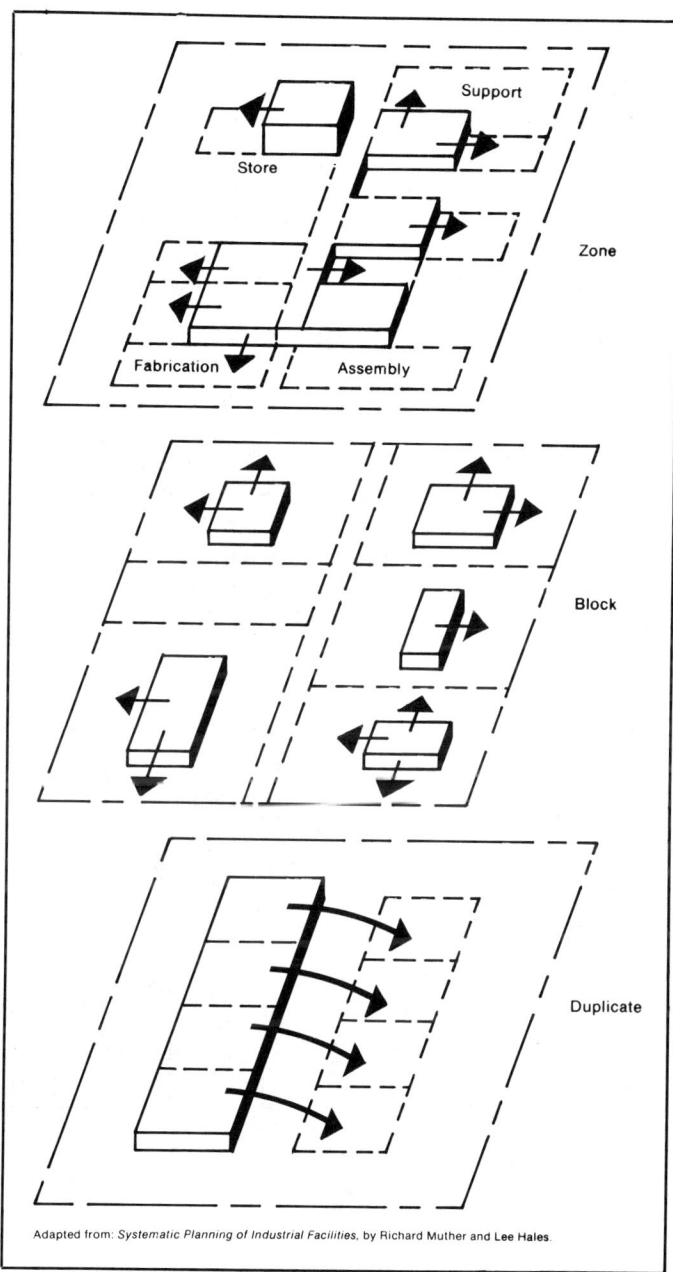

Figure 4. *Master site concepts.*

Again, two or three alternative facilities plans should be developed for each. When this is done, one or two growth plans and master site concepts typically reveal themselves to be workable across several futures. Costs and other factors being equal, these will be the best long-range plans, since they have the best chance of working no matter what comes to pass.

As can be seen in Figure 5, this approach requires the construction of perhaps a dozen alternative plans. This can be done only if the plans are kept brief and conceptual, and time-consuming details are avoided. Remember, the long-range plan need give only directions, not dimensions.

A STRONGER FACILITIES FUNCTION

The techniques described here can improve the quality of our planning—and, indeed, of our facilities themselves. But the key to lasting improvement will be found in a stronger facilities function—one that is represented at the corporate and business levels of planning, not just the site or shop floor.

There is a growing recognition in the electronics industry that product and process engineering must work more closely together, especially in times of rapid growth and change. This is the essence of IBM's "early manufacturing involvement." To this we should add *"early facilities involvement"* in product

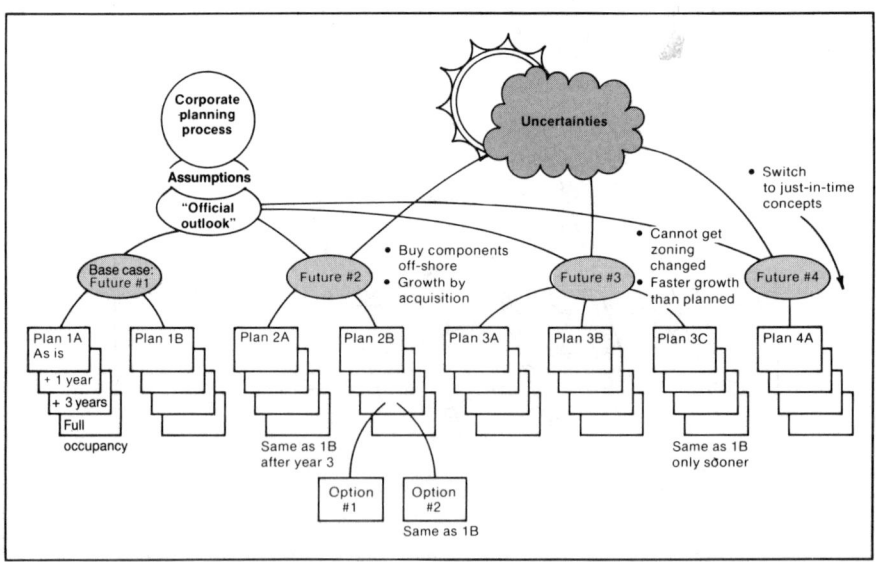

Figure 5. *Planning for multiple futures.*

planning and manufacturing decisions. As is suggested in Figure 6, it's time for facilities planners to forge stronger ties with the rest of the organization. No invitation is required.

Suggested Readings

Hax, A. C., and N. S. Majluf. 1984. *Strategic Management: An Integrative Perspective.* Prentice-Hall, Inc. Englewood Cliffs, NJ.

Hayes, R. H., and S. C. Wheelwright. 1984. *Restoring Our Competitive Edge.* John Wiley & Sons, Inc. New York.

Muther, R., and L. Hales. 1980. *Systematic planning of industrial facilities* (Vol. I and II). Management and Industrial Research Publications Inc. Kansas City, MO.

Reprinted from April 1985 issue of *Industrial Engineering.*

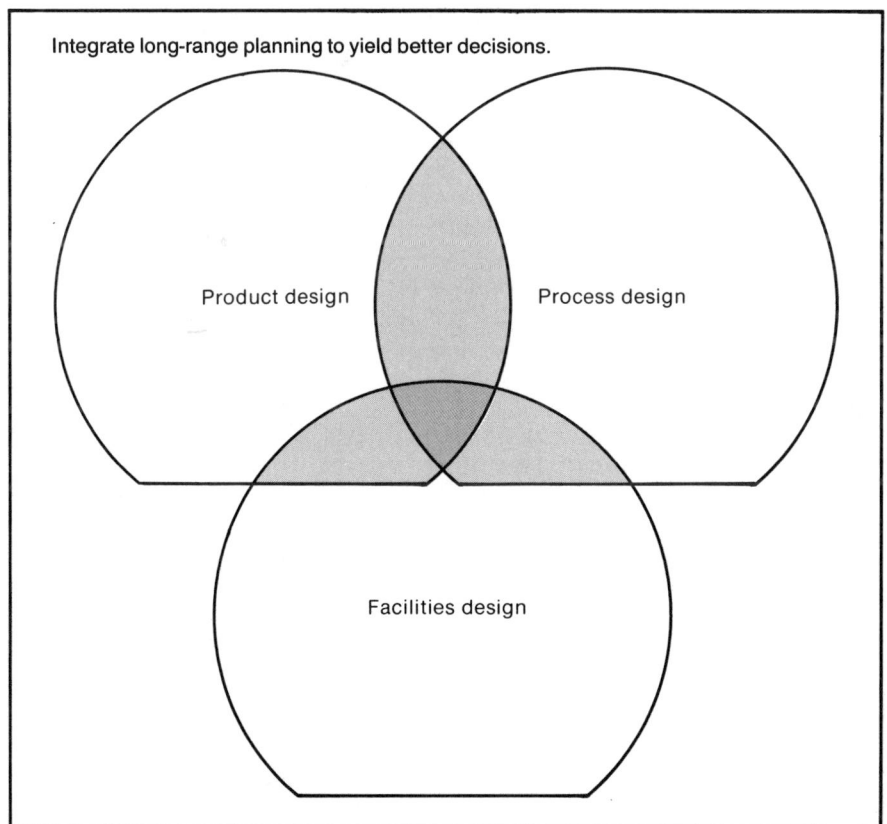

Figure 6. *Cooperative facilities design.*

24

Conceptual Model Aids in Development Of Manufacturing System for PC Assembly Shop

Arn McKenna

This chapter discusses the evolution of a manufacturing system for a printed circuit assembly (PCA) shop. The business environment is characterized by more than 100 types of PCAs with mixed volumes and rapidly changing product and process technologies. The product line is portable personal computers and associated peripheral products.

A conceptual model is used to evaluate the abilities of different manufacturing systems to meet organizational objectives. The need to apply different manufacturing technologies appropriate to particular product groups is demonstrated.

THE MANUFACTURING SYSTEM

A method is needed for reducing a manufacturing system to its fundamental elements in order to evaluate and compare systems that operate in different business environments. A vector diagram is shown in Figure 1 that models asset utilization, service level and product quality as the forces that determine the type of manufacturing system.

When the organizational objectives are met, the forces are correctly balanced, and the resultant vector is zero or a "bull's eye." Although the elements of the

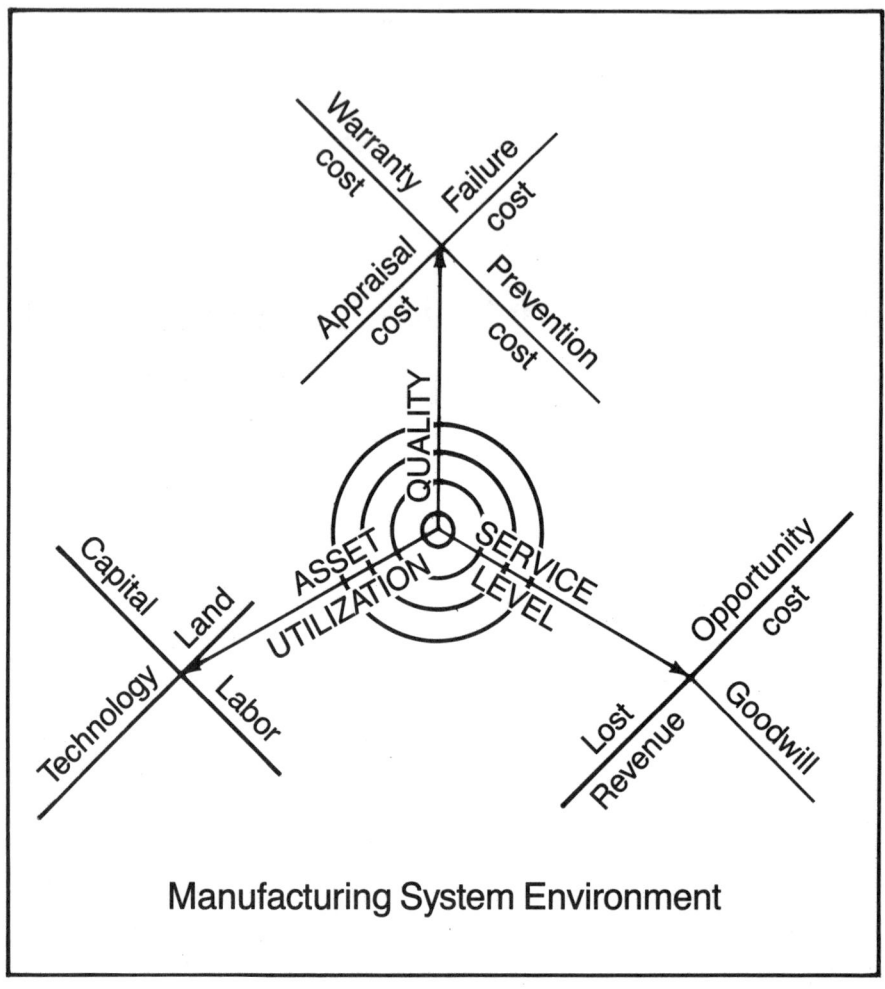

Manufacturing System Environment

Figure 1. *Forces that determine the type of manufacturing system.*

model are not quantified in terms of magnitude or direction, the interactiveness of the elements and the need for balance are demonstrated.

PCA SHOP ENVIRONMENT

In general, PCA fabrication is carried out in a job shop environment. A single product may include from one to several PCAs. A basic PCA shop process flow is shown in Figure 2. In cases of high volume finished products that use few PCAs, the shop environment may be characterized as a hybrid flow/job shop.

As the PCA product mix increases, the different PCAs compete for the shop resources. In some cases, there are processes that can be completed only by matching PCAs of different types to form functional sets. Other processes such as temperature cycling may require large periods of time to complete and add a paced time element to the manufacturing system design. Such a process flow is shown in Figure 3.

High technology portable computer PCAs are characterized by rapid change and relatively short product life. The major cost element for the PCA is the material used. A major product contains from one to five PCAs and usually requires several peripheral products (that use PCAs) to extend the range of application of the major product.

The PCA shop must have the flexibility to produce new product PCA sets and peripheral product PCAs in volume while maintaining moderate production levels of PCAs for mature products.

New technologies are likely to use more complex process steps, which requires the manufacturing system to support a wide product mix with complex process flows and to provide short manufacturing times to reduce materials costs. Additionally, new technologies are usually less controllable than mature technologies and need process steps designed to assure an appropriate level of product quality.

The process flow shown in Figure 3 is used to increase the reliability of products that make extensive use of CMOS integrated circuit technology. The "infant mortality" or early failure mechanisms are excited in the factory by

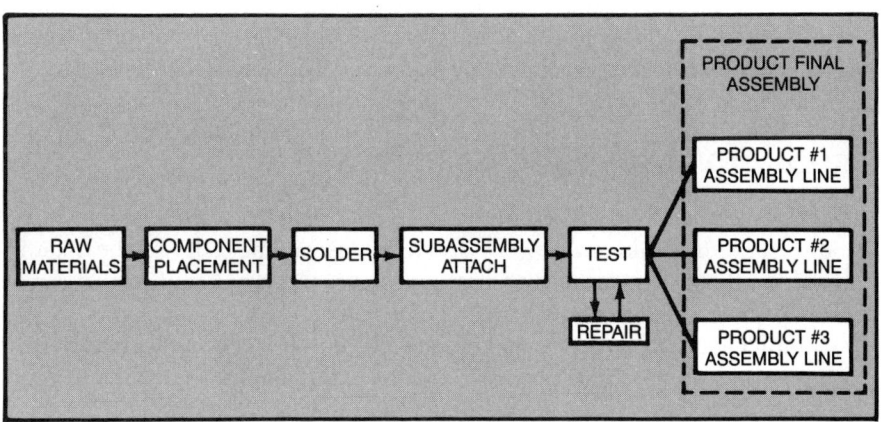

Figure 2. *Basic process flow for a PCA fabrication.*

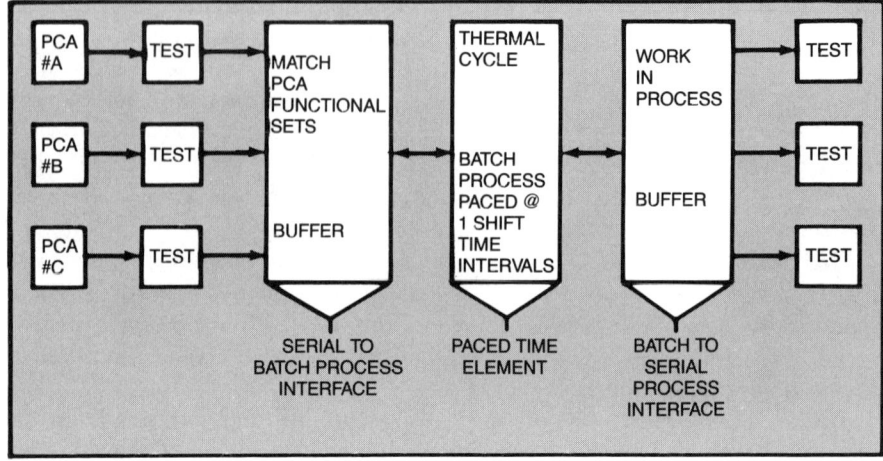

Figure 3. *Thermal cycle process step increases product reliability.*

stressing the PCA with thermal cycling while the circuitry is functioning. Referring to the bull's eye diagram in Figure 1, the warranty cost is reduced while the in-house failure and repair expense is increased.

Material efficiencies are affected in four ways by this process requirement. First, the manufacturing time is increased by the length of the thermal cycling process. Second, because the PCAs must be matched in functional sets to operate, additional material scheduling is required.

Third, the chamber adds a batch element to the process. Fourth, a paced time element is added to allow loading and unloading of the temperature cycling chamber during regular business hours. These losses are offset by the decrease in field failures.

PRODUCT VERSUS PROCESS DESIGNS

The PCA shop can be analyzed from two perspectives, as a process shop or as a product shop. By ranking the process steps in descending order by the quantity of products that use a particular process, work centers are identified that can be set up as product generic processes.

Multiplying by the product volumes and the number of minutes required for each process step identifies processes that have the potential to achieve economies of scale if they are set up as product specific (see Figure 4). When each process step's potential is identified as being product generic or product specific, other factors are considered to determine the final configuration.

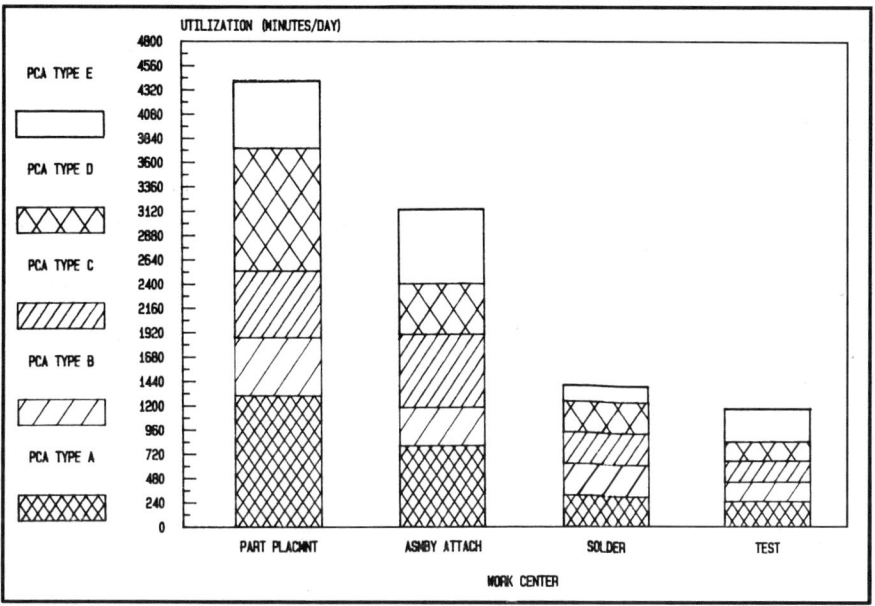

Figure 4. *Work center utilization by PCA type.*

The products can be sorted into groups of similar technological sequences. These groups yield information as to which process steps have the potential to be combined to form manufacturing cells. By reducing the manufacturing system to product specific processes, product generic processes and manufacturing cells, the system can be simplified into work centers designed for expected product volumes and an anticipated product mix.

The product versus process manufacturing system configuration criterion described above is based on asset utilization. This method gives a good start, but it doesn't tell the whole story.

In using the time needed by each process step as a factor, the assumption is made that floor space, labor and machinery utilization are the predominant criteria for the manufacturing system design. No allowances are made for factors that affect product quality, level of customer service or efficient use of materials.

Additionally, the assumption of maximizing asset utilization depends on the realization of a predetermined product mix and volume.

These assumptions can be faulty when you are dealing with a product line that is volatile in both volume and mix. Two strategies are employed to add flexibility to the manufacturing system, a method of linking the work centers to minimize work in process (WIP) and a generic work cell design to assure high asset utilization.

LINKING THE WORK CENTERS

Linking the work centers is the most difficult task in completing the manufacturing design. Because some of the work centers achieve the greatest economies by producing high volumes of a small group of products (product specific processes) and others achieve economies by serving a large group of products (product generic processes), WIP inventory accumulates between work centers. The large product mix and rapid change of products make classical work measurement-based line balancing impractical.

Three constraints prove to be a challenge in achieving economies in materials by reducing the WIP inventory when linking the work centers: (1) the prioritizing of PCAs that are competing for a common work center; (2) the need to match one PCA with another at a particular place in the process sequence; (3) the requirement that a particular quantity of matched sets of PCAs must be at a particular work center at a particular time of day.

The linkage between work centers takes on a hybrid form that complements the asset utilization of the work centers. An MRP-based system is used to deal with raw material vendor lead times and assures that raw materials are available to meet the expected product demand. New product designs use parts that are already used in other products where possible to add flexibility to the raw materials inventory.

At the beginning of each shift, the work center supervisors discuss the current WIP configuration in the shop, the service level expectations and product quality. The result of this meeting is a common set of process expectations for the shift that form the basis for the individual work center supervisor's judgments.

Each work center then uses a process method appropriate to the process type. Generally, the product specific work centers will produce batches, the product generic work centers will act as flow processes and the manufacturing cells will set up and process batches.

The work centers use a "demand-pull" card system to communicate the material needs to the preceding work center for serial flow processes. For more complex process flows that require PCA set matching or critical timing, the judgment of the individual work center supervisor is relied upon to "stack the deck" or prioritize the material needs from the preceding process.

The linkage system allows each work center to use a technology that is appropriate to achieve economies for the particular product mix and product volumes that are processed. A large amount of latitude is given to the work center supervisors in meeting service level expectations. The result is a manufacturing system that is very responsive to changes in product mix, product volume and process technology, yet maintains high service level and asset utilization.

DESIGNING PRODUCT GENERIC PROCESSES

When the work centers have been identified as being product specific, product generic or cellular with the expected product mix, the ability to quickly alter the work center configuration to support unexpected mixes and advanced technologies must be considered.

In some cases automation may be employed for a basic process where technological advancement is unlikely for the economic life of the process. In other cases, labor intensive processes provide economies in materials by virtue of their ability to deal with nonuniformities in raw materials and through the application of human judgment to correct unanticipated situations. Few automated processes match the flexibility of a motivated worker.

In both cases, the work center is set up to serve the most products possible for high asset utilization over a broad range of product mix.

The application of these strategies throughout the PCA shop has resulted in a manufacturing system that produces on a regular basis over 100 PCA types in an area of 12,000 sq ft. Roughly 120,000 components are loaded per day at a cost that is very competitive with that of a product specific shop. First-time PCA designs are routinely fabricated within a two-day period, coincident with normal production.

Reprinted from April 1985 issue of *Industrial Engineering*.

25

Developing a Material Handling Plan For Electronics Assembly

Thomas P. Cullinane

During the past two decades there has been tremendous progress in designing and developing hardware, systems, and procedures for automating and improving electronics assembly operations. The most outstanding gains have been realized in automatic insertion, automatic testing, and automatic inspection. Although electronics assembly systems today are far superior to the systems of two decades ago, they very often fall short of the industry's productivity goals. One area in which opportunities for improvement are abundant throughout the industry is material handling.

Material handling is critical within any electronics assembly system. Observations within a wide range of organizations have demonstrated that material handling can be the factor that determines the success or failure of an assembly system. Material handling can account for between 20% and 80% of the total operating costs of a firm. As electronics firms modernize, automate, and update their operations and equipment to achieve quality and production goals, effective material handling will very often be the key element in the success or failure of these systems.

Material handling has been defined in several different ways. In an assembly system environment, it should be considered as an extensive, not totally scientific activity that adds the very important values of time and place use to materials. Material handling adds value or utility to a component or product in that this activity or action places the product where it is needed when it is needed. Material handling can consume resources, create high levels of work in process inventories, be responsible for lost products, and create production delays if not properly planned. When planning for material handling activities in most of today's assembly type operations, only the facilities and operators that come into direct contact with the material handling equipment are considered. This approach is a component approach and is sometimes called bottom-up design. Most current-day planners do not recognize that this component approach is incorrect, because it appears to satisfy the material handling needs of the system being designed. The component approach views material handling only in terms of the current system needs. It does not take into account long-term goals of the assembly operations, the impact of the overall organization needs, the integration of new technologies with mature technologies, and the need for flexibility. The component approach increases the chances of developing a material handling system that cannot adjust to even minor changes in the requirements placed on it. From a corporate perspective, a preferred approach to this bottom-up method is one that is driven by life cycle concepts and reduces the chances of suboptimization.

A planning process which is based on a systems development life cycle specifies that the analyst look beyond the system under study. This approach requires the analyst to consider corporate goals and to recognize future company needs as part of the material handling system planning process. It ensures that the most significant opportunities, as defined by the corporate goals, are taken into account in a manner which will provide for integrating new technologies into the system.

When using the systems development life cycle approach to planning a material handling system for electronics assembly, the corporate goals and needs become drivers and are used to recognize candidate opportunities for improvement and technology needs. The main advantage of planning a system from a "top down" systems development life cycle approach is that new technologies are evaluated from the perspective of the system's long-term goals, not from the short-term perspective of "let's just get this thing working."

PLANNING A MATERIAL HANDLING SYSTEM*

The purpose of this chapter is to describe a 10-step process for planning material handling systems for an electronics assembly operation. The foundation for the 10 steps is derived from the system development life cycle. The life cycle consists of four phases:

1. understand the problem;
2. design the system;
3. build the system; and
4. install and use the system.

*See Cullinane (1984).

The 10 step process addresses the "understand the problem" phase of the life cycle. After the 10 steps have been completed, the product of the work will be a well thought out master plan for material handling system design, analysis, and control.

Step 1: Define Material Handling Goals

The first step of this planning process involves determining a clear understanding of the company's material handling goals and documenting those goals. This requires that the organization's goals be studied very closely such that the material handling goals can be defined at the highest levels of the organization. This first step also involves developing an understanding of the degree to which the organization's goals are understood by everyone. Ensuring that the goals of each of the organizational elements are congruent with the overall goals will reduce the probability of developing a counterproductive material handling system.

Step 2: Develop "As Is" Material Handling Model

Before recommending the future designs and operation of material handling systems, it is important to understand and analyze "what is" or the "as is" condition. This involves understanding all of the important functional, informational, and dynamic characteristics that constitute the current assembly operation. Graphic modeling tools, such as flow diagrams and systems flow charts, are most useful in this work. Another useful modeling tool is the Integrated Computer Aided Manufacturing Definitions Language (IDEF$_0$) developed by the U.S. Air Force Materials Laboratory at Wright Patterson Air Force Base in Dayton, Ohio.

Once the initial definition of the assembly environment has been established, the material handling system can be partitioned into several functional areas.

These functional pieces, when taken together, represent the whole material handling and storage environment and are referred to as tech areas. Some tech areas might be receiving operations, automatic identification, horizontal transport, vertical transport, palletizing, material sortation, and staging.

Step 3: Examine and Prioritize Tech Areas

Each of the tech areas must be studied to determine if problems currently exist or if they need to be modernized. Material handling problems are not easily definable. When a tech area is first examined there may be a general feeling that a material handling problem exists, but where it is or what it is may be unknown. In situations such as these, checklists such as those shown in the *Advanced Basics of Material Handling* (1984) will be most useful. The "Principles of Material Handling" that appear in *Advanced Basics of Material Handling* can also be very helpful when analyzing technical areas for improvement. Problems must be fully understood, and for a specific assembly system many questions will need to be asked. *Basics of Material Handling* (MHI 1977) provides a discussion of the types of questions that need to be asked.

Each tech area is evaluated to determine its potential for improving system operating effectiveness. Criteria which are developed from the goals are used to rank and evaluate the tech areas. It is at this point in the planning process that time and available resources will determine how much effort is put into the next seven steps.

Step 4: Develop Material Handling Tech Area "As Is"

Step 4 is similar to Step 2 and may not require any extensive effort. The goal of this step is to detail the functional, informational, and dynamic characteristics of each one of the material handling tech areas. At this point the management information system is examined to determine the degree to which it supports the tech area of interest. While examining each tech area, it is very important to realize that the flow of information is as important as the flow of materials within an automated material handling environment. The interface between the management information system and the tech areas must be clearly defined.

Step 5: Identify Material Handling Performance Measure

A performance measure is a variable that is monitored to determine if a particular activity or set of activities in the tech area is performing acceptably relative to the stated goals. Some typical performance measures include the following:

1. volume of throughput;
2. operator utilization levels;
3. size of unit load;
4. speed of handling; and
5. volume of breakage.

Step 6: Analyze Tech Areas

Analyzing a tech area requires three activities. First, the defined area is examined to ensure all relevant data has been recognized. Second, performance measures are evaluated to determine if they actually do the job they were intended to perform. Third, ongoing or planned company projects are reviewed to determine which, if any, will impact the tech area of interest.

Step 7: Assign Priority to Problems

The tech areas that have problems of considerable importance are identified at this stage. To accomplish this task the performance measures found to be most important are used to examine the elements of the technical areas that "appeared" to be a problem during step 6. The cause of poor performance should be documented at this stage. A clear problem statement should be written, and if a problem statement cannot be easily written, the data is incomplete, or the problem is not fully understood and should be examined further.

Step 8: Conceptualize Solutions

The design process begins in this step. With the problem areas isolated, possible direct and supporting technology should be researched to establish a baseline from which designs can emerge. The design of each improvement concept should include the following:

1. an initial description of major design considerations and constraints;
2. a description of one or more candidate design concepts;
3. the identification of off-the-shelf technology which might be used and that technology which must be developed (enabling technology); and
4. some preliminary assessments of how well the proposed design concepts will fit in the current design/manufacturing environment.

Step 9: Assess Impact

Each potential design must be examined in terms of payoff, risk, payback period, and other relevant management criteria.

Step 10: Specify the Material Handling Plan

The material handling plan will be an integrated set of technical plans and management plans for design, development, and implementation of each material handling improvement concept. Those design concepts that have surfaced during the evaluation process are the ones around which the material handling is designed.

SUMMARY

The results of applying the 10-step material handling system planning process have been excellent. This procedure allows the analyst to establish the facts, identify opportunities for improvement, define performance measures, examine possible solutions to material handling problems, and create a plan.

Suggested Readings

Cullinane, T. P. 1984. Planning material handling systems. *Proceedings of NAVSUP-MHI Seminar*. February. Houston, Texas.

The Material Handling Institute. 1984. *Advanced Basics of Material Handling*. The Material Handling Institute. Pittsburgh, Pennsylvania.

The Material Handling Institute. 1977. *Basics of Material Handling*. The Material Handling Institute. Pittsburgh, Pennsylvania.

Reprinted from *1986 International Electronics Assembly Conference Proceedings*.

Thomas P. Cullinane is Professor and Chairperson of Industrial Engineering and Information Systems at Northeastern University in Boston, Massachusetts.

26

Material Handling Equipment Alternatives Examined for Progressive Build in Light Assembly Operations

Gary B. Orr
Scott M. Sopher
James M. Apple, Jr.

Light assembly operations, which can be found in several different industries, such as electronics, small appliances and apparel, currently use different material handling equipment alternatives.

Normally, the assembly requires multiple operations performed at several work stations (e.g., assembly, test and repair).

When the total assembly task is divided into small, sequential elements, it is referred to as *progressive build*. The movement and control of parts between work centers where the successive elements are prepared is the material handling opportunity in these light industries.

MATERIAL TYPES

There are three types of material that must be handled in progressive build:

- Base units.
- Material added to the base.
- Bench material.

The handling unit most commonly used for moving assemblies in progress is a tote box size container. Other types of containers include slave pallets, trays and fixtures.

Base units

Base units are the fundamental building block and originate at the first assembly station. The base unit will go through all the assembly stations. All other material is added to the base unit to form the final product.

Material added to the base

Kitted or stock parts

These are parts sent to the assembly work station either in kits or in periodic issues. The parts are often handled in the same container as the base units. However, these parts are usually too large, too expensive or replenished too frequently to be stored at the work station.

Bench material

Bench material consists of hardware (e.g., nuts, bolts, washers) and other material that is bulk issued to the floor (e.g., post, wires). It is stored at the work bench or in the assembly area.

Subassemblies

Subassemblies are similar to kits in that they are issued to assembly as required. However, they are built up at another work station, placed in storage and issued for assembly to a base unit.

The handling problem typically involves moving material from storage to the work station and back to storage. The objective is to get the material to the work station to support a predetermined schedule. The type of handling system required hinges on several factors to be discussed later.

In progressive build, the handling alternatives are:

- Bench assembly.
- Paced assembly.
- Transporters.
- Assembly carousels.
- Assembly automated storage/retrieval system.

BENCH ASSEMBLY

This assembly operation is generally used in lieu of automating the material handling process. The system consists of operation to operation transfers of material, where by the operator passes the finished unit to the next station. This system utilizes manual material handling between processes. Typically, balancing between operations is not closely controlled and material queues are provided between each station. Consequently, material control becomes difficult without the aid of an extensive material tracking system.

The number of operators working on each assembly is flexible to an upper limit, but rapid changes in the work load will necessitate a high degree of workload balancing in order to utilize the workers. The workplace is most commonly individual benches or long work tables.

PACED LINED

Paced line assembly represents an increment of automation beyond bench assembly in that material is transferred between stations either continuously or intermittently. Operations must be balanced to successfully use this assembly concept. Consequently, the line is not amenable to rapid changes in work content. Paced lines minimize the work-in-process queues between the work stations.

Apple (see "For further reading") lists 10 prerequisites to using a paced line concept:

- Sufficient volume.
- Operation can be broken into small time elements.
- Consistency among station fixtures.
- Line cannot be totally unflexible to model changes.
- Material must be supplied to each station.
- Enough operations to be done to warrant the line.
- Low maintenance of fixtures.
- Set-up time must be minimized.
- Standard product design.
- Parts must be interchangeable.

Some of the material handling equipment used for paced assembly include:

- Conveyors (chain, belt).
- Manual slide line.
- Single level carousel.

The paced assembly concept cannot be easily used in an individual incentive program; however, group incentives can be applied. Also, management must maintain the line balance in order to minimize internal delays.

TRANSPORTER

A transporter is most commonly a two-level, bi-directional belt conveyor which dispatches totes to work stations on the top level and automatically returns totes on the bottom level. A dispatcher controls the part flow from a control panel located at the head of the transporter. A divert arm is located at each work station to divert incoming totes off the conveyor. After each operation, the outgoing tote is placed on the bottom level queue station and returned to work-in-process (WIP) storage.

The conventional transporter is limited to routing from a dispatch station. To accommodate sequential and balanced operations, a system has been developed with operator controlled downstream routing. In addition, another company has developed a system which provides both downstream and upstream routing from any work station.

A flow rack has been the conventional WIP storage mode for the transporter installation. In addition, carousels have been used to store totes of WIP and raw material. Robots have been used in conjunction with carousels to automatically feed totes onto the transporter, thereby eliminating the need for a dispatcher.

DIRECTED CARRIER

Another transportation system is a directed carrier. The carrier system consists of a cart on a rail. The rail supplies power and direction to the cart. The cart moves at work height from the WIP storage area to the production equipment.

The system offers precise positioning capability to accommodate the requirements of robots, and the cart is used to deliver material to and return products from the work station. The acceleration and deceleration of the cart can be regulated to meet most any special handling constraints.

ASSEMBLY CAROUSEL

The assembly carousel is a special application of chain conveyor with fixed carriers that move in an elliptical pattern. Work-in-process material is stored in a horizontal, multi-level bin that is constantly in motion. Assembly stations are located adjacent to the carousel. Raw material must be delivered to each station, this is done either by the carousel or by an external handling system.

Operators pick totes from either a specific shelf or a specific position within a shelf. After each operation, totes (or the base unit) are placed onto a designated shelf (or position within a shelf) for the next operation. The spread of the carousel is variable, which allows adjustments for work delivery requirements.

ASSEMBLY ASRS

A new concept for light assembly is full integration of the storage and handling systems. A system has been developed for the ASRS to deliver raw materials and in-process material to individual work stations located adjacent to the ASRS. The system offers random storage locations for each operation and varying queue lengths at each work station.

This system offers extensive feedback on inventory levels and operator production reporting. It is commonly used in conjunction with a mezzanine to utilize the building height and to increase the number of work stations served by a single SR machine.

Several parameters impact the decision on which type of material handling equipment is appropriate. These parameters are:

- **Material type**—Quantity, size and weight of the base units, kitted parts and bench material.
- **Work content/assembly**—Amount of time to complete an assembly.
- **Assemblies/tote**—Number of assemblies per tote or container.
- **Number of operations (work stations)/assembly**—Number of work stations required to complete an assembly.
- **Product mix**—Number of different products built or number of different products run simultaneously.
- **Throughput**—Number of assemblies processed at each operation per dispatching unit.
- **Accurate positioning**—Placing the part in a prescribed position to facilitate automation.
- **Life of product or process**—Expected life of a product or an assembly process.

An initial screening of parameters and constraints is the first step in evaluating material handling equipment alternatives. The alternatives should be compared with the parameters to reject the infeasible opportunities (see Tables 1 and 2). For example, a very high throughput rate requirement may preclude transporters, or a complex product mix may eliminate a paced line alternative.

In some cases, adjustments can be made in the parameters so that they match the characteristics of a particular system. For example, combining operations will reduce transactions and may make an otherwise unacceptable alternative a feasible solution.

Ranking a list of criteria based on their importance to your application is the second step in evaluating equipment alternatives. These criteria are:

- **Space:** The amount of floor space required per work station and associated material handling equipment.
- **Work-in-process:** All the material handled between raw material and purchased parts storage and finished goods storage.
- **Manufacturing cycle time:** The elapsed time to manufacture a product.

TABLE 1 – V³ Analysis For Assembly Alternatives

	Assembly ASRS Transporter	Assembly ASRS Transporter	Assembly ASRS Transporter	H
HIGH	Assembly ASRS Transporter	Assembly ASRS Transporter	Assembly Carousel Transporter	M
	Assembly ASRS Transporter	Paced Line Assembly Carousel Transporter	Assembly Carousel Transporter	L
	Assembly ASRS Transporter Paced Line	Assembly ASRS Transporter	Assembly Carousel Transporter	H
VOLUME MEDIUM	Assembly ASRS Transporter Paced Line	Paced Line Assembly Carousel Transporter	Bench Assembly	M
	Paced Line Assembly Carousel Transporter	Bench Assembly	Bench Assembly	L
	Assembly ASRS Transporter	Transporter Paced Line Assembly Carousel	Assembly Carousel Transporter	H
LOW	Paced Line Assembly Carousel	Bench Assembly	Bench Assembly	M
	Bench Assembly	Bench Assembly	Bench Assembly	L
	LOW	MEDIUM	HIGH	VALUE

TABLE 2 – Constraints Between Parameters And Alternatives

Parameters	Alternatives Bench	Paced	Assembly Carousel	Transporter	Assembly ASRS
Material type	Flexible	Usually small although systems are available for large parts	Usually tote size	Tote size less than 3 lb	Tote size
Work Content/ Assembly	Material delivery can be complex.	Must be balanced	Tote size	Must be within the tote	Tote size
Assemblies/Unit load	Flexible	Usually more than 3 and less than 10	Tote size	Dispatcher capacity	Dispatching
Number of Operations	Floor Space	Interfacing assembly lines	Carousel size	Flexible	Length of the storage aisle or building height.
Product Mix	Flexible	Flexible among product families	Slow to change out.	Control/dispatch system	Flexible
Throughput	Size and number of WIP stations	Slowest operation	Slowest operation	Dispatcher	Dispatching
Accurate Positioning	Requires fixtures	Flexible	Requires fixtures	Flexible on some systems	Flexible
Product Life	Flexible	Set-up	Flexible	Flexible	Flexible

- **Control:** The opportunity to receive and regulate feedback from the process.
- **Operating cost:** Sum of the labor and associated overhead costs to operate the system.
- **Equipment cost:** Cost of the material handling equipment.
- **Line feeding:** External handling required to support the operators.
- **ROI:** Required return-on-investment for capital expenditures.
- **Flexibility:** The ability to react to changes in requirements (e.g. volume, mix, process).
- **Manageability:** Management per worker ratio.
- **Worker's attitude:** Social and/or technical changes acceptable to the worker.

After the criteria are ranked, a weight should then be given to each criterion.

Given the feasible solutions, the third step in the evaluation process is ranking the alternatives for each criterion. For some criteria, like ROI, use classical engineering economy techniques as part of the evaluation. However, for other criteria it is more difficult to evaluate the alternatives. In Table 3 one such criterion is ranked based on the authors' experience in light assembly operations.

The final step in the evaluation process is to measure the feasible solutions against the list of criteria and select the best alternative.

TABLE 3 – Criterion Alternatives Ranked

	Amount of WIP
Lowest	Paced line
	Assembly ASRS Transporter
	Assembly carousel
Highest	Bench assembly

Suggested Readings

Apple, J. M. 1977. *Plant Layout and Materials Handling.* 3rd Edition. John Wiley & Sons. New York.

Reprinted from April 1985 issue of *Industrial Engineering.*

27

Automatic Guided Vehicles Move PCBs Between Assembly Stations

Donald H. Daebler

Electronic manufacturers have for years been automating their printed circuit assembly operations. These automated operations, however, were individual assembly and test areas, or "islands of automation." It is the materials handling field that provides methods of physically tying these islands of automation together in the factory. One means of doing this today is with wire-guided vehicle systems.

An automatic, wire-guided vehicle system usually consists of:

- Vehicles designed to move materials on pallets and in PC board magazines and totes.
- Guide-path wires on the factory floor for the vehicles to follow.
- A control system for automatic routing of vehicles, positioning vehicles at assembly machines and conveyors, and assuring safe, collision-free operation.
- Battery charging systems.
- Custom-designed conveyors that function as links between vehicles and workstations for material delivery and pick up.
- Pallets, PC board magazines, and totes in which material is moved.

The difference between vehicle systems and conveyor systems is the sophistication in system controls available with guided vehicles. Complex conveyor systems do not provide the flexibility and programmability found in vehicle systems with high levels of control.

GUIDE PATH

A wire-guided vehicle moves through a factory by following a wire embedded in the floor. The guide-path wire, laid out in the form of a loop, is usually a 1.5 mm² PVC-insulated electric cable. Guide-path wires carry alternating currents at various frequencies generated by loop power units. A guide-path wire must not be installed too close to solid-iron constructions or cross-iron structures since this will greatly affect the magnetic field from the loop, which is sensed by the vehicle for guidance control (see "Autocarriers—This Is How They Work," Volvo Corporation of America).

Most wire-guided vehicle systems are very flexible in their operation, as the system's software can be reprogrammed. A system can also be expanded to accommodate more vehicles and more stops, and the scope of existing stops can be expanded.

Future expansion is important to consider when designing the system. It may initially include optional pick up and delivery points that could be used in the future. These optional points will permit the factory to adapt to new product and production schedules without any major changes required in the wire-guided vehicle system.

The guide-path system may be a single loop or several loops (Figure 1). In a single-loop system, a vehicle's on-board microprocessor is programmed to make the necessary stops along the guide path. Programming using a key pad or bar code reader on the vehicle is common in the smaller, basic systems. More complex systems have more than one guide-path loop. Each guide-path loop is assigned a separate frequency that a vehicle can recognize and follow.

A vehicle must have the capability to follow a specific frequency to a certain position, then switch to another steering control, guide-path frequency. The vehicle may have all of the necessary intelligence in its on-board microprocessor to make the necessary changes, or a system-control computer may make the guide-path changes through a means of communication with the vehicle.

Communication between vehicles, or between vehicles and the central-control computer in larger systems, handles the following important information:

- What material movement has high priority for pick up and delivery.
- What guide-path frequencies the vehicle will follow to get to its destination.
- When the vehicle is to travel in a forward or in a reverse direction.
- When to start, stop, or change speeds to avoid a collision with another vehicle.
- Where the vehicle is located and what problems exist on the factory floor.

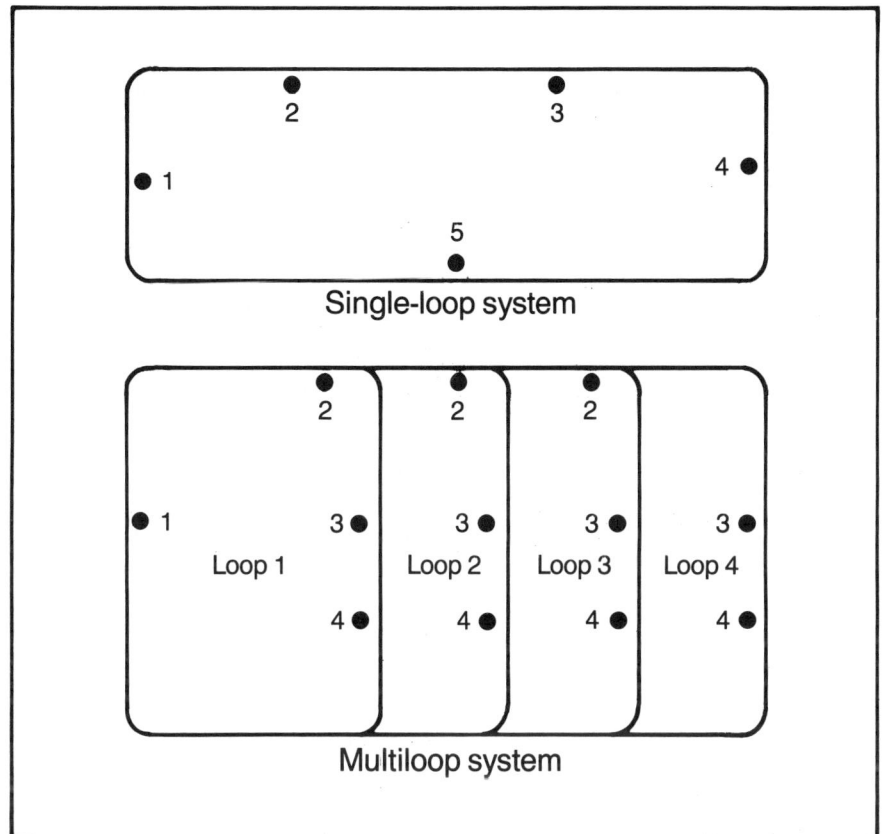

Figure 1. *Guide paths for vehicles are made with electrical wires formed as single or multiple loops in the factory floor.*

LOADING THE VEHICLE

Vehicles must be designed to handle the types of loads they will have to move. For example, in stockroom operations the vehicle should be sized for the maximum pallet size when handling pallets. In PC board assembly operations the vehicle should be designed to handle PC board magazines and totes. Designing a vehicle to handle all types of loads is usually difficult and could impair the reliability of the system.

A vehicle may be required to pick up and deliver loads at different elevations. An example of this would be PC board magazine feed systems used on component insertion equipment. Also, being able to operate at various elevations provides the system designer with greater flexibility in interfacing other assembly equipment with a vehicle.

For loading and unloading, a vehicle must position itself accurately in front of a conveyor. Positioning accuracy of plus or minus one-quarter inch will usually assure reliable material transfer and is in the range of normal vehicle capability.

Floor conveyors, like the vehicle, should be designed to handle specific load types. Conveyors designed and used for dedicated types of loads, in contrast to general-purpose conveyors, will enhance the reliability of the guided-vehicle system.

Powered zone conveyors should be the type used. They will move pallets, PC board magazines, and totes, while keeping the different loads of material from bumping or touching each other—unlike gravity delivery conveyors.

Single-level conveyors as well as bi-level conveyors are suited for handling PC board magazines and totes. The bi-level design, a type of zoned conveyor, uses an elevator to move the load between levels. This type of conveyor lends itself to operations that require an operator to be present. A single-level conveyor, however, will be the type used most often for handling large pallets.

Standardization of PC board magazine and tote dimensions is recommended. Totes and magazines similar in length and width will work on the same vehicles and conveyors; their height only becomes a consideration when bi-level conveyors are used.

DELIVERY TO AUTOINSERTERS

An automatic, wire-guided vehicle system can move PC boards and components from the stockroom to automatic component insertion and placement machines in printed circuit assembly. Components, and in some cases the PC boards, are delivered in totes. Circuit boards, however, are more apt to be in magazines that are designed for the type of automatic board-handling systems used on component insertion and placement machines.

Component delivery from the stockroom to insertion machines is accomplished by placing pick up and delivery conveyors in strategic locations throughout the assembly area. These conveyors should be of the type that facilitate vehicle delivery of full totes and the pick up of empty totes for a return to the stockroom.

PC boards are usually loaded into magazines or totes in operations prior to assembly. An ideal way to load boards in magazines or totes is to make the loading operation an integral part of a bareboard testing or stockroom operation.

Assembly and test equipment may be designed to operate with automatic board loaders and unloaders. Then, PC board magazines must be positioned in these loaders with a specific orientation for properly feeding boards to the assembly and test equipment. If magazine orientation is required, it must start

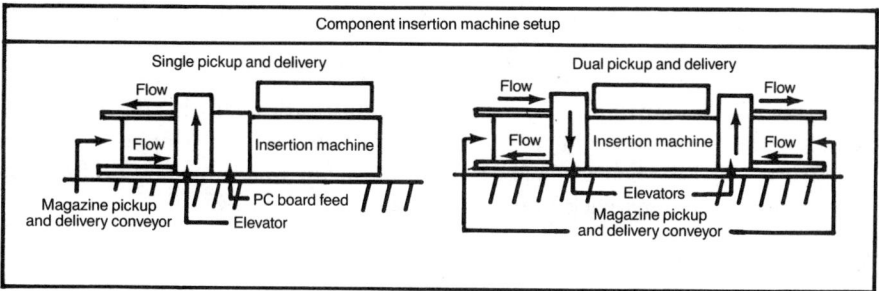

Figure 2. *Single or dual PC board loaders may be used with automatic insertion machines.*

at the magazine loading operation and be maintained by the vehicle system throughout assembly.

When automatic loaders are not used on the insertion or placement machines, a powered zone conveyor is used to provide the material-handling interface between the assembly machine and the vehicle system.

Assembly machines equipped with automatic PC board loading equipment may have a loading system using either single or dual loaders (Figure 2). Both systems are unique, and the guided-vehicle system must be designed to accommodate whichever system is used.

Machines designed with a single board loader will accept magazine deliveries from the vehicle system on the lower conveyor level. When all boards in the magazine have gone through the assembly operation, the magazine is moved to the upper conveyor level for pick up by the vehicle.

Assembly machines designed with dual PC board loaders have one loader at the input side and another loader located at the output side of the machine. The loader on the input side of the assembly machine accepts magazines of PC boards that require component insertion or placement. The vehicle system places these magazines on the lower conveyor level. When the magazine is empty, the loader places it on the upper conveyor level for pick up by the vehicle.

On the output side of the assembly machine, the loader accepts empty PC board magazines from the vehicle system on the upper conveyor level. When the magazine is full of PC boards, the loader places the magazine on the lower conveyor level for vehicle pick up. Using assembly machines with dual loaders will require additional handling of empty magazines, and the vehicle system must be sized for the added workload.

Loaders used on component-insertion and placement machines are usually designed with buffer zones on both conveyor levels to prevent machine downtime caused by untimely vehicle pick up and deliveries.

PC board assembly equipment not having automatic board-loading capability needs to be equipped with a vehicle service conveyor. An operator is also required to load the boards on and off of the assembly machine and position the completed assemblies back in the magazine.

SOLDERING AND CLEANING

Machine soldering and cleaning operations in printed circuit assembly is the next area where the guided vehicle system must be designed to provide the required material flow. This area is usually combined with a hand-assembly line located in front of the soldering machine and a solder touch up and PC board repair line located at the rear of the cleaning machine. These operations are normally set up to form one long, in-line flow.

The vehicle system is designed to deliver magazines or totes of PC boards for the hand-assembly and soldering operations at the front of this line. Then the system must pick up, at the rear of the line, magazines or totes of PC boards after they have been assembled, soldered, cleaned, and repaired.

A production line designed for soldering and cleaning operations can either be equipped with dual board loaders used on insertion machines, or with zoned service conveyors to provide the required vehicle interface.

Reprinted from *Electronic Packaging and Production*, March, 1987. © 1987 by Cahners Publishing Company.

28

Design for Line Flexibility

We-Min Chow

Designers of manufacturing lines often face the problem of determining appropriate line capacity to satisfy production demand. The demand is usually derived from a market forecast, subject to change from time to time. Since there is a significant amount of lead time to install a manufacturing line and the actual demand may change in a much shorter period of time, it is impractical to adjust the line so that the capacity is consistent with the demand. Theoretically, we may consider that the market demand fluctuates around a time-dependent mean value, or trend, which is a smooth function over time. It is therefore more reasonable to plan the line capacity according to the trend.

Unless the demand trend is constant, the problem of planning line capacity still exists. In many industries, particularly in the high-technology area, a product lifetime can be short, e.g., 3 to 5 years. The demand trend is typically a concave function. When a new product is first introduced to the market, one can expect to see an increasing demand trend because of better performance or lower cost. A few years later, the product may be phased out. If the designer plans for the peak demand, an overinvestment and underutilization situation occurs at both early and late stages. If the design is based on the average demand, the production line will not have enough capacity to support the peak period. Consequently, it is desirable to be able to adjust the line capacity dynamically.

Line adjustment needs careful planning. When the line is in a capacity-expansion mode, the action of adding more work stations may disrupt the existing line operations. The new work stations might be placed in an undesirable area, which could degrade the line performance. To reduce the capacity by removing work stations, on the other hand, could also disrupt the line and leave waste space that might otherwise be used for another purpose

This paper discusses a simple method that can resolve the problems just described. The main point in this method is to construct a manufacturing line of high flexibility with a direct-access material-handling system.

It is assumed throughout this paper that the manufacturing line is composed of a number of work stations, connected by a material-handling system. The material-handling performance is a function of the working environment defined by a number of factors such as material attributes (size, weight, etc.), workstation size, line layout, and traffic flow. Our study is limited to lines comparable to magnetic-disk assembly lines. We assume that the work stations are about 10 × 10 square feet and work units are about 2 × 2 × 2 cubic feet and less than 200 pounds. A work unit can be a semifinished product, a part container, or a kit.

In the following sections, we shall discuss a material-handling system and its operation characteristics, a line-configuration procedure, and a method for expanding or contracting the line.

DIRECT-ACCESS HANDLER (DAH)

When the line capacity is adjusted by adding or deleting work stations, problems may arise from line disruption. There are two types of disruption. To install or remove a work station might disrupt operators working in the same area. Furthermore, since the number of work stations is changed, an uneven material-flow pattern and accompanying traffic congestion might develop.

Whether this potential disruption results in a real problem is mainly contingent on the line layout. If a direct-access material-handling system is introduced, it is possible to have a flexible layout such that problems from disruption are negligible. We shall consider one specific type of material-handling system, namely the direct-access handler (DAH).

A DAH is similar to a robotic device riding on a carriage that can move along a linear track. The DAH is equipped with a rotation arm and a picking device at the end of the arm. The picking device is used to pick and to place materials. Work stations may be on one side or on both sides of the track.

Suppose that a number of assembly work stations are served by a DAH. The work stations are adjacent to each other, either placed in one side of the carriage track or evenly distributed into both sides of the track. The DAH picks up parts trays or subassemblies at an input/output port (I/O port) located at one end of the track and delivers them to the work stations. Empty trays and completed assemblies are returned to the same I/O port. If there is more than one type of work station, subassemblies may have to be moved from one station to another by the same DAH. Figure 1 illustrates a layout where a DAH serves a number of work stations and the I/O port is a section of a conveyor.

Each work station has a pick-and-place port (P/D port) which serves as a local I/O port to the station. In addition, a multi-level rack may be installed in front of work stations. When the DAH attempts to deliver an object and finds the port occupied, the object may be placed in the rack. The object is retrieved later when the port becomes available.

The DAH can be considered as a single-server queueing system. The arrival rate is equivalent to the rate of service requests generated from the work stations. A service request may be delayed due to congestion (that is, the DAH is busy in serving earlier requests). Once the DAH becomes available, it moves from the last delivery point to the sending station, picks up a work unit, moves again to the receiving station, and completes the delivery. The time interval between the first move and the completion of delivery constitutes the service time.

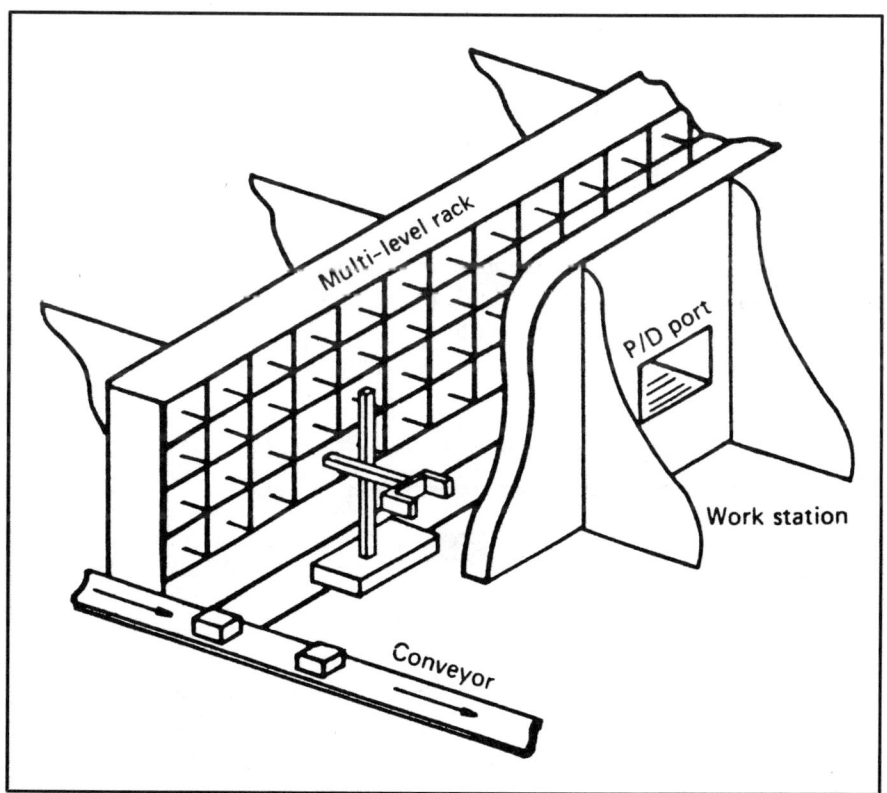

Figure 1. *A DAH layout.*

If the distance between the work stations and the traffic flow pattern are known, it is easy to compute the service time, as follows. Let

a = carriage acceleration and deceleration

b = carriage top speed

Θ = arm rotation speed

g = pick time

h = place time

d_{ij} = distance between stations i and j

r_{ij} = degree of arm rotation between stations i and j.

Using a, b, and d_{ij}, it is easy to compute the travel time t_{ij}. Suppose that the current DAH position is at station k. In order to move an item from station i to station j, the service time will be

$$s(k,i,j) = max\ (t_{ki},\ r_{ki}/\Theta) + g + max\ (t_{ij},\ r_{ij}/\Theta) + h. \tag{1}$$

If the service request rate from one station to another is constant, the probability, $P\ [S = s(k,i,j)]$, is directly proportional to the product of the receiving rate at station k and the request rate from i to j.

Let

f_{ij} = traffic volume from i to j
(This does not include repositioning moves.)

$$F = \Sigma\ \Sigma\ f_{ij}$$

$$P_{ij} = f_{if}/F$$

$$P_k = \Sigma \frac{f_{lk}}{1\ F}.$$

Then

$$P\ [S = s(k,i,j)] = p_k p_{ij}.$$

We assume that the request arrival process is Poisson, we then have an M/G/1 queue system. (Intuitively, the Poisson assumption is justified for a large number of work stations with asynchronized cycle times.) Thus the average waiting time under the first-come-first-served queueing discipline is (see Cox and Smith [1]).

$$E\,[W] = E\,[S] + \frac{E[S]\,\rho}{2\,(1-\rho)}(1+C^2) \tag{2}$$

where S = service time
   W = waiting time (delay plus service time)
   ρ = traffic intensity or DAH utilization
   = product of arrival rate and $E[S]$
   C = coefficient of variation of S.

Using equation (2), together with Little's formula (see Little [2]), L = λW, the average queue length can be evaluated.

The DAH performance has been previously studied and reported (in Chow [3]). The following observations have been made:

1. The average waiting time, as compared to the average queue length and utilization, is a more crucial and sensitive performance measure in the sense that any change in arrival rate or service time will affect the average waiting time most. (This is easy to see by differentiating the average waiting time and the average queue length with respect to ρ.)
2. It is important to keep the service time small. This can be done by providing advanced hardware, a good work station layout, or an efficient dispatching algorithm.
3. Under the current technology and the assumed working conditions, pick and place times contribute a major portion of the service time. Therefore, a reduction in pick and place times would result in greater performance improvement than would come from increasing the top speed or the acceleration and deceleration. This implies that the DAH travel distance is not as important as the traffic volume and service time has a small coefficient of variation.
4. Placing work stations on both sides of the carriage track is a better layout than placing them on only one side, because the saving in travel time usually outweighs the arm-rotation delay.
5. When the traffic is not heavy or the travel distance is not important, the first-come-first-served rule is just as good as other rules.

6. As compared with a continuous-flow material-handling system, such as a conveyor, the DAH is less sensitive to the layout, but more sensitive to traffic volume.

The characteristics of DAH performance can be understood best by looking at the relationship between the average waiting time and the utilization. Consider two different DAH's:

	DAH 1	DAH 2
Top speed (ft/sec)	3.3	6.6
Acceleration/deceleration (ft/sec/sec)	1.6	3.3
Arm rotation speed (degree/sec)	0.0	60.0
Pick time (sec)	7.5	3.8
Place time (sec)	9.0	3.8
Vertical travel time (sec/ft)	1.3	1.0

The first DAH does not offer arm-rotation capability and can only be used to serve work stations on one side of the track. Both DAH's allow their picking arms to move vertically. This is an important feature from a space-saving point of view, since excess work in process can be placed in vertical storage. Since the work-station size is much larger than the work-unit size, the horizontal travel time is usually longer than the vertical travel time. If both motions occur simultaneously, we may assume that the vertical motion never affects the DAH service time and that equation (1) is still valid.

Applying these two DAH's separately to a disk assembly line under different layout and material flow assumptions, we obtain two sets of data derived from the M/G/1 queueing model. Data are fitted separately by two smooth curves as shown in Figure 2. These curves are called operating-chararacteristic (OC) curves. In general, an OC curve is an increasing convex function. A high-performance DAH has a small service time and the corresponding OC curve is close to the lower right corner. For the same average waiting time, the high-performance DAH can tolerate more traffic volume and therefore offer more useful capacity.

It can be seen that both curves in Figure 2 are bending upward drastically when the average waiting time goes beyond about 60 seconds. (This value corresponds to a utilization of 0.67 for DAH 1 and 0.83 for DAH 2.) Obviously this will be a desirable operating point.

If the average waiting time is less than this value, we can use the DAH more aggressively by increasing the workload. If on the other hand the average waiting time is greater than 60 seconds, the DAH performance becomes very unstable. Any small increment in workload will seriously impact the material-handling performance.

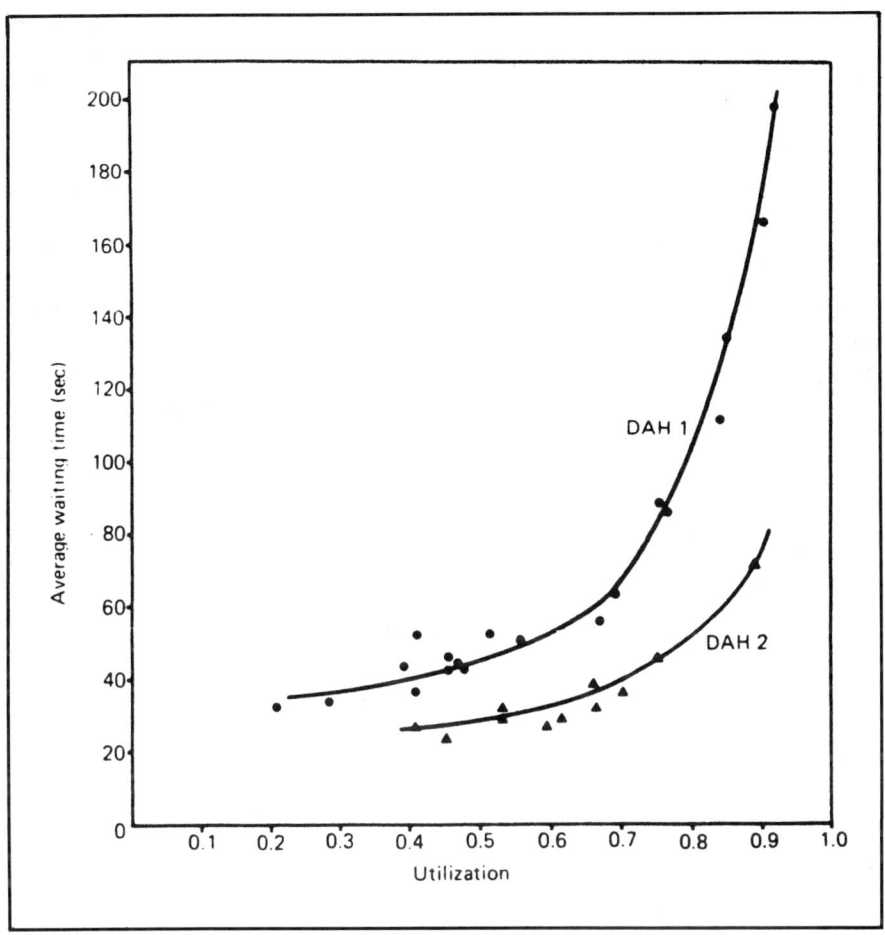

Figure 2. *DAH operating-characteristic curves.*

Note that this 60-second threshold is also valid for a family of DAH's whose OC curves lie in between the two curves shown in Figure 2.

LINE CONFIGURATION

The optimal solution for layout of a manufacturing line has many parameters: the building shape, the number of work stations and their sizes, material flows between stations, the material-handling system and its capacity, and so forth. This problem has not been totally solved yet due to its complexity. For a general discussion, we are going to simplify the problem by ignoring the building constraint. We are only interested in how to configure the line so that any capacity change can be made without significantly disrupting the existing line operation.

The direct-access capability of the DAH is a useful property for a flexible line. One does not have to lay out the work stations strictly according to the material-flow directions. For a given number of work stations and the material-flow pattern, work-station modules can be created, with each module composed of a number of work stations served by only one DAH. The module-design criterion is such that the average waiting time of a work unit inside the module can never exceed 60 seconds. The material-handling device connecting modules is not our concern in this study, for simplicity's sake. No matter what kind of device is used, we assume that it always has sufficient capacity to avoid traffic congestion in this area.

The modular approach has at least three major advantages:

a. Since the number of the modules is much smaller than the number of work stations, the line-layout problem is simplified.
b. Each module provides a natural partition for local traffic control. Such an arrangement will facilitate the entire control system. In a computer-controlled environment, this means a distributed control system in which each local controller deals with only one module. Logically, all local controllers are identical and subject to the same design.
c. The line can easily be expanded or contracted.

Since our primary objective is to construct a flexible line, we shall concentrate our discussion on the last item.

The procedure that leads us to a flexible modular line is rather simple. First, we determine the number of work stations, based on work-station cycle times, yield factors, station availabilities, and production demands. Next, the material-flow volumes between work stations are computed. Finally, we "pack" work stations into modules such that the average waiting time in each module is no more than 60 seconds. Since the number of modules is identical to the number of DAH's, the minimal number of modules can be the secondary objective. Two algorithms are developed below to achieve these objectives. We need the following notation:

m = number of work-station types

n_i = number of type i work stations

s_i = average cycle time of a type i station

Q = production demand

a_i = availability of a type i station

p_{ij} = proportion of finished work from type i to type j stations

r_i = start factor of type i stations (i.e., expected number of times a product is submitted to a type i station)

f = expected amount of products that must be initiated to produce one good product unit

T = effective working time during a day.

The yield factor of a type i station can be interpreted by the routing probability, p_{ij}. For instance, if type j is the successor of type i, then p_{ij} is the yield factor for type i stations. With a probability $1 - p_{ij}$, the semi-finished product either goes to a rework station or to the scrap when rework is not permissible. The start factors are determined by a set of simultaneous linear equations. Assume that the initial work station is type 1 and the last one is type m. We have

$$r_1 = f + \sum_{i=1}^{m} p_{i1} r_i$$

and $\qquad\qquad\qquad\qquad\qquad\qquad\qquad\qquad\qquad\qquad$ (3)

$$r_j = \sum_{i=1}^{m} p_{ij} r_i \qquad j = 2, \ldots, m.$$

Let y be the yield factor of station type m. Set $r_m = 1/y$. Then f and $r_j, j = 1, 2, \ldots, m - 1$, can be evaluated.

Since the available time of a type i station during a day is $a_i T$, the total quantity that all type i stations can produce each day will be

$$q_i = \frac{a_i T n_i}{s_i r_i}.$$ (4)

The following algorithm determines the number of work stations.

Algorithm 1

(i)   Let $n_i = 1$, $i = 1, 2,\ldots, m$.

(ii)   Compute q_i, $i = 1, 2,\ldots, m$, by using equation (4).

(iii) Find $Q = \min (q_i \mid i = 1, 2, \ldots, m)$.

(iv)   Record Q and k such that $q_k = Q$.

(v)   *If* Q exceeds the peak demand, stop the algorithm.
       Otherwise, replace the value of n_k by $n_k + 1$ and go to step (ii).

The outcomes observed in step (iii) defines a sequential process for line expansion. Let (Q_1, k_1), (Q_2, k_2), $\ldots$, (Q_v, k_v) be the sequence that has been recorded. The initial line capacity will be Q_1 and each station type consists of exactly one station. After having included a type k_1 station, we obtain a line that is capable of producing Q_2 product units. If we add an extra type k_2 station, the line capacity becomes Q_3, and so on.

The utilization of each type i station is defined by the product of the job arrival rate and the average cycle time:

$$u_i = \frac{Qr_i}{Tn_i}s_i. \tag{5}$$

A line configuration, flexible for dynamic expansion and contraction, can be obtained by packing work stations into modules. Meanwhile, the queueing model is invoked to make sure that the 60-second condition is satisfied in the course of each expansion or contraction. This procedure is presented by the following algorithm.

Algorithm 2

(i)   Apply algorithm 1 and define a set of linear sequential lists for different line capacities:

For Q_1, $L_1 = (1, 2, \ldots, m)$

For Q_2, $L_2 = (1, 2, \ldots, m, k_1,)$

. . . .

For Q_v, L_v = (1, 2, ..., m, k_1, k_1, k_2, ..., k_{v-1})

where each element designates a work-station type.

(ii) Determine the traffic volumes between work stations and from parts storage or kitting areas to each individual work station, for Q_1, Q_2, ..., Q_v.

(iii) Let i = 1.

(iv) If L_i is empty, skip to step (ix).
If L_i is not empty, let X_i be an empty list.

(v) Copy an element from L_i into X_i according to the sequential order in L_i.

(vi) Let all the work stations corresponding to the elements in X_i be in the same module and served by a single DAH. Taking the flow information developed for Q_i at step (ii) and work-station sizes as the previous section, compute the average waiting time.

(vii) If average waiting time does not exceed 60 seconds, then go to step (v). (This will place more stations into the module.) If the average waiting time is greater than 60 seconds, then delete the last element from X_i and go to step (viii). (We have defined a tentative work station module for Q_i.)

(viii) If X_i is empty, no feasible solution exists. Stop. (This means that the DAH does not have enough capacity.) If X_i is not empty, record X_i for future reference.

(ix) If $i < v$, then increase i by 1 and go to step (iv). (An additional tentative module must be defined.) If $i = v$, go to step (x). (All tentative modules have been found.)

(x) Determine the shortest list among all available X's. Denote this list by Z, and destroy all X's. (A common module for all Q's has been defined.)

(xi) Delete elements from L_1, L_2, ..., L_v, if these elements also appear in list Z. Record the content of Z.

(xii) If L_i is empty, for all i, stop. (The algorithm is completed.) If L_i is not empty, go to step (iii). (Define modules for the remaining work stations.)

At the end of the algorithm, we have a sequence of records about the content of each module. The total number of DAH's to be installed is equal to the number of times the list Z has been recorded in step (xi).

We can now illustrate this procedure with a numerical example. Consider an assembly line composed of eight different types of work stations. The assembly sequence is $1 \rightarrow 2 \rightarrow 3 \rightarrow 4 \rightarrow 5 \rightarrow 6 \rightarrow 7 \rightarrow 8$. The estimated average cycle times (in hours) for the station types are 0.05, 0.10, 0.15, 0.15, 0.15, 0.35, 0.15, 0.10. We assume that (i) $a_i = 0.95$ and $r_i = 1$, for $i = 1, \ldots, 8$, (ii) $T = 20$ hours, and (iii) the peak demand is 200 pieces a day. Results from Algorithm 1 are summarized below in Table 1. If the daily demand is less than or equal to 54, a total of eight work stations is needed, one of each type. When the demand just exceeds 54, a type 6 work station should be added to the line. This is done by adding the work station to the end of the last module or creating a new module for the station. The line capacity is 108 pieces per day. For the capacity between 109 and 126, four additional stations are included in the line: types 3, 4, 5, and 6. This procedure is carried out until the capacity exceeds the peak demand.

The utilization of each work station is computed by using equation (5) with $Q = 200$. We have

$$(u_1, \ldots, u_8) =$$

$$(0.54, 0.54, 0.81, 0.81, 0.81, 0.95, 0.81, 0.54).$$

If no buffer is allowed at work stations, utilization can be interpreted as the probability that a work unit cannot be fed directly into the work station (because of the Poisson arrival assumption). In this case, the unit is placed into the vertical storage area and retrieved later on demand.

Work stations 10 feet wide will be used. Each station, regardless of its function, consumes exactly one part container to make one product. Thus for every scheduled product, the DAH has to move three work units: one full container, one empty container, and the product itself. Each module has an

TABLE 1 – Line Capacity Plan

Capacity	Type of work station
54	1 2 3 4 5 6 7 8
108	6
126	3 4 5 6
162	7
189	2 6
217	8

input-and-output (I/O) port, located at one end of the carriage track. Work units that arrive and leave the module are handled through the port. The modules are connected by a conveyor, and each I/O port contains an accumulator so that a DAH can always successfully make a delivery to its corresponding I/O port.

We assume the same hardware features as DAH 2 (see the first section), except that the pick and place times are 4.5 seconds. The module contents and the DAH performance derived from Algorithm 2 are shown in Table 2. A layout of the full capacity line is illustrated in Figure 3 where work stations are distributed into both sides of the DAH aisle.

Two numbers are associated with each work station. The one at the upper right corner is the work station type, and the other (in parentheses) is a demand threshold. When the production demand reaches this threshold, the corresponding work station must be installed.

Initially, a line capacity of 54 is installed with eight work stations (one of each type) and two modules. The first module consists of work station types 1 through 7, and the second contains only one type 8. When the demand is increased, more work stations are added to the line, and the boundary line of the second DAH module is moved downward (see Figure 3 and Table 2). This process continues until the capacity reaches 162 pieces per day. For a capacity between 163 and 189, we create the third module that contains a single type 6 station. Adding another type 8 station, the line is capable of producing 217 pieces per day, which already exceeds the peak demand.

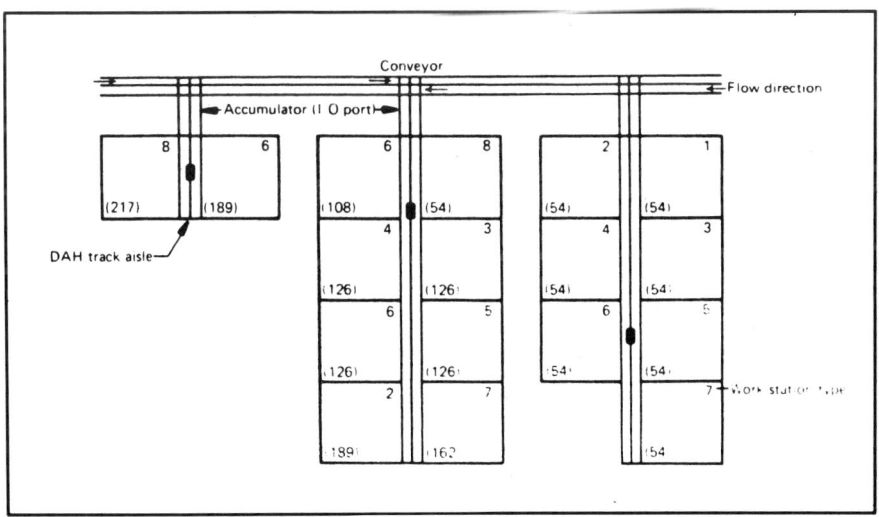

Figure 3. *A flexible line.*

TABLE 2 – DAH Performance Under Different Demands

Demand	Module	Module content	Average waiting time	Utilization
54	1	(1,2,3,4,5,6,7)	21 seconds	0.39
	2	(8)	13	0.03
108	1	(1,2,3,4,5,6,7)	40	0.74
	2	(8,6)	13	0.12
126	1	(1,2,3,4,5,6,7)	32	0.65
	2	(8,6,3,4,5,6)	20	0.38
162	1	(1,2,3,4,5,6,7)	36	0.71
	2	(8,6,3,4,5,6,7)	28	0.59
189	1	(1,2,3,4,5,6,7)	40	0.74
	2	(8,6,3,4,5,6,7,2)	47	0.78
	3	(6)	13	0.05
200	1	(1,2,3,4,5,6,7)	46	0.79
	2	(8,6,3,4,5,6,7,2)	40	0.71
	3	(6,8)	13	0.11

The procedure for contracting a line works in the opposite direction. When work stations are eliminated from the existing line, more space is created and can be used for another purpose.

The above numerical example assumes a simplified assembly sequence. In reality an assembly process may have a tree structure, if there are subassemblies, or a directed graph with cycles, if rework is necessary. Our procedure is not limited to any specific structure.

DISCUSSION

We have presented a flexible line concept and a procedure for line expansion and contraction. This is done by arranging work stations in a modular fashion. The line capacity can be dynamically adjusted by extending or compressing existing modules, and/or by adding or deleting modules. Since these actions can only take place at the end of the manufacturing line, line-disruption problems will not occur.

It is possible to reduce the number of direct-access handlers by rearranging the work-station layout. However, there are two problems: (a) One is not very likely to find an efficient method for the minimal solution, and (b) even if the minimal solution can be found, it does not guarantee the flexibility for nondisruptive expansion or contraction. Assume that the travel time of the DAH is small, as compared with the pick and place times, so that the DAH performance can be approximately determined by the service request rate and the pick and place times. For $C \doteq 0$, equation (2) becomes

$$E[W] \doteq E[S] + \frac{\lambda \ (E[S])^2}{2 \ (1 - \lambda \ E[S])}$$

where λ is the service request rate and $E[S] = g + h$.
Solving for λ, we have

$$\lambda = \frac{2 \ (E[W] - E[S])}{E[S] \ (2 \ E[W] - E \ [S])}$$

This expression defines the maximal service request rate that a DAH can support with an average waiting time less than or equal to $E[W]$. There is a service request rate associated with each work station. We want to pack the values of the service request rates into identical bins of capacity λ so that the number of bins we packed is minimal.

If the work stations are independent (i.e., no material flows between work stations), as a special case, the problem is reduced to the so-called "bin-packing problem" (see Graham [4]). No efficient algorithms for the exact solution are known. The problem is further complicated by work-station interdependency due to the existence of material flows between work stations. (Intuitively, we prefer to have two work stations placed in the same module if the traffice between them is heavy.) This verifies our first statement. When a heuristic method is tried for the minimal solution, we obtain a different layout (shown in Figure 4) where only two DAH's are used. The average waiting times are 55 and 54 seconds, respectively. If we start with a line capacity equal to 54 pieces per day, any future expansion must take place from within the existing line skeleton. This means possible line disruption. Furthermore, this layout creates a number of waste spaces before the line capacity reaches 200. Whether one should design for flexibility or for the minimal material-handling cost is certainly dependent on the cost structure. And the question cannot be answered unless the total manufacturing cost is evaluated. In the magnetic disk assembly area, a work station usually costs much more than a handler does. The former varies from the order of $100,000 to the order of $1,000,000, while the latter is between $50,000 and $150,000. Since the number of work stations can be an

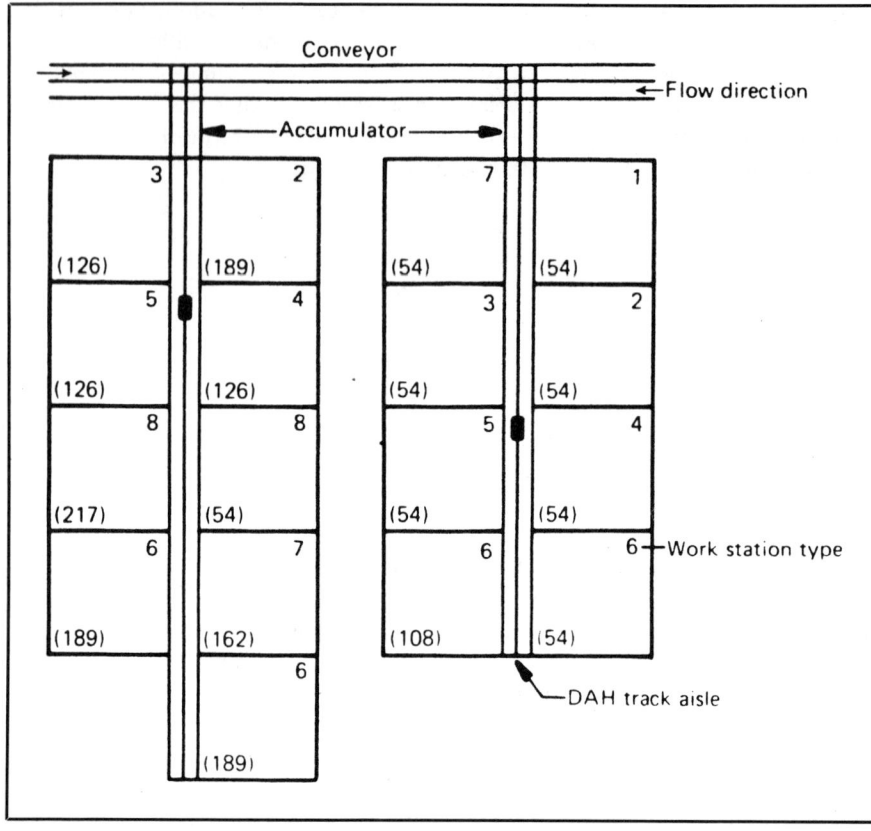

Figure 4. *A layout with minimal number of DAH's.*

order of magnitude higher than the number of handlers, it is justified for the flexible line concept.

During a line expansion (or contraction) process, one could also try to redo the layout of the entire assembly line for the minimal number of DAH's. Whether or not this is a correct approach depends on the new layout cost. This cost should include work-station rearrangement, reconstruction, possible line damage (machines and work units), operation slowdown and even temporary line shutdown. Every time an additional DAH is needed, as demand increased, we simply compare the DAH installation cost with the new layout cost and adopt the less expensive solution. If we anticipate that the peak demand can be reached rapidly, then it is possible to have a "minimal" number of DAH's without a new layout. For example, we may rearrange the second module in

Figure 4 such that the work stations for high demand thresholds are near the bottom of the module. Two DAH's (the minimal number for this example) are installed for the initial production line. As demand increases, work stations are merely added to the bottom of modules and thus a new layout is unnecessary. As a general procedure, this approach requires multiple DAH's for the initial line. Since the demand forecast is not always accurate, the minimal number of DAH's is not guaranteed.

In the second section it is assumed that some kind of material-handling device is used to connect the DAH modules. For a discrete type of device such as a DAH, the I/O port of each module can be regarded as a work station and the previous analysis of DAH's may be applied. For a continuous type of such as a conveyor, the important factor is the maximal processing rate, or equivalently the maximal number of requests processed per a unit of time. This rate is the ratio of traveling speed to the minimal distance headway on the conveyor. It is imperative that the rate of service requests for using the conveyor at each I/O port be below the maximal processing rate. Chow [3] compares a conveyor with a DAH and concludes that the conveyor, as a continuous type of device, has a much higher processing rate than the DAH. This implies that a conveyor, used as the module-connecting device, cannot be the bottleneck. On the other hand, since the arrivals at an I/O usually follow a random pattern, the receiving DAH may temporarily become a bottle-neck. In this case we may increase the conveyor length for the purpose of accumulation. This will increase the travel time between modules and consequently the work-in-process level.

The flexible line concept does not optimize the material flow. Instead, it attempts to optimize capital investment and space utilization by adding and deleting work stations with minimal line disruption. The proposed algorithm may force the same type of work stations to be distributed into different DAH modules. Suppose that there is more than one assembly operation. A completed work unit-from the first operation may be sent to a second type of work station either in the same module or in a different one. Selection of work stations should be carried out to balance work loads. When a remote work station is selected, the work unit will be involved in multiple material handling, e.g., DAH (sender)⟶ conveyor (module connector)⟶DAH (receiver). If this travel time is too long, we must allow extra work-in-process to keep the work stations from starving. The important thing is to make sure that the processing rate of each material handling device is higher than the service request rate.

Another technique for dynamic adjustment in line capacity is called "line replication." We install a low-capacity line. When the demand rises, we simply replicate the line. For instance, we may install four identical lines, each with a capacity of 54 pieces per day. In such a case, the line size had a great impact on the manufacturing system. If the size is too large, the system may not be adaptive to demand changes. If the size is small, the system would have a low utilization. The key problem is to determine the optimal single line size. This problem is further complicated by the uncertainty inherited from the market forecast. A study of a magnetic disk assembly line showed that under a fixed demand pattern a flexible line design, as compared to the line replication approach, can save about 16% of total manufacturing cost, including work station and space costs plus work-in-process, direct labor and operating expenses.

Finally, we point out that the module length can be adjusted to meet a building constraint. This can be done easily by imposing restrictions on the module lengths. If we have to install two or more independent lines, the optimal line sizes can be computed by the same technique developed for the well-known knapsack problem (see Gilmore and Gomory [5]). We use Algorithm 1 to determine the costs for different capacities and invoke the knapsack algorithm to find the minimal cost such that the total capacity is greater than or equal to the peak demand. The flexible line concept can be applied to each independent line.

Suggested Readings

D. R. Cox and W. L. Smith. 1961. *Queues.* Methuen.

J.D.C. Little. 1961. *A Proof of the Queueing Formula: L = λW.* Operations Research. Vol. 9. pp. 383-387.

W. Chow. 1983. *An Analysis of Automated Storage and Retrieval Systems in Manufacturing Assembly Lines.* IBM Technical Report TR 02.1082. IBM General Products Division. San Jose, California. (also to appear in IIE Transactions)

R. L. Graham. 1976. *Computer and Job-Shop Scheduling Theory.* edited by E. G. Coffman. Chapter 5. John Wiley.

P. C. Gilmore and R. E. Gomory. 1966. *The Theory of Computation of Knapsack Functions.* Operations Research. Vol. 14 pp. 1045-1074.

Reprinted from the March 1986 issue of *IIE Transactions.* Industrial Engineering Research and Development.

29

Process Approach to Planning a Successful Kitting System Is Outlined

Scott Conrad
Raymond Pukanic

The essence of successful manufacturing is having the right parts in the right quantity, in the right place, at the right time. The objective of a successful kitting system is to make this happen.

In today's electronics market success depends not only on having a good product (R&D) and good marketing, but also upon a low manufacturing cost. A large part of controlling the manufacturing cost is controlling the work-in-process (WIP). Controlling WIP is the objective of a kitting system.

A good kitting system is a manual or computer controlled means of tracking where material has been, where it is, where it is going, and how much of what got lost or damaged along the way. This article describes a systems or process approach to planning an effective kitting system (see Figure 1).

Like all systems, the kitting system must interface with other systems, primarily the production system (where the parts become a product), the stores or warehouse system (where the "raw" parts are stored) and the delivery system (the means of transporting material between stores and the production area, and between production areas).

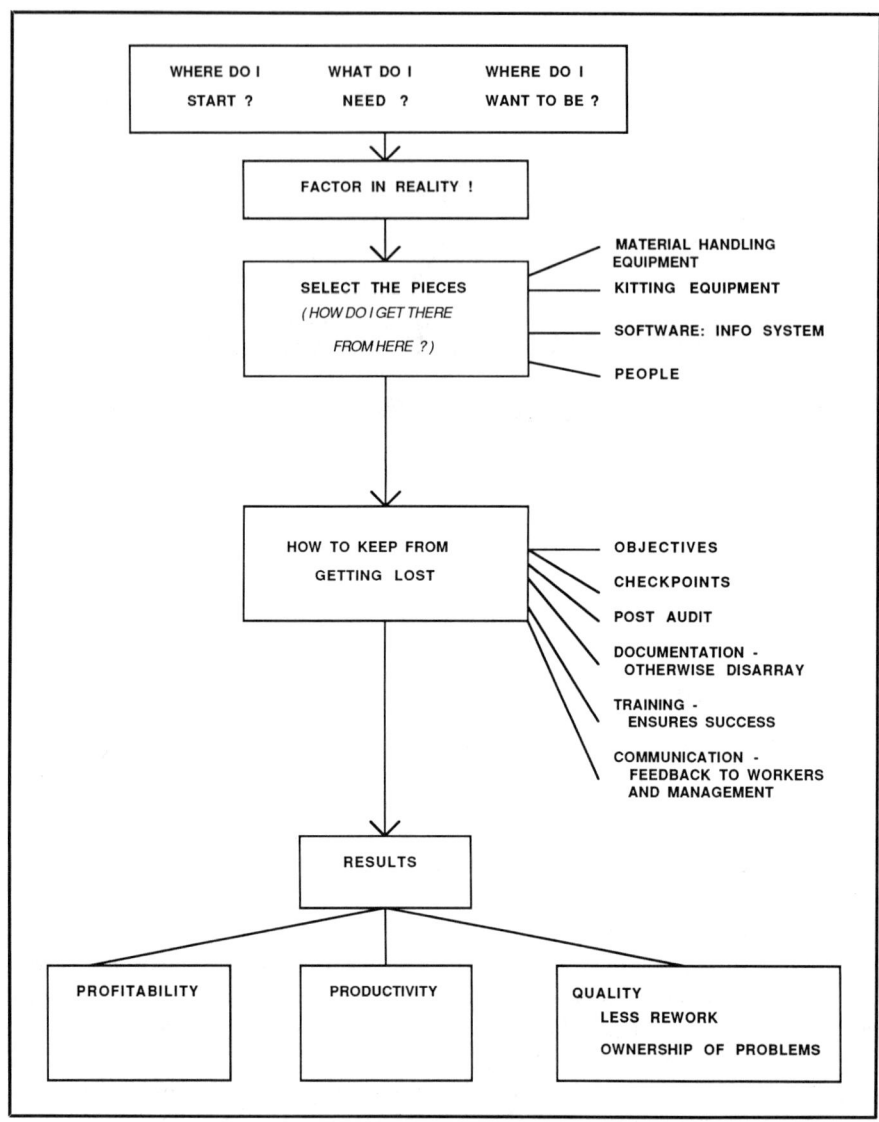

Figure 1. *A process approach to planning a kitting system.*

Inventory management is one of the key elements of survival in the electronics industry. Having a simple, easy to use and maintain system for moving material through the manufacturing process yields:

- *Minimum WIP*—Kit sizes are set to minimize WIP.
- *High visibility of WIP* kits = A predetermined quantity of parts, no guessing.
- *Minimum handling of material* (Stores puts a part in the kit and production picks it up only once to place it into a product).
- *Better methods in stores and production:* A planned and engineered approach for moving material versus an evolutionary approach.

SURVIVAL TEAM

To develop an efficient material flow process, a "survival team" composed of production management, materials management and industrial engineering must be formed. (See Figure 2.) This team focuses on better material handling and control. The place to start is at the beginning of the manufacturing/assembly process, with kitting. However, kitting does not end with the parts picking process. There has to be a close link between stores and all of the production/ process steps. The IE typically plays the key role of survival team facilitator in this interaction process from the up-front planning stages through implementation and post-implementation audit.

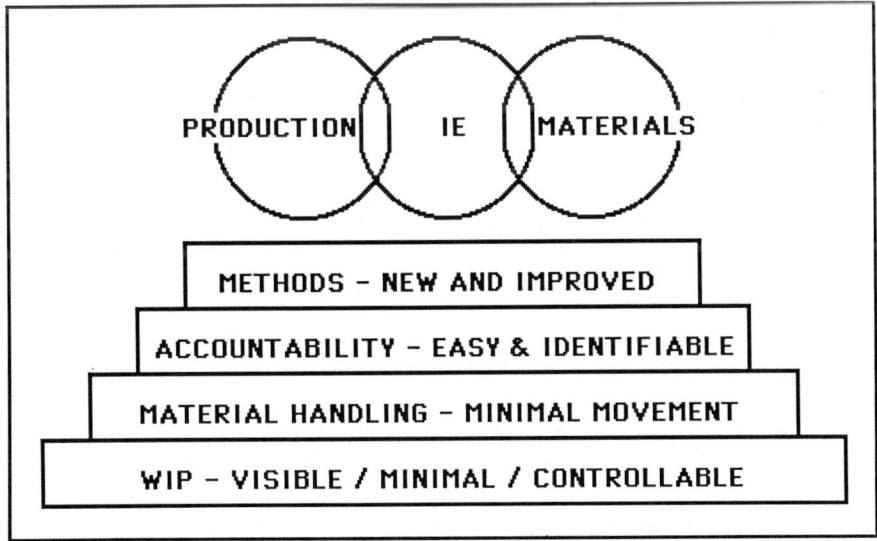

Figure 2. *Survival team success pyramid.*

When properly implemented, the following critical elements can favorably affect material handling costs in your facility:

- Of primary importance is *visible WIP*. Staging areas on the production floor need to be minimal, close to where they will be used and easy to locate and access. Many inventory dollars are tied up in redundant inventories that are "lost" on the production floor because there are too many places where parts can be hidden. This emphasized the need to also have control of WIP.

Ideally, the production manager should have only the materials needed for the moment, with the materials that are needed next ready to be delivered. There will, however, be some level of realistic material buffer (WIP), whether it's a few hours or a few days. A good control system is imperative for survival, and its accuracy and response become more critical as the WIP level decreases.

- *Minimal WIP levels*—minimal movement of materials and material handling protection with the proper equipment will minimize material shrinkage (material yield and process yield). The more material there is "lying around," the greater the chance it is going to be handled and/or rehandled. The more frequently materials are handled, the more susceptible they are to damage, particularly if they are not protected in appropriate material handling containers. "Materials typically spend 1.5% of their time in process and 98.5% of their time in handling"—Jim Tompkins
- *The control system and material handling logistics* allow easy accountability for both stores and production. A simple, but effective tracking system should replace reams of paperwork and multiple tags on materials. When you identify clearly the material handling and control responsibilities of stores and production personnel, their responsibilities become easier to carry out.
- *Better methods* of material handling and control are also supported by the survivor team. Existing methods should be periodically reviewed for ways of improving them or replacing them with new methods.

KITTING ISSUES

The physical constraints are some of the first items to address. The sizes of the materials will determine storage and picking methods in addition to stores layout.

Depending on the particular requirements, an optimal stores kitting area could consist of picking zones that included a combination of static shelving (where picking densities were low or materials bulky) and vertical or horizontal carousels (with higher picking densities). Pickers would be assigned to a zone rather than dedicated to a particular kit. This would eliminate queueing problems when more than one picker had to use the same carousel zone. The kits would move through the appropriate zones until they were completed (see Figure 3).

The electrical or electrostatic discharge (ESD) protection requirements can be critical for preventing damage to electronic component parts in storage and retrieval. The U.S. Department of Defense, a major contractor of electronic components and equipment, has set ESD protection criteria based on the classification of electronic device sensitivity. The ESD control handbook listed under "Suggested Readings" is helpful in identifying critical electronic components that will require special storage and picking areas or zones, and associated protection during transport.

WHERE DO YOU BEGIN?

First, let's compare where you probably are now to where you could be with an efficient kitting system. If you're like most electronics manufacturers, you are probably doing things the "same old way" (SOW). You build your product in large batches; i.e., monthly run quantities. The material is pulled from stores by part number and placed in bags and boxes stacked on pallets, then delivered to the largest opening in the aisle next to the line.

Production resorts the material on the pallets, making sure the right parts and quantities are available. Shortages get reported; overage usually gets thrown away. The assemblers sort through the pallets every day, picking the parts needed to build product for each day. The most visible aspect of SOW is a lot of

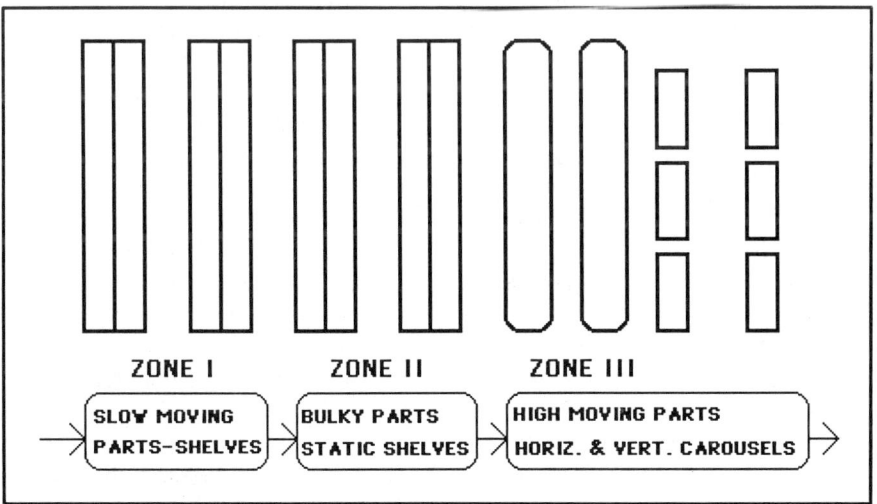

Figure 3. *"Optimal" kitting area.*

WIP sitting out in the open with a lot of people sorting through it—not an efficient way to build a high quality product.

With an efficient kitting system, one finds that manufacturing lot sizes tend to be much smaller; e.g., one day, one week or one at a time. The parts will probably still be pulled by part number, but since the kits are labeled and the quantity of each part small, wrong part and quantity errors are rare. The assembly people get the kits delivered right to the work station and can at a glance determine if a part is missing or in the wrong quantity. Shortages still get reported, and overage tends to stay in the kit so stores can spot it when the kit is returned and look into possible pulling or documentation problems. The real beauty of the system is that a part gets touched by human hands only twice—once going into the kit and once going into the product. A good kitting system is simple and efficient.

The next question is how to develop a good kitting system.

Setting up a material kitting plan requires resources: people, time and money. People are needed to plan, set up and implement the kitting system. Time is required to prototype and phase in a kitting system. Typically, it takes six months to a year to evolve to the right kit layouts, quantity and recycling system. Money is required to buy the totes, labels, carts, subcontainers and any other kitting hardware.

To get the money and the commitment of people to set up kitting, top management support is needed. If management does not recognize the need and potential savings involved with a kitting project, the project is not likely to succeed.

So how do you convince top management of the need for the project?

First, walk before you run. Start by outlining a plan for a small, well defined kitting project on a new product (if possible), or an existing product where kitting will offer a high payback. New projects are usually easier since they suffer less from "same old way" thinking on how material should be handled.

The following six steps are key to establishing a sound foundation for this project:

1. When you plan, set specific, realistic goals; e.g.:
- Maximum of three kits per assembly at all times: one in stores, one on the line and one in transit (maximum WIP = three times the number of assemblies).
- No kit will be delivered to the line unless all parts are included (partial deliveries end up as WIP waiting for the last part).
- Reducing assembly time at least 10% by setting up kits so that parts are in assembly order and the assembler does not need to sort the parts to check for right part and quantity.

2. Estimate the expected savings from the project; e.g.:
- Reducing WIP from two weeks to one week = $100,000 WIP reduction first year. Ongoing savings for eliminating holding costs of 25% or $25,000.
- Reducing assembly labor time 10% = $50,000 per year.
3. Estimate the expected costs of the project; e.g.:
- Additional stores labor to kit parts $35,000 per year including overhead = one full-time person.
- Engineering time: two man months = $12,000.
- Labor to set up kits: $5,000.
- Kitting equipment: carts/totes/bins will require an initial outlay of $50,000. The ongoing replacement or update of equipment will cost $1,000 per year.
4. Calculate the return on investment (ROI). In this example a simplified approach was used to calculate the payback and internal rate of return (IRR) (refer to the cash flow model in Figure 4). The following assumptions were used:
- A 50% tax rate was used for expense, savings and depreciation.
- A 10% tax credit was used.
- The capital was depreciated over five years using the straight line method.
5. Set up a gantt chart of what needs to be done by whom to successfully implement the project.

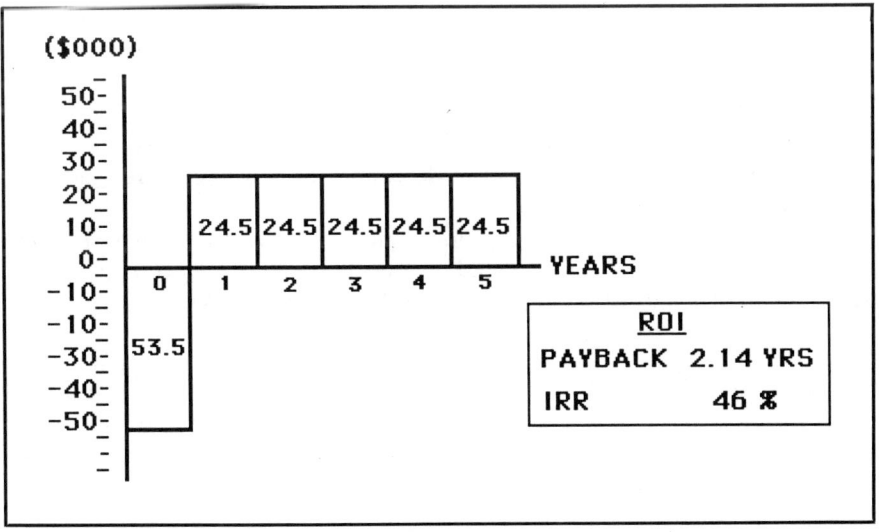

Figure 4. Cash flow model.

6. Finally, make a formal presentation of the plan to top management and all of the key people involved in the project. Communicate to all what the objectives of the kitting program are, their roles and the expected results. At the end of your presentations the support of the following groups is needed:
- Production line management.
- Materials/stores management.
- Champion—manufacturing manager or company president.

Once the key players are convinced the program is sound, you can proceed.

IDENTIFYING KITTING NEEDS

Define existing system

Now, let's step back for a minute and talk about how to define just what your kitting needs are. First, define the existing system. Flowchart the physical flow of material from receiving to finished product. Do not just flowchart what it should be or what people "think" it is. *Select a random sample of parts and physically follow them through the process.* As you go along, make notes on how the process can be improved.

Brainstorming session

Next, sit down with the materials and production people and brainstorm what the ideal material flow and delivery system would be. Be creative and allow crazy ideas; the evaluation comes later. The involvement you get among the end users through this process is key to the success of the project. People will take ownership of their own ideas and are more likely to make them work.

Even if you can't use all the ideas, at least everyone had the opportunity to participate in the planning process. This is particularly important with groups who may be resisting the change or in which the "not invented here syndrome" prevails.

Flowchart as best as you can the "ideal" material flow system. Now compare the existing to the ideal and begin asking questions about how you move closer to the ideal. The "ideal" system might include the following:

- Kit sizes of one.
- Zero WIP.
- No wrong parts.
- No paperwork.
- No counting of parts.
- No cycle counting required.
- Handling parts only once.

What are the realities of your materials/production situation?

- Multiple stock locations or central warehouse?
- Electrostatic discharge (ESD) sensitive parts with special handling and packaging requirements?
- Lot sizes that are fixed by production and material handling equipment? (e.g., auto-inserters, transport vehicle size, line fixtures, modular tote size.)
- Do existing information and accounting systems allow only batch transactions? (e.g.: if the system previously was set up for runs of 10 to 20 and lot sizes approach one, the computer can be overloaded with the increased number of transactions.)
- Kit sizes—What sizes and weights of materials need to be moved? (Human lifting ergonomic considerations versus what is needed.)
- Assembly sequences?
- Work station interface—Where does the "kit" or material get staged in the work area?
- Part storage by sequential part number (versus assembly sequence, FIFO, type/size of part, volume used, value, ESD sensitivity)?
- Part storage on steel shelving, carousels or a minitrieve system (horizontal or vertical carousel, aisle savers or other automated equipment)?
- Detrashing of vendor packaging—at stores or on line?
- Is vendor packaging compatible with the kits?
- Can the information system do "phantom pulls"?

The picture that comes to the stores manager's mind is one of a lot of partially filled kits in the warehouse. To prevent this, the information systems must be able to do "phantom pulls" to see if all the parts for a kit are available *before* the parts are physically pulled. This minimizes the probability of a kit's not being completed on the first pass.

SELECTING THE PIECES

At this point, one might wonder if kitting is possible and if so, if it is worth the effort. Don't get discouraged! It *is* possible and worth the time and effort to do it right. Give it your best shot!

Once a good understanding is achieved of what the product to be kitted is, how it is assembled, how the parts must be handled (weight, size, quantity, packaging) and how the stores operation works, one can begin to design the kitting hardware (totes, carts, subcontainers, etc.).

In designing the cart or transport method, the following questions need to be answered:

- Can it be loaded and unloaded safely?
- Does it fit down the aisles of the plant?
- Must it be secured in a truck for intersite transfer?
- Does it protect the parts from theft and the elements?
- How large are the carts?
- Can they be linked together in a chain?
- How much storage space do they require?
- Must the cart be grounded for ESD reasons?
- How many kits can a cart hold?
- How heavy are the parts?
- Will the carts be pulled by a fork truck, an AGV (automatic guided vehicle) or a person?

Transport systems can be as simple as a stack of totes on a pallet moved with a pallet jack or as complex as an automatic pickup and delivery AGV.

Totes come in almost as many shapes, sizes, materials and colors as one can imagine. A good rule of thumb to keep in mind when selecting totes is that form follows function. The function of the tote is to hold a subset of parts. This could be a set of small ESD sensitive components to be loaded onto a pc board, in which case a conductive tote with a lid would provide faraday cage protection. A well designed modular tote can be subdivided or partitioned to accommodate different sizes of parts, particularly small components.

Before buying a tote, think about how it can or will be labeled. Stores needs to identify what parts go into which totes, and the line needs to know what assembly is in each tote to build the product.

A big area of concern when looking for a tote is how big it should be. Some of the major items to consider when selecting a tote are:

- When full, can it be safely lifted?
- Does it fit well into carts and work stations?
- Are multiple depths of tote available with the same "footprint" to accommodate a larger range of part sizes?
- Can the tote be divided? (e.g., does it have ribs in which dividers can be inserted?)
- Do the totes "stack" so that parts inside are not damaged from another tote "nesting" on top?
- Do the totes have handles for easy manual lifting or for grabbing by a mechanical extractor?
- Do the totes have a lip for hanging by at work stations?

- Can the totes be collapsed for easier storage and transport when empty?
- Is the material the totes are made of static generative? (Many plastics hold a static charge that can destroy electronic components.) A quick check with a static meter will tell whether or not a tote is "hot".
- What colors are desired? Different colors for each assembly area? The best guideline regarding color is to keep it simple. To paraphrase Henry Ford: "Use any colors you want as long as they are all the same." The use of one color tote with different colored labels to distinguish between products or assembly types provides more flexibility. Also, labels are easy and inexpensive to change.

SUBCONTAINERS

Subcontainers are the "little boxes" or totes within the larger tote. Subcontainers can be simple cardboard boxes or specially designed boxes to hold fragile or sensitive parts. The selection of subcontainers should include consideration of the following:

- How will they be labeled for easy identification by stores and production personnel?
- Sizes—form follows function, and they have to fit in the tote, too.
- Materials—foam to protect fragile parts, conductive or antistatic material for ESD protection, or custom dunnage.
- Storage—Do they nest or stack? How are they kept in the tote?

INFORMATION SYSTEM INTERFACES

The kitting system is not just carts and totes; it is the right parts in the right quantity, in the right place, at the right time. The kit and its pieces must be clearly labeled to tell stores which parts go in which boxes, in what quantity, and when the kit must be delivered to production. The kit must also be labeled so that production can tell at a glance what the kit is (what product) and which parts are which, in assembly order, for fast, error-free assembly.

Production also needs to know where to send the empty kit. Meanwhile, accounting wants to know what parts are where and for how long. The information system needs to provide all this information, all the time.

Kits can be labeled with computer printed, typewritten or neatly printed labels. A key requirement is that the correct information be on the label in a font and character size that those who have to read it can easily see and understand. The labels must also be placed on the kits in a consistent fashion so that no time is wasted looking for the label.

A note here on automatic identification systems is appropriate. To track kits in and out of production areas, bar codes, magnetic stripes or optical character recognition may be used. Our suggestion regarding automated data entry is to start out manually, evolve exactly what information is needed for tracking purposes and then evaluate automatic data entry systems as a separate project.

One role of the information system is generation of the labels; the other role is keeping the system up to date. The system as we define it is a listing of what carts, totes and subcontainers are required for each job and what parts go into them. This is the documentation of what goes where. If this information is not documented, maintaining a kitting system will be impossible. Sooner or later, kits get mixed up, and if there is no documentation to get them back in order, they stay mixed up.

POST-IMPLEMENTATION AUDIT AND MAINTENANCE

Often people think the job is done once the system is up and running. This is seldom true. Someone needs to make sure the system is still running well and achieving the objectives that were set at the beginning of the project.

At three and six month intervals the process performance should be documented. Items such as inventory levels, inventory turns, WIP levels and production throughput should be tracked over time and related to goals. The good and the bad should be taken seriously and necessary changes or improvements made.

An ongoing maintenance plan is also important. The production line needs to take responsibility and ownership to ensure success. An ongoing production reporting system will provide feedback to production management and also point out successes in improving overall productivity.

CONCLUSION

The role of the IE and the "survival team" does not end once the initial system is set up. It must also be maintained and improved upon. For example, vendors and subcontractors could package parts in standard subcontainers for faster placement in kits; the transportation and data entry systems could be automated. A system is always changing, and if we do not keep up with the changes and improve the system, it quickly falls apart. A kitting system is no exception to the law of entropy, according to which an unattended system has a tendency to gravitate toward disorganization.

Kitting parts for efficient material flow is fundamental to survival in the competitive world of electronic manufacturing. The engineering of a material flow system helps control WIP by making it visible, minimizing handling and contributing to better methods throughout the plant. Designing an effective system requires a "process" approach (see Figure 1) involving production, materials, information systems and industrial engineering to get the right parts, in the right quantity, in the right place at the right time *all the time*.

Suggested Readings

_____. 1985. AGVS for assembly: flexible layout, easy expansion. *Material Handling Engineering*. September. pp. 58-62.

_____. 1985. Bar codes speed shipping, achieve 99.7% accuracy. *Modern Materials Handling*. September. pp. 88-89.

Davies, A. L., M. C. Gabbard, and E. F. Reinholdt, 1982. Storing warehouse stock for efficient selection. *The Western Electric Engineer*. Winter. p. 43-47.

Emerson, E. R. and D. S. Schmatz. 1981. Results of modeling an automated warehouse system. *Industrial Engineering*. August. pp. 28-32 & 90.

_____. 1985. *ESD Control Handbook for Protection of Electrical and Electronic Parts, Assemblies and Equipment*. U.S. Department of Defense Handbook No. 263. May.

Fitzgerald, K. R. Vertical carousels boost picking, conserve space. 1985. *Modern Materials Handling*. September. pp. 86-87.

_____. 1985. Ford takes a new material handling route to board testing. *Material Handling Engineering*. September. pp. 76-80.

_____. 1985. Just-in-time leads to better quality. *Manufacturing Guidebook*. Modern Materials Handling. p. 29.

_____. 1985. Pointers on preparing conveyor specifications. *Modern Materials Handling*. September. pp. 89-90.

_____. 1985. Storage technology trims the manufacturing fat. *Manufacturing Guidebook*. Modern Materials Handling. pp. 32-35.

Reprinted from the February 1986 issue of *Industrial Engineering*.

30

Key Problems and Solutions in Electronics Testing

Miguel Garcia-Rubio

The increasing significance of product testing on total product costs is a matter of much interest in the electronics manufacturing industries (Goel 1980). Testing costs of integrated circuits and circuit packs (printed wire board assemblies) are approaching 50% of total manufacturing costs. Fault detection and isolation in the field account for 35% of total maintenance time (Shemeta and Spillman 1983). Testing is a constant concern throughout the product development and manufacturing process that gets major attention when crises arise and is "forgotten" when the process becomes manageable. The driving forces or problems behind these crises are the high pace of growth of product complexity, the reduction of already short product life cycles, and the poor recognition, industrywide, that the highest leverage point in addressing testing concerns is early in the design stage. These problems account for high costs in manufacturing test system installation, and system maintenance. The ability to adequately test the system is a prime requisite for rapid fault isolation and correction. Thus, testability becomes a major consideration when developing any product. Design for testability (DFT) in electronics manufacturing has gained much attention (Williams and Parker 1983). However, because of a lack of a corporate testing strategy, many companies are having difficulty in committing to, and implementing, a specific DFT technique. I will present a more detailed view of these problems and discuss those solutions that I think are the most important.

PROBLEMS AND CHALLENGES

Manufacturing

The notion that product testing is primarily a manufacturing function in the electronics manufacturing industries is becoming obsolete (Riezenman 1987). Products are now too complex both in terms of the functionality they must provide and the development and manufacturing process they must go through to continue having designers not accountable for design-related manufacturing and field-support problems. Electronics technology is constantly and rapidly evolving, resulting in a product complexity growth rate that is outpacing the capabilities of modern test equipment. Functional testers require extensive programming for accurate fault diagnosis and usually take a long time to get running. Also, they provide a very limited input-output access to the device or circuit pack under test. On the other hand, in-circuit testers overcome the programming requirements and test point access limitations by using a "bed of nails" test fixture. The latter allows for direct probing of components on a circuit pack and independently tests them. However, automatic test equipment (ATE) will not meet an increasing percentage of testing needs for future products. The reasons ATE in-circuit test equipment is expected to decline in usefulness for future products are several. The difficulty in mechanically fixturing dense boards (i.e., double-sided surface mount), extremely fast circuit speeds, and excessive demands on memory to store device vectors are among the most troublesome, especially considering that ATE is the most expensive piece of equipment on the factory floor. However, in-circuit ATE has many alluring qualities that make it difficult to abandon. For example, test programs are relatively easy to "fix" when design changes occur by simply adding and subtracting test vectors for particular components, assuming no fixture changes are required.

An ideal time to gather information on the status of a circuit pack assembly process is when the pack is being tested. Most ATE have automatic data collection and networking capabilities. With the proper data reduction techniques, much can be learned about the quality of the product design as well (Yav 1987). To differentiate between design and manufacturing problems, the data collection, processing, and reduction to information has to be done on a factorywide basis. These require a factorywide networking strategy that becomes difficult to implement in the face of short product and technology life cycles which render many ATE on the factory floor obsolete. In general, device and circuit pack testing are primarily done to "test" the process, and system testing is primarily done to "test" for product performance and compliance to specifications (Davis 1982).

Proliferation of ATE is a symptom, not the problem. The main reason behind this is the lack of a corporate testing strategy. Without a "big picture" perspective, individuals and organizations within most companies have no clear understanding of the link between the end-to-end test process and material flow (and yields) on the factory floor. Poorly defined standards for fault coverage, no management encouragement of designers to make test a design function, and efforts to automate the factory floor without a corporate commitment to simplify the test process keep fueling the high costs of testing. If manufacturing continues to be given the largest share of product test concerns, symptoms like proliferation of expensive ATE, long idle times, scrap of (possibly good) boards that become "untestable," and use of production test stations for time-consuming diagnostics and repair will continue to manifest.

Design

The other side of the coin is design, where test program generation is currently the main problem (Goel 1980). Long test development times can result in late product introductions with corresponding loss of profits. The problem gets even more complex when test development times become, in duration, comparable to the already short product life cycles. Long test program development times result from not addressing testability early in the design process (Shemeta and Spillman 1983). There are few constraints on design engineers to consider the economics of test development and manufacturing test when they design their products. In many cases, design for testability (DFT) only occurs after the design is complete and about to be introduced to manufacturing (Riezenman 1987). The outcome is that at this late point, testability becomes very expensive to implement. There are many success stories in which DFT has simplified and reduced test generation efforts (Timoc 1984). Long test development times are very costly and seem to be growing exponentially. At issue here is fault coverage: if the required level must be close to 100%, then resulting test development costs may be excessive and test program execution times on the factory floor may be too long. On the other hand, if fault coverage is relaxed, there is the possibility that some products (mostly devices and circuit packs) may pass the test and fail during system test or malfunction in the field, with the corresponding capital expenditures in diagnosis and repair. A vicious cycle may even develop in which a circuit pack passes an in-circuit test and fails the system test repeatedly. Defect isolation is very time-consuming at best and often represents a substantial percent of installation and repair costs on large systems (Shemeta and Spillman 1983).

Complicating the aforementioned problems is the inadequate link between design and test. Asking designers to think about a particular test set when "building" testability into their designs can be a nightmare in face of the great diversity of ATE. Once designers have successfully introduced a design to manufacturing, a new wave of ATE technology comes in, forcing manufacturing to buy different and incompatible models and brands. Even a different configuration of the same model can prove troublesome. That different model forces the designer to "get acquainted" with a new piece of test equipment and develop specialized test programs which will not be useful the next time around, lengthening the test development process and requiring manufacturing to debug previously correct programs. Translating files of simulation results into test programs is a nonstandard and nontrivial process that is gaining attention.

SOLUTIONS

The following paragraphs present what I consider the most important solutions proposed in industry.

1. Companies must understand the "end-to-end test process" as a set of operations that span the entire concept-to-customer process. Quoting Lewis E. Platt from Hewlett-Packard (1986), *"The same old way doesn't work any more. No amount of automation will work if our basic assumptions and expectations are obsolete...the 'find it and fix it' approach to quality won't work."* This understanding has to be followed by a long-term objective and commitment to a more disciplined and simplified test process, where design, manufacturing, test, and material flow procedures are integrated.
2. Recognize that the biggest payoff results from addressing product test concerns at the beginning. This recognition requires an up-front testing strategy for all levels of product integration that addresses design, manufacturing, test, materials, installation, and maintenance before the product is designed. Borrowing from Vishwani Agrawal of AT&T-Bell Laboratories (personal communication), we must stop being *"device-wise and system-foolish."* In addition, there must be a corporate commitment to testing standards like fault coverage, testability measures, test equipment, networking, surface mount technology, and documentation.
3. Form a product development team that incorporates members from the above functional organizations where testability becomes a major design responsibility. Testability must be recognized as a design function with DFT instead of expensive ATE, as the most effective vehicle to control test generation and manufacturing test costs. This implies methodology constraints at design that focus on manufacturing test needs in a way which does not hinder the creative process. For example, the level sensitive scan design (LSSD) technique developed by E. Eichelberger at IBM became an industry success (Eichelberger and Williams 1977). It restricted the design style but allowed for easier testing and diagnosis (Radke 1986). Integrating design with test must also be addressed at the tool level and requires CAE/CAD tools that support DFT in a form transparent to designers, minimizing

nonvalue—added functions, such as translating data from one data base to another, error correction, and audits. Of particular importance is automating test program generation and simplifying testability analysis, which currently require extended manual involvement (Williams 1987). The process of converting design data into test programs must be automated. Companies must strive to integrate their CAE/CAD tools with the test equipment to be used on the factory floor.

4. Commit to built-in self-test (BIST). One of the best approaches to system testability is BIST. According to Vishwani Agrawal (personal communication), *"One nice thing about BIST is that it describes the future of testing in one word."* BIST is the internal capability of a device, circuit pack, or a system to test itself. It is a technology and design discipline that "builds" the test set into the product and allows for a test capability growth which can keep pace with future technologies. Although software-based BIST is now a common feature at the system level, implementing it at device and circuit pack levels is very difficult. To guarantee adequate fault coverage, the latter requires test sequences that are too long, thus demanding more memory than available on ATE. It also lengthens the test process, which can result in major material velocity reductions on the factory floor. Hardware-based BIST overcomes these problems by using additional circuitry that eases the testing of the device (board) and interdevice (inter board) connections using internal device (board) capabilities. This self-testing function is initiated and controlled externally.

Although BIST has been successfully applied in many cases (Timoc 1984), a standard technique for testing general sequential logic present in most VLSI device is still lacking. Also, to accommodate the additional circuitry, BIST imposes a real-estate overhead on the device or circuit pack floors that translates into lower yields. Thus corporate commitment to BIST standards is very important. Companies must develop BIST strategies that address their internal needs and those of their suppliers and customers. Developing CAD tools that support BIST design is particularly important and requires major test standards efforts, such as the IEEE Testability Bus Standardization Committee, where several standards are being considered. Among the most important are proposals from the Joint Test Action Group, with a boundary scan test-bus aimed at commercial applications, and the very high speed integrated circuits test-bus, developed mostly for military applications (Pradhan et al. 1987). Currently an effort is under way to combine both standards (private communication).

Finally, an integrated BIST strategy requires reusing device-BIST at circuit pack and system levels (see Point 2). In the long run, this will be the best way to reduce costs in system installation, field maintenance, and repair.

5. Develop a strategy and implementation plan for computer integration of the manufacturing (CIM) process. Although the scope of CIM is much broader than just testing on the manufacturing floor, it nevertheless is heavily dependent on test information flow. Manufacturing testing can be seen as a production control mechanism in which test information is used to correct manufacturing process problems in real-time. For that purpose, companies must have a CIM strategy that integrates the different test stations in terms of the following:

- a clearly defined test strategy (see Points 1 and 2);
- automatic data logging and reporting capabilities;
- data base standards for test data;
- application programs for reducing test data into useful information (this implies developing both statistical and knowledge-based methods to improve test diagnostics);
- networking strategy that defines protocols and communication standards; and
- a link to the design process in which test process information from the factory floor can be fed back to correct design problems (see Point 3) and improve product reliability.

CONCLUSIONS

In their effort to increase productivity, companies must develop and commit to a long-term, "cradle-to-grave" test strategy. Integrating the different disciplines to develop and produce products is, in my opinion, the only alternative. Understanding how design mistakes, test, and repair affect both material flow in the factory and customer support in the field leads to simplified and thus harnessed test processes. According to Dr. Charles M. Savage from D. Appleton Company, Inc. (1985), . . . *"it makes little sense to computerize the contradictions and confusion of existing processes. Instead, it is necessary to 'simplify the complex'." This simplification requires an up-front testing strategy that has BIST as its most important element. Any automation effort will have to come in the form of CAD/CAM tools that support a simplified test process implementing the previously mentioned strategy.*

Suggested Readings

Davis, B. 1982. *The Economics of Automatic Testing.* McGraw-Hill Book Company (UK) Limited. London.

Eichelberger, E. B., and T. W. Williams. 1977. *A logic design structure for LSI testability. Proceedings of the 14th Design Automation Conference.* pp. 462-468.

Goel, P. 1980. Test generation cost analysis and projections. *Proceedings of the 17th Design Automation Conference.* pp. 77-84.

Platt, L. E. 1986. The same old way doesn't work any more. *IEEE Design and Test of Computers.* February. pp. 82-83.

Pradhan, M. M., R. E., Tulloss, H. Beenker, and F.P.M. Beenker. 1987. Developing a standard for boundary scan implementation. *Proceedings of the IEEE International Conference on Computer Design.* Rye Brook, NY. October. pp. 462-466.

Radke, C. E. 1986. Experiences in VLSI testing. *IEEE Design and Test of Computers.* p. 83.

Riezenman, M. 1987. Linking design to test. *Electronic Engineering Times.* p. T6.

Savage, C. M., Editor. 1985. *A Program Guide for CIM Implementation.* The Computer and Automated Systems Association of the SME. Dearborn, MI.

Shemeta, E., and R. Spillman. 1983. Computer-aided testability design analysis. *RADC-TR-83-257 Final Technical Report.* Boeing Aerospace Company.

Timoc, C. C., Editor. 1984. *Logic Design For Testability.* IEEE Computer Society Press. Silver Springs, MD.

Williams, T. W., and K. P., Parker. 1983. Design for testability — a survey. *Proceedings of the IEEE.* 71, No. 1, pp. 98-112.

Williams, W. 1987. An automatic test generator for programmable logic devices. *Proceedings of the 1987 IEEE International Test Conference.* pp. 658-667.

Yau, C. W. 1987. *ILIAD: a computer aided diagnosis & repair system.* Proceedings of the 1987 IEEE International Test Conference. pp. 890-898.

Miguel Garcia-Rubio is a systems engineer in productivity and quality improvement for AT&T Bell Laboratories, Holmdel, New Jersey.

31

Electronics Parts Handling in Receiving Inspection

Myron F. Wilson

The need for improved productivity at all levels in all companies is well accepted. Competition, particularly with foreign countries, has become very keen during the last several years. World distribution capabilities have given foreign competitors direct access to American markets. These foreign countries usually have significantly lower labor costs due to lower standards of living. At the same time, due in large measure to governmental efforts, many of these countries have significantly improved the quality of their delivered products to be on par and often better than comparable American products. At best, the difference in labor costs will only change over a long period of time. Therefore, American companies must direct additional effort toward productivity issues.

In attacking the productivity issue, one traditional approach is by means of capital investments—automating manual tasks and those that require high skill levels of the individual operators involved. Great improvement are possible and, in fact, are being accomplished. Robots of all types with ever-increasing capabilities are being placed in the production process. Although initial applications were limited to undesirable environment and repetitive boring tasks, more recently much broader applications are being used. Electronics techniques incidentally have had a significant impact in this automation improvement with the increasing complexities and capabilities of microprocessors at ever-decreasing costs. Capital investments are certainly and excellent approach to the productivity improvement requirement. However, capital available is limited to any enterprise.

Therefore, other techniques are needed which will result in improved productivity without the need for cash. Significant productivity improvement can be made with small capital investments, permitting the available money to be spent where the capital expenditure approach is necessary.

Innovative approaches usually entail a measure of risk—the risk of eventual cost and the risk of customer concerns. Risk must be traded with opportunity. However, with careful consideration of the various facets, a proper analysis can occur. In these types of approaches, a total system analysis is usually necessary, rather than merely changing an individual function.

This chapter describes such a program. When the program was initiated, it was recognized to have some risk. However, experience has shown that the increased productivity and the savings accrued were an excellent trade-off. There still remains some risk—one being customer acceptance of the technique— when the customer feels the need to specify a given process due to his perceived need to control some element in the final product, such as quality and reliability.

Individuals and companies involved must be willing to assume such risks to experience the important yield of increased productivity. Productivity can be improved while actually conserving capital and other cash expenditures.

The approach described here is a change in the way electronic piece parts are handled in the incoming inspection and test function. For many years, as quality control techniques have evolved, incoming parts inspection has been refined and increased. The availability of complex, fast test equipment has resulted in many companies changing from sampling plans to 100% testing. At the same time, many piece parts are becoming more complex, extremely fragile, and susceptible to damage from handling due to electrostatic discharges.

As we observed these increasing test requirements and projected implied future needs, it became apparent that even more complex testing equipment would be required to continue the traditional inspection approach. The software requirements to service the test equipment and to tailor the tests for individual piece parts would also become an expensive item. At the same time, it was recognized that this effort was duplicating what the supplier of the parts should be doing. Conceivably, this position would be multiplied by each of the thousands of companies receiving a given part from a single supplier.

The question was asked, "Why not place the responsibility directly on the supplier? Why not set up a comprehensive program, tell the supplier what is expected, and arrange to furnish information the supplier needs to deliver only good parts?" Give the supplier the responsibility and expect the supplier to perform!

INVESTIGATION AND ANALYSIS

We also realized that the capabilities of suppliers vary widely, as well as their desires to participate in a program such as we were considering. Some complex parts are very difficult to test for all parameters and even some suppliers have limited capabilities to perform total comprehensive tests. The first step before involving the various suppliers was to fully understand our initial situation of the quality level of the parts we were using in our production environment.

We started by analyzing the parts that were removed during the production testing process. We found that three types of parts caused more than 65% of all removals. These were microcircuits, capacitors, and transistors. (Figure 1.)

The number of removals, of course, depends on the number of parts used. Consequently, we analyzed the parts installed over a period of one year and specifically observed the three commodities identified. We found these three part types comprised only 51% of all parts used. The individual part types had removal rates of 1.0% for microcircuits, 0.2% for capacitors, and 0.7% for transistors. The total of 24 million parts used experienced a removal rate of 0.35%. (Table 1.)

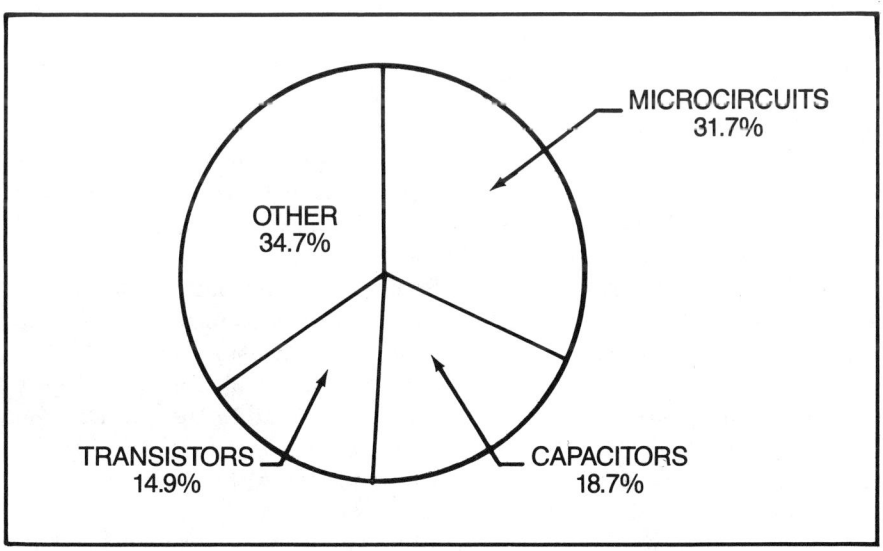

Figure 1. *Part removal distribution*

TABLE 1 – Removal Rates

Commodity	Removal Rate	Usage
Microcircuits	1.0%	3.5 million
Capacitors	0.2%	7.2 million
Transistors	0.7%	1.6 million
All Parts	0.35%	24.0 million

During the year, 84,000 removals were performed. Costs per removal, considering many factors, may vary from $50 to $100 per removal. Thus, the calculated cost for these actions range from 4.2 million to 8.4 million dollars—a significant cost item. These were parts that had been tested during incoming inspection.

Parts are removed for various reasons and are generally regarded as defective parts if the replacement process results in the product performing correctly after the change. However, the removed parts are not always defective. Other reasons for removal include design margin or misapplication, improper test and diagnosis, and workmanship errors, such as poor solder joints. To determine what portion of these production removals were actually defective parts (those parts that fail to meet the specified requirements), random samples of parts were subjected to extensive failure analysis.

This verification process is difficult and time-consuming. Often when the parts are removed, the person doing the removing is not careful and breaks the leads or otherwise damages the part in rough handling. In retesting the part, all specified parameters must be examined, not just those observed in the normal incoming test activity. The parts must be carefully examined in a laboratory environment by a knowledgeable technician. Sample reexaminations have been performed several times over a period of two years. The data shown in Table 2 indicate that only about 25% of the removals were actually defective. These results are also supported by Hewlett-Packard's Loveland Instrument Division findings, where more than 75% of the replaced parts met their specified requirements (Sanders 1984). Detailed analysis indicated 36% of the removals were directly related to the misapplication of the devices.

These data, recognized to be limited, show the removal rate due to defective parts is approximately one fourth the total removal rate. Therefore, if the total overall removal rate was 0.35% as shown in Table 1, the rate of defective parts was approximately 0.09% or 900 parts per million (ppm).

TABLE 2 – Defective Parts in Production Removal

	Sample #A	Sample #B	Sample #C
Sample Size	67	159	215
% Defective	21	35	26

These defect rates occurred during a period of time when our standard incoming inspection process was 100% testing of active components and a 1% AQL sampling of each lot of passive components. Typical of most incoming inspection activities, only a relatively small number of possible parameters were tested from each lot. Examining the history of lot rejections during this time, the reject rate of the three predominately rejected commodities was significant. However, most of the rejections were due to administrative problems, such as wrong parts received or wrong markings rather than functional defects (Table 3).

TABLE 3 – Incoming Inspection Reject Rates

Commodity	Total	Administrative	Functional
Microcircuits	5%	4%	1%
Capacitors	4%	3%	1%
Transistors	4%	3%	1%

As a part of the supplier quality investigation, a program was developed and implemented in production to obtain data necessary to measure a supplier's performance concerning the quality of his parts. This program included reporting all part removals, supplier identification, reason for removal, and other production-related information.

Based on the reason for removal, the actual defective parts were separated from the removals for other reasons. Periodically, the accuracy was verified by comparing the results obtained from a thorough failure analysis to the reason stated by the production technicians. This ongoing analysis compared very favorably with the 900 ppm previously calculated.

Reject rates of 4% and 5% experienced in receiving inspection would appear to be significant in preventing grossly bad parts from reaching the factory floor.

However, the data indicate that predominant reasons for lot rejections were administrative and not functional. The functional reasons were only about 1%. To quantify the true benefits of the incoming inspection that was being performed, several detailed investigations were conducted. The most significant investigation involved 27 selected microcircuit part numbers. For a period of three months, electrical attribute testing was discontinued for these devices. The subsequent removal rate in production was very closely monitored using the program initiated in production and descibed earlier. During this period, there was no significant change in the production removal rates compared with those experienced before, the investigation. Parallel investigations during this same period of time involving other commodities also revealed minimal benefits from incoming tests of functional attributes.

SUPPLIER QUALITY IMPROVEMENT PROGRAM

Thus, we decided to initiate a comprehensive supplier quality improvement program. Supplier responsibility for product quality was emphasized reducing dependence on us, his customer, to perform any required sorting operations. The program was established with four main elements.

The first, a key element, was to establish a production data program. (Getting production personnel to properly document and retain adequate records concerning their actions has always been difficult.) Primary issues covered in this data improvement program were as follows:

1. complete and accurate information covering all removal actions with the reason for removal specifically noted;
2. vendor identification for each part removed to cover situations where multiple vendors may be furnishing a given part number;
3. detail noting the location with an electronic circuit where the part was found to be performing inadequately; and
4. retention of removed parts with the special data noted previously, thereby permitting retest, review of the detailed circuit applications, and return to the supplier if desired.

The second element of our supplier quality improvement program was a planned, documented program to improve the relationship with each of our major vendors. The production data, described previously, were used to communicate to the various suppliers the detail performance of their piece parts in our specific production applications. This information was conveyed to each supplier during periodic meetings in which the program was explained and detail information exchanged. During these meetings it became apparent which suppliers were interested in our program and willing to assume full responsibility for the quality for their parts. Over the last two years we have reduced the number of suppliers by 22%.

The third element in our program was reduced testing of piece parts in the incoming inspection process. Due to the high number of administrative problems occurring, all incoming lots continue to be observed for count, correct part, and proper markings. This continues to be a situation that suppliers seem to have difficulty solving. Although all lots are visually examined, only about 2% of the parts are subjected to a functional test. These part numbers, which are continually changing, are generally parts in critical applications where removal is difficult or very expensive or initiated by production line complaints. The actual parameters tested are limited to those necessary for the particular applications.

The fourth element is a result of the third. The resources of quality engineers that were formerly assigned to the functional testing aspects of incoming inspection were reassigned to the production areas to follow up and take corrective actions when parts were removed. Thus, many more application situations can be reviewed in production than would be possible without the additional resources. No organization can have all the quality specialists they would like, and resources must be given priority. Remember, only 25% of the removals were bad parts. Thus, it is better to concentrate on the 75% segment that is involved with other than bad parts.

RESULTS

By modifying the incoming inspection activity to only examine for administrative errors as noted earlier, the productivity of the individual inspectors improved significantly. As displayed in Figure 2, the number of lots per inspector increased from 10 lots per day to 34 lots per day—an improvement of 240%. This reduction in inspectors is estimated to save about $1.5 million per year.

The cost savings due to the reduced testing level is obvious. However, other savings resulted that are not so obvious. If the change in philosphy had not occurred, we would have had to invest in test equipment to test ever-increasing complex microcircuits. The cost avoidance would not only include the initial equipment investment, but also the ongoing costs of programming tests to be performed. Software programming will be more and more difficult in the future; both in terms of cost as well as securing the type of people capable of performing this function.

Another saving resulting from the reduced testing concerns the reduced inventory level of piece parts. Our experience had been, and is usual in the industry, to have between one week and two weeks of parts between the dock door and the stock room. It is common to have dock-to-stock computer systems

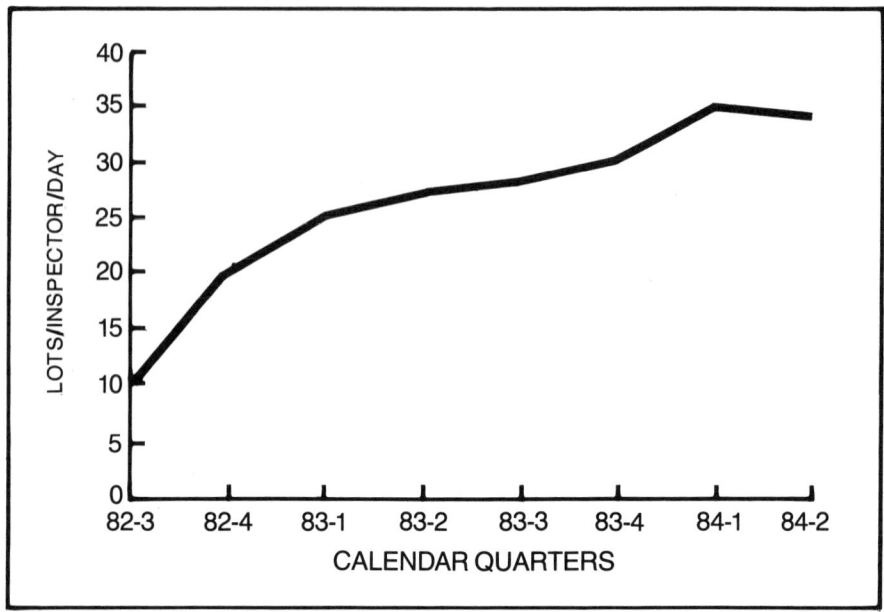

Figure 2. *Incoming inspection productivity*

to keep track of this volume of parts to know the location at any time and permit expediting particular parts whenever necessary.

The results of the inventory reduction are shown in Figure 3. The number of lots received per month is on the left scale and the number of lots in backlog at any time is on the right scale.

The total lots received have been reduced over the period due to general inventory reduction actions in conjunction with a new production control system installed in the factory. Figure 3 indicates that at the present time about 8,000 lots per month are being received, but only about 40 lots remain in backlog. These 40 lots are primarily rejected lots held for material review board action. Physically, parts are received by a truck in the morning, unloaded, unpacked, inspected, repacked, and sent to stock before the end of the day. This reduction in parts inventory from 1 to 2 weeks to 1 day is a real savings. No longer is it necessary to expedite parts within the incoming inspection process to handle production line shortages. Consequently, a dock-to-stock computer system is not necessary.

When reviewing the savings by reduced incoming test cost, the obvious question is what has been the effect of the program on the production activity.

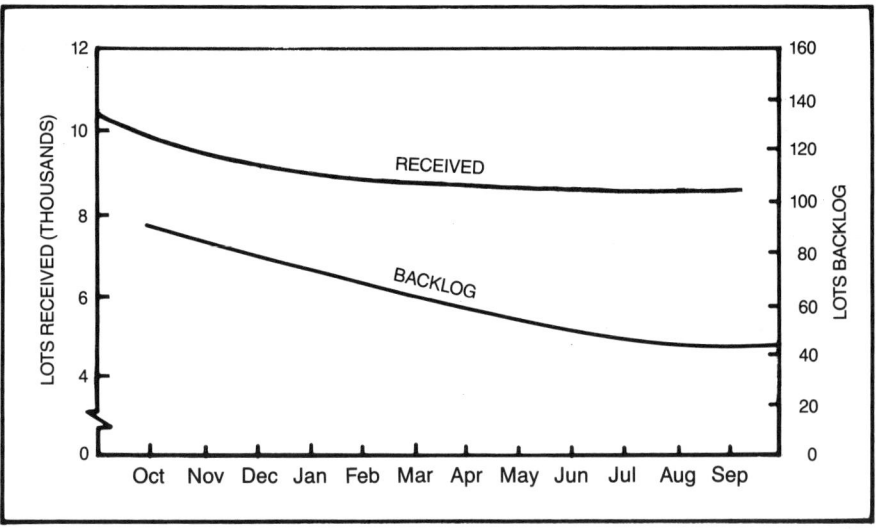

Figure 3. *Receiving inspection*

Obviously problems uncovered during the production process are much more expensive to find and fix than those identified during a piece part inspection/test process.

The resulting test repair experience is shown in Figure 4. During this time, production test repairs have been reduced from 8.5 per unit to about 4 per unit, a reduction of 52%.

The reduction in test repairs is the result of more than just better parts from the suppliers. As pointed out earlier, the quality engineers were reassigned to work problems in production and to attack the 75% segment of the reasons for parts removals. These results would certainly indicate that the elimination of parts testing did not cause any increase in problems later in the process. This reduction in test repairs per unit is estimated to have resulted in about 24,000 fewer part removals per year. Using an average cost of a repair action mentioned earlier, this savings would be between 1.2 million and 2.4 million dollars per year.

Reviewing the benefits of our supplier improvement program, we have experienced increased productivity, cost savings, and cost avoidance and improved piece part quality. Some of the economic benefits are listed in Table 4.

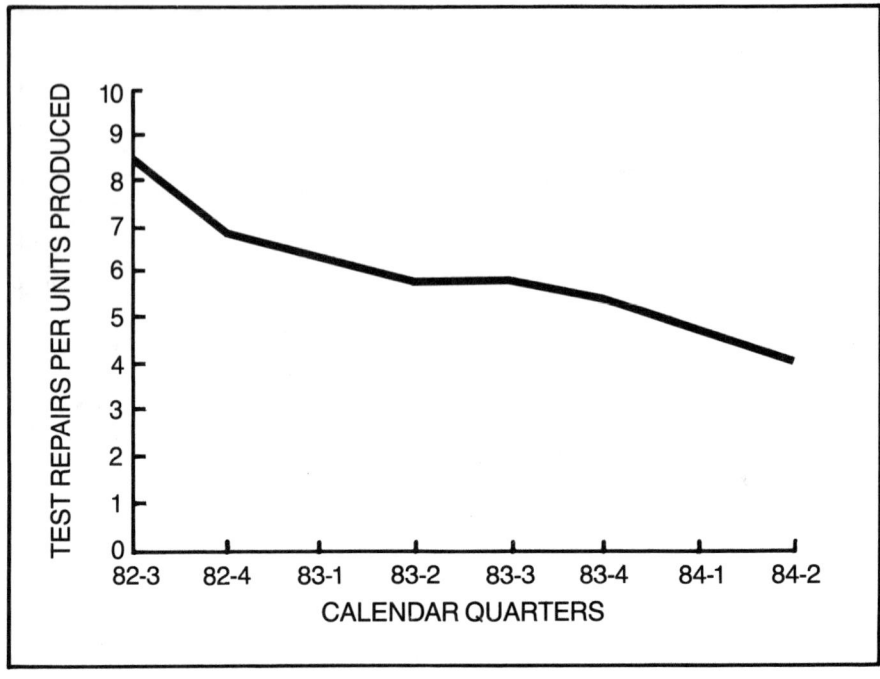

Figure 4. *Production test repairs*

TABLE 4 – Economic Benefits

Factor	Improvement
Incoming inspection productivity	240% increase
Incoming cost savings	$1.5M per year
Reduced inventory in process	8 days to 1 day
Production test repairs	52% reduction
Active supplier base	22% reduction

FIELD PERFORMANCE

For any endeavor the effect on the customer must be considered. When significant productivity gains are attained, the customer must be closely monitored to ensure no adverse reaction. We have for more than 20 years maintained a program of customer communication in the area of equipment reliability performance. The program, call "RECAP" (reliability evaluation and corrective action program), collects actual airline reliability performance data. Presently some 20 different carriers are currently participating in the RECAP program.

Production of a new commerical avionics system was started at the same time the supplier quality improvement program was implemented. This provided an excellent opportunity to measure the true effectiveness of the program and to recognize the effect on the removal rates of the new equipments.

Using the RECAP program described earlier, we are able to conclusively monitor the overall program. Although the data is still being analyzed, more than 9 billion part hours of data have already been collected. Initial results indicate that some part types are experiencing failure rates five times better than predicted. This significant improvement may be the result of digital circuit techniques in the particular equipments.

The field performance of nine new products has been compared with predicted results. These nine products include computers, as well as avionics senors such as radios. All these product use digital circuitry. Field experience of more than 300,000 aircraft flight hours from seven different operators was included in this study. The MTBF predictions used were computed using traditional parts count reliability prediction techniques (as described in MIL-HDBK-217.) The average field performance of these nine products, assembled with parts not functionally tested in receiving inspection, is shown in Figure 5. The predicted reliability performance is also shown as a reference.

From Figure 5, the actual MTBF performance is observed to be at a level approximately 2.3 times better than predicted. The rate of growth is very close to what was expected.

The results of our supplier quality improvement program have been excellent. Productivity has been increased throughout the production process. The suppliers have done a good job in improving the quality of their products, and the production quality engineers have improved the removal/bad part ratio. The improvement in removal rates and the quality of the piece parts is shown in Table 5.

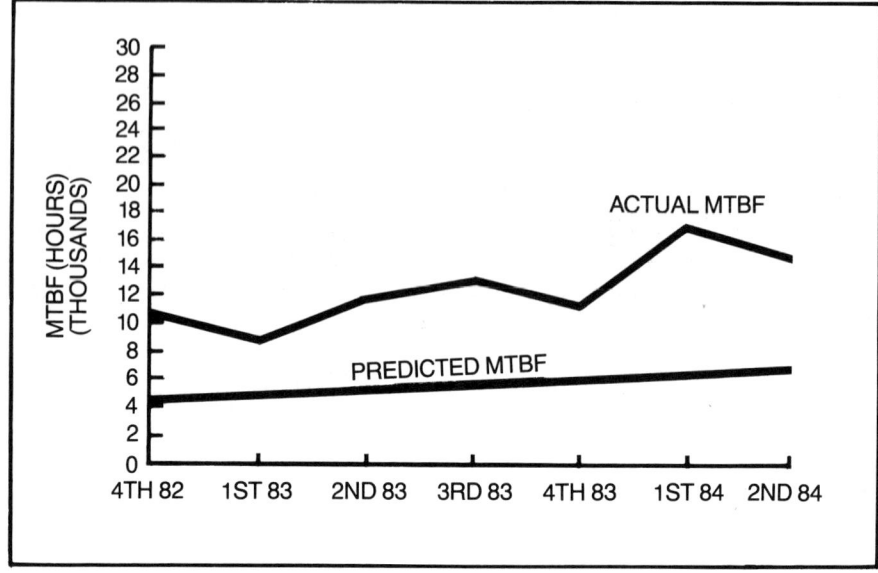

Figure 5. *MTBF performance*

TABLE 5 – Piece Part Quality

	Removals	Bad Parts	Ratio
Original	3600 ppm	900 ppm	4:1
Latest	2300 ppm	650 ppm	3.5:1

SUMMARY

Summarizing the benefits of the supplier quality improvement program, we have experienced the following:

1. cost savings by not performing functional testing during incoming inspection;
2. inventory-in-process reduction from eight days to one day;
3. fewer removals during production test process; and
4. field reliability significantly better than predicted.

The program has worked well. When we started in 1982, there was some concern with eliminating functional testing of incoming parts. But the results to date are all positive lower costs, lower inventory, and better field performance.

Productivity improvements can be accomplished without significant capital expenditures.

Suggested Readings

Sanders, S. 1984. Eliminating traditional incoming inspection. *Evaluation Engineering*. 23(6): 92-94.

Reprinted from *1985 Annual International Industrial Engineering Conference Proceedings.*

Myron F. Wilson is group director of Avionics Operations at Rockwell International in Cedar Rapids, Iowa.

32

Ten-Step Guide Provides an Orderly Procedure for Solving Packaging Problems

Roger W. Hixson

Many industrial engineers have occasion to tackle a packaging problem now and then. Whether you go it alone or have the expertise of a packaging engineer within your company, the following may help. Note that this information applies to products that have been manufactured or processed (filled, wrapped, etc.) and are ready for the final pack operation. Process packaging is another topic.

The 10 steps to improved packaging are a guide to solving a packaging problem in an orderly manner. They are:

1. Identify current packaging costs.
2. Categorize costs/select product.
3. Decide if outside help is needed.
4. Determine product characteristics.
5. Evaluate packaging material/equipment.
6. Consider other factors.
7. Select best options/final proposals.
8. Document.
9. Implement.
10. Follow up.

Step 1

Identify the total costs associated with packaging the current products. This will involve identifying at least the following:

- Purchase price of packaging material.
- Storage (warehousing) of package material and supplies.
- Labor and overhead required to package product and unitize it on a pallet.
- Packaging equipment.
- Additional freight on pack material.
- Customer costs due to defective pack material.
- Costs of returning and repairing damaged product due to defective packaging.

Step 2

Categorize these products by total packaging costs, ratio packaging cost to billings, and low or high volume product; then pick a product from the top half of your product package cost analysis with a moderately high volume. This will assure that you are spending your time where it will pay off the most.

Step 3

Based on the potential cost savings, decide whether to go it alone or seek outside help:

- Develop a good relationship with several local packaging material representatives. They can provide free consultation (usually within the limits of their own products) and samples of material for alternatives.
- There are several good packaging system representatives who can advise on individual equipment items or entire turnkey systems. They normally represent different competitive equipment, and therefore are not limited to one company's products.
- Don't overlook paid consultants who have the expertise to give a broad range of advice on engineering and even implementation. Their fee may be paid back many times on a large project.

Step 4

Determine the characteristics of the product:

- Fragility: How sensitive is the product to damage due to dropping or vibration? This can be determined by testing the product in-house with appropriate drop and vibration equipment, or it can be done by an outside lab. Materials representatives sometimes provide that service free.

The sensitivity (G-factor) is listed in Table 1 for several products. This G-factor, along with packaging material manufacturers' curves, can help determine the type of cushioning required to protect a product.

- Weight. The height a packaged product should be dropped for test purposes has been determined to be as shown in Table 2. Drop on all faces, and three leading edges, then open and examine contents visually and functionally. Repeat on loose cargo vibration test.

(Reference: National Safe Transit Association, 6022 W. Touhy, Chicago, IL 60648, for pre-shipment test procedures.)

Some packaging materials provide good shock protection on the first drop, but may become deformed and lose their protective ability on subsequent drops. Also, consider using a pallet on larger items to encourage the use of mechanical handling, which normally is less dangerous to the product. Don't overlook the fact, however, that many palletized loads are stacked or can tip over in handling.

- Individual or bulk shipped product. A light object alone may be relatively easy to protect, but add several in a bulk shipment and the results may be entirely different.

TABLE 1 – Sensitivity (G-Factor)

Very sensitive	20-40g	Electronic instruments, precision
Sensitive	40-60g	equipment
Moderately sensitive	60-85g	Electromechanical equipment
Moderately rugged	85-110g	TV, optical equipment
		Refrigerator, appliances

TABLE 2 – Heights for Test Drops

Lbs	Drop (Inches)	Type of Handling
1-20	30	One-man throw
21-40	24	One-man carry
41-60	18	Two-man carry
61-100	12	Light equipment
Over 100	Incline impact	Heavy equipment

- Method of shipment. Air and parcel usually prohibit palletization. Shippers often have a limit on the overall dimensions of a shipping container. Some may require a premium for too much cube per weight, while others may go strictly by weight. Check your local transportation to be certain.
- Package cosmetics. The marketability of most products is enhanced by a sharp package. Even the inner pack has an effect on the customer's opinion about the product just purchased and the company it represents. Insist on good visual quality from your vendor. Printing and logos add very little to the price of a carton. The cosmetic factor is a point that needs to be covered with marketing and management early in the analysis.
- Other factors. Static electricity, humidity, temperature and storage time are factors that may require handling solutions designed into the package. For example, some sensitive electronic components must be shielded with static dissipating pack material to assure safe transit.

Step 5

Evaluate the product needs against all the different types of packaging material:

- Expanded polystyrene (EPS) is one of the most popular and least expensive pack materials. If volume can justify the mold ($1,000-$10,000), EPS makes a clean pack. It can also be hot wire fabricated for smaller volume packs.

Shock protection may be designed into collapsible "rib" structure. Even this may not be sufficient for a product that is moderately heavy and fragile.

- Polyethylene and polyurethane foams are a possible alternative to EPS for fragile products. They are more resilient than EPS, but also more expensive. A lightweight product will do well with a light density foam, but heavier products will require a more dense foam to prevent "bottoming out" when they are dropped. PE foams and PE/EPS copolymer foams have recently been molded, which opens up a new option for fragile products.
- Thermofoam (as in hamburger trays) has also expanded into the product packaging field. This low-cost foam can be designed into an attractive, lightweight and shock-protected pack for moderately light and fragile products. The mold is also less expensive than EPS.
- Foam-in-place is a very versatile, light, resilient and inexpensive pack. A small investment in hose assembly and two-part chemicals is all that is required. For low volumes, the product is wrapped in a protective plastic wrap and a small amount of chemical is squirted into an open carton.

As the chemical expands into a foam, the wrapped product is laid onto the foam bed, which quickly sets up. A small amount of chemical is again squirted onto the product, and the carton is closed as the foam expands to fill all voids to provide a snug, cushioned pack. Note that the hose assembly requires periodic cleaning, and the chemical operation may require ventilation.

This method may also be molded on semi- or fully automatic equipment. Its advantages are the small storage requirement for raw materials and instant pack material availability.

- Peanut pack, bubble wrap and foam sheets provide moderate protection for low volume or infrequently shipped products.
- Cell dividers are good for shipping smaller items. Either corrugated (which must be hand assembled) or chipboard (which comes pre-assembled and collapsed from the factory) may be designed for surprising strength and product protection. It may also be used in conjunction with individual product protection such as bubble-wrap in a corrugated carton.
- Die cut corrugated inserts are a traditional and economic pack material.
- Corrugated cartons are used on a majority of product pack or overpack. The printed circle on most corrugated cartons is a manufacturer's certification of the strength of the carton as printed on the seal.

I advise you to get acquainted with a good corrugated representative. There are usually two distinct types: one geared to carload type shipments, the other to smaller custom type products. Be sure you have the right one for your needs. Some basic corrugated terminology:

- **Wall strength:** bursting protection.
- **Type of flute:** inner corrugation style.
- **RSC:** regular slotted carton, flaps meet.
- **FOL:** flaps overlap.
- **Telescopic:** two-piece carton.
- **Multi-wall:** two or more laminated corrugated walls.
- Heat shrink or bubble pack provides a means of placing a product into a clear protective dome, which not only gives moderate pack protection, but also is valuable for display.

Step 6

Other things to consider:

- Cartons may be secured with glue, tape, strap or staples. Material and labor costs and the availability of application equipment should be evaluated.

- Labeling may be pre-printed, applied in-process using an auto-coder or applied as a gummed sticker. You should incorporate a standard bar code on the label since most industries are using it for inventory control.
- Use cut-outs in carton to display contents where needed.
- Special carton markings such as "fragile contents," "this side up" and "do not drop" are good ideas, but do not count on these instructions being followed at a busy commercial dock.
- Safety of package design is important. Heavy items should be designed for mechanical handling. Bulky items should have built-in handles.
- Automate the packaging process as much as can be afforded. Auto carton setup, carton seal and palletizer are simple examples. Auto packaging of product sometimes requires expensive, specially built equipment that requires volume for justification.
- Where possible, use pack material for in-process handling.

Step 7

Determine the best pack options as outlined in Step 1 and test each. At least two different types of packs should be tried until you get more familiar with packaging. The inner pack should be designed first, with the carton being added as a shell (unless you have already established standard cartons).

Give your packaging representatives all the product information available, and they will assist you in the pack design. Test your product in its sample pack by simulating loose cargo vibration and/or drop tests as outlined in Step 4.

Also, before the final method is established, ship a product by its probable mode of transportation to its probable destination. Have the product returned unopened, and evaluate the results. Many good laboratory designed packs have failed this "real world" test.

Step 8

Documentation: Once you have defined and evaluated the proposed pack, it is imperative that you document as many of the requirements as possible. Depending on the structure of the organization, a detailed drawing with appropriate notes is needed with sign-off by all who will be affected by the change. A pack process sheet should also be generated to inform those concerned of what materials and methods are required in the final package.

Step 9

Implementation: If all has gone well up to this point, implementation of the proposed pack will be a natural step with no surprises for anyone. If you have unique customers, they may require one or more trial shipments. If tooling is required, you may have to wait 2-12 weeks. Purchasing should buy only minimal initial quantities of materials if changes might need to be made after implementation.

Step 10

Follow-up: It is imperative to follow up the pack implementation to tally the positive and negative aspects of the new method. Talk to the pack operators, warehouse people, shipping clerks, traffic department, carriers and customers if possible. See if any material is being returned due to defective packs. Any changes should be made early.

COST SAVINGS REPORT

It is probably the right time here to generate a cost savings report on the installed method, and obtain a little recognition for a job well done.

Each new packaging problem will be easier as you gain more experience. After a few projects, you will probably notice that a standard carton size or interchangeable inner pack may generate more volume for added savings and make life in general much easier.

Once you have generated a few cartons, you may want to design the inner pack to fit a standard carton. Also, inner packs designed to fit more than one product may cost a minimal amount more per unit, but also may reduce overall costs through volume, storage and orders. Also consider standardizing cartons that fit neatly on a pallet.

For more packaging information, I recommend the McGraw-Hill *Handbook of Package Engineering*, by Joseph F. Hanlon; referring to a monthly publication such as *Packaging*; and contacting a good packaging school like the one at Michigan State University for a list of related seminars.

Reprinted from February 1985 issue of *Industrial Engineering*.

33

How to Implement MRP II in the Electronics Industry

Larry K. Warren

Most companies in the electronics industry, as well as in other industries, operate in a dynamic business environment. Product designs change as technology changes. New product lines are added, and mature ones are dropped. Improvements in manufacturing technologies reduce costs, increase output rates, and reduce production lead times.

Manufacturing resource planning (MRP II) is a systems approach to dealing with these business changes as they affect production planning, scheduling, materials management, and performance reporting. This chapter describes MRP II concepts, its benefits, considerations important to its successful implementation, and a case example of a company using MRP II.

THE CONCEPTS

MRP II has evolved over the past several years from other production and materials control system approaches. First, there was the reorder point technique. It was used to determine the point in time (schedule) at which a new production or purchase order should be placed to meet anticipated demand. The technique considered demand over the replenishment lead time, safety stock, and economic order quantities. Its primary disadvantage, however, is that it treats all items as independent demand items. Reorder point does not permit linking component part requirements to finished goods requirements.

To address this problem, material requirements planning (MRP) systems were developed. These are computer-based systems that tie component requirements to higher level requirements through a product bill of material structure. Using material requirements planning concepts, a finished good production forecast can be used to directly define production component and purchased component requirements. This is a significant improvement over the reorder point technique, but there is still a weakness: a delay on the shop floor or in purchasing means a potential missed finished good completion date. Material requirements planning provides no follow-up, or control loop, to ensure that material plans are executed properly.

Thus, MRP II was established. MRP II is a system that incorporates MRP concepts, but goes beyond that by providing a control loop to monitor production and purchasing execution of those plans, so that corrective action can be taken early if problems arise.

To perform its function, MRP II consists of several subsystems. It normally includes the following:

1. An engineering data base that contains the component material (bill of material) and production routing definition of each finished good and component;
2. Master production scheduling which defines a finished good and replacement (spares) part production forecast;
3. Material requirements planning that converts the finished good forecast into a component parts forecast using product bills of material;
4. Inventory control which tracks on-hand and on-order inventory balances used in determining the timing and quantity of new component orders;
5. Capacity planning that determines the production resources necessary to build required components;
6. Scheduling and shop floor control which develop detailed production schedules and monitor progress against those schedules;
7. Purchasing that defines purchase order placement dates and monitors progress against those dates; and
8. Performance reporting which calculates performance factors such as on-time completion statistics, efficiencies, and actual cost versus standard.

Although all of these system functions are important, their relative importance varies depending on specific company needs. Many electronics companies, for example, have a large number of components, many purchased, with varying lead-times and incoming acceptance rates. These same companies may have relatively limited production facility capabilities, such as only board assembly, mechanical assembly, final assembly, and test facilities. Their needs are heavily oriented toward material planning and monitoring purchasing and incoming inventory control and are less oriented toward production capacity planning and follow-up. For these companies, the material planning and control loop is much more important within MRP II than the production planning and control loop.

Similarly, a company that builds to customer order, such as a defense avionics manufacturer, has different needs than one which produces to a forecast, such as a consumer products manufacturer. The first company must generally tie component production and purchase orders to a specific customer finished good order. The latter company does not do this, but instead can summarize component requirements for all finished good orders and build or purchase components to this summarized requirements schedule without specific end-item identification.

There are a variety of MRP II software packages available commerically, and almost all recent MRP II installations have made or are planning to make use of off-the-shelf software. Most software packages have sound material planning capabilities, fewer have good purchasing control features, and fewer still have good shop floor control and reporting capabilities. Almost all software packages offer material control over build to forecast (no need to directly tie component orders to end-item orders) products. A few offer control over made-to-customer-order items. When choosing a software package, company needs and requirements must be defined, and software packages can be evaluated against those requirements. Each package's strengths and weaknesses relative to those requirements must be fully understood before software is selected.

If the software is selected properly, and if it is implemented properly, the benefit of MRP II can be substantial. Component inventories will decrease by ordering only those components needed to produce needed end items. At the same time, component shortages will decline by ensuring that needed components are ordered. Production productivity will improve due to improved scheduling and reduced time spent waiting for components on shortage status. Vendor performance will improve due to more timely purchase order release procedures and improved performance monitoring capabilities. And there may be other benefits in areas of reduced lead time and improved customer service. One antenna manufacturer, for example, realized a 15% reduction in inventories, a 35% reduction in lead-time, and almost a 10% reduction in overall manufacturing costs resulting from MRP II. MRP II can be an effective tool in improving production performance.

Implementation Considerations

These benefits can be derived, however, only if MRP II is implemented properly. No more than a third of MRP II users can be considered Class A (i.e., highly effective) users. The rest range from moderately successful users (i.e., portions of MRP II are being used in day-to-day production operations) to completely unsuccessful users, where considerable resources were spent with no resulting use of or benefit from MRP II. The differences among these companies is almost always in the level of management commitment and the implementation approach taken. The key factors in effective implementation of MRP II are as follows:

1. Top management must be committed to expending the necessary resources and making the necessary interdepartment policy decisions to ensure MRP II's success—a policy of instilling discipline in all reporting functions must be enforced by management;
2. The engineering data base (bill of material and routing) must be fully and accurately developed and procedures must be in place to process changes and updates in a timely fashion;
3. The high transaction volume systems, inventory and shop floor control, must be designed to assist with and control user input—automated data entry should be used where possible and cycle counts and production count reconciliations should be made;
4. Procedures must be developed to provide a reasonable master production schedule;
5. All MRP II subsystems must be understandable and straight forward to use, and special purposes codes and formats should be avoided where possible;
6. User training must be thorough; and
7. Post implementation follow-up must be exhaustive and all problems identified must be corrected. If proper implementation procedures are not followed, the probability of MRP II success is low to nonexistent.

MRP II TIE WITH COMPUTER-INTEGRATED MANUFACTURING AND JUST IN TIME CONCEPT

MRP II is an important part of computer-integrated manufacturing (CIM) and its concepts tie directly with just in time (JIT) concepts. CIM is an approach to manufacturing that automates as much of the process as possible. In CIM, computer-aided engineering analysis of product functional requirements and preliminary designs feed a computer-aided design (CAD) system, which performs detailed electrical, board, and mechanical design. CAD detailed designs drive the automated development of computer-aided process planning and machine and robot operating instructions in computer-aided manufacturing (CAM) system. These CAM operating programs drive automated precision graphics and artwork preparation, board fabrication and component assembly, mechanical fabrication and assembly, (and component and systems level test cells.) To function in a truly automated environment, these manufacturing centers must be scheduled and controlled automatically. MRP II is the planning, scheduling, and control component of CIM. While at this point CIM is more of a concept than a functioning system, it is being aggressively developed piece by piece by many companies in the spacecraft, communications, avionics, defense, and computer manufacturing industries. MRP II plays a key role in these companies' CIM plans.

Similary, MRP II concepts, in many ways, mirror JIT concepts, although JIT affects a company in a broader sense than does MRP II. Simply stated, JIT philosophy is to eliminate or minimize all unproductive activity, such as set-up, material handling, queue, and inventory. This is accomplished through cell-oriented arrangements that cut travel distance and queues and provide improved physical control over material. And, by reducing set-up time to near zero through automation and software controls, lot sizes of one part become feasible, thus reducing finished inventory and lead time. MRP II supports JIT by providing a scheduling and control function. MRP II, then, is an important concept within broader strategies—CIM and JIT—that many companies believe represent the future of state-of-the-art manufacturing.

MRP II CASE EXAMPLE: XYZ AVIONICS CO.

Although many companies have experienced difficulties in implementing MRP II because of poor software selection or installation practices, others have been successful. XYZ Avionics Co. is a $50 million (annual sales) avionics and black box manufacturer that was successful. Its operations consist of PCB assembly and test, cable and harness assembly, mechanical fabrication and assembly, final assembly, and various component, board, and system level test processes. It builds its products to customer (generally Navy and Air Force) order. XYZ's approach to implementing MRP II is described as follows.

XYZ's first step in implementing MRP II was to define its objectives and requirements in detail. For example, a very general objective of reducing components shortages was detailed into a specific objective of reducing incoming component reject rates through better monitoring of vendor performance. Another objective was to reduce missed component receipt dates through improved control over long lead items. From this detailed set of objectives, specific user requirements were defined. Major requirements included make-to-order tracking of component orders relative to finished customer orders, maintaining multiple on-hand inventory locations, realtime updating of inventory records, logical kitting (allocation) of components to production orders, and detailed PO tracking and vendor performance reporting. These and other requirements were summarized into a checklist, and five MRP II software packages were evaluated against that requirements list. A technique for scoring each vendor proposal against user requirements was developed, and the software selection decision was based primarily on its ability to satisfy user requirements.

The package that was selected had make-to-order component tracking and on-line update capabilities and it met most other major requirements. Where essential requirements were not met, software modifications were made. Users were a key part of the design of all software modifications and new program development.

Concurrent with software modification and development, the engineering data base (bills of material and routings) was built. A key part of this activity was developing and "as manufactured" as well as an "as designed" bill of material. Design engineers were responsible for creating the data base and keeping it current. Strict procedures for processing engineering changes were developed and applied, even as the data base was created, so that it would be current when completed. A data base audit program helped to identify logical errors, such as a manufactured item without identified components.

XYZ developed and installed their system in modules. The engineering data base was first, followed by inventory control, material requirements planning, purchasing, shop floor control, and then performance reporting. A new module was not installed until all existing modules were functioning well in production use. This is a critical point that many MRP II users miss. Installing a new module, before an existing one is working effectively, will almost guarantee failure. Even if the implementation schedule must be slipped, a new module should never be implemented until all supporting modules are at full effectiveness.

And before installing a new module, XYZ reaudited its data base to ensure accuracy and trained users thoroughly, including simulating use of the system with live production data. After installation, XYZ followed up with users by auditing the system. Input error rates were tracked, and if improvement was not shown over time, data input procedures were analyzed to ensure ease of use and user understanding. In some cases, both procedural and system changes were required to reduce input errors. For example, the original design of an inventory location-to-location transfer transaction called for two transactions to be entered—one to reduce on-hand inventory at the "from" location and another to increase on-hand inventory at the "to" location. In use, this design provided two opportunities for error. Based on postinstallation discussions with users, this transaction was redesigned into one that contained a single part number and quantity along with a "from" and "to" location designation. Error rates were reduced accordingly.

In short, XYZ was successful not only because it made a reasoned software decision and followed proper, user-oriented installation procedures, but also because top management committed resources to the project. Quality of results was placed before meeting schedules, and people and system resources were committed to ensure that input data and data bases were controlled, accurate, and useful. Without such a commitment, MRP II cannot succeed.

With the commitment, it can be a strategic tool in ensuring manufacturing competitiveness.

Reprinted from *1986 International Electronics Assembly Conference Proceedings*.

Larry K. Warren is president of LMS Warren, Inc. in Olympia Fields, Illinois.

Part 4

Automating Electronics Assembly Processes

This section provides chapters that outline methodologies and techniques for automating electronics assembly processes. Fu et al. (1987) present a methodology that combines traditional economic analysis with systems modeling techniques to evaluate various quality strategies for a printed circuit board manufacturing line. They describe an application of their methodology to estimate effects of various incoming component quality levels and scrap rates at test stations on total annual costs.

The use of minilines in electronics assembly processes is presented by Barr et al. (1986). The authors outline how the miniline concept or philosophy can be used to solve manufacturing problems. Falter and Choobineh (1986) provide a step-by-step procedure for justifying CAM in electronic industry. A study dealing with CAM justification in a printed circuit board is also presented. A procedure for bridging islands of automation in the factory of the future is presented by White (1982). White points out that what appears to be lacking in today's environment are economies that would make automating all factory operations cost-effective. Pukanic (1985) provides a framework for incorporating flexibility and integration in materials handling requirements in electronics assembly. Material handling interface diagram and tools are provided to aid systems designers involved in electronics assembly. A case study on how to implement a factory control system (FCS) in a surface mount line is provided by Israni and Lujack (1987), who present the levels at which FCS applications are feasible. A procedure for implementing group technology (GT) concepts in electronics assembly is offered by Styslinger and Melkanoff (1985). Outlined is the development of a classification and coding system that describes manufacturing attributed to circuit card assemblies.

Maes (1987) presents a case study that outlines how a surface-mount assembly was automated in a European plant. The various levels of automation for printed circuit boards are also presented. The case study also discusses how the Kanban method was used to achieve just-in-time delivery. A methodology for designing and implementing integrated systems is provided by White (1986). Cross (1987) discusses 10 work-flow principles to improve performance with regard to cost, quality, delivery, and flexibility. Cross demonstrates how the 10 work-flow principles can be applied in the work environment. A classification of the papers presented in this chapter is shown in Part 4 matrix.

Part 4 Matrix

Chapter Number	Title	CONTENT CLASSIFICATION						ELECTRONICS ASSEMBLY PROCESSES						
		Model	Quantitative Technique	Strategy/Methodology	Case Study	General Theory	Design	Receiving & Store	Inspection	Kitting	CP Assembly Mfg.	Equip. Assembly	Packing	Mat'l Handling
34	Combining Economic Analysis and System Modeling to Evaluate Test Strategies for Circuit Board Assembly Lines	X	X						X		X			
35	Miniline: Research Applied to Manufacturing				X	X	X				X			
36	CAM Economic Justification - A Case Study in Electronics Industry			X	X						X	X		
37	Factory of Future Will Need Bridges Between Its Islands of Automation					X	X				X	X		
38	Flexibility and Integration Are Key Material Handling Concepts in Electronics Assembly Equipment			X		X		X			X	X	X	X
39	Factory Control Systems in Electronics Manufacturing	X		X	X	X	X	X	X	X	X	X	X	X
40	Line Scan Vision System			X	X	X			X		X			
41	Group Technology for Electronics Assembly	X			X	X	X	X			X	X		
42	Automating Surface-mount Assembly in a European Plant				X	X	X				X			
43	Becoming the Systems Integrator Before the Year 2020			X										
44	Making Manufacturing More Effective by Reducing Throughput Time				X	X	X				X	X		

34

Combining Economic Analysis and System Modeling to Evaluate Test Strategies for Circuit Board Assembly Lines

Bor-Ruey Fu
Charles Falkner
Rajan Suri
Scott Garlid

ABSTRACT

This chapter presents a methodology that combines traditional economic analysis with system modeling to provide a better management tool for manufacturing decisions. In particular, we focus on quality strategies, and a network of queues model is used to generate system performance estimates which drive the economic analysis. The important issue considered is how the approach can be used to evaluate the economic ramifications of various quality strategies in proposed manufacturing systems. We demonstrate the usefulness of the methodology as a basis for planning a system configuration for a printed circuit board (PCB) manufacturing line with different quality strategies. The key features of our approach are the integration of system and economic analysis, the ability to rapidly explore many different alternatives, the use of emerging analytic modeling tools instead of more time-consuming simulations. All the analysis described here was accomplished using a PC/AT computer and PC-based software such as LOTUS 1-2-3.

A major problem a potential user of new manufacturing technology may face is in justifying a large investment. New technologies are often very capital-intensive and may be difficult to justify using traditional economic analysis (Blank 1985; Kaplan 1986; Primrose and Leonard 1986; Suresh 1985). The user must extend his analysis to include strategies, plus subjective and objective factors which typically have not been included in traditional justification approaches. Many of these factors arise from the operational complexity of manufacturing systems. Thus, a user must evaluate many operational trade-offs to ensure the most economical alternative is selected before beginning the formal justification process (Blank 1985; Falkner and Garlid 1986).

The complexity of manufacturing systems is largely due to many interactions between components of the system and the dynamic and uncertain manufacturing environment. Because of this and the staggering number of factors that can affect system performance, operational decision making often becomes an issue of too much information rather than too little. For example, reducing batch sizes to reduce in-process inventories and decreasing system flow times often must be traded off with more machine set-ups, reduced throughput rates, and potential material handling overload. The comparison for these trade-offs should be based on the goal of the manufacturing facility—profits.

Several modeling techniques are available for analyzing manufacturing systems. These techniques include computer simulation, queueing network theory, static allocation, perturbation analysis, and Petri nets (Suri 1985). Although each of these methods is valid for analyzing and evaluating manufacturing systems, none has more than partial acceptance by industry. This is due primarily to a lack of experience in system modeling techniques and insufficient understanding and confidence in the results and implications of such models.

We use an analytical modeling tool, MANUPLAN II, for the manufacturing system analysis. The choice of this tool was based on several factors. For decisions at the planning level it is easy to set up models, and each "what if" is evaluated in seconds. It runs in a PC environment, making it accessible to most engineers, and it uses LOTUS 1-2-3 as an interface, which makes integrating economic and system modeling straightforward in a single model. This modeling approach is gaining acceptance in industry for planning and justifying manufacturing systems (Haider et al. 1986; Mills 1986; *CIM Strategies*). For an overview of MANUPLAN II see the Appendix.

QUALITY IN PRINTED CIRCUIT BOARD MANUFACTURING

As electronic products become more complex and vertical integration increases, product testing and testing strategies become a vital link in the success of both the product and its manufacture. Indeed, some industrial studies show that testing costs can exceed 50% of the product cost (Miczo 1986).

The cost of testing depends on many factors, but a simple rule of thumb states that testing costs increase by an order of magnitude for every level of integration (Davis 1982). Thus, the cost of correcting a fault may be $1 at the component level, $10 after the component has been inserted into a board, $100 after the board is inserted into a product, and $1,000 if the product is in the field. Obviously, improving quality and correcting faults in early stages of manufacturing is of utmost importance. This chapter describes two main aspects of quality strategies.

Product Quality Strategy

We can divide the economics of quality strategies into two categories: 1) the cost of manufacturing better quality products (including the cost of purchasing better quality raw material and the cost of implementing improved manufacturing processes and procedures); and 2) the cost of testing and repair (including capital and operating costs). This chapter will focus on the costs of testing and repair. Thus, we will look into the benefits (in terms of reduced total annual cost of test/repair) of improved product quality, without considering the increased cost of manufacturing such a product. This will allow us to answer the following question with hard monetary numbers. What is the value of increased quality of product, in terms of overall reduction in system costs? The answer to this question enables the manufacturing manager to justify the purchase of better quality raw material or to implement a better manufacturing process, albeit at a higher cost.

Scrap or Rework Strategy

A second quality strategy is the trade-off between scrap and rework. Decreasing scrap will save on material costs, but will require more resources for rework and also add to the system work in process. Which factor will dominate the other depends on the system data. Hence, another question relevant to quality strategies is presented. For this system is it better (in terms of overall economics) to scrap parts or rework them? Actually, this question needs to be asked for each test point in the system. We will simultaneously increase/decrease the scrap rate at all test points in the system, according to equations given later. This is done only for simplicity to give main insight without a large number of output graphs.

MANUFACTURING SYSTEM MODEL

System Description

We consider a manufacturing line that produces three types of PCBs. Raw components and boards arrive at the line from the receiving dock. The first set of operations performed on the boards are a series of assembly processes (Figure 1) consisting of dual, single, axial, and radial insertion, followed by wave soldering. After they are assembled, the boards go through in-circuit tests with a repair and retest loop, followed by functional and preburn-in-tests with corresponding repair and retest loops, then a burn-in-process followed by memory and system tests with repair and retest loops.

Modeling the Quality Parameters

The primary purpose of this particular model is to estimate the effects of various quality strategies on the economics of system performance. By quality we focus on the product yield after a test procedure is conducted. The yield—the probability that a PCB passes the test is given by (Davis 1982)

$$P_{PASS} = \text{Yield} = (1 - PFB)^{(FC \times TL)}. \tag{1a}$$

where probability of faulty board (PFB) is the fraction of parts which theoretically would not meet the requirements of a given test if that test were perfect. This is the fraction that actually contains one or more faults, but these faults may or may not be detected by the test. Test level (TL) is the fraction of PCBs that are actually tested. A test level of 0.10, for example, indicates a sampling procedure where one PCB in 10 is tested. Fault coverage (FC) is the fraction of all possible faults that a given test and software can detect. This parameter is particularly applicable in the automatic testing of PCBs. In such an environment, software is written for each test, and the more complete and complex the test software, the higher the fault coverage. We assume FC = 1.0 and TL = 1.0— that is, faults are always detected in the relevant test procedure. With this assumption, we focus on other quality-related parameters as discussed. A more detailed analysis would involve looking at FC and TL values below 1.0. However, this would also involve making more detailed probability models of the circuit board failure characteristics that we wish to avoid.

Next, we define a decision parameter scrap rate (SR), which is the fraction of failed boards that we will scrap instead of attempting to repair them (in practice, this could correspond to setting a threshold level based on the particular subtest that the board failed). Thus, of the boards that arrive at the test station, the proportion which is scrapped is

$$P_{SCRAP} = SR \bullet (1 - P_{PASS}). \tag{1b}$$

Hence, the proportion of boards that come out of the tester and are sent for rework is

$$P_{REWORK} = 1 - P_{PASS} - P_{SCRAP}. \tag{1c}$$

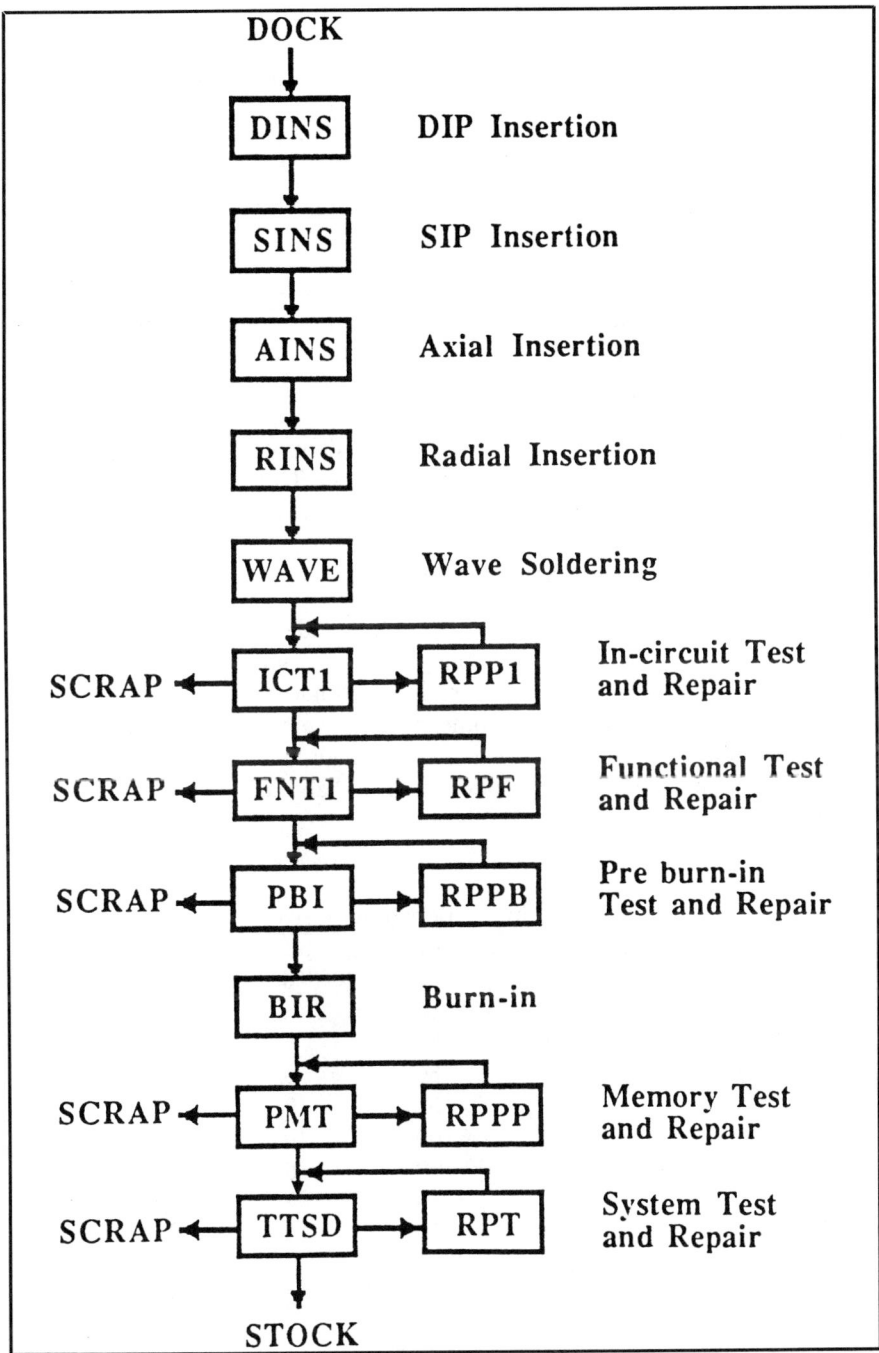

Figure 1. *Process routing for PCB-A.*

For simplicity, we assume that repaired boards have, on average, the same quality as boards arriving at the tester from the previous operation. This can be generalized to accommodate other situations (Network Dynamics Inc., 1987).

We now describe certain details of the system and their representations as inputs to the model.

Details of the MANUPLAN II Model

The data items displayed are sections of the LOTUS 1-2-3 worksheet used by MANUPLAN II for input. We begin by specifying the amount of time the equipment will operate—that is, the available time. In this model, the time is one shift (8 hours) per day for 260 days per year.

*Opern. *Unit HR	Flow time Unit DAY	Demand Period YEAR	HRs worked per DAY 8	DAYs worked per YEAR 260

Description of the Equipment Groups

The equipment in the system is specified to the model by giving the equipment group name, the number of machines in the group, and their reliability (expressed as the mean time to failure [MTTF] and the mean time to repair [MTTR] in hours per piece of equipment). The overtime factor can be used to make individual equipment groups run for a longer or shorter time period than the available time. We do not model material-handling capacity or delays in this model. However, this aspect is easily added (Network Dynamics Inc. 1987).

* Equip. * Name	No. in Group	Reliability-(HRs) mttf	mttr	Overtime Factor
DIP	7	80	8	1
SIP	7	80	8	1
AXIAL	6	80	8	1
RADIAL	7	80	8	1
SLDR	2	360	12	1
GR70	1	160	16	1
GR71	1	160	16	1
RP1	3	960	8	1
GR95	1	160	12	1
ADAR	3	160	12	1
PBO1	2	160	12	1
BIR1	−1	160	12	1

* Equip.	No. in	Reliability-(HRs)		Overtime
* Name	Group	mttf	mttr	Factor
PMT1	3	720	9	1
TT1	4	160	12	1
S018	1	240	20	1
S087	·1	240	20	1

BIR1 is a burn-in station. In MANUPLAN II, the value of "−1" denotes a special type of station called a "Delay Station". For the meaning and uses of such a station in manufacturing see Network Dynamics Inc. (1987).

Part Description

The first part in the system (PCB-A) has an annual demand of 15,000 good pieces and moves through the system in lots of 35 boards. Similar data follow for the other boards.

* Part	Demand	Lot
* Number	per YEAR	Size
PCB-A	15000	35
PCB-B	15000	35
PCB-C	15000	35

Part Routing

The next set of input data describe the routing of each part. As an example, see Figure 1, which shows the routing for PCB-A. In MANUPLAN II, all material must flow into the system from an entry point called DOCK and exit the system into either STOK (stock, for good boards) or SCRP (for scrap). Figure 1 also shows the operation description. This routing is specified as three data items: an operation, the next operation or operations, and the routing proportion between operations. The routing proportion is the result of the quality characteristics. For example, the following data show that at the in-circuit test (ICT1) 80% of PCB-A boards pass the test, 8% are sent for repair, and 12% are scrapped. These proportions are programmed into the LOTUS spreadsheet as Equations 1−3, which depend on the parameters such as PFB and SR. By this we combine the quality calculations with the system model.

* Routing for Item (Part)		PCB-A
* From	To	Proportion
* Opern	Opern	
DOCK	DINS	1
DINS	SINS	1

* Routing for Item (Part)		PCB-A
* From	To	Proportion
* Opern	Opern	
SINS	AINS	1
AINS	RINS	1
RINS	WAVE	1
WAVE	ICT1	1
ICT1	FNT1	0.80
ICT1	RPP1	0.08
ICT1	SCRP	0.12
RPP1	ICT1	1
FNT1	PBI	0.80
FNT1	RPF	0.08
FNT1	SCRP	0.12
RPF	FNT1	1
PBI	BIR	0.80
PBI	RPPB	0.12
PBI	SCRP	0.08
RPPB	PBI	1
BIR	PMT	1
PMT	TTSD	0.80
PMT	RPPP	0.12
PMT	SCRP	0.08
RPPB	PMT	1
TTSD	STOK	0.80
TTSD	RPT	0.12
TTSD	SCRP	0.08
RPT	TTSD	1

Similarly, routing data are specified for the other two parts.

Operation Assignment

These data include the assignment of operation(s) to equipment groups and specify the operation set-up and process times. If an operation is always performed on only one equipment group, the proportion assigned is 1 (this is always the case in our model, but see [Network Dynamics Inc. (1987)] or [Suri et al. (1986)] for other situations). The assignment for PCB-A is shown as follows.

* OPERATION ASSIGNMENT for ITEM (part)
  PCB-A

* Opern. * Name	Equip. Name	Proportion Assigned	time/lot (setup)	time/pc. (run)
DINS	DIP	1	0.20	0.100
SINS	SIP	1	0.33	0.120
AINS	AXIAL	1	0.25	0.130
RINS	RADIAL	1	0.22	0.150
WAVE	SLDR	1	0	0.033
ICT1	GR70	1	0.15	0.033
RPP1	RP1	1	0	0.300
FNT1	GR95	1	0.10	0.020
RPF	RP1	1	0	0.300
PBI	PBO1	1	0.10	0.100
RPPB	RP1	1	0	0.300
BIR	BIR1	1	24.00	0
PMT	PMT1	1	0.15	0.200
RPPP	RP1	1	0	0.300
TTSD	TT1	1	0.15	0.300
RPT	RP1	1	0	0.100

Similar operation data are specified for the other two parts. This completes the description of the manufacturing model.

Outputs from MANUPLAN II

With the above data specified in the LOTUS spreadsheet, we can now run MANUPLAN II to evaluate the system. To give the reader an overview of our analysis, we will briefly describe the outputs from MANUPLAN II without the economic analysis (which will follow).

A typical run of MANUPLAN II for this system takes about 100 seconds on a PC/AT class computer. Outputs from MANUPLAN II are placed in the LOTUS 1-2-3 spreadsheet and include equipment utilization in setup, run and down time, work in process, and flow time. A typical summary of the equipment utilization is as follows:

	EQUIPMENT UTILIZATION SUMMARY						
	% of capacity required				WORK IN PROCESS (in lots)		
Equipment Name	for SETUP	for RUN	for REPAIR	TOTAL UTILIZATION	at EQUIPMENT	in QUEUE	TOTAL
DIP	2.6	83.5	8.6	94.7	6.03	5.19	11.2
SIP	3.7	73.5	7.7	84.9	5.40	2.48	7.9

EQUIPMENT UTILIZATION SUMMARY

Equipment Name	% of capacity required for SETUP	for RUN	for REPAIR	TOTAL UTILIZATION	WORK IN PROCESS (in lots) at EQUIPMENT	in QUEUE	TOTAL
AXIAL	3.9	73.6	7.7	85.2	4.65	2.54	7.2
RADIAL	2.6	74.9	7.8	85.3	5.43	2.44	7.9
SLDR	0	52.4	1.7	54.2	1.05	0.39	1.4
GR70	5.9	45.5	5.1	56.6	0.51	1.17	1.7
GR71	4.4	41.5	4.6	50.5	0.46	0.75	1.2
RP1	0	82.6	0.7	83.3	2.48	2.14	4.6
GR95	9.0	53.1	4.7	66.7	0.62	2.32	2.9
ADAR	0.9	60.1	4.6	65.5	1.83	0.77	2.6
PBO1	1.6	54.5	4.2	60.3	1.12	0.69	1.8
BIR1	0	0	1.2	1.2	10.93	0.19	11.1
PMT1	1.4	66.1	5.1	72.6	2.03	1.11	3.1
TT1	1.0	67.6	5.1	73.7	2.74	1.00	3.7
S018	0	72.1	6.0	78.1	0.72	2.71	3.4
S087	0	72.1	6.0	78.1	0.72	2.91	3.6

and a typical production summary is:

Part Number	Good Prodn (pieces)	Scrap Prodn (pieces)	Desired production can be achieved WORK IN PROCESS (pieces)	FLOW TIME dock-stok in DAYs
PCB-A	15000	11403.7	1142.3	14.23
PCB-B	15000	4837.5	819.1	12.29
PCB-C	15000	4837.5	682.1	9.64
TOTAL PIECES:			2643.6	

The scrap production is the amount scrapped *in addition* to the good production. The work in process is the average number of pieces of a particular board in the system at any time. The flow time shows the average time for a good board to go through the system. All these numbers will be used in our economic model (see *Details of the Economic Model*).

MODELING THE COST STRUCTURE

Including economic parameters in the system model allows decision makers to calculate the cash flow to evaluate trade-offs. For example, total annual cost, average cost per PCB, and net present value can be determined. The economic results can also be broken down both by system components and cost components to aid in identifying the factors for improvement. For example, by using an economic-based system model, a manager can make a decision based on the output that estimates cost per PCB and net present value over the system life. On the other hand, using traditional system modeling alone, the manager would have to make a decision based on operational parameters, such as throughput rates, flow times, work-in-process inventories, and equipment utilizations. If the manager was very ambitious, total cost estimation for inventory, power consumption, maintenance, labor, etc., could be assigned and calculated after the model had been run (Suresh 1985). On the other hand, the manager could apply traditional economic analysis without a system model, but cost estimates for many variables and overhead costs, such as inventories, tooling, power consumption, and material handling, would be less accurate because the complex interactions and dynamic nature of the manufacturing system would not be analyzed. When using these other approaches there is a tendency to analyze a much smaller number of alternative designs. An important factor in our approach is the ease with which many alternatives can be studied for the combined economic and system model.

Once costs are associated with the system components and operational parameters, the engineers, the supervisors, or foreman can perform sensitivity analysis to determine where the largest potential for improvement exists. This type of analysis would be valuable both during the initial planning and justification of the system and during the ongoing system operation.

Overview of Economic Model

For the economic analysis to be most useful, costs must be shown on both a global and local basis. Global cost, or total annual cost per PCB produced, is most useful for evaluating and comparing several test strategies in a macroscopic way. It is difficult, however, for such a measure to indicate where cost improvements could be made. Localized costs may, therefore, be collected by machines, by PCB types, and by cost components. In this model, total cost is based on the cost components: materials, equipment, labor, overhead, work in process, and maintenance. All costs will be presented in terms of annual costs. This requires reducing capital costs to annual equivalents. We use a simplified approach explained as follows. However, our basic framework is easily extended to more sophisticated accounting frameworks (Smith 1983).

Material Cost

This is the total cost of purchasing all raw materials necessary to meet production. This cost includes the increased input of material caused by scrap in the system (this increase is calculated by MANUPLAN II).

Equipment cost

The cost of equipment is amortized over the expected life of the investment and converted to an annual cost based on the minimum attractive rate of return (MARR) [13]. Taxes and depreciation are ignored, but could easily be included.

Labor cost

This consists of the cost of the operators responsible for supervising, operating, or setting up each machine. A machine can have any number of operations assigned to it depending on the physical situation. The total labor hours are calculated from the amount of machine hours obtained as outputs from MANUPLAN II.

Overhead cost

Overhead cost includes energy, indirect manufacturing expenses, allocated administration costs (but not WIP), equipment amortization, or maintenance, which are separately considered.

Work-in-process (WIP) cost

This cost reflects the capital tied up in WIP that could generate a return if invested elsewhere. The quantity of WIP, by product, at each point in the system is calculated by MANUPLAN II. The difficulty in determining WIP cost is estimating an accurate value per unit at each production stage. In a station with no branching or repair loop, the value of a PCB after an operation is completed is the value before the operation plus the costs incurred during the operation. In a station with a repair loop the calculation requires some additional analysis. As shown in Figure 2, let

$X \quad$: Average cost of incoming PCB.
$X' \quad$: Average cost of a PCB in WIP at test station.
$X_1 \quad$: Average cost of a PCB leaving test station.
$T \quad$: Test cost per PCB.
$R \quad$: Rework cost per PCB.

(Note: p, q, r are the routing probabilities at the test station, where $p + q + r = 1$). Then

$$X' = \text{Prob(Part arriving at TEST comes from Route [1])} \bullet \$X$$

$$+ \text{Prob(Part arriving at TEST comes from Route [2])} \bullet (\$X_1 + \$R)$$

$$= (1 - q) \bullet \$X + q \bullet (\$X_1 + \$R). \tag{2a}$$

$$\$X_1 = \text{Prob(Part Leaving TEST comes from Route [1])} \bullet (\$X + \$T)$$

$$+ \text{Prob(Part Leaving TEST comes from Route [2])} \bullet (\$X_1 + \$T + \$R)$$

$$= (1 - q) \bullet (\$X + \$T) + q \bullet (\$X_1 + \$T + \$R). \tag{2b}$$

Solving Equations 2a and 2b we have

$$\$X' = \$X + (\$T + \$R) \bullet \frac{q}{1 - q} \tag{3a}$$

$$\$X_1 = \$X + \$T + (\$T + \$R) \bullet \frac{q}{1 - q} \cdot \tag{3b}$$

Maintenance cost

This is the cost to repair machines based on machine down time predicted by the system modeling tool (MANUPLAN II). Indirect costs incurred, such as the increase of WIP caused by machine failure, are covered by the MANUPLAN II outputs that predict this increase.

Details of the Economic Model

We will describe how all the costs in our economic model are calculated. The Lotus 1-2-3 spreadsheet contains an area for some economic data to be input by

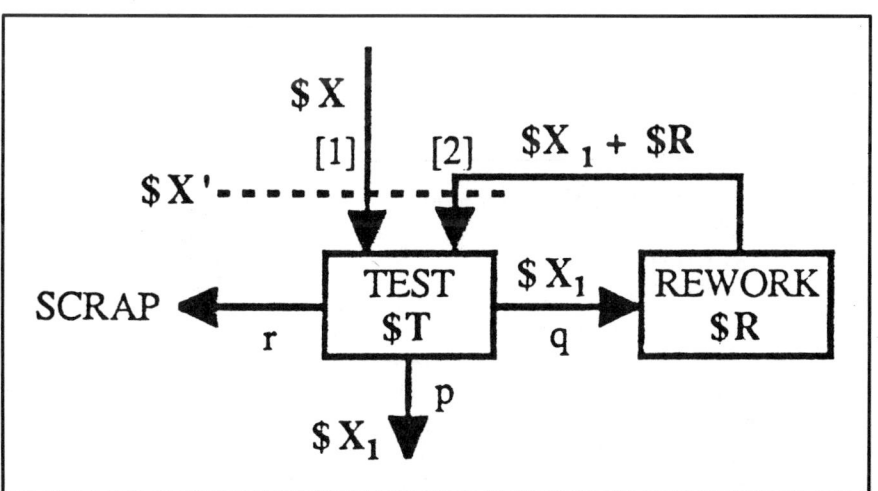

Figure 2. WIP cost at test station with repair loop.

the user. Based on this data and the MANUPLAN II inputs and outputs, we then calculate total annual costs according to Equations 4–21 below.

Table 1 summarizes the basic economic input data. The raw material cost per unit is the sum of the purchase cost of components plus any value added by manufacturing operations preceding the assembly process shown in Figure 1. The WIP holding interest rate is the interest rate to be charged for the value of the inventory; it is the opportunity cost for the capital tied up in WIP. We assume a uniform labor cost per hour that includes both wages and benefits. If different classes of labor are required for different machine groups, these can be easily incorporated into the calculations. The maintenance cost per hour is an aggregate average covering labor, supplies, and spare parts and can be estimated from maintenance records. If required, the maintenance cost per hour can also be varied by machine groups. The minimum attractive rate of return (MARR) is the effective annual percentage return on investment that just meets the investor's threshold of acceptability (Smith 1983). Based on this input data, as well as MANUPLAN II inputs, we calculate two quantities that will be used in other equations

$$\text{(Operating Hours/Year)} = \text{(Days/Year)} \bullet \text{(Hours/Day)}. \qquad (4)$$

$$\text{(Capital Recovery Factor)} = \frac{(\text{MARR}) \bullet (1 + \text{MARR})^{\text{Equipment Life}}}{(1 + \text{MARR})^{\text{Equipment Life}} - 1} \qquad (5)$$

The MARR value of 0.15 leaves a value of 0.20 for the capital recovery factor in this study.

TABLE 1. – Economic Model Input Data

Cost	Inputs
Raw Materials Cost per Unit:	
PCB-A	$100.00
PCB-B	$200.00
PCB-C	$150.00
WIP Holding Interest Rate	25.00%
Labor Cost per Machine Hour	$20.00
Overhead Cost per Machine Hour	$5.00
Maintenance Cost per Machine Hour	$23.00
Equipment Life in Years	10
Minimum Attractive Rate of Return (MARR)	0.15

Table 2 shows the cost by equipment groups. The first four columns are input. The equipment name and number in the group correspond to the system description provided in MANUPLAN II. Column 2 shows the equipment purchase and installation cost, and Column 4 gives the equipment manning in terms of average number of labor hours per machine hour. We assume that any fractional part of a man not assigned to equipment within the system is productively used elsewhere. If set-up is performed by specialists at a different rate, the calculations could be reorganized to model this situation. Columns 5 and 6 are calculated as follows:

$$\text{(Equipment Cost/Year)} = \text{(Equipment Cost/Unit)} \bullet \text{(No. of Machines)} \bullet \text{(Capital Recovery Factor)} \tag{6}$$

$$\text{(Labor Cost/Year)} = \text{(Labor Cost/Hour)} \bullet \text{(Labor Hour/Machine Hour)} \bullet \text{(Machine Hours/Year)}, \tag{7}$$

where the machine hours per year (shown in column 10) is calculated from machine utilization for set-up and run, using MANUPLAN II output, by

$$\text{(Machine Hours/Year)} = \text{(Machine Utilization for Setup \& Run)} \bullet \text{(No. of Machines in Group)} \bullet \text{(Operating Hours/Year)}. \tag{8}$$

Columns 7 and 8 are determined from

$$\text{(Overhead Cost/Year)} = \text{(Overhead Cost/Machine Hour)} \bullet \text{(Machine Hours/Year)}, \tag{9}$$

$$\text{(Maintenance Cost/Year)} = \text{(Maintenance Cost/Hour)} \bullet \text{(Machine Utilization for Repair)} \bullet \text{(No. of Machines)} \bullet \text{(Operating Hours/Year)}, \tag{10}$$

where the machine utilization for repair is obtained from MANUPLAN II output.

The ninth column is the operating cost per machine hour, which is an average for operating a machine in the group. This value is used to calculate the value added to WIP inventory (refer to the details in Table 4). It is calculated as follows:

$$\text{(Operating Cost/MC Hr.)} = \frac{\text{(Equip. Cost/Yr.} + \text{Labor Cost/Yr.} + \text{Overhead Cost/Yr.} + \text{Maint. Cost/Yr.)}}{\text{(Machine Hours/Year)}} \tag{11}$$

The last row of Table 2 shows the totals of the annual equipment amortization cost, labor cost, overhead cost, and maintenance cost that are used again in Table 4 to show the annual costs by products.

TABLE 2. – Annual Costs Shown By Equipment Groups.
(Dollar values are in thousands, except operation cost/hr.)

= = = = = =INPUTS = = = = = = ————————————OUTPUTS————————————

Equip. Name (1)	Equip. Cost (2)	No. of MCs (3)	Lbr. Hr. /MC Hr. (4)	Equip. Cost/Yr. (5)	Labor Cost/Yr. (6)	Overhead Cost/Yr. (7)	Maint. Cost/Yr. (8)	Op. Cost Cost/Hr. (9)	Annual MC Hrs. (10)
DIP	$250	7	0.5	$349	$125	$63	$29	$45.12	12,534
SIP	$200	7	0.5	$279	$112	$56	$26	$42.11	11,239
AXIAL	$150	6	0.5	$179	$97	$48	$22	$35.83	9,668
RADIAL	$225	7	0.5	$314	$113	$56	$26	$45.12	11,285
SLDR	$225	2	1	$90	$44	$11	$2	$66.87	2,180
GR70	$600	1	1	$120	$21	$5	$2	$139.02	1,070
GR71	$600	1	1	$120	$19	$5	$2	$152.50	955
RP1	$300	1	1	$120	$103	$26	$1	$48.38	5,157
GR95	$600	1	1	$120	$26	$6	$2	$119.39	1,290
ADAR	$700	3	1	$418	$76	$19	$6	$136.74	3,804
PBO1	$500	2	1	$199	$47	$12	$4	$112.11	2,334
BIR1	$100	1	1	$20	$0	$0	$1	$9.86	0
PMT1	$750	3	1	$448	$84	$21	$7	$133.14	4,213
TT1	$800	4	1	$637	$114	$28	$10	$138.47	5,705
S018	$800	1	1	$159	$30	$8	$3	$133.17	1,500
S087	$800	1	1	$159	$30	$8	$3	$133.17	1,500
Total		48		$3731	$1041	$372	$146		

Table 3 shows the WIP holding cost of PCB-A for the year (similar calculations are done for PCB-B and C). Column 3 is the added dollar per lot by each operation. This cost is obtained by multiplying the operation time for a lot (which is computed from three MANUPLAN II inputs: the operation times for set-up and run and the lot size) by the operation cost per machine hour obtained from column 9 in Table 2. This calculation is

(Added Dollar/Lot) = [(Operation Time for Run/pc.) • (Lot Size) + (Operation Time for Setup/Lot)] • (Operation Cost/Machine Hour).   (12)

Column 4 is WIP cost per lot in a given operation. For example, in DINS (DIP insertion) the incoming WIP has a value that is the cost of raw material. After DINS is completed the cost for a lot is the cost of raw material plus the added dollar per lot (i.e., operation cost for a lot for DINS). This sum is then the WIP cost per lot for the next operation [SINS (SIP insertion)]. Although this procedure holds for sequential operations, the calculation for test operations is more complicated: the WIP costs per lot for operations with branching and rework loops are calculated according to Equations 3a and 3b (the branching probability q can be found in *Details of the MANUPLAN II Model*, e.g., q equals to 0.08 on the route from ICT1 to RPP1).

Column 5 shows the WIP (in lots) for each operation as calculated by MANUPLAN II. The cost of WIP inventory, in column 6, is the product of columns 4 and 5,

$$(\text{WIP Cost}) = (\text{WIP Cost/Lot}) \bullet (\text{WIP in Lots}), \qquad (13)$$

and the WIP holding cost (column 7) is then

$$(\text{Holding Cost/Year}) = (\text{WIP Cost/Year}) \bullet (\text{Annual WIP Holding Factor}). \qquad (14)$$

where the annual WIP holding factor is obtained by continuously compounding the WIP holding interest rate, that is

$$(\text{Annual WIP Holding Factor}) = e^{(\text{Interest Rate})} - 1.$$

In this case, with interest rate = 22.31%, we get annual WIP holding factor = 0.25.

TABLE 3 – WIP Cost Shown By Operations of PCB-A

Opern. Name (1)	Equip. Name (2)	Added Dollar per Lot (3)	WIP Cost per Lot (4)	X WIP (in Lots) (5)	= WIP Cost (6)	Annual Holding Cost (7)
DINS	DIP	$166.93	$3,500	3.31	$11,580	$2,895
SINS	SIP	$190.78	$3,667	2.58	$9,471	$2,368
AINS	AXIAL	$172.00	$3,858	2.75	$10,592	$2,648
RINS	RADIAL	$246.83	$4,030	2.94	$11,859	$2,965
WAVE	SLDR	$77.23	$4,277	0.58	$2,468	$617
ICT1	GR70	$181.42	$4,798	1.69	$8,087	$2,022
RPP1	RP1	$507.94	$4,995	0.60	$3,011	$753
FNT1	GR95	$95.51	$5,486	1.16	$6,380	$1,595
RPF	RP1	$507.94	$5,590	0.52	$2,923	$731
PBI	PBO1	$403.59	$6,493	1.81	$11,762	$2,941
RPPB	RP1	$507.94	$6,952	0.71	$4,966	$1,241

TABLE 3 – WIP Cost Shown By Operations of PCB-A

Opern. Name (1)	Equip. Name (2)	Added Dollar per Lot (3)	WIP Cost per Lot (4)	X	WIP (in Lots) (5)	=	WIP Cost (6)	Annual Holding Cost (7)
BIR	BIR1	$236.53	$6,952		6.09		$42,308	$10,577
PMT	PMT1	$951.94	$8,395		3.13		$26,312	$6,578
RPPP	RP1	$507.94	$9,477		0.65		$6,146	$1,537
TTSD	TT1	$1,474.69	$11,024		3.74		$41,229	$10,307
RPT	RP1	$169.31	$12,700		0.37		$4,717	$1,179
			Total		32.64		$203,812	$50,953

Table 4 details the total system costs by part and major category: material cost, equipment amortization cost, labor cost, overhead cost, maintenance cost, and WIP cost. In Table 4, the material cost per year is calculated from raw material cost per unit (from Table 1) and the number of PCBs used per year. The number of PCBs used per year consists of the production goal (from MANUPLAN II input data) plus the annual number of scrapped PCBs calculated from "MANUPLAN II." Therefore,

$$\text{(Material Cost/Year)} = \text{(Raw Material Cost/Unit)}$$
$$\bullet \text{(Good \& Scrap Products/Year)}. \tag{15}$$

The equipment cost per year (for part x) is obtained from the totals in Table 2 by prorating on the amount each part uses that equipment:

$$\text{(Total Equipment Cost/Yr. for Part } x) = \sum_i \text{(Equipment Group } i$$
$$\text{Cost/Yr. allocated to Part } x) \tag{16}$$

where

(Equipment Group i Cost/Yr. allocated for Part x)

$$= \text{(Equipment Group } i \text{ Cost/Yr.)} \bullet \frac{\text{(Machine Utilization for Part } x \text{ at Equipment Group } i)}{\text{(Sum of Machine Utilization for all Parts at Equipment Group } i)} \tag{17}$$

the first term is the corresponding equipment group i in column 5 of Table 2. The numerator and denominator of the second term are from MANUPLAN II outputs, which include detailed machine utilization allocated to all parts.

Column 3 is the sum of labor costs over all operations, where labor cost for a given operation is calculated as follows:

$$\text{(Labor Cost for a Given Operation)} = \text{(Labor Cost/Hour)} \bullet \text{(Labor Hour/MC Hour)} \bullet \text{(Annual Machine Hours for Setup \& Run} \quad (18)$$
$$\text{for that Operation)}.$$

The first two terms on the RHS have been explained previously, and the last term is a MANUPLAN II output.

Column 4, the overhead cost per year (by part), is the sum of overhead costs over all operations for the part which is given by:

$$\text{(Overhead Cost for a Given Operation)} = \text{(Overhead Cost/Hr.)}$$
$$\bullet \text{(Annual Machine Hrs. for Setup \& Run for that Operation)}. \quad (19)$$

Column 5 is the maintenance costs per year (by part) that comes from the totals in Table 2 prorated by the fraction a part is to total production:

$$\text{(Maintenance Cost/Yr.)} = \text{(Total Maintenance Cost/Yr.)} \bullet \text{(Fraction of Maintenance Cost for a Part)}. \quad (20)$$

where

$$\text{(Fraction of Maintenance Cost for a Part)} = \frac{\text{(Sum of Repair Utilization for a Part over all Equipment)}}{\text{(Sum of Repair Utilization over all Equipment \& Part)}} \quad (21)$$

The repair use in Equation 21 is obtained from MANUPLAN II output.

The last column shows the cost per shipped product. This is the total annual cost divided by the production goal. Because each of the components of this total contain the costs of scrap, rework, and system congestion, the calculation allocates these costs to each product shipped.

TABLE 4 – Total System Costs Shown By Part
(Dollars shown are in thousands, except cost per good unit)

Prod. Type	Material Cost/Yr. (1)	Equip-ment Cost/Yr. (2)	Labor Cost/Yr. (3)	Over-head Cost/Yr. (4)	Mainte-nance Cost/Yr. (5)	WIP Cost/Yr. (6)	Total Cost/Yr.	Cost per Shipped Unit
PCB-A	$2,641	$1,919	$496	$159	$58	$51	$5,324	$354.93
PCB-B	$3,967	$999	$280	$102	$41	$53	$5,442	$362.85
PCB-C	$2,976	$813	$265	$111	$47	$32	$4,244	$282.92
Total	$9,584	$3,731	$1,041	$372	$146	$136	$15,010	

NUMERICAL RESULTS

The data used in this example are based on information from a corporation for whom a similar study was performed. For the purpose of illustrating the capacity of this approach, the number of products has been reduced to three. Even with three products, there are nine automatic testing equipment (ATE) stations each requiring a unique PFB estimate, which could be varied. To minimize the combinatorial analysis to answer the questions posed, each ATE station was assumed to have a representative PFB and SR. For the analyses these were uniformly varied through a common multiplier. We investigate several quality strategies that involve changing the PFB and SR data to analyze how estimates of total annual cost from the model change. We use total annual cost as a measure of system performance to keep the ideas simple. The method for considering life cycle cost described in Equation 4 is easily incorporated into the model.

The PFB is assumed to be the characteristic quality level of the incoming components, either as purchased raw material or as output from fabrication processes elsewhere in the factory. The quality strategies tested are represented by four levels of PFB varying from 10% to 25%. A reduction in the PFB represents improvements in the manufacturing system, such as renegotiating purchase contracts to obtain improved quality of raw materials or purchasing a more sophisticated manufacturing process guaranteed to improve output quality. Table 5 illustrates the cost estimates obtained on model runs for levels of PFB studied. Note that the percentages of material, labor, and burden (equipment, WIP, maintenance, and overhead) correspond to typical industry reports.

Generally, we found that WIP, maintenance, and overhead were relatively minor contributors to total cost and exhibited relatively small variation over the quality parameters studied. Therefore, these cost components are omitted from our basic results shown in figure 3–5. This does not mean that improvements

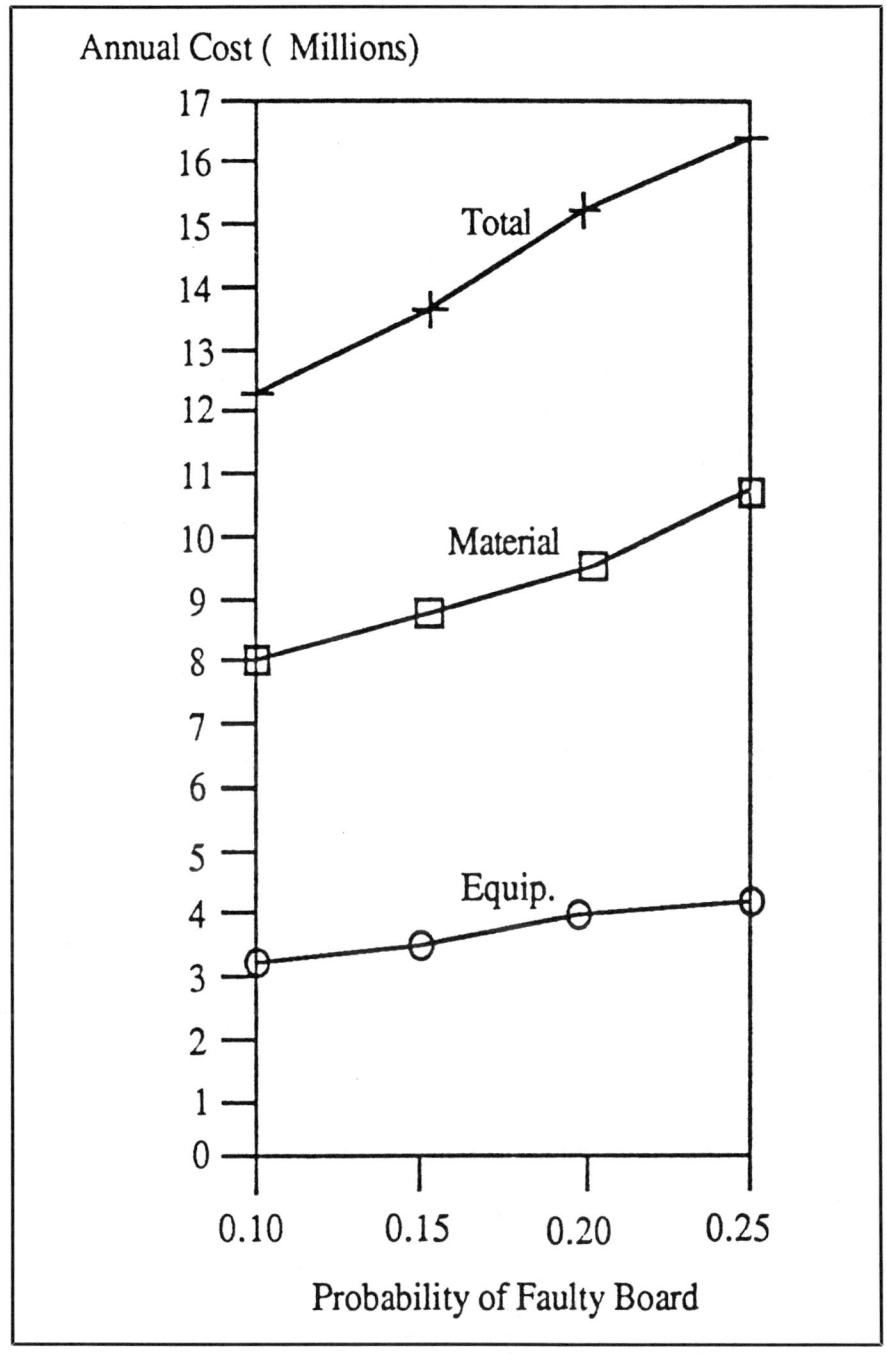

Figure 3. *Effect of component quality on costs.*
(When SR Factor = 0.6).

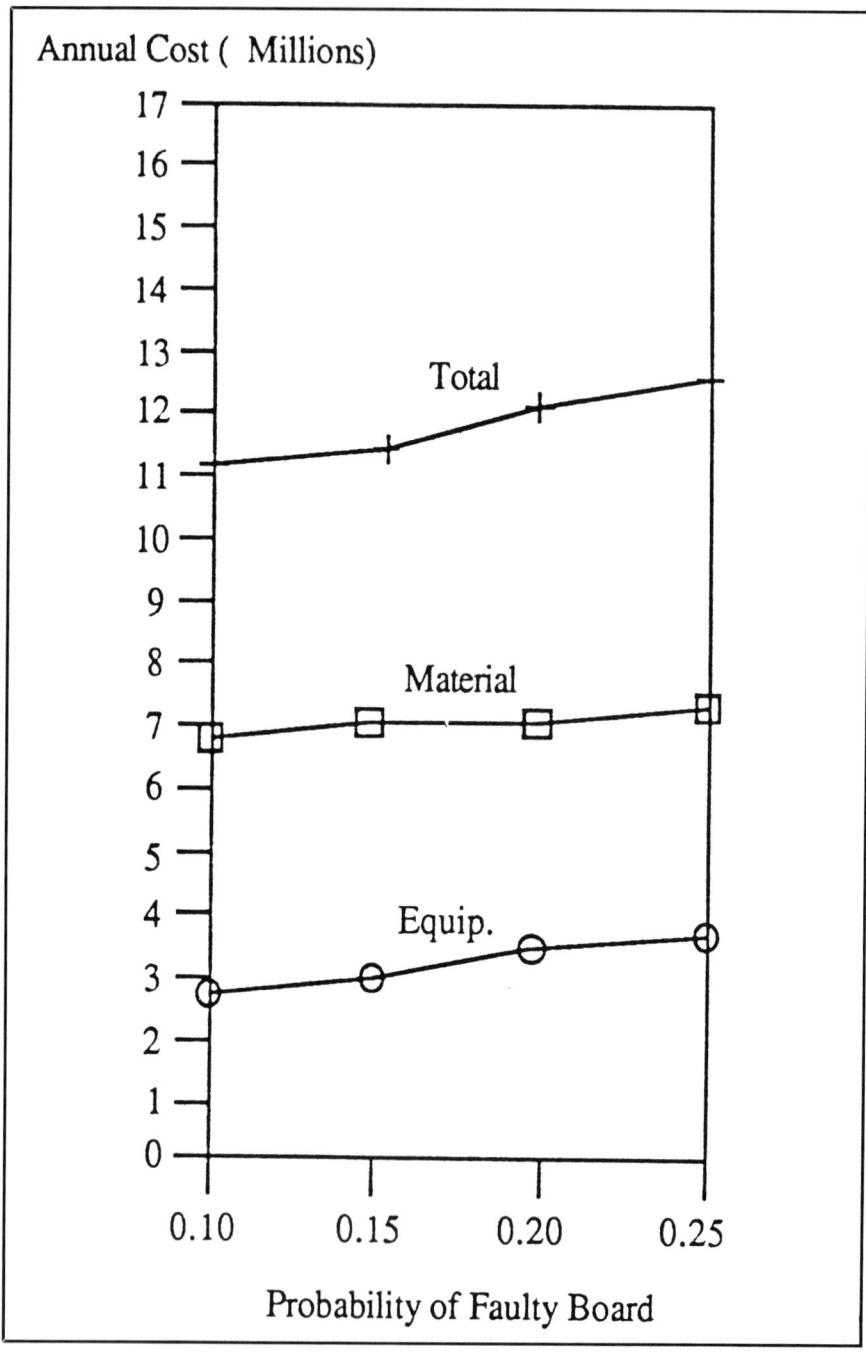

Figure 4. *Effect of component quality on costs.*
(When SR Factor = 0.06).

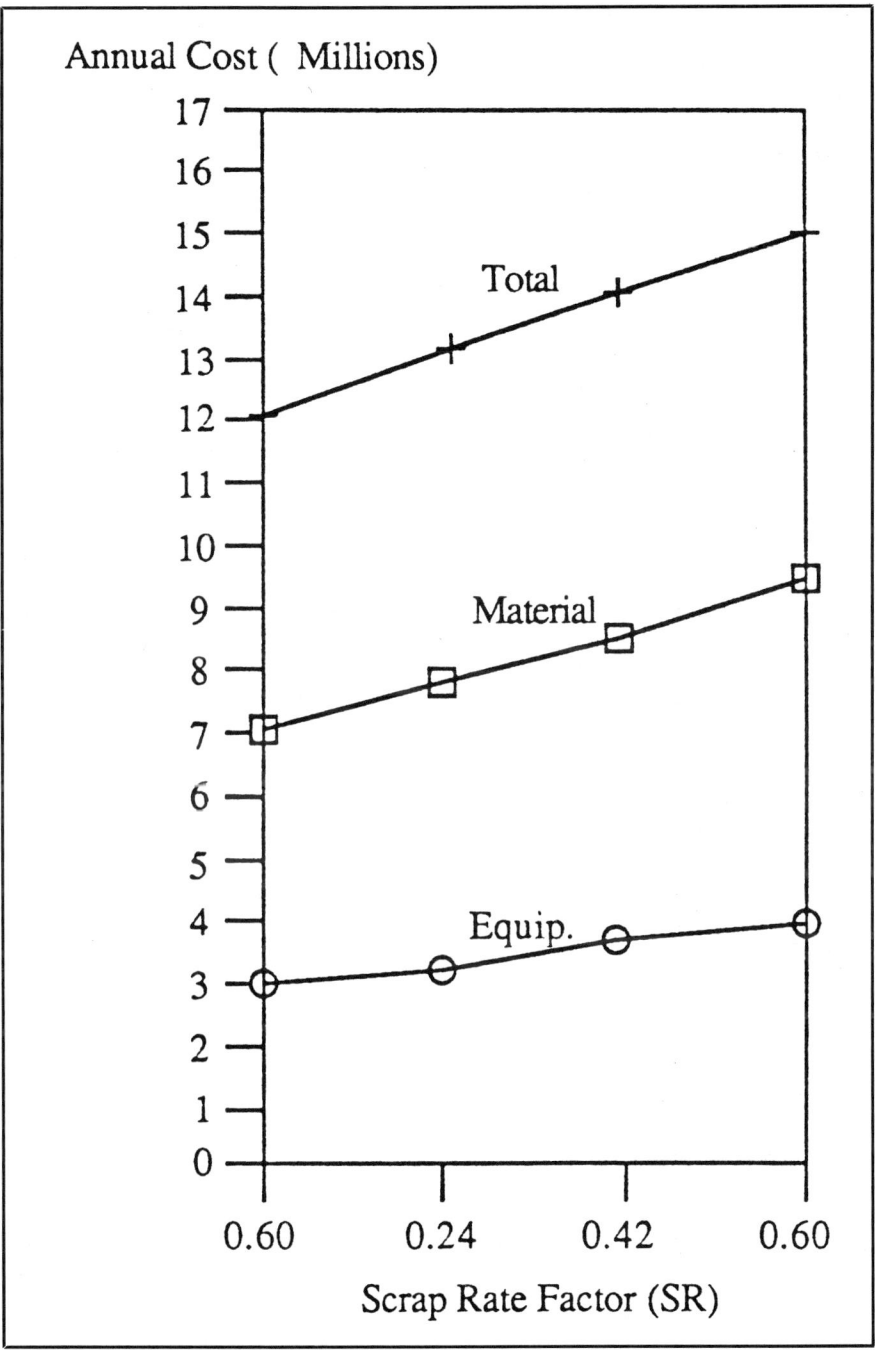

Figure 5. *Effect of scrap rate on costs.*
(When (PFB) = 0.2).

are insignificant. For instance, if the PFB can be reduced from 20% to 10%, WIP, maintenance, and overhead costs will be reduced 25%, 15%, and 18%, respectively (see Table 5).

TABLE 5 – Itemized Annual Cost for Varying Levels of Component Quality (Dollars shown are in thousands)

Item	Probability of Faulty Board				Percentage of Total Cost
	0.10	**0.15**	**0.20**	**0.25**	
Total	$12,371	$13,492	$15,011	$16,529	100
Material	$7,918	$8,672	$9,583	$10,704	63-65
Equip.	$3,078	$3,278	$3,731	$3,865	23-25
Labor	$845	$940	$1,041	$1,172	6-7
Overhead	$304	$337	$372	$417	2-2
Maint.	$124	$135	$146	$162	1
WIP	$102	$131	$136	$208	1

The second quality strategy studied involved various levels of SR. The SR is the percentage of boards not passing the test at an ATE station that are then scrapped. By scrapping a high percentage of the boards which fail, less rework is required. We studied four SR factors from 0.6 to 0.06, corresponding to a net scrap rate of 5% – 15% at a given test, which appeared realistic from our knowledge of industry data.

All of our estimates of total annual cost for a particular quality strategy are based on the optimum manufacturing system—that is, the one with the lowest annual cost. Thus, for each quality parameter studied, the optimum lot size and equipment configuration were determined. This prevented confounding our results from other factors of system performance.

Figures 3 and 4 illustrate the variation of total annual cost with PFB. Figure 3 and Table 5 are based on a high SR, whereas Figure 4 is for a low SR. The significant conclusion from comparing these graphs is that the variation in total cost is driven by material cost at high scrap rates and by equipment costs at low SRs. In either case, the results indicate that if the initial estimates of PFB are high, large amounts of money can still be invested toward lowering the PFB to obtain significant improvements in overall system performance.

Figure 5 illustrates the effect of SR at a fixed level of PFB. For our example, a policy of minimizing SR produces the best economic performance of the system. For all of our ATE the ratio of the value of a part to be scrapped to the rework cost per unit was high, varying from about 3 to 40-50. For other situations where this ratio is closer to one, a policy of high scrap may have some benefits.

Finally, figure 6 summarizes our analysis showing the effects of both PFB and SR on total annual cost.

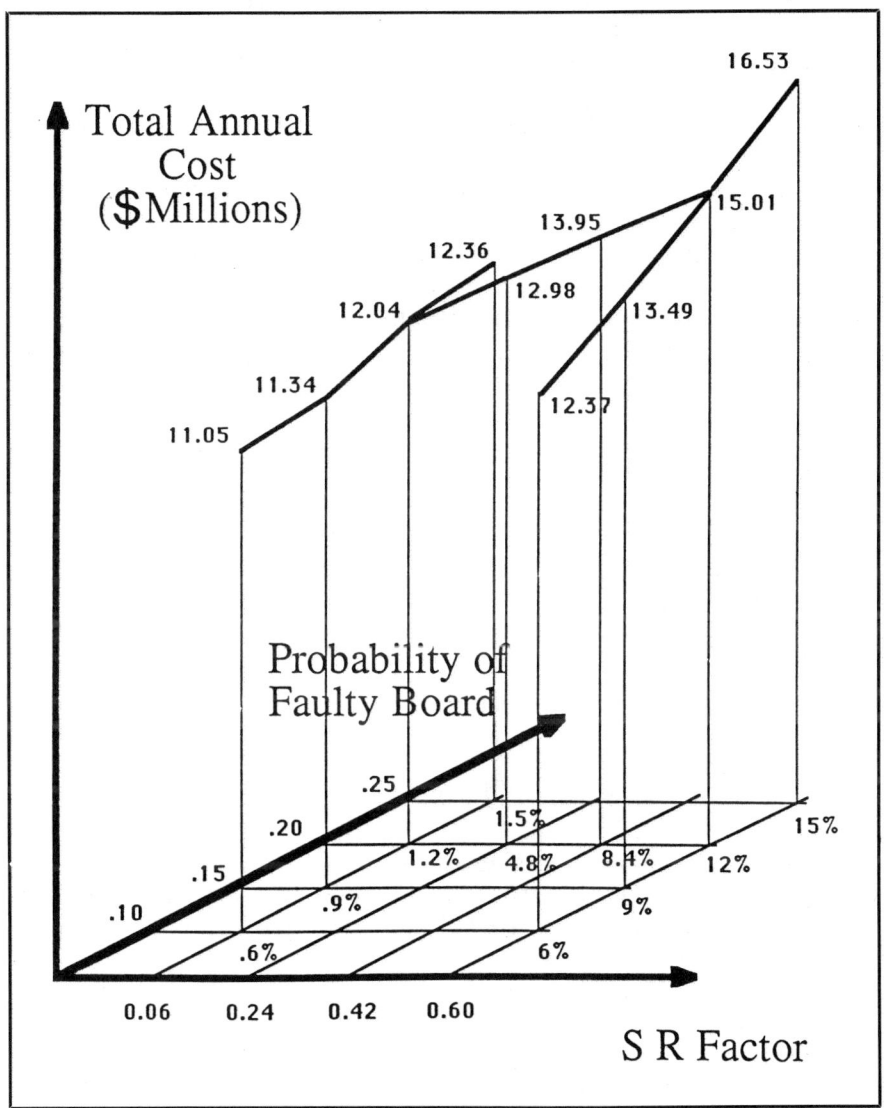

Figure 6. *Joint effects of component quality and scrap rate on total costs.*

CONCLUSIONS

We have demonstrated the feasibility of combining system modeling and economic modeling to explore the economic impact of different quality strategies. The particular approach used here can rapidly explore many different strategies, in a matter of minutes, on a microcomputer such as a PC/AT.

Although every manufacturing manager will admit that improved raw material quality is beneficial, quantifying that benefit in economic figures has been difficult. Our approach allows this benefit to be clearly quantified in monetary figures.

The analysis in this model covered only the variation in the SR and PFB. Changes in fault coverage and test level are not discussed. We did not attempt to characterize the cost of improved manufacturing processes. In addition, we have assumed that PFB is independent of the ATE sequence. An entirely different set of quality models arise when the probability of detecting a fault at an ATE depends on the preceding test results for that board. However, our aim is to demonstrate our approach's feasibility. For more realistic analysis the model can be extended to include such features.

ACKNOWLEDGEMENT

We thank Ken Anderson of Siemens Corporation for his guidance in helping us understand the principal concepts in PCB testing.

Suggested Readings

Blank, L. 1985. The changing scene of economic analysis for the evaluation of manufacturing system design and operation. *The Engineering Economist.* 30(3): 161-168.

Box, G. E. P., W. G. Hunter and J. S. Hunter. 1978. *Statistics for Experimenters.* John Wiley and Sons. New York.

Davis, B. 1982. *The Economics of Automatic Testing.* McGraw Hill. New York.

Falkner, C. H. and S. Garlid. 1986. Simulation for the justification of an FMS. *1986 Fall Industrial Engineering Conference Proceedings.* Boston, Massachusetts. pp. 99-108.

Garlid, S., C. H. Falkner, B. R. Fu and R. Suri. 1987. Evaluating Quality Strategies for CIM Systems. *1987 Fall Industrial Engineering Conference Proceedings.* November. pp. 30-36.

Haider, S. W., D. G. Noller and T. B. Robey. 1986. Experiences with analytic and simulation modelling for a factory of the future project at IBM. *Proceedings of the 1986 Winter Simulation Conference.* Washington, D.C. December. pp. 641-648.

Kaplan, R. S. 1986. Must CIM be justified by faith alone. *Harvard Business Review.* March-April. pp. 87-95.

Miczo, A. 1986. *Digital Logic Testing and Simulation.* Harper and Row. New York.

Mills, M. C. 1986. Using group technology, simulation and analytic modelling in the design of a cellular manufacturing facility. *Proceedings of the 1986 Winter Simulation Conference.* Washington, D.C. December 1986. pp. 657-660.

Network Dynamics Inc. 1987. *MANUPLAN II User's Manual.* Network Dynamics Inc. Cambridge, MA.

Primrose, P. L. and R. Leonard. 1986. The use of a conceptual model to evaluate financially flexible manufacturing system projects. In *Flexible Manufacturing Systems: Current Issues and Models,* F. Choobineh and R. Suri, eds. Industrial Engineering and Management Press. Norcross, GA.

Primrose, P. L. and R. Leonard. 1986. Evaluating the intangible benefits of flexible manufacturing systems by use of discounted cash flow algorithms within a comprehensive computer program. In *Flexible Manufacturing Systems: Current Issues and Models,* F. Choobineh and R. Suri, eds. Industrial Engineering and Management Press. Norcross, GA.

Smith, G. W., 1983. *Engineering Economy: Analysis of Capital Expenditures. First Edition.* The Iowa State University Press. Ames, Iowa.

Suresh, N. C. 1985. Justifying multimedia systems: an integrated strategic approach. *Journal of Manufacturing Systems.* 4(2): 117-134.

Suri, R., 1985. An overview of evaluative models for flexible manufacturing systems. *Annals of Operations Research.* 3: 13-21.

Suri, R., G. W. Diehl and R. Dean. 1986. Quick and Easy Manufacturing Systems Analysis Using MANUPLAN. *1986 Annual International Industrial Engineering Conference Proceedings.* pp. 195-205.

Suri, R. and G. W. Diehl. 1987. Rough cut modelling: an alternative to simulation. *CIM Review.* 3(3): 25-32.

CIM Strategies. 1985. Zero in on feasible plant designs before simulating.

APPENDIX: OVERVIEW OF MANUPLAN II

MANUPLAN II is a rough-cut analysis software tool for modeling manufacturing systems such as FMS, group technology cells, and job shops, which enable system designers to evaluate alternative manufacturing systems quickly and easily (Network Dynamics Inc. 1987; Suri et al. 1986; Suri and Diehl 1987). Often a system that would take several weeks to evaluate using simulation can be evaluated in days with MANUPLAN II. Analyzing a system performance takes only a few seconds. By changing the system design data, such as number of machines, lot sizes, and operation process times, manufacturing designers can quickly evaluate different systems and obtain the trade-off between inventory levels, production requirements, flow times, and equipment reliability.

Unlike most other tools that evaluate a manufacturing system, MANUPLAN II is not a simulation—it uses an analytic model. The mathematical model used combines network-of-queues and reliability modeling to estimate the dynamics of the interaction between the machines and the workpieces in the system. The model includes the effect of equipment reliability (machine failures and repairs) on manufacturing system performance. Inputs to MANUPLAN II and outputs from MANUPLAN II have been illustrated in the text.

A particular advantage of using MANUPLAN II is that the input and output data are contained in a LOTUS 1-2-3 spread sheet and are available to the analyst in a flexible environment to provide any type of summary which is useful to the particular problem being studied. Our emphasis on economically estimating the impact of various quality strategies illustrates this flexibility.

A preliminary version of this work, titled "Evaluating Quality Strategies for CIM System," by Scott Garlid, Charles Falkner, Bor-Ruey Fu, and Rajan Suri, was presented in the Fall *Institute of Industrial Engineers Conference Proceedings*, November 1987, 30-36. This paper contains a more detailed description of the economic equations and system model, as well as a more extensive analysis of outputs of the model for various quality strategies.

Bor-Ruey Fu is a Ph.D. student in Industrial Engineering at the University of Wisconsin-Madison.

Charles H. Falkner is professor of Industrial Engineering at the University of Wisconsin-Madison.

Rajan Suri is associate professor of Industrial Engineering at the University of Wisconsin-Madison.

Scott Garlid is an Industrial Engineer in Advanced Manufacturing Systems with Intel Corporation, Chandler, Arizona.

35

Miniline: Research Applied To Manufacturing

Donald E. Barr
William T. Chen
Robert Rosenberg
Donald P. Seraphim
Patrick A. Toole

The rapid growth of the electronics industry fueled by very large-scale integration (VLSI) is making great demands on engineers for production of complex miniaturized printed circuit products. The processes used to make printed circuits include physical and chemical phenomena governed by fundamental laws of material science, physics, and chemistry, which heretofore have been studied only in university laboratories and research institutions. Now we need to know how to apply the tools and techniques of pure science to the hustle and bustle of the production line.

Our studies show what fundamental laws are affecting which aspects of the production process both qualitatively and quantitatively, and we have found that they have had a profound effect on manufacturing processes, tools, and controls. While the examples and experiences reported in this paper are associated with advanced complex electronic systems, similar approaches can be used in other industries.

Many of todays high-tech production lines a) have numerous and interactive process steps, b) require large capital investment, and c) often depend on materials and process inputs from other sites or suppliers. Nevertheless product quality is approaching zero defect levels while performance is continuing to improve. Printed circuit board production lines illustrate such high technology industry where evolution is proceeding rapidly.

THE MINILINE CONCEPT

The "miniline" is a concept or philosophy for solving manufacturing problems and for scaling-up new products or processes, as well as solving manufacturing "line-down" situations. It is extremely useful in resolving reliability quality, and yield improvement issues. The concept brings together interdisciplinary skills from manufacturing engineering, quality engineering, development, research, universities, consultants, etc. This group of key people focus on a given problem, then attempt to hypothesize or model the problem from all perspectives. The traditional method is to run matrix experiments on the product line to zero in on the problem. The approach in the miniline is to duplicate, under laboratory conditions, the manufacturing process on a small scale, often in sophisticated analytical instruments. It is the task of the specialist team to identify experiments and the tools with which the experiments can be performed. Key to the success of this approach is the ability to duplicate the manufacturing process in the laboratory, where it can be tightly controlled while being closely monitored. The creation of minilines is described in this paper with concrete examples for a sophisticated printed circuit facility.

One type of miniline speeds up development and manufacturing scale-up for high technology products. During development and manufacturing scale-up, it is often the custom to use pilot lines. Pilot lines are usually designed to process one or a few pieces of a batch, i.e., one printed circuit board or a small batch at a time. However a complete full-size tool set is still used. Therefore almost all of the drawbacks of a production line still prevail: complexity, capital intensiveness. dependence on input.

In recent publications [1]-[5], we described the circuit boards and processes used in the IBM 3081 Processor Unit and the IBM 4381 Processor Unit computer systems. In this chapter we will describe how the miniline was used to problem-solve, develop new processes, and enhance controls for large complex production lines like those used for production of the 3081 and 4381 printed circuit boards.

An important consideration to begin with is that there is an immense breadth of discipline in printed circuit packaging technology. Even one set of processes and materials (photo-processing, for example) requires a range of interdisciplinary skills. An example of just one of the wet lines used in the process set is shown in Figure 1. The sequence of tanks in Figure 1 illustrates the large variety of chemicals used and hints at the complexity of such processes.

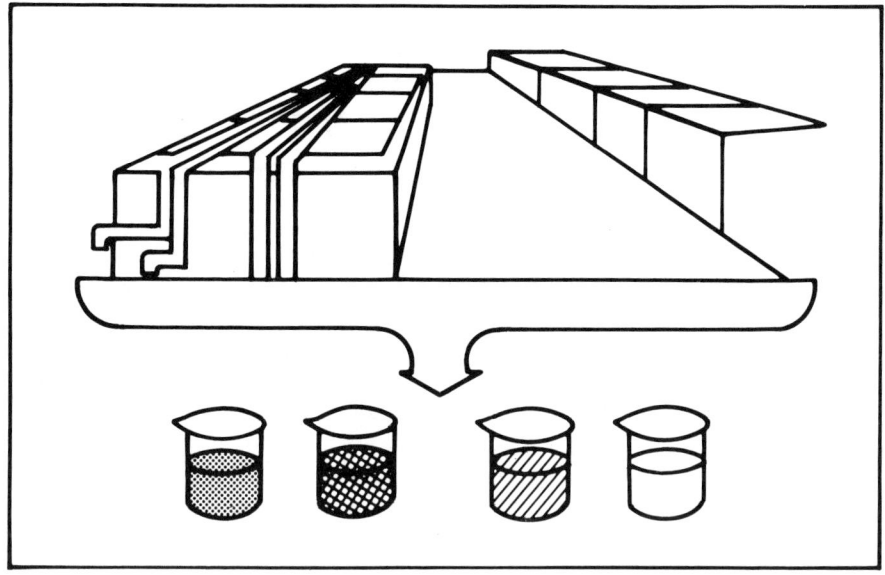

Figure 1. Miniline applied to plating.

Matrix methods are sometimes used to study the interaction of process variables. However, with a production line which may be 200 process steps long, it will often take too long to run an effective matrix experiment. Therefore it is important to consider taking some new approaches:

- test tubes and beakers;
- microcalorimeters;
- Fourier transform infrared spectroscopy;
- Rutherford backscattering;
- Ph.D.'s; and
- professors.

Usually industry research departments and universities are not known for practical solutions in the sense of being immediately implementable. The miniline brings industry, research, and universities much closer to manufacturing reality by solving problems using fundamentals such as analysis and simple experimental simulation. By using the miniline approach we can often solve a problem in a very short period of time. A key to problem solving is to realize that, whatever changes are suggested by the experiments, they must be implementable immediately to be practical. This generally negates major capital tooling changes or major material changes. Therefore industry research and the university consultant must have an overview of the production process tooling limitations to be able to make an effective recommendation.

PRINTED CIRCUIT BOARD PROCESSES AND MINILINES

Epoxy Glass

The starting material glass cloth and resins can be our first example. The cloth is immersed in resin with a high solvent content which is evaporated to a dry gel state in a tower with a sequence of heaters (Figure 2). One key to a reliable process is to stage the heating so that the solvent has the opportunity to diffuse to the surface of the resin rather than to create gas bubbles internally on the cloth yarn surfaces. Thus the fundamental approach is to make precision measurements of solvent weight loss at various temperatures. This allows the diffusion process to be modeled [6], and calculations can be made on how fast the drying or treater tower can be run at various temperatures without a concern for gas bubble formation.

In fact a hot stage microscope with a programmed heating stage can also be used to follow the process visually. Such experiments are, in fact, just as well

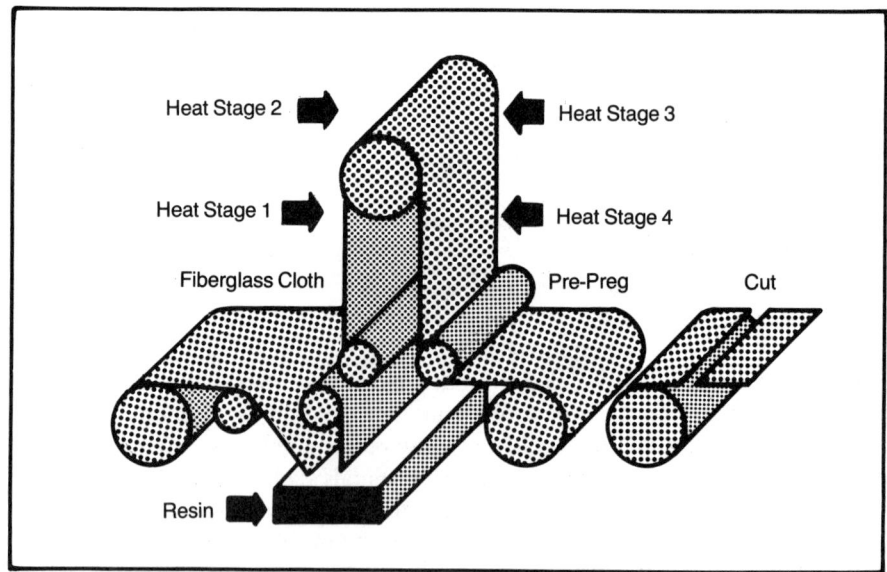

Figure 2. *Glass fabric impregnation (treater tower).*

done in a university environment as in industry. The key is to achieve fundamental data, solubility constants, diffusion coefficients, etc., as a function of temperature. The miniline in this case is an accurate temperature controller, scale balance, and a hot stage microscope to verify the bubble formation. We have sponsored this type of research at a university.

A few years ago, a microcalorimeter experiment took days to set up and run. Now with the latest instruments, we can simulate the entire resin cure cycle within hours. The full enthalpy exotherm for cross-linking can be followed, exactly simulating a lamination cycle. However the global linkages are the last to form and may not be accurately detectable with the calorimeter since very little of the exotherm is left at this stage, Fortunately, on cooling through the glass transition, the transition shows up as a specific heat anomaly. Also since the transition temperature is quite sensitive to the final globe linkages, the calorimeter is an ideal miniline and quantitative monitor of the effectiveness of the production process (see Figure 3).

The process can be followed by other techniques as well. For example, flow measurements of the resin-impregnated glass cloth (prepreg) can be accomplished in a small press with time, temperature, and pressure control. During pressing, the total mass flow out of a specified area can be measured accurately. This flow is, in fact, a parameter of definite interest in determining whether the features in a printed circuit laminate will be adequately filled with resin, coated, and insulated. This type of flow measurement combines in one measurement

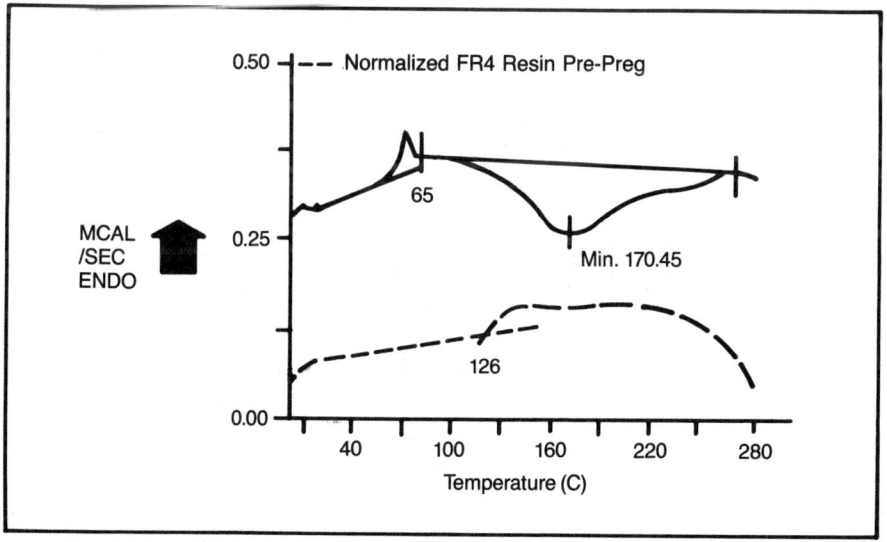

Figure 3. *Calorimeter scan.*

both the effect of temperature on cure state and the effect of viscosity decrease due to increasing temperature. The rate of heating must fully simulate that of the production press. Rheological measurements can also be made *in situ* during a temperature excursion simulating a press cycle. The data from such an experiment, in fact, shows that the viscosity reaches a minimum and then increases rapidly as cure increases. Indeed there is only a narrow operating range of time during the press heat-up cycle where flow will occur adequately.

This flow measurement experiment together with other cure-rheological studies (Figure 4) gave a full picture of the effects of viscosity, cure, and temperature on prepreg production [7].

PHOTOPROCESSING

The printed circuit boards for the IBM 3081 Processor and the IBM 4381 Processor contain approximately a mile of circuitry. There are 6 signal planes, 12 patterned voltage planes, and 2 surface planes. In the early stages of manufacturing scale-up, a good deal of confusion arose due to the strongly interacting features of the circuitization process. Photoresist adhesion varied considerably from panel to panel and within a panel (Figure 5).

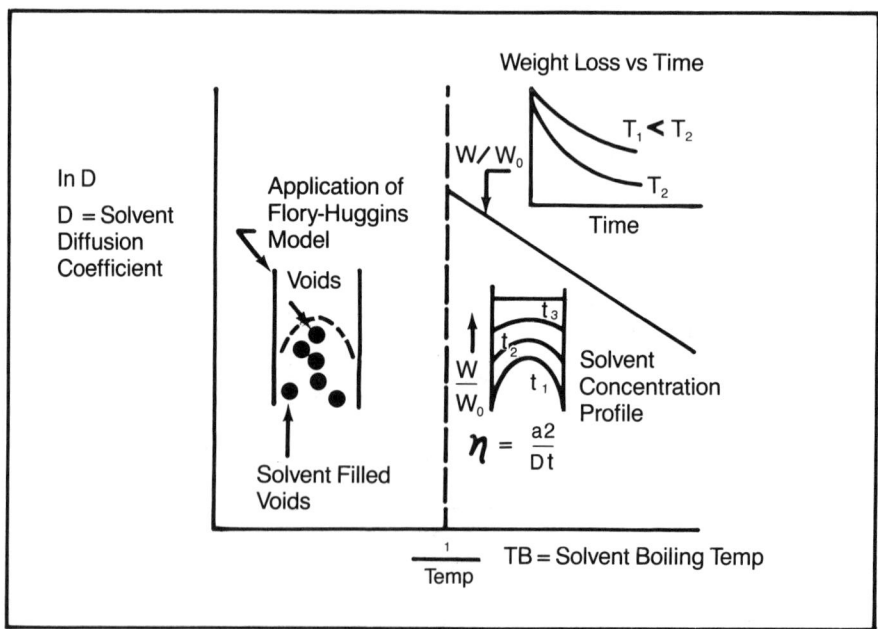

Figure 4. *Flow measurement and cure-rheological studies.*

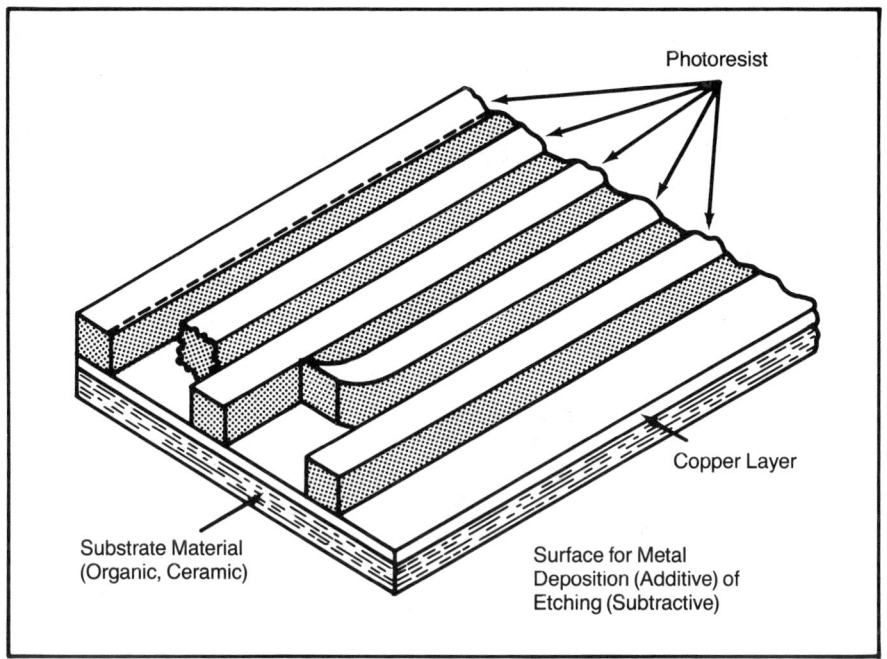

Figure 5. *Schematic of poor resist adhesion in a panel.*

Since there is a variety of failure modes for photoresist this may be one of the best examples we can cite for the value of an interdisciplinary team with a wide variety of scientific characterization skills. For example, the mechanics expert can develop techniques to measure adhesion or surface roughness accurately and assure that the mechanical phenomena (stress and strain due to swelling of the photoresist) are understood.

The photochemist will focus on the degree of cross-linking and assure that the bottom layer of the resist is exposed to the required photon flux with the correct spectrum.

The surface analyst will concentrate on the incoming surface characterization. What are the contaminates? Is the surface oxidized? If a thin conversion coating is used, how thick is it and how uniform is it? How does it vary with the concentration of the coating and temperature of the conversion coating process? What impact do the development materials have on the conversion coating?

The polymer chemist may focus on development and swelling. Is the resist able to withstand the chemicals? How fast do the chemicals diffuse into the resist? How much expansion takes place?

Finally the electrochemist may be consulted to understand the potential for corrosion and blistering (cathodic delamination) in the process environment. Electrochemical [8] measurements like open cell potential can be made after surface preparation treatments. A Pourbaix diagram can be constructed to show where the stable surface phases are relative to the chemical potentials, etc.

A wide variety of minilines is, therefore, appropriate. Measurement techniques are easily set up to monitor the chemical effects *in situ*. For example, cyclic voltammetry is used to determine the potentials for the oxidation boundary at various conditions. A sequence of beakers easily reproduces the production line chemicals, and scanning Auger can be used to characterize the surface after various treatments.

The turnaround time of experiments is orders of magnitude faster than the time required on a pilot or production line. The key variables are rapidly brought to focus and their effects measured. Correlations can be achieved with production line parts analysis. Better still, chemicals from the production line can be tested in the miniline to verify their goodness. Many times, contaminants from exposure to solutions (drag in) from earliers tanks in the sequence have an effect when large batches of parts concentrate the contamination. This can be easily uncovered in miniline experiments.

One of the first miniline experiments in the circuitization process involved the design of an experimental apparatus in which the hot alkaline plating solution was pumped through a windowed cell containing an actual part of a resist imaged circuit board. With this plating cell, it was possible to see and photograph the plating process and to observe the formation of undesirable or slow plating. In this particular miniline setup, we were able to determine not only the defect sites which caused uneven plating, but much to our surprise, also the fact that even low-level tungsten filament lighting exhibited a strongly undesirable effect on the processed photoresist.

The model for the cause for shorts and open defects is based on photoresist adhesion failure. This general situation, of course, is common to most electronic packaging production, whether it be circuit boards or modules and whether the process be additive or subtractive. Additional minilines were established to provide fundamental information on the entire system starting with the photoresist and initial metal surface prior to and during all subsequent processing steps. In order to follow the enthalpy of reaction during exposure of the photoresist and to obtain the associated kinetics, a differential scanning calorimeter (DSC) was modified [9] to include a variable monochromatic ultraviolet light source. With this miniline apparatus, we could follow the cross-linking reaction in the photoresist at various wavelengths. Subsequently we did obtain a correlation between exposure and resist performance during the plating process.

In the case of the metal surface miniline, we were able to follow the anodic and cathodic nature of the metal surface using electrochemical monitoring during all the initial metal cleaning. Here we made direct potentiometric open cell measurements while moving the metal through all the processing solutions involved in the manufacturing line. At the same time, similar experiments were being conducted in our research area to characterize the surface using the scanning Auger technique. In these experiments the metal surface preparation processes were simulated in a set of beakers. This provided a fundamental understanding of the metal surface during the process changes and allowed us to move quickly to implement improved process changes with confidence.

These resist-plating minilines utilized actual parts which were monitored during simulated processing in the laboratory. This provided fundamental understanding of the numerous complex interrelationships which exist in these complex fabrication processes. In addition the miniline experiments were conducted in the laboratory using sensitive analytical instruments without disturbing the manufacturing line until the improved process modifications were defined and the process windows were clearly defined.

Localized debonding of the photoresist is one good example of a problem which requires an interdisciplinary collaboration of scientists and engineers to model (Table 1).

Plating

The plating process consists of a whole series of steps, starting from treatment for surface cleaning, the various photoresist process steps, surface activation, seeding, and finally plating in the plating bath. The quality of the circuitry on the surface and in the plated-through holes depends critically on the pretreatment steps, as well as on the plating bath itself. Laboratory beaker lines with laboratory prepared solutions and the manufacturing line solutions are ideal for following the step-by-step surface conditions utilizing Auger, ESCA, and voltammetry. Considering the size of the usual plating bath, one cannot help but ask whether the chemistry, kinetics, and surface conditions in a 1000-gal scale bath can be reproduced in a 100-cm^3 beaker.

However specially prepared coupons prepared in beakers under controlled conditions can be placed in the bath to evaluate the potential impact of the process steps on a real operating bath. The bath is in a pseudosteady state since the chemicals are continuously consumed and added, and reaction products accumulate. Smaller lab baths are run based upon the manufacturing bath solution to study the effects of dynamic operating conditions such as bath potential and plating rate, as a function of added chemicals and product loading. The plated circuit lines can be evaluated for ductility and cyclic fatigue by using standard testing systems (e.g., Instron Corp. or MTS Systems Corp. instruments) on copper lines removed from plated coupons.

TABLE 1 – Resist Performance Variables Miniline Tools

	Polymer/photochemistry	Micromechanics	Surface science	Electrochemistry
Materials & process variables	Resist thickness Expose intensity Development time Post bake Surface reflectivity Adhesion promoters	Resist swelling Flexural rigidity Surface smoothness mechanical interlock Temperature variation case II diffusion Handling dynamics	Environmental contamination Oxides Process contamination solution permeation	Solutions chemistry Temperature pH
Physical parameter	Cross-link density distribution Resis cross-section profile	Interface stress concentration Buckling stress case II diffusion mechanical adhesion	Local debonding adhesive/cohesion delamination	Surface potential in different process solutions
Miniline tools	Photo DSC TMA FTIR NMR DSC	Instron peel Scratch test Holography Finite element Two-material bend test	SCM ESCA Auger Optical Microscope	Voltammetry Open cell Potential

As there are 40,000 plated-through hole connections in the most advanced printed circuit boards, the plated-through hole receives a great deal of attention through miniline modeling. One consideration is the magnitudes of stresses on the plated-through hole [10]. A second factor is the ductility of the copper. Still another factor is the adhesion to the hole wall [10], [11]. Assuring an adequate uniform thickness of copper in the hole is one of the most difficult tasks.

Electroless copper deposition from a plating bath driven by formaldehyde oxidation has proven to be an excellent method for producing equal copper thickness at the center of the hole and its edge. The ductility of the copper can be characterized by thermal cycle or by flexure fatigue, the latter being much more rapid. Thus any changes in the chemistry of the bath can be monitored by mechanics testing to assure that the product quality is met. Liquid circulation in the bath can be monitored by dye techniques. The more difficult task of showing fluid flow in a hole can be scaled. The major concern is whether bubbles are trapped in the hole and whether the agitation motion is sufficient for the fluid front (bubbles) to pass through the full length of the hole.

Soldering, Flux Chemistry, and Intermetallics

The solder joint is one of the most critical interconnections in electronic packaging. As a typical processor has hundreds of thousands of solder connections, reliability of a connection must be excellent. The Fourier transform infrared instrument (FTIR) is one of the most used minilines. Fortunately flux chemical action, which is very complex, appears to be amendable to simulation with this instrument.

Solder-joining minilines can be utilized to gain the fundamental information required to establish processes with clearly defined process windows. Two such minilines described here can be used to follow the depletion of tin from the solder-joint during the joining process. The first one involves the flux-metal oxide dissolution process. The heat cycle parallels the conventional manufacturing process and the rate of dissolution is monitored by the FTIR spectra changes (Figure 6). For abietic acid flux, the loss of the free carboxylic acid peaks and simultaneous formation of the ionized carboxylate peaks in the spectra can be used to obtain the desired kinetics as a function of temperature. Indeed experiments may be run to determine which contaminants the flux will attack and which will cause nonwetting of the solder.

Differential scanning calorimetry (DSC) is especially convenient for following the loss of tin from the solder to copper. In this case there is a copper-tin intermetallic layer which forms when the solder wets the copper surface at elevated temperatures. The depletion of tin from the solder can be conveniently monitored by the change in melting point of the mixture. It is also

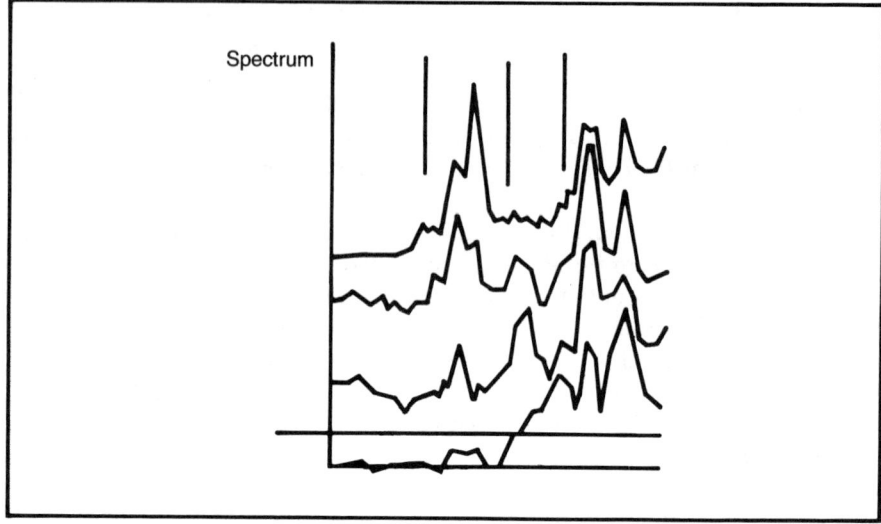

Figure 6. *FTIR spectra of acid flux reactions.*

convenient to program the DSC to simulate the solder-joining furnace and then to simulate the entire process

MECHANICAL DRILL MINILINE

Mechanical drilling is also important because in each IBM 3081 Processor printed circuit board there are about 40,000 0.40-mm (16-mil) diameter drilled holes. Any broken drills or poorly drilled holes would be detrimental to the production yield and reliability. Scanning electron microscopy (SEM) is the miniline tool used for in-depth study of the drill fracture surface from drills broken during manufacturing processes. This establishes how the fracture originates in the carbide material and propagates across the drill cross section and also any defect in the drill that could cause the stress concentration for premature failure. A laboratory robot with identical drive (r/min), insertion and retraction rate (1/min), and control system as the production tool is used to measure the load on the drill. A strain gauge measurement system of torque and thrust is developed on the robot stand so that the torque and thrust on the drill as it drills any circuit board composite section can be measured continuously.

From this data, one can determine the dynamic loading (force and torque) on a drill as a function of its design (whether new or worn) going through a board section with different materials (resin, glass, copper) and material thicknesses and as a function of the various drill parameters (r/min, dwell, and insertion rate). From this combination of dynamic loadings as input to a finite element model, one can calculate the stress distribution in the drill at the region of fracture. Interestingly, since fracture analysis indicates that most drill failures occur in flexure superimposed on torsion, it becomes important to incorporate a buckling mode of deformation in the finite element model to understand the fracture mechanism. An important consideration is to develop a practical simple bending device to screen out any defective drills that may cause premature failures and damage to the boards during production.

The carbide materials are selected for proper balance of strength and wear. Our investigations revealed that as the drill wears, the thrust and torque gradually increase accompanied by the increased heating of the drill itself. Then, the temperature of the drill exceeds the glass transition temperature of the resin materials and results in so called "resin smear" on the copper plane in the drilled hole. This results in a subsequent process for smear removal being required prior to plating of the plated-through hole. The amount of smear and the quality of cutting over the drilled surface are both affected by the amount of wear on the drill.

An important question in the plated-through hole is the adhesion between the plating and the drilled hole wall. This can be measured by establishing a peel test (Figure 7) across sections of the 0.40-mm diameter hole wall. This is another tool for quantifying and controlling the reliability and integrity of a very critical feature in the printed circuit board [10]. While mechanical drilling is a relatively stable process, subtle changes in drill material and design (controlled by drill vendors) and in the board materials and process design could significantly affect plated-through hole drilling either in drill breakage or hole quality.

CONCLUSION

The interaction of the scientists with manufacturing engineers is synergistic. The versatility of moving the miniline to an industry research or a university laboratory expands the expertise and analytical tools which can be applied. Some of these tools are really unique, such as Rutherford backscattering equipment [12] which is more often associated with nuclear science than with the factory floor.

Minilines can serve both to generate initial data bases and diagnose and prevent detrimental changes in the manufacturing line.

- A miniline is a tool for simulating physical-chemical reactions in a manufacturing process or product.
- A miniline can be a set of beakers or test tubes allowing collapse of an acre of tanks into an experimental setup on a tabletop.

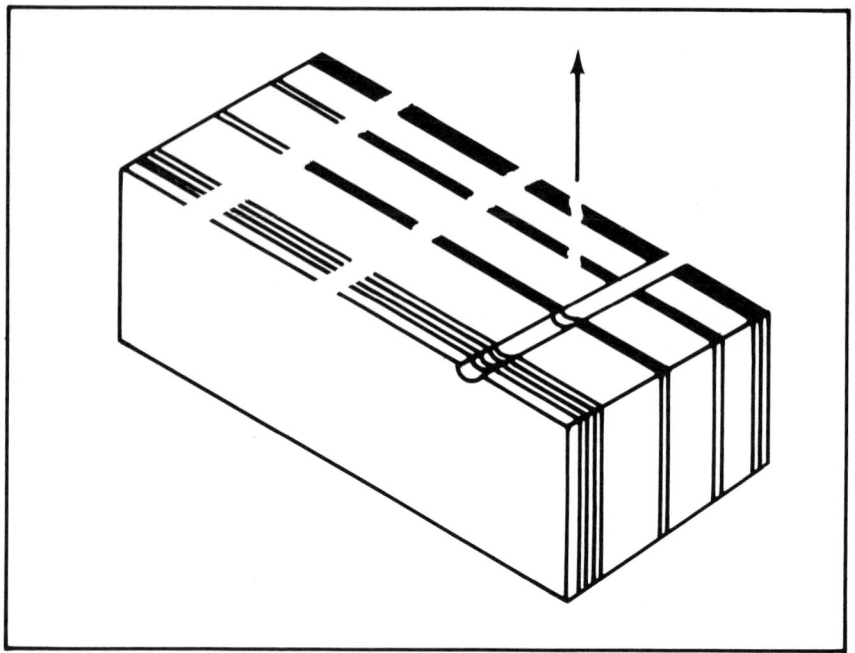

Figure 7. Plated-through hole peel test.

- A miniline can be a set of instruments for making a variety of surface, bulk, or mechanical measurements. Examples are electrochemical equipment, FTIR Auger, photocalorimeter, Instron, etc.
- A miniline may require access to sophisticated equipment at universities: Rutherford backscattering, neutron diffraction scanning Auger, etc.
- A miniline can be an observation setup: hot stage microscope, time delay photography, high speed movie, etc.

Finally the miniline is a means for an *interdisciplinary* team of research scientists and manufacturing engineers to collaborate. Production remains intact while the problem is tackled off-line on the miniline. Characterization instruments and interdisciplinary investigations lead to intimate knowledge of the production process and product.

- Quality is improved;
- Process optimizing results in improved yields and increased productivity;
- Problems are solved rapidly at low cost;
- Invention is accelerated; and
- Science advances.

Suggested Readings

1. D. P. Seraphim. 1982. A new set of printed circuit technologies for the IBM 3081 Processor Unit. *IBM Journal of Research Development.* Vol. 26, pp. 37-44.
2. R. F. Bonner, J. A. Asselta, and F. W. Haining. 1982. Advanced printer circuit board design for high performance computer applications. *IBM Journal of Research Development.* Vol. 26, pp. 297-305.
3. D. P. Seraphim and I. Feinberg 1981. Electronic packaging evolution in IBM. *IBM Journal of Research Development.* Vol. 25, pp. 617-629.
4. J. R. Bupp, L. N. Chellis, R. E. Ruane, and J. P. Wiley. 1982. High density board fabrication techniques. *IBM Journal of Research Development.* Vol. 26, pp. 306-317.
5. G. G. Werbizky, P. E. Winkler, and F. W. Haining. 1979. Making 100,000 circuits fit where at most 6,000 fit before. *Electronics.* pp. 109-114, August 2.
6. M. J. Colucci 1984. Void formation in glass-reinforced epoxy resin prepregs. thesis. University of Texas. Austin, TX, December.
7. B. K. Appelt, J. D. Gotro, T. L. Ellis, G. P. Schmitt, and J. P. Wiley. 1985. Composite lamination-analysis and modeling. *Proc. 43rd Annual Technical Conference.* Society of Plastics Engineers, Inc. (ANTEC '85), p. 289.
8. J. M. Atkinson, R. D. Granata, H. Leidheiser, Jr., and D. G. McBride. 1985. *IBM Journal of Research Development.* Vol. 29, p. 27.
9. B. K. Appelt and M. J. M. Abadie. 1984. *Society of Plastics Engineers.* Vol. 30, p. 257.
10. W. T. Chen, L. C. Lee, C. K. Lim, and D. P. Seraphim 1984. Mechanical modelling for printed circuit boards. In *Proc. Tech. Program*, Paper No. 41. Printed Circuit World Convention III. Washington, DC. May 22-25.
11. D. P. Seraphim, L. C. Lee, B. K. Appelt, and L. L. Marsh. An overview of materials science in printed circuit packaging. E. A. Giess, K. N. Tu, D. R. Uhlmann, Eds. Materials Research Society Symposia Proceedings. Vol. 40. Electronic Packaging Materials Science.
12. E. J. Kramer. Polymer-polymer interdiffusion. E. A. Giess, K. N. Tu, D. R. Ulhmann. Eds. Materials Research Society Symposia Proceedings. Vol. 40. Electronic Packaging Materials Science.

36

CAM Economic Justification—A Case Study in Electronics Industry

Barry Falter
Fred Choobineh

Manufacturing industries within the United States now can use computer-aided manufacturing (CAM). CAM can potentially improve manufacturing productivity and thus improve the ability of a company to compete effectively in the marketplace. Many manufacturing companies' competitive secret weapon is superior manufacturing capability—not marketing or product design (Wheelwright and Hayes 1985). Also, maintaining the status quo or base case on manufacturing technologies is no longer viewed as riskless, because failing to implement CAM results in a less competitive position in the market environment (Gold 1982).

Traditionally, manufacturing equipment justification was viewed as a tactical investment decision. A tactical decision is concerned with accomplishing internal operating plans and activities at a detailed level. Tactical decisions are made by middle- and lower-level management, with each manager viewing decisions through his or her narrow domain. Most of the tactical decisions in a manufacturing environment are done by engineers at the bottom level of a manufacturing operation. This creates a bottom-up process of improving manufacturing operations with engineering economy techniques. Because of the narrow view of the analysis, the bottom-up process is primarily concerned with cost reduction of direct labor, materials, and energy. This approach causes problems in justifying equipment that has a broad and longer term effect.

Adopting CAM requires a broader scope of justification than is normally applied to purchasing equipment and facilities. CAM justification must account for the traditional effects on direct labor, materials, and energy. In addition, its effects on inventory, flexibility, reduced uncertainty, manufacturing overhead, and the quality of the product must be considered (Kaplan 1983; 1984). Thus, justifying CAM projects should be done in the context of the total manufacturing system. This means that the justification process not only should consider the CAM application, but also its interactions with the production elements around it.

Considering the interactions of a CAM project with its surroundings in a justification process requires a good cost data base. However, traditional cost-accounting procedures distort costs through overhead buckets allocated to direct labor and they are unable to measure the key aspects of manufacturing, such as quality and flexibility (Blank 1985, Kaplan 1984; Suresh and Meredith 1984).

In addition to justifying CAM projects in the context of the total manufacturing system, the projects should be justified in the context of the strategic goals of the company. The strategic goals are achieved through a strategic plan that fosters the competitive strategy of the company.

The key aspects of a firm's competitive strategy are demonstrated by the "Wheel of Competitive Strategy" (Figure 1). The wheel hub represents the firm's goals, and the spokes are key operating policies directing the functional areas of the company. Top management specifies the key policies based on the strategic plan formulated to accomplish the firm's goals. Formulating a competitive strategy involves two internal and two external factors. The factors internal to the company are the personal values of the key implementers and the company strengths and weaknesses (a profile of assets and skills relative to competitors). The external factors are industry opportunities and threats (economic and technical) and societal expectations such as government policy. These factors determine the limits of what a company can successfully accomplish (Porter 1980).

A sustainable competitive advantage is the fundamental basis of long-term above-average industry performance (Porter 1985). Implementing CAM affects the key factors of company strengths and weaknesses, thus increasing or decreasing the limits of what a company can successfully accomplish for maintaining a competitive advantage.

A strategic planning viewpoint is necessary to integrate with CAM justification the external environmental issues of a company's future financial, competitive, technological, and growth market positions (Blank 1985). Top management cannot only specify a minimum return on investment (ROI) or asset productivity level to justify CAM equipment. Top management must include strategic technological factors in developing the firm's strategic plan.

Figure 1. *The wheel of competitive strategy (Porter 1980).*

These factors must then be translated into the operational policies that form the framework for each functional area's decision-making process. The strategic technological factors affecting the firm's long-term growth can then be integrated with the traditional economic analysis to justify CAM investments.

As the implementation cost increases and the overall effect of a CAM application broadens, the required involvement of top management in the decision-making process increases. A large integrated CAM application requiring large capital expenditures and creating broad long-term benefits requires a top-down decision-making process. A stand-alone CAM application is less capital intensive, thus requiring less top management involvement, but it still may have a broad impact throughout the whole manufacturing system and other functional areas, such as marketing. This broad impact of a stand-alone CAM application still requires linking tactical decision making by middle management with the strategic objectives of the company.

Middle management should be pushing for innovation and modernization because they form the communication backbone of a company (Settles and

Mize 1985). Middle managers are close enough to the operating or bottom level to understand the technological aspects of a CAM project. Middle managers can analyze the broad impact and cost-benefits of a CAM project and, therefore, can understand and communicate the link between the company's tactical and strategic objectives. Top management's function is to integrate the strategic technological factors into the overall tactical or operational policies, enabling middle management to fully exercise their unique position during the techno-logical decision-making process.

This chapter's objective is to demonstrate a proper economic justification of a CAM project. The project involves acquiring a robot for loading printed circuit boards with the specified components. The justification procedure accounts for the factors within the strategic framework of the company, as well as the direct and indirect cost and savings impact on the manufacturing system. A factor analysis or scoring method is used to explicitly account for the company's strategic objectives, and the net present value is used for analyzing the cash flows.

We will review the benefits of CAM applications, discuss the competitive strategy-related concepts followed at Hewlett-Packard, present a case study, and provide some concluding remarks.

A REVIEW OF BENEFITS OF CAM PROJECTS

The benefits of CAM projects can be categorized as tactical and strategic benefits. Strategic benefits are generally more difficult to identify and quantify than the tactical benefits. An example of strategic benefits of CAM projects is their ability to create a technological competitive advantage aiding a company in sustaining long-term, above-average performance within the industry.

Among the tactical benefits of CAM often mentioned (Gold 1982, Kaplan 1983, Kaplan 1984) are lower unit costs, reduced throughput time, reduced raw material inventory, reduced work-in-process inventory, reduced finished good inventory, reduced material handling costs, reduced storage requirements, and improved scheduling. Depending on the type of CAM application, some or all of the mentioned tactical benefits may be realized. Some CAM benefits may be tactical in nature, but with strong strategic implications. Examples of these tactical/strategic benefits are quality and flexibility.

The zero-defect approach is an emerging concept in quality that has been made possible by automation. Advocates of the zero-defect approach claim that total long-term manufacturing costs decrease as defects decrease (Kaplan 1983). The cost savings from producing high quality parts are obtained from reduced inspection, rework, scrap, material handling, cost of warranty, and field service and effective increased production capacity from less rework and less produc-tion time allocated to scrapped products. Some other effects of quality improve-ments are increases in profitability from the ability to raise the price to the customer and/or to an increase in market share due to a better quality product and image (Kaplan 1983). Also, the increase in morale and well-being of the workers—not being subjected to boring/dangerous work—will have a positive effect on work-life and increase quality (Sullivan 1984).

Increased manufacturing system flexibility may be the most important benefit of implementing CAM. This system flexibility is an amalgamation of the ability to process smaller lot sizes, improved ability to process a mixture of different parts in a serial manner, and flexibility in scheduling. Improved scheduling and control of manufacturing would result in producing the required quantity and prevent reaching production capacity prematurely during high demand periods.

Implementing CAM may allow combining different operations. This ability to combine operations would simplify material flow, maintenance, and indirect labor efforts. Also, the risk of machine obsolescence would be reduced due to the higher level of flexibility to produce varied parts (Sullivan 1984).

New products require a highly flexible manufacturing environment to employ technological innovation and advanced features after initial introduction to manufacturing, whereas a mature product does not require a very flexible manufacturing environment (Kaplan 1983). The flexibility provided by CAM would allow new products to make the transition to a mature and stable product with ease.

The strategic benefits of CAM and the benefits derived from improved quality and higher degree of flexibility are not generally included in the bottom-up economic justification process. This exclusion may result in not implementing the CAM project and, thus, maintaining the status quo that is risky in today's world market.

COMPETITIVE STRATEGY-RELATED CONCEPTS AT HEWLETT-PACKARD

Since the competitive forces within a market change with time, company executives should adjust the competitive strategy of their company accordingly. Under these circumstances the strategic plan that has been successful in the past may cause the demise of the company in the future. Thus, strategic planning may be viewed as a dynamic, rather than a static, process.

In the past few years Hewlett-Packard (HP) has shown that it is gradually changing its business strategies. HP has been known as a manufacturer of high quality instruments and calculating machines. However, at present, computer products generate more than half of its revenues, and HP is competing with Digital Equipment, IBM, and Wang Laboratories. The transition from an instrument maker to a computer maker has been guided by a careful strategic plan. However, elements of strategic management have changed to accommodate the new market environment.

The previous approach to strategic management followed by HP is referred to as the value-based incremental approach (Wheelwright 1984). Under this approach the company's long-term direction and strategic plan are given by a set of values and beliefs held by top management. This set of values and beliefs permiates across the organization and, thus, becomes generic business strategies. Companies using this approach of strategic management may reject business opportunities that do not satisfy its values and beliefs and generally support projects which have been well thought out and planned (Wheelwright 1984).

HP previously followed a business strategy that sought competitive advantage in selected small markets based on high value and multiple-features products. Its marketing strategy advocated controlled growth for high value/high price products. Its manufacturing strategy emphasized high quality products and on-time delivery dates. HP's research and development strategy emphasized features and quality or products that are designed for performance. HP followed a conservative financial strategy with no debt policy, believed in no worker layoffs, and has had a companywide incentive program.

Now HP closely follows the previously followed strategies with the following exceptions. First, HP is striving toward marketing low-priced products without sacrificing the quality (e.g., ink-jet printer and System 37) (Catalano 1985). This strategic change is necessary, because HP is determined to be competitive with giant companies such as IBM. Second, HP has moved away from its product-based decentralized corporate structure and has reorganized into three large marketing groups. This more centralized structure allows the chief executive to set strategy for each group and to take advantage of the economy of scale.

A CASE STUDY

The existing process under study is the preparation and loading of discrete axial lead components (resistors, capacitors, etc.) and integrated circuits (IC) on a printed circuit board (PCB). The existing base case method is direct labor-intensive, and the CAM alternative being considered is using computer-controlled robots to do the preparation and loading.

The existing operations are batch-oriented and consist of inserting ICs on a PCB using a semiautomatic insertion machine, performing discrete axial lead components, and hand loading the axial lead components on the PCB. All of these operations require intensive set-up, thus requiring large batch sizes and resulting in excessive raw, work-in-process, and finished goods inventory. The direct labor-intensive operations are error prone, thus quality problems exist creating extensive material scrap, inspection, and rework of the PCBs after soldering. The excessive work in process, direct-labor interface, and multiple operations require extensive overhead operations to control and handle material and to manage and schedule the operations.

The proposed CAM system will combine the performing of the leads of axial components and semiautomated and manual steps of loading a PCB. It consists of three stations and two robots. The stations are part feed/performing, straightening, and load. The robots function as a feeder robot and a load robot. The two, three-axis robots are used to interface the three PCB stations into an integrated system. The part feed/performing station supplies the feeder robot with the necessary parts from reels and tubes. A straightening station is necessary to straighten and determine lead orientation on the part. The load station allows fixturing a PCB within the load robot work envelope. The basic process consists of the feeder robot picking a part from the part feed/performing station and placing it in the performing fixture for automatic performing. The feeder robot places the part in the straightening fixture. In parallel the loading robot picks up a part from the straightening station and loads it on the PCB fixtured in the load station. A vision system inspects for defects throughout the process.

The proposed automated process will lower cycle time by combining operations, lowering operation time, set-up time, material handling, and batch size. This will result in significantly reduced overhead in the areas of management, scheduling, and inventory. It will show a significant increase in quality, with reduced downtime, resulting in less rework, increased output, and significant overall savings.

STRATEGIC AND ECONOMIC ANALYSIS

The economic analysis is summarized and shown in Tables 1 and 2. In Table 1, the savings from direct labor, scrap, and energy have been classified under the traditional savings, and all other relevant quantifiable savings have been classified under the nontraditional savings. This classification has been done to emphasize the impact of the product on the manufacturing system. The net-present-value (NPV) of the proposed CAM project was calculated to be $55,980 and is shown in Table 2. Based on the sign of the NPV, the project should be accepted on a purely economic basis.

The strategic factor analysis of the CAM project versus maintaining the status quo is shown in Table 3. The factors stated in Table 3 are viewed to be among the most important strategic factors that have an impact on the long-term health of a business in the electronics industry. The scoring method has been used to strategically analyze the project. Among the strategic factors considered is the rate of return on investment (ROR), which has been determined from the engineering economic analysis. The scoring method allows the weight factor of ROR to be set according to its impact on the company's survival.

TABLE 1 – Economic Analysis

		ANNUAL BEFORE-TAX CASH FLOWS	
	YEAR:	ZERO	ONE THROUGH TEN
INVESTMENT			
CAPITAL FACILITIES:			
Vendor-supplied System *		$-700,000	
Accessories		-25,000	
PROCESS DEVELOPMENT:			
Engineering		-114,000	
Installation		-15,000	
Tooling/Fixtures, Other		-15,000	
TRADITIONAL SAVINGS			
Direct labor			$ 125,402
Scrap (material)			8,386
Energy			-313
NONTRADITIONAL SAVINGS			
OVERHEAD REDUCTION:			
Management			7,747
Scheduling			8,587
Production Floorspace			3,564
Maintenance			-1,416
Sustaining Engineering			-2,952
INVENTORY REDUCTION:			
Raw Materials			11,500
Work-In-Process			20,500
Finished Goods			3,500
W.O. Material Preparation			35,039
Inventory Floorspace			35,424
QUALITY IMPROVEMENTS:			
Rework Time			10,292
Inspection Time			37,639
Material Handling (replacement parts)			39,123
Inspection/Rework Floorspace			3,600
Rework-produced Scrap			1,921
Management of Inspection/Rework			6,864
BEFORE-TAX CASH FLOW PER YEAR:		$-869,000	354,407

* Cost represents three of the proposed CAM systems to meet capacity requirements.

TABLE 2 – Cash Flow and NPV Analysis of the Proposed CAM Process

	YEAR	
	ZERO	**ONE THROUGH TEN**
Before-tax Cash Flow	$−869,000	$ 354,407
Straight Line Depreciation		86,900
Taxable Income		267,507
Income Taxes *		133,754
After-Tax Cash Flow		220,653
Uniform Series Present-Worth Factor		4.192
Present Value	$−869,000	$ 924,980

NET PRESENT VALUE = $−869,000 + $924,980 = $ 55,980

NOTE: Before Tax Cash Flow (BTCF) - see Table 1
Depreciation = (869,000/10) = 86,900 per year.
Taxable Income = BTCF − (Depreciation per year).
Income Taxes = Taxable Income x 0.50.
After Tax Cash Flow (ATCF) − BTCF − Income Taxes.
Uniform Series Present-Worth Factor = (P/A, i, n)
  with i = 20% per year, n = 10 years.
Present Value (PV) of Uniform Series (years 1 to 10) =
  ATCF x (P/A, 20, 10).
NPV = PV(year 0) + PV of Uniform Series (years 1 to 10).
* Investment tax credit was ignored as its political outcome is unknown as of 12/85.

Integrating the economic and strategic analysis enables proposals, traditionally rejected on economic grounds, to be strategically evaluated. If a proposal is both economically and strategically favorable, the proposal is given the highest priority level for capital funding. Proposals that are neither economically nor strategically sound should be rejected.

If the CAM alternative is economically unfavorable and strategically favorable, a sensitivity analysis must be applied to the economic analysis of the proposal.

TABLE 3 – Strategic Factor Analysis of the CAM Versus Status Quo Alternatives

		ALTERNATIVES	
WEIGHT W(i)	TECHNOLOGICAL STRATEGIC FACTORS	CAM, SCORE X(1)	STATUS QUO, SCORE X(2)
1.0	Quality Contribution To Market Share/Profitability	0.7	0.6
1.0	New Product Introduction On Time/High Quality	0.8	0.4
0.9	High ROR	1.0	0.9
0.9	Development of Engr./Mgmt. Expertise	1.0	0.0
0.8	Low Risk	0.4	1.0
0.7	Ability To Meet Customer Due Dates	1.0	0.8
0.6	Quality of Work Life	0.8	0.6
0.5	Expandability To Other Products	0.8	0.0
0.4	Ease Transition Of New Product To Mature State	0.5	0.3
0.3	Process Flexibility	0.7	0.5

Overall Strategic Value:
Sum of W(i) * X(i); i = 1 to 10:   5.52   3.80

Steps: (1) Management must set the weight of the highest attribute(s) equal to 1.0 and the other attributes less than 1.0.

(2) Estimate the X(i) score that reflects each alternative's level of satisfaction for each strategic factor. A score of 1.0 means that the alternative maximizes performance on a strategic factor. A zero signifies minimum performance. For example, the X(i) score of the CAM alternative for the low risk strategic factor is 0.4.

(4) Determine the overall strategic value = sum of W(i) * X(i); i = 1 to 10, j = 1 to 2.

(5) The alternative with the highest strategic value is the most desirable strategic alternative.

The sensitivity analysis consists of using different variations of the alternative and/or reconsidering the accuracy of the estimated cash flows used in the economic analysis. The sensitivity analysis will attempt to find a favorable economic condition while maintaining the favorable strategic value of the CAM alternative. After sensitivity analysis, if the proposal is strategically favorable but not economically favorable, funding for the proposal should be considered only after higher priority proposals are funded.

If the CAM alternative is strategically unfavorable and economically favorable, sensitivity analysis must be applied to the strategic analysis of the proposal. The sensitivity analysis consists of reconsidering the accuracy of the scores given to each strategic factor under each alternative. The strategic sensitivity analysis will attempt to find a favorable strategic value while maintaining the favorable economic analysis of the CAM alternative. After the sensitivity analysis, if the proposal is economically sound but strategically unsound, funding for the proposal should be given a low priority, because the proposal doesn't support the company's desired long-term market position. Giving priority to the funding of proposals, as described, will ensure investments in CAM and support the company financially and strategically over the long term.

Table 3 shows the CAM proposal strategically outweighing the status quo. The CAM proposal earned a strategic value score of 5.52 versus 3.80 for the present method. If the CAM proposal had scored a zero on the high ROR strategic factor, the CAM proposal still would have strategically outweighed the present method. In this situation sensitivity analysis would be applied to the economic analysis in an attempt to reach the ideal situation of an alternative being both economically and strategically favorable.

CONCLUSION

The existing equipment justification techniques must be modified so that they are suitable for justifying CAM. Traditional justification of equipment involves engineering economic analysis at the organization's lowest level. Thus, a limited view of only the lowest level is taken. This creates an internal or tactical viewpoint and, therefore, is primarily concerned with the traditional cost reduction areas of direct labor, materials, and energy.

CAM's effects are too broad to be analyzed traditionally. CAM affects the integrated elements in the environment around actual production. Therefore, CAM's effects on the nontraditional cost-benefit areas such as inventory, quality, flexibility, and manufacturing overhead systems must also be considered during CAM equipment justification.

CAM's broad effects reach into the external environment, affecting a company's strategic position. CAM affects strategic factors, such as increased market share from improved product quality, improved ability to meet customer due dates, and decreased time required for new product introductions to the market. Therefore, the base case of maintaining the status quo and not implementing CAM results in a less competitive position for the company in the market. CAM requires that tactical decisions be integrated with the strategic decisions to account for the broad impact of CAM on the environment. This requires that top management include the strategic technological factors in developing the firm's strategic plan. The strategic plan must then be translated into operational policies for each functional area. These policies enable middle management to exploit their unique position or expertise when justifying stand-alone CAM.

The CAM alternative demonstrated in the case study wouldn't have been economically justified using only the traditional areas of savings. Even if the CAM alternative would have demonstrated minimum performance in the nontraditional area of economic analysis, it still would have been strategically favorable. Therefore, the case study presented demonstrated the value of including the nontraditional areas affected by CAM and the value of strategically analyzing a CAM alternative.

Existing equipment justification techniques must be modified to allow a company to justify and exploit CAM, which is economically and strategically sound. Exploiting CAM will allow a company to increase productivity and compete effectively in the marketplace.

Suggested Readings

Blank, L. 1985. The changing scene of economic analysis for the evaluation of manufacturing system design and operation. *The Engineering Economist*, 30 (3): 227-244.

Catalano, F. J. 1985. Hewlett-Packard focuses on the office spectrum. *Electronic Business*. August. 46-55.

Gold, B. 1982. CAM sets new rules for production. *Harvard Business Review*. November-December. 89-94.

Harvey, D. F. 1982. *Strategic Management*. Charles E. Merrill Publishing Co. Columbus.

Hodder, J. E., and H. E. Riggs. 1985. Pitfalls in evaluating risky projects. *Harvard Business Review*. January-February. 128-135.

Kaplan, R. S. 1983. Measuring manufacturing performance: a new challenge for managerial accounting researh. *The Accounting Review*. 58: 687-705.

Kaplan, R. S. 1984. Yesterday's accounting undermines production. *Harvard Business Review*. July-August. 95-101.

Michael, G. J., and R. A. Millen. 1984. Economic justification of modern computer-based factory automation equipment: a status report. *Proceedings of the First ORSA/TIMS Special Interest Conference on Flexible Manufacturing Systems: Operations Research Models and Applications*. Ann Arbor, Michigan. August. 30-35.

Porter, M. E. 1980. *Competitive Strategy-Techniques for Analyzing Industries and Competitors.* The Free Press.

Porter, M. E. 1985. *Competitive Advantage-Creating and Sustaining Superior Performance.* The Free Press. New York.

Settles, F. S., and J. H. Mize. 1985. Managing change from the middle. *Industrial Engineering.* (9): 14-16.

Sullivan, W. G. 1984. Replacement decisions in high technology industries — where are those models when you need them? *1984 Annual International Industrial Engineering Conference Proceedings.* 119-128.

Suresh, N. C., and J. R. Meredith. 1984. A generic approach to justifying flexible manufacturing systems. *Proceedings of the First ORSA/TIMS Special Interest Conference on Flexible Manufacturing Systems: Operations Research Models and Applications.* Ann Arbor, Michigan. August. 36-42.

Wheelwright, S. C., and R. H. Hayes. 1985. Competing through manufacturing. *Harvard Business Review.* January-February. 99-109.

Wheelwright, S.C. 1984. Strategy, management, and strategic planning approaches. *Readings on Strategic Management.* Ballinger Publishing Co. Cambridge, MA.

Reprinted from *1986 Annual International Industrial Engineering Conference Proceedings.*

Barry Falter is a Process Engineer for Hewlett-Packard's Logic Systems Division in Colorado Springs, Colorado.

Fred Choobineh is associate professor of Industrial and Management Systems Engineering at the University of Nebraska-Lincoln.

37

Factory of Future Will Need Bridges Between Its Islands of Automation

John A. White

The automated factory, also known as the factory of the future, has captured the attention of both managers and engineers. The business press, computer magazines and engineering publications have jumped on the automated factory bandwagon.

Also, several major corporations are attempting to position themselves to be suppliers of automated factories; others are assessing the role of the automated factory in their own manufacturing strategy.

It is interesting to observe the increasing list of characters in the developing "automated factory" drama. Among those who seem to want to play leading roles are manufacturing equipment suppliers, material handling equipment suppliers and computer system suppliers.

At a recent industrial engineering conference, a representative of a major machine tool supplier described the factory of the future as consisting of three major components: the manufacturing equipment, the material handling system and the overall control system.

Interestingly, he went on to state that, of the three, the one that could be specified most arbitrarily was the handling system.

Alternately, some material handling system suppliers seem to believe their equipment is the best solution for the automated factory, regardless of the manufacturing equipment to be used.

Not to be outdone, some computer system suppliers are promoting real time hardware and software packages for controlling the automated factory—independent of the manufacturing and material handling technology to be used.

Recognizing my obvious bias, I believe a change of script is in order for the automated factory. Namely, the industrial engineer should be called on to design the automated factory to ensure that it is an "integrated system of people, equipment, materials and energy."

AUTOMATED VS. AUTOMATIC

Before proceeding further with a discussion of the automated factory, it is important to define what we mean by the term. The *automated factory* is not the same thing as the *automatic factory*. The automatic factory is a peopleless factory. In the automated factory, automation and mechanization dominate, but people are still needed to perform a limited number of direct tasks and a greater number of indirect tasks.

People are also needed in the automated factory to deal with unusual situations. For example, it is seldom cost-effective to design an automated system to handle exceptions. Instead, exceptions should be treated as exceptions.

Fujitsu Fanuc's new facility located in Japan near Mt. Fuji is a near-automatic factory. On the third shift, robots are used to assemble robots.

The Yamazaki Machinery Works in Nagoya, Japan, recently announced the opening of an automated factory. On the night shift, only a night watchman is present, while 18 machining centers continue to operate in the $18 million flexible manufacturing facility. On the day shift, people are used in the receiving area to operate the programmed hoists and cranes used to load castings on wire-guided trucks and material transporters. People also perform tool sharpening and computer programming tasks.

This facility only performs machining; it does not assemble the machine tools. A close look at the limited operations performed and the narrowly defined product line provides a clue to what is needed in the automatic factory.

Based on our best assessment of the factory of the future, it appears that a hierarchical factory system will best meet the needs of the future. From decision points located strategically throughout the *automatic* factory, parts and sub-assemblies failing to meet stringent inspection standards will be routed to an *automated* factory designed to handle the exceptions.

The automatic factory will be designed for either high volume/low variety or high value/low variety production. A key to the automatic factory is low variety. The automated factory will be designed for a wider variety of production requirements.

MATERIAL HANDLING OBJECTIVES

In designing material handling systems for the automated factory, the following objectives should be considered:

- Create an environment that results in the production of high quality products.
- Provide planned and orderly flows of material, equipment, people and information.
- Design systems that can be easily adapted to changes in product mix and production volumes.
- Design a layout that accommodates expansions in product mix and production volumes.
- Reduce volume of work-in-process.
- Provide controlled flow and storage of materials.
- Integrate processing, inspection, handling, storage and control of materials.
- Eliminate manual material handling at work stations.
- Eliminate manual material handling between work stations.
- Utilize the capabilities people have from the neck up, not the neck down.
- Deliver parts to work stations in predetermined quantities and physically positioned to allow automatic transfer and automatic parts feeding to machines.
- Deliver tooling to machines in a controlled position to allow automatic unloading and automatic tool change.
- Utilize space most effectively, considering overhead space and impediments to cross traffic.

The term *islands of automation* is frequently used to describe the transition from conventional or mechanized manufacturing to the automated factory. Interestingly, some people use the term as though it were a worthy objective to create islands of automation. On the contrary, the creation of such islands can be a major impediment to the integrated factory.

Typical islands of automation include numerically controlled machine tools, robots, automated storage/retrieval systems and flexible machining systems. In some cases, the islands are very small (e.g., an individual machine or work station); in other cases they are department-sized.

As an example of the creation of relatively small islands of automation, consider an appliance manufacturer who installed a number of robots along an existing assembly line. The resulting labor reduction generated a cost savings; the robots were certainly justified economically.

However, an opportunity was missed to increase productivity for the total system. Materials were delivered to and removed from the robots using the existing material handling system. Because the production rates for the robots differed from the manual rates, materials were stacked on the floor around the machines and in the aisles.

From a myopic point of view, the robot was impressive; but from a systems point of view an island of automation had been created.

An example of a relatively large island of automation was observed in a gear-box plant for a truck manufacturer. Castings were delivered by lift truck to a machining center. The castings were then manually placed one at a time into a magazine which fed castings one at a time to a robot.

The robot subsequently fed the castings to each of three machines and then placed the semi-finished part on a conveyor. The conveyor delivered the part to a second robot, which fed the part to each of the three additional machines and then placed the semi-finished part on a second conveyor.

The part was then delivered to a third robot, which fed the part to each of three more machines and placed the finished part on a peg-rack dolly for pickup and delivery by a manned tugger to a storage area. Three islands of automation had been linked together to form a much larger island.

As noted, the castings were delivered by lift truck, in a wire basket, to the first robot. The castings had come from the foundry in a nearby building.

At one point, the castings were on a belt conveyor positioned and spaced in such a way that they could have been automatically placed in the delivery container in a controlled fashion. However, they were simply dumped into the wire basket. Consequently, at the gear-box building, someone had to reach into the wire basket, grasp a casting, orient it properly and place it in the magazine that fed the first robot.

Why didn't the foundry workers place the castings in the basket so that they could be removed automatically? "It wasn't their problem."

Many management systems represent major impediments to the design of integrated systems. Evaluating managers strictly on their cost center performance discourages them from incurring small costs to generate big savings downstream.

From a systems viewpoint, islands of automation are not necessarily bad as long as they are considered interim objectives in a phased implementation of an automated system. However, to obtain an integrated factory system, the islands of automation must be tied together or linked.

An obvious vehicle that can be used to physically "build bridges that join together the islands of automation" is the material handling system. Likewise, information bridges can be provided through the control system.

MATERIAL TRACKING

A key ingredient of the automated factory is the shop floor control system. One element of a total shop floor control system is the material tracking system, which passes information about the material to such automated equipment as machine tools, robots, storage/retrieval systems, sortation conveyors, palletizers and guided vehicles.

Two approaches used to perform material tracking are *continuous tracking* and *interrupted tracking*. With continuous tracking a single input external to the system is required. The input could be keyed in manually or read automatically from magnetic or optical codes on the material.

Following the initialization of the tracking system, the material is tracked continuously based on feedback from the equipment/material interfaces rather than the material itself.

With interrupted tracking, "snapshots" of the material are taken periodically by automatic identification equipment. Between "reads" by the automatic identification system, the material can be considered to be in a tunnel and is invisible to the control system. However, its status is known and available in real time.

Feedback is not provided to the control system from the equipment/ material interface. Rather, information is transmitted to the automated equipment by the control system based on information received from the material.

In both continuous and interrupted tracking systems, the computer system serves as the courier of information from the material and/ or the equipment/ material interfaces "upstream" to the automated equipment "downstream." Large data bases and relatively sophisticated computer systems are required for continuous tracking; alternately, interrupted tracking typically places fewer demands on the computer system.

A recent development in interrupted tracking has resulted in the material's serving as the courier of data from the computer system and the equipment/ material interfaces upstream to the automated equipment downstream. Referred to as PREMID (Programmable REMote IDentification), the Swedish-developed product utilizes a "smart badge" on the material for dynamic storage of information.

At strategically located data transmission points, microwave transmitters transmit data to the badges, which can receive, store and transmit information. In turn, when the material arrives at an equipment/material interface it relays information to the automated equipment. The material handling network becomes a part of the information network with this method of material tracking.

MATERIAL HANDLING EQUIPMENT

Which material handling equipment will be included in the automated factory? The answer to the question obviously depends on the material characteristics, flow requirements and constraints imposed by the facility and manufacturing equipment. A case can be made for practically all unmanned material handling equipment's playing a role in the automated factory.

At this time the leading candidates for transporting material between specified points appear to be belt, chain and roller conveyors, towline and trolley conveyors, monorails and automated guided vehicles.

Because of the desirability of keeping material under control physically by having it properly positioned and oriented for automatic loading/unloading, specially designed fixtures, tote boxes, containers and/or slave pallets will be used throughout the system.

Work in process will likely be stored either in miniload and carousel storage/retrieval systems or in unit load storage/retrieval systems, depending on its size. However, computer controlled lift trucks also will be used to perform storage/retrieval operations in aisle-to-aisle applications.

Monorail systems will be used for both storage and material transport. The monorail system in the automated factory will involve microprocessor-controlled carriers that operate much like the automated guided vehicle system (AGVS). The primary distinction between the monorail system of the future and the AGVS is that the former will be installed overhead.

Because of its ability to be installed in three dimensions, the monorail can represent a highly flexible alternative to the AGVS. However, the installation cost for the monorail will be an economic factor to contend with.

A monorail system will consist of both powered and unpowered carriers. Using computer-controlled people-mover systems as a model, the monorail system can provide automatic switching and traffic control to allow a high degree of activity on relatively short paths. Automatic transfers of material will be possible between carriers on the monorail just as packages can be automatically transferred between trains at specified transfer points.

The automated factory will include improved recognition systems based on vision, sonar, laser and microwave technology. Additionally, voice encoding will play an important role in the factory of the future. Robots and robot-like devices will perform most of the material handling at the work station.

A number of improvements in robot technology are needed for the robot to realize its potential in the automated factory. A universal language is needed to allow robots to communicate with one another as well as with manufacturing equipment and material handling equipment. The language need not be sophisticated.

Perhaps the best analogy is our own communication system. Industrial engineers have a language of their own, as do computer scientists, managers and accountants. Yet industrial engineers, computer scientists, managers and accountants can communicate with one another using a common, general language.

Just as there will exist a hierarchy of robots by skill levels, there will also exist a communication hierarchy. Not all robots will be the same, just as not all people are the same. Robots will not have the same skills and capabilities, just as people do not have the same skills and capabilities.

For example, some robots will be blind and relatively "stupid"; they will perform simple pick-and-place functions. Other robots will be quite "smart" and perhaps have sophisticated vision capability.

SYSTEMS INTEGRATION

The heterogeneous human work force can function as an integrated system because of the ability to communicate. Similarly, a heterogeneous work force of robots can function as an integrated system only if there exists a common communication system—one that is simple to teach.

In describing the early generation of robots in use in Japan, Yoshitaki Kitao listed for an English-speaking television audience the following limitations:

- *Target blindness*, the inability to accurately find and acquire control of randomly oriented and positioned materials.
- *Material blindness*, the inability to discriminate between different materials.
- *Environmental blindness*, the inability to adapt to a changing environment.
- *Blindness to other machines*, the inability to communicate with and to sense the locations of other machines.
- *Blindness to people*, inability to communicate with and to sense the locations of people.
- *Blindness to deterioration*, the inability to detect deteriorating quality in its performance, as well as its own deterioration, and an inability to "heal thyself, physician."

Recent robotics research has concentrated on the limitations listed. Such concepts as high-speed micro-manipulators, electromechanical muscles, environmental sensing, voice encoding, tactical sensors, mobility and remote data transmission are being developed in a number of research laboratories.

SUMMARY

In summary, the hardware required for the automated factory exists today for many applications. The same appears to be true for the automatic factory. The missing ingredients do not appear to be hardware components; rather, what appears to be lacking is an economic environment that would make it cost-effective to automate all factory operations. As hardware and software costs decrease relative to humanware costs, the economic viability of the automatic factory improves.

For the automatic factory to become a reality, product designers and process designers must be concerned about the material handling process. Concern for material handling cannot be an afterthought.

Products must be designed for both manufacturability and handleability. Specifically, the shapes and sizes of materials, parts, tooling, subassemblies and assemblies must be carefully considered to ensure that automatic transfers, loading and unloading can be performed.

To facilitate the consideration of material handling in designing the automated factory, product designers, process designers and material handling systems designers must work together.

As high technology areas emerge, many will find applications in the handling, storage and control of material. Europe has led in the development of hardware technology, and the United States has led in the development of controls technology. However, the Japanese appear to be the leaders in applying new technology to factory systems. Their systems discipline is such that they "make it work."

Material handling is an important key to the factory of the future, and the material handling system can be the factory integrator. However, for it to be a contributor to the automated factory, it must be interfaced with the manufacturing system and the control system. Despite what some may claim, material handling should not be specified arbitrarily.

Reprinted from April 1982 issue of *Industrial Engineering*.

38

Flexibility and Integration Are Key Material Handling Concepts in Electronic Assembly Environment

Ray Pukanic

There is one thing that is constant about the electronics manufacturing environment—change! The high level of competition caused by rapidly advancing technology and the need to satisfy customer demands for higher quality at the lowest possible cost have resulted in short product life cycles and product demand curves that are difficult to predict.

This chapter addresses some of the problems of change and how they can be handled through the use of modular work stations and modular tote material handling systems that can be reconfigured many times over to respond to the changing needs in the electronics assembly environment. A standard footprint in the tote makes it easy for material handling equipment and people to handle different loads. In addition, there is a "subsystem" of subcontainers that fit inside the tote in different configurations.

The workplace must be well integrated with material handling; otherwise excessive material handling events, potential damage due to handling and excessive work in process inventories can result.

I like to think of material handling at the micro and macro levels. Macro material handling encompasses material movement throughout the facility, particularly between assembly, test and inspection functions.

Micro material handling deals with handling of materials at the work station level. This involves the totes coming in, being worked on and leaving and the support materials, tools and assembly aids at the work station itself that the worker deals with in carrying out the assembly, test or inspection function.

When we think of improving material handling and increasing productivity, typically automation comes to mind—automatic insertion, robots assembling and testing circuit boards, islands of automation tied together with conveyors, AS/RS systems or automated guided vehicles, etc.

Care must be taken to choose the right level of automation for the amount of material handling needed. It is imperative to identify the type and amount of flexibility required to adapt to these variables:

- Various sizes and methods of inserting components and how they are soldered.
- Different sizes of printed circuit boards and how they are handled, inserted into and soldered.
- How materials will be handled:
  Kit versus bulk issue.
  Kit size.
  Delivery method.
- Production volumes:
  Product mix.
  Demand curves.
  Expected life cycles.
- Are there product "families" that can be scheduled together?

As you can see, the list of criteria for flexibility requires careful consideration. Other items could be added to the list as well. The key in identifying flexibility requirements is to ask the question, "What do I expect to change or not to change in this process?"

The volume/complexity graph in Figure 1 provides a conceptual guideline for material handling evaluation. Zones I through IV represent the spectrum of material handling requirements from low volume/low complexity to high volume/high complexity.

Zone I is typically a "manual delivery/ manual return" environment. The rate of material movement is low, and forcing any high level of material handling automation would not improve the process, nor would the equipment justify it.

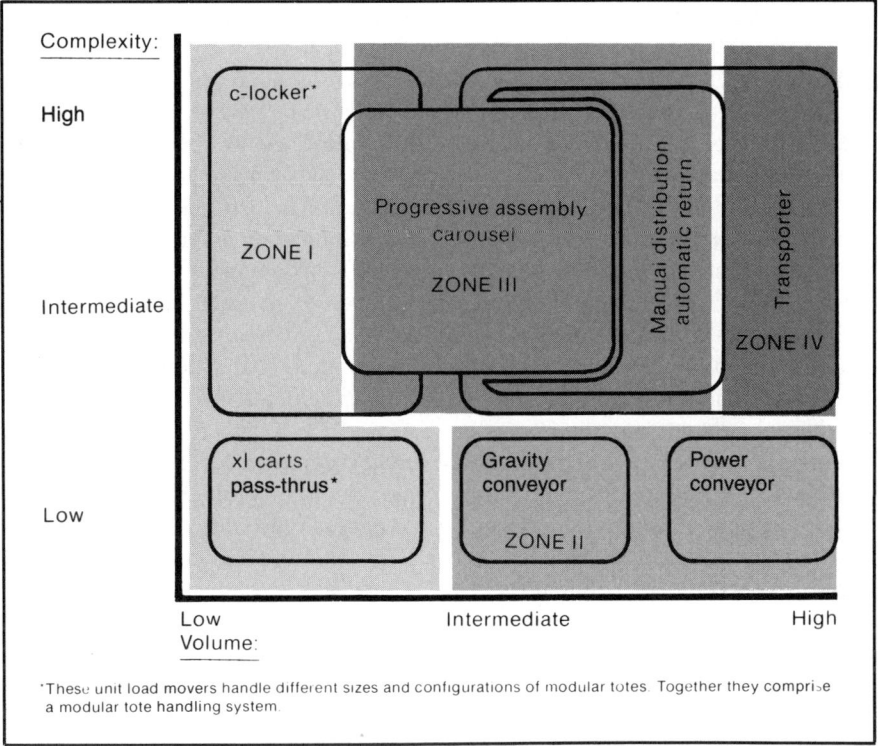

Figure 1. *Material handling evaluation example.*

Manual material handling can be a very efficient method of moving materials if macro and micro material handling are well planned using an integrated systems product. Unit load movers that accommodate modular totes and subcontainers can make manual delivery and return quite efficient.

The material handling strategy in Zone I can be carried out by a dispatcher moving a unit load mover through the manufacturing process to provide both delivery and return functions.

A workload study on the dispatching function should be performed as part of the material handling strategy. It may yield some surprising results regarding the size operation that can be handled manually. If dispatching requires more than one person, you usually move into Zone II.

In Zone II, since the complexity is still low, but the material handling volume increases, gravity and power conveyors are warranted to replace the frequent manual trips that would be required. The dispatching delivery function can still be performed manually, but conveyors can be used to move and queue materials to the next station or return them to the dispatcher.

We start getting into some sophisticated material handling in Zone III. The progressive assembly carousel stores and moves material in the same place. The assembler never needs to leave the workplace to get materials. The materials, in totes, are easily obtained from the shelf on the carousel slated for assembly, and are then returned to the carousel. The assembly progresses up the shelves on the carousel as it progresses through the assembly process.

Another material handling strategy in Zone III is manual delivery/conveyor return. The dispatcher concentrates on delivering modular totes and stages them for the assembler at the work station. The conveyor return simplifies the dispatcher's job by accumulating totes in the random order they are completed in, rather than requiring the dispatcher to anticipate which station will be completed next.

Zone IV covers the higher volume/higher complexity levels of material handling. Typically, automation is easier to justify at this point. More importantly, automation almost becomes a necessity for domestic manufacturers seeking to be competitive with overseas manufacturers. There are many ways to automate, but they all still boil down to the basic material handling interfaces shown in Figure 2.

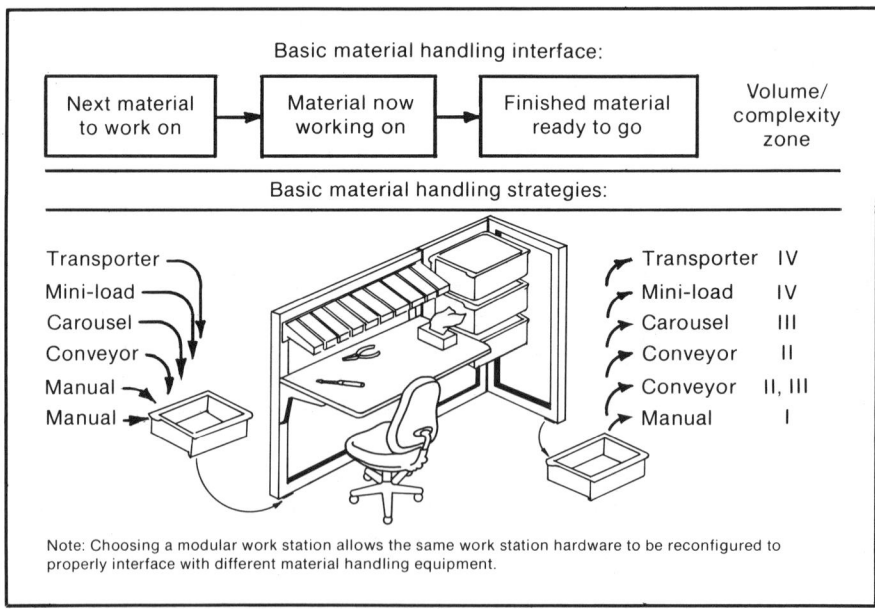

Figure 2. *Basic material handling interface diagram.*

The main difference automation makes is that materials are moved to, from and between the assembly operations faster and under controlled tracking (e.g., CIMS—MRP-II with bar code control linking the manufacturing operations). Modular totes are brought to the assembler faster by increasing the delivery rate through shorter distances or faster delivery methods, or both.

Transporters deliver materials quickly via belt conveyor, and delivery priorities are controlled by a dispatcher who has a light panel that reflects the dynamic activity of material needs. "Material need lights" are activated by the assembler when the "next material" tote is moved to the "material now working on" position.

This movement activates a micro-switch which lights up the dispatcher's panel boards. "Finished material ready to go" is placed in a tote and then onto a conveyor belt, which returns the material to the dispatcher for distribution to another assembly or test operation.

A mini-load AS/RS can also be configured to deliver and retrieve modular totes. The computer-controlled mini-load would automatically do the job of the dispatcher by acting on "material need signals" from the work stations. It would automatically deliver totes of materials to and where needed; it would automatically retrieve finished material from work stations; and it would automatically store material (WIP) between operations within the mini-load itself.

PEOPLE AND AUTOMATION

As we keep learning over and over again, we don't eliminate all people because of automation. Someone needs to keep an eye on how materials are actually moving and be able to respond to changing dynamics.

At the same time, as we start creating "islands of automation," it is critical that systems, both hardware and software, be well integrated; otherwise, we can be stuck with a beautiful mess that has a high price tag.

What we are faced with is people assembling and testing. We must consider the human anatomy and its limitations when choosing the size and weight capability of our modular tote handling system. People have to physically interact at the assembly, test and/or dispatching levels. Somewhere, people will be moving those materials, and it is necessary to plan for this.

When choosing a modular tote and the materials to put inside, keep in mind that the tote should be no wider than a person and the weight should not exceed 35-50 lb. This will make material handling workable (i.e., humanly possible) at the work station level and, perhaps, eliminate the need for another person on the line just to help move totes (what an embarrassment in an automated factory!).

ELIMINATING "CART-ITIS"

Electronics manufacturers are starting to take a hard look at productivity improvements and manufacturing costs. Just-in-time philosophies and MRP-II systems are bringing close attention to reducing WIP inventories. But "cart-itis" may be the main culprit in excessive WIP inventories.

One of the easiest ways of moving materials without a huge investment is by using carts. They have a lot of advantages:

- They are easy to purchase (typically a catalog stock item).
- They move (are relocated) easily.
- You can increase your storage capacity easily.

Their disadvantages include the following:

- If not designed for, they can be a hindrance to material flow and people traffic.
- It is hard to discipline staging zones for materials.
- When empty, carts are usually discarded in aisles or in the way of intended material flow.
- It is too easy to build up excess WIP inventories. This habit perpetuates rework which perpetuates damage which perpetuates yield losses.

The disadvantages are what I call "cart-itis"—you can easily get "hooked" on carts to solve your short-term problems, but in turn, create greater long-term problems.

If we look at the basic material handling steps shown in Figure 3, we see that we could use two carts—one before and one after each operation-and between each operation, a "pool" of carts for "surging" material.

If these operations can be located next to each other with "pass through" shelves between them, only *two* carts are required per string of work stations. The number of carts is reduced by combining material movement with material staging. Then:

- WIP inventories are reduced by eliminating the cart-itis between operations.
- Material handling events are eliminated.

HEAVY-DUTY MATERIAL HANDLING

Some electronics equipment gets quite heavy (over 35 lb) for safe, easy lifting—particularly when large power supplies are involved. The transformer alone can weigh 20 lb or more.

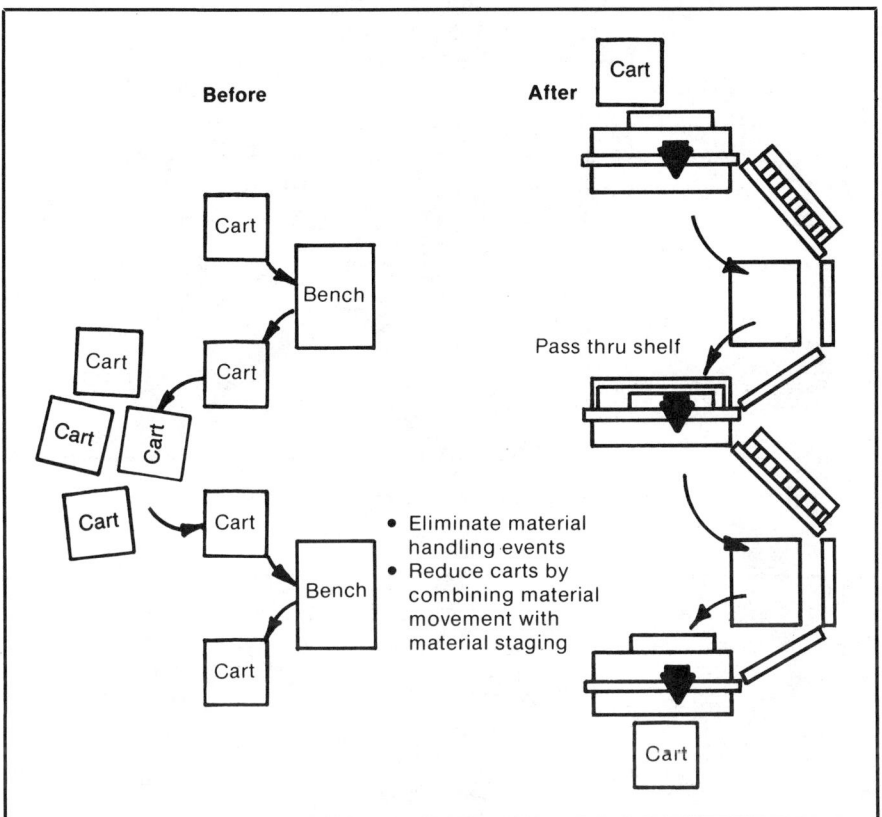

Figure 3. *Example of "cart-itis".*

At the same time, there can be a lot of small parts and pieces that have to be accessed for the assembly. A work station with an "arena" design can put parts and tools within easy reach.

Figure 4 shows an assembly area for a power supply. The power supply was assembled on a modular tray slid onto a conveyor at the same height as the work station and then moved to inspection and test without the workers having to lift the unit, which weighed 65 lb.

After test, the power supply can then be loaded into the final unit using a work manipulator (see Figure 5), which reduces the effective load to 10 lb for the person to lift.

To system test
Assembly power supply
Material
Inspection
Sub-system test

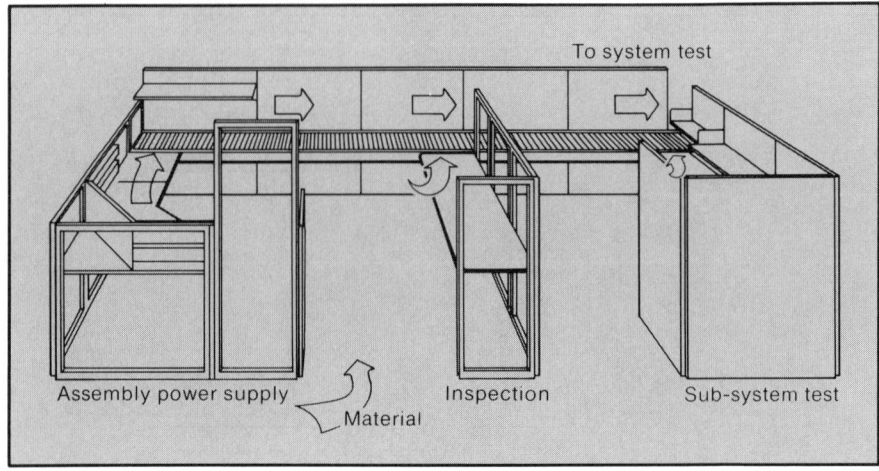

Figure 4. *Heavy-duty material handling.*

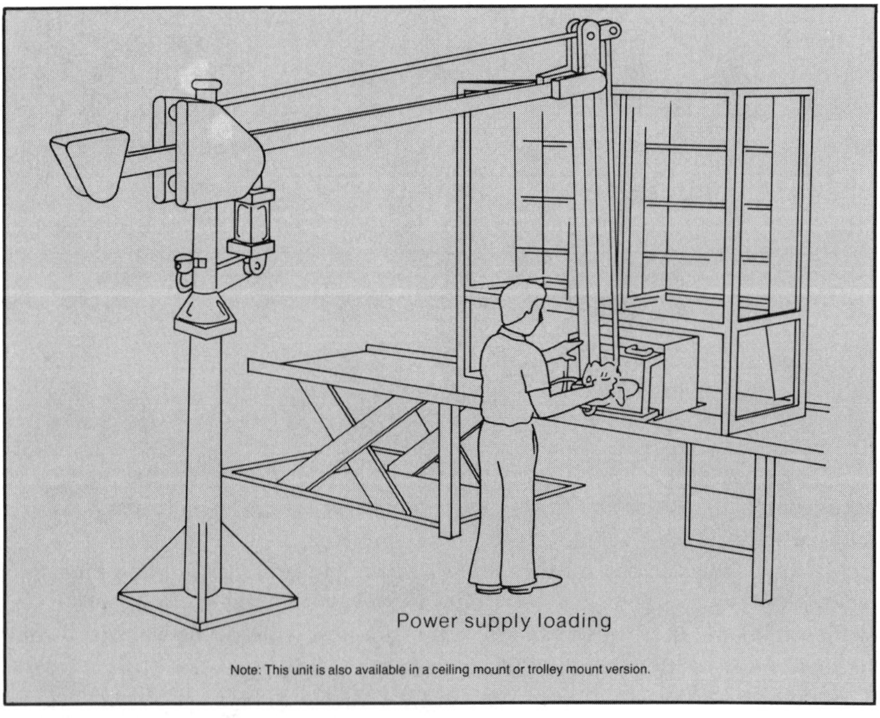

Power supply loading
Note: This unit is also available in a ceiling mount or trolley mount version.

Figure 5. *Part supply loading using conco balance master with parallel linkage.*
(Figure courtesy of Conco-Tellus)

Suggested Readings

Tompkins, J. A. 1983. Options given transport systems. *Industrial Engineering.* January. p. 24.

Reprinted from April 1985 issue of *Industrial Engineering*.

39

Factory Control Systems in Electronics Manufacturing

Sharat S. Israni
Jeff Lujack

In electronics manufacturing, an integrated hardware and software Computer Integrated Manufacturing (CIM) solution is used in monitoring and controlling floor-level machines. A factory control system (FCS) provides this functionality. The FCS translates the plant's strategic production decisions into specific operations on the floor and at the same time provides feedback to the plant-level systems on the status of work orders and machines in the factory. In this manner, it is a key part of a closed-loop feedback system for production control.

In this chapter, the nature of electronics manufacturing is characterized and the requirements of an FCS are described. The generic FCS has to provide a range of functions, including links to Computer Aided Design (CAD), Materials Requirement Planning (MRP), tool tracking, scheduling, Work-in-process (WIP) tracking, and machine operator instructions. In specific applications, a subset of these functions is usually sufficient. As a case study, a surface mount line is examined that is used in producing high density printed circuit (PC) boards. Some particular implementation problems are discussed.

WHAT IS AN FCS?

A factory control system (FCS) is an integrated hardware and software CIM system used in monitoring and controlling factory floor machines. Figure 1 shows a production control hierarchy found in several manufacturing environments. In this chapter, we will discuss electronics manufacturing—Israni (1987a) offers more detail on other environments. The National Bureau of Standards has described a five level hierarchy in use in the aerospace industry (Jones and McLean 1986). In addition, Katajamaki (1984) provides a generalized hierarchy and puts it in a larger organizational context.

The center level controller is sometimes referred to as an area controller.

As factory automation in electronics manufacturing has advanced, so has the degree of integration. Traditionally, floor level machines (e.g., wave soldering, autoinsertion) have become increasingly sophisticated, with elaborate communications and control systems built in. These machines have become islands of automation. At the same time, systems that provide plant level functions (such as order entry and master scheduling) have become well established. The objective of an FCS is to provide the link between these two levels. FCS helps translate the strategic production decisions (such as which orders to produce) into operational commands for the floor machines. In the other direction, it provides the plant level systems with feedback on the actual status of workorders and machines at all times.

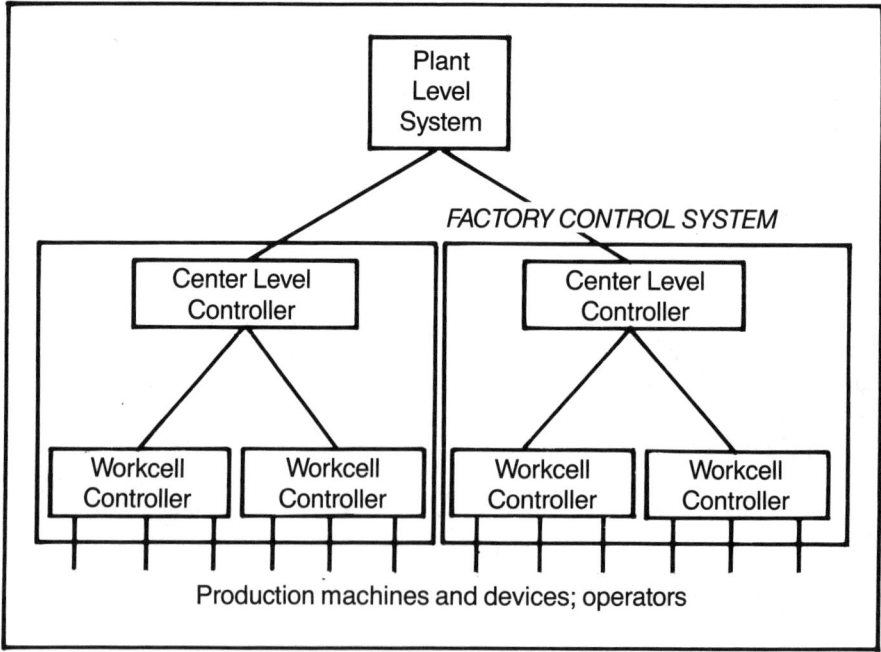

Figure 1. *A production control hierarchy.*

FUNCTIONS OF AN FCS

To achieve the objectives described previously, the FCS has to perform some or all of the following functions.

Scheduling and Dispatching. Finite scheduling is performed on workorders that are allocated for production to an area (*line balancing* is an implicit function performed at this time). This prepares the FCS for production. When an operator or machine signals a request for a new job, the dispatcher selects the job to be started at that machine. At the same time, the dispatcher downloads the *operator instructions* and/or *recipes* for that job, which are appropriately utilized or displayed. (*Recipe generation* is the creation of machine control and automated test programs.)

Material Flow Control. When a particular workorder is to be started at a machine, the raw or in-process materials required for it are directed for this module. Automatic commands may be issued to the material handling equipment (AGVs, AS/RS's) or manual commands may be issued to the operator. When any material movement takes place, this module is updated with the status of the move.

Machine and Operator Interface. This interface involves monitoring the status and production counts of machines and communicating to them (i.e., to their controllers). This module often has intelligence built in that allows for arbitration, control, and arithmetic computation. Also, all *operator data entry* is typically managed by this module. Downtime-cause data are a typical example of what the operator enters.

WIP Tracking. The location of all workorders on the factory floor, as well as other information, such as the count of parts produced, is tracked by the FCS. This module relies on the data collected by the monitoring module.

Historical Production and Quality Data. These data are preserved in a long-term data base and used to generate *reports*. Examples of such reports are pareto diagrams of machine downtime by cause and parts rejects by cause.

Not all of these functions are required in each production FCS. A subset is usually picked in most stepwise implementations of CIM on the factory floor and may even be sufficient for the medium term at that factory. At this time, there are no industry-standard FCSs.

For more detail on FCS functions, see Israni (1987a,b). For more information on the theory behind some of these functions, see McClain and Thomas (1980).

DATA FLOWS IN AN ELECTRONICS FCS

Figure 2 shows the flow of data in a largely integrated FCS used in electronics manufacturing. The flows pertain to the framework within this article—obviously, there are several alternate configurations in practice.

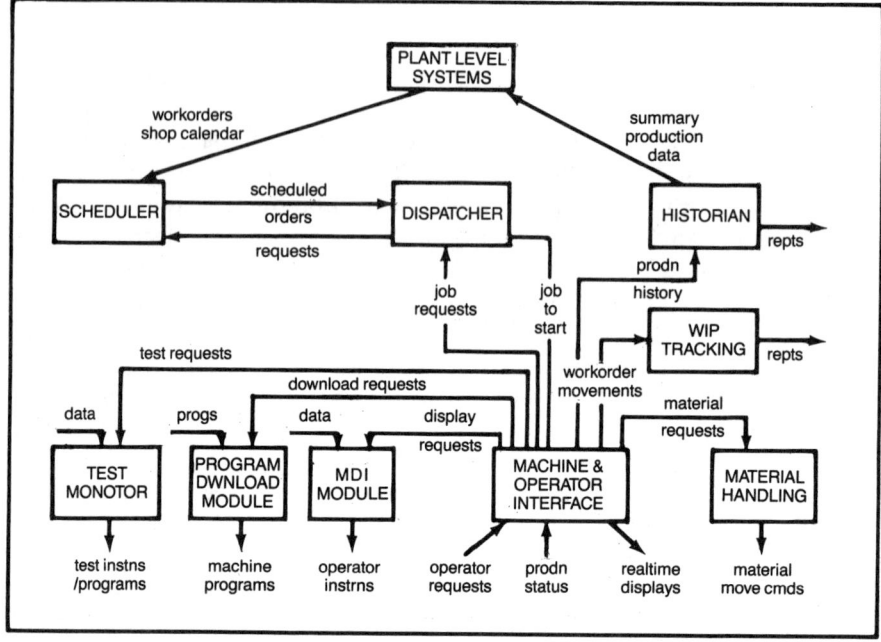

Figure 2. *Data flows in an electronics FCS.*

Generally, the flows are as follows. Workorder requirements (what parts should be produced, how many, when due, etc.) are downloaded from the plant level systems to the scheduler. The scheduler generates the finite schedule for the next shift (or few shifts). When the shift schedule, along with the necessary machine programs and operator instructions, is ready, the system is ready for operation.

The operator requests the next job scheduled at his or her machine. The operator interface module passes on this request to the dispatcher, which interprets the shift schedule to describe the job to be started. The machine interface module then directs requests for data, recipes, material movement, and/or operator instructions to the appropriate module, as in Figure 2.

Machine status information is collected by the machine interface module. Workorder movements are reported to the WIP tracking module, and production history is reported to the Historian. Summary production data are passed up to the plant level, thus closing the loop.

CASE STUDY: IMPLEMENTING A SURFACE MOUNT LINE

The case presented is about a new surface mount line implemented by Hewlett-Packard in Sunnyvale, California. The problems faced and priorities selected by the engineers are described.

In the quest to lower manufacturing costs and increase component density on PC boards, manufacturers are going to surface mount technology in ever-increasing numbers. To maximize potential efficiencies inherent in the process, a new installation needed to be designed to be monitored and controlled by a set of workcell controllers. These computers would link the machines on factory floor with plant level systems currently used for traditional MRP functions.

Implementing a relatively new manufacturing process brings with it a novel set of problems for manufacturing management and engineering. Simultaneously designing and implementing new automation adds another layer of complexity. Because a large part of the software investment would be in the workcell controller, an "off-the-shelf" software product was desirable. However, this was not possible, because the process is relatively new. Therefore, a significant amount of new software had to be developed.

To fine-tune new designs while meeting early production schedules, software systems had to be put in place as early as possible. These software systems had to be both flexible and maintainable. This inherent conflict was addressed by initially identifying the most critical requirements and implementing them in a manner that would permit further enhancement as we gained experience with the new process.

A "software communications environment" was defined to develop custom programs required for each workcell. This facilitated timely implementation of critical functionality while providing a well-documented environment for future growth. Programmers could rely on a well-defined set of data communication and program management methods while developing custom software. This communications environment was built entirely on standard components from HP-UX; Hewlett-Packard's real-time implementation of UNIX™ System V with Berkeley extensions.

Initial Implementation

Initially, the workcell controllers would handle expedient chores on the factory floor. Plant level systems would play a more passive role, collecting data in the "background" for the longer term. This helped speed implementation through simultaneous development by programmers working on the general plant level computer system. Later, more time-critical functionality would be introduced.

A board handling system was designed that could operate autonomously yet still be integrated with the workcell controllers. This dual requirement allowed the earliest possible exercise of the line mechanically and alleviated some of the schedule pressures from the workcell controller software implementors. Furthermore, an autonomous board handler provided additional reliability during line operation.

Workcell Controller Functionality

For initial designs, first priority was given to functionality that was required by all workcells in the line. First and foremost, the workcell controller had to be able to coordinate material flow between upstream and downstream workcells. A board identification system, using a barcode reader, was installed so that the workcell controller could identify each board as it arrived at the workcell and download programs and commands necessary for workcell operation.

In addition, the workcell controller was designed to act as a virtual workcell so that board handlers, barcode readers, and machines could be controlled by the operator from a common interface. Critical machine status displays using text and graphics were required. The operator could start up and shut down the workcell and manually intervene in workcell operation (such as pause and restart) for maintenance, change-over, and fault correction. These capabilities had to be robust and easy to use.

The workcell controllers were required to provide many "off-line" functions as well. They had to manage machine programs or instructions (recipes) to shorten machine set-up and changeover times. Steps required to generate these recipes were simplified to minimize expertise required to configure and tune the workcell. For some workcells, recipe optimization was necessary to minimize build time. Furthermore, the workcell controller had to be able to upload and store recipes from machines that could generate them locally—through a learn mode, for example.

Plant Level Functionality

Initially, plant level machines were configured to allow engineers to manually down load data about manufactured products to the workcell controllers. Material lists and CAD data were acquired and translated on workcell controller systems to reduce recipe generation and set up times. Defect data from the test area were automatically loaded back up to the plant level computer for correlation and reporting. Later, plant level functionality would be expanded to increase overall line efficiency. Line balancing, scheduling, and more "real-time" statistical reporting would be developed on a center level controller between the plant host and the workcell controllers as line operation matured.

Results of Implementation

As a result of this implementation, changeover times have been reduced. More importantly, the amount of time and effort required to introduce a new product into the line has been significantly reduced. This reduction helps fight the tendency to build too much manufacturing capacity in response to the high rate of new product introductions experienced in the computer business.

In addition, manufacturing rates have increased due to high-speed part placement rates and lowering of manual intervention in the process. Board handling systems and a reduction of manually placed parts have made this possible. The process is still too new to evaluate long-run product quality statistically. However, preliminary indications are that the line is producing products of excellent quality.

Future Directions

As we gain more experience with surface mount manufacturing, our ability to improve its flexibility and efficiency will increase. Our initial installation of workcell controllers between the workcells and existing plant level computers give us a platform for further enhancement, and we see this as an evolutionary cycle of implementation. First designs address most-critical issues with an eye to future enhancement and improvement as the manufacturing process matures.

Future challenges include better selection of process parameters to track over the short and long term to maximize efficiency and minimize defects. Improved exception handling and error recovery and a general robustness to improper or inadvertent operation are challenges of equal or greater magnitude to the core functional implementation. Finally, scheduling functions such as line balancing will be incorporated to further eliminate tedious manual tasks and improve overall line efficiency.

CONCLUSION

We have described a comprehensive hierarchy of integrated software functions used in monitoring and controlling an electronics manufacturing factory. In practice, there are few installations that can currently be considered integrated to that extent. Most electronics manufacturers are working toward integrating their islands of automation at the workcell level and their present-day plant level systems into an effective factorywide network. The case study presented is one such effort.

Suggested Readings

Israni, S. S. 1987. Factory floor monitoring and control needs in various manufacturing environments. *IIE Integrated Systems Conference Proceedings.* November 5-7, pp 348-351.

Israni, S. S. 1987. The workcell controller—a key link in machine control. *IE News: Manufacturing Systems.* 22(1): 1-3; 21(4): 1, 4.

Jones, A. T. and C. R. McLean. 1986. A proposed hierarchical control model for automated manufacturing systems. *Journal of Manufacturing Systems.* 6(1): 15-25.

Katajamaki, M. 1984. CAD/CAM in robotic applications design and simulation. *Proceedings of Autofact 6.*

McClain, J. O. and L. J. Thomas. 1980. *Operations Management.* Prentice-Hall. Englewood Cliffs, N.J.

Dr. Sharat Israni is a Member of the technical staff in the Industrial Applications Center of Hewlett-Packard Company.

Mr. Jeff Lujack is a Member of the technical staff at Hewlett-Packard's Computer Manufacturing Group in Palo Alto, CA.

40

Line Scan Vision System

Norman R. Brunelle
Frank P. Higgins

The automation of inspection tasks has always been of prime concern to process engineers. Until recently, the technology was not available to automate anything but the simplest tasks. With advances in machine vision and computer technologies, many visual inspection tasks can now be automated. This paper describes an AT&T development that can be used as an efficient tool for automating the inspection of printed circuit board assemblies. Significant increases in reliability are achievable as compared to current methods, namely human inspectors.

BACKGROUND-CUSTOMER NEED

Since the advent of printed circuit boards (PCBs) in the late 1950s and early 1960s, the verification of completed assemblies has been and continues to be a major concern in the electronics industry.

As the technology progressed through the 1970s and now into the 1980s, we could assemble more circuits and, therefore, more components per square inch of PCB than ever before. But smaller components and, in many instances, increased lead densities have brought assembly verification to the point where human inspection is less than 80 percent reliable.

Operator fatigue also decreases the reliability of any inspection, and leads to human discomfort and dissatisfaction with the work environment.

All these factors have led to the development of expensive electronic testers for verifying PCB assemblies. Although the tester are effective, their operation is costly—the initial expense of equipment and fixtures, labor to operate the equipment, and labor to repair defects. Therefore, improved quality of the assembly operation and reduced costs of verification remain primary objectives of the electronic industry.

Assembly defects can be classified into three major categories:

- *Wrong or missing components*—These defects can be attributed to operator error or malfunctions of insertion equipment. As more automation is used in assembly, these defects tend to be minimized. But a new defect type is emerging: a crumbled lead, or a lead that is not inserted correctly into its hole in the PCB (Figure 1). Often, electronic testers do not find this type of defect because the lead may still make contact with the circuit but is not soldered to it. This defect will eventually cause an electrical failure days and sometimes months after the unit is in the field.
- *Wrong polarity*—This type of defect usually results from operator error, but can also result from mixed orientation within the component packaging.

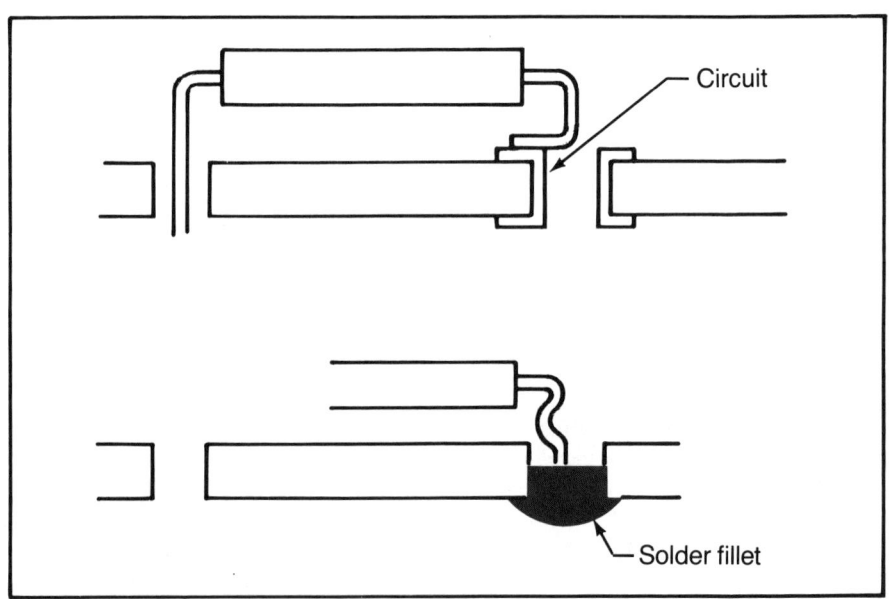

Figure 1. *A new defect type emerging from automated assembly of printed circuit boards. Top, a folded-under lead; bottom, a hiked or crumbled lead.*

Once again, assembly automation will tend to minimize this type of defect. But operator error in loading the components and random defects in packaging will always exist. Thus, the need for verification will always be present.

• *Solder defects*—These defects are generally classified as shorts and opens. They are mainly caused by variability of the automated soldering process, cleanliness of metallic components, and the PCB's design. As the technology advances, these defects are also being eliminated and, as some engineers feel, may eventually become negligible. However, most processes are not yet at this level.

For at least the last 20 years, the assembly verification process has been a target for improved methods.

Visual aids and overlays have been and continue to be used with some success. But these technologies are slow and subject to human inefficiencies. Optical comparator processes have also been attempted with similar results. For example, Teradyne, Hewlett-Packard, and Fairchild, among others, have developed post-solder electrical verifiers to address these problems.

These electrical testers generally will find almost all the typical assembly defects, except the curled-up lead or lead partially through the hole. Although electrical verifiers are efficient in identifying defects, they are extremely expensive (hundreds of thousands of dollars each). They also require intricate fixturing and programming to isolate and evaluate complicated circuits.

Because testing includes electrical measurements, the assembly must be soldered before test. When a defect is found, it generally requires analysis before being verified and repaired. The cost of this procedure, including repair, can average anywhere from $100 to $500 a day for every one percent of defectiveness.

The industry yield for first pass testing of electronic assemblies averages 75 percent, with half the defective assemblies failing because of workmanship or noncomponent-failure defects. Detection and repair of these defects can cost from $250,000 to several million dollars a year, depending on the volume and yields of the assembly process. Improved yields also reduce retest requirements, which means a savings in test equipment.

As these examples show, there is a strong force to increase test yields, and improved inspection reliability is one way to achieve this.

AT&T is expending much effort to improve performance in this area. Increased automation and modernization of our processes are the first priority in most of our factories. Improved test yields—which translates into shorter intervals and, therefore, reduced inventories—is a driving force for modernization.

Assembly processes are being designed to move the product forward constantly in an *in-line* fashion. Processes are also being networked with a shop flow computer to provide a flexible manufacturing information database—e.g., computer-aided design and manufacture (CAD/CAM)—and a mechanism for implementing process control.

Many companies are using machine vision, a developing technology, to address the reliability problem of verification and provide the database needed to evaluate the process and control those areas that generate the problems.

SYSTEM DESCRIPTION

Based on the above requirements, a set of image-acquisition specifications for a machine vision system can be developed.

The system's resolution must be at least as small as the smallest feature of interest. For presolder lead inspection, this is about 5 mils. Because a typical or practical scanning area for PCB inspection is about 10 inches by 10 inches, this implies that the system's linear resolution must be better than one part in 2000. The system must also be able to inspect a PCB (including acquiring and processing the image) in about 10 seconds, the dwell time of typical in-line assembly processes.

The Sensor

The critical element of such a system is the sensor; in our case, a line scan camera was selected.

A line scan camera consists of a single row of image sensing elements (pixels) and associated electronics, integrated with an appropriate lens system. The analog charge proportional to the intensity of the incident light that strikes each element is amplified and strobed serially out of the camera.

Line scan cameras are available commercially, and have from 256 elements to 2048 elements. One with 2048 elements meets the linear resolution requirement described above.

However, the image must be built up by acquiring line images. This is done by moving the camera relative to the object in a direction perpendicular to the linear array. The speed of this motion depends on several factors; the most critical is the amount of incident illumination.

If we assume maximum available illumination, the data rate of a 2048-element line scan camera can be as high as 9000 lines/second. For a scanning resolution of 5 mils/line, this equates to 45 inches/second, well over ten times more than the requirement for this application.

Therefore, a line scan camera meets all the requirements for the sensing element of a PCB inspection system. As will be illustrated later, the camera has some additional advantages when used for presolder lead inspection.

Camera Interface

While the sensor is an important component in the system, careful attention must be given to system integration to meet the processing time requirement.

Figure 2 is a block diagram of the line scan vision system. The line scan camera is connected to the rest of the system through a custom camera interface. In addition, minimum system has an image memory, a central processing unit (CPU) with program memory, and input/output (I/O) interfaces, all interconnected via a bus.

We selected a commercial, nonproprietary bus for this application called the VME bus. The CPU controls and synchronizes the camera interface with the translation stage through an I/O interface and translation stage controller. The CPU also has access to the shop flow computer where PCB code and CAD/CAM information reside.

The operation of the camera interface is the key to efficient image acquisition and processing.

First, it is not practical or necessary to store each frame of the entire PCB image. For presolder lead inspection, the locations of all leads are available

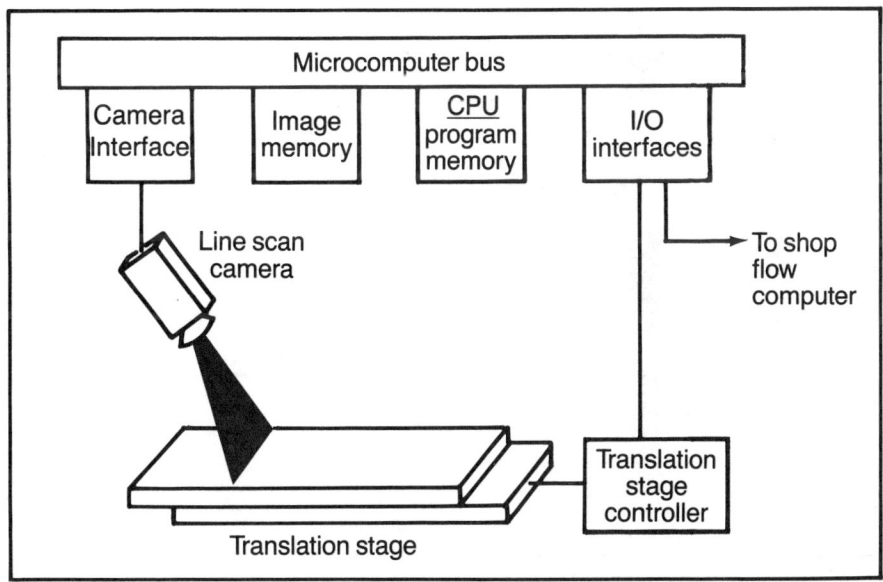

Figure 2. *The line scan vision system. A custom interface connects the line scan camera to the rest of the system.*

from the CAD/CAM database. Therefore, a more effective technique is to use this CAD/CAM database to define *windows*, or small regions of interest, that contain the pertinent image data of the PCB. Each window would include a small area surrounding each PCB hole that has a component lead inserted through it.

We designed the camera interface with these concepts in mind. This interface is a programmable switch that writes the serial image data from the line scan camera to the image memory, The program stored in this interface is derived from the CAD/CAM database for a particular PCB code. Because the program contains the information required to operate the switch, only image data from these predefined windows are stored in the image memory (Figure 3). The data are then available to the CPU for later processing.

The major advantage of this technique is it significantly reduces the image memory size required to achieve high resolution over a large area. However, two conditions must be met for its successful implementation.

First, we must have information available about *where to look*. This is usually found in a CAD/CAM database. Second, the board must be placed accurately

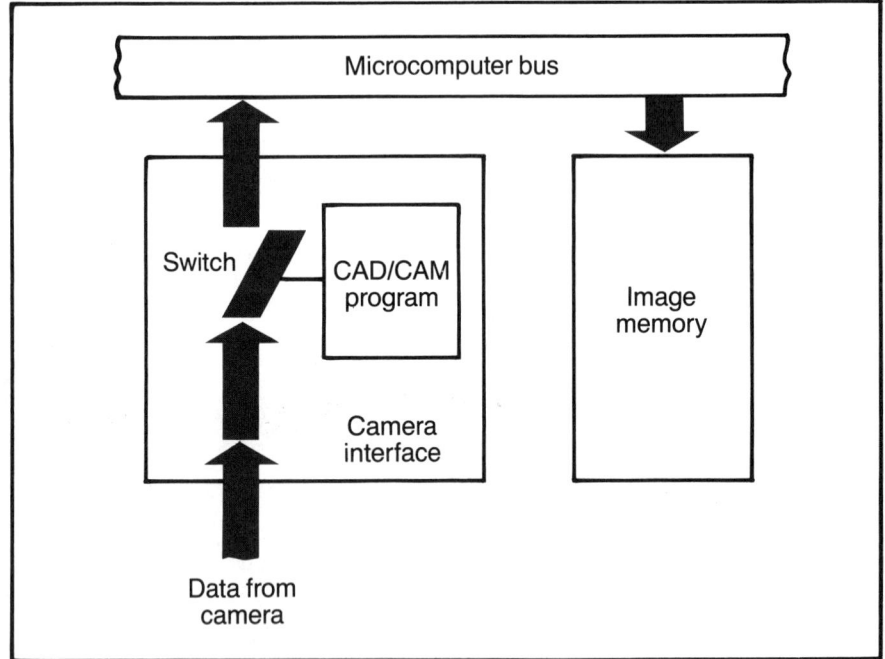

Figure 3. *The camera interface for the line scan vision system. Information in the CAD/CAM program operates the switch that enables image data from the camera to reach memory.*

with respect to the start point of the camera. This is not a problem for PCB inspection, because tooling holes are available and are used to register a PCB accurately in other steps of the manufacturing process. Typical placement accuracy for PCB inspection might be plus or minus 5 mils (one pixel).

Thus, the two conditions that would limit the use of this technique in other inspection applications are easily overcome for circuit board inspection.

Processors and Memory

In integrating the camera interface into the system, the concepts of multiple processors and shared memory are important to overall system performance.

When acquiring an image, the camera interface accesses the bus and image memory as just another CPU on the bus (Figure 2). This access occurs only when the preprogrammed switch is *on*; i.e., when regions of interest are being scanned by the camera. Therefore, the CPU can read and process data already stored in image memory during the *off* times of the scan. On a typical PCB, this is a large percentage of the scan time.

Concurrent processing is an important, but not obvious, advantage of using a windowing technique to acquire the images. If one large image is stored as a single frame, processing cannot begin until the camera interface relinquishes the bus and image memory, i.e., at the end of the scan.

Using multiple processors in this architecture increases system performance at minimum cost. For example, assume that an inspection algorithm for presolder lead inspection operates at a speed of 100 leads/second. (An inspection algorithm is a subroutine or subprogram that processes the image data and returns a numeric value to identify an acceptable or defective condition.) If the dwell time of the manufacturing process is 10 seconds, a single processor inspects 1000 leads/dwell time at this rate.

Adding an additional processor doubles the throughput, which permits inspection of PCB codes with up to 2000 leads/dwell time. The cost of the hardware for this increase in speed is about 5 percent of the overall system cost.

It is easy to visualize how these processors operate in parallel. Each processes successive, alterate images (windows of data), so they are relatively independent of each other. Almost double the speed is achieved because only a small percentage of system processing overhead is used for this approach.

PRESOLDER LEAD INSPECTION

We selected presolder lead inspection as an initial application of the system. Some technical issues regarding this application will be reviewed here to point out the advantages of using a line scan system.

Many components inserted in PCBs have leads that are left unclinched as part of the normal manufacturing process. The most common example of this is a multileaded component with only its corner leads clinched. Because this is an acceptable condition, an automated inspection system must be capable of reliably detecting the lead both in the clinched and unclinched state.

For example, typical leads on an integrated circuit are 5 mils by 15 mils and protrude 50 mils from the lower surface of the board.

Because a clinched lead presents a significant cross section to the camera, the lead's image contains a large amount of useful information. However, an unclinched lead presents a small, almost undetectable cross section to the camera when the viewing angle is normal (perpendicular) to the PCB. To increase the cross section of an unclinched lead, we can use a higher resolution (<5 mils/pixel) or view the lead at an angle.

Viewing at an angle is a simpler solution and, when set up properly with a line scan camera, does not have the imaging disadvantages associated with a two-dimensional camera.

Figure 4 illustrates the differences of angular viewing with a line scan camera and a two-dimensional camera. Two effects are avoided because the image is *a line* and the angle of view is rotated about this line.

First, when viewing at an angle with a two-dimensional camera, the image focal plane is now at this angle with respect to the object. Unless the depth of focus is very large, different areas of the image are in and out of focus. Rather than image a *plane* (the case with the two-dimensional camera), a line scan camera images only a line. Because the two-dimensional image is built up through relative object-camera motion, this line is always in focus thus avoiding the depth of focus problem.

Second, when a two-dimensional camera images an object while viewing at an angle, the image has perspective distortion (Figure 5). A line scan camera avoids this distortion because each line is the same size in the two-dimensional image that is built up from successive one-dimensional scans.

With angular viewing, if the lead is properly illuminated, two pieces of information about the lead can be detected in the image: the lead itself, and its shadow. Figure 6 illustrates this effect.

The angle of incident illumination is set equal to the viewing angle. Thus, the conductive pad that surrounds the hole reflects the incident illumination. If a lead is present, it blocks the light coming from the upper part of the pad and casts a shadow on the lower part of the pad.

Figure 7a, a gray scale plot of an image of an unclinched lead through a hole, illustrates this effect. Figure 7b is an image of a 45° clinched lead under the same conditions, while Figure 7c shows a defective condition, a missing lead.

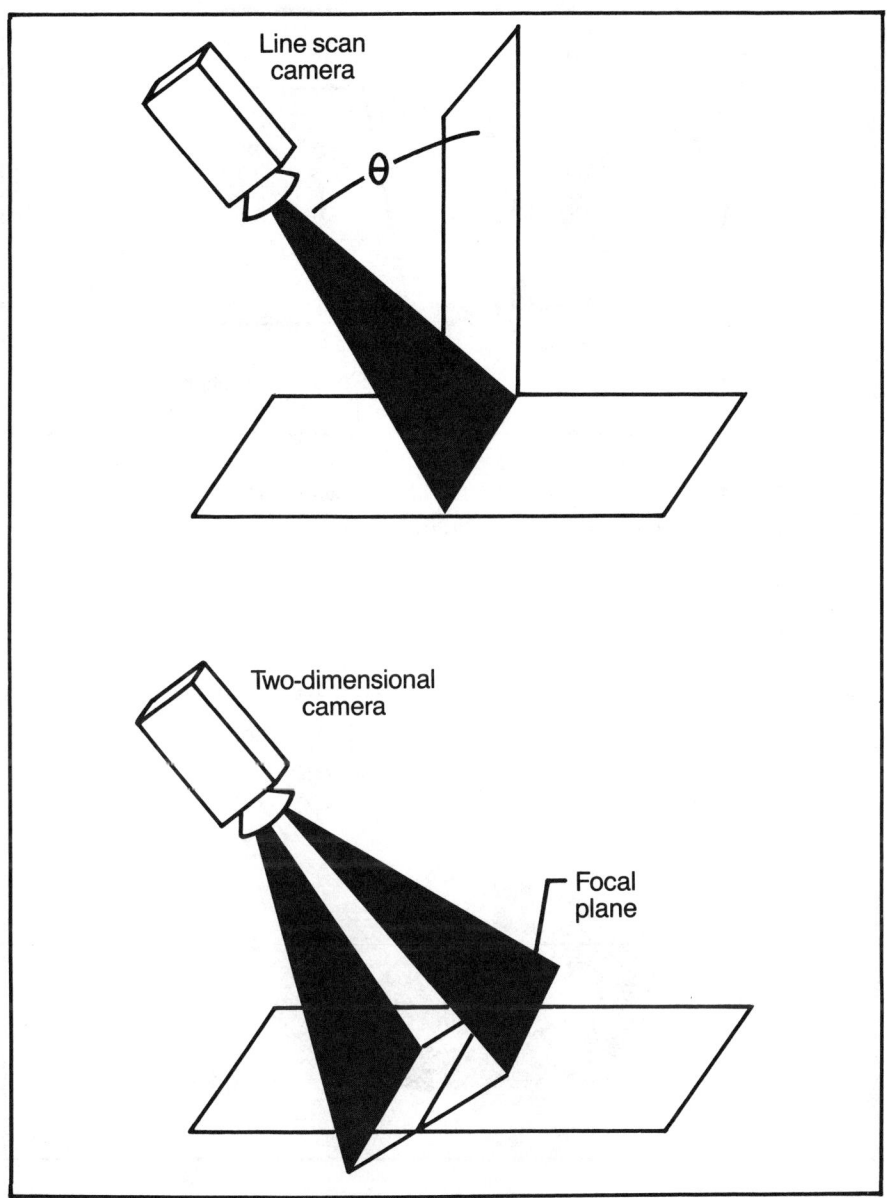

Figure 4. *Angular viewing with a line scan camera and a two-dimensional camera. The line scan camera's image is a line, and the viewing angle rotates about this line. A two-dimensional camera's image is a plane, which when rotated, can create focusing and distortion problems.*

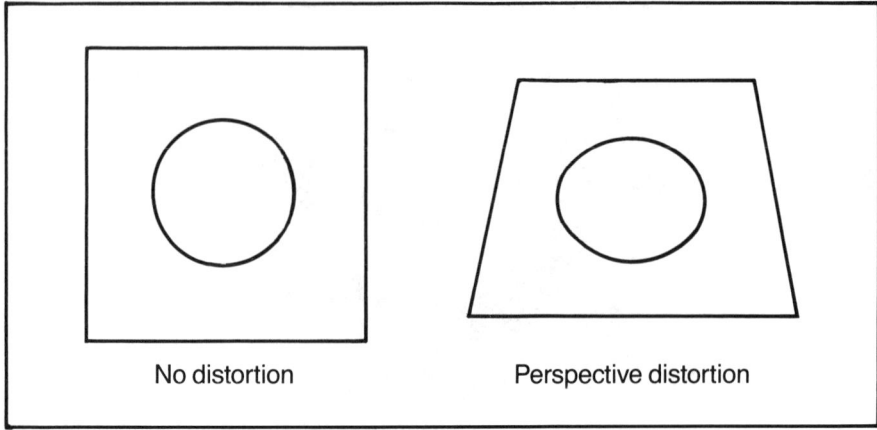

Figure 5. *Perspective distortion (right) encountered when viewing at an angle with a two-dimensional camera.*

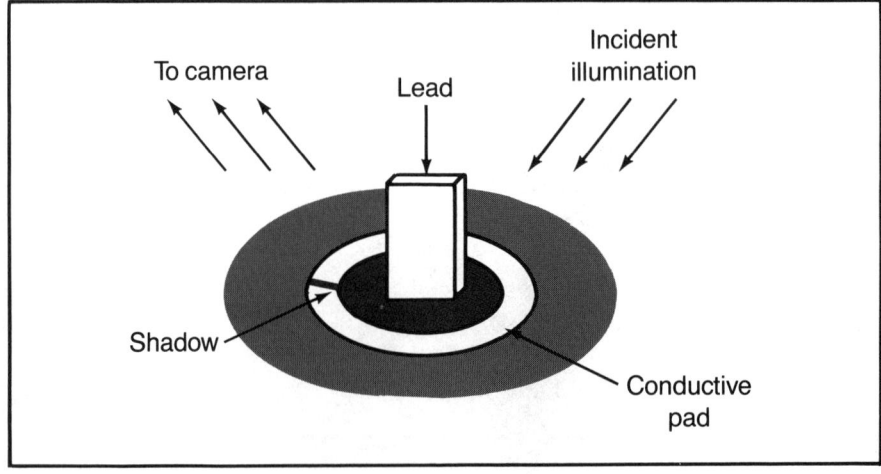

Figure 6. *Illumination and viewing conditions for an unclinched lead. When properly illuminated, the lead blocks the incident light and casts a shadow on the conductive pad.*

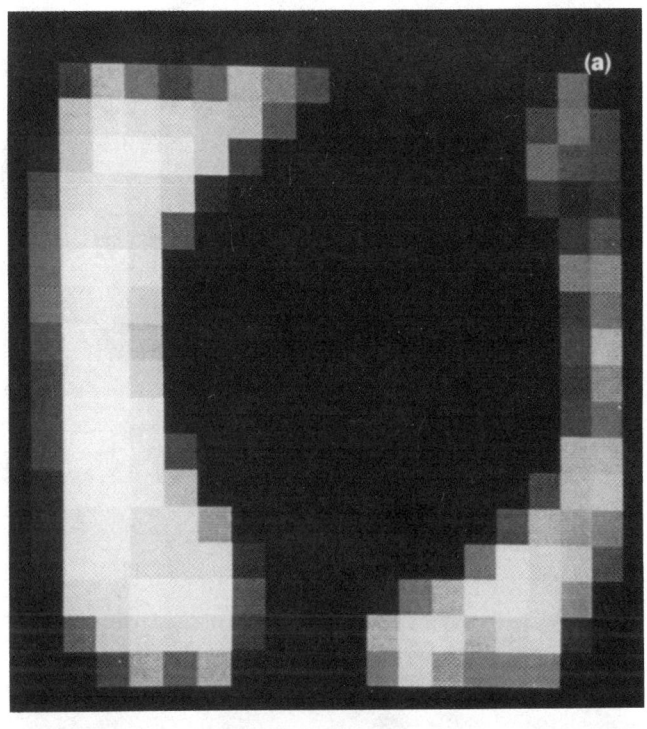

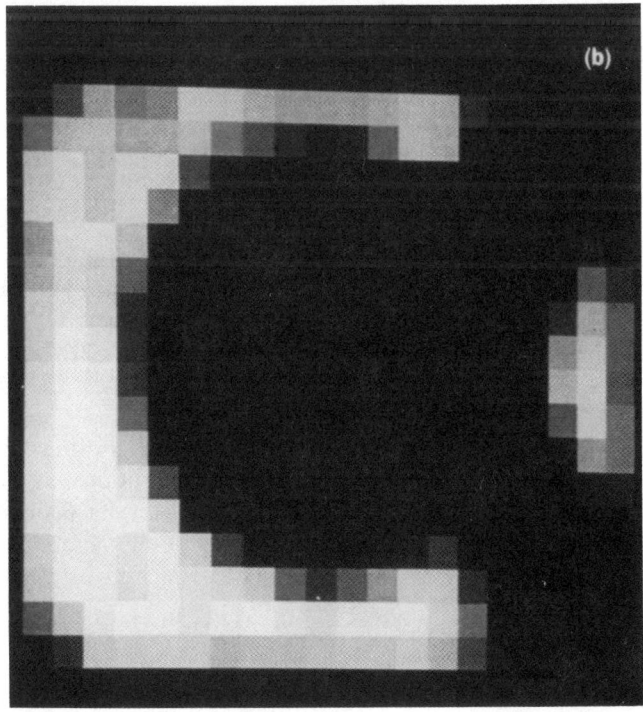

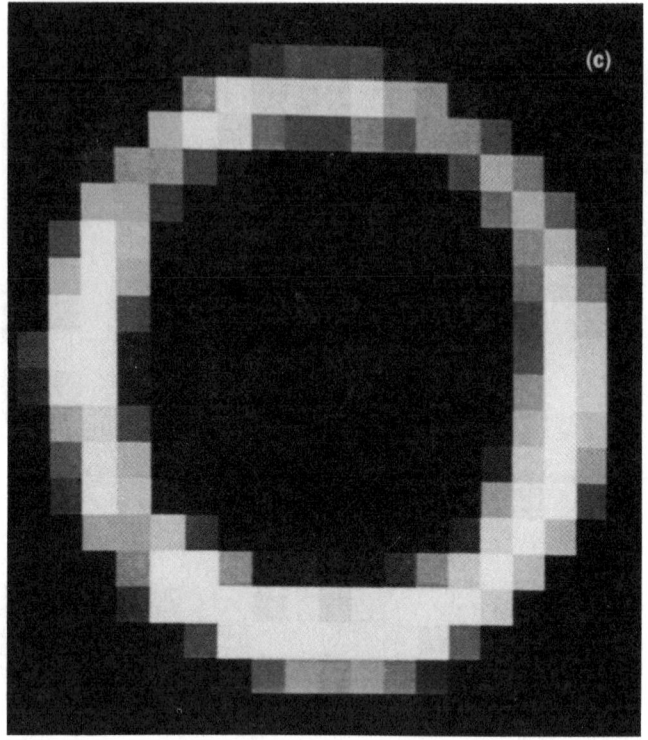

Figure 7. *Gray scale images of a PCB hole illustrating the effect of angular viewing: (a) an unclinched lead, (b) a 45° clinched lead, and (c) a hole with a missing lead.*

Algorithms

It is beyond the scope of this paper to discuss the algorithms for determining passing versus defective conditions for this application. However, some general statements should be made using presolder lead inspection as an example.

The lead count of a typical PCB code is easily about 1000 per PCB. Also, there can be a large variation in the images of the leads, because of their different shapes and clinch angles.

If we use the architecture described here, practical algorithms must run at a speed of 100 inspections/second. In addition, the algorithms must be robust.

Generally these two conditions conflict; i.e., the faster the algorithm, the less robust it is, and vice versa. However, if the front end illumination and viewing techniques just described are used, simple, fast but effective algorithms can be applied.

A histogram of the probability that acceptable and defective sites will occur versus the quantitative result of an algorithm provides insight to the meaning of robust. Figure 8 illustrates two probability densities, one from a fully populated board and another from a defective or bare board. Robustness is related to the separation of these curves. The more separated they are, the easier it is to establish a threshold that differentiates between acceptable and defective sites.

However, the two curves generally overlap at some point. Establishing a threshold for this condition leads to the terms false accept and false reject rates. The *false accept rate* is the percentage of defective sites indited as good. This is represented as the tail of the defective curve that crosses the threshold into the acceptable regime. The *false reject rate* is the percentage of acceptable sites indited as defective, and has a similar, but opposite representation on the acceptable curve.

Let us quantify what we mean by robust in terms of these rates. Assume that there are 2000 leads per PCB with a miss insertion rate of one in 10,000 leads (a

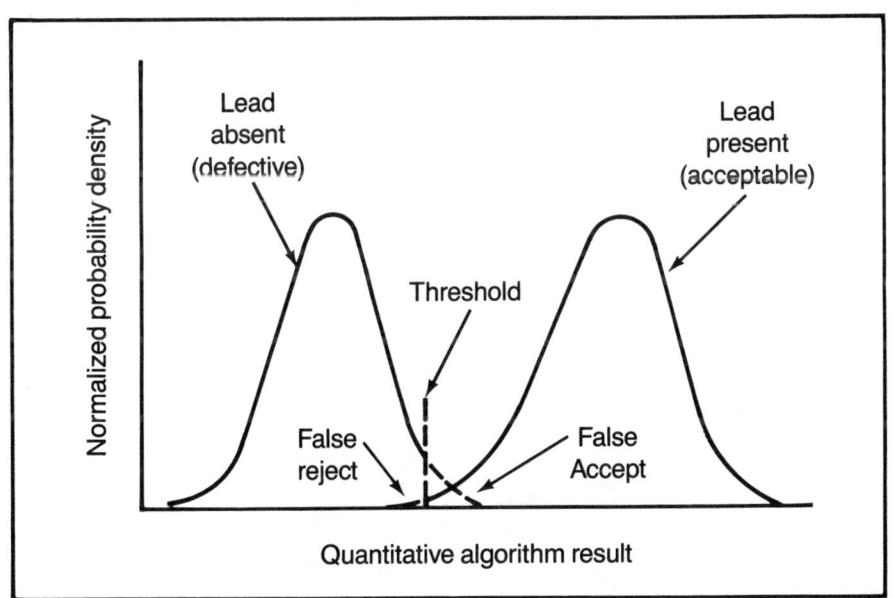

Figure 8. *Normalized probability density functions for acceptable and defective sites. The false reject or false accept rate represents where the tail of the curve crosses the threshold into the other regime.*

typical performance of an insertion process). Figure 9 is a simplified plot of the average number of PCBs that pass through the repair loop of an automated lead inspection system as a function of false reject rate.

If we assume perfect operation (zero false rejects), 20 percent of the PCBs would pass through the repair loop (have real defects). However, if the inspection system is not perfect, it falsely indites other PCBs, which loads the repair loop.

At a false reject rate of just over 0.04 percent, the repair loop transforms from a repair loop to a reinspect line because all the product is going through it. Therefore, a false reject rate of less than 1/10000 (out of 10,000 acceptable sites, one is falsely rejected) quantifies a robust in-line inspection system. Typical commercial two-dimensional systems falsely indite one defect out of 1000 acceptable sites.

If Figure 8's curves overlap, there is obviously a tradeoff between the false accept and reject rates. Initially, one might assume that the false accept rate should be about zero, or the same order of magnitude as the false reject rate. However, this may not be practical or even necessary.

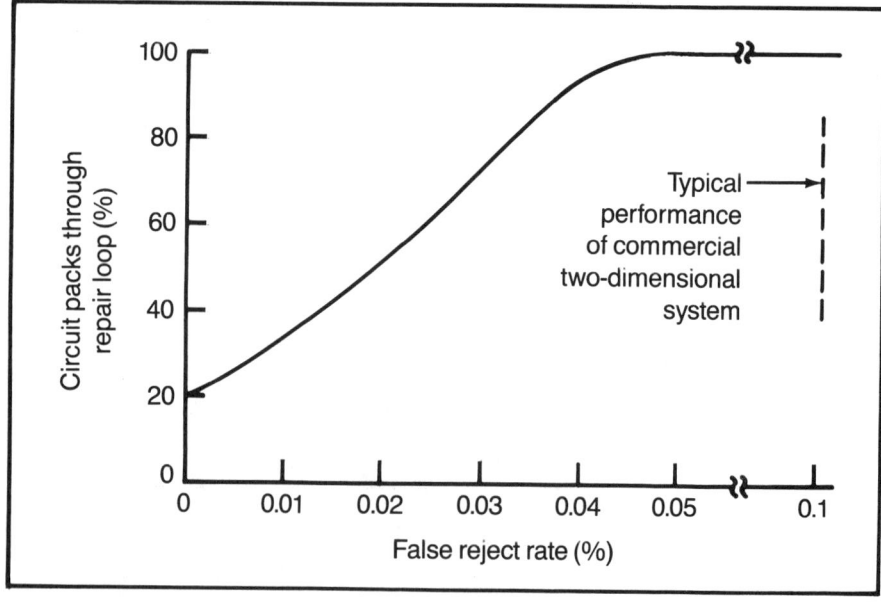

Figure 9. *Average number of PCBs through repair loop (with 2000 leads/ pack) versus false reject rate. Commercial two-dimensional inspection systems typically exhibit a 0.1-percent false reject rate with mixed-technology circuit packs.*

For the example given above, a simple calculation shows that for a false accept rate as high as 1/50, only one out of 250 PCBs will have a defect that is not detected. This seems to be a reasonable false accept rate for an in-line inspection system.

A line scan vision system with the architecture described here can be effective at achieving these goals for a robust inspection. The major reason for this is the large amount of information used in the analysis. Most of it comes from the large number of pixels or data available for the algorithm to process.

For a given scan time, a line scan system acquires over ten times more pixels than two-dimensional systems. For presolder lead inspection, additional information (the shadow) is available about the presence of a lead. Furthermore, the CAD/CAM database also identifies the type of lead (flat IC lead, axial lead, etc.) and its clinch angle.

The more significant the information available to the algorithm, whether image data or CAD/CAM data, the more separable are the curves in Figure 8. Thus, this results in a more robust inspection.

CONCLUSIONS

The complexity of today's electronic designs is pushing the limits of performance of the manufacturing process. Machine vision can provide the technology to monitor this performance effectively by automating assembly verification.

We described a machine vision system specifically designed to address efficiently the problem of PCB inspection. It has several features that provide improved performance over current methods. Although we only described the use of the system for presolder lead inspection, it is not limited to this function. Future uses of the system will involve other PCB inspection applications.

Reprinted with permission from the *AT&T Technical Journal*.
Vol. 65, No. 4. pp. 58-65. © 1986 AT&T.

41

Group Technology for Electronics Assembly

Thomas P. Styslinger
Michel A. Melkanoff

Before discussing how group technology can be used in electronics manufacturing it is first necessary to define what group technology (GT) is and how it has been used. The definition of group technology used in this paper is the following.

Group technology is a manufacturing philosophy for producing small to medium lot sizes of parts wherein the parts are grouped together to take advantage of their similarities in manufacturing and/or design.

Group technology was first developed in the Soviet Union when S. P. Mitrofanov showed that setup times could be reduced from one hour to just a few seconds by taking advantage of similar parts using similar fixtures.[1] Mitrofanov based his studies on the premise first postulated by A. P. Sokolovski who suggested that parts of similar configuration and features, all other factors being equal, should be manufactured in the same way by using a standard technical process. GT then spread into Europe and Great Britain in the mid 1950s and into the United States and Japan in the early 1960s.

Group technology involves the development of part families and Group technology based on cell layouts (see Figure 1). A part family is a collection of parts which can be produced using the same set of resources.

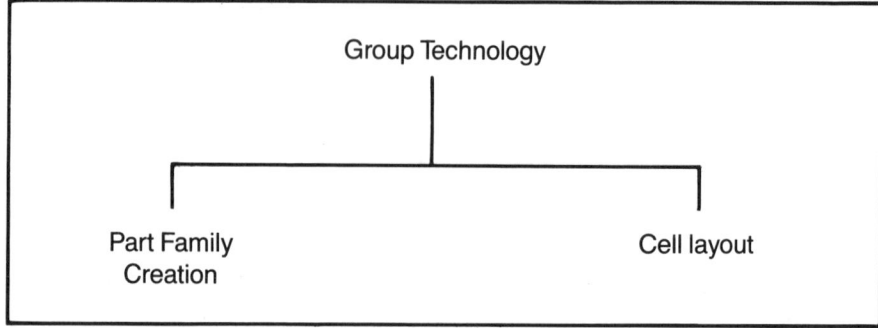

Figure 1.

There are two types of factory layouts in widespread use today: line and functional. The line layout is typical of the mass production industries, such as the automobile industry where single type high volume items flow down a dedicated assembly line which has little flexiblity. The functional layout is typical of the job shop environment where the machines are arranged according to function, lathes with lathes, mills with mills, etc. The functional layout is used in the low volume environment where a high degree of flexibility is required. The GT cell leyout consists of grouping personnel and machines in one area because they contain the necessary capabilities for producing a family of parts. These layouts are illustrated in Figure 2.

There are many benefits to be gained by using the group technology philosophy. Typically, parts pass through numerous departments, single operations, supervisors, machine operators and production control personnel. All of this adds up to lost parts, confusion and inefficiency. GT exploits product similarity in design, tooling, fixturing, and routing to form part families. The GT cell layout because it specializes in the manufacturing of a particular family causes a reduction in part handling, setup time, work-in-process and through-put times. Supervisory responsibility is increased because the supervisor is not merely accountable for an operation but for producing a product. Assemblers and machine operators derive increased job satisfaction because they see the results of their efforts. Reported GT implementations have seen 85% reductions in setup times and over 100% increases in throughput rates.[2]

CLASSIFICATION AND CODING

Classification and coding are often used interchangeably with group technology, but all three terms are distinct and different.

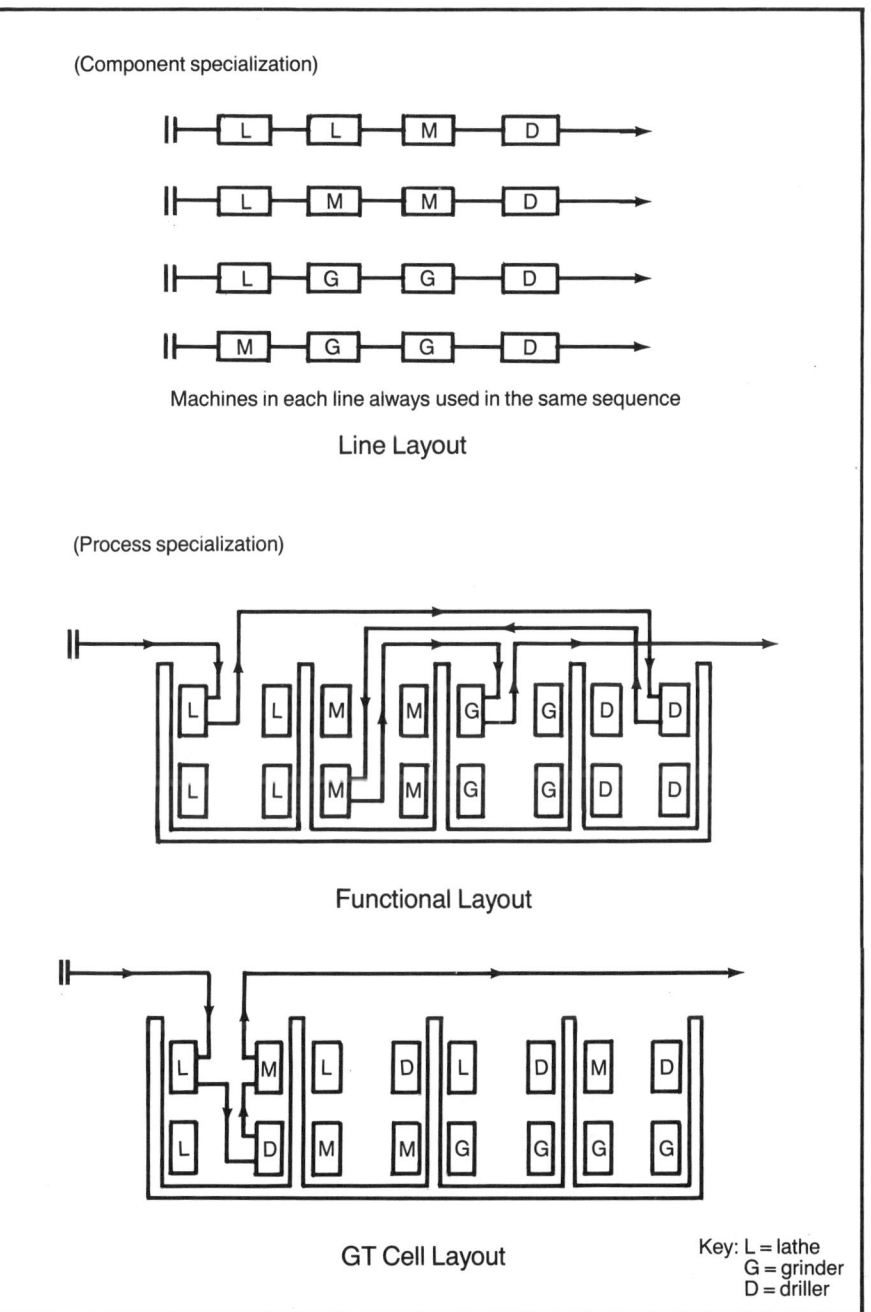

Figure 2. *Factory layouts.*

Classification is a technique to organize related data into a logical and systematic order using a specific set of criterion that categorizes similar items together. Coding on the other hand is the arbitrary assignment of one or more symbols to a part, which when deciphered communicates specific meaning or intelligence. Neither classification nor coding are defined as a manufacturing philosophy and therefore should not be confused with group technology.

There are three methods for grouping parts into families:

1. Visual inspection;
2. Production flow analysis; and
3. Classification and coding.

As its name implies, visual inspection consists of grouping parts into families according to their appearance. The method is inexpensive but also rather arbitrary and inaccurate. Production flow analysis is a technique where all the route sheets for the parts being classified are collected, sorted and analyzed to determine part families. The weakness of the production flow analysis is that the routing sheets have been generated by different process planners who have their own styles in developing a process plan. Classification and coding is the most accurate method but also the most expensive.

A classification and coding system to be used for the development of families for group technology is based on the type of item being studied. Since most of the GT systems have been implemented in the metal fabrication area the majority of classification and coding systems describe the design and manufacturing attributes of metal parts. The classification code, can be one of three types:

1. Monocode;
2. Polycode; and
3. Multicode.

The monocode is hierarchical in structure where each digit's meaning is determined by the previous digit. In the polycode each digit's meaning is determined by its position on the code. The multicode is a hybrid of the previous two. The majority of the codes in use today are of the multicode variety.

A classification and coding system can have other uses besides developing part families for GT. In metal part fabrication classification codes are being used to retrieve designs in order to reduce design duplication. Classification codes are being utilized to retrieve previously developed process plans which are copied and modified for the current part. The classification code can be used as an estimating tool. When a new design is created, the estimator codes the new design and retrieves a similar design with a cost history.

GROUP TECHNOLOGY AND ELECTRONICS MANUFACTURING

Electronics manufacturing can be broken down into four major manufacturing areas:

1. Component fabrication;
2. Printed wiring board fabrication;
3. Circuit card assembly; and
4. Product assembly.

This chapter will focus on group technology as it applies to circuit card assemblies, although it appears from a cursory evaluation that GT could be applied to the other areas as well. Group technology for circuit card assemblies (CCA's) involves the development of CCA families and the utilization of those families in the design of GT cells. Once those cells have been developed a material transport system and control system can be used to link the cells together, forming the base of a flexibility manufacturing system (FMS).

The construction of CCA families requires using one of the three family formation methods. Visual inspection is inexpensive but highly dependent on the arbitrary decisions of the person performing the categorization. the confidence level in results of a visual inspection based study are usually not sufficient to promote investment in an FMS. The routing method is better but depends on the planner's style who generated the routing. The best method is to classify and code the CCA's attributes, develop a set of family criteria and then have the computer decide to which family the CCA belongs.

Once the families have been developed and quantified the analysis then proceeds on to determining the resources required to produce each particular family of CCA's. These resources are translated into personnel and equipment which determine the floor space required for each GT cell. The required interfaces between cells can then be studied to determine the overall layout of the FMS.

After the cell or FMS has been designed, a scheduling system must be developed which will determine which cell or cells are required to produce the CCA. The CCA's classification code and cell family parameters can be used when it is scheduled for production to determine which cell or cells are required.

ELECTRONICS CLASSIFICATION AND CODING

The classification and coding system used for group technology must be capable of describing the circuit card assembly manufacturing attributes. Those attributes have to be described in sufficient detail to make part family formation possible. The manufacturing attributes used for part family formation are selected for classification and coding on the basis of their utilization in determining the processes and process setups required in production of circuit card assemblies.

The classification and coding systems designed for metal part fabrication have included both design and manufacturing information. Unlike metal parts, the physical design of a circuit card assembly does not determine the assembly's function. Many different functions may be performed on a single CCA, such as clocking, counting and arithmetic (ALU). On the other hand a CCA may only perform part of a total function, such as RF signal generation which is shared between several CCA's. The classification of electrical functions performed on circuit cards is beyond the scope of this chapter, but would be an interesting and valuable topic for further investigation. The classification system developed for circuit card assemblies focuses strictly on those attributes used in the manufacturing process.

The producibility of a circuit card assembly is determined by the physical attributes of the printed wiring board and the components attached during assembly. It is this type of information which must be classified and coded for use in electronics group technology. The following paragraphs discuss some of the attributes required for part family formation.

The physical description of the circuit card assembly must contain dimensional information including the following:

1. Shape;
2. Length;
3. Width; and
4. Thickness.

The length and width dimensions in conjunction with shape define many of the fixtures and material handling devices used in the assembly process. The length and width also have a moderate effect on the wave solder parameters. Circuit card thickness plays a critical role in the auto insertion and wave soldering processes. For automatic insertion the thickness of the printed wiring board is an important parameter in setting the height of the insertion head. Experience has shown that when board thicknesses vary more than 0.010 of an inch the insertion head must be manually adjusted to insert the component at the correct height. If the component is inserted too high it will be loose after clinching. On the other hand, if the component is inserted too low the impact of the insertion head may damage the printed wiring board. The printed wiring board thickness is also a factor in the setup of the wave solder process. An increase or decrease in thickness changes the thermal conductivity of the PWB and in turn affects the temperature setting of the wave solder preheaters.

Other printed wiring attributes must also be considered when setting the wave solder process controls; they include the following:

1. Base material;
2. Minimum and maximum conductor width; and
3. Conductor spacing.

These attributes are used in determining the settings for the preheat temperatures, conveyor speed, solder temperature and solder height. If the process attributes of the circuit card assemblies waiting for wave soldering are know those with similar attributes may be grouped together decreasing the number of process control changes.

The number and types of components used on the circuit card assembly is a major factor in determining whether or not automatic insertion is either feasible or economical. Components can be classified into the following categories:

1. SIP, Single In-line Package;
2. DIP, Dual In-line Package;
3. Axial;
4. Radial;
5. Can;
6. Flat Pack;
7. Pin Grid Module; and
8. Special Package.

Automatic component insertion equipment typically specializes in the insertion of a particular component package type. Notice that the components were classified by package type and not function, such as, integrated circuit, resistor or transistor. Automatic component insertion is dependent on package type not function.

In the production of military hardware the assembly specification to which the assembly must be built may impose special processing requirements. A specification which requires no exposed copper on leads forces a manufacturer to either trim all leads to the required length before wave solder or to double wave solder the circuit board. The classification system should separate those CCA's which require special processing from those which do no.

LIMITATIONS OF CLASSIFICATION AND CODING

There are many other circuit card manufacturing attributes which need to be classified and coded besides those discussed above. These include mechanical components used, percent of components auto-insertable and testing requirements. Most of the software packages designed for group technology place limitations on the types of codes and code lengths allowed. For example, one common GT package allows only numeric codes and a maximum code length of thirty digits. Considering the amount of information required to develop part families, using a limited code length system presents problems. For example, when coding component packages used in the circuit card assembly, an additive or binary code is necessary to describe combinations of items. If all of the component packages are used on an assembly this would be the result.

Code	Package Type
001	SIP
002	DIP
004	Axial
008	Radial
016	Can
032	Flat Pack
064	LCC
128	Pin Grid Module
256	Special Package

Total   511

It takes three digits to uniquely encode one single attribute such as the types of component packages used. If every attribute required three digits and a thirty digit maximum code length was used, a maximum of just ten attributes could be coded.

There are two solutions to the limitations of classification coding, multi-linked codes and flexible coding. In the multi-linked coding system space is allowed in the record for a unique number which will serve as a link between one code and another. In the case of a circuit card assembly the linking element would be the printed wiring board part number from which the circuit card assembly is made. The circuit card assembly code would contain information such as the electrical and mechanical components used and the printed wiring board code would contain information on the base material and conductor information. Figure 3 presents an example of this procedure.

Flexible coding is a method of exchanging attributes into and out of a fixed length coding system. It involves having a user defined coding system and a data base of attribute information. A user defined coding system is one in which the user can specify which attributes will appear in which digits. Figure 4 illustrates flexible coding using a six digit fixed length coding system. In Case 1 the user is initially interested in fixutre information so he selects the shape, length and width attributes from the data base and specifies the digits in which they will appear.

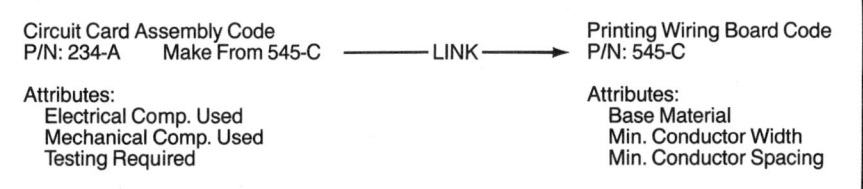

Figure 3. Multi-linked code.

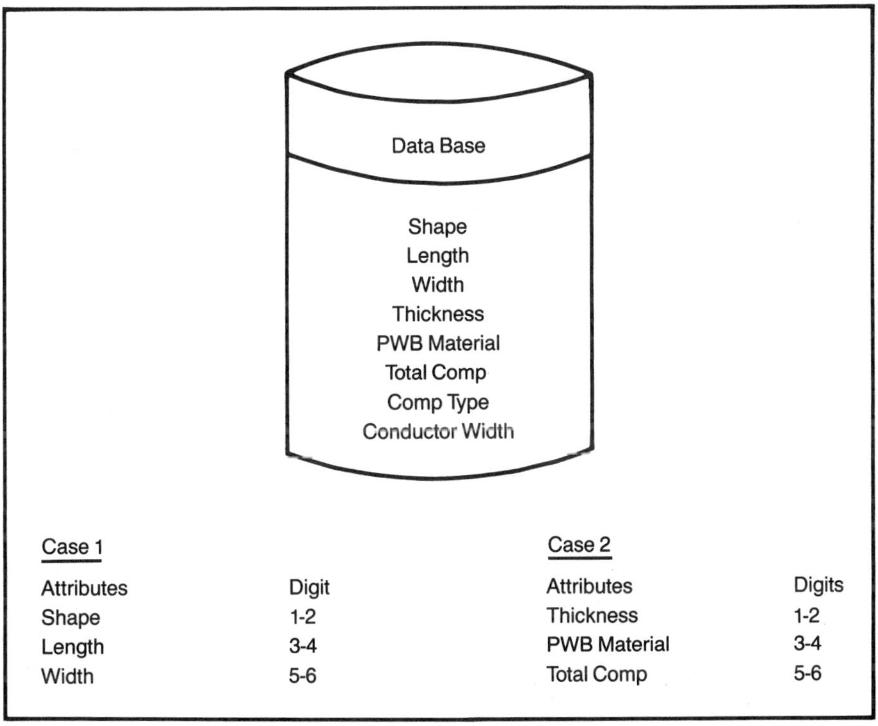

Data Base

Shape
Length
Width
Thickness
PWB Material
Total Comp
Comp Type
Conductor Width

Case 1			Case 2	
Attributes	Digit		Attributes	Digits
Shape	1-2		Thickness	1-2
Length	3-4		PWB Material	3-4
Width	5-6		Total Comp	5-6

Figure 4. Flexible coding system.

In Case 2 the user is still limited to the fixed code length of six digits, but he is interested in attributes which may affect his wave soldering process. Therefore he selects the thickness, printed wiring board material and the total number of components used on a circuit card assembly from the data base. The data base contains, for each circuit card assembly, attributes which may be selectively entered into a fixed length code. This ability gives the user flexibility in performing a group technology analysis on general circuit card attributes or focusing on those specific attributes which may affect a particular process. In the multi-linked coding system the user does not have this flexibility because he is forced to select one of the predefined coding systems. The flexible coding system based on a manufacturing attribute data base is a major step forward in the integration of computer aided manufacturing.

The flexible coding system requires additional software support compared to the multi-code system. It also requires the development and maintenance of an attribute data base. As a result the initial cost is higher for flexible coding if an attribute data base has to be designed and loaded.

CLASSIFICATION AND CODING SYSTEMS AND DATA BASE INTEGRATION

Integration and the ability to quickly and accurately handle information are key elements in today's production environment. An isolated stand alone classification and coding system used for group technology is not a step forward, but a step backward into the past. A stand alone system has both data entry and maintenance problems. The initial loading of a classification and coding system requires many man-months of effort. If a database containing the attributes required in the manufacturing process has been established, the loading of the coding system becomes a much simpler task. A computer program can be written to automatically translate attribute information into the classification codes required. Maintenance also becomes simpler because as the manufacturing data base is updated, the files used by the classification and coding system are also updated.

There has been a great deal of discussion concerning the integration of GT and process planning and how GT can drive process planning. In the mechanical fabrication environment, classification and coding systems have been used to retrieve previously written process plans, commonly called variant processing planning. In a variant process planning system the planner codes the part being planned, then using some computer software searches the previously coded parts file for a similar part. If a match is found, the planner modifies the similar part's planning to fit the part currently being planned. The variant process planning system approach works well for fabricated parts but appears unsuitable for assembly planning.

Assembly planning describes the discrete steps required to take a conglomeration of parts and turn them into a single assembly. Considerable information on the individual parts being used and the processes required is necesary to create an assembly process plan. Consider for example the placement of an axial leaded resistor on the PWB; the resistor hole spacing must be known in order to determine the distance out from the resistor body to bend the leads to fit the holes. A classification code of fixed length is incapable of storing this type of detailed information in the quantity required.

The group technology system will interface with the circuit card planning sytsem even if a classification and coding system is not used to drive the planning system. A group technology layout will decrease the multitude of routings used in the circuit card assembly area. This in turn will decrease the routings available to the process planner. The GT system informs the planner that given the attributes of the circuit card being planned it belongs to a particular production family. The planner will then have the option of utilizing the GT assembly cell or using a less efficient method of assembling the circuit card. If a generative or semi-generative planning system approach is used for assembly planning, both it and the classification and coding system utilized for GT will interface with a common manufacturing data base. The data base will contain the information necessary to drive both the GT system and the assembly planning system.

The manufacturing resource planning (MRP) system will probably also require an interface to the common manufacturing data base. This interface will act as the medium for the GT system to automatically access information on total annual production quantity, releases, routings and time standards. This interface will provide the GT system user with timely and accurate information while also decreasing the amount of manual system updating required.

CASE STUDY

In late 1983 Hughes Aircraft's radar systems group and ground systems group set out to determine the feasibility of GT as applied to the manufacturing of circuit card assemblies. To Hughes' knowledge GT has never heretofore been applied to the manufacturing of electronic products. The feasibility study was conducted at Hughes' ground systems group, a manufacturer of radar, sonar, data systems and communications products. Hughes Ground Systems Group produces approximately 70,000 circuit card assemblies in 1,200 different configurations.

The purpose of the feasibility study was:

1. To determine if hard benefits of group technology electronics applications exist;
2. To clarify and quantify the circuit card assembly environment; and
3. To demonstrate to management the value of group technology and the potential applications using actual production data.

Data Collection and Classification Code Formation

The first step in the feasibility study was to select a representative sample of production circuit cards. The circuit card assemblies selected were a representative cross section of the products being manufactured and technologies used in the products. It was determined that a 100 card sample would provide sufficient information to draw conclusions. The group technology software used required the following information on each assembly.

1. Part number;
2. Classification code;
3. Annual quantity produced;
4. Annual number of releases;
5. Work center routing;
6. Cost center routing; and
7. Setup and run time applied to each work center.

The work center and cost center routings along with the annual quantity and release information were obtained from the manufacturing resource planning system. The run times were obtained manually from time standard data sheets.

Classification Code

A classification code was developed by studying the circuit card drawings and process plans. Also, experts in each of the assembly areas were interviewed to determine the circuit card attributes considered when setting put a process. A total of twelve attributes were selected for classification and coding (see Figure 5).

Classification Attributes
1. Type (digital, analog, ...)
2. Geometrical shape
3. Size (length, width, thickness)
4. Quantity of components
5. Maximum component height
6. Lead length
7. Hand assembly requirements
8. % Auto-DIP insertable
9. % Auto-axial insertable
10. Connector types
11. Mechanical components
12. Special features

Figure 5. *Feasibility study classification attributes.*

Each circuit card was coded by referencing the assembly drawing and the process plan. The GT software used was originally developed for metal fabrication and had fixed fields with a total length of thirty digits. The classification coding system used was force fitted into the fixed fields and length.

Special Note

The classification code used in the feasibility study bears only a remote resemblance to the classification code developed for planned production use.

Analysis

After all the information for each of the 100 circuit cards was tediously entered into the GT data files the analysis began. The first step in the analysis was to perform a routing analysis (see Figure 6). The results showed that 20 different routings were required to produce the 100 different circuit card configurations. The highest single routing usage (No. 8) consisted of 34 CCA configurations requiring 9252 hrs., (43%) of the total shop capacity. The second highest routing (No. 14) consisted of 12 CCA configurations requiring 2497 hrs. (12%) of the shop capacity. These two routings only differ by one operation, hand assembly (see Figure 7).

The next step in the analysis was to check if the circuit card assemblies which used routings No. 8 and 14 had similar manufacturing attributes. The classification codes of the circuit card which used these two routings were then analyzed. The results showed that 40 of 46 circuit cards had the following attributes:

Classification = Attribute
Function Type = Digital
Shape = Rectangular
Dimensions = 5 in. × 5 in.
% Auto-Dip Insertable > = 80%
Connector Type = Riveted and Eyeleted Connector
Mechanical Components = Thermal Mounting Plate
= Card Guides
= Ground Tabs
= Buss Bars
Special Features = None

The two routings were then separately analyzed by classification code yielding surprising results. Eleven of the twelve circuit cards using routing No. 12, the routing which had no hand assembly operation called out, did in fact require hand assembly, only one was totally auto insertable. The explanation for this

ROUTING REPORT
TOTAL PART NUMBERS IN FILE 100
TOTAL UNIQUE ROUTINGS 20
TOTAL REQUIRED CAPACITY 21425 Hrs.

ROUTING NO.	FLOW USAGE	TOTAL CAPACITY	USAGE HISTOGRAM
1	6	817.5	******
2	2	388.1	**
3	5	1398.0	*****
4	2	62.2	**
5	1	257.7	*
6	6	496.2	******
7	1	4.6	*
8	34	9252.4	**********************************
9	8	1895.1	********
10	1	31.8	*
11	8	3076.6	********
12	2	94.0	**
13	2	125.9	**
14	12	2497.2	************
15	1	304.3	*
16	2	379.4	**
17	2	34.7	**
18	3	152.2	***
19	1	33.3	*
20	1	123.6	*

Figure 6. *Routing report.*

surprising result is that the process planner has the option to call out for a small amount of hand assembly to be performed in the auto insertion area.

The auto-DIP insertion classification attribute was then raised to 90% and using the same attributes as above, the results showed that 35 of the 46 had the required attributes and required 39% of the total shop capacity. It was decided that this set of parameters which separated out the 35 circuit cards would define a production family of parts around which an auto-DIP assembly cell would be built. The auto-DIP assembly cell would include all of the operations performed before wave solder. Wave solder was not included because circuit cards other than those which would be assembled in the cell would need access to the wave solder machine. A block diagram of the cell is shown in Figure 8.

ROUTING MATRIX

FLOW NO.	COMP PREP	MECH ASSY	AUTO DIP	AUTO AXIAL	HAND ASSY	WAVE SOLDR	REFLOW SOLDR	2ND ASSY	COMF COAT	FINAL ASSY
14	X	X	X			X			X	
8	X	X	X		X	X			X	
13	X	X	X	X	X	X			X	
2	X	X		X	X	X			X	
3	X	X		X	X	X		X	X	
12	X	X				X			X	X
4	X	X	X		X	X			X	
9	X	X	X		X	X	X		X	
10	X	X			X	X	X		X	
11	X	X			X	X	X		X	
18	X				X	X	X	X	X	
19	X				X				X	
6	X	X			X	X		X	X	
5	X	X			X	X			X	
1	X	X			X				X	
15	X	X			X			X	X	
17	X	X			X			X	X	
16	X	X							X	X
20	X	X							X	X
7	X	X							X	X

Figure 7. Routing matrix.

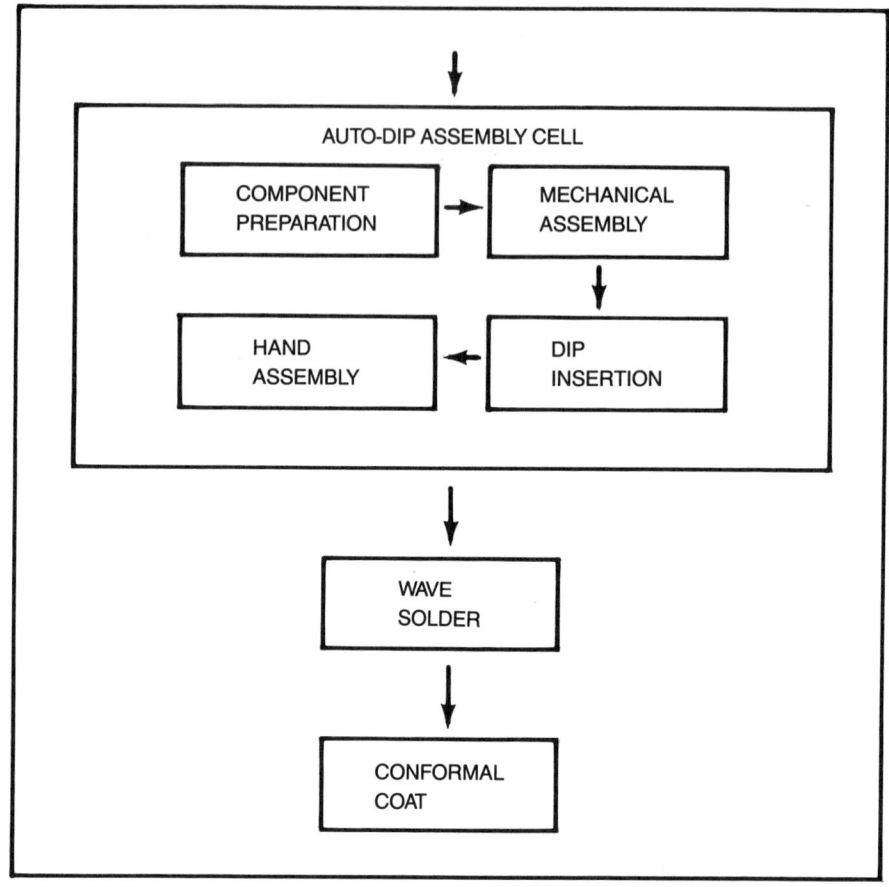

Figure 8. *Auto-DIP assembly cell.*

Using the GT software a machine and personnel load analysis was performed to determine the machine and personnel requirements for the auto-DIP insertion cell. The results showed that one auto-DIP insertion machine would be required working two shifts, which was fortunate because one of the two auto-DIP insertion machines used in the shop could be dedicated to the cell. Personnel who normally only performed a single monotonous operation would perform a variety of assembly operations in the cell. For example, a person performing buss bar installation would install card guides, connectors and buss bars, thus achieving fuller personnel utilization.

A methods engineering analysis was performed to estimate the benefits of installing the auto-DIP assembly cell. A 26 percent improvement in personnel and machine utilization was forecast, with an estimated value of $275K. Also the implementation of the cell would reduce material handling by 40 percent saving an estimated $75K. A 25 percent work-in-process reduction was forecast for those circuit card assemblies using the cell. The benefits of implementing the cell totaled an estimated $435K. The cost of implementing the cell is minimal because no major equipment purchases are necessary.

Study Conclusions

The results of the feasibility study showed that group technology applied to the production of electronic assemblies is both possible and profitable. Group technology proved to be a viable tool in the development of assembly cells which will in turn be the basis of an electronics flexible manufacturing system. The implementation of an auto-DIP assembly cell will improve personnel and machine utilization, reduce material handling and WIP, thereby improving the product throughput rate.

FINAL CONCLUSIONS

Group technology is an excellent tool for use in the development and scheduling of electronic flexible manufacturing systems. It need not be limited to metal fabrication applications. The major problem in applying GT to electronic products is the inclusion of enough manufacturing attributes in a fixed length code to accurately describe the assembly. Of the multi-linked and flexible coding options available for solving the problem, the flexible coding system seems to be the best choice because of its integration with a common manufacturing data base. The case study proved that group technology is a viable tool in the development of circuit card assembly cells. The utilization of group technology will prove to be a key element in the efficient and accurate design and implementation of flexible manufacturing systems in the electronics industry.

References

U.S. Air Force. 1977. Scientific organization of batch production. Integrated Computer-Aided Manufacturing (ICAM) Task II-Final Report. Vol. III. December.

Hyde, W. F. 1981. *Improving Productivity by Classification, Coding, and Data Base Standardization.* Marcel Dekker Inc. New York.

Suggested Readings

Burbidge, J. L. 1975. *The Introduction of Group Technology.* John Wiley & Sons Inc. New York

Carter, C. F. 1984. Look! the tools are in your hands. *Production Engineering.* September. pp. 84-89.

Coombs, C. F. 1979. *Printed Circuits Handbook.* McGraw-Hill Book Company. New York.

Groover, M. P. 1980. *Automation, Production Systems, and Computer-Aided Manufacturing.* Prentice-Hall Inc. Englewood Cliffs, NJ.

Kohen, S. I. 1982. *Electronic Manufacturing.* Restion Publishing Company Inc. Reston, VA.

Maller, R. J. 1984. Integrated manufacturing—the concepts, the structure: simplicity and focus or collapse. *Production Engineering.* September. pp. 62-65.

Reprinted courtesy of the authors, Thomas P. Stylinger and Michel A. Melkanoff, and the Society of Manufacturing Engineers. © 1985 from the FMS for Electronics Conference, February 25-27, 1985, Cambridge, Massachusetts. SME Technical Paper EE85-132.

42

Automating Surface-Mount Assembly in a European Plant

M. Maes

N.V. Philips Industrie, located in Leuven, Belgium, 25 km due west of Brussels, is part of the Philips Consumer Electronics Division. The facility produces a wide range of electronic products, including hi-fi audio systems, hybrid customized electronic systems, and general consumer electronics goods.

The Leuven facility also makes most of what it consumes itself, including basic plastics, metal components, special machines and tools, as well as printed circuits and hybrids. Although automation is widely used throughout, nowhere is it put to better advantage than in the production of printed circuit board assemblies, where it now constitutes some 85 percent of the production capacity.

This level permits an increase in product reliability, quality, and throughput, as well as reducing manual content and helping to combat "foreign" competition. The goal is to eventually boost the level of automation to around the 90 percent mark, including automation of the present manual transport of materials between workstations.

The automation program at Leuven started about 10 years ago, at which time the initial concentration was on the placement of bridge-wired components. It was thought best to solve this problem before considering how to automate the assembly of horizontally and vertically mounted ones. The present (85 percent) level of chip-mounting automation has been in effect about 18 months and took two years to implement. Future automation will be limited solely by production needs.

It was desirable to be able to quote a definite amortization period, and at first it was felt that this would be a simple matter. It was quickly realized, however, that this was not possible since fluctuations in the cost of board components rendered applicable calculations unusable. Nevertheless, board components have subsequently become less expensive and assembly times much faster, to the point where such calculations are possible again.

AUTOMATING ODD-FORMS

One of the main targets for increased automation in the future will be the assembly of exotic components—the so-called "odd-form" components, such as switches, plugs, connectors, and transformers. These, because of their size and/or configuration, cannot currently be mounted by conventional (e.g., vertical, horizontal or SMD) placement machines. Successful automation in this area would offer a manufacturer a strong technological advantage in the marketplace, and a great deal of developmental work is being carried out toward this goal, notably in the Far East, but especially in Europe. The Philips Leuven facility hopes to be one of the first companies to enter this challenging new area.

Although the aim is to push odd-form insertion automation to its ultimate limit, a 100 percent level is hardly likely, simply because it will never be possible to automate certain of the more difficult odd-forms (Figure 1). Perhaps the best that one can hope for is a 95 percent level.

At present, there is some misunderstanding about working with odd-form components, given that they can range widely in shape, from trimmers which are fairly easy to mount, through the switches, transformers, and others mentioned earlier, which do not allow easy assembly using automatic placement machines.

Although a high degree of automation has been achieved, placement of the more complex-type odd-form components will demand manual assembly. The total automation of assembly operations including the "difficult" odd-forms (perhaps by using robotic devices) is unlikely on the grounds of economy and effort required. Odd-form components and their placement areas are extremely small, with the result that some 5 to 7 percent of the entire assembly program must therefore be a manual operation.

CONSTRUCTION AND DESIGN

Much attention has been paid in the past to standardizing board dimensions to reduce waste to an absolute minimum. With production currently running at about 9,000 boards per 24-hour working day, the incentive is a considerable savings in money and materials.

Thus, board widths are controlled at Leuven to 200.835 mm. Board lengths, on the other hand, are currently of three sizes, but the aim is to reduce these to

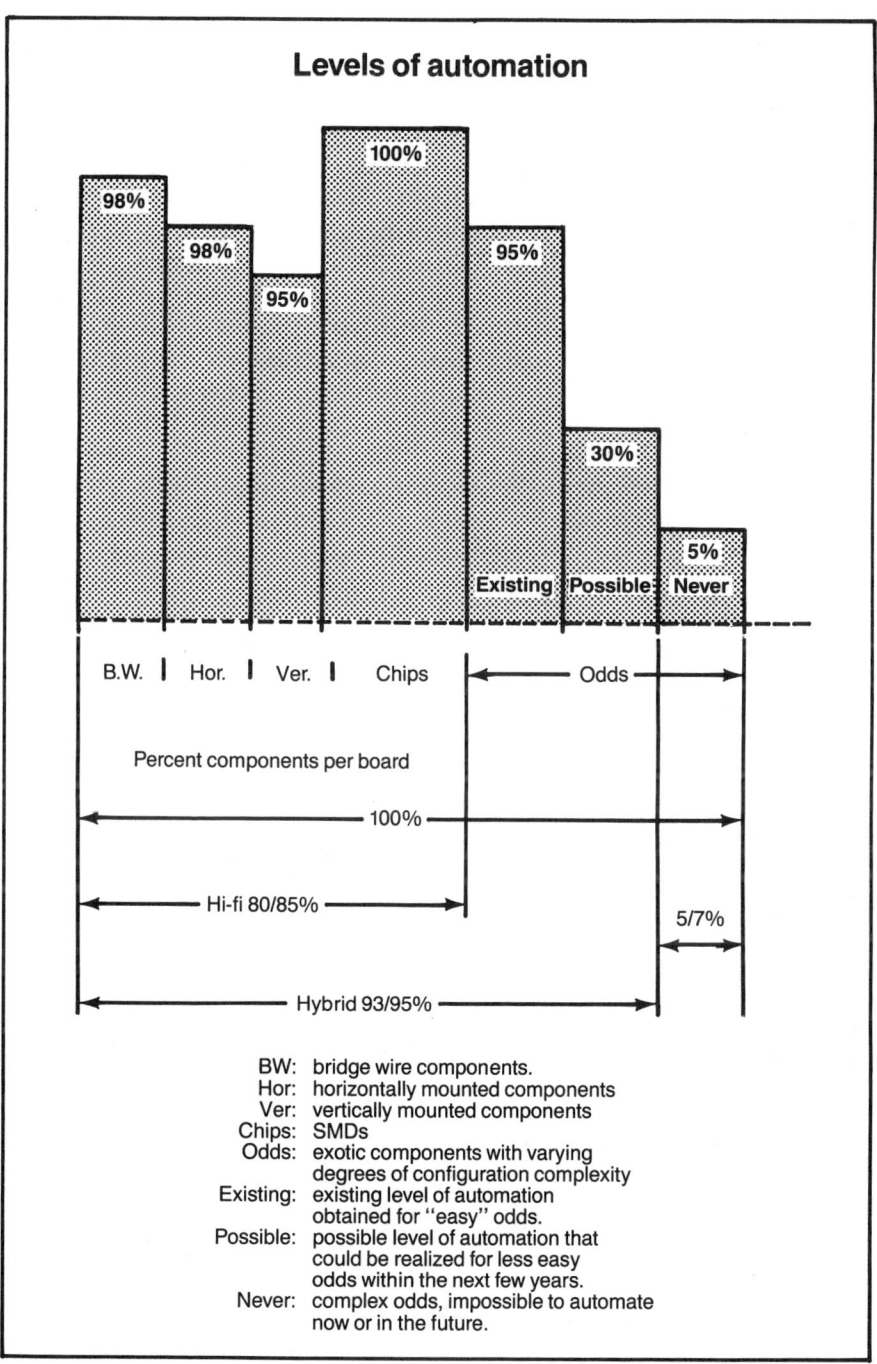

Figure 1. *Actual and anticipated levels of assembly automation attainable for hi-fi PCBs. Odd-forms rule out 100 percent automation.*

two sizes (300 and 420 mm) in the near future. Incidentally, the automatic placement systems operate non-stop around the clock, but for manual insertion two shifts per day are operated.

Since thermal problems are rarely encountered with the boards assembled at Leuven it is possible to standardize on two basic types: phenolic for hi-fi boards, and a (50 mm by 50 mm) ceramic substrate for hybrid circuits.

At the beginning of the SMD placement venture, a mass of evidence supporting the need for special board materials and layouts, etc., confronted engineers. Fortunately, most proved to be unfounded. By the same token, extensive use of SMDs was expected to reduce overall board size. However, in practice it was found that with the greater-space surface mounting made available, users were demanding even more circuit complexity, requiring a higher number of actual devices. Thus, any potential board size reduction was largely negated.

Board design is implemented on a system supplied by ComputerVision, comprising a host computer, five graphics workstations and a central plotter which not only allows complex modeling and related board-design procedures, but provides the necessary photomask for the in-house bare-board factory. This system enables a so-called paperless production facility, which ensures the highest possible accuracy and precision.

In line with present convention, the standard boards comprise several subboards, each of which is a complete entity in its own right and separated during the assembly of the end-product itself. The possibility of automating the board-separating part of the process is being studied. Some possibilities are the use of waterjet or laser cutting, though the choice of laser cutting may not be economical since it presents the added problem of removing the resulting accumulations of carbon.

Although a great deal of double-sided mounting is done at Leuven (both simple and complex configurations on both the top and bottom side), generally a sandwich construction is used, the "meat" being the copper itself. For some boards (those with decoder and digital functions as used on certain Philips end products) a double copper sandwich is used.

ASSEMBLY AND SOLDERING

Two main hi-fi board production lines serve automatic component placement. Each line, depending on the process stage, uses either Universal or Philips-built MCM machines. Between each workstation are buffer stores where boards are placed in special transport containers, each capable of holding 25 boards. Container routes are determined by cassette-stored software according to the type of end-product. The kanban method is used throughout to achieve just-in-time (JIT) delivery.

The MCM machines can be fitted with alternative software programs according to board type. Each has four software-controlled heads with automatic loading/unloading. A hardware-programmable adhesive-placement module is also employed, which simultaneously places all adhesive dots. Used in conjunction with other MCM-system modules, it is also possible to place components simultaneously. Thus, overall throughput is considerably increased.

An automatic board-assembly procedure is as follows: the bridge-wired components, as well as wire-wrapped horizontal and vertical components, are placed by Universal machines. Then SMDs are placed using Philips' MCM machines, followed by manual mounting of the odd-forms. Philips has worked closely with Universal for a number of years and joint working committees exist between both concerns.

Philips adopted the use of a fully automated in-line production soldering system at Leuven, the so-called "solder-cut-solder" process. Stuffed hi-fi boards are fluxed, pre-heated and solder dipped, after which leads are cut and the board goes on to final fluxing, pre-heat, and dual-wave soldering.

The use of infrared reflow is reserved for hybrids. For SMD assemblies, the dual-wave method is used. Soldering is so closely controlled throughout the process that cleaning is unnecessary. Although initially cleaning was done, extensive studies showed there was little to be gained from the process within production requirements.

Boards enter the ovens for the pre-heat state of Leuven's solder-cut-solder process. When pre-trimmed terminations are adopted, this process will be modified. Total throughput time for a board is about two minutes; solder con tact time is about five seconds. After soldering comes the first "touch up" process (e.g., checking items like solder pads). This is followed by the first of several automatic tests. A reject rate between 15 and 20 percent has been achieved, due mainly to the very tight board specification.

From bare-board "bed-of-nails" testing to final assembled PCB, stage-by-stage testing is performed visually, manually, semiautomatically and fully automatically. Inspection and test runs the gamut from a visual comparison of the circuit (after chip mounting and glue curing) with the layout mask, to full use—of ATE (automatic test equipment). Automatic testing is first done on an in-circuit basis, after which functional testing and alignment is performed.

All boards are mechanically and visually checked, in-circuit tested and all adjustable components are trimmed to their required parameters or signal values. They are then marked with a tracking code, which allows day-to-day production to be adjusted as required.

The boards are then put into containers with all the other component parts of a given end-product. Each container bears a bar code indicating the particular end-product the kits are destined for, and automatically routed to the required on-line manual assembly station on the floor above. Throughout the entire production cycle, extensive use is made of daily batch orders derived from master production schedules through Philips' own proprietary (COPICS) communications-oriented production and information control system.

Reprinted from *Electronic Packaging and Production*, March 1987. © 1987 by Cahners Publishing Company.

43

Becoming the Systems Integrator Before the Year 2020

John A. White

The need for integrated systems in manufacturing and distribution has long been recognized (White 1981; 1982). However, few truly integrated systems have been installed. Instead, organization, information, and automation islands have been created.

Aristotle is credited with having said, "The whole is greater than the sum of its parts." Others have made similar claims, yet little evidence exists to support such claims. Actions taken in the past have involved applying economic justification techniques incrementally. The costs and benefits of the whole are accounted for by summing the costs and benefits of the component parts of the system.

Many disciplines have claimed integrated systems to be within their purview, including industrial engineering. As an example, consider the revised definition of industrial engineering adopted in March 1985 by the Institute of Industrial Engineer's (IIE's) Board of Trustees:

"Industrial engineering is concerned with the design, improvement, and installation of *integrated systems* of people, materials, information, equipment, and energy. It draws upon specialized knowledge and skills in the mathematical, physical, and social sciences together with the principles and methods of engineering analysis and design to specify, predict, and evaluate the results to be obtained from such systems."

The industrial engineer's role in becoming the systems integrator is the central theme of this chapter.

A number of issues related to integrated systems are addressed. Reasons for and against designing and implementing integrated systems are considered. The scope of the systems integrator and the role of the industrial engineer in designing and implementing integrated systems are explored. Hardware and software issues, as well as the challenges involved in designing, selling, specifying, and implementing integrated systems, are examined.

WHAT ARE INTEGRATED SYSTEMS?

Among the various issues that must be addressed regarding integrated systems, the most important one is the definition of the subject. Industrial engineers have been using the term for decades, yet considerable confusion exists regarding what is and what is not an integrated system. At the heart of the issue is the definition of a system. A general definition of a system is as follows: a set of objects or elements, with relationships, between them and/or their attributes, organized in such a way as to achieve a predetermined objective as they interact within their environment.

The elments of the system can be loosely or tightly connected and can operate independently, so long as they collectively achieve a predetermined objective. In general, an integrated system is one that is tightly connected—the individual elements in an integrated system are synchronized.

Examples of integrated systems include an integrated material handling system, an integrated energy distribution system, an integrated monetary system, an integrated accounting system, an integrated personnel system, an integrated manufacturing system, and an integrated transportation system. This chapter will focus on integrated systems in the context of manufacturing, warehousing, distribution, and/or material handling.

Integrated systems have received increased attention because of the emergence of computer integrated manufacturing (CIM). The popularity of CIM has increased to the point that it is in danger of becoming just another three-letter acronym (White 1986).

Unfortunately, too much hype and not enough delivery is being associated with CIM, and its promise, though great, is not being realized. Some organizations are advertising they sell CIM—others promote themselves as turn-key CIM suppliers. The numbers involved in playing the CIM game have grown to the point that one needs a program to distinguish the players. To complicate the matter further, each company plays a different game, by different rules. The only thing the players have in common is they use the same label (CIM) to describe what they are doing.

A term-by-term examination of CIM, in reverse order, is useful. CIM applies to manufacturing—however, manufacturing is defined sufficiently broadly to include production and distribution activities. All values added and support functions of manufacturing are included. Likewise, both direct and indirect labor activities are embraced by CIM (the thrust of CIM is directed more toward indirect labor activities than toward direct labor activities).

The scope of CIM in manufacturing is both horizontal and vertical. The latter includes the functions of product, process, and schedule design, whereas the former includes production, assembly, materials handling, packaging, quality control, production control, maintenance, etc. CIM includes both planning (CAD and CAPP) and execution.

Second, CIM depends on integration. The synergistic benefits of systems integration that result in two plus two being greater than four is one of the promises of CIM. However, the term integration must be defined very broadly—it includes the interface and coordination of functions, the linkages of physical components, and the information handoffs (both vertically and horizontally) throughout the entire organization.

Also, the breadth of integration requires more than integrating two or three machines or workstations—it requires more than integrating two or three departments. CIM, when implemented to its fullest, will incorporate the total manufacturing enterprise, including multiple production plants, suppliers, and customers through procurement and accounting processes.

The integrating aspect of CIM is necessary, but not sufficient. As an example, the just-in-time system used by Toyota relies on KANBANS to achieve integration. A JIT manufacturing system can be highly integrated and not involve the computer at all (although most do rely on the computer, at least behind the scenes).

Third, CIM depends critically on the use of the computer to perform the integrating function. Although CIM includes the physical integration of hardware components, integrating the information subsystems is essential to CIM. For example, one can achieve CIM without automating production or materials handling. The production processes can be labor-intensive in a CIM environment, so long as they are integrated via the computer.

Unfortunately, many feel it is necessary for computer integration to occur using a highly centralized software architecture. Alternatively, one can network a number of microcomputers to achieve computer integration. Also, many fail to consider using a combination of people and computers to integrate manufacturing. CIM does not require that all information linkages be performed via computer. One should design the system such that certain linkages can be performed either manually or by computer (White 1985).

CIM is not a panacea and is not the right answer for everyone. It requires a major commitment of resources and considerable patience from management.

WHY HAVE INTEGRATED SYSTEMS?

Despite all of the rhetoric concerning the need for integrated systems, few shining examples exist. There appear to be many reasons for failing to design and implement integrated systems. Some of the more common reasons are as follows (White 1986).

First, designing and implementing integrated systems is not easy. A detailed understanding must exist of systems requirements and interactions. Many details must be considered—nothing can be left to chance. System complexity grows exponentially with the number of operations.

Second, designing and implementing integrated systems is a radical departure from tradition. Since the early 1900s, the approach has been to break the system into its basic components, analyze the system in terms of its operations, inspections, transportations, storages, and delays, use division of labor, design hierarchical organizations, and form cost centers. An emphasis on compartmental or segmentalist thinking has dominated.

Third, there are few apparent rewards in designing and implementing integrated systems. Managers are responsible for managing individual organizational units "by the bottom line." Few, if any, incentives exist to use team or systems objectives. Individual, rather than team, performance is rewarded, despite claims to the contrary.

Fourth, "insurance" is preferred, often in the form of inventories, additional space, redundant equipment, and excess personnel. Many believe a highly integrated system has no cushion for error; they believe a finely tuned "integrated engine" will experience considerable downtime. Also, if something goes wrong, the person to blame can be easily identified (White 1986).

However, insurance should not cost more than the "rare event" one is seeking to avoid. Often the cost of the "insurance" cannot be justified.

Minimizing risks seems to be preferred to maximizing gains. As an example, high inventory costs are preferred to costs of stopping production. As a result, large inventories are used as an "insurance policy" under "just-in-case" management.

Fifth, many organizational barriers must be overcome. The organization chart tends to create boundary lines. Individual managers have difficulty sacrificing for the benefit of the whole.

Sixth, the concept of integrated systems is not well understood. It has different meanings for different people, depending on their background and experience. For example, to a hardware supplier, manufacturing systems are integrated if they fit together physically. To a computer supplier, information systems are integrated if they share a common data base and provide real time control. To a design engineer, integrated design systems require functional integration of product, process, and schedule design (White 1985).

Each person views systems integration differently and each believes that systems integration has been achieved by satisfying their view. Furthermore, each tends to think of an integrated system as a pipeline, rather than as a network of pipes. Consequently, integrated systems are viewed as tightly connected, inflexible, and very risky.

Seventh, a leadership void exists. In those cases where systems integration has been attempted successfully, a champion existed. A strong leader emerged and made a strong commitment to an integrated approach. Without a leader, there is little support or commitment for systems integration.

Eighth, few success stories and numerous horror stories exist. As Skinner noted,

"The new, computer-based 'total systems' approaches to production management offer the promise of new and valuable concepts and techniques, but these approaches have not overcome the tendency of top management to remove itself from manufacturing. Years of development of 'the factory of the future' have left us each year with the promise of a great new age in production management that lies just ahead. The promise never seems to be realized. Stories of computer-integrated manufacturing (CIM) and new automated equipment disasters are legion; these failures are always expensive, and in almost every case management has delegated the work to experts (Skinner 1985)."

Despite the highly publicized systems that have been successfully implemented by some leading firms, the belief continues that systems integration is expensive, risky, and complex.

Ninth, resources are limited. In addition to a scarcity of capital, there exists a critical shortage of qualified people. The qualifications of a systems integrator should include not only technical skills, but also management and financial skills and an understanding of operational requirements. Among the technical skills needed is competency in developing control systems, for it is the control that tends to determine success versus failure.

Tenth, some individuals are threatened by the use of integrated systems. In many firms direct labor cost represents less than 15% of manufacturing cost. Material cost, on the other hand, accounts for 50%, and indirect costs account for the remaining 35% of total manufacturing cost. Instead of targeting direct labor costs for reduction, integrated systems are aimed at reducing indirect costs. Because indirect labor is used to fill the "gaps" between operations, and integrated systems eliminate gaps, they will have an impact on indirect jobs.

Many excuses are given for not designing and implementing integrated systems, but there are three good reasons for doing so. First, the promise is great. Although delivery has been poor, the promise of integrated systems has not diminished.

Second, designing and implementing integrated systems is feasible. The development of computer hardware and software allows information to be provided accurately and in a timely fashion for decision making. As a result, management's span of control has been expanded. A segmentalist structure is no longer required.

Third, the survival of manufacturing in some countries may well depend on the use of integrated systems. Eliminating the "fat" in today's manufacturing and distribution systems is essential for some nations to be competitive.

ATTITUDES CONCERNING AUTOMATION

Automation has been applied in a variety of industrial settings for more than 20 years. For example, automated storage and retrieval systems, robots, automated guided vehicles, automatic identification, and numerically controlled machine tools have been used extensively in production environments.

Yet, in spite of a very lengthy record of "proven" successes, "horror stories" about implementing automation continue to arise. Reports are not uncommon of schedule slippages of months and years, coupled with costs doubling and tripling estimates. Further, "success stories" exist involving the use of fairly conventional systems. Claims and counterclaims abound as to which is the "best" approach—automated or conventional systems. In embracing a "back-to-the-basics" approach, some strongly oppose making investments in automation. Why is that? Ten observations about the current situation follow (White 1986).

First, automation technology should not be blamed. Rather, the design, implementation, and use of the overall system should be examined before judging culpability. With exceptions, seldom will the technology not work. Usually, either the scope of the application was too broad, the wrong technology was applied, the technology was applied incorrectly, or the technology was not being used properly.

Second, too many automated systems are designed "for show, not for dough." Too little attention is paid to the tangible benefits derived from capital investments in automation. Well-designed automated systems can be justified economically. However, rather than invest in front-end design to ensure that the resulting investment will pay dividends, many buy more hardware or software instead. Automation is often overpromised and underdelivered (White 1984).

Third, rather than being designed on the basis of requirements, such systems are often designed on the basis of the state-of-the-art solutions in existence or, in some rare cases, in anticipation of advances in the state of the art. They are "solution-driven" solutions, instead of "requirements-driven" solutions.

Fourth, the real system requirements are not well defined. The supplier is often asked to do "free design" and quote a price for satisfying a user's needs, without those needs being specified quantitatively in terms of performance requirements. In a cost-competitive environment, less attention to detail results, and critical flaws creep into the design.

Fifth, the most significant lesson learned is that the primary benefit of automation is the systems discipline it imposes. The AS/AR is a highly disciplined technology in comparison with the usual conventional system: aisle-captive S/R machines are served by an input/output system. Because of the computer control, the machines can perform dual command cycles created by matching storage locations and retrieval locations for maximum throughput. Slave pallets are used to achieve uniform "footprints" in the close tolerance environment. To the extent that the same level of discipline is obtained using more conventional approaches, justifying automation is very difficult. It is very difficult for automation to compete against a well managed, highly motivated work force, but very few well managed, highly motivated work forces exist.

Sixth, the "bottleneck" is due to software, not hardware. The control system is often designed with an excess of control. Rather than err by undercontrolling the system, the designer tends to overcontrol. In this case, the KISS principle means "Keep Insisting on Simple Software!"

Seventh, the system being procured becomes the vehicle for addressing requirements that are tangential to the primary system. Examples include adding to the AS/RS control system, purchasing, receiving, shipping, accounts payable, accounts receivable, inventory control, and other functions. The result is a major software job undertaken by a firm whose area of expertise represents less than half the effort. In general, suppliers without the requisite software experiences and qualifications are trying to satisfy users' gluttonous appetites for software sophistication. "Software-smart suppliers" generally are not "material handling hardware-smart" and vice-versa.

Eighth, the user lacks the discipline to "freeze the design" following the procurement decision. As a result, requirements continue to be defined during the system design effort. Design changes are both expensive and disruptive to the delivery schedule. Also, changes made during the design and implemention process are "spur of the moment" responses to an exception condition.

Ninth, the computer hardware to be used is generally specified before the computing requirements are defined. As a result, the computer is often undersized in the initial quotation, but the error is not discovered until the system is installed and fails to handle the transaction requirement. Suppliers are required to "nail down" the computer to be used too soon in the design process.

Tenth, unreasonable, unnecessary, and expensive requirements are placed on the system to "never fail." Reliability, availability, and maintainability values are specified arbitrarily, rather than by considering the cost impact of purchasing redundant computers. The need for "non-stop" and "hot backup" capabilities must be considered carefully. The cost of having "full system recovery" within x minutes or seconds of any unscheduled computer stoppage should be assessed before arbitrarily specifying a value for x.

DESIGNING INTEGRATED SYSTEMS

In designing integrated systems, a delicate balance must be maintained in terms of the degree of risk one is willing to take with respect to leading-edge technologies versus proven technologies. Because the system is being designed for the future, incorporating yesterday's proven technology might be more risky than incorporating tomorrow's unproven technology.

A trade-off is required between the possibility of technological obsolescence of yesterday's technology and the possibility that tomorrow's technology might not be available on time, it might not function as required, and it might cost more than predicted. Perhaps because of the risks of the unknowns associated with future technology very few integrated systems have relied on unproven technology.

Top-down design and bottom-up implementation are important. Designing from the top will ensure that the system is compatible with and supportive of the firm's business objectives. Bottom-up implementation ensures that individual components can operate independently. However, because of the top-down design, their operations will be synchronized. It is this "loose, tight" coupling of integrated systems that seems to be essential in today's environment.

The systems designer must think "big," yet deal with the "little." It is a challenge to develop simple solutions that will stand the test of time. The work simplification philosophy (eliminate, combine, and simplify) must be applied at all levels: operations, functions, departments, and the total system.

A number of challenges face the designer of integrated systems. Among those suggested by the previous discussion are the following: keeping the solution simple and "requirements driven"; involving the ultimate user in the design process; integrating the various design functions; designing for automation, especially when automation is not immediately justifiable; avoiding paying for excess insurance; achieving systems discipline; and maintaining a long-term focus while being responsive to short-term needs. The designer must also avoid the "not invented here" syndrome.

SELLING INTEGRATED SYSTEMS

In selling integrated systems to upper management, a number of lessons have been learned. Among those reported previously are the following: design the whole, sell the whole, and implement the parts; be sure you know what you are doing; do your homework; thoroughly analyze the "do nothing" alternative; involve the user; analyze, analyze, analyze; sell, sell, sell; be realistic and thorough in estimating costs and benefits; ensure accountability; and find a champion (White 1985).

SPECIFYING INTEGRATED SYSTEMS

For many, a significant change is needed in the way in which integrated systems are procured. In particular, many are trying to use detailed design specifications in procuring integrated systems. At present, performance or functional specifications are needed. When a firm purchases an industrial truck, a conveyor, or a palletizer, for example, detailed design specifications are used routinely and written at the component level. However, when an integrated system is purchased, the rules are quite different. Specifically, when procuring an integrated system, the specification must emphasize on *what, when, where,* and *why*, rather than *how, who,* and *which*.

As firms obtain experience in implementing integrated systems and as increased standardization occurs in data bases, protocols, and communication interfaces, detailed design specifications will be used to procure integrated systems. However, until that time occurs, functional specifications will continue to be the preferred procurement instrument.

IMPLEMENTING INTEGRATED SYSTEMS

Successful integrated systems must be planned. Implementation includes installing and debugging the system, training employees, and auditing the system to ensure that it meets requirements and is used properly.

As noted previously, a phased implementation plan should be developed. As with most major changes, a transition from a segmentalist approach to an integrated approach will meet with opposition. Consequently, one must protect against premature rejection of the system from start-up problems that might arise.

Few people can maintain a long-term focus and simultaneously avoid making mistakes in the short-term. Yet that is exactly what is required, because a "greenfield" system is seldom feasible. Rather, more commonly an existing system must continue to function while an improved system is being designed and installed. A phased implementation plan is required to avoid stopping production.

THE SYSTEMS INTEGRATOR

The scope of the systems integrator varies considerably, depending on one's perspective. To be comprehensive, a full systems integration must represent a union of the various views. An integrated design process must be used, the hardware must "fit together," and the information requirements must be met.

Because of the complexity of systems integration, a team approach, rather than an individual approach, is needed. Furthermore, many different functions should be represented on the team. No individual can be the systems integrator.

However, one individual must be responsible for seeing that the team performs effectively.

For a manufacturing system, the following functions should be represented on the design and implementation team: manufacturing, assembly, product design, process design, production planning, materials management, distribution, quality control, information processing, and industrial engineering. The manufacturing representation should include direct labor representatives to benefit from the experience and perspective of the ultimate system user.

THE INDUSTRIAL ENGINEER'S ROLE

By the year 2020, the industrial engineer will probably not be the systems integrator. However, the industrial engineer will play a critical role in designing and implementing integrated systems. We do not expect that any one person will be the systems integrator, but rather a team approach will be required. Furthermore, there will be instances in which the likelihood of success will be increased by having some other discipline serve as the team leader.

The role of the industrial engineer in designing and implementing integrated systems depends on both the abilities of the individual and the context in which systems integration occurs. If a depth of understanding of the underlying manufacturing processes is required, then it might be more appropriate to have as the team leader someone with that knowledge. However, because of the breadth of exposure to the organization common to the industrial engineer, the team leader probably would often be an industrial engineer.

Depending on the scope of practice for industrial engineering in a firm, it might not be possible to convince management that the industrial engineer is qualified to be a member of the systems integration team, much less the leader. Furthermore, such individuals may not be qualified to perform systems integration. This particular concern is due to the differences that exist in the job title, industrial engineer, and industrial engineering curricula in colleges and universities (White 1985).

Suggested Readings

Skinner, W. 1985. *Manufacturing: The Formidable Competitive Weapon.* John Wiley & Sons, Inc. New York.

White, J. A. 1981. Material handling: its vital role in the '80s. *Proceedings of the 3rd IE Managers Seminar.* April. Institute of Industrial Engineers. New Orleans, LA.

White, J. A. 1982. The automated factory and integrated systems in the '80s. *Proceedings of the 4th IE Managers Seminar.* March. Institute of Industrial Engineers. Chicago, IL.

White, J. A. 1984. Design for automation. *Modern Materials Handling.* January: 29.

White, J. A. 1985. CIM provides the missing link. *Modern Materials Handling.* January: 29.

White, J. A. 1985. Selling integrated systems. *Modern Materials Handling,* March: 29.

White, J. A. 1985. Stregthening the role of industrial engineering in education and society. *Industrial Engineering.* December: 46-53.

White, J. A. 1986. The industrial engineer as the systems integrator. *Proceedings of the 8th IE Managers Seminar.* March. Institute of Industrial Engineers. Washington, D.C.

White, J. A. 1986. Time to reevaluate your insurance. *Modern Materials Handling.* May: 25.

White, J. A. 1986. Impediments to systems integration. *Modern Materials Handling.* July: 23.

White, J. A. 1986. CIM: another TLA? *Modern Materials Handling.* September: 25.

White J. A. 1986. Warehousing in the New Age. *Proceedings of the 7th International Conference on Automation in Warehousing.* October. Institute of Industrial Engineers. San Francisco, CA.

White, J. A., and J. M. Apple, Jr. 1985. Material handling requirements are altered dramatically by CIM information links. *Industrial Engineering.* 17(2): 36-41.

Reprinted from *1986 Fall Industrial Engineering Conference Proceedings.*

John A. White is Regents' Professor of Industrial and Systems Engineering, Georgia Institute of Technology, Atlanta.

44

Making Manufacturing More Effective By Reducing Throughput Time

Kelvin F. Cross

Manufacturing is "in." There are numerous articles and texts about symptoms, problems, and solutions in manufacturing. Special emphasis has been placed on strategy, automation, and various Japanese techniques. Yet the question remains, "Specifically, what should my company do?" The right answer lies in doing what's right for each individual business, not because it's in vogue, fun to do, or because engineering likes it.

However, all manufacturers share some similar characteristics that provide the foundation for immediate process improvement. The objective of every manufacturing operation is to help the business make money. Manufacturing helps the business accomplish this by supporting sales through flexibility, delivery, quality, and cost. The company's market niche and/or the niche of specific products determine how the manufacturing operation can best help the business make money. For instance, the flexibility to customize products could be more beneficial to sales than the ability to manufacture the cheapest products. Accordingly, the manufacturing process must be designed to be flexible enough to customize products.

Also, it must be recognized that, over time, the way to make money changes. Products mature, markets mature, and new niches are created. As the business evolves, so must the manufacturing process. One force in particular drastically impacts the nature of the business and its manufacturing operation. That force is the customer, and it becomes a more powerful force as the market increasingly becomes a buyer's market, as it does for every product.

THE SIGNIFICANCE OF A BUYER'S MARKET

Essentially, in a seller's market the customer may accept a high price for a long or erratic lead time for delivery of a standard product of marginal quality. In other words, the company can sell whatever it builds. The short-term manufacturing strategy is simple: increase output. Nothing else matters.

Most manufacturers are not so fortunate as to be in a seller's market, and when they are, it isn't for long. Inevitably, a seller's market is transformed into a buyer's market since every business wants to be in a seller's market. When such a market develops, numerous businesses go into it. Eventually, more product is produced than is demanded by the market (witness calculators, microcomputers, video games, etc.). Worldwide competition in most industries is decreasing the longevity of the seller's market phase and intensifying the buyer's market phase (as in steel and automobiles). In a buyer's market everything matters: cost, quality, delivery, and flexibility, although one of these four strategic attributes of manufacturing is usually emphasized to support the company's market niche.

Also, in a buyer's market, process innovation may become more important than product innovation.[1] As a product matures, the rate of product innovation declines and is surpassed by the rate of process innovation. More engineering effort may go into designing the process than into further redesigns of the product. For instance, as the product life cycle matures and product design and production volumes stabilize, it may become possible to automate. Again, the manufacturing process must do more than simply make the product. It must do it effectively in terms of cost, quality, delivery, and flexibility.

AUTOMATION IN PERSPECTIVE

Recognizing the need for process innovation, manufacturing management may desperately grasp for the "automation solution." However, automation in and of itself does not necessarily help the business to make money. In a situation where product customization sells, any automation that would inhibit flexibility could hurt the business. Manufacturing might be highly efficient, but not effective. The key is to determine what is effective.

Too often automation is applied as the solution to problems that either don't exist or aren't understood. An effective manufacturing operation results from an understanding of manufacturing's role in the company and from organizing the process properly *prior* to automation. This point was made recently by James

Harbour, an auto industry consultant, in the *Wall Street Journal* (May 13, 1986):

> New technologies haven't made any massive improvement in [the auto industry's] productivity. So far, they have turned out to be more show than substance. Auto makers could make bigger gains by scrapping outmoded work rules, managing their workforces better and handling their parts inventories more efficiently.

Even in the introduction of information systems, the same principle applies. An article in *Fortune* (May 26, 1986) by William Bowen, called "The Puny Payoff From Office Computers," stated:

> Ideally, you should change the way work is done before you put in new equipment. Nancy Bancroft, manager of office systems consulting at Digital Equipment Corp., advises prospective customers to scrutinize their procedures before deciding what to buy. "If people are doing the wrong things when you automate," she says, "you get them to do the wrong things faster." Paul Strassmann, former vice president of the information products group at Xerox and now a consultant, advises: "Automate only after you simplify."

Again, the focus must be on the design and operation of the process for overall effectiveness. Only after the process is simplified and functions well manually should automation be introduced. If this procedure is followed, automation can be pulled in where it makes sense rather than pushed in because it's the "latest and greatest."

MANUFACTURING STRATEGY IN PERSPECTIVE

Development and implementation of a sound manufacturing strategy can be critical to the success of any manufacturer. However, a manufacturing strategy is not essential for beginning many basic improvements. Most manufacturers have the potential for strategic process improvement without relying on detailed strategic direction.

For any manufacturer it makes sense to build only what will be sold. Unfortunately, the customer may expect delivery in less time than it takes to build and deliver it. In a hypothetical scenario (see Figure 1, top illustration), the customer expects one month delivery, and it takes a month to deliver an order from finished goods inventory. Therefore, all purchasing and build time must be done prior to receiving real customer orders. This means that the purchase/build activities must of necessity be done to an internally generated forecast.

In an ideal situation (see Figure 1, center illustration), the purchase, build, and delivery time would all be accomplished within the lead time dictated by the marketplace. Nothing would be done unless there is a customer. Job shops and aircraft and ship manufacturing may include purchase time in their quoted

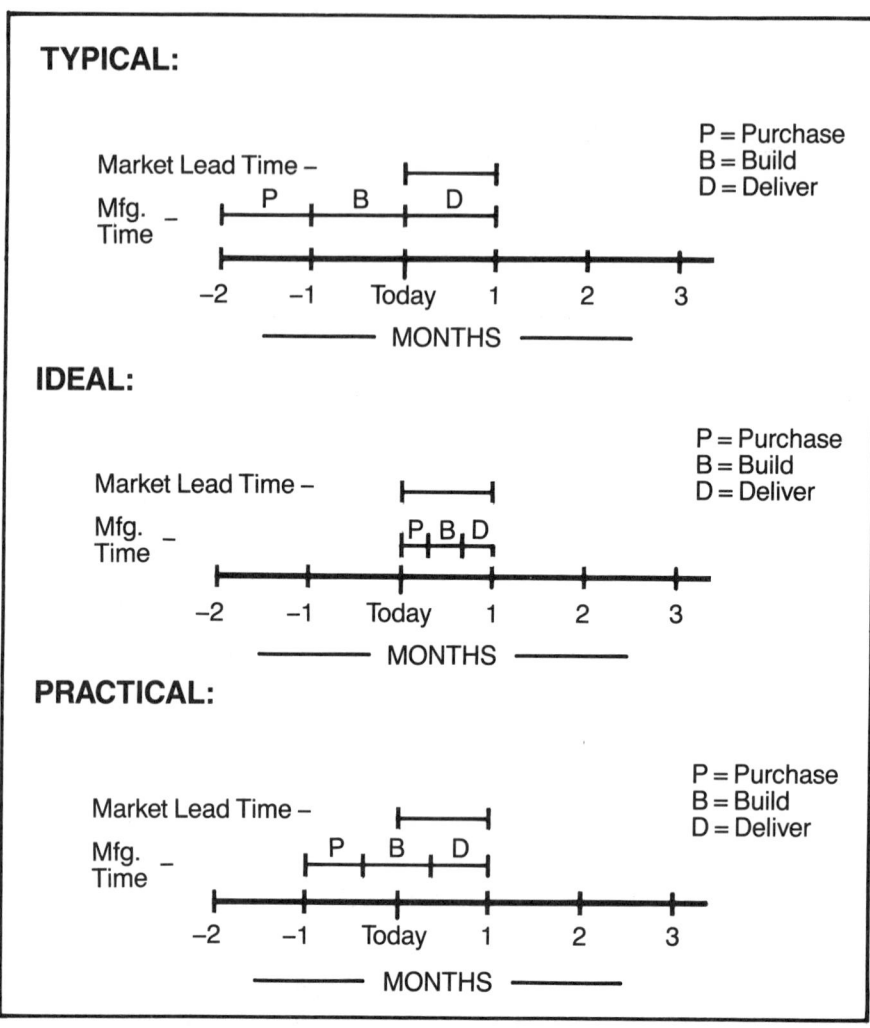

Figure 1. *Three purchase, build, and delivery time scenarios.*

delivery times, reflecting a true build-to-order environment. But for most repetitive manufacturers, this scenario is impossible.

Regardless of industry, the purchase lead time is, in many instances, beyond the control of the company. However, the build and deliver time is within the company's control. Therefore, the practical objective for repetitive manufacturers is to keep the purchased materials in the raw state as long as possible so that work-in-process (WIP) and finished goods inventories are not tying up raw material that could be used to build and deliver a customer order. The practical objective (see Figure 1, bottom illustration) is to shorten the build and deliver cycle as much as possible. Products can then be built and delivered to a customer order within the market-demanded lead time.

Although seasonal industries may have difficulty with this objective, a trade-off must be recognized. Does it make more sense to carry capacity or finished goods inventory? Regardless of capacity/inventory strategy, it does make sense to shorten the manufacturing cycle time. Typically, only 5 percent of the total cycle or throughput time is devoted to adding value. In many cases, the product is waiting to be worked on 95 percent of the time.

An unnecessarily long throughput time contributes to poor strategic performance regarding cost, quality, delivery, and flexibility.

Cost is directly affected by inventory. Rapid throughput results in minimized WIP and finished goods inventories. Carrying costs are reduced and cash flow is improved. Also, the cost of reworking WIP and finished goods to comply with an engineering change order or a specific customer's requirements is eliminated.

Quality is enhanced by the rapid identification and correction of problems before too many defective subassemblies and/or finished products are built. Rapid throughput means that the emphasis is placed on building the product right the first time.

Queues and waiting times are eliminated by an emphasis on rapid throughput time. As the build cycle approaches the value-added time, it becomes more predictable and dependable. Delivery is then improved by ensuring both timeliness and reliability.

The definition of "flexibility" isn't always clear. Does flexibility mean the ability to handle product changes? process changes? product volumes changes? product mix changes? Or does it mean the ability to customize products? In some cases flexibility is merely an excuse for chaos on the manufacturing floor. In any case, the dramatic reduction of throughput time can only help. Flexibility, by most definitions, is achieved by the ability to build to order. And this is attainable only with a rapid and predictable build/deliver time. Rapid throughput permits flexibility and eliminates chaos.

It follows that basic strategic improvements can begin from one simple tenet: *reduce throughput time.* In fact, flexibility, through build-to-order manufacturing, *requires* rapid throughput time.

WORK-FLOW PRINCIPLES

Rapid throughput can be achieved by the integration of group technology/work cells, job enrichment, optimized production technology (OPT) principles,[2] and Just-in-Time manufacturing techniques. Without elaborating on these general approaches, there are specific related work-flow principles that provide clear direction for reduced throughput time and immediate process improvement. The remainder of this article is devoted to examining those principles.

Establish a Product Orientation

A production department can be organized by process or by product. In a process-oriented department each step of the process is consolidated, usually as a work center. For instance, all of the cutting machines are grouped together as a work center, feeding all of the drilling machines in a subsequent work center. The underlying principle is the maximization of equipment utilization. Unfortunately, this is accomplished at the expense of throughput time.

As depicted in the top illustration of Figure 2, a process orientation typically creates a convoluted network of unmanageable flows. The material can flow through any one of the available machines at each processing step, creating a bewildering network of possible paths. Therefore, in a process-oriented flow it is not unusual to have large lot sizes, buffers of material, and significant work-in-process. Sophisticated control and tracking systems are established to deal with these problems, usually with questionable results.

An effective alternative is to establish a product-oriented flow. "Product" does not necessarily mean the end product of the company, but rather the product of a particular process. For example, within a computer manufacturing plant, a circuit board is the "product" of the circuit board assembly department. Likewise, a cable is the product of the cables assembly department. However, within those departments it is advantageous to establish "production modules" as depicted in the bottom illustration of Figure 2. The number of possible paths are dramatically reduced by segmenting the operation into three "minifactories."

The production module concept derives from the integration of three concepts in manufacturing: (1) group technology, (2) the work-cell concept, and (3) job enrichment.

Group technology is defined as the identification and bringing together of related or similar components and processes in order to take advantage of their similarities in design and/or manufacturing. In other words, products/assemblies are classified according to their similarities in manufacturability.

A work cell, or production module, is then designed to specialize in the production of a particular family of products/assemblies. This specialization provides a product focus and enhances the capability to respond to product revisions and new products. Product changes are facilitated by the ability to communicate easily within a small group. Also, new product introduction problems and other changes only affect one group and not the entire operation as in a process-oriented layout.

The production module establishes a job design that is desirable from the standpoint of human factors. A semiautonomous group of workers is assigned to

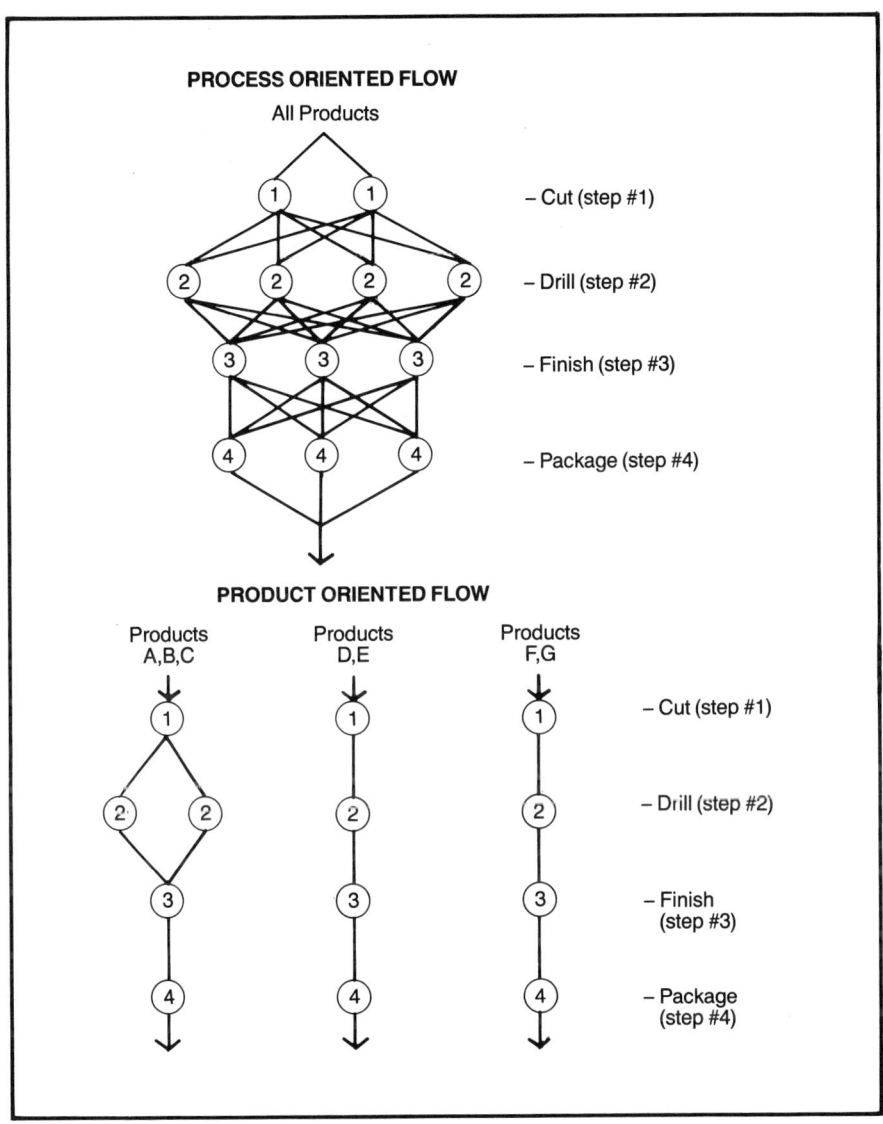

Figure 2. *Process and product orientation.*

a production module, and the group members are measured and rewarded as a group. This creates a teamwork situation in which the team is responsible and accountable for its production quality and quantity. People's jobs become enriched as they acquire proficiency at more than one job and thereby become capable of building a whole piece or product, not just a small portion of it. The result is that there is a more tangible and meaningful relationship between their work and both their product and their product's impact on the company.

Eliminate Buffers

In many cases throughput time is slowed by the means used physically to move the product. Whether intentionally or not, the operation of some means of material transport depends on buffers. The addressable conveyor, or transporter system, is an example. When the material flow is examined carefully, it becomes obvious that throughput time is lengthened.

Each movement of the product from one step to the next involves loading onto the conveyor, unloading to the flow rack, unloading the flow rack back to the conveyor, and unloading the conveyor onto the workbench. Even assuming that the load/unload and conveyance times are negligible, the fact that the product must be held means that time will be lost. As illustrated in Figure 3, a simple three-step, twenty-minute operation could easily take hours in actual throughput time.

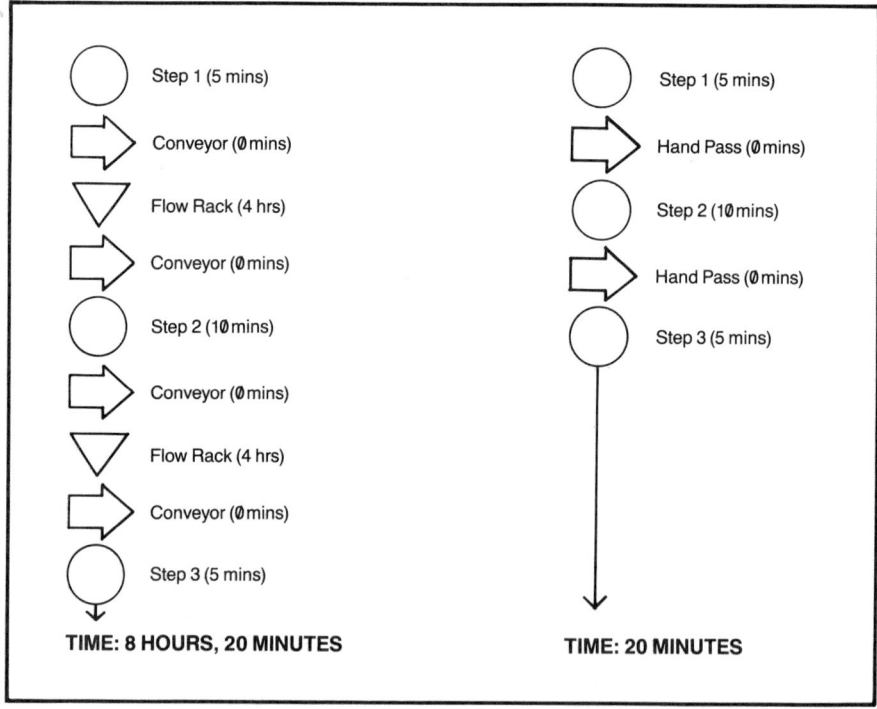

Figure 3. Eliminating buffers.

However, if the operation were structured to allow a handoff from one step to the next, throughput time would nearly equal the hands-on, value-added time. In addition, dramatic reductions in floor space and a reduction in confusion could be achieved.

Establish One-At-A-Time Processing

It is advantageous to process work as expeditiously as possible. Achieving this objective might dictate that work units be processed individually rather in batches.

For instance, assume (see Figure 4) that the first processing step receives 25 units that take 5 minutes each to process. (A unit could be product in a factory or paperwork in an office.) Assume the second step also requires 5 minutes per unit. If the units are processed as a batch of 25 at both steps (i.e., all 25 units are completed before they are sent to the next operation), it will take 250 minutes, or just over 4 hours, until the units reach step 3. Processing the units individually, the first unit would reach step 3 in 10 minutes, and all 25 units would reach step 3 in 130 minutes, or just over 2 hours.

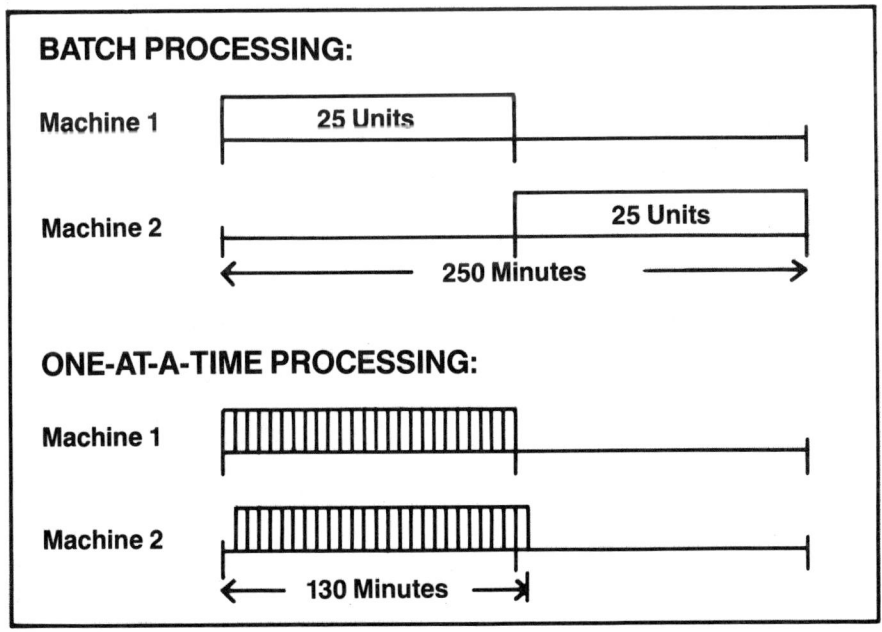

Figure 4. *Batch processing and one-at-a-time processing.*

Balance the Flow to the Bottleneck

Understanding the bottleneck, and balancing the flow through it, is one of the major OPT principles defined by Eliyahu Goldratt.[3] It makes no sense to pass more material through any operation than the subsequent operation can handle.

For instance, Figure 5 displays a capacity to output 125 units/day at the first step. However the fifth step can only handle 80 units/day. While every other step can handle more than 80 units, it is ineffective to do more than 80. Doing more than 80 will result in an accumulation of inventory prior to the fifth operation and, therefore, a reduction in throughput time. The output of the

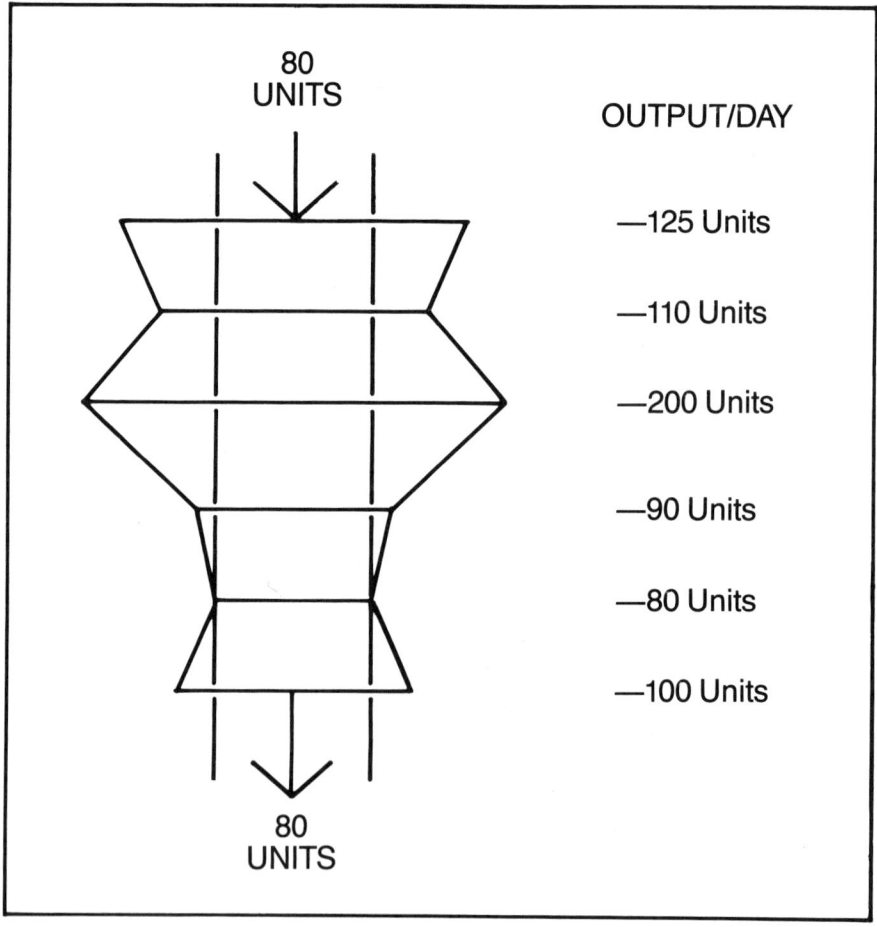

Figure 5. *The impact of bottlenecks.*

department as a whole is gated by the bottleneck, so the department is only capable of an 80 units/day output.

In real situations, bottlenecks are not usually fixed. They are typically dynamic, constantly changing in magnitude and location. The mix of product, the lot sizes, and the set-up times between lots all have a major impact on the bottleneck. The key is to be aware of the importance of the bottleneck and recognize that the process as a whole is regulated by it.

Minimize Sequential Processing

Sequential processing can create two problems that lengthen throughput time. First, the operations are dependent on each other and therefore gated by the slowest step. Second, no one person is responsible for the whole assembly or has a whole job.

Assume an assembly requires 20 minutes of labor to be put together. (See Figure 6.) In the sequential process, each individual gets one-quarter of the work, or 5 minutes per unit. But let's assume that the third person always takes 6 minutes per unit. It is irrelevant whether the reason is that he or she is slow or that the process is not properly balanced by engineering. In either case, the slowest step becomes the gate and limits production to 10 units per hour. In real situations the problem might not always be a particular person but rather a floating bottleneck created by fatigue, product mix, etc. Sequential processing permits this kind of problem to occur.

While sequential processing is an inevitable part of manufacturing, in many cases it is not essential. Sequential steps can at times be replaced with parallel processing. In the example, if each of the four people do the whole assembly, output per hour could be increased by 15 percent. The slowest person could still work at the same pace and take 20 percent more time, but this would not affect the others' productivity.

For some manufacturers, material prep and/or kitting is a part of the manufacturing process that unnecessarily creates an additional sequential step. Often it is possible to store components at the workbench and give the assembler the instructions to pick (and prep) the parts as required.

From a human factors and quality standpoint, parallel processing is beneficial. Each worker can take pride in building a whole assembly and therefore will accept responsibility for productivity and quality. It then becomes possible to eliminate another sequential step: inspection.

Schedule Effectively

Throughput time is directly related to scheduling (or the lack thereof). On production lines that handle more than one assembly, the order in which assemblies are produced is significant.

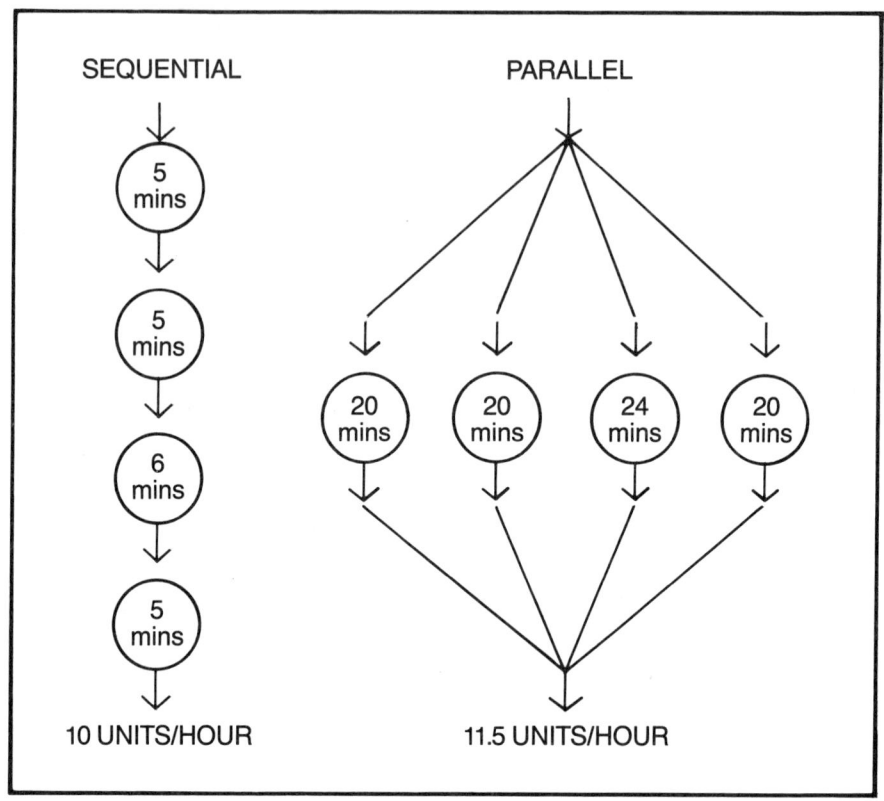

Figure 6. *Sequential and parallel processing.*

Figure 7 assumes that product A takes 10 minutes on machine 1 prior to being processed for 5 minutes on machine 2, and that product B requires only 5 minutes on machine 1 prior to taking 10 minutes on machine 2.

Schedule 1 depicts starting job A first. Job A takes 10 minutes before transfer to machine 2, which is when job B can begin. The completion of both jobs takes 25 minutes.

Schedule 2 illustrates what happens if job B is started first. Both jobs are completed in 20 minutes. Just by sequencing the jobs properly, throughput time is reduced by 20 percent.

In real situations, scheduling is not simple. Lot sizing and set-up times vary. There are numerous jobs on numerous machines. All the more reason why scheduling should not be ignored.

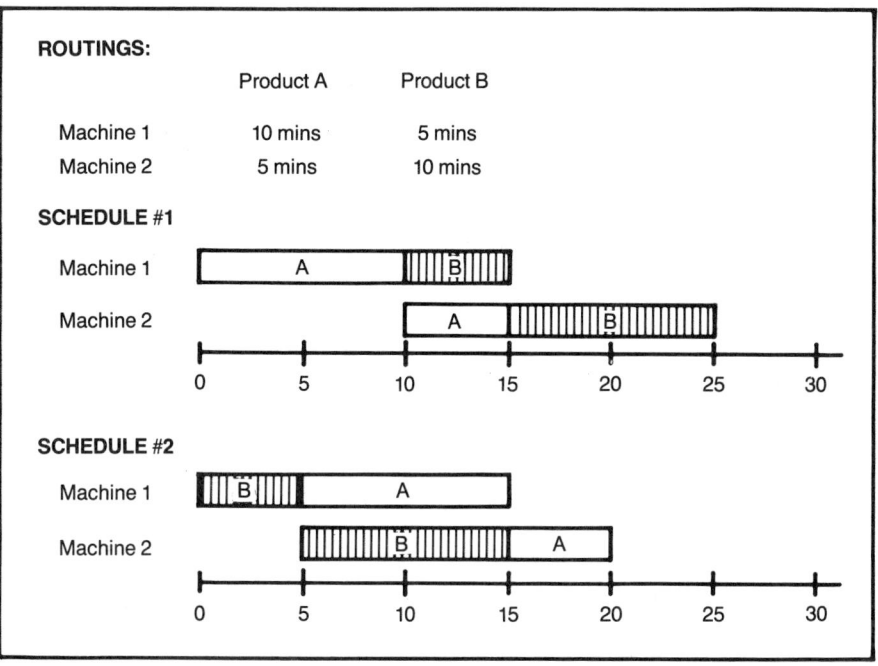

Figure 7. *Effective scheduling.*

Minimize Multiple Paths

Multiple paths create confusion and a balancing/scheduling nightmare. Buffer inventories are typically employed to relieve the symptoms and alleviate the need to address the problem. Unfortunately, buffer inventories needlessly add to throughput time.

In the illustration on the left in Figure 8, it is easy to see why inventories might be used to eliminate the complexity of scheduling and balancing. The operation depicted produces an electromechanical assembly with a mechanical assembler, an electrical assembler, an inspector, and a tester. The 100 units entering the operation do not all follow the same path. At the very beginning, there is a 60-40 split, with 40 of the 100 bypassing mechanical assembly. Those 40 units go to electrical assembly, along with 30 units from mechanical assembly. Subsequently, 50 units bypass inspection and go directly to test, while the other 50 go through inspection. In actual situations, the quantity or percentage of product that travels each route could change hourly.

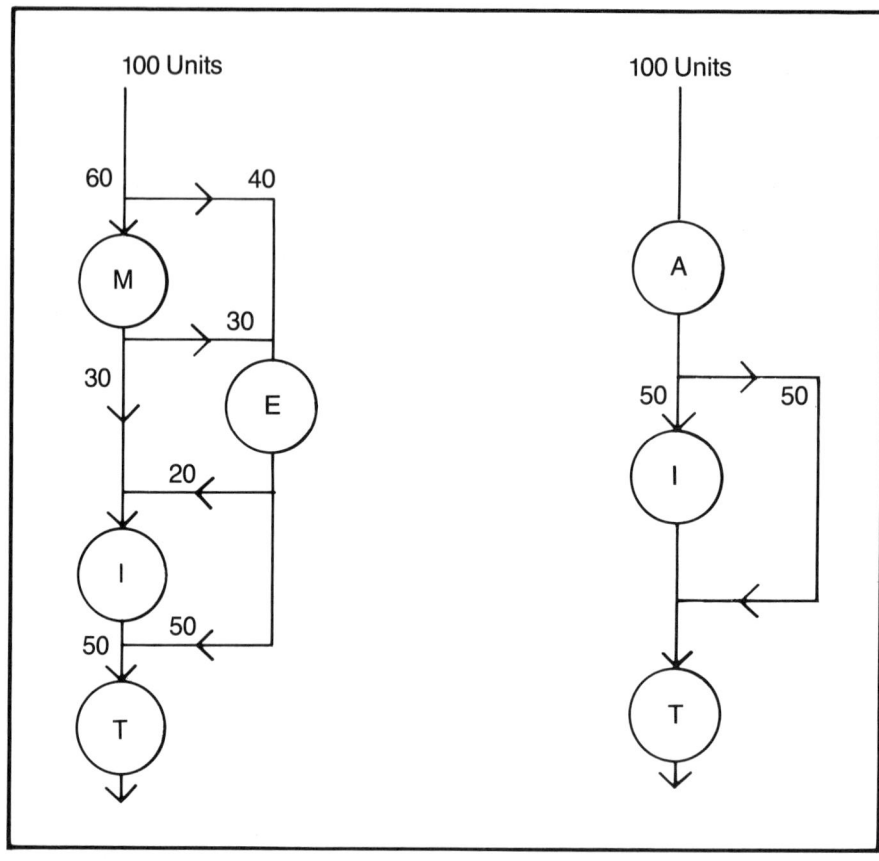

Figure 8. *Minimizing multiple paths.*

The process on the right in Figure 8 does exactly the same job in a greatly simplified manner. The simplification is accomplished by using assemblers who are responsible for both mechanical and electrical assembly. Elimination of the distinction between mechanical and electrical assemblers immediately makes the process more manageable. There is only one bypass loop to be reckoned with. The need for inventories or scheduling is minimized, and the need for tracking systems is eliminated.

At the same time, jobs are enriched since each individual is given the whole assembly job. In many cases this would make it possible to eliminate inspection completely. It might also be to make each assembler responsible for his or her own testing.

Build It Right the First Time

The exhortation to "build it right the first time" sounds too obviously true to warrant stating. But if taken seriously, it has implications of current manufacturing practices, especially with regard to inspections, testing, and repair operations. Their primary objective must be to identify quickly process problems and initiate corrective action so subsequent product will, in fact, be built right the first time.

This will have important consequences, since even a small improvement in yield can be significant. An improvement in yield from 90 percent to 95 percent means a 50 percent reduction in rejects. As a consequence, repair and retest time and effort are cut in half.

Minimize Transactions

Transactions related to tracking and control add no value to the product. If the process itself is greatly simplified in the various ways suggested in this article, it then becomes possible to eliminate various transactions and the effort required to perform and monitor them. For instance, sophisticated shop-floor control systems, work order systems, and material trackings are often not required within a simplified work environment.

Eliminate Randomness in the Process

Randomness in the process creates confusion and impairs productivity and throughput time. Although it can never be eliminated entirely, random occurrences such as material shortages, equipment breakdowns, human error, etc., can be dramatically curtailed and their impact minimized.

Sound material planning is essential, as is the availability of raw materials at the point of use rather than in the warehouse, stockroom, or kitting area.

Adequate preventive maintenance is required. If avoiding machine downtime is given a high priority, then it certainly can be reduced by a significant degree. If preventive maintenance personnel can't make the machines run reliably, perhaps they should be replaced.

Jobs can be designed with the objective of minimizing the potential for error. Providing process sheets and visual aids can also reduce the chances of mistakes. Ergonomic principles must be incorporated both into the task and into the tools provided to complete it. These considerations, combined with the responsibility to do the job right the first time, will reduce the likelihood of the operator making random human errors.

CONCLUSION

In summary, significant improvements in throughput time can be achieved by integrating the work-flow principles into the design and management of the manufacturing processes. These principles are:

1. Establish a product orientation;
2. Eliminate buffers;
3. Establish one-at-a-time processing;
4. Balance the flow to the bottleneck;
5. Minimize sequential processing;
6. Schedule effectively;
7. Minimize multiple paths;
8. Build it right the first time;
9. Minimize transactions; and
10. Eliminate randomness in the process.

It should also be noted that in this article, the creation of whole jobs has been recommended in connection with a number of these principles. That is because these principles are more easily achieved through this approach to job design. While these principles can be designed into the process, it is the workers who will make them effective. Their flexibility to handle multiple tasks and to contribute mentally as well as physically can make possible dramatic reductions in throughput time.

Suggested Readings

Abernathy W. J. and Utterback J. 1975. Dynamic Model of Process and Product Innovation. *Omega* 3(6):639-57.
For a discussion of OPT principles, see E. Goldratt and J. Cox. 1986. *The Goal: A Process of Ongoing Improvement.* North River Press. Croton-on-Hudson, NY. See *ibid.*

Reprinted with permission from *National Productivity Review*, Vol. 6, No. 1, Winter 1986-87.
© 1986 Executive Enterprises, Inc., New York, NY.

Part 5

Product and Process Design for Electronics Assembly

This section provides articles that examine how product and process design issues play a major role in improving productivity and quality in electronics assembly. The consideration of product and process design issues seeks not only to adapt the person to the environment, but also to focus on creating product designs that are easy to manufacture. In the electronics industry, product and process design issues are gaining more and more attention as methods for improving both individual and organizational productivity are developed. Gatenby (1987) provides an overview of product design (Design For "X") issues to improve corporate productivity and quality. Human factor issues in product and process design are detailed later in this summary.

An approach for designing electronics assembly and test stations is provided by Pukanic and Morelli (1985). Daetz (1987) examines in detail the effect of product design on quality and product cost. The author provides design guidelines for quality improvement. Design measure and cost function matrix are also provided as an aid to practitioners. Whitney (1987) provides a comprehensive approach for ensuring optimal process design in electronics assembly processes. Whitney further presents specific examples from IBM, Digital, and other major companies.

The electrostatic discharge (ESD) process poses a major challenge for electronics firms. Pukanic (1985) presents methods for controlling ESD problems. One area Pukanic covers is how to deal with clean room contamination. Holmgren (1984) offers useful techniques for managing vision problems related to tasks. The electronics assembly processes require substantial use of lighting fixtures especially for visual inspection. Holmgren provides a comprehensive checklist for work places with light tables in graphic work. Hwang and Salvendy (1984) provide a methodology for obtaining optimal allocation of tasks between human supervision and automated systems. Microscopes are used extensively in electronics assembly inspection. Ostberg and Moss (1984) provide measures and guidelines to aid microscope users. The authors further provide specific recommendations that will help minimize workers' fatigue levels due to microscope use. Product design implications for electronics assembly processes are examined in detail by Domas and Helander (1984), Mosley (1987), Hales (1985), and White (1987). Part 5 matrix summarizes the papers presented in this section.

OVERVIEW OF HUMAN FACTORS IN DESIGN OF ELECTRONICS ASSEMBLY

Human factors is a field that is becoming increasingly important in the electronics industry as a method of improving productivity, health, and safety. Human factors needs to be incorporated into the proper design of the workspace, the equipment and tooling, and the actual manufacturing process flow.

In our modern world, manufacturing technologies are rapidly changing, and the impact of the rapid changes in technologies and processes can adversely affect employees. This impact can range from a minor irritation, causing additional fatigue, to a severe life hazard that can affect the surrounding populace as well as the company's employees, such as the Chernobyl nuclear reactor accident.

To better understand human factors, a definition is needed before proceeding further. Human factors, sometimes call ergonomics (especially in Europe), is the study of the human being in relationship to the environment in which the individual must operate or interface throughout life. This environment can range from one's kitchen or workshop to a jet fighter cockpit or a data entry workstation. However, this article shall focus on some of the concerns and applications found in the electronics industry.

There are a variety of specialized disciplines that contribute to the study of human factors. Some examples are architectural design, biology, biomechanics, cultural anthropology, engineering, industrial design, medicine, psychology, sociology, statistics, and others. There also appear to be two general groups involved in studying human factors. The one group consists of the specialists, while the second group consists of the generalists. The specialists are usually those professionals who are engaged in the scientific study of specific aspects of human factors, such as the human eye's response to visual flicker rates of a visual display terminal (VDT). The generalists are usually those who have studied human factors and/or applied several aspects of human factors to some form of an integrated system, such as an industrial engineer who is designing a new production facility. Both of these groups are complementary and necessary for the successful design of human factors systems.

Human Factors in Manual Assembly Operations

When human factors principles are applied to manual assembly operations, the level of productivity for many different types of operations can be significantly raised.

In manual operations, an operator or assembler is usually required to perform repetitive tasks on daily basis. Some of these repetitive tasks may be hazardous to the employee's long-term safety and health. An illustration of this would be in the area of repetitive hand motions. The particular types of occupational diseases that can be caused by these repetitive motions are carpal tunnel syndrome, tenosynovitis, and white finger disease. The cause of these diseases are often related to the design of the tools or the lack of adequate job design.

One very common example of a cause of carpal tunnel syndrome is the repetitive use of lead-cutting pliers. Pliers are often inadvertently designed in such a manner that they place high pressure at the base of the palm causing the carpal tunnel to become inflamed. This results in pain and the inability to effectively use one's hand. Once a person has acquired this disease, the likelihood of a recurrence is very high, especially if the disease has seriously progressed. In that event, the employee may be permanently disabled. This type of case is very costly to employers in terms of increases in workers' compensation premiums, a significant loss of productivity on the part of the injured employee, and most importantly, a valuable employee has sustained a permanent disability due to improper tool selection and job design.

The solution to this problem is to develop a better product design that minimizes the amount of repetitive motions required to assemble the product and to select the most ergonomically correct tooling for the assembly process.

Workstation design is another concern that needs to be addressed in electronics assembly. Although most curricula include facility layout and design, their greatest error is the continued inclusion of the 50-percentile human model as the basic criterion for design. When it comes to workstation design, the 50% model should be used only when there is absolutely no other way to design and/or select a workstation system. The workstation of the future is probably going to have to be completely adjustable so as to fit not only the "average" person but also to adapt to fit an entire range of the population (95-99 percentile models). The workstation must be able to fit to the person—this includes seating and the worksurface or envelope.

Lighting is an important concern that needs to be considered by facility designers. Typically, the greatest concerns are lighting output as well as energy efficiency. However, there are also important issues that affect quality and productivity in regard to lighting. These are visual acuity, visual fatigue, seasonal depression, and hormonal regulation. The quality, quantity, and wave length of lighting have been shown to have an effect on the previously mentioned issues. Although some of the scientific studies are controversial, other aspects of lighting in addition to lumen output and energy savings need to be considered. There could possibly even be long-term negative health implications from conventional artificial lighting.

Other concerns for assembly areas can include safety requirements for electrical, mechanical, and chemical systems, temperature and humidity controls, and room and furnishing color.

Some words of caution to the professional who is evaluating the selection and purchase of the ergonomically designed workstation are in order. First, several good handbooks on ergonomics and human factors should be read (some good primers are listed in the suggested readings). Secondly, the design of the equipment should be thoroughly investigated. Many products are advertised to be ergonomic, when actually they may incorporate as little as a single aspect of ergonomics. Products that meet as many ergonomic design requirements as possible should be selected. Using chairs as an example, a good ergonomic chair will have the following easy-to-adjust features: seatpan height, backrest height and angle, seatpan tilt (forward and rearward), and a tip-resistant safety feature such as a five-star pedestal base.

A good point to remember when evaluating existing systems is that if a method or design is working and it does not impose on the health, safety, or well-being (comfort—both physically and psychologically) of an individual over a working lifetime, then that particular method or design is most likely realistically sound. The professional should concentrate efforts on other areas where the returns on the time and money invested will be greater.

Many studies have shown that significant productivity increases are possible by having more ergonomically and work-flow correct workstations. The highest productivity gains that we have seen publicly documented have been in the 50-62% ranges. However, more realistic improvements have been in the 20% ranges.

At a major medical electronics manufacturer, the following criteria were incorporated in the redesign of an assembly area: flexible and adjustable workstations, ergonomic seating (chairs included forward tilt capability), full-spectrum task lighting, and an integrated manual material handling system which could be used at the workstation. During the first month, the productivity rose by over 35%. After the first month, the productivity increase remained in the 16-20% range. The process flow and headcount remained the same. Therefore, the aforementioned factors accounted for the productivity improvements.

When one calculates the investment rate of return on ergonomic workstations, the payback period will usually satisfy even the tightest of capitalization requirements.

Automation

Human factors in automation requires even more planning and forethought than with manual assembly operations. While automation may take many forms, this section will be concerned with the basic automation that can be found in the electronics industry, such as robotics, autoinsertion, or material handling equipment.

In automation, human factors must address the role of designing the automation control station [which will most likely include a VDT and/or a personal computer (PC) as the main control interface], as well as job enrichment, safety, maintenance, and training.

With the VDT and/or PC-type controller workstations, postural and fatigue issues now become a much larger part of the system dynamics. There is currently an American National Standards Institute standard pending for VDT workstation design criteria. This standard could well affect the future legislated requirements that will specify minimum standards which must be adhered to for equipment and workplace design where VDT and/or other types of computers would be used. Such standards, when available, are an invaluable aid to designing this type of workstation. Again, adjustability will become a byword for such systems.

From a safety standpoint, automation is a real human factors challenge. Now the professional must look at risk analysis and understand how people will think when they intentionally bypass safety interlocks to make the job "easier." There have already been instances in which robotic homicide has occurred involving both authorized maintenance and unauthorized personnel. Physical guards, interlocks, and signs to protect personnel have to be designed, but also software control devices which will not allow safety equipment to be overridden. Further, that protection must remain in effect for programmers and maintenance personnel while still allowing them to perform their functions quickly and efficiently.

Because of the rapid introduction of automation within the electronics industry, training programs will need to become an important aspect of human factors. Training programs must not only be developed to train operators and maintenance technicians in the operation, safety, and maintenance of the equipment, but also to retrain those employees who become displaced by the new technologies. These programs must be human-centered because people are still the most valuable resource a company has, regardless of the level of automation incorporation into the electronics assembly operation.

Suggested Readings

Handbooks

Alexander, D. C. and B. M. Pulat. 1985. *Industrial Ergonomics: A Practitioner's Guide.* Industrial Engineering and Management Press. Norcross, Georgia.

Eastman Kodak Company. Human Factors Section. 1983. *Ergonomic Design for People at Work.* Volume 1. Lifetime Learning Publications. Belmont, California.

Grandjean, E. 1980. *Fitting the Task to the Man, An Ergonomic Approach.* Taylor and Francis Inc. London.

IBM Corporation, Human Factors Center. 1984. *Human Factors of Workstations With Visual Displays.* IBM Corp. San Jose, California.

Kleeman, W. B. 1981. *The Challenge of Interior Design.* Van Nostrand Reinhold Company. New York.

Mandal, A. C. 1985. *The Seated Man, Homo Sedens.* Dafnia Publications. Klampenborg, Denmark.

McCormick, E. T., and M. S. Sanders. 1981. *Human Factors in Engineering and Design.* McGraw-Hill. New York.

Wodson, W. E. 1981. *Human Factors Design Handbook.* McGraw-Hill. New York.

Articles and Papers

Brown, J. 1983. Idustrial zoning: factory cubicles boost productivity. *Electronic Business.* September 1983. pp. 176-177.

Hasselquist, R. J. 1981. Increasing manufacturing productivity using human factors principles. *Proceedings of the Human Factors Society 25th Annual Meeting.* Human Factors Society. Santa Monica, California.

Hooper, R. L. Jr. 1985. The use of human factors to improve productivity in electronic assembly facilities. *Proceedings of the 1985 International Electronic Assembly Conference.* October 7-9. Institute of Industrial Engineers. Atlanta, Georgia.

Hooper, R. L. Jr. 1986. Human factors and safety in the manufacturing environment. *Proceedings of the Third International Conference on Human Factors in Manufacturing.* November 4-6. IFS (Conferences) Ltd.

Priest, J. W. 1985. Ergonomic changes in workplace can improve the productivity of product operations. *Industrial Engineering.* 17(7): 40-43.

Tichauer, E. R., and H. Gage. 1977. Ergonomic principles basic to hand tool design. *American Industrial Hygiene Association Journal* (38), November 1977. pp. 622-634.

Wurtman, R. J. 1975. The effects of light on the human body. *Scientific American.* 233: 68-77.

* Material presented in this section was provided by R. L. Hooper, Jr., (IBM Rolm); D. L. Morelli (Applied Risk Management Inc.) in collaboration with the Editors.

Part 5 Matrix

Chapter Number	Title	Model	Quanti-tative Tech-nique	Strategy/ Method-ology	Case Study	General Theory	Design	Receiv-ing & Store	Inspec-tion	Kit-ting	CP Assem-bly Mfg.	Equip. Assem-bly	Packing	Mat'l Hand-ling
45	Design for "X" (DFX): Downstream Key to Efficient, Profitable Product Realization	X		X	X		X	X	X	X	X	X	X	X
46	A Systems Approach to Ergonomically Sound Design of Electronics Assembly/Test Stations					X			X		X	X		
47	The Effect of Product Design on Product Quality and Product Cost					X	X				X	X		
48	Manufacturing and Design: A Symbiosis				X	X	X	X			X	X		
49	ESD: Understanding the Problems and Methods for Controlling			X	X	X					X	X		
50	A Total Ergonomic Investigation of Working Places with Light Tables				X	X			X	X	X	X	X	
51	Human Supervisory Performance in Flexible Manufacturing Systems		X		X			X	X		X	X		
52	Microscope Work - Ergonomic Problems and Remedies	X				X			X					
53	Manual Versus Robotic Assembly: Some Implications of Product Design				X	X	X	X	X	X	X	X	X	X
54	Artificial-Intelligence Techniques Boost Utility of CAE/CAD Tools				X	X	X				X	X		
55	Designing for Assembly: A Computer-Based Approach				X	X	X				X	X		
56	Tapping CAD Data for PC Board Panelization			X	X	X	X	X	X		X			X

45

Design for "X" (DFX): Key to Efficient, Profitable Product Realization

David A. Gatenby

THE DFX CONCEPT

Electronics design encompasses product design and process design of individual electronic components (e.g., silicon devices) and of electronic assemblies (e.g., printed circuit boards and system-level packaging and interconnection). Historically, designers have primarily focused on functionality and features of electronic systems during the design phase. Increasing competitive pressures have now forced companies to pay more attention to the effects that designs have on downstream operations (e.g., manufacturing and installation), because the cost and quality of manufacturing and delivering an electronic product is primarily determined by its design.

Design For "X" (DFX) is an approach to design products and processes for cost-effective, high-quality downstream operations from manufacture (including fabrication, assembly, and test) through end-customer usage. Figure 1 shows a high-level model of DFX where the "Xs" include major downstream processes (manufacturing, distribution and installation, service and maintenance) and end-customer requirements, as well as specific activities and requirements within each major "X" category. Several of the more specific activities shown in the model occur in more than one downstream operation—for example, assembly, testing, and material logistics are design concerns for manufacturing, distribution, and installation.

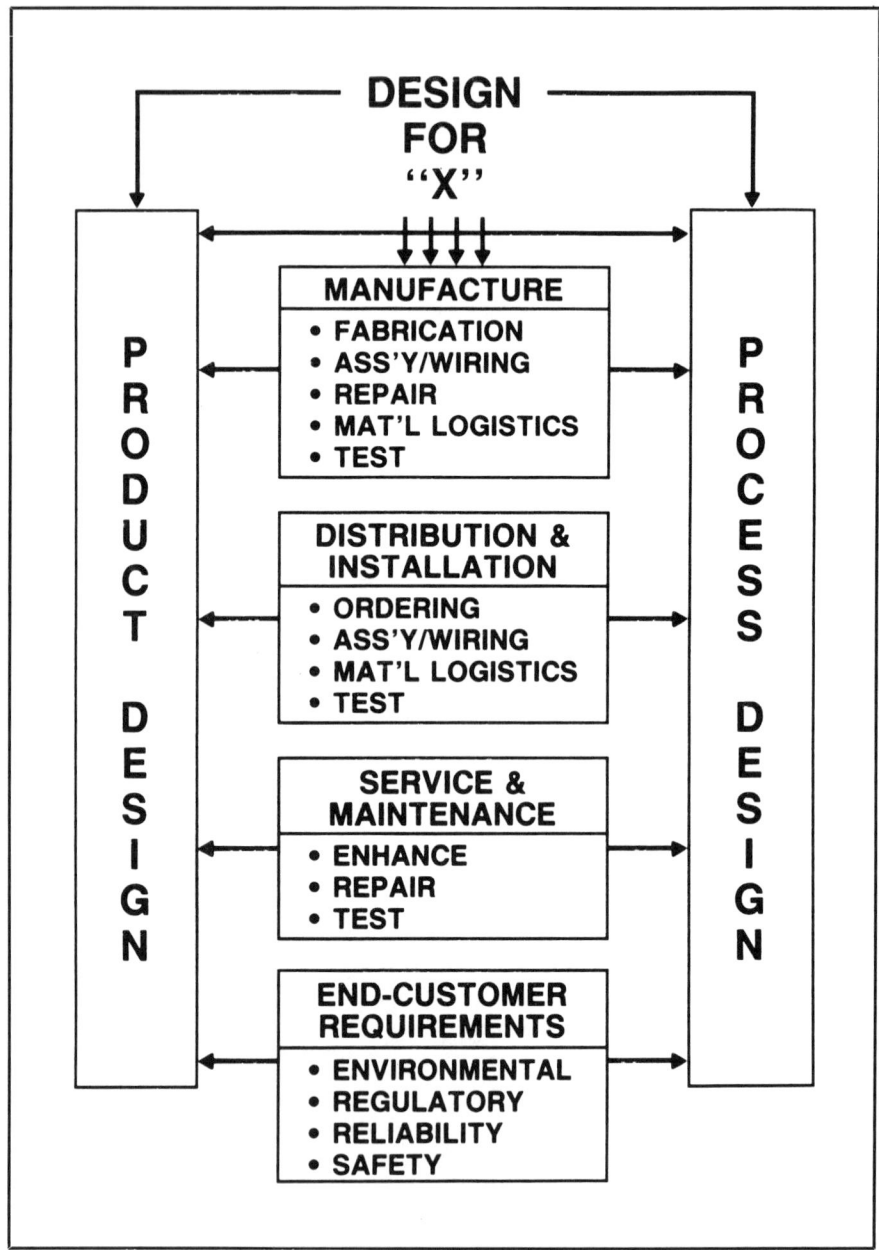

Figure 1. *Design for "X" (DFX) model.*

Designers need to understand how a design's attributes affect all of the "Xs" so that one DFX concern is not optimized at the expense of another. For example, service, repair, material logistics, and fabrication processes need to be considered when combining or integrating parts to improve assembly operations (for both mechanical piece parts and housings, as well as silicon devices). These types of DFX relationships and trade-offs, their inherent interdisciplinary scope, and rapid changes in many technologies make DFX complex. A systematic approach to understanding DFX and its complexities is described later.

DFX must be an integral part of product realization to achieve high quality, rapid product introduction and profitability. This integration must be driven into eduational programs, product realization processes, information systems, organizational structures, and reward systems. DFX-related techniques in each of these areas are reviewed later.

A SYSTEMATIC APPROACH TO UNDERSTANDING DFX

DFX for electronic products can be complex, because it requires an understanding of the relationships between design and associated downstream operations and requirements. This interdisciplinary scope, along with changes in electronics technology, amplify the complexity of effective DFX. A systematic approach to understanding DFX, a prerequisite for effectively supporting it during product realization is presented as follows.

DFX Linkages & Printed Circuit Board Design for Manufacturability

DFX linkages show the relationship between downstream processes (the Xs), design attributes, and design activities—this DFX linkage concept is shown in Figure 2. A printed circuit board (PCB) design for manufacturability (DMF) linkage example is shown in Figure 3. In this example, the fluid mechanics of the wave-solder *process* dictate that rows of in-line through-hole device (THD) pins be parallel to the direction that a PCB passes over the wave to avoid solder bridging. The corresponding *design attribute* is the orientation of multileaded THD components. Component orientation is determined in the component placement *design activity* of PCB layout. This DFX example can be viewed at three different levels of specificity:

1. DFM—because the process is a manufacturing process (top X box in Figure 1);
2. Design for assembly (DFA)—because the process is part of the overall assembly process (subcategory of manufacture in Figure 1); and
3. Design for soldering (DFS)—because the specific process being designed for is soldering.

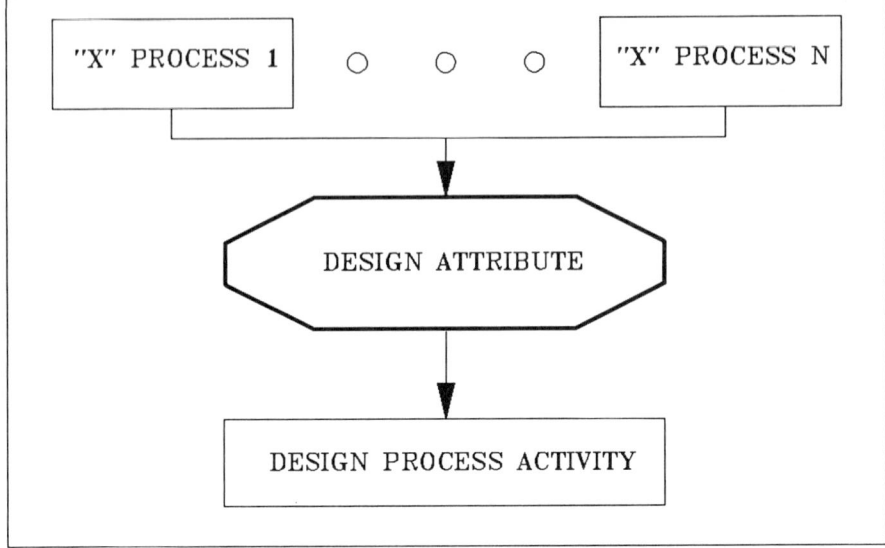

Figure 2. *DFX linkage concept.*

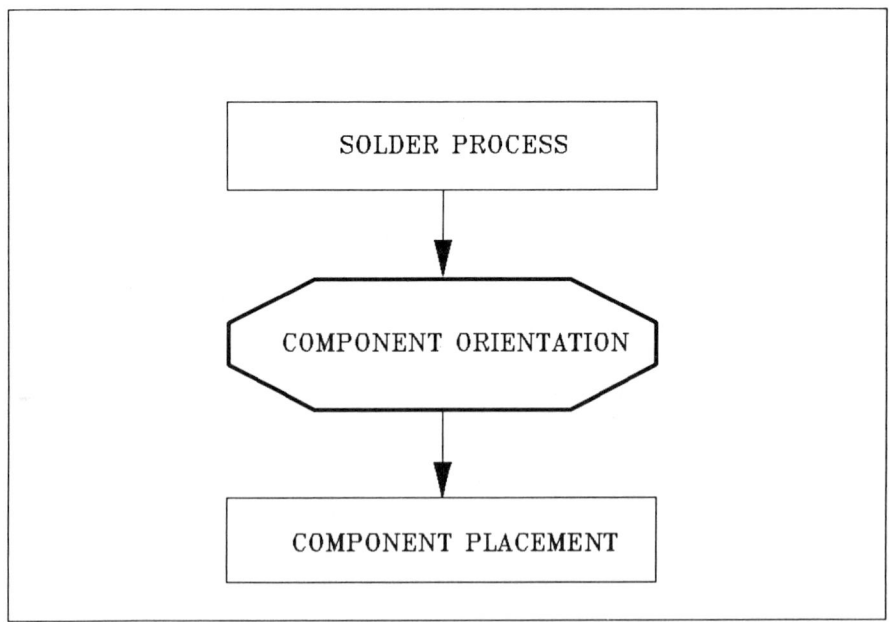

Figure 3. *PCB DFM linkage example.*

Often, more than one process determines a given design attribute. For example, THD component insertion and assembly-process inspections can also dictate preferred component orientations to increase operation throughput and accuracy. Figure 4 shows these two additional PCB DFM linkages, as well as the one shown in Figure 3.

DFX linkages become more complex when there are conflicting goals between processes that are affected by a single design attribute. For example, PCB holes that are significantly larger than their associated THD leads can improve component insertion yields, whereas PCB holes which are tighter (slightly larger than THD leads) can improve wave-solder yields. Also, lead-to-hole ratios typically cannot be customized for each lead size because PCB fabrication operations can be improved by minimizing the number of different hole sizes (to reduce drill bit inventories and to avoid multiple drilling machine set-ups). Figure 5 shows the PCB DFM linkages for this more complex example. Note that the drilling process has a different style box to distinguish the PCB fabrication process from the PCB assembly processes.

Different processes in a DFX linkage (labeled "1" through "N" in Figure 2) may or may not have conflicting goals and requirements and may or may not be

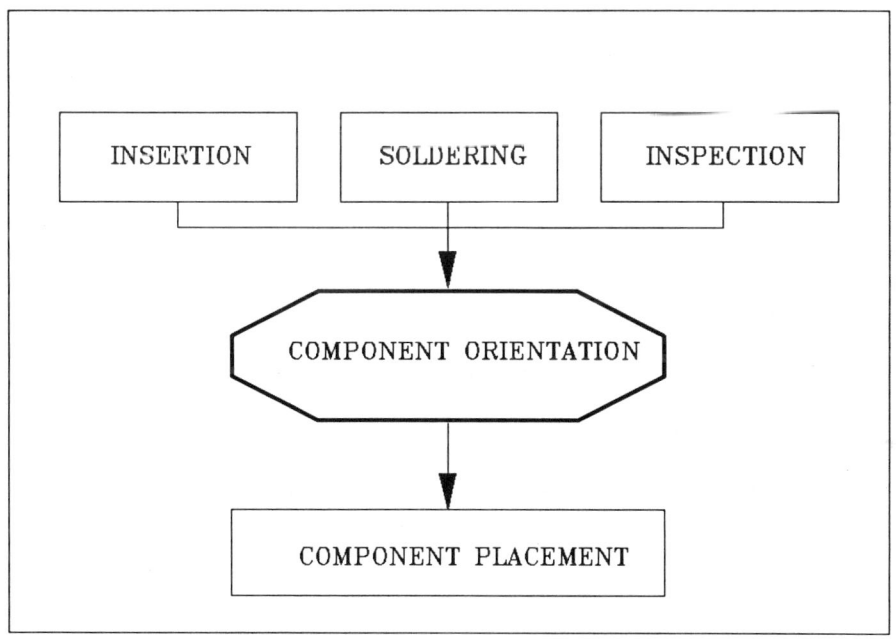

Figure 4. *PCB DFM linkage example.*

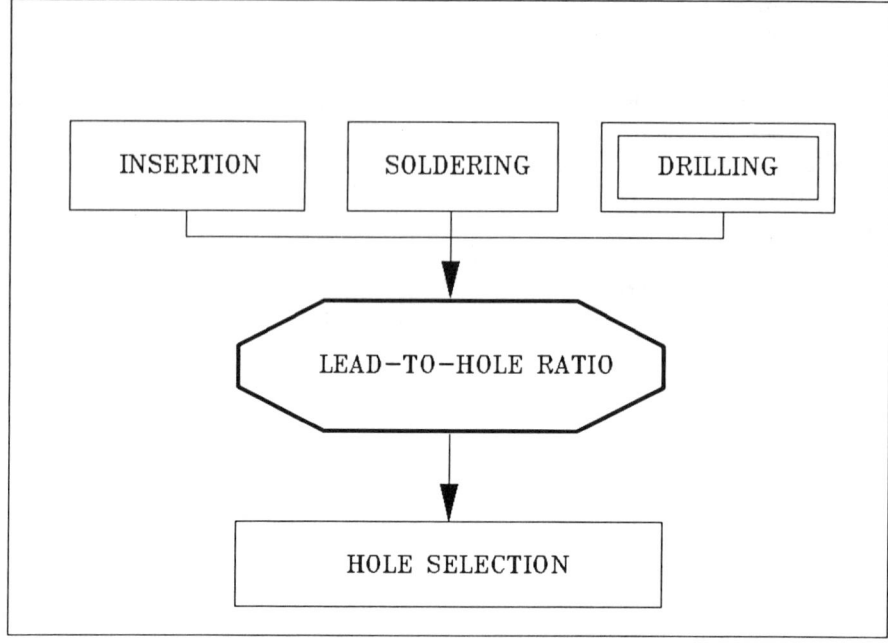

Figure 5. PCB DFM linkage example.

part of the same overall downstream activity (e.g., fabrication vs. assembly, or manufacture vs. installation). Recognizing and resolving these DFX trade-offs is probably one of the biggest challenges for designers in today's competitive environment.

DFX Taxonomy Example: PCB DFM

An important step in systematically understanding DFX is to organize the DFX linkages that were described previously. A taxonomy—that is, a classification scheme—that groups design attributes by design activities has proven effective for organizing DFX linkages.

Figure 4 shows the PCB DFM linkages for the component orientation design attribute. The following list identifies manufacturability—affecting design attributes that are determined in the component placement activity:

1. clearances (adjacent components, PCB edges, tooling holes);
2. grid locations;
3. lead spans;
4. mounting side; and
5. orientation.

Figure 6 summarizes some of the undesirable effects that these placement design attributes can have on the steps in the PCB assembly process. Similar groupings of design attributes and their relationships to PCB fabrication and assembly have been analyzed for each of the following design activities:

1. circuit and test design;
2. component selection;
3. reserved space allocation (e.g., for labeling during manufacture);
4. component placement;
5. hole size selection and layout;
6. conductor design (including electrically nonfunctional conductors such as patterns to improve plating processes);
7. solder mask design; and
8. nomenclature design (e.g., silkscreen marking).

ASSEMBLY STEP	UNDESIRABLE EFFECTS
MATERIAL PROCUREMENT	Ordering Extra Material
COMPONENT PREP	Add Insulation, Manual Assembly
INSERTION/PLACEMENT	More Set-Ups, Manual Assembly
SOLDERING	Insufficients, Shadowing
CLEANING	Component Damage
INSPECTION	Fillet Visibility, Slow Throughput
SHEARING/ROUTING	Interferences to Blade/Holder
TESTING	Special Fixturing
REPAIR	Tooling Interference
HANDLING/TOOLING	Bending, Fall-Out, Jam Transport
TRACKING	Label/Scanner Interference
NEXT-LEVEL ASSEMBLY	Interferences to Housing

Figure 6. PCB DFM relationships matrix.

The breadth of relationships between these design activities and their corresponding design attributes and the PCB fabrication and assembly processes is shown in Figure 7. Note that the third design activity column in Figure 7 corresponds to the placement DFM relationships listed in Figure 6.

DFX and the Electronics Packaging Hierarchy

DFX linkages and taxonomies must be developed for all of the downstream considerations in the DFX model (Figure 1) for all levels in the electronics

MANUFACTURING STEP	DESIGN ACTIVITY							
PCB Fabrication:								
DRILLING			•					
PLOTTING				•	•			
PLATING			•	•	•			
SOLDER MASKING				•	•			
TESTING (Bare-Board)		•	•	•	•			
ROUTING/BLANKING				•	•			•
NOMENCLATURE APPLICATION		•					•	•
REPAIR				•				
HANDLING/TOOLING	•			•	•		•	
PCB Assembly:								
MATERIAL PROCUREMENT	•	•	•					
COMPONENT PREPARATION	•	•	•		•			
COMPONENT INSERTION/PLACEMENT	•		•	•	•	•	•	
SOLDERING	•		•	•	•	•	•	
CLEANING	•		•					
INSPECTION			•	•	•		•	
SHEARING/ROUTING			•		•	•		
TESTING (Assembled-Board)	•	•	•		•	•		
REPAIR	•	•	•	•	•	•	•	•
HANDLING/TOOLING	•		•	•	•			
TRACKING	•		•			•	•	•
NEXT-LEVEL ASSEMBLY	•		•			•		•

Column headers (diagonal): COMPONENT SELECTION, CIRCUIT & TEST, PLACEMENT, HOLE SELECTION, CONDUCTORS, SOLDER MASK, NOMENCLATURE, RESERVED SPACE

Figure 7. *PCB DFM relationships matrix.*

packaging hierarchy. Previous examples included DFM linkages and taxonomies for a single level in the electronic packaging hierarchy: PCBs. Although some relationships with adjacent packaging levels are implicitly addressed (e.g., via component selection and next-level assembly criteria), DFX linkages and taxonomies are necessary for all levels in the packaging hierarchy. A common example of this is design for testability (DFT) where individual devices, PCBs, and units and systems must be testable. Designing for electrostatic discharge (ESD) protection and electromagnetic compatibility (EMC) also require DFX for every packaging level. For example, ESD must be designed for at the device level (e.g., NMOS and CMOS protection circuit design rules), at the PCB level (e.g., selecting devices with the highest possible ESD thresholds), and at the unit and system levels (e.g., grounding and circuit isolation arrangements). Similarly, designing for EMC should be considered at the device level (e.g., minimizing integrated circuit (IC) emissions), PCB level (e.g., filtering and grounding techniques in layout), and at the unit and system levels (e.g., shielded enclosures and cabling).

DFX linkages and taxonomies must be examined for every electronics packaging level to fully understand and build in DFX considerations. This systematic understanding of DFX paves the way for supporting DFX throughout product realization.

SUPPORTING DFX IN PRODUCT REALIZATION

DFX considerations should be built into every design early in the product realization process (PRP). Approaches for supporting DFX in the PRP include establishing a knowledge base and providing DFX education and training, supporting DFX in development processes and information systems, sharing information and experiences, and organizational approaches and incentives for achieving DFX.

Knowledge Base and Education

DFX concepts and their relationship to corporate productivity and quality improvement should be understood by personnel at every level. Top management support of DFX is essential because DFX activities typically require increased efforts earlier in the PRP with paybacks reaped downstream. Middle managers need to understand the reasons for changing their operations and allocating people for DFX-related activities. Personnel at the operational levels need to understand DFX and its benefits to help motivate them to implement DFX in their work. Policy statements related to DFX (e.g., corporate quality and safety policies) and management commitment to implementing DFX should be well-advertised through courses, documentation, and executive plans and presentations.

DFX knowledge (e.g., DFX linkages and trade-offs) is developed through teamwork and interdisciplinary education. Rotations and internships (across organizations in the DFX model (Figure 1) are excellent ways to accelerate DFX knowledge; acquisition and appreciation. Universities can also pave the way for enhancing DFX by emphasizing multiple disciplines in engineering curricula and emphasizing cooperative (versus competitive) learning. Teaching team attitudes should be coupled with teaching specific team methods (e.g., inspections) to lay a solid foundation for DFX.

DFX knowledge can be communicated in several forms:

1. verbal (e.g., in dialogs with DFX experts);
2. written (e.g., in design guidelines, standards, practices, or rules); and
3. digitally encoded (e.g., in computerized data bases of components and design rules).

Clear communication of DFX knowledge is essential throughout the PRP and in education and training programs. DFX taxonomies can provide a framework for writing DFX documentation as well as digitally representing DFX rules and guidelines. DFX knowledge is used throughout the development process using manual and automated techniques; DFX support in processes and tools is described next.

Development Process

A carefully formulated, documented PRP description or methodology is necessary to achieve productivity and quality improvements. Process definitions should include DFX-related tasks at key points in the development process, including the following:

1. establishing DFX rules, guidelines, and targets;
2. initiating team design or redesign activities;
3. performing team inspections, evaluations, and reviews of designs for DFX; and
4. tracking of DFX performance (e.g., conforming rules and results in target X processes).

Teamwork is essential to achieve DFX. Selecting the right team members for each activity is crucial to an effective process. It is also important to integrate information systems into the process. For example, linking of on-line design capabilities at several locations can enable simultaneous viewing of designs and joint usage of DFX capabilities (e.g., design rule audits and reports) by designers and manufacturers. CAE/CAD (computer-aided engineering/computer-aided design) support of DFX will be discussed next.

CAE/CAD

DFX can be supported in a CAE/CAD environment in several ways:

1. managing DFX knowledge on-line (e.g., in component or DFX rule data bases;
2. automatically building in DFX considerations (e.g., per on-line rules); and
3. auditing designs against DFX rules.

   Figure 8 shows a simple architecture for supporting DFX with information systems. Figure 9 illustrates how a DFX taxonomy can be applied to planning information systems support of DFX. CAE/CAD support of DFX for each design attribute can be assessed with respect to DFX rules and their usage in a given design activity. In figure 9, for example, component orientation design rules are managed on-line, they are used by interactive and automatic placement functions, and they can be audited against the PCB design. The taxonomy can help CAE/CAD developers and their customers to systematically identify where additional DFX support is needed. CAE/CAD support of DFX should

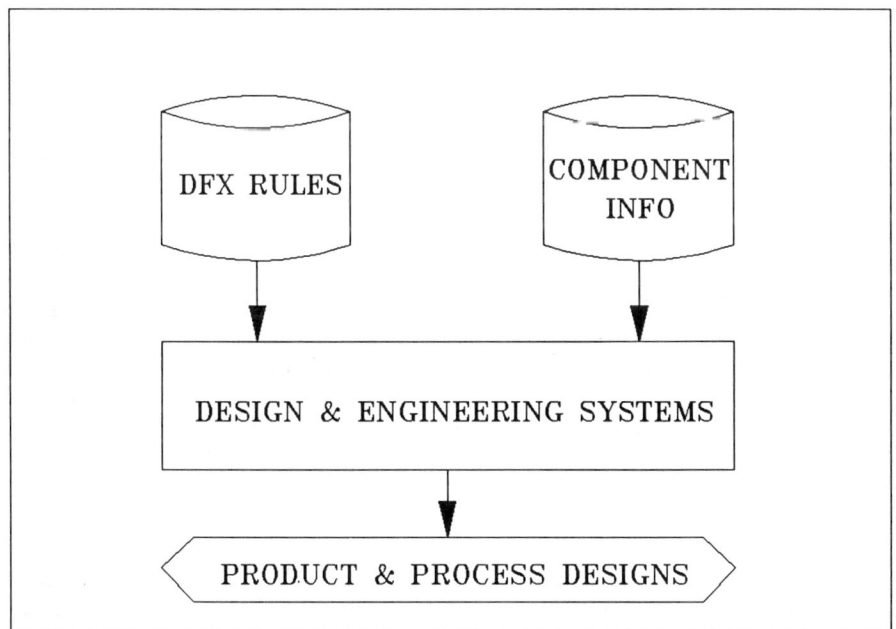

Figure 8. *Information systems support of DFX.*

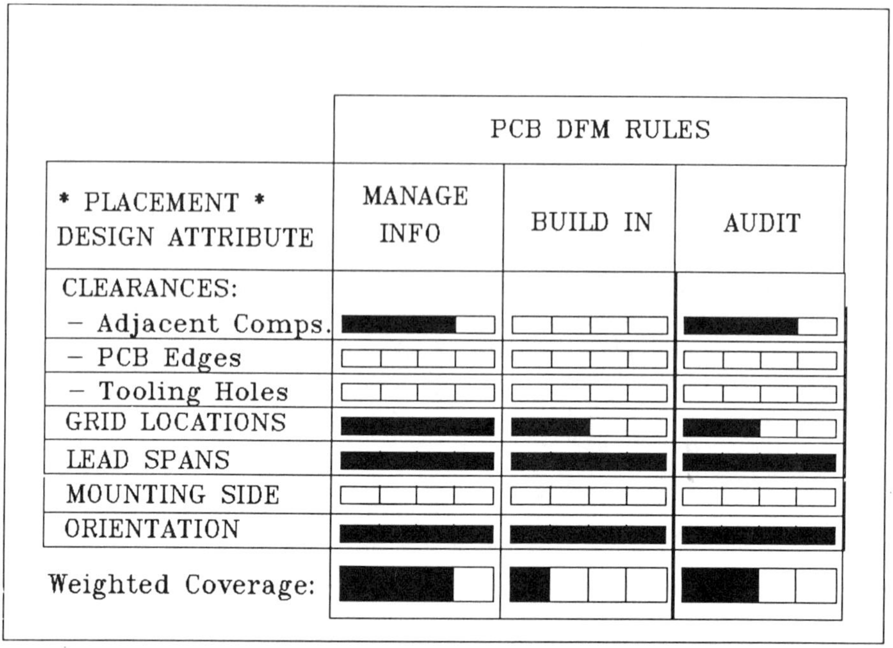

	PCB DFM RULES		
* PLACEMENT * DESIGN ATTRIBUTE	MANAGE INFO	BUILD IN	AUDIT
CLEARANCES: – Adjacent Comps.	▰▰▰▱	▱▱▱▱	▰▰▰▱
– PCB Edges	▱▱▱▱	▱▱▱▱	▱▱▱▱
– Tooling Holes	▱▱▱▱	▱▱▱▱	▱▱▱▱
GRID LOCATIONS	▰▰▰▰	▰▰▰▱	▰▰▱▱
LEAD SPANS	▰▰▰▰	▰▰▰▰	▰▰▰▰
MOUNTING SIDE	▱▱▱▱	▱▱▱▱	▱▱▱▱
ORIENTATION	▰▰▰▰	▰▰▰▰	▰▰▰▰
Weighted Coverage:	▰▰▱	▰▱▱	▰▱▱

Figure 9. *DFM coverage analysis: Information systems.*

be closely coupled with the development methodologies and PRP definitions discussed previously. In cases where DFX criteria is more subjective or qualitative—for example, designing for foreseeable misuse (to enhance product safety and prevent liability)—processes are the dominant means of DFX support.

Simulation and modeling techniques (e.g., for circuit and mechanical designs) are increasingly important to ensure proper design margins and robustness of designs (e.g., in conjunction with Taguchi's methods). Emerging design automation technologies (e.g., expert systems and synthesis tools) are beginning to incorporate more DFX considerations into their on-line knowledge bases and corresponding design solutions.

Information and Experience Sharing

DFX technology, trials, tribulations, and triumphs should be shared throughout a corporation. Conferences and forums dedicated to DFX should be run periodically to share experiences (e.g., in product development, manufacturing, installation, etc.) and information on productivity and quality methods, tools, and education and training.

Industrywide conferences on DFX (e.g., on DFM) can also be valuable sources of information for DFX experiences and techniques. Representatives from different engineering disciplines should take advantage of these conferences to make contacts, develop perspective, and possibly procure DFX technologies from other companies.

Organizational Approaches and Reward Systems

Many successful companies have achieved DFX through organizational and management structures. For example, one approach to ensuring the cooperation between product designers and manufacturing engineers is to combine them into a single organization reporting to a common manager. Manufacturability becomes a common, high-priority goal that designers and manufacturers work on jointly from design concept through production. The Japanese have been particularly successful at this approach — their manufacturers are involved at product conception, and their designers are involved through production ramp up.

Rewards and incentives are important ingredients for fully supporting DFX. Teams and individuals should be recognized and rewarded for DFX accomplishments. For example, DFX should have equal or higher weight than firefighting in determining team and individual performance. DFX innovations should be highly valued, just as functional innovations have been in the past. These values should be reflected in reward systems, because they both can translate into profits — the bottom line.

SUMMARY, CONCLUSIONS, AND DIRECTIONS

This chapter has presented a broad view of DFX — designing for downstream considerations from manufacture through end-customer operation. A systematic approach to understanding DFX has been demonstrated; it includes identifying and organizing DFX linkages between downstream processes, design attributes, and design activities. Education, development processes, CAE/CAD information systems, information exchanges, and managerial approaches to supporting DFX in product realization were discussed.

The level of DFX support in product realization and PRP support technologies need to increase for U.S. electronics companies to become more competitive. DFX specialists for example, in designing for EMC, ESD, safety and liability-prevention, reliability and maintainability, need to be better integrated into mainstream PRP activities, as well as PRP support technology development, such as in CAE/CAD. DFX knowledge bases and educational programs must be extended, DFX tools and processes must be improved, and DFX technologies and experiences must be shared and leveraged. Those who focus on prevention in the PRP, synonymous with DFX and quality by design, will survive and succeed.

Suggested Readings

1. _____. 1987. Product realization. *AT & T Technical Journal.* 66(5).
2. _____. 1986. Quality: theory and practice. *AT & T Technical Journal.* 65(2).
3. Boothroyd, P. and G. Dewhurst. 1987. Design for assembly in action. *Assembly Engineering.* 30(1): 64-68.
4. _____. 1987. Quality. *Business Week.* June 8.
5. _____. 1987. Design for manufacturability: getting it right the first time. *Conference Proceedings.* Chicago, IL. Management Roundtable, Inc. Chestnut Hill, MA.
6. _____. 1987. Revitalization of U.S. electronics manufacturing. *Electronic Business.* 13(2).
7. _____. 1987. Special report: on good design. *IEEE Spectrum.* 24(5).
8. IEEE International Symposium on Electromagnetic Compatibility. San Diego, CA. September 1986. *Catalog # 86CH2294-7.* Institute of Electrical and Electronics Engineers. New York.
9. Imai, M. 1986. *The Key to Japan's Competitive Success.* Random House, Inc. New York.
10. *Proceedings of the 2nd International Conference on Product Design for Manufacture and Assembly.* Newport, RI. April 1987. Troy Conferences. Rochester, MI.
11. *Proceedings of the 23rd Design Automation Conference.* Las Vegas, NV. June 1986. Institute of Electrical and Electronics Engineers, Inc. New York.
12. *Quality by Design: A Quality Manual for the AT & T R & D Community.* Order # 500-021. August 1986. Issue 1.1. AT & T Customer Information Center. Indianapolis, IN.
13. *Reliability and Maintainability Symposium Proceedings.* Las Vegas, NV. January 1986. Catalog # 86CH2247-5. Institute of Electrical and Electronics Engineers. New York.
14. Thorpe, J. F. and W. H. Middendorf. 1979. *What Every Engineer Should Know About Product Liability.* Dekker Publishing. New York.
15. Walsh, S. 1987. Service: the missing link in electronics industry automation. *Electronics Test.* 10(9): 37-43.

Dave Gatenby is a DFX and CAE/CAD planner for AT & T, Berkeley Heights, New Jersey.

46

A Systems Approach to Ergonomically Sound Design of Electronics Assembly/Test Stations

Ray Pukanic
Donald L. Morelli

Recognition has been growing that poor attention to ergonomic factors in the design of electronics work stations can contribute to a wide variety of problems. These problems include lower productivity and performance as well as higher absenteeism rates and workers compensation claims.

Ergonomic deficiencies are best identified, evaluated and corrected in a "systems" or multidisciplinary manner, so that all pertinent factors become part of the solution. For any industry, these factors would include, but not be limited to:

- Production demands.
- Quality.
- Profit.
- Optimum utilization of human resources.

Several methods of identifying ergonomic deficiencies are available, including:

- Review of existing records.
- Job station observation.
- Worker surveys/questionnaires.

RECORDS REVIEW

Almost all the records commonly kept by businesses can be helpful in the efforts to locate ergonomic deficiencies. If production or quality suddenly dips downward, or if these items never reach expected levels after a process change, work place and/or work method changes may be indicated.

Higher than normal absenteeism from *particular* jobs or work stations is a clear "red flag" for ergonomics evaluation. More obvious may be high numbers of injuries associated with specific jobs.

Unfortunately, all too often the indicators in these records are not used as effectively as they should be for meaningful problem solving. When such indicators appear, the records should be analyzed in an effort to identify the ergonomically undesirabe *task demands* or *work elements* that probably exist.

Many times these task demands or work elements are induced by the hardware of the work station. The more common deficiencies of the electronics work place include:

- Chairs or work surfaces at inappropriate heights.
- Poor lighting.
- Hand tools which force hands/wrists into extreme positions.
- Material handling requiring twisting or a large range of motion.
- The use of one hand as a "biologic jig" to hold IC boards.

Since the hardware of the work place is often involved, records on tool or equipment damage as well as product scrappage may also be useful in identifying work stations with ergonomic deficiencies.

Whenever undesired events can be traced to specific jobs, work stations or work elements, ergonomic deficiencies may be indicated, and efforts should be directed at identification and correction. A guide to help identify ergonomic deficiencies is provided in Table 1.

JOB STATION OBSERVATION

Observation of the work stations and the tasks themselves is probably the best method for identifying ergonomic shortcomings. Options for structuring these observations include:

- The "classical" work sample, tempered to identify ergonomic factors.
- Task analysis.

The work sample is useful in gathering information on absenteeism (in the sense of people not spending the desired time at the work station), the frequency of performance of various tasks, the frequency of tool usage and the

TABLE 1 – Ergonomic Target Guide

IF YOUR OPERATIONAL PROBLEMS INCUDE:
- ☐ *PRODUCT DAMAGE/HIGH SCRAPPAGE*

MOST LIKELY WORKER INJURIES:
STRAINS/SPRAINS OF THE BACK AND SHOULDERS.

LOOK FOR:
- STATIC OR AWKWARD POSTURES INDUCED BY THE WORK PLACE.
- EXCESSIVE MATERIAL HANDLING.
- MANUAL MATERIAL HANDLING OVER A LARGE RANGE OF MOTION.

☐ *POOR PRODUCT ASSEMBLY/HIGH REWORK*

MOST LIKELY WORKER INJURIES:
CARPAL TUNNEL SYNDROME, FINGER CUTS AND LACERATIONS, SHOULDER COMPLAINTS.

LOOK FOR:
- WORKER MODIFICATIONS TO CHAIRS (PADDING).
- WORK SURFACES AT INAPPROPRIATE HEIGHTS.
- EXTREME POSITIONS OF THE WRISTS (FLEXION, EXTENSION AND DEVIATION).
- REPETITIVE PINCHING/GRIPPING REQUIREMENTS.

☐ *PRODUCTION UPSETS*

MOST LIKELY WORKER INJURIES:
STRAINS/SPRAINS OF THE BACK AND SHOULDERS, CUTS AND CONTUSIONS OF THE FINGERS,

LOOK FOR:
- IMPROPER QUEUING OF PARTS.
- EXCESSIVE WALKING ALONG ASSEMBLY LINES.
- ABSENCE OF LINE-STOP AUTHORITY.
- MANUAL MATERIAL HANDLING OVER A LARGE RANGE OF MOTION.

☐ *WAREHOUSING/ORDER FILLING DEFICIENCIES*

MOST LIKELY WORKER INJURIES:
STRAINS/SPRAINS OF THE BACK, CONTUSIONS TO THE UPPER EXTREMITIES.

LOOK FOR:
- HIGH TURNOVER OF WORKERS.
- EXCESSIVE REHANDLING (PALLETIZING/DEPALLETIZING).
- POOR SCHEME OF ASSIGNING WAREHOUSE ADDRESSES.

Note: Table material © Donald L. Morelli.

amount of material handling performed by the worker. All such information assists in evaluating the nature and severity of any suspected ergonomic problem.

The task analysis is similar to any work or time study in that a list of the sequence of work elements is generated. However, more specific information is gathered on factors such as:

- Body postures required.
- The range of motion necessary.
- Hand and wrist positions and actions.
- The distances required for material handling.

The findings of either a work sample or a task analysis should indicate to the designer or industrial engineer where and what changes should be made. The various indicators for change need to be evaluated according to a systems approach so that the *overall* objectives of the work requirements are achieved. This can be done by checking the present design and proposed changes against certain criteria:

- Importance—The importance or vital nature of a component *relative to completing the task* will indicate its proper location.
- Frequency of use—The locations of components are prioritized according to their frequency of use, minimizing long reaches.
- Function—In addition to the adage, "form follows function," this implies the desirability of grouping components with similar functions.
- Sequence—Tied to function; a logical progression of activities is vital to motion economy.

A list of typical situations and task demands that present invitations to improve work place ergonomics is presented in Table 2.

Although the value of worker input has been downplayed or not taken seriously in the past, quality circles, design review teams and participative decision-making approaches have begun to tap this veritable fountain of *good* information and ideas. From an ergonomic viewpoint, things that the workers *do* can speak as loudly as what they would be likely to say.

Workers know the job as only they can. Their perspective is one of "getting the job done," despite what many may think. The workers can be an excellent source for identifying specific problems as well as concepts for solutions.

Modifications to the work station by the workers are usually instant indicators of poor fit to the worker and further invitations to improve work place ergonomics. The most prevalent modification one finds in the electronics industry is the addition of large amounts of padding to chairs, not just to soften the seat, but to adjust the worker's position relative to the work surface. Adjustability of chairs or of the work surface may be a worthwhile option.

TABLE 2 – Invitations to Improve Work Place Ergonomics

- IS ABSENTEEISM HIGH ON A PARTICULAR JOB?
- IS TURNOVER HIGHER THAN IN SIMILAR JOBS?
- IS PRODUCTION EFFICIENCY LOWER THAN EXPECTED/ PREDICTED?
- IS PRODUCT QUALITY LOW?
- DOES THE PROCESS RESULT IN TOO MUCH MATERIAL WASTE?
- IS THERE HIGH EQUIPMENT OR TOOL DAMAGE ON SPECIFIC JOBS?
- IS THE WORKER FREQUENTLY AWAY FROM THE WORK PLACE?
- IS THE WORK PACE RAPID AND BEYOND THE WORKER'S CONTROL?
- DOES THE JOB REQUIRE FREQUENT USE OR MANIPULA- TION OF HAND TOOLS?
- IS THE WORKER REQUIRED TO MAINTAIN ANY SINGLE POSTURE FOR LONG PERIODS?
- DO WORKERS SIT ON THE FRONT EDGE OF THEIR CHAIRS?
- ARE WORKCHAIRS MODIFIED BY ADDITIONAL CUSHIONS?
- ARE THE WORKERS REQUIRED TO HAND-HOLD PARTS THAT COULD BE POSITIONED IN JIGS, CLAMPS OR FIXTURES?
- ARE DIALS AND CONTROLS DIFFICULT TO READ AND IDENTIFY?
- DOES THE JOB REQUIRE SPECIAL LIGHTING?
- DO HAND TOOLS OR OTHER EQUIPMENT TRANSMIT VIBRATION TO THE WORKER'S HANDS, ARMS OR WHOLE BODY?

Note: Table material ©Donald L. Morelli.

RISKS OF POOR ERGONOMICS

To illustrate the types of information available from the three methods, let's consider what each indicates about electronics work stations.

Review of injury histories indicates that workers in electronics assembly and test jobs have a high incidence of disorders linked to ergonomic deficiencies of the work place. These include:

- Repetitive motion disorders involving the hands and wrists, the shoulder and the elbow.

- Complaints involving the neck and shoulder.
- Injuries involving the lower back.

Such incidents indicate that deficiencies may primarily exist in:

- Hand tools.
- Relative heights of the work surface or work piece.
- Static loading of the muscles of the upper torso to hold the head in a forward flexed posture.
- Material handling at the work station that may require repetitive twisting or a large range of motion.

Observations of many electronics work stations support the indications of the injury data. Hand tools, combined with nonadjustable fixtures or hand holding of components, frequently induce a combination of ulnar deviation, flexion or extension while performing forceful gripping activities. Such requirements have been linked to occurrences of tenosynovitis, tendonitis and carpal tunnel syndrome.

Because of the need to bring the work into focus for detail work, the head many times is held forward in positions in excess of 30 degrees. Such postures will certainly lead to fatigue of the muscles of the upper torso and neck.

In situations where the work surface may be too high relative to the worker, undesirable abduction (raising) of the upper arm occurs. This can also lead to seemingly questionable claims of injury to the shoulders. In fact, irritation of the tendons of the rotator cuff as well as entrapment of the nerves which run through the shoulders and down the arms can occur.

The location of tool holders, parts totes and/or material carts many times induces repetitive twisting or reaching to perform routine tasks. Where such requirements exist, the incidence of lower back injuries tends to be high.

All of these undesirable elements can be improved by application of the criteria mentioned previously (importance, frequency, function and sequence), as well as adherence to basic ergonomic principles such as:

- Keep the wrists straight.
- Keep elbows down.
- Minimize moments on the spine.
- Minimize twisting and bending.
- Provide adjustable chairs and/or work surfaces.

It is not just coincidence that both informal conversations with electronics workers and formal surveys often yield comments to the effect that work stations should be:

- Versatile.
- Adjustable.
- Flexible.
- Adaptable.

Such characteristics are among those routinely recommended for ergonomic improvements to work stations. This only reinforces the desirability of gathering information from workers prior to major work place changes.

WORK STATION DESIGN

It is important to keep these ergonomic principles in mind when designing work stations. It also helps if you think of yourself doing the job. Sit down in the existing station that is being used. Take careful note of where work items are placed and ask yourself: How easy are they to reach? How easy are they to see? What percentile are you in the anthropometric tables?

The person using the work station daily probably has it set up the best possible way within the fixed constraints of the inflexible work bench. You may also find that some make-shift alterations have been made for easier use. Seek out these red flags! They are typically areas where a better ergonomic solution is needed.

Most workers subconsciously adapt to the "undesigned" work place. They may not even be aware that there are better ways of performing their work. However, after years of adapting to the work place, other physical problems tend to creep up which can lead to high medical and health care costs to the manufacturer in the long term.

MATERIAL HANDLING INTERFACE

Figure 1 shows a form used for capturing the information necessary to properly design a work station. Its title, "Micro Material Handling Work Station Design," emphasizes the focus on the work process. Materials and equipment are handled in the locality of the work station. Focusing on the needs and equipment constraints will lead to better ergonomic solutions. When addressing equipment location we must consider:

- Size.
- Weight.
- Frequency of use.

work station no.			work station name *INSPECTION/FUNCTIONAL TEST*		stand-up X	combination
description of work performed *ASSEMBLED UNITS ARE USUALLY INSPECTED*					sit down	esd protection X
TO MAKE SURE EVERYTHING IS THERE AND THEY ARE COSMETICALLY					panel height	
ACCEPTABLE. THE UNIT IS CONNECTED TO A MONITOR TO MAKE					work surface size *2 X 4 FT.*	
SURE IT FUNCTIONS TO SPECIFICATIONS.					w.s. width	other
*tool & equipment requirements					size depth	see comments
item	size	weight	best location			used often
TERMINAL 1	*25 X 18 X 12*	*30 LBS*				
TERMINAL 2	*19 X 16 X 7*	*15 LBS*				
MONITOR	*22 X 15 X 17*					
① TOOL BOX						
AIR TOOL			*OVERHEAD*			
SHELF (MANUAL STORAGE)	*4 FT*					
② TURNTABLE						

*utility requirements	item	yes	no	best location		used often
general task lighting						
critical task lighting		X		*2 REQUIRED*		
electrical outlet strip #outlets		X		*12 ON 2 STRIPS*		
air		X				
vacuum						
other						
comments *ESP PROTECTION ON WORK SURFACE, FLOOR MAT AND WRIST STRAP.*						
① FUTURE INSTALLATION						
② TO HELP MANEUVER UNIT DURING INSPECTION						
*note: indicate preferred or ergonomic placement of these items						

Figure 1. *Micro material handling work station design.*

This also applies to location of utilities so they can support rather than hinder the work process.

When ergonomic principles are used, the work place is designed around the person, his or her equipment and the work he is asked to perform. If we use the 5th-95th percentile from the anthropometric data in the *Handbook of Industrial Engineering* (see Salvendy, "Suggested Readings") as design parameters, the solution should work out well. But be careful when using the data, particularly if the population in your manufacturing facility is skewed toward short or tall people.

The population of production workers on the West Coast, where there are many Orientals and Hispanics, is typically skewed toward the shorter end.

It is interesting to note, from the anthropometric tables for design parameters of the 5th to 95th percentile, that the spherical reach for male and female is obviously different (i.e., minimum 23.5 in. to 26 in. horizontal radius); however, the following items are the same for male and female:

• Sight arcs of 60 degrees each side (120 degrees total) are met with an easy head rotation of 15 degrees.

- Sight arcs of 15 degrees above horizon and 30 degrees below give a range of 45 degrees.
- A work surface height range of 5 in. will satisfy the entire range of males and females.

A work station that is well designed with all attention given to ergonomic detail may prove to be an "ergonomic problem" if it is too rigid. Flexibility and adjustability allow the work station to adapt to the range of different heights and sizes of people. Designing for flexible, adjustable and adaptable work places provides the forgiving qualities demanded by change. Some situations that call for change are:

- The operation changes to require stand-up versus sit-down assembly (or vice versa).
- Additional equipment is needed as part of the test procedure.
- People on different shifts use the same work station.

Systems furniture, with built-in flexibility and adaptability, has been around for open offices since 1968. In recent years, manufacturers have finally provided a systems furniture solution for manufacturing assembly and test applications. When planned properly, systems products consisting of support panels, work surfaces and work support components can provide ergonomic solutions. Planning and installing the work station can still be "fine tuned" to height adjustments or equipment changes that fit the needs of the person or the work process. A total reconfiguration of the work station can be easily accomplished within minutes if major changes in the process occur between planning and installation.

TEST STATION DESIGN

Systems product solutions for test station design can accommodate many of the needs in the environment. Some of these needs typically fall into the following categories:

- Need to access test equipment and computers/terminals.
- Location of equipment and controls within easy reach.
- Comfortable visual angle.
- Elimination of glare on CRT screens from overhead and surrounding lights.
- Better display of information for work or reference while using a personal computer or terminal.

Figure 2 illustrates a systems product solution for a test station design. (Circled letters on the figure refer to those below.)

A) The corner work station provides an easily reached arena design.

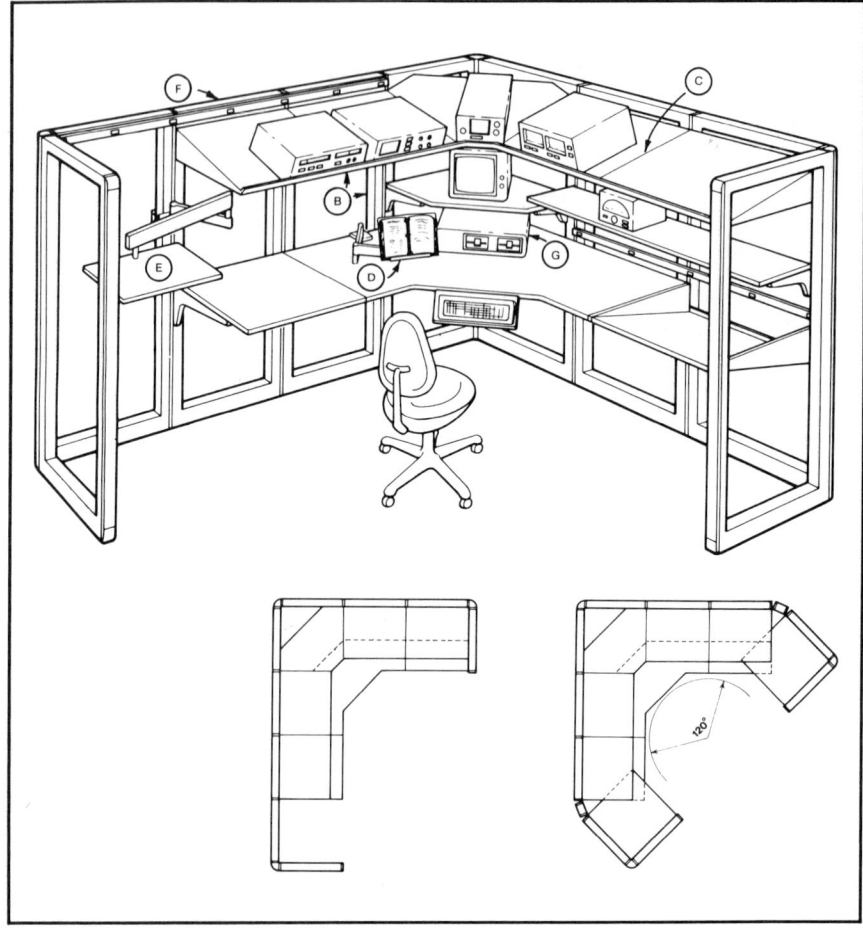

Figure 2. *Systems product solution for a test station design.*

B) Above the work surface are two levels of shelving for test equipment resembling a cockpit design. The first level would contain the more frequently used equipment at eye level or slightly below.

C) The shelves at the next higher level are at a 15 degree angle for easier viewing.

D) A reader stand on a pivot arm is provided to put reference material within at an easy view angle. It also presents information on the same viewing plane as the CRT screen.

E) A heavy duty pivot arm allows movement of equipment in three dimensions for infinite adjustment to particular needs. When positioned properly, it can be used to share equipment between stations.

F) Power strips can be mounted where needed.

G) The depth of the corner work surface easily accommodates a computer/ terminal with enough space in front for additional work. The keyboard tray is adjustable for different heights and can be retracted under the work surface when not in use.

Figure 3 is a plan view of the test station, where additional equipment is still within the 120 degree sight arc.

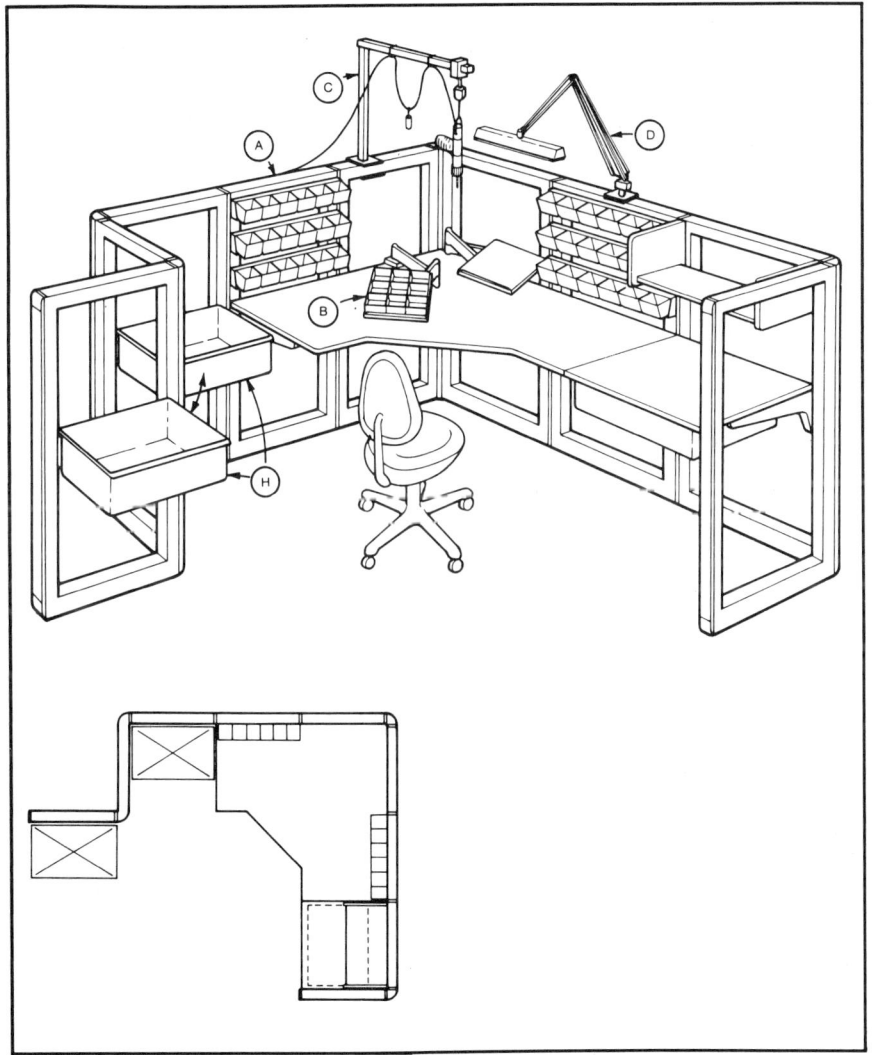

Figure 3. *Plan view of the test station.*

In both Figure 2 and Figure 3, more frequently used equipment would be located in the center of the work station. The less frequently used equipment would be located on the higher sloped shelf and near the ends of the sight arcs.

ASSEMBLY STATIONS

The corner work station is also applicable to work-intensive assembly stations. In Figure 3, the arena design provides the following ergonomic advantages (refer to circled letters on drawing):

A) Access to many small parts within easy reach to the front and sides of the assembler. Rail-mounted subcontainers are at a "friendly" 15 degree angle.
B) Additional small parts can be dispensed on trays mounted to pivot arms. A person with short arms can bring them in closer to the work. A person with longer arms can move them a comfortable distance away.
C) A pivoting tool arm supports the weight of a power driver. The unique design of the handle allows both vertical and horizontal supported use of the tool.
D) Adjustable critical task lighting allows individuals to adjust according to their work needs and eliminate their glare angle.
E) Adjustable height of the work surface at one-inch intervals allows the work surface to accommodate the height of the work piece.
F) An adjustable slant board can be used for PCB parts loading. Each person can adjust the angle as needed.
G) Electrical outlet strips can be mounted at work surface height where tools are used.
H) Proper material handling interface with the station eliminates awkward lifts and reaches.

ERGONOMIC SEATING

If there is anyplace in an electronics facility where ergonomic seating is a necessity, it is the manufacturing assembly area. Assembly and test workers spend most of their day sitting in one place doing their work. Over time, improper seating can lead to serious medical problems. Good ergonomic seating provides:

• Adjustable lower lumbar support in seat back.
• A waterfall edge on the front seat to prevent cutoff of blood circulation.
• Forward/backward seat back adjustment/movement.
• A recessed seat to give even support.
• Pneumatic height adjustment.
• Five-star bases for better stability.

LIGHTING

Glare is a big problem. Typical overhead lighting causes glare on CRT screens. When designing a facility, use a minimal amount of indirect lighting (eliminates glare) to illuminate workways and utilize critical task lighting at the work place. Care should be taken when using nonadjustable task lighting for assembly areas. The reflection from the solder and wire on PCBs can cause irritating glare.

Many companies have designed a lot of exterior windows into the building to humanize the environment. However, too much outside light can also be a problem. The ability to control the amount of outside light with adjustable slatted windowblinds would provide a better working environment.

Electronics assembly environments need ergonomic attention in their design. The hardware and tools are available with systems designed for flexibility and adjustability in manufacturing environments. With creative thinking and problem solving, most work places can be designed with people in mind... ergonomically.

Suggested Readings

Grandjean, E. 1982. *Fitting the Task to the Man, an Ergonomic Approach.* Taylor & Francis Ltd. London.

Hasselquist,R. J. 1981. Increasing manufacturing productivity using human factors principles. Proceedings of the Human Factors Society 25th Annual Meeting.

Hutchingson, R. D. 1981. *New Horizons for Human Factors in Design.* McGraw-Hill. New York.

Salvendy, G. (editor). 1982. *Handbook of Industrial Engineering.* John Wiley & Sons. New York. chapter 6.9.

Smith, L. A., and J. L. Smith. 1982. How can an IE justify a human factors activities program to management. *Industrial Engineering.* February. pp. 39-43.

Tichauer, E. R. 1978. *The Biomechanical Basis of Ergonomics.* John Wiley & Sons. New York.

Reprinted from July 1985 issue of *Industrial Engineering.*

47

The Effect of Product Design on Product Quality and Product Cost

Douglas Daetz

Product design involves all those activities undertaken to implement the functionality specified for a product. It is assumed that the product's specification was developed with the participation of marketing personnel, for only by using feedback from customers and marketing personnel can a company be reasonably confident that a newly designed product will find enough buyers to be profitable.

Rather than list all the steps and project control mechanisms in an "ideal" process of product design, this section will simply point out two cost-effective augmentations to the processes typically used:

1. Assign one or more manufacturing engineers to the design team from the start of the project.
2. Provide a means for design and manufacturing engineers to concretely evaluate and improve the manufacturability of a design.

Currently, the assertion that these measures are cost-effective is based more on the principle of "prevention costs less than repair" than on extensive data from real cases. However, the combination of theoretical arguments and anecdotal evidence has convinced the management of most of Hewlett-Packard's product divisions to couple manufacturing/process/product engineers into design teams and to teach their design and manufacturing engineers a methodology for analyzing and scoring the "assemblability" of a mechanical design.

Involving manufacturing engineers from the beginning of a product development project has benefits for both the product itself and for the manufacturing system that will produce the product. By asking the designer questions like "Can't you combine those two parts?" or "Can you use just one common screw type instead of three different types of screw?", the manufacturing engineers will help improve the product's inherent manufacturability. The design engineers, in expressing to the manufacturing engineers what will be needed in the manufacturing processes to achieve the desired functionality of the product, will be providing lead time for the development and debugging of any new manufacturing processes or capacity enhancement needed. At HP, this concept is called *concurrent product and process development.*

The first step to achieving consistent quality is being able to measure attributes of interest. Until a few years ago, there were no quantitative tools and systematic methodologies for evaluating a design's manufacturability. In 1980, however, Geoffrey Boothroyd, a professor at the University of Rhode Island, published "Design for Assembly," a report detailing a methodology he developed at the University of Massachusetts in Amherst. The Boothroyd approach analyzes each part in an assembly from the standpoints of 1) necessity of existence as a separate part; 2) ease of handling, feeding, and orienting; and 3) ease of assembly. Results of the analysis include an estimate of the assembly time (including handling, feeding, and orienting time) and a rating for design efficiency.

By 1983 a second systematic and quantitative methodology for examining the design of a mechanical assembly had emerged, confirming the necessity of measuring assembly ease (or assembly simplicity). The "Assemblability Evaluation Method" was developed in Japan by Hitachi's Production Engineering Research Laboratory (PERL) over a three-year period after one of PERL's engineers attended a Boothroyd "Design for Assembly" seminar. Hitachi still considers its methodology to be proprietary. General Electric, which obtained a license from Hitachi to use the methodology, improved it and developed English-language training materials. (GE has made its improved methodology and training materials available to selected U.S. companies under its license to Hitachi.)

Both the Hitachi/GE and Boothroyd methodologies now count major corporations among their users. HP opted to emphasize the Hitachi/GE methodology because it appeared to meet HP's requirements for product variety and transferability. Being simpler than the Boothroyd approach, the Hitachi/GE method required less time to train engineers to use it effectively. In less than two years—starting with two facilitators who taught two-day courses and trained additional HP division facilitators—over 1,300 HP design and manufacturing/process/product engineers have learned this relatively objective, quantitative methodology that shows them how well they are doing with respect to *assembly simplicity* and clearly points to the parts of an assembly that need improvement. From the quality assurance viewpoint, using design reviews (a fairly common practice) and assigning a quality/reliability engineer early in a design project (a less common practice) should be included in a good design process. In addition, the use of an "experimental design" approach like the Taguchi method would undoubtedly be of great benefit in the development of more robust (less marginal) designs for a given cost or less costly designs for a given robustness. However, given both the limited background in experimental design of most U.S. engineers and the scarcity of people in the United States who know enough about the Taguchi method to train others in its use, uptake of the Taguchi method into U.S. firms has scarcely begun. Within HP, for example, the Taguchi method is just beginning to be used.

THE EFFECT OF PRODUCT DESIGN ON PRODUCT QUALITY

What limits 100% quality of conformance, i.e., what keeps some units of a product from conforming to the design specification? In many cases the problem is poor design—of both the parts themselves and how they fit together. Some parts may be designed with features that are difficult to fabricate repeatedly or with tolerances that are unnecessarily and unbuildably tight (often due to a "worst case" rather than a "statistical" analysis of tolerance stackup). Some parts may lack details for self-alignment or features that prevent insertion in the wrong orientation. In other cases, the design parts may be so fragile or so susceptible to corrosion or contamination that a fraction of the parts will inevitably be damaged in shipping or internal handling (particularly true if effort has not been made to design appropriate packaging for such delicate parts). Sometimes a design, due to lack of refinement, simply has more parts than are really needed to perform the desired function(s), so there is a greater chance of assembly error. Thus, problems of poor design may show up as errors, poor yield, damage, or functional failure in fabrication, assembly, test, transport, and end use. As Figure 1 shows, there are three arenas in which a product's design affects quality and cost: at the supplier's plant, in the manufacturer's own plant, and at customer locations.

How might a design induce quality problems at a supplier? A common cause of problems is incomplete or inaccurate specification of the item to be provided by the supplier. The supplier provides items that conform to the specification it received, but unfortunately, the parts do not satisfy the design intent because of some specification error of commission, omission, or transmission. This problem often occurs with custom parts (as opposed to stock parts) due to either weakness in the design process (e.g., requiring designers to do their own drawings of new parts but not enforcing their use of common drawing standards), engineers who do not follow set procedures, or sloppiness in the procurement

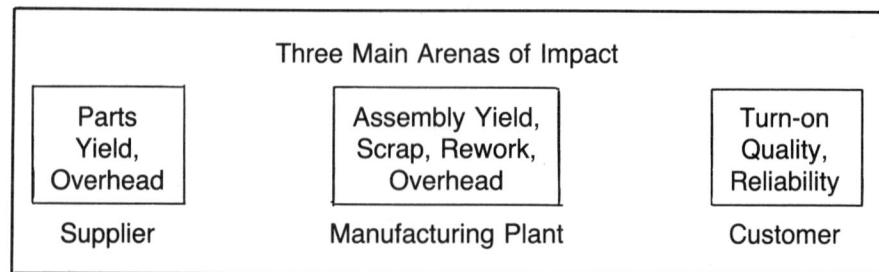

Figure 1. *The effect of product design on product quality.*

and purchasing process. In any case, the greater the number of different parts in a design and the more suppliers involved, the more likely it is that a supplier will receive an incomplete or inaccurate part specification. Obvious counter-measures include designing a product as much as possible around *preferred parts* (i.e., parts already approved based on their reliability and qualified source of supply); *minimizing the number of part numbers* in the design; and procuring parts from *a minimum number of vendors.*

Another common design flaw that leads to supplier quality problems is a specification that is too tight relative to the inherent variability in the supplier's manufacturing process, causing some items to fall out of the specification tolerance band. If the supplier must inspect all parts to separate conforming from nonconforming units, the purchaser has two problems: it must pay, directly or indirectly, for the supplier's yield losses, and it must defend against errors the supplier may make in screening out the nonconforming items that the process will continually generate.

How might a product design cause quality problems in the manufacturing plant where the product is assembled? If part fabrication is done in the plant, then many of the same problems as described above for a supplier can occur. In addition, there will be problems in the area of assembly and test. For example:

- Designs with numerous parts may cause part mix-ups, missing parts (because among so many parts, a missing part is not so noticeable), and more test failures.
- If some parts are very similar but not identical, the chances of an assembler using the wrong part in a given location are increased.
- Parts without details to prevent insertion in the wrong orientation may be assembled improperly.
- Complicated assembly steps and/or tricky joining processes may lead to incorrect, incomplete, unreliable, or otherwise faulty assemblies.

- Designs that require adjustments during or after assembly increase the chance of errors.
- The designer's failure to consider the conditions that parts will be exposed to in the assembly process, e.g., temperature, humidity, vibration, static electricity, finger oils, and dust, may lead to subtle weaknesses in some units or unit failures during testing.

Figure 2 summarizes and augments the foregoing points concerning the ways that design can directly affect assembly yields, the amount of scrap and rework, and manufacturing overhead expense. Implementing the guidelines in Figure 2 can improve quality and cut costs. For example, if told to *minimize the number of parts* in a design, a designer can typically find a way to cut parts by 20% to 40%. Eliminating engineering changes on released products (the last guideline in Figure 2) reduces in-plant and service errors and lowers direct manufacturing and overhead costs. A few companies, including some of HP's divisions, are already using the number of engineering changes after product release—the lower the better—as a measure of the effectiveness of R&D groups.

The final arena in which a product's design affects quality is at the customer location. The turn-on quality and reliability of units received by customers depend cumulatively on what happened at the suppliers and in the manufacturing plant. The survivability of units during shipment to customers also depends somewhat on the design of the product, and to a great degree on its packaging. Two additional design considerations that affect customer satisfaction—the perceived quality of the product—are the product's ease of use and the number of options of the product, i.e., the number of varieties in the model family. From the quality standpoint, a design should be so simple that correct assembly and use of the product are foolproof (which may be aided by concise, clear documentation) and should have as few options as possible.

THE EFFECT OF PRODUCT DESIGN ON PRODUCT COST

There is an obvious correlation between product quality and product cost. Product cost includes the manufacturing cost, the expenses associated with product warranties, and engineering redesign costs. Furthermore, short-run loss of orders due to specific product problems and long-run loss of orders due to a general lowering of reputation, although these costs are hard to estimate, represent product costs due to poor quality.

Many of the characteristics of a product's design that affect conformance quality also have an associated effect on manufacturing cost. Table 1 lists the principal components of manufacturing cost.[1] The items are grouped by three main categories:

1. Labor directly and indirectly applied to actual production of products.

Minimize Number of Parts
- Fewer part & assembly drawings —► Less volume of drawings & instructions to control
- Less complicated assemblies —► Lower assembly error rate
- Fewer parts to hold to required quality characteristics —► Higher consistency of part quality
- Fewer parts to fail —► Higher reliability

Minimize Number of Part Numbers
- Fewer variations of like parts —► Lower assembly error rate

Design for Robustness (Taguchi method)
- Low sensitivity to component variability —► Higher first-pass yield
—► Less degradation of performance with time

Eliminate Adjustments
- No assembly adjustment errors —► Higher first-pass yield
- Eliminates adjustable components with high failure rates —► Lower failure rate

Make Assembly Easy and Foolproof
- Parts cannot be assembled wrong —► Lower assembly error rate
- Obvious when parts are missing —► Lower assembly error rate
- Assy. tooling designed into part —► Lower assembly error rate
- Parts are self-securing —► Lower assembly error rate
- No "force fitting" of parts —► Less damage to parts, better serviceability

Use Repeatable, Well-Understood Processes
- Part quality easy to control —► Higher part yield
- Assembly quality easy to control —► Higher assembly yield

Choose Parts that Can Survive Process Operations
- Less damage to parts —► Higher yield
- Less degradation of parts —► Higher reliability

Design for Efficient and Adequate Testing
- Less mistaking "good" for "bad" product and vice versa —► Truer assessment of quality, less unnecessary rework

Lay Out Parts for Reliable Process Completion
- Less damage to parts during handling and assembly —► Higher yield, higher reliability

Eliminate Engineering Changes on Released Products
- Fewer errors due to changeovers & multiple revisions/versions —► Lower assembly error rate

Figure 2. *Design guidelines for quality improvement.*

2. Materials for products and manufacturing processes.
3. Overhead chargeable to manufacturing operations.

In many companies, all but direct manufacturing labor and product material cost are counted as overhead. For most U.S. produced electronic and computer products, direct material cost represents 50% to 80% of total manufacturing cost, while direct production labor cost ranges from 2% to 15%. Manufacturing overhead cost for these products would then be in the 15% to 45% range. Material cost is the biggest slice of the manufacturing cost pie, with manufacturing overhead cost the next largest; the smallest piece is direct production labor cost.

Members of GE's "Design for Assembly" training staff point out that about 75% of manufacturing costs are determined by the design of the product. With products in which parts alone are 65% to 80% of the manufacturing cost, the design may account for 90% or more of the total manufacturing cost. Thus, no matter how clever manufacturing engineers, quality engineers, production managers, assemblers, or production control specialists are, they won't be able to affect more than 10% to 25% of the manufacturing cost of a given design. Therefore, design engineers and their managers must be given more information on direct and indirect costs of the alternative designs they conceive.

Figure 3 was developed to help design engineers and their managers see the costs associated with their design decisions. This tableau qualitatively rates the effect of nine design measures on direct material cost, assembly labor cost, and 16 different overhead cost elements. For example, design changes are attributed "major driver" status for all cost elements except facilities/space, where they are still considered to have "significant impact."

To date there are not many precise, quantitative data nor verified functional relationships between product design and product cost. Preliminary findings, however, suggest the following:

- Assembly time (assembly cost) is roughly proportional to the number of parts assembled, given the same type of assembly environment.
- Material costs per unit can usually be reduced by reducing the number of parts in each unit.
- The system cost of carrying a part number in a manufacturing division might range from $500 to $2,500 annually, not counting the costs of the parts themselves.
- Establishing and qualifying a new vendor for a new part may cost almost $5,000.
- On average, in a large design and manufacturing facility (more than 500 employees) the system cost of making design changes after a product has been released to manufacturing may be $5,000 to $10,000 per design change.

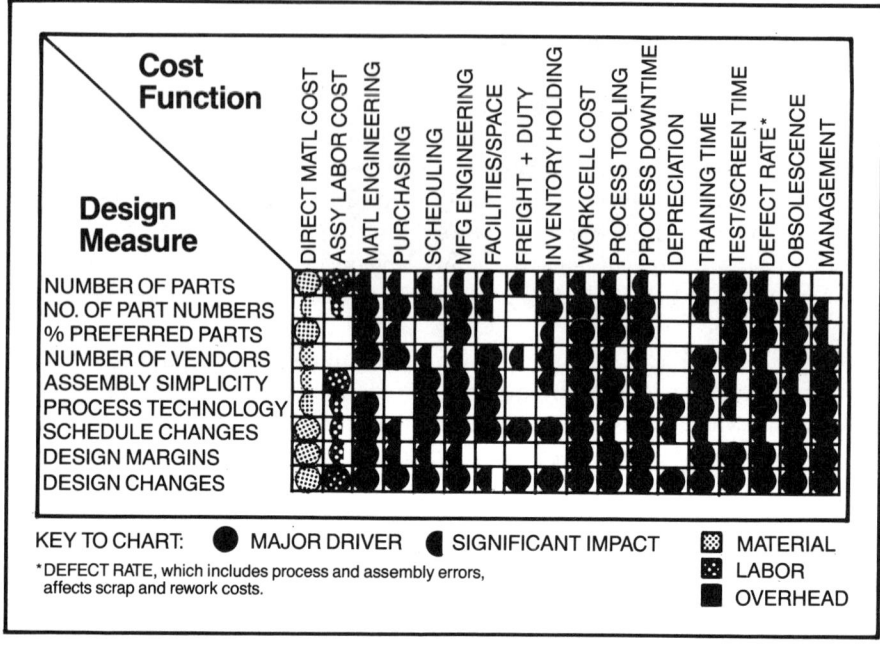

Figure 3. *Manufacturing costs driven by design.*

The above discussion focused on reducing manufacturing cost through design, but using the guidelines in Figure 2 is also expected to reduce product development costs. For example, if products and manufacturing processes are designed concurrently and designs involve fewer parts and part numbers than before, then expenses for engineering labor, drawings, redesign, and tooling should be reduced. Another bonus may be earlier product introduction.

TABLE 1 – Principal Components of Manufacturing Cost

Labor
- Direct Manufacturing
- Indirect Manufacturing (includes supervisory and manufacturing-related engineering support labor)

Materials
- Product
- Process

TABLE 1 — Principal Components of Manufacturing Cost (Cont)

— Consumables
— Equipment (Expensed, Capitalized)

Overhead
- Labor (includes purchasing, accounting, facilities, etc.)
- Nonproduction materials
- Equipment
  — Engineering (Expensed, Capitalized)
  — Office (Expensed, Capitalized)
- Utilities
- Facilities (includes buildings, land, and taxes thereon)
- Cost of capital
- Corporate charges (e.g., allocations of some corporate headquarters expenses to the manufacturing operation)

CONCLUSION

Focusing on quality and manufacturability during the development phase of a product's life cycle is crucial. The level of conformance quality that may be achieved in production and the product's cost are largely determined by the design of the product.

At this time, some of the key product design measures for achieving competitive quality and competitive product cost appear to be:

- designing product and process concurrently.
- measuring and striving for assembly simplicity.
- minimizing the number of parts.
- minimizing the number of part numbers.
- using as high a percentage of preferred parts as possible.
- minimizing the number of vendors.

These measures are not completely independent of each other, and as yet there has not been a systematic analysis to quantify their individual and joint contributions to both product quality and product cost. To a certain extent they may even seem to product development engineers to compound their problem of making tradeoffs in their designs in order to satisfactorily meet functionality, schedule, and other constraints. However, giving product development engineers explicit guidelines for making their design choices should make their job easier. The six measures listed above represent a start for focusing attention on the effect of product design on product quality and product cost.

Acknowledgment

I am grateful to the Hewlett-Packard Co. for allowing me to share some of what I have learned during the past two years through my work in the Manufacturing Research Center (MRC). More specifically, I would like to thank Bob Grimm, recently retired director of MRC, and Mike Lee, manager of the Control Systems Department of MRC, for reviewing and suggesting improvements to this article, and for approving its publication.

Footnote

1. Neither "General and Administrative" (G&A) and "Selling" expenses, nor "Profit" and "Taxes" are included in this tabulation of the elements of manufacturing cost.

Suggested Readings

Andreasen, M. M., S. Kahler, and T. Lund. 1983. *Design for Assembly*. IFS Publications Ltd. U.K. and Springer-Verlag. Berlin/Heidelberg/New York.

Bolz, R. W. 1981. *Production Processes: The Productivity Handbook*, 5th edition. Industrial Press Inc. New York. [See first three chapters.]

Boothroyd, G., and P. Dewhurst. 1983. *Design for Assembly*. University of Massachusetts, Dept of Mechanical Engineering. Amherst, MA.

Schreiber, R. R. 1985. Design for Assembly. *Robotics Today*. June. 7(3): 45-46.

48

Manufacturing and Design: A Symbiosis

Daniel E. Whitney

Not so long ago a product's designers seldom worried about how it would have to be built. Automation engineers took the new design and did their best to devise an assembly method, either by hand or by machine. Manufacturers tried where they could to use inexpensive parts and to push them into assembly as fast as possible. The result: problems arising out of the design or fabrication, or caused by parts being out of tolerance, had to be solved by the assembler's ingenuity—usually in undocumented and unappreciated ways.

This "hidden agenda" made it at least difficult, sometimes impossible, to introduce robots and other advanced assembly technology. Manufacturers balked at the cost of equipment that robots needed but that people did not—such as special pallets, parts feeders, and control computers. Manufacturers quickly found that robots could not work around such problems as parts out of tolerance; feeders jammed and the robotic system came to a standstill.

High technology, it seemed, was inadequate. Few companies wanted to spend money for better-made parts just so robots could put them together. Instead, they demanded better and brighter robots that could adapt to and solve the problems.

However, in the last few years attitudes have begun to change. Manufacturers are realizing that better parts not only help the robots; they actually make for a better product.

Manual assembly is rapidly disappearing as an option in high-tech products because mere people are inadequate in terms of quality, documentation, and cleanliness. With increasing international competition, there is little time to put manufacturing mistakes right. Designers now have to work with manufacturing engineers to juggle a welter of options, realizing that no matter what the technology, a product's manufacturing costs are determined from the moment it is designed.

How can product and process engineers improve design and choose the best manufacturing techniques? Three examples point the way:

- Designers of an IBM lap-top computer were faced with a problem that taxed their ingenuity, as well as their manufacturing and purchasing skills. The power-supply board and the disk drive had to be connected by a 4-inch ribbon cable. The task; plug in the two ends of the cable automatically.

Such a task presents no problem to a human worker, who easily sees and handles the plugs as they flop around on the ends of the cable; but it is out of the question for automatic insertion machines, and even for today's most advanced robots. To get away from doing the job by hand, the IBM team redesigned the plugs.

They had a vendor make each cable so that its two plugs would stack, one on top of the other, to form a sort of block with a loop of cable out to one side. The connected plugs, packed in trays, are shipped in that position. During assembly, a robot grips the two-plug block, using a simple vision system, and inserts the whole block into the power-supply board. Then it detaches the block's upper half (the other plug) and inserts it into the disk drive; finally, the robot unkinks the cable.

- Japanese shipbuilding companies like Ishikawajima-Harima Heavy Industries Company and Mitsubishi Heavy Industries have brought efficiency to their complex manufacturing processes by making the design of the ship a part of the construction planning. Once the ship's overall shape and characteristics have been determined, designers plan the size and shape of its sections by the order, method, and location in which they will be made.

First, the size and shape of the subassemblies are decided. Then planners identify individual pieces of hull plate, pipe, deck, and so on for each subassembly. Next they draw up schedules for the ordering of raw materials and joining together of parts into subassemblies. Once the subassemblies have been built and connected, little final assembly work is needed.

- In Volkswagen's Hall 54 plant in Wolfsburg, West Germany, some 25 percent of a car's final assembly is by robots or special machines, up from 5 percent. Among significant changes in car-design practice was the front-end configuration. By adding just one part to the frame, the front end can be left open for the engine to be installed by hydraulic arms in one straight upward push. What once took several men at least a minute is now done by a machine in 26 seconds.

Another telling example involved screws. Volkswagen convinced its purchasing department to pay 18 percent more for screws with cone-shaped tips. Robots could then insert the screws, even if the sheet metal or plastic parts were out of alignment. So many West German companies adopted cone-point screws over the next two years that their price fell to that of the ordinary, flat-tipped kind. Everyone from manufacturing to purchasing was happy.

PLANNING FOR PRODUCT AND PROCESS

Intense competition leaves little time for mistakes or learning from experience at either the design or the manufacturing stage. Engineers at both levels need to work together to create the product and its assembly system more or less at the same time.

A design team's agenda comes in five parts:

- Making the design work—the realm of the design engineer.
- Selecting components and identifying those that have to be designed and made from scratch.
- Choosing fabrication and assembly techniques—for example, manual assembly, robots, or special high-speed machinery.
- Selecting mounting, joining, and fastening methods—for example, screws against adhesives.
- Understanding the interactions between the choice of components, manufacturing methods, assembly sequence, and so on.

When working with manufacturers, Charles Stark Draper Laboratory always asks product designers to rethink their product: Why three springs instead of two? Why use screws to put it together if it will never need to be taken apart?

The Polaroid SX-70 camera series, introduced in the 1970s, is an excellent example of snap-together products with no fasteners; in the IBM Proprinter, screws have been eliminated to ease assembly [see "Matrix Printer: No Pulleys, Belts, or Screws."].

In general, the team must identify critical design and manufacturing decisions and make them in the correct sequence to avoid backtracking. Planning for manufacturing involves creating a suitable assembly sequence, identifying subassemblies, deciding at what stage to test subassemblies, and designing each part so that its functional and tooling tolerances (gripping and jigging surfaces) are compatible with the method of assembly.

One technique to aid planning is to recognize similarities among parts so they can be grouped and worked on together, with similar parts sharing a machine or a measuring method. A factory making a variety of analog and digital circuit boards, for example, will have standard insertion machines that rapidly install inexpensive digital components. But analog boards contain many expensive, odd-shaped parts that are hard to install.

Robots and human workers are the only choices for the assembly line, and because labor is relatively expensive in some western countries, boards with odd-shaped parts may have to be built elsewhere. On the other hand, boards with only standard-shaped parts can be built economically in those countries. Designers must appreciate those factors and avoid combining the two types of component on one board.

When boards have to be assembled in countries with low labor costs, long distances between builder and user may cause logistical problems if small batches are ordered, or if the design is changing rapidly—one reason why companies are turning to robots to load circuit boards. A highly simplified logic diagram outlines some of the choices a design team faces as it works its way toward the best combination of components, insertion techniques, and logistics [Figure 1].

Keeping Suppliers Current

Manufacturing techniques, especially with more mature products, can be the heart of a company's competitive edge. A key example is the so-called just-in-time (JIT) production technique—where inventory is cut back so that no more than the appropriate amount of materials is ready at each point in the manufacturing process. The JIT manufacturer takes control, ordering what he needs only when he needs it. A typical order to a supplier might be: "I want 4337 of model A, 143 533 of model B, 6 of model C, and 507 of model D—by 7 a.m. tomorrow."

But suppose the items being ordered are themselves assemblies. How does the supplier fill the order? Not by stocking a huge warehouse and drawing out such orders each day—models and quantities ordered fluctuate greatly, and inventory costs are high. Instead the supplier, using the combinatoric method of product design, may also be a JIT manufacturer. In this method, generic parts are combined into the required specific product, whose market has been analyzed to determine the range of models needed—a range reflected in the various types of each part.

This analysis for an electronic subassembly, for example, might show the need for three voltages (5, 12, and 24 volts), two ranges (zero to peak and minus to plus peak), and three output pins. The design team ensures that any of the 18 combinations (3 by 2 by 3) will fit mechanically and function electrically.

Along with the parts themselves, manufacturing engineers design equipment for part feeding, assembly, and testing, so that batches of any size and sequence

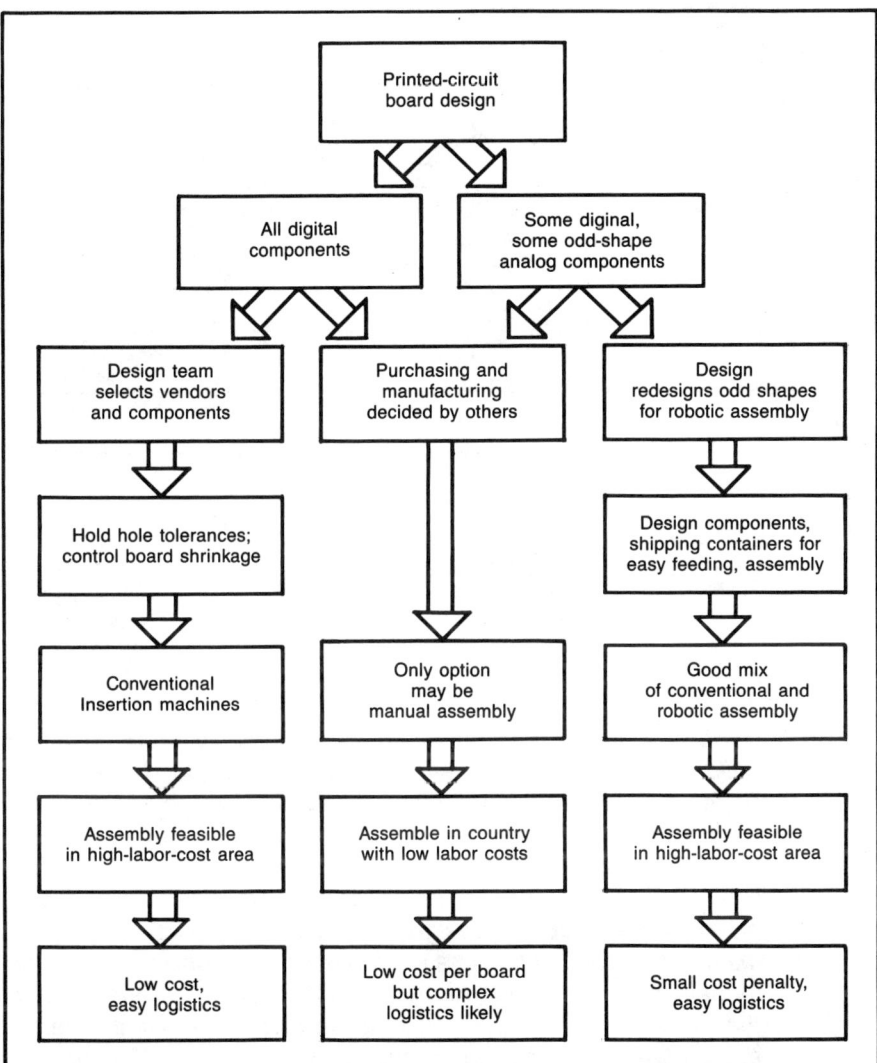

Figure 1. *A design team has various options to consider when preparing to manufacture a PC board. The team must decide which path yields the lowest cost, the easiest inventory management, the greatest manufacturing flexibility, and the widest options for redesign—all according to the circumstances. But if the designer and manufacturing engineer do not work together, some options may be closed off.*

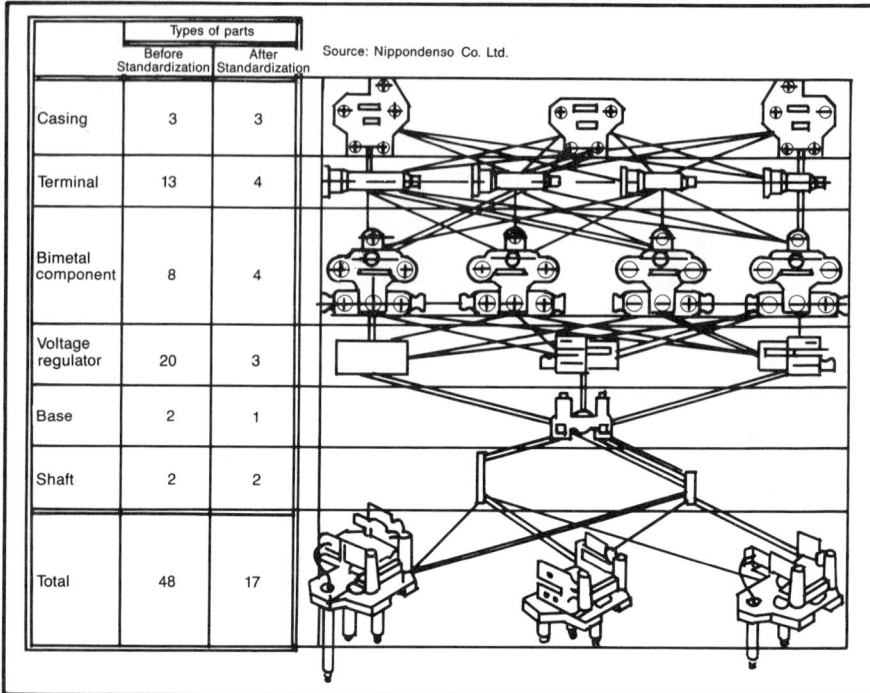

Types of parts		Source: Nippondenso Co. Ltd.
	Before Standardization	After Standardization
Casing	3	3
Terminal	13	4
Bimetal component	8	4
Voltage regulator	20	3
Base	2	1
Shaft	2	2
Total	48	17

Figure 2. *Depending on what assembly path is selected, Nippondenso can build any one of 288 variations on a panel gage. Three of the gages built for Toyota are shown in this artibrary route selection, which illustrates the combinatoric method of design and assembly. By combining variations of six generic parts, the company can make any of the variations in any quantity. A multistation assembly machine, specially designed to take just-in-time orders from main contractors, turns out a gage every 0.9 second and changes models, also in 0.9 second, by passing a dummy casing through the machine. To facilitate making the gages, the Nippondenso design team reduced the number of basic parts to 17 from 48.*

can be built with minimal changeover time. In Japan, Nippondenso Company Ltd. manufacturers automotive gages for Toyota by combinatoric assembly. Changeover from one type of gage to another takes no more than 0.9 second (see Figure 2).

The Best Assembly Method

The way things used to be, assembly methods were not chosen by an industrial engineer until well after a product's design was complete. In addition, the assembly lines were manual, and the most economical process gave each worker about the same amount of work to do.

But the design team must now consider automatic assembly and its effect on the way parts are designed and tested. A manufacturing engineer has many choices of automatic-assembly equipment. Some are slow and cheap, others fast and costly. One manufacturer uses the following rule of thumb; if the basic cost per piece for the fastest variable-center distance-insertion machine for integrated circuits is equivalent to 1, then the approximate cost for a conventional radial-lead part inserter, or an insertion machine for dual in-line packages of integrated circuits, would be between 1 and 2. A robot would cost roughly 50 to 75 times, and a human worker in the United States 150 times, the basic cost per piece.

Some assembly equipment can only insert one part, and must immediately pass the work to another station. A robot, on the other hand, can perform several steps in a row and becomes more economical the more steps it is asked to do. However, if steps that a robot can do are separated by steps that it cannot do, two robots may be needed, which could be too costly. Many sequences should be analyzed to see which affords the least expensive assembly system.

In many products, success of automated assembly depends on how predictable is the positioning of mating surfaces, which depends on tolerances on the surfaces that mate, as well as on those that are gripped by the robot.

A product's assembly can usually be broken down into several sequences, each presenting different surfaces to the fixtures and grippers. Each sequence must therefore be analyzed to identify which surfaces are to be gripped. Since all tolerances are determined during design, it is essential that the assembly sequence be established at this time, not months after.

Sensors, of course, can be used to help steer parts into the mating position, but the equipment can be slow and expensive. In making ICs, for example, robots must typically be able to insert electric parts in less than 2 seconds. Machines for inserting one type of IC package—conventional dual in-line packages—can be anything from 10 to 50 times as fast as robotic insertion. In such cases there is not enough time for sensing: chamfers on the parts and compliance in the insertion tooling together make a practical solution.

The design team must also decide the best stages at which assemblies should be inspected. It is usually costly, and often too late, to find a bad part after it has been buried under other parts or permanently fixed into its place in the product. Each testable fault must be classified as to when it can first occur, when it can first be detected, and when it can no longer be easily detected and fixed. The cost of testing and repairs must also be determined. Even for the same fault, the results of such an analysis may vary with each assembly sequence.

In one case, a manufacturer was designing a robot assembly line for videocassette-recorder chassis. The design team decided that a test was necessary to verify that the chassis motor and linkages were correctly assembled. But the power terminals to conduct such a test were not connected until several assembly steps had obscured the mechanism. The solution; a specially designed robot tester with its own power leads and touch sensors to see that the mechanism worked. After the test, the robot left the mechanism ready for the next step in the assembly process.

The Surface-Mounting Challenge: Access To Circuits

Jon L. Turino

New manufacturing methods can create headaches in design and testing, especially when intricate microchips are involved. A case in point is the surface-mounting technology that over the past few years has become popular for producing printed-circuit boards.

Effective circuit design for surface mounting parts on boards means making sure circuit designs can be built and tested as economically and as quickly as possible. Built-in, on-board testability should thus be integral to any board design employing surface-mounting technology.

Surface mounting, which has been prominent for several years now, differs from traditional insertion mounting of dual in-line packages. With insertion mounting, the package leads are inserted through holes in the board and fastened to the back by a wave-soldering process, where liquid solder flows across the board's bottom surface and attaches to the leads and walls of the hole. This method of mounting restricts the packages to the one surface of the board.

Surface-mounted packages are put on flat patches of solder already on the board, allowing packages in both surfaces. What's more, the surface-mounted parts are smaller and can be placed closer together than the older kinds. Surface-mounting technology gives the designer two to four times the functional circuit space available with insertion fabrication.

Some 25 percent of all PC boards built in the United States are found to have either manufacturing defects or electrical interaction and functional faults. A good design for a surface-mounted board must therefore include access points for automatic testing. Such tests during manufacturing can not only correct faults, but also improve the process and verify the quality of the assembly work.

However, the automatic equipment must have access to all internal circuit nodes—physical access for in-circuit testing, electrical access for functional testing. Such access requires exclusive space on the board. Indeed, physical or mechanical access may require anything from 5 percent to 50 percent of the board space, depending on the design. The least expensive and most reliable in-circuit test fixtures gain access to circuits only at the board's bottom surface, through spring probes spaced on 0.100-inch centers.

Such a board design usually shuts the bottom surface off for components. At very least, it means making sure that all circuit noes are brought to the bottom of the board through via holes, as well as ensuring properly sized and spaced test pads on the bottom surface, with any components placed away from the pads.

The Surface Mount Technology Association, along with most makers and sellers of in-circuit testers, recommends that test pads be at least 0.035 inch in diameter to give probes a target of reasonable size. The pads should be placed at least 0.020 inch away from the soldered component lead, since surface-mounted components have a tendency to drift during soldering that can result in broken probes or broken parts.

More expensive fixtures, with 0.050-inch center probes, can be used if 0.100-inch test pads will not work, although cost tradeoffs can vary by several thousands of dollars. Components can be placed on both surfaces of the board, with fixtures able to access both surfaces. But dual-sided fixtures cost as much as 10 times the standard kind, and are sometimes unreliable.

Providing electrical access to the board's internal circuit nodes takes up much less board space, and supports both in-circuit cluster testing, as well as performance or functional testing. Electrical access can be converted to physical access in a number of ways. One technique, used by such manufacturers as General Electric Company is to put multiplexers on the board so multiple internal nodes can be monitored by only a few physically accessible pins; another method is to add shift registers. During testing, internal nodal data is latched into the shift registers and then serially scanned by only four or so pins from easily reached pads, or from the normal board-edge connector or a test-only connection scheme.

An even more elegant method, pioneered by Logical Solutions Technology, is to use a combination of selectable (by addressing) and serial scan techniques to interface with a testability bus. The bus can be brought either to a set of inexpensive pads on 0.100-centers on the board's bottom surface, or to some other physical interface.

Either in-circuit or functional, automatic test equipment sends data in real time to any circuit node, and observes data from any node in a similar manner. The latching capability of some testability circuits lets relatively slow automatic test equipment diagnose "fast" unit-under-test circuits. The combination, while requiring extra space on the board and adding a little to the cost of parts, needs less space than some other approaches. The method also supports either functional or performance testing, the latter now the method of choice for high-yield surface-mounting technology.

Matrix Printer:
No Pulleys, Belts,
or Screws

John Newman
Morris Krakinowski

A new design trend in industry—developing products with few parts so they can be manufactured easily on an automated assembly line—is embodied in IBM Corporation's Proprinter. The tabletop printer, a wire-matrix type for the company's personal computer line does not have the usual array of screws, pulleys, belts, and brackets. The parts simply snap together.

As a result, the printer can be assembled by robots, rather than manually. It is being produced at an IBM plant in Lexington; Kentucky, and has proved popular with consumers. The new design and the sparing use of parts have translated into lower manufacturing costs, higher reliability, and even increased capability compared with the IBM 5152 printer.

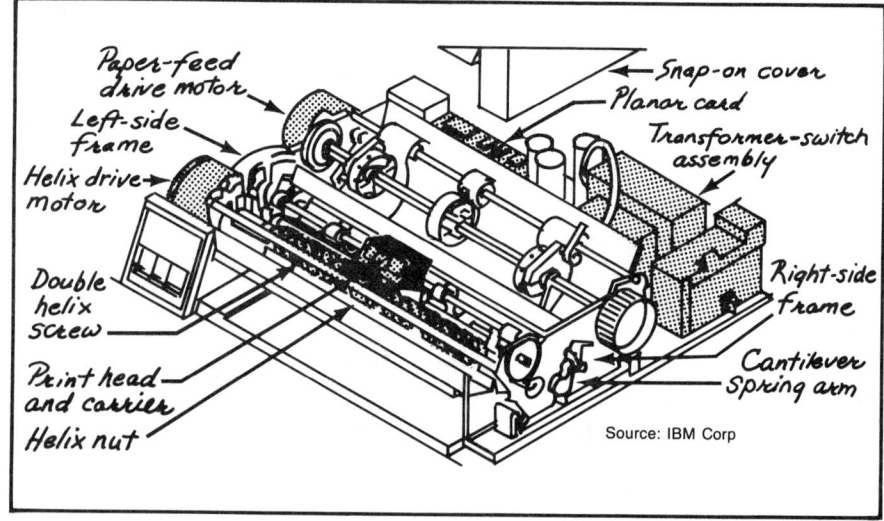

Figure 3. *The Proprinter uses 60 percent fewer parts than its predecessor because of a stringent design strategy: use no screws, bolts, or brackets and—wherever possible—have single parts carry out multiple functions. The product is not only cheaper to assemble but is also more reliable and more capable than the old IBM printer.*

To develop the Proprinter, IBM first created a set of "design for manufacturability" guidelines, in early 1983. Second, manufacturing engineers, tool designers, test engineers, and purchasing personnel were drawn into the design process at an early stage; and third, the product and the manufacturing process were developed simultaneously. The many manufacturing requirements that are reflected in the product design contributed to the primary goal of the Proprinter project—cutting production costs while maklng a high-quality product.

Fewer is Better

Because the Proprinter was to be assembled on a completely automated line, the design team aimed for a product with the fewest parts. A reduction in the number of parts cuts down on the number of handling operations, manufacturing fixtures, assembly operations, and robots.

The net result of the design approach can best be illustrated by comparing the Proprinter with the IBM 5152 printer, designed by a Japanese company. Like the Proprinter, the IBM 5152 is a dot-matrix variety. But it prints at only one rate—80 characters per second—and can handle only one type of paper.

The Proprinter, by contrast, has three print rates, ranging from 200 CPS to a near-letter-quality 40 CPS. Envelopes or individual sheets can be fed through the front of the printer while the usual tractor-feed paper is hooked up.

These extra functions were achieved with 60 percent fewer parts. The Proprinter has 61 parts, with no discrete fasteners, while the IBM 5152 uses 152 parts—only half of them functional, since 74 are fasteners. Although common fasteners cost only a few cents apiece, assembly of individual fasteners manually or by a robot is very time-consuming and as a rule increases the cost and reduces productivity.

Parts for Multiple Use

Two major strategies helped hold down the number of parts; wherever possible a single part performs multiple functions. At the same time a new fastening technique—elastic "snap action" fasteners—replaces discrete screws nuts, and washers.

The final design contains 13 moving parts and subassemblies, including two motors, gears, and pressure rollers, all held in place without screws or external hardware. Both motor assemblies, for instance are held in place by tabs in the left side frame. The motor is clamped in place by inserting its flange into the printer's side frame with a twist. Ordinarily the two motors might have been mounted with as many as 24 screws, nuts, and washers.

Wide use was made of six types of thermoplastic molded materials to obtain the required wear, rigidity, and stability for complex parts. The Proprinter's cantilever spring arm, incorporated in the printer side frames, is an example of a multifunction part. The cantilever springs replace conventional coil springs and apply load on the paper pressure roller. They also retain the pressure roller and serve as bearings for the rotating part.

The print-head drive is another instance of a single piece replacing an array of parts. A double-helix screw and molded nut drive the print-head and ribbon. The screw rotates while the nut moves axially, causing the print-head, which is attached to the nut, to traverse a horizontal print line. The helix screw and coupling are pressed directly onto the motor shaft. The preassembled helix drive then snaps into the side frame—a reliable robot assembly operation.

The double-helix print-head drive replaces the belt-and-pulley arrangement—with its multiplicity of small gears, snap rings and screws—found in many table-top printers. The flexible belt, for one thing, is difficult, if not impossible, for a robot to handle.

Electromechanical Package

The fewest interconnections and assemblies possible, as well as low cost, were the guidelines for the design of the electromechanical package. It consists of two subassemblies; the power transformer, and the logic and power circuitry on the planar card.

Five connectors are mounted on the planar card for a like number of electro-mechanical components. To eliminate discrete wire cables and to permit auto-matic assembly, the connectors on the planar card are mounted near the components. The transformer, two motors, and the print-head have pigtail connectors that plug into the board. The keyboard panel plugs directly into a connector on the board.

Originally the ac receptacle, the filter, and the power switch were mounted on the planar card. The design team later decided to put these components within the stamping that houses the transformer—a less expensive and more reliable solution.

Automatic Assembly Steps

Proprinters are assembled on an automated line at IBM's Lexington, Kentucky, plant (the original line described here, was in Charlotte, North Carolina). Eight robots put together 41 parts and subassemblies. Such an assembly operation usually consists of several robot arm movements. IBM engineers eliminated non-productive turning or orienting operations of each robot in the Proprinter line. The reprogrammable machines execute operations only to pick up and insert parts. Beginning with a base plate, the robots add parts sequentially layer by layer to build the assembly. Each part is firmly attached right away, leaving none loose to be held down for the next assembly operation.

The design of the print and paper-advance mechanism takes into account automated assembly considerations. Product design engineers proposed to hold the paper guide shafts and the print platen between two simple side plates. However, an elaborate fixture was needed to hold all the elements in alignment to slip on the side plates axially.

Automation process engineers, however, preferred a smooth sequential assembly on a single machine base without auxiliary fixturing. The result? A design in which the Proprinters side frames separately snap into the base. The paper feed roll is then installed and the print platen is snapped into slots on the side frames, forming a study structure that supports the rest of the print mechanism.

The exceptions to on-line assembly are the feed roll, pressure roll, and the tractor subassemblies. They are assembled off-line to reduce the time of final assembly. For example the tractor, which has 10 parts is assembled automatically in a special assembly station, with parts fed by vibratory bowl feeders.

Designing the Line

IBM engineers developed the automated manufacturing line for the printer in parallel with the product. The printer assembly line consists of eight robotic workstations linked by a conveyor system. Each station on average handles five to six parts and all of them complete their operations in the same amount of time.

Large parts are presented to the assembly robots in specially designed pallets. These parts, like the injection-molded pieces are accurately placed on the pallet by a parts handling robot. They are removed from the press without losing orientation. The parts handling robot holds the part while secondary finishing operations are performed. It then places the part accurately in a matrix array on a pallet. No human operator is involved.

The printer's rigid molded plastic base was carefully designed to withstand the stresses of automated transfer during assembly. First a laminated copper ground plate is installed on the base, followed by the planar electronics board. Molded guide pins in the base with generous tapers assure proper alignment and retention of the two broad flat parts.

The multifunction base plate has a series of lips and vertical hooks and slots that serve as fasteners for the large subassemblies and the cover. The heavy transformer and the power switch subassembly hook into a lip and engage two snaps. Two elastic rubber disks inserted in the base preload the subassembly against the lip and snaps, and serve as shock absorbers.

The lighter paper guides, each of which has snaps, are inserted directly into the slots in the base. The paper guides form a typical layered assembly that defines the paper path in the printer. The lower paper guide is inserted from above, followed by the upper guide.

Other parts such as shafts are fed from hoppers; small parts are delivered by vibratory bowl feeders.

Most parts have no special features to make gripping easy for robots. Rather, each robot uses custom grippers to handle a specific group of parts; no time is lost in changing grippers.

The dexterity of the robots, with seven degrees of freedom, is fully employed during assembly. The insertion and securing of most printer parts require the robot to execute motions along multiple axes.

Consider, for instance the assembly of the side plates. The side plate hooks into a lip on the base plate and is angled in until it snaps into place. The single-sided snap-action joint is a reliable fastener and can withstand heavy impact stresses. A Scara-type robot, which can execute only a vertical up and down motion, could not be used to insert the side plates. The design of the fastening method dictated a more complex robot. Even so, the parts were specifically designed to aid robotic assembly. The hook on the printer's base plate has a generous lead-in chamfer; the side plate has an angled catch for easy automatic mating.

The design team developed new techniques to manipulate and plug electric connectors. On the paper drive and print-head motors, prewired connectors are temporarily "parked" at fixed locations. After the motors are installed, the robot swings to the predetermined spot to grip the connector and plug it into the planar board.

Plugging the control keyboard into its socket on the planar logic board is a more complex operation. On the back surface of the preassembled keyboard there are an array of pins and four aligned buttons. Four keyhole-shaped slots on the paper guide plate act as guides during insertion. The semirigid connection of the keyboard to the planar board can withstand a 50-G drop test—an IBM requirement for reliability. A judicious distribution of mass in the printer assembly prevents severe dislocations that could damage the connection.

Digital Scopes: Assembly in Record Time

Kenneth I. Werner

Many of today's engineers and scientists first saw a Tektronix oscilloscope in an undergraduate college laboratory. They feel they have grown up with Tek scopes. But in doing their own growing up, the laboratory oscilloscopes made by nearly all instrument companies have become complex and, as a result, time-consuming and very expensive to assemble.

One of the exceptions to that trend is the 11400 line of digitizing oscilloscopes from Tektronix Incorporated, Beaverton, Oregon. It is complex by virtue of being fast and having many advanced functions, but it can be assembled in 45 minutes. By contrast, a predecessor, the 7854 digitizing scope, took 9 hours to assemble.

Simplicity Through Complexity

Tom Rousseau, project manager for the 11400 series and his design engineers faced a difficult problem. Their job was to design scopes that would be even more complex than the 7854, with its more than 200 knobs and buttons to control 250 functions. To make the new line easy for engineers to use, the 11401 and 11402 were to be programmable instruments with soft buttons, touch screens, and extensive onboard diagnostics. All told, the approach mandated 0.75 megabyte of firmware. Yet the time to assemble the oscilloscopes had to be cut drastically.

The biggest savings lie in the designs of the printed-circuit boards and the way they are connected. The complexity of the 11400 series circuit boards requires dense component placement, so surface-mounted components were called for wherever possible. To keep assembly time down, the design called for maximum use of machine insertion. As a result, boards were laid out with components oriented in a common direction to facilitate machine insertion, and they were spaced to accommodate the insertion heads.

Where surface-mounted components are not used, the designers selected components housed in DIP, TO-5, and other packages with leads exiting from the bottom, which favors automatic insertion. Components were further selected and arranged to minimize the number of insertion heads required and the number of times the heads had to be changed.

Another concern was the extensive cabling between boards on the 7854, which made cable installation a time-consuming and error-prone task. On the 11400 series most cables are replaced by card cages and back planes; the few cables that remain are indexed to prevent incorrect connection.

The digital design of the 11400 scopes helps speed final factory calibration. All but one of the 200 calibration constants in the 11400 series are done automatically. Laser trimming of resistors has largely replaced the classic trim pot, and most of the calibrations that remain are adjustments for the CRT.

Before the design of the 11400 scopes got under way, Al Peecher, then manufacturing manager (now retired), and Greg Rogers, then engineering manager for the Laboratory Instruments Division, decided to bridge the traditional separation of engineering and manufacturing.

Previous experience had indicated that designers often used components and board layouts that defied machine insertion. "Engineering would design it so it worked," said scope codesigner Walter C. Ventgen, "and Manufacturing would spend two years redesigning it so it could be made at a profit." Peecher and Rogers reversed this trend by stationing a group of manufacturing engineers in the design engineering area to make sure manufacturing requirements were incorporated.

One major change in design practice came at the board prototyping stage. The standard procedure at Tektronix, as in most of the electronics industry, had been to hand-assemble prototype boards in the engineering department. A working prototype verified the design but it usually looked nothing like the production board. The 11400 team made it a rule that all prototype boards would be assembled by the manufacturing department. As a result manufacturing was able to catch many potential assembly problems at an early stage.

Phone Chip: Shared Circuits Save Space

Paul Wallich

For nearly 10 years, engineers at AT&T's Consumer Products Laboratory wanted to put the basic circuitry of a telephone onto a single chip, but manufacturing-cost estimates held them back. In 1984 they started work on a chip that incorporates ringer, dialing, and power-conditioning circuitry on a single bipolar die. The chip was first used in AT&T's Model 1600 Featurephone, introduced in August 1985.

To Larry Strelt, an engineer in AT&T's components department in Indianapolis, Ind., the design job looked simple enough; the one chip had to fit into a telephone and replace the multiple chips then commonplace. But in fact the task was not all that simple. The telephone system is standardized throughout North America, and the chip had to fit perfectly—everywhere.

Streit and his fellow engineers—including one from AT&T Bell Laboratories and one from AT&T Technologies (formerly Western Electric, now merged into AT&T), both in Reading, Pennsylvania—chose a set of tools widely used by IC designers. They employed schematic-capture software for the initial circuit design and an AT&T version of the Spice analog-simulation program to verify its operation. Finally, they had proprietary AT&T chip-layout tools to put the circuit on silicon.

The biggest challenge in the design, Streit said, was reducing the number of gates in the chip's digital circuitry while still maintaining the required functions. To do this, Streit combined the circuits that generate dual-tone multifrequency signals for dialing, with those that produce the telephone's ring. The combined dialing-and-ringer circuitry has just over half the gates needed by the two separate circuits. An off- or on-hook control determines whether the circuit is in the dialing or ringing mode. Another challenge Streit said, was designing the chip so that ringer and voice-transmission circuitry shared a power supply.

The designers had to accept one major tradeoff in making their telephone chip a working product. Peter Schuw, head of the Consumer Products Laboratory's coded-products department, recalled that to make the chip as versatile as possible, the designers had to sacrifice their goal of a completely self-sufficient IC. A separate microprocessor was added to control such functions as stored-number dialing. "In the best of all possible worlds, you would have put the dialing microprocessor on the chip," Schuw said, but 4-bit microprocessors are so cheap and so readily available that the two-chip combination was much more cost-effective.

Furthermore, Schuw said, the micro-processor is made in CMOS and so consumes little power, while the new telephone chip is bipolar, for easy connection to telephone lines. Making the ringer, dialing, and power-conditioning circuitry in CMOS would have required elaborate protection against line surges and other hazards, while a bipolar microprocessor would have consumed much more power. The telephone chip can also be used with different microprocessors in many different systems.

The telephone chip's design has been changed somewhat since it went into production, with some additional analog circuitry moved on-chip (the original design required some discrete analog components on the circuit board). The added functions fit into the same die area, so the chip's cost is unchanged. Streit noted that any design, however well it works, can usually be improved by another pass.

49

ESD:
Understanding the Problems and
Methods for Controlling

Raymond L. Pukanic

Despite recent attention in the electronics industry, static electricity is an old and common phenomena. Walking across a carpet and touching a door knob elicits a shock—it is electrostatic discharge (ESD).

A static discharge or spark is created when an object of sufficient charge attempts to neutralize itself by releasing its charge onto an oppositely charged recipient. For electronics manufacturers this static discharge can destroy components, or, even worse, degrade them so they fail at a later date. ESD is common and is a problem.

The common strategy for reducing static damage is to prevent the charges from building up. Although this sounds simple the methods of implementation can be very involved and frustrating. A major problem is that ESD is intangible and cannot be detected until damage has already occurred.

Why are we concerned with ESD? Not only can a damaged device be undetected in an unprotected environment but the negative impact to manufacturing costs can be staggering. Analytical Chemical Labs, after extensive research over several years, reported losses from ESD within four separate groups as shown in Figure 1.

An average manufacturer's cost for a 1% loss is $2,000. If just one disk drive per day is lost, the cost may be $3,500.

GROUP	STATIC MIN.	LOSSES MAX.	REPORTED AVG.
COMPONENT MFG	4%	96%	16-20%
SUBCONTRACTOR	3%	70%	9-15%
CONTRACTOR	2%	35%	8-14%
USER	5%	70%	27-33%

Figure 1.

UNDERSTANDING THE PROBLEMS

Fundamentals of Static Electricity

The basics of static electricity are easy to understand. Static electricity occurs because charges on one material are transferred to another, usually by friction. When a material has gained a charge, it will try to neutralize itself by giving its charge to a willing recipient.

However, to more fully understand how static electricity forms and dissipates, it must be viewed from the atomic level. An atom contains a neucleus made up of neutrons (no charge) and protons (positively charged). The electrons (negatively charged) move in different orbits around the neucleus. In an electrically balanced atom there are an equivalent amount of protons and electrons. When friction is applied to this balanced atom, one or more electrons may leave. In this situation, a positive ion remains, because there are more protons than neutrons. Conversely, the recipient of the electrons now become negative ions.

A material composed of these positive or negative ions will emit a corresponding electrostatic field around itself. Because nature strives for equilibrium, electrostatic fields of opposite charges will naturally attract each other.

Triboelectric Charging

Triboelectric charging is actually the result of friction that was explained previously. The amount of charge generated by friction depends on the types of materials brought together and their polarity (positive or negative). Scientists have organized lists based on the atomic structures of various materials called a *triboelectric series* (Figure 2).

The more readily a material assumes a charge, the farther it is from the center, or "0" point, of the list.

From the triboelectric series (Figure 2) polarity of the charge a given material will take on can be determined.

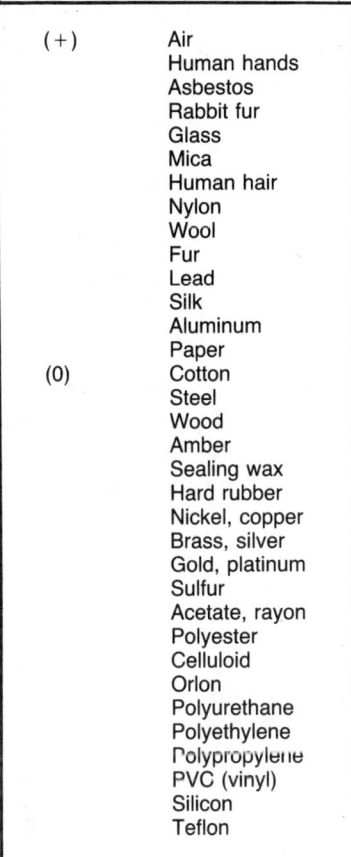

Figure 2. *Triboelectric series.*

Human hands and the materials they typically come into contact within electronics assembly are at opposite ends of the chart. This means that there is a greater chance for stronger electrostatic fields to build up in this environment.

One way to eliminate triboelectric charging is to have everything in a static-sensitive area made of the same material. But, unfortunately, this is not practical.

Inductive Charging

Static electricity can also be transferred without physical contact of materials. This happens through a process called induction that behaves in a similiar manner to a magnetic field. A magnet (with a strong magnetic field) may actually pull a nearby steel bar into contact with itself. The further away the steel bar, the less the magnetic influence. In the same manner, a charged materials' electrostatic field can have an inductive influence on a nearby neutral material (Figure 3).

When a neutral object is brought into a charged object's electrostatic field, its own protons and electrons move, polarizing to the charge. The originally neutral object is then charged by induction.

This method of inductive charging is often difficult to detect because the materials do not have to make physical contact.

Classifying Material

The U.S. Department of Defense has identified classes of materials based on their surface resistivity. When comparing different materials, a common denominator—ohms per square—is used. This measurement gives the same surface resistivity reading anywhere on the same surface material. The material classifications are listed in Figure 4.

Static Damage

Devices are damaged when a sufficient amount of static electricity goes through the sensitive circuit. The static discharge could short or open the micro circuit, leaving "catastrophic" damage. This is the more desired type of damage, because it is easily detected in product performance during test.

A worse situation is when the static discharge "degrades" the device. The unit appears to be working fine during test—however, the degraded device is subject to latent damage in the field. At this point repair costs can be $600 to

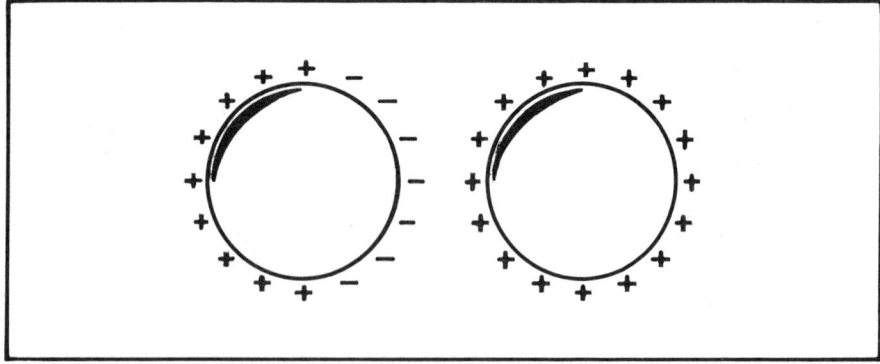

Figure 3. *Inductive charging.*

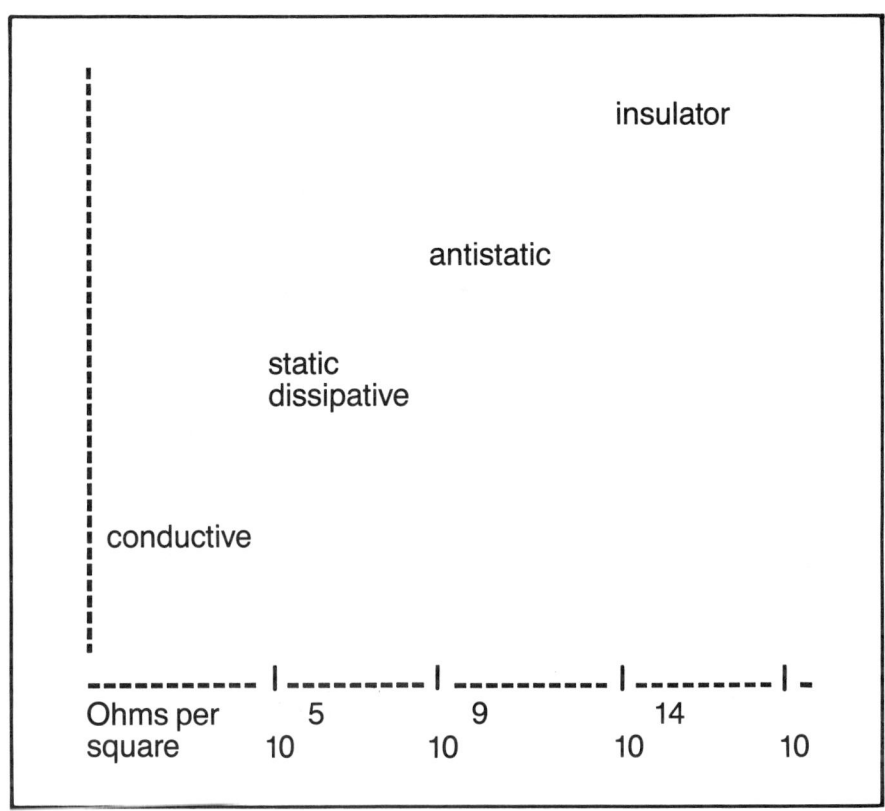

Figure 4. *Material classification.*

$1,000, which is about 100 times the cost of the failed device. Cost avoidance of an effective ESD program can contribute to the bottom line (profits) in this instance.

Classification of Device Sensitivity

The U.S. Department of Defense has categorized electronic devices into three classes based on their sensitivity to static electricity (Figure 5).

In 1981, most devices were in Class III. Today most devices are in Class II, with a trend toward Class I. Interestingly, the human threshold of sensitivity is about 4,000 volts. If we depended on our senses to detect high electrostatic fields, we would destroy most of the electronics devices used today.

Device Classification	Static Threshold
CLASS I	up to 1,000 volts
CLASS II	1,000–4,000 volts
CLASS III	4,000–15,000 volts

Figure 5.

Clean Room Contamination

In clean rooms, great care is taken to keep particle counts at very low levels. A few particles only 5 microns in diameter could cause a high speed disk drive to crash. This could result in a very costly loss of data to the end-user (particularly if it is a bank or financial institution).

Build up of static electricity in the clean room could cause high concentrations of particulate around the electrostatic charge areas.

METHODS FOR CONTROLLING STATIC ELECTRICITY

Ionizers

The principle behind ionization is to shower the selected manufacturing environment with positive and negative ions. The ions would seek out any opposite charges in area and, thus, neutralize them.

Applying ionizers can be difficult. A tote moving on a conveyor could take 60 seconds to neutralize. By that time the static damage may already have occurred.

Antistatic, Static Dissipative, and Conductive Materials

Materials have been developed and manufactured into tote bins, work surfaces, floor mats, and many other items to provide electrostatic protection. Depending on the sensitivity of the electronic device in question the appropriate material should be chosen. The more sensitive the device is to electrostatic fields, the more conductive the protecting material should be. However, moving the charge too fast could create sparking, which could damage the device.

A clear understanding of how the material functions in carrying out the static protection is very important. Some antistatic materials are filed with a surfactant (a detergent) that is moisture-dependent. As long as there is moisture in the air the surfactant leeches toward the surface, drawing the moisture toward it.

However, there is a finite amount of surfactant in the antistatic material and it is not obvious when the surfactant is depleted. Therefore, you do not know when your static protection has turned into a static generator.

Other Control Methods

Additional items, such as wrist/ground straps, conductive mats and floors, shoe grounding straps, gowns and clothing, conductive bags, and conductive chairs are important elements in an effective static protective manufacturing environment.

APPLYING CONTROL METHODS

Electronics assembly stations should be equipped with static-protective work surfaces and the operator should be grounded with a wrist strap. A conductive chair should also be used to eliminate the chair as a static generator. The conductive chair will only work effectively on a conductive mat or flooring.

The other major area for protection is material handling. Conductive carts with drag chains to a conductive floor work very well. Conductive flooring (a ground path for the cart) is important throughout the manufacturing area where the cart could travel.

If the manufacturing area does not have conductive flooring throughout, then conductive totes with lids (providing a faraday cage for ESD shielding) should provide adequate protection.

Implementing an effective ESD program requires a commitment to discipline. An ESD task force should be formed including engineering and production supervision. Regular training of operators and assemblers is a must. The task force should be the ongoing monitoring program.

Selected Readings

U.S. Department of Defense. DOD-HDBK-263. May 2, 1980.
Testone, A. Q. Static electricity in the electronics industry. Testone Enterprises. Lee, MA

Reprinted from *1985 Annual International Industrial Engineering Conference Proceedings*.

Raymond L. Pukanic is a senior facilities consultant with the professional services group at Herman Miller, Inc., Santa Clara, California.

50

A Total Ergonomic Investigation of Working Places with Light Tables

Dag Holmgren

VISION

The proclamation of the Swedish Work Environment Association about regulations regarding working positions, movements and stress reads as follows: "Vision environment ought to be, if possible, of such kind that there is no cause for straining work positions . . ."

The term "work position" is often used as a synonym for "vision distance" and consequently we have the definition for accommodation conditions.

Accomodation effort and its effect on organisms is a field in which the Finnish ophthamologist, Kaisu Vikari, has specialized for a long time.

Dr. Viikari has found the same relationship between the organism of the human being and the accommodation as the English ophthamologist Sir Duke-Elder. The important conclusion in their research is that, probably, there is a greater connection between what is normally called poor vision environment and the human body.

A poor vision environment does not only cause eye problems but also neck and back problems.

In her book *Panacea, Daughter of Tetralogi* Dr. Viikari accounts for the results of clinical cases, over many years, where accommodation has been the direct or indirect cause of individual complaints.

BRANCHWISE INVESTIGATION

Photography

There are special demands on light tables in the photographic branch due to the need for exact color estimation and comparisons of look-down and look-through pictures. The lighting problem consists in the eye not being able to recognize directly transmitted light (through a slide) in the same way as reflected light (illuminated paper copy) even if the "basic picture" is the same. The reason for this is that the slide functions as a filter, i.e., certain parts of the light is being filtered away, whereas the paper picture reflects the light, as certain parts of the spectra are being absorbed by the colored grains in the paper surface.

For a perfect result, the producer of photographic materials must use exactly the same pigmentation for proper pictures and slides.

The reason why perfect pictures are being produced is that those who work in this field have developed a skill in compensating light flows. Tests with a prototype with the same kind of light source as that of a light table and an over-light source have shown that the pictures were being differently understood although the intensity of the two light sources could be adjusted separately.

Another reason is that light tubes have a light spectrum composed with the purpose of rendering an impression as colorful as possible. The light is an approximation of a perfect light. The so called "triple strip light tubes" are definitely not suitable for color estimation.

The Light Table

The requirements for light tables for photographic work are as follows.

Light: color temperature 5000°K ±300 color rendering Ra-index larger than 92. Luminance max. 1500 cd/m². Adjustable. Flicker-free.

Masks: Picture surface adjusted to picture; neutral in order to adjust surrounding luminous area to picture. The mask must not be too dark.

Work position: Possibilities to adjust angle and height of luminous surface. Support for pictures under inspection.

Comment to color temperature: The temperature should be 3000°K when producing pictures for home projectors as the color temperature of the projector lamp has about this temperature.

Graphics

The Graphic Health Board of Stockholm have made a thorough study of work places equipped with light tables with the intention of an improvement to these and, above all, putting together specifications for new purchases of light tables or for new design of work places with light tables.

Work at light tables has been studied from the point of view of weight and light. Different factors have been appraised according to a check list.

Most of the criticism that has come forward in these studies concerns the design of the light table and the stand, and also reflections caused by general lighting.

Alterations suggested on already existing equipment are:

- Use of already existing adjustability (height and angle)
- Reducing of table thickness
- Paper masks for luminous surface
- Dimming device required
- Inspection of general lighting.

Before purchasing a light table it ought to be considered whether there really is a need for under-light in performing expected tasks or if a normal work table is sufficient.

When purchasing light tables the following points should be taken into consideration.

- Size of light table in relation to type of work
- Flicker-free, uniform light
- Dimming device
- Suitable masks available
- Minimal heat production
- Compliance with safety regulations
- Correct work position
- Stand adjustable in height and angle.

When designing a new work place it is important to consider the space around the light table, the flow of work material and the position of the light table in relation to daylight.

Annoying reflections in the working surfaces or direct blinding can be avoided by careful planning of the general lighting. In choosing work lamps with adjustable light, each work place can be illuminated according to individual demand.

The checklist at the end of the paper was developed by The Graphic Health Board.

Drawing and Office

Within this sector, the total drawing work needs a closer investigation. Light tables are being used to a certain extent as some tasks demand a luminous surface.

In Sweden, the company "3 Ergonomer" has started a project where they intend to study draughting in general. Whenever they come across work places with light tables the result from this project will be used. ITAB Design AB is a part in this project responsible for light table questions.

Drawing and office work at light tables consists of many different tasks with different demands on illumination. Generally speaking, the tasks are eyestraining and the studies that have been made show the following ergonomical problems.

1. Vision problem related to individuals
2. Vision problem related to task
3. Vision problem related to illumination
4. Unsuitable work position due to vision circumstances
5. Unsuitable work position due to furnishing and technical equipment

Vision Problems Related to Tasks

Draughting is precision work with high demands on vision. Very often the vision distance is more than 15-25 cm which makes the eye work at a highly tensed accommodation level. This permanent muscular tension already leads to symptoms of fatigue after no more than one hour's work.

Vision Problems Related to the Individual

Uncorrected defect of vision at draught work can lead to an extra effort of vision. Oversighted people without vision aid often use part of their accommodation ability to correct this vision handicap, which results in additional strain on the accommodation musculature. The brain compensates for smaller defects of the ocular muscles by an extra adjusting movement of the eyes. In certain cases the ocular mechanism gets tired out which leads to vision problems.

By vision at close range a certain dislocation of the optical axis takes place. The centering of glasses, adjusted to greater vision distances, does not accord, thus causing a picture distortion.

Vision Problems Related to Illumination

Draught work means both look-down and look-through. Consequently there must be two counteracting light systems, light table and general lighting. If there is a great unbalance between these different light system the vision work becomes more and more difficult through contrast reduction.

The eye adapts itself to the medium luminance of the vision field. A light table with unshielded light sections causes an incorrect adaption and this could have a blinding effect.

If the drawing film is shiny or the light table has a shiny surface there is danger of uncomfortable reflections and these would render the vision work more difficult.

Bad Working Positions

The draughtsman leans forward while working because of the short vision distance thus increasing the pressure on his back.

An obvious improvement of his working position would be to use magnifying glasses. However, at the same time, there is a decrease of the field depth which renders the drawing task more difficult to perform.

The work surface is usually very large, forcing the draughtsman to lean forward across the table while drawing. Without any extra support this puts an even greater strain on his back. Inclining tables are no solution since the engraving tools then would drop to the floor.

Measures Suggested

1. People who are to work at light tables ought to have an eye check-up first.
2. People working at light tables ought to have eye check-ups every 2-5 years depending upon age.
3. People who need special glasses because of their work ought to obtain these through their employer.
4. General lighting should follow the same principles as lighting for viewing screen terminals. Fittings should be placed in such a way that reflections can be avoided.
5. Working places with high demands on "look-on"should have adjustable light with dimming device.
6. Light tables should have adjustable light with luminance up to 1500 cd/m^2.
7. Light tables should have a matted work surface with an even light.
8. Light tables should have some kind of built-in mask system.
9. Lighting of light table should be even and flicker-free.

X-Ray

It is very difficult to take any measures within this sector especially regarding body postures. Most hospitals are equipped with alternators, auto-alternators not meant for sitting work and yet this kind of work is being performed to a large extent in a sitting and consequently twisted position.

Principally, these alternators could be rebuilt but considering the economical consequences these suggested re-designs would probably not be met with sympathy.

In a study of four Swedish hospitals equipped with altogether 19 alternators, light intensity varied from 2000 to 12000 lux. which is probably a representative figure in general. This result also explains why so many x-ray pictures have to be retaken because they are too dark.

The equipment is very often not correctly located and there is a lack of service and information.

The possibilities of masking off non-used luminous surfaces are ignored, leading to a blinding effect where the image under inspection appears to be too dark.

A check-up by Spri, The Institute of Hospital Planning and Rationalisation, showed that rejected, dark pictures had been inspected on light tables with insufficient lighting due to light tubes not being in good working order.

Practical Examples

A study was made at the x-ray department of Mora Hospital. Some of the findings show that the uneven luminance distribution in the room is probably one of the reasons why the personnel have an uncomfortable impression of the light. Our tests proved that the light table showed a luminance 50 times stronger than the wall behind.

The large contrasts between dark and light surfaces should be diminished by improving on luminance distribution and elimination of all reflections. The medium luminance of the walls should be about the same as that of the picture.

Measures taken were: Information about how to mask off the luminous surface and only use the space necessary for inspection. Rearranging the alternators by turning them away from the walls at a 90° angle in order to create more space and eliminating irritating disturbances among the personnel. In turning the alternators this way, the luminance of the walls increased from 40 cd/m² to 350 cd/m².

Ceiling fittings were disconnected and a light tube, in a bright red, with a low luminance mirror screen was put over the working table. As a positive side effect, the bright red color of the lamp added an extra pleasing touch to an otherwise rather sterile looking room.

This rearrangement of an x-ray department proves that an uncomfortable working place very often can be improved upon by rearranging already existing equipment.

Dental X-ray

At the Institute for Post Graduate Education in Jönköping a number of proto-type work places have been designed and tested.

Because of the smaller size of the equipment it has been possible to test new ideas. The result of these tests has been most significant. Physical troubles have practically all ceased after the introduction of the new work places.

These tests are continuing and will be assessed at a later stage of this project. This work is presented in a paper by Olof Eckerdal DMD, Ph. D.: "Ergonomics in Oral Radiology."

Industrial X-ray

The Swedish Technical Control Institute (STK) in Jönköping works with industrial quality control. They inspect welded seams, surface finish, materials etc. and they are also responsible for the registration and control of welders.

Very often X-ray photography is used for inspection work, and at the Jönköping office they both work with filming photography and image inspection. Mostly the work takes place at the actual working location.

The film has an emulsion on both sides and consequently it has a very high density, requiring high intensity light tables.

- The light tables had 3-5 halogen lamps for a light source.
- The light could be adjusted with a wheel, and mini-light and maxi-light could be obtained by pressing a foot pedal.
- Light tables with halogen lamps mini-light—10000 cd/m^2 and maxi-light—50000 cd/m^2.
- Light tables with 5 halogen lamps mini-light—500 cd/m^2 and maxi-light—45000 cd/m^2. In the US there is a hygienic exposure limit for the vision center of the eye bottom (macula lutea): Persons are not allowed to look directly at surfaces with luminances exceeding 10000 cd/m^2 for more than 10 seconds per 8 hour day.

According to STK the films should have this light intensity. Each light table was equipped with differently sized masks according to film being used.

Working Place—Work

Light tables were arranged in such a way that reflections in films could be avoided. One of the light tables was put in a corner with curtains as a shielding device.

Inspection work is very strictly organized and each image is to be inspected for a certain amount of time, there are norms for density etc.

At certain inspection places there are special film inspectors in order to keep inspection quality even.

Since lighting intensity is very high it is necessary to shield off all stray light. This problem had been quite successfully solved with metal masks of different dimensions according to films being inspected and a magnifying glass was used as an optical aid.

PC-Board Inspection

Studies show that PCB-inspectors were suffering from pain in the shoulder and back due to certain defects in the working environment.

Working positions were unfavourable due to unsuitable working tools and insufficient blinding lighting.

Solutions to some of these problems have been introduced in a model which can be used for practical tests of different ideas and further demands on design.

The test model comprises:

• Inspection fittings with shielding device
• Table top
• Light table

Work position becomes more erect than before and the strain on the back lessened by making it possible for the elbows to rest on the table.

The PCB is leaning against "steps" on the light table. The board can be inspected through a shielding frame which creates a calm background and increases concentration. The shielding frame also has the function of a distance indicator eliminating bending movements of the neck.

Work Light

By studying different work places in the electronic industries it has been found that there is a great need of a special type of work light.

Most work lights are not very well suited for the small distance between the light, the workpiece, and the worker's eyes which is very often necessary when examining minute components or at workplaces for handicapped people. The lamp usually consists of either a normal light bulb with a shade giving off a lot of heat or a light tube fitting with a flickering light.

In order to solve these eyesight problems we have designed a new type of work light using neon tubes and the same driving technique as in light tables.

By using the unlimited design possibilities of the neon tube, this lamp has been shaped into a U-form with a very thin profile allowing a light output from three sides which gives a completely shadowfree work area.

CHECKLIST FOR WORK PLACES WITH LIGHT TABLES IN GRAPHIC WORK

• Work position: non-fixed, non-outer position.
• Adjustable sitting and standing position.

- Prolonged standing position: negative.
- Working height: well balanced standing position for eyestraining work.
- Body support: Support for body against table, rounded table edges, possibly arm support.
- Chair design: Adjustability: Easily adjusted seat depth, placing of back-rest and height stable. If equipped with castors; non-slippery floor.
- Combination chair—work table: Well adjusted sitting position for eyestraining work.
- Space for feet and legs: a) Sitting position: 70 cm on foot level, 50 cm on knee level, 70 cm laterally. b) Standing position: room for feet under work table.
- Feet support: Well supported when in sitting position.
- Design of light table: Thickness: 10 cm maximum, matted glass, stable, tapered leaning edge, base plate white inside.
- Adjustability of light table: Variable transmission (between 68-88 cm by sitting down and 95-110 cm by standing up) of table height and angle, shielding possibilities of pale green paper 60% reflection.
- Dimming device: Variable dimming of light, placed within comfortable reach.
- Working space: Adequate loading surfaces; be able to move freely around work table.
- Layout of accessories, most frequently used: Manuscript holder.
- Accessories, design: Easily adjustable rulers, handy cutting tools, suitable height of loading table.
- Light system of light table: Uniform illumination of surface (10% variation max.), no disturbing sound or heat, flicker free, adjustable from 100-1500 cd/m^2. Luminous surface should be 10-25% shorter than length of light tubes (conventional type), well earthed base plate. Light tubes: universal white or neon tubes.
- General lighting: Lamps with direct/indirect light distribution (300-500 lux), glare shield (K 15 DST), 1 or 2 × 40 W, dimming device. Placing: No disturbing reflections. Light tube: Universal white (soft white on large premises). Wall fittings might be needed for acceptable luminance contrasts (light table—wall surface).
- Working place light: Turnable incandescent lamp, light tube fittings might be needed in certain work routines (reading of ruler scales).
- Placing of work place: No disturbing light.
- Color scheme of room: Harmonious colors. White ceiling. Factor reflection 80-85% in ceiling and 60-65% on walls.
- Reflections: No disturbing light from window or fittings.
- Light maintenance: Cleaning of fittings and change of tubes after 7-8000 hours burning time.

- Noise: Sound level of max. 45 dB (A) (when irritating noise).
- Climate: Appr. 20°C, 30% rel. humidity.

A Swedish Work Environment Fund Project.

51

Human Supervisory Performance in Flexible Manufacturing Systems

Sheue-Ling Hwang
Gavriel Salvendy

OBJECTIVE AND SIGNIFICANCE

In the developing computer aided manufacturing systems, the human is no longer directly controlling the system processes. Instead the operator interacts primarily with the computer which directly controls the processes. The human becomes less involved in the manual control of the inner loops of the manufacturing system and more concerned with supervisory control from outer loops of the system (Edwards & Lees, 1972). In terms of the tasks, in computer aided manufacturing systems, the operator deals with information processing and decision making more than materials handling and manipulation.

The first problem is how many machines should an operator be responsible for in an FSM. Most FMSs use a teamwork concept to keep the system functioning. Manpower stability is a desirable objective since a few key people command a major manufacturing resource (Klahorst, 1983). In general, an FMS has 4 to 12 machines which are monitored by one or two operators through computer interaction. Periodically the computer will notify the operator to change machine tools if necessary (the average tool life is 30 to 90 min.). After the operator changes the tool, he inputs this data into the computer in a statement, such as "tool A in machine 2," so that the computer can start to calculate the life time of this new tool. The computer calculates or acts upon the data entered by the operator in order to control the machine processes. The whole system may malfunction if the operator inputs the wrong data into the computer. Borrowing the concept of the arousal-performance function (Yerkes and Dodson, 1908) where performance is maximum under the median arousal situation, it may be possible to obtain a similar function between number of machines and performance. If we assign too many machines for an operator to monitor, he may suffer from mental overload; if we assign too few, he may become bored and therefore less vigilant. Both overload and underload decrease the operator's performance. What will be the optimum number of machines for an operator to monitor? This problem is of great concern to the manufacturing community.

The second interesting problem is how many tasks should be performed or allocated to the computer and how many should be allocated to the human. The implementation of an FMS has changed the tasks the human performs in a manufacturing system. While the overall quantity of tasks the human performs has decreased in an FMS, the level of decision making and information processing has increased. This is because the human must match in information processing and decision making the tremendous speed at which computers present and process information. Therefore, one of the most challenging problems for human factors specialists is how to design jobs for people who must interact with computers which control FMSs. Rosenbrock (1982) suggests that we should not eliminate the human qualities of skill and judgement when designing such systems. In fact, these qualities should be made more productive by the interaction between the human and computer. Thus, it seems reasonable to propose that the optimal relationship between the human and computer in an FMS is one of shared responsibility and decision making. Responsibilities for decision making in the operation and control of an FMS should be assigned to the intelligence, either human or computer, most able to perform decision making at that instant. Hwang et al. (1983) in an attempt to address this point have proposed two experiments for examining the human-computer relationship in an FMS. One experiment aims at providing a systematic structured task analysis of how the human supervisor controls an FMS. The second experiment addresses issues of the optimal allocation of functions between computer and human. People tend to think that an "ideal" FMS makes the computer responsible for almost every task while the human monitors the minimum. From the point of view of safety, this kind of system is risky. The system may break down only by pushing a wrong button. The reasonable strategy is to let the human do what he is good at, i.e., flexible decision making and monitoring, while letting the computer perform the routine calculations, control and monitoring. Therefore it is necessary to understand the nature of the human and the computer so that we can assign the appropriate amount of work to the human and computer.

The objective of the present study are:

1. to find the optimum number of machines for an operator to monitor in an FMS,
2. to find the appropriate allocation of functions between the operator and computer.

These objectives can be transferred to the following hypothesis:

1. Hypothesis 1—It is assumed that there is an interaction effect between number of machines and allocation level. As the number of machines increases, the performance may decrease on high allocation level while increase on low allocation level.
2. Hypothesis 2—It is assumed that there exists an optimal number of machines for the operator to monitor. The optimal number of machines also depends on the allocation level. The relationship between number of machines and performance is assumed to be in the form of an inverted-U function. The operator who is monitoring the medium number of machines may show the maximum level of performance.
3. Hypothesis 3—It is assumed that subjective stress increases as the number of machines and allocation level increase.
4. Hypothesis 4—It is assumed that physiological stress increase as the number of machines and allocation level increase.

METHODOLOGY

Subjects

The subjects were senior or first year graduate students in the School of Industrial Engineering at Purdue University. Thirty students participated as paid subjects in the experiment. All of them had previous experience in manufacturing system engineering concepts. These subjects consisted of nine females (mean age = 21.4 years, SD = .73), and 21 males (mean age = 22.1 years, SD - 1.20).

Experimental Task

The task of the human operator was the operation and control of a simulated FMS. The task was simulated by an IBM Personal Computer. There were four kinds of machining centers in the displayed FMS—1) Vertical Machining Center (VMC), 2) Horizontal Maching Center (HMC), 3) Lathe (LAT), and 4) Inspection Center (ISP). The purpose of this FMS was to produce six different kinds of parts. Each part was required to go through various machining centers where different tools were used to process the part. The subject's task was to monitor the system by checking the machines which had a tendency to break down. The subject also moved parts to the normal machining centers when necessary. In the experiment, two types of FMSs, System "A" and System "B", were simulated. It was designed in such a way that System "A" was more automatic than system "B". When moving parts in system "B", the subject had to check whether the parts in the other machine queue were overloaded or not (the maximum parts in queue were 5 parts). Also the subject had to check to see whether the tools for the parts were available. In System "A" the part status and tool status were automatically controlled. Therefore, it was not necessary to check the part status or tool status before moving the parts.

Operational Definition of Variables

The independent and dependent variables are defined in the following sections.

Independent Variables. There were two independent variables in the experimental task. The first one represented the number of machines. The number of machines on each level was decided by examining the size of existing FMSs. The second independent variable was task allocation. The levels of each independent variables were described as follows:

1. *Number of Machines.* The number of machines which were assigned to each subject consisted of the following five levels:

Level 1 .... 4 machines
Level 2 .... 8 machines
Level 3 .... 12 machines
Level 4 .... 16 machines
Level 5 .... 20 machines

2. *Task Allocation.* Task allocation contains two levels:

Level a .. low level of task allocation where the subject monitors the machines and moves the parts without considering either part status or tool status.
Level b .. high level of task allocation where the subject monitors the machines and moves the parts after checking the part status and tool status.

Dependent Variables: Three dependent variables are defined as follows—

1. *Performance.* The performance time and the number of errors made by a subject while monitoring the system were collected. The computer recorded the time automatically from when the question first appeared on the screen until the subject responded. This elapsed time is the performance measure

for each subject. An error means an incorrect response, such as moving a part to a "down" machine.

2. *Physiological Measures.* Physiological measures of arousal were collected from a pair of electrodes for EKG and respiratory rate. The task difficulty level was manipulated in the experiment, theorized to result in a change of arousal level. Physiological measures, on the other hand, are direct index of arousal level. The higher the respiratory rate and heart rate or the lower the value of sinus arrhythmia, the higher the arousal levels. From EKG recordings, one can obtain mean heart rate (MHR) and sinus arrhythmia (SA) which has been previously used as a measure of information load. The standard variance of the time between heart beats (S^2) and Mean-Square successive difference (MSSD) were used to quantify SA. It has been previously discussed that MSSD is more appropriate for measuring mental workload than S^2 (Sharit and Salvendy, 1982). The present study considered respiratory rate, MHR, SD, and MSSD as physiological measures of workload.

$$\text{(where MSSD} = \sum_{i=1}^{n-1} (X_{i+1} - X_i)^2/(n-1)).$$

3. *Subjective Measures.* The subject's score from the subjective questionnaire was the third dependent variable used in the study. Two questionnaires were used following each experimental session to evaluate how the subject felt about the task. One of the questionnaires was derived from Pearson's Feeling Tone Checklist (Pearson, 1957). Ten questions from Pearson and seven additional computer-work related questions formed a stress checklist (Barfield et al., 1983). Another questionnaire was developed to evaluate task complexity and stress.

Experimental Design

Form of the Design: In order to arrive at a best estimation for any difference under the constraints of a limited number of subjects, a cross over design was used.

Two sequences were necessary to each level of machine number in this counter-balanced experimental design. As least two subjects per sequence were necessary to calculate error terms. There were three subjects in each sequence. Therefore the total number of subjects was 30. The subjects were randomly assigned to each sequence. Each subject had to complete two sessions of the experiment. For every subject, both sessions had the same number of machines but only one allocation level (high of low) was used in each session.

Experimental Apparatus

An IBM Personal Computer presented the experimental task to the subject. The subject entered his answer by using the keyboard, and the answers and response time were automatically recorded.

The Beckman Biomedical recorder (Type R-411 Dynograph Recorder) was used to measure the EKG and respiratory rate of each subject through an Impedance Pneumograph coupler (Type 7212, P/N 704-0080) which was connected to two electrodes on the subject's thorax.

A four-channel FM recorder recorded the physiological signals from the output of the Beckman. Two channels of the FM recorder recorded respiratory signals and EKG signals while the third channel recorded the "break" signal sent from the IBM PC. This indicated the beginning or ending of the experimental session.

Procedure

The subject was asked to read the manual describing the experimental task and procedure and practice the questions following the manual one or two days before the experiment.

Upon arrival, the experimenter described the system briefly and allowed the subject to ask questions.

The subject placed two electrodes between his/her fifth and sixth rib interspace and one ground electrode on the back of the hand.

The subject was randomly assigned to one of the machine levels and sequences. Each subject participated in two experimental sessions with one allocation level in each session (high allocation represents system "B" while low allocation represents system "A").

In the first experimental session, the subject practiced two to three trials. The experimenter gave feedback to the subject whenever he/she made a response. The practicing trials lasted until the subject was familiar with the task.

The program was run with "seed" equal to 10 (on low allocation level) or 20 (on high allocation level). At the start of the experiment, at least one machine would be "down" in each trial. The subject had to answer 9 to 12 questions in order to deal with the down machine. Meanwhile the subject's answers and response time to each question were recorded by the printer. There were five trials for each experimental session.

After the end of trials, the subject completed the subjective stress and task complexity questionnaires.

Between the experimental sessions, the subject rested for five minutes in order to reach a physiological baseline level.

The procedures for the second experimental session were identical to the first session.

RESULTS

The following dependent variables were collected to test the hypotheses.

1. Performance. This included both the duration of performance and errors.
2. Physiological measures. This included mean heart rate and sinus arrhythmia.
3. Subjective questionnaire responses. These included evaluation of mental work load, stress and task difficulty.

This section contained the results from the following analyses:

1. MANOVA—Due to the multiplicity of dependent variables, it was necessary to apply a MANOVA to the performance measures, subjective questionnaire, and physiological measures.
2. Correlation between dependent variables to discover whether performance correlates with either subjective response or physiological measures, and how much we can predict subjects' performance by their subjective response and physiological measures.

   Performance data included number of errors and response time. Since the range of the number of errors was very small (0, 1, or 2) nonparametric statistical analysis was applied to the number of errors. The results from Freedman's two-way analysis did not show any significant effect of either machine level or allocation level on the number of errors. The interaction effect was not available from nonparametric analysis. Thus, response time which represented performance measures in the following section, along with subjective questionnaire and physiological measures were analyzed in parametric statistics.

MANOVA

Performance (Response Time)

The Bartlett-Box test revealed significant heterogeneity of variance among cells for the response time of question 1 and question 8, 17. Thus, the data on questions and 8, 17 were transformed by taking the square root of each measure. The MANOVA summary (Table 1) for performance (response time) shows the Wilks' criterion values for all independent effects. The MANOVA demonstrated that the main effect for machine level and the main effect for task allocation were significant.

Hypothesis 1 which assumed that there was an interaction effect between machine level and allocation level was rejected because the interaction effect between machine level and task allocation was not significant. Since the main effect of sequence was not significant, the response time of these eight questions was not affected by presenting the easy task first followed by the difficult task, or by presenting the difficult task first followed by the easy task.

TABLE 1 – MANOVA Summary Table for Response Time

Source	K	dfH	dfE	U	P
Between Subjects					
M (Machine)	8	32	49.54	.068	.052
Q (Sequence)	8	8	13	.650	.560
M × Q	8	32	49.54	.084	.106
Error Term					
Within Subject					
T (Task allocation)	8	8	13	.245	.005
M × T	8	32	49.54	.156	.476
Q × T	8	8	13	.535	.279
M × Q × T	8	32	49.54	.119	.269
Error Term					
where k = number of dependent measures					
dfH = degrees of freedom for treatment effect					
dfE = degrees of freedom for error effect					
U = Wilks' likelihood ratio statistic					

Furthermore, the univariate analysis of variance tested the significant effects on each individual question. Both main effects of machine level and task allocation levels were very significant on the square root for response time of question 1. The square root of response time increased when the number of machines in the system increased. Also the response time was longer on the high allocation level (t = 4.26, df = 20, p < .005).

Subjective Questionnaire

It was evident from the Homogeneity of Variance tests that the scores of the stress scale had unequal variances. To stabilize the variances, the scores of the stress scale were transformed by the square root transformation before proceeding with the MANOVA for the stress checklist and task complexity variables.

The results of the MANOVA revealed a significant interaction effect between sequence and task allocation, a main effect for sequence, and also a main effect for task allocation. An ANOVA was then calculated to check whether the interaction effect and main effect were significant for each scale of the subjective questionnaires.

From the results of MANOVA and ANOVA, hypothesis 3, which assumed that the subjective response of stress and task complexity would increase as the machine level and allocation level increased, was partially accepted. The machine level did not affect subjective response of stress on task complexity. However, the subjects subjectively felt more stressful and the task more complicated on high allocation level compared to the low allocation level.

Physiological Measures

The physiological data, i.e., MHR, SD of the time between heart beats, and Mean Square Successive Difference of the time between heart beats, had a wide range of individual differences. Thus the ratio between the original data in each session and the individual's baseline response were used as an index of physiological reaction. Although the index has reduced the between subject differences, the homogeneity of variance tests (Bartlett-Box) revealed that all the ratio of physiological data were non-homogeneous and hence need to be transformed before proceeding with MANOVA or ANOVA calculations. Therefore the reciprocal transformation of the ratio of MHR was taken. Also the square root transformation was applied to the ratio of MSSD. Hypothesis 4, which assumed that as machine level and allocation level increased, MHR would increase while SA decreased, was rejected since neither task allocation nor machine level showed significant effects on phsyiological reaction.

Correlation

Correlation Between Performance and Subjective Response

The stress checklist and stress scale correlated to the response time of questions 2 and 3 on the low allocation level. The task complexity scale also related to the response time of question 3 on the low allocation level. More significant correlations were found on the high allocation level. The stress checklist correlated with RT of question 2. The task complexity scale correlated with RT of question 1 and question 16. In addition, the stress scale related to the RT of question 1 (= .39), 16 (= .40). and 17 (= .36).

Multiple R: Regression analysis was utilized to predict performance. Subjective response and physiological measures were used as predictors of performance. When combining the subjective questionnaire and physiological measures, the multiple R was only significant on question 12 (R = .65). The greatest variability which any response time could be predicted was less than 43 percent of the variance accounting for total performance.

CONCLUSIONS AND IMPLICATIONS

The results of this study implied that an Inverted-U theory may not always be supported. For the simple tasks which require very small amounts of mental workload, the performance may not be affected by arousal level at all. For the difficult tasks which require complicated mental processes, the high arousal level will deteriorate performance significantly.

The lack of interaction between the number of machines and allocation level may have implications for the job design of FMSs. Since allocation levels are more sensitive (than the number of machines constituting an FMS system) to the stress level and the change of performance, one may assign the operator any number of machines (between 4 machines to 20 machines) without adverse effects as long as they are associated with an "appropriate" allocation level. The study suggests that it is possible to assign the appropriate level of task allocation according to the individual's stress level.

The subjective stress questionnaire may be used as a criterion to predict the performance of the supervisor in an FMS. The results of the present study indicated that the subjects were more stressful in the high task allocation system (where the decision making is simple). And thus the subject's performance decreased as the task allocation level increased. These results imply that the system should be so designed as to alleviate stressful human-system interactions. This can be accomplished by the following alternatives:

1. Allocate to the computer more tasks. The supervisory tasks of the operator in an FMS should consist only of simple decisions.
2. For more complex decisions, an expert system or decision support system should be developed which will work in conjunction with the supervisor.

Suggested Readings

Barfield, W., B. M. West, F. Robertson, L. Taylor, and N. Tamplin. 1983. Stress as a function of the rate at which information is presented on a video display terminal. *Proceedings of the Human Factors Society 27th Annual Meeting.* Norfolk, Virginia. October 10-14. pp. 516-520.

Casali, J. G., and W. W. Wierwille. 1983. A comparison of rating scale, secondary-task, physiological, and primary-task workload estimation techniques in a simulation flight task emphasizing communications load. *Human Factors.* 25 (6): 623-641.

Edsell, R. D. 1976. Noise and social interaction as simultaneous stressors. *Perceptual and Motor Skills. 42.* pp. 1123-1129.

Edwards, E., and F. P. Lees. 1972. *Man and Computer in Process Control.* Huddersfield: H. Charlesworth.

Fantion, E., D. Kasdon, and N. Stringer. 1970. The Yerkes-Dodson Law and alimentary motivation. *Canadian Journal of Psychology. 24* (2): 77-84.

Hochhauser, M., and H. Fowler. 1975. Are effects of drive and reward as a function of discrimination difficulty: evidence against the Yerkes-Dodson Law. *Journal of Experimental Psychology: Animal Behavior Processes. 104* (3): 261-269.

Hokanson, J. E., and M. Burgess. 1964. Effects of physiological arousal level, frustration, and task complexity of performance. *Journal of Abnormal and Social Psychology.* Vol. 68. pp. 698-702.

Hwang, S. L., J. Sharit, and G. Salvendy. 1983. Management strategies for the design, control, and operation of flexible manufacturing systems. *Proceedings of the Human Factors Society, 27th Annual Meeting.* Norfolk, Virginia. October 10-14. pp. 297-301.

Keppel, G. 1973. *Design and Analysis: A Researcher's Handbook.* Prentice-Hall, Inc. Englewood Cliffs, NJ.

Klahorst, H. T. 1983. How to plan your MS. *Manufacturing Engineering.* September. pp. 52-54.

Lens, W. 1980. Achievement motivation and intelligence test scores: a test of the Yerkes-Dodson Hypothesis. *Psychologica Belgica. 20* (1): 49-49.

Murphy, L. E. 1966. Muscular effort, activation level, and reaction time. *Proceedings of the 74th Annual Convention of American Psychological Association.* pp. 1-2.

Pearson, R. G. 1957. Scale analysis of a fatigue checklist. *Journal of Applied Psychology. 41.* pp. 186-191.

Pinneo, L. R. 1961. The effects of induced muscle tension during tracking on level of activation and on performance. *Journal of Experimental Psychology. 62.* pp. 523-531.

Rosenbrock, H. H. 1982. *A flexible manufacturing system in which operators are not subordinate to machines.* Unpublished paper. The U. of Manchester Institute of Science and Technology. England.

Royce, J. R., and S. R. Diamond. 1980. A multifactor-system dynamics theory of emotion cognitive-affective interation. *Motivation and Emotion. 4.* (4): 263-298.

Sharit, J., and G. Salvendy. 1982. External and internal attentional environments II. Reconsideration of the relationship between sinus arrhythmia and information load. *Ergonomics. 25* (2): 121-132.

Yerkes, R. M. and J. D. Dodson. The relation of strength of stimulus to rapidity of habit formation. *Journal of Comparative Neurology and Psychology.* 18. pp. 459-482.

Reprinted from *Proceedings of The Human Factors Society 28th Annual Meeting,* pp. 664-669, 1984. © 1984, by the Human Factors Society, Inc. and reproduced by permission.

52

Microscope Work—Ergonomic Problems And Remedies

Olov Ostberg
C. Eugene Moss

In assessing the potential for health problems in the electronics industry in Asia, Lin (1982) concluded that microscope work was an important occupational health problem. The microscope operators were typically women between 15-25 years of age, and the symptoms of microscope work included headaches, tenosynovitis due to repetitive motions, asthma due to soldering fumes, and possible visual impairment due to prolonged microscope viewing. Simons et al. (1942) reported that sustained voluntary or involuntary contraction of the ocular and/or neck muscles, such as occurs in using microscopes, can give rise to headaches, and pain and stiffness in the neck. In fact, a certain type of musculoskeletal neck pain accompanied by numbness and weakness of an arm, when sitting at a conventional microscope, has been termed "microscopist neuralgia" (Robinowitz et. al., 1981). Emmanuel and Glonek (1975) found that among microscopists at a major U.S. company, 80% were troubled by headaches or neckaches, 75% by backaches or stiff shoulders, and 75% by eyestrain. A study of hospital's cytology laboratories within the greater Stockholm county revealed that 84% of the microscopists suffered from job related musculoskeletal pain and 82% from eyestrain (Karlqvist and Tapio, 1980).

Dickerson (1976) examined the sources of complaints from a significant number of employees at a worldwide microscope operations company. Management believed that the complaints were closely linked to production problems, and initiated an action plan involving six vision criteria for visual acuity, eye-muscle balance, and depth perception, and eight eye protection codes for working time, microscope maintenance, and eye rest. Using these criteria, Dickerson found that only 73% of the employed microscopists were visually suited for full-time work and another 13% were capable of performing low-power microscope work on a part-time basis. The employment of the remaining 14% had to be terminated based on their inability to meet the minimum visual requirements.

AREAS OF CONCERN

Microscope Myopia

Myopia is a spherical error of refraction. Essentially, it is too much refractive power or, more often, too long an eye for the image of a distant object to be exactly focused on the retina. A temporary, functional or operational myopia can be defined as a situation specific over-accommodation.

One of the first accounts of operational microscope myopia is Druault's (1946) discussion of the visual problems experienced by physicians working at the microscope. He hypothesized that near-sightedness and double vision were the result of undue accommodation and convergence caused by the use of monocular microscopes. Schober et al. (1970) have shown that operational myopia is indeed more marked with monocular than binocular microscopes. Furthermore, these authors showed that converging ocular tubes gave rise to a substantially higher degree of operational myopia than parallel tubes. According to Baker (1966), inexperienced microscopists have a higher degree of microscope myopia than experienced microscopists.

From a report by Hennessy (1975), it is known that the microscope viewer typically works with an accommodation between 1-3 diopters, which is equivalent to a viewing distance of between 33-100 cm. Furthermore, he showed that the variation in microscope myopia is explained by an individual's involuntary accommodation in darkness or in viewing situations where the eyes are deprived of feedback for depth adjustment: night myopia, instrument myopia, open loop myopia, bright sky myopia, etc. This involuntary response, to become myopic when visual cues are missing, occurs because the eyes adopt an intermediate point of focus. This does not mean, however, that the eyes are at rest when the daytime viewing distance corresponds to the involuntary night/dark focus.

MacLeod and Bannon (1973) suggested that the microscopist should consider the object viewed through the microscope as being very far away in order to avoid viewing with active accommodation. Operators were also recommended to look up from the microscope often to avert a pattern of nearpoint overfocusing. They asserted that the proper use of microscopes should be no more fatiguing than any other normal visual task. However, such a statement is of limited value as "proper use" alludes to the viewing without involuntary accommodation (which is impossible), and as "other normal visual tasks" falsely allude to the microscope task as being a normal visual task. Furthermore, cytodiagnostic microscopy prohibits taking rest breaks by requiring long periods of concentration, which in combination with an anxiety for making detection errors in pathological determinations, sometimes makes the microscope work extremely fatiguing (Johansson, 1981).

It should also be pointed out that many microscopes are designed with converging eyepiece tubes. Whether this is due to optomechanical design solutions or to help the eyes get a fused image of the viewed object, the outcome is that the microscope causes the eyes to converge and accommodate. This is also a factor contributing to microscope myopia.

The Use-Abuse Theory of Visual Impairments

An early article in Illuminating Engineering (1967) gives a colorful description of management's frustration about low productivity due to the microscopists' unwillingness to tolerate such high visual demands and resulting discomfort. Microscopists naturally tend to relate their visual problems to working conditions. Employers, however, typically argue that if the employee meets the visual job criteria, takes good care of the microscope, and follows instruction on how to operate the instrument, then "The only fatigue you should experience is the fatigue felt by everyone at the end of a productive, satisfying day" (Fairchild, 1980). An ophthalmologic research team in Austria found that the accommodation during visually demanding work resulted in a small temporary myopia aftereffect, which could be used as an abuse measure. This led, in turn, to the recommendation that during continuous microscope operations "there shall after each hour's work be at least a one hour switch to a task not involving high visual demands" (Holler et al., 1975).

Similar to Dickerson (1976), Soderberg et al. (1983) were able to define several ophthalmologic factors contributing to microscopists' visual strain at an electronic plant. It was found that 80% of the full-time microscopists experienced various symptoms of visual strain. A statistical relationship was found between these symptoms and uncorrected astigmatism, fusion insufficiency and microscope use time. The authors were convinced that these factors, as well as uncorrected spherical refraction insufficiencies, had a bearing on the development of visual strain, especially in older microscopists.

Zoz et al. (1972) examined 593 microscope workers and found several indicators of marked visual fatigue, particularly among hyperopes and astigmatics. One indicator was that the eye became progressively more myopic during the work day. Visual screening examinations revealed that after six months of microscope work the percentage of persons classified as myopics had increased 37%.

Microscope work seems to be the most demanding work in terms of resultant temporary myopia, but the same consequences can occur for other types of close visual work such as watchmaking, industrial inspection, and cartography (Koitcheva, 1983). Although it is becoming accepted that prolonged work at demanding visual tasks may result in a temporary functional myopia, it is still debated within the ophthalmological community whether this foreshadows a permanent dysfunction (Taylor, 1981). Lanyon and Giddings (1974) and Young (1977) concluded that prolonged exposure to nearpoint viewing was a significant factor in the etiology of myopia. Based on large scale statistical data from the U.S. Public Health Services, Angle and Wissman (1980) convincingly showed that the use-abuse theory *is* valid, and that visual nearpoint work *is* a causative factor in the development of myopia. Long-term work at a microscope may thus increase the liklihood that the operator becomes myopic, especially if the microscope work is started at a younger age.

Lighting Aspects

Microscope work often is performed in laboratory-type rooms, with bright and reflecting surfaces, and with a high level of illumination. One method used to solve these lighting problems has been to install tailor-made microscope working cabinets with built-in lighting (Bergkvist et al., 1981; Dickerson, 1980; *Illuminating Engineering*, 1967). Microscope work also means that the attention is concentrated on a bright field of limited angular extension in an otehwise dark surround, as set by the microscope's built-in field stop and the design of the eyepieces. This means that the eyes are exposed to a field of view with extremely high contrasts between the center and the surround. This contrast may be many hundred times higher than the 3:1 ratio usually recommended for strain free viewing.

Of greater concern, however, is the hazard of having too much light and non-visible optical radiation impinging on the retina. One handbook recommends that older persons "should use the microscope with plenty of light", but later a warning is given that "when a microscope is too bright, a distinct pulling sensation in the front of the eye can be felt" (Burrells, 1977). Another warning comes from a production engineering treatise where it is concluded that microscope operators usually use excessively intense light: "The effects are not noticeable at first, but often prolonged viewing will cause severe eyestrain" (Froot and Dunkel, 1975). A light intensity high enough to cause a "pulling sensation" may also be high enough to cause damage to the retina. A general guideline is that the viewed light is a potential hazard for retinal damage if the luminance exceeds $10,000 \text{ cd/m}^2$. Field measurements indicate that the luminance as seen by microscopists frequently are on the order of 1,000-5,000 cd/m² (Bergkvist et al., 1981; Karlqvist and Tapio, 1980; Soderberg, 1978). These luminance levels may still be too high if the optical radiation is rich in short wavelengths (blue light). Fortunately, the predominant wavelengths transmitted through microscopes are in the red or even infrared region and thus have relatively long wavelengths.

The bright but seemingly non-hazardous microscope light could still be of concern because of the fact that microscopy necessitates excessive accommodation. It is not known what a retinal stretch means in terms of altered vulnerability to heat producing light—and accommodation does cause retinal stretch (Enoch, 1975). Secondly, myopia appears to enhance the possibility of retinal detachment (Ogino and Hashimoto, 1978), and light exposure may accelerate the condition of a potential retinal detachment (Lanum, 1978). Taken together, these factors indicate that microscope work could predispose workers for the development of detached retina. This is also the conclusion in a report evaluating a cluster of retinal detachments in workers who welded platinum and gold fine wire under a microscope (NIOSH, 1980).

Posture

Emanuel and Glonek (1975) and Karlqvist and Tapio (1980) found that approximately 80% of full-time microscopists suffered from aches/pain/stiffness of the neck, back and shoulders. These problems arise because the positioning of the eyes, and hence the posture of the whole body, is closely controlled by the microscope's focal length and the position and inclination of the eyepiece. That these problems can be quite severe is indicated in a report by Kumar and Scaife (1979), who carried out an ergonomic investigation at a plant employing 444 microscope operators. On a typical day, no less than 5% of the workers were found to receive medical attention for work-related musculoskeletal ailments.

Soderberg (1978) studied the postural problems at microscope operations in an electronic plant. She found that 45% of the operators suffered from work-related musculoskeletal ailments, primarily in the neck and shoulder region. The causes of the operator's postural problem was traced to one or more of the following factors (in order of significance): (1) unsuitable location of controls, (2) lack of workable supports for hands and underarms, (3) unsuitable height and angle of eyepiece tubes, (4) a general lack of space and work surfaces, and (5) static posture and/or very small repetitive work movements.

In order to adjust stage position, focus, and light influx, the clinical microscope operator's hands will have to be at the controls more or less continuously. And in industrial operation the hands are likewise involved. Unfortunately, microscope controls are invariably located in positions which force the operator to adopt awkward hand and arm positions, especially if the microscope has been elevated to match the seated operator's eye height.

POSSIBLE ERGONOMIC CONTROL MEASURES

By means of automation, many microscope jobs in industry and health care have already been eliminated; some altogether (blood cell count; micro soldering) and some via the introduction of projection screens (which poses a new set of ergonomic problems). The present review of the problems and remedies of microscope work only looks at the topic from an occupational safety and health perspective.

On Lighting and Microscope Design Recommendations

There is a need for improvements regarding the microscope lighting and the microscope viewing characteristics, but final recommendations cannot be given without further research.

Accommodation is induced by the lack of depth cues, by convergence, and by looking downward. Perhaps microscopists' involuntary accommodation could be diminished by horizontal and parallel eye piece tubes with built-in artificial depth cues. The optics of the microscope could also be designed with the understanding that instrument myopia has to be accepted (Leibowitz and Owens, 1975). Fox and Bahr (1981) suggested that the detail visibility in certain applications could be enhanced merely by using a white stage (increased contrasts). High quality microscope optics may reduce the need for high lighting levels. Microscopes could be equipped with a filter device to cut out the infrared and near-infrared portion of the optical rays exiting the oculars, and a detector device could warn the operator when the visible light is too bright for long term viewing.

On Work Postures Recommendations

Musculoskeletal injuries rank second in the current NIOSH (1983) list of areas for remedial action. From the Swedish 'NIOSH', Soderberg (1978) has suggested that some of the work posture problems in microscopy could be reduced if the ocular tubes would permit a more horizontal line of sight. This idea was adopted by Bergkvist et al. (1981), who created ergonomic accessories for existing microscopes. A favorable report on the benefits of such redesigned microscope workstations has been given by a group of pathologists (Robinowitz et al., 1981). Interestingly enough, another physician suffering severe back/neck/shoulder pain during microscope work has designed a complete "sitting machine" with pushbutton operated adjustments of the table and the chair (see Dickerson, 1980).

Of considerable interest is the recent Swedish *Ordinance Concerning Work Postures and Working Movements* (1983), which entered into force on January 1, 1984. This ordinance is of interest because of its generic nature (applicable to any type of work) and because its Commentary specifically mentions the need for a good chair for microscope work and also the need for adjustability of the microscope workstation. The Ordinance spells out compulsory principles for workplace designs, and when these principles, for practical reasons, cannot be met "the person doing the work must be given suitably disposed breaks".

On Work-Rest Break Recommendations

When ergonomic approaches involving redesign of work tasks or equipment are not feasible, then alternatives such as work-rest breaks and shortened working hours must be considered. Ideally, the relief opportunities should be integrated into the work tasks and work procedures, thereby making the work well-balanced (Soderberg, et al., 1983).

A USSR study has classified microscope work in a category requiring reduction in working time (Zoz et al., 1972). Hasselstrom (1976) reported that the USSR limits work time for cytodiagnostic microscopy to 24 hours per week. One of the major Swedish electronics industries recommends that employees carry-out non-microscope work during the last half-hour of a microscope working day, and encourages the use of natural work-rest breaks throughout the day for musculoskeletal relief exercises (Ericsson, 1982). Emanuel and Glonek (1975) recommended the insertion of ten minutes work-rest breaks two hours before and after lunch respectively. During the breaks the employees were advised to spend some time at a relief program consisting of both musculoskeletal and visual/ocular exercises.

Work-rest breaks are equally important where the microscopes have been equipped with projection screens, especially when the operator is required to view the screen for prolonged periods of time. NIOSH (1981) has acknowledged needs for work-rest breaks for operators of visual display termials (VDTs). These recommendations may be applicable for projection screen microscopes, and could also have a bearing on regular microscope work.

CONCLUDING REMARKS

This review of ergonomic issues in microscope work has taken account of the available research and technical literature dealing with such tasks as well as related visually demanding jobs. Evidence of visual dysfunctions, musculoskeletal strains and general fatigue are significantly recurring themes to dictate concerns and remedial actions. Microscope redesign to offset factors contributing to eyestrain and workstation modifications to ease postural demands are in order. This was also the conclusion of a French workshop on the safety and health apsects of microscope work (ANACT, 1980), the summarizing statement of which was "The operation of binocular instruments is highly fatiguing and merits the highest attention".

Suggested Readings

ANACT report: 1980. *Work with High Visual Demand. The Case of Work with Microscopes and Binocular Magnifiers.* (In French). Montrouge, France: Agence Nationale pour l/Amelioration des Conditions de Travail.

Angle, J. & D. A. Wissman. 1980. The epidemiology of myopia. *American Journal of Epidemiology.* pp. 220-228. *III.*

Baker, J. R. 1966. Experiments on the function of the eye in light microscopy. *Journal of the Royal Microscopical Society.* pp. 231-254, 85.

Bergkvist, H., L. Carlsson, & M. Stott. 1981. *Terminal and Microscope Work. Design of Workplaces, Work Equipment and Lighting.* (In Swedish). Gruppen. Stockholm: Ergonomi Design.

Burrells, W. *Microscope Technique. 1977. A Comprehensive Handbook for General and Applied Microscopy.* John Wiley & Sons. New York, NY.

Dickerson, O. B. 1976. Visual criteria aid in job placement. *International Journal of Occupational Safety & Health.* pp. 39-41, May/June.

Dickerson, O. B. 1980. Practical ergonomics. In C. Zenz (Ed.) *Developments in Occupational Medicine.* Year Book Medical Publisher. Chicago, IL. pp. 159-168.

Druault, A. 1946. Visual problems following microscope use. (In French). *Annals Oculistique.* pp. 138-142, 179.

Emanuel, J. T. & R. J. Glonek. 1975. Ergonomic approach to productivity improvement for microscope work. *Proceedings of the AIIE Systems Engineering Conference.*

Enoch, J. M. 1975. Marked accommodation, retinal stretch, monocular space perception and retinal receptor orientation. *American Journal of Optometry & Physiological Optics.* pp. 376-392, 52.

Ericsson, L.M. 1981. Ericsson booklet: *Advice and Directions Concerning Microscope Work.* (In Swedish). Molndal, Sweden.

———. 1980. *More than Meets the Eye* (booklet). Fairchild Camera and Instrument Corp. Mountain View, CA.

Fox, C. H. & G. F. Bahr. 1981. Relieving muscle fatigue and eyestrain in microscope. *Acta Cytologica.* pp. 195-196, 25.

Froot, M. A. & W. E. Dunkel. 1975. Visual inspection of integrated circuits: A case study. In C. G. Drury & J. G. Fox (Eds.) *Human Reliabilty in Quality Control.* Taylor & Francis. London. pp. 289-292.

Hennessy, R. T. 1975. Instrument myopia. *Journal of the Optical Society of America.* pp. 1114-1120, 65.

Hesselstrom, K. 1976. The working conditions of laboratory technicians engaged mainly in microscopy. (In Swedish). *Laboratoriet.* (Stockholm). pp. 250-265. No. 6

Holler, H., M. Kundi, H. Schmid, H. G. Stidl, A. Thaler, & N. Winter. 1975. *Work Load and Visual Demands in Display Screen Work.* (In German). Vienna: Gewerkschaft der Privatangestellten. (Verlag des OGB).

———. 1967. Lighting Keeps abreast of the microelectronic (editorial article). *Illuminating Engineering.* pp. 32-37, 62.

Johansson, C. R. 1981. Cytodiagnostic microscope work. (In Swedish). Department of Psychology, University of Lund.

Karlqvist, L. & S. Tapio. 1980. The working environment of cytology laboratory technicians. (In Swedish). Stockholm: Stockholms Lans Landsting.

Koitcheva, V. 1983. Visual functions and work. (In French). *Le Travail Humain,* pp. 93-111, 46.

Kumar, S. & W.G.S. Scaife. 1979. A precision task, posture, and strain. *Journal of Safety Research.* pp. 28-36, 11.

Lanum, J. 1978. The damaging effects of light on the retine. Empirical findings, theoretical and practical implications. *Survey of Ophthalmology.* pp. 221-250, 22.

Lanyon, R. I. & J. W. Giddings. 1974. Psychological approaches to myopia: A review. *American Journal of Optometry & Physiological Optics.* pp. 271-281, 51.

Leibowitz, H. W. & D. A. Ownes. 1975. Anomalous myopias and the intermediate dark focus of accommodation. *Science.* pp. 646-648, 189.

Lin, V. 1982. Health and the international divisions of labor: Issues in studying occupational health in less developed countries. *Proceedings of the American Public Health Association Meeting.* Montreal, Canada..

MacLeod, D. & R. E. Bannon. 1973. Microscopes and eye fatigue. *Industrial Medicine and Surgery.* pp. 7-9. 42(2).

———. 1980. *NIOSH Health Hazard Evaluation Report, TA 80-023-865.* NIOSH. Cincinnati, OH.

———. 1981. *NIOSH Research Report: Potential Health Hazards of Video Display Terminals.* NIOSH. Cincinnati, OH.

———. 1983. NIOSH: Leading work-related diseases and injuries—United States. *Morbidity and Mortality Weekly Report.* 24-26 & 32(2).

Ogino, N. & M. Hashimoto. 1978. Myopia and ocular fundus changes. (In Japanese). *Nichi Gan Kai Shi.* pp. 90-94, 82.

———. 1983. *Ordinance Concerning Work Postures and Working Movements.* (In Swedish). National Board of Occupational Safety and Health. Stockholm.

Robinowitz, M., C. F. Bahr, & C. H. Fox. 1981. Relieving muscle fatigue and eyestrain in microscopy. *Acta Cytologica.* pp. 585-586, 25.

Schober, H. A. W., H. Dehler, & R. Kassel. 1970. Accommodation during observations with optical instruments. *Journal of the Optical Society of America.* pp. 103-107, 60.

Simons, D. J., E. Day, H. Goodell, & H. C. Wolff. 1942. Experimental studies on headache: muscles of the scalp and neck as sources of pain. *Research*

Publications of the Association for Research in Nervous and Mental Disease. pp. 228-244, 23.

Soderberg, I. 1978. Microscope work II. An ergonomic study of microscope work at an electronic plant. (In Swedish). *Arbetarskyddsstyrelsens Undersokningsrapport.* Stockholm. No. 40.

Soderberg, I., B. Calissendorff, B. Elofsson, B. Knave, & K. G. Nyman. 1983. Investigation of visual strain experienced by microscope operators at an electronic plant. *Applied Ergonomics.* pp. 297-305, 14.

Soderberg, I., E. Gunnarsson, B. Calissendorff, S. Elofsson, & K. G. Nyman. 1981. Microscope work III. Investigation of visual strain and eye changes in histological technicians during two different microscope-intensive working routines. (In Swedish). *Arbetarskyddsstyrelsens Undersokningsrapport,* Stockholm. No. 25.

Taylor, H. R. 1981. Racial variations in vision. *American Journal of Epidemiology.* pp. 62-80, 113. (Followed by a debate: 138-142, 115, 1982.)

Young, F. A. 1977. The nature and control of myopia. *Journal of the American Optometric Association.* pp. 451-456, 48.

Zoz, N. I., A. Y. Kuznetsov, M. V. Lavrova, & V. A. Taubkina. 1972. Visual hygiene in the use of microscopes. (In Russia). *Gigiena Truda i Professional' nye Zabolevanija.* pp. 5-9, 16(2).

53

Manual Versus Robotic Assembly: Some Implications of Product Design

Karl Domas
Martin Helander

Many examples of successful robot applications abound in Japan, the United States, and especially Sweden, which has the highest number of robots per capita. Typically the robots are used for tasks such as welding and spray painting. Robots are ideal for such duties, since the work environment is often hazardous to humans. In the last couple of years, robot capabilities have improved due to increased dexterity and movement accuracy. It is now possible to use robots for more complex industrial tasks, including production involving flexible manufacturing and automatic assembly. For such tasks it is not self-evident who should perform the job—robots or humans. Models are needed in order to allocate tasks between humans and robots.

Ever since the time of the industrial revolution and the introduction of machinery, people have been concerned with the allocation of tasks between humans and machines. In the beginning there was a relatively high concern for the human operator. Early inventions such as the "Spinning Jenny" and the "Spinning Mule" were designed so that rather than eliminating human skills they required an extension of the operator's skill beyond what was required in order to perform the previous task (Rosenbrock, 1982). Later developments in automation have usually gone the other way; they have substituted for human skills. In many cases, the only tasks left for human operators were those difficult to automate or those required for supervision of the system.

With the introduction of robots in the working environment, another level of complexity of automation is added. There are still left-over tasks that are difficult to automate. A typical example is industrial inspection which cannot yet be performed proficiently by robots. These tasks are often monotonous and boring, because they rely primarily on the operator's ability to perform visual discriminations. There are no cognitive elements, nor any social skills required, ingredients which are imperative for job satisfaction. But more complex and satisfying tasks are also created, such as robot programming and maintenance.

In the past, we have been preoccupied with the effects of the introduction of robots on the production of the direct labor, since this is where the effects are immediate and most visible. This chapter also brings into focus the effects of the introduction of robots on the indirect labor, for example, product designers and financial managers. In this case, the involvement is more intangible and intrinsic, but the impact of automation may be even more dramatic than for the blue collar workers.

Description of Robots

According to the Robot Institute of America, a robot is "a reprogrammable, multi-function manipulator designed to move material, parts, tools, or specialized devices, through variable programmed motions for the performance of a variety of tasks." This definition makes robots seem like just another piece of hardware. But robots are more! Most of us perceive robots as machines incorporating a mechanical interpretation of human motor functions, a representation of anthropomorphic qualities. The computer is an additional part of a robot providing means of control, while sensors, actuators, wires, and other devices represent the electronic gluebinding between the mechanical manipulator and the computer. The software is the soul breathed in by the programmer to make the robot perform its magic.

One example of a robot is the IBM 7565 Robotic System shown in Figure 1. It consists of a mechanical manipulator which, with seven degrees of freedom: X, Y, Z, for the linear motions, roll, pitch, yaw for the horizontal motions, and the opening and closing of the gripper. The Manipulator is controlled by the IBM Series/1 computer by programming the various motions. In addition, there is sensory feedback to the computer from a variety of sensors. Linear displacement sensors provide information on positions of the linear components; rotary potentiometers provide information on rotary displacements of the wrist. The highest concentration of sensors is, however, found in the gripper. A linear potentiometer is connected to the linkages that move the fingers of the gripper and thus, provide information on the degree of opening of the gripper. A light emitting diode with a phototransistor serves to detect the presence or absence of objects between the fingers. An arrangement of strain gauges provides information on force exerted on the fingers in three directions. In addition to these

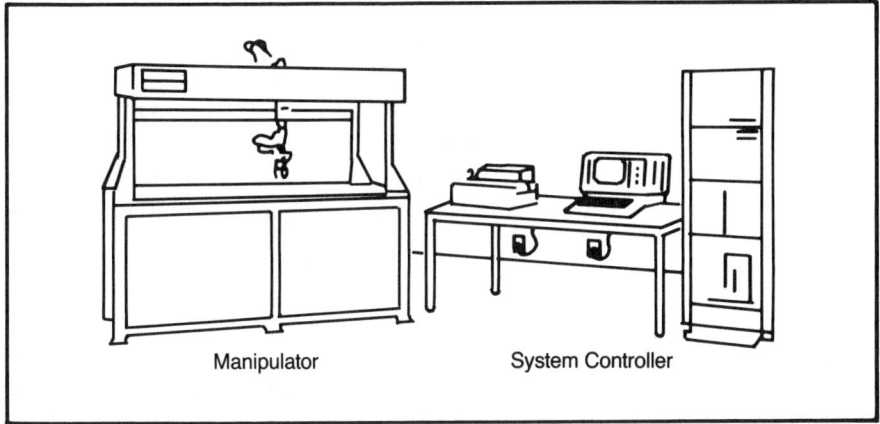

| Manipulator | System Controller |

Figure 1. *IBM 7565 robot with system controller.*

sensors, the Series/1 computer provides a means to connect up to 64 additional sensors via the digital input ports, and up to 64 output devices via the digital output ports. The Series/1 computer is also capable of communicating with other computers, thus providing networking capability and computer control of several machines and robots linked together.

Similar to the robots of older generations, the 7565 can be "taught" to perform a job via a teaching pendant. The operator then guides the manipulator by pressing appropriate buttons on the pendant. When the desired position is reached, the computer reads the signals from the sensors and stores values corresponding to the positions of all the joints. Later, the operator can command the robot to replay these sequences.

The real power of the 7565 lies in that it can be "programmed" to perform a job via a high level programming language called AML (A Manufacturing Language). Similar to other high level languages like Fortran, PL/1, and APL, AML is very powerful and requires trained programmers to use it efficiently.

A program is first written that can drive the robot to perform the desired actions. Then, before the robot is installed in the manufacturing environment, this initial program must be augmented by software that handles error recovery and other external signals indicating missing parts, misaligned or wrong parts, and so forth. All of these eventualities must be considered by the system designers and programmed a priori. These programs typically take more time to program and require more memory space than the programs that direct the robot motions. During the system design it is not possible to predict all eventualities. The programs must, therefore, be written to allow inclusion of new programs, as new requirements are defined. This is where robot programming is touching on artificial intelligence. Ideally a robot programming system should be self-teaching so that it can gather information from the manufacturing processes, analyze it, and produce programs to allow the robot to adjust to new conditions or events. This is the real meaning of flexible automation. It is not so much the ability to reprogram the robot to perform another job in the afternoon than it was doing in the morning. Flexibility of automation refers to the adaptability of the robot to various factors in the external environment. This is presently the area of greatest challenge in the development of robotic systems.

Assembly Lines

A fairly recent technique of assembling products is by using "continuous flow manufacturing." A computer network initiates the transportation of materials, subassemblies, and components, usually utilizing different types of conveyors. Along the conveyors are several work cells, where assembly or test operations are performed. The work cells can be either manual or automated to varied degrees. In most cases, fully automated production is desirable; at least in the long term. However, in many cases, it is impossible to bring the entire manufacturing system into operation all at once. A common reason is that the products were not designed so that robots can perform all of the steps in the assembly process. It will then take time to implement the necessary changes in product design so that robots can be used. Other reasons are that the times for delivery of automated equipment is longer that the product development. The production may then have to start before the automated assembly line is operational. Using CAD/CAM equipment product development may take only 6 months, much less that the 1-1.5 years necessary for implementing an automated assembly. Since the product life time may be as short as 2-3 years, it is often difficult to justify the costs necessary for automatic assembly. Such considerations are now putting pressure on industrial engineers to implement assembly lines of a more generic nature that can be used to produce any product fitting a certain class.

In any case, during a transition period, a mixture of manual and automated work cells is found coexisting on an assembly line. As the product designs improve and corresponding manufacturing process changes are introduced, the manual work cells may be reproduced by robot work cells. A modern assembly line, therefore, has the flexibility to adapt to new products and processes on a timely basis.

Allocation of Tasks

The allocation of tasks between humans and robots depends largely on the design of the product to be assembled. In the past, products were designed predominantly for manual assembly. Recently, a new methodology of design has emerged, the intention of which is to facilitate automatic or robotic assembly.

This methodology is variously referred to as design for manufacturability, design for ease of assembly, or design for automation (DFA). The principles of DFA first emerged in industry in the form of compilations of empirically developed rules and guidelines. In recent years, new analytical methods have been developed and there are now several principles for DFA; the most common are listed in Table 1 (Boothroyd, 1982).

It has come as a surprise to most design engineers that once the product has been designed for automatic assembly, it is also easier to assemble manually. In other words, most of the design principles that make it easier to assemble a product by a robot also facilitate manual assembly. The choice of robotic or manual assembly is, therefore, nor always straightforward. In many instances, when the product has been redesigned for automation, the assembly work is simplified and the assembly time is minimized to the point that capital spending on automation can no longer be justified.

There are several examples of this situation we would like to share with you. One is a paper pick mechanism for a small printer (Bailey, 1983). In this case, the original product has 27 parts; far too many to be assembled by a robot. After redesign, the number of parts was reduced to 14, 13 of which could be assembled by a robot, see Figure 2. The remaining part had to be inserted by a

TABLE 1 – Principles of Design for Automation

Design for unidirectional assembly, preferably top-down
Eliminate or reduce the number of screws
Design for insert and snap assembly
Design chambers for self-adjustment
Eliminate parts that are difficult to feed automatically, such as springs, washers, fragile parts, etc.
Eliminate parts requiring extremely tight tolerances
Eliminate parts that are difficult to orient
Eliminate parts that are difficult to handle; either too bulky or too small
Combine parts to reduce the number of assembly steps
Eliminate cables, wires, and other flexible parts

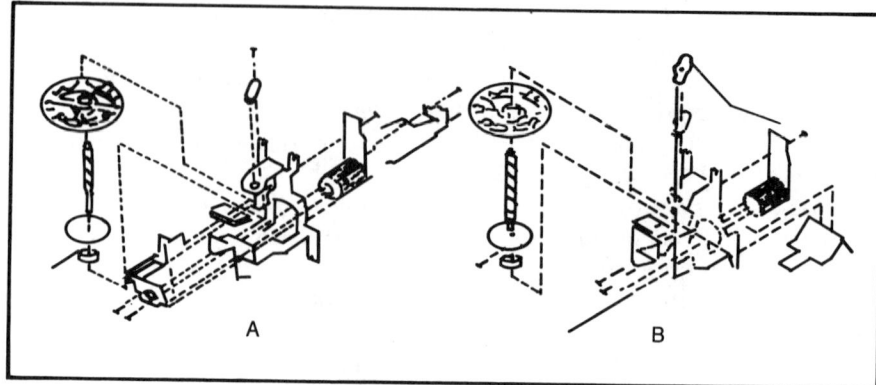

Figure 2. *Design of paperfeeding mechanism for an IBM printer. (A) Before redesign for automation the product had 27 parts. (B) After redesign the product has 14 parts, 13 of which could be assembled by a robot.*

human operator. This is then a case of left-over tasks, and there would be difficulties in designing the job so that the human operator would be satisfied. However, the surprising outcome of this study was that the assembly of the product had been so much simplified that it did not pay off to use a robot. This product is presently assembled manually.

Another example is a computer terminal product in which all internal cables were eliminated and the parts simply snap together, see Figure 3. The assembly time is so short that automation could not be justified. While the family automated process may still be desirable, economics and expediency substantiates the manual operations. Within this environment, the selection of automated robotic jobs is governed more by the technical feasibility and economics than by disciplined analysis of operations determining what jobs are best performed by people and what jobs are best performed by people and what jobs are best performed by robots.

Discussion

Robots will be playing increasingly important roles in our lives in the future. For people involved directly with production, the effects are immediate and already visible. While changing many jobs, sometimes into extinction, new jobs are being created, thus providing career paths for many. Here, education and training play a crucial role. However, robots are also indirectly affecting lives and jobs of many other people such as product designers, product and business

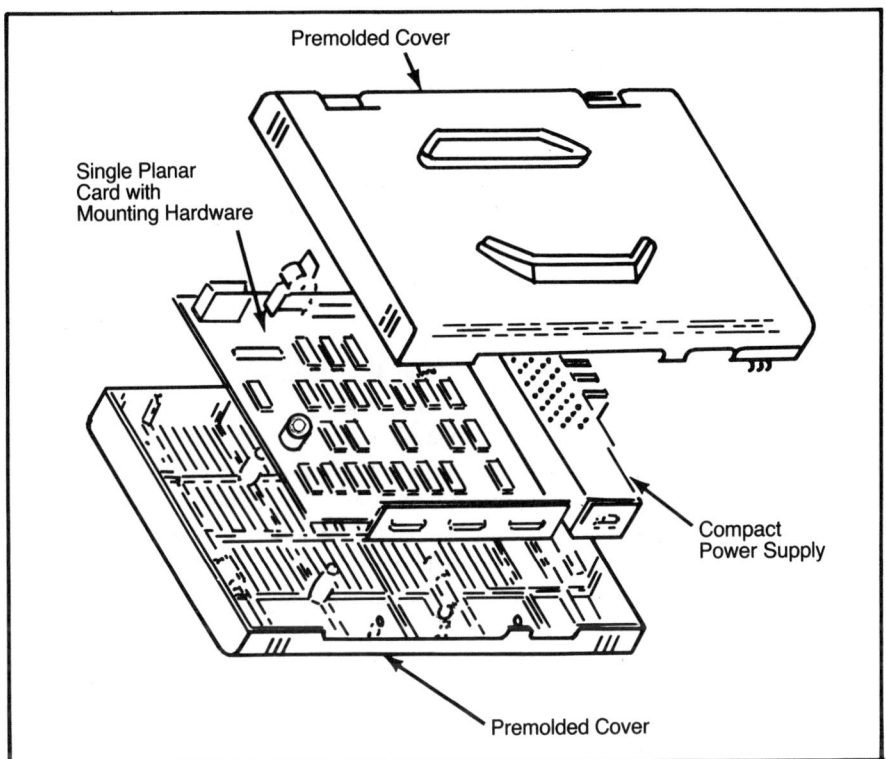

Figure 3. *A product designed for ease of assembly and automation.*

planners, and financial specialists. Most of us will be affected by the introduction of robots into production, in many different ways and to varying degrees. We need to understand robots in order to be able to determine not so much what robots can do better than people, but to determine what jobs are interesting for humans and when we want to exploit the robots to perform the uninteresting, unhealthy, and dangerous jobs.

For example, compared to robots, human operators excel in perceptual discriminations; it is, therefore, tempting to use humans for inspection tasks and leave the assembly to robots. However, one must keep in mind that inspection tasks are often inherently monotonous and boring; often more so than assembly tasks. From a job satisfaction point of view, it would be better to leave the inspection task to the robot and the assembly task to the human. It is difficult to propose general guidelines; much depends on the characteristics of the workplace and the people performing the task. There are, however, several principles for job satisfaction in industrial environments dealing with psychological issues such as avoiding pacing and fragmentation of work, and increasing work autonomy and worker responsibility.

In most cases, human assembly is simplified by the same principles as robotic assembly. For example, the use of one size of screw rather than several simplifies robotic assembly considerably. Humans may be less impositioned by the use of several sizes of screws, although it may be predicted that the reaction time for choosing a screw would be reduced if there were only one size.

There are, however, important differences between humans and robots that must be considered. For example, robots prefer to insert screws vertically; horizontal insertion is much more difficult since there must be special end effectors to prevent screws from falling. In contrast, humans prefer horizontal insertion which offers biomechanical advantages over vertical insertion.

Ideally, products should be designed so that they can be assembled by humans or robots. Thereby, it is possible to maintain full flexibility in the choice of manufacturing method. There are already models which can predict robotic assembly times (Boothroyd, 1982). Research is now needed to develop models that can predict human assembly times. Although methods engineering and principles for economizing human motion have a long tradition within industrial engineering (e.g., Barnes, 1963), implications of product design on ease of assembly are virtually unknown. There is a substantial amount of knowledge in the human factors literature on laboratory studies of reaction times and perceptual-motor skills. Surprisingly, this body of knowledge has never been interpreted in the more practical context of industrial assembly. It seems ironic that the introduction of robots has opened our eyes to such important human issues.

Suggested Readings

Bailey, R. J. 1983. Product Design for Robotic Assembly. *Proceedings of Robot 7.* Society of Manufacturing Engineers. Dearborn, MI.

Barnes, R. M. 1963. *Motion and Time Study.* Wiley. New York.

Boothroyd, G. 1982. *Design for Assembly Handbook.* Department of Mechanical Engineering. University of Massachusetts. Amherst, MA.

Nof, S. Y., J. L. Knight, & G. Salvendy, 1980. Effective Utilization of Industrial Robots—A Job and Skill Analysis Approach. *AIIE Transactions.* 12. pp. 216-225.

Paul, R. P. & S. Y. Nof, 1979. Work Methods—A Comparison Between Robots and Human Task Performance. *International Journal of Production Research.* 17. 277-303.

Rosenbrock, H. H. 1983. Seeking an Appropriate Technology. *Proceedings for IFAC Symposium on Systems Approach to Appropriate Technology Transfer.* Austria.

Reprinted from *Proceedings of the Human Factors Society—28th Annual Meeting—1984.* © *1984,* by the Human Factors Society, Inc. and reproduced by permission.

54

Artificial-Intelligence Techniques Boost Utility of CAE/CAD Tools

J. D. Mosley

Artificial intelligence has found a valuable niche in computer-aided engineering and computer-aided design. CAE/CAD tasks frequently require the use of heuristics, and the application of AI techniques can reduce uncertainty by offering software-based inferential reasoning capability.

Yet the effects of AI in the engineer's work place are currently more evolutionary in nature than they are revolutionary. For example, no full-fledged intelligent CAE package has yet entered the market—you'll find only traditional CAE software that borrows techniques developed from AI research. However, as CAE software developers increase their AI expertise, you can expect increasingly more sophisticated CAE/CAD packages and, eventually, an expert system that provides innovative CAE/CAD solutions.

IDENTIFY APPLICABLE TASKS

Although it is often difficult to distinguish between software that does and doesn't incorporate AI techniques, it is easy to identify engineering tasks that would lend themselves to AI solutions. Brian Horsley, senior engineer and AI systems specialist at PA Technology, cites pc-board routing as a classic problem for AI to tackle. Routing involves sifting through several outcomes, all of which are technically correct, in search of the best solution.

Instead of yielding to the clearcut answers that computers generate when using analytical techniques, routing problems must be solved by comparing degrees of efficiency. An AI approach employs certain heuristics and facts to reduce the number of likely routes, thus streamlining the computer's search for optimal routing.

Significantly, the strategy (that is, inference engine) used to guide the computer through selection of the various heuristics and facts (often embodied in a set of design rules) must be independent of the knowledge base that contains the heuristics, rules, and facts. In this way, a change in circumstance or parameters will result in a revised application of the contents of the knowledge base. This flexibility differentiates an intelligent system from an algorithmic one.

Other focal areas for intelligent CAE packages include component layout, troubleshooting, and logic simulation: key areas where the benefits of balancing heuristic and theoretical knowledge significantly reduce the amount of time an engineer must devote to a design effort. But the state of the art at most major CAE companies has not yet spawned a classically defined, commercially available AI-based CAE system.

EXPERT SYSTEM ROUTES PC BOARDS

Nevertheless, Telesis Systems Corp has announced a rule-based expert-system pc-board router, called Insight, that analyzes the physical data and design rules necessary to develop and optimize routing strategy. After you enter English descriptions of the board's physical characteristics and design rules regarding such aspects as line width, spacing, and pad description, Insight accesses its database of routing techniques to create the unique set of routing instructions that yields the highest possible completion rates for the board you've described. Insight considers nearly 30 costing factors and automatically routes boards with as many as 14 signal layers.

Insight was developed in Prolog, a programming language used extensively in AI research but has been ported to C for increased speed and portability. The manufacturer claims that Insight's multipass, multialgorithm system outperforms rip-up routes by a factor of 10:1. You can use Insight for any design technology, including digital, analog, SMD, fine-line, and ECL. Moreover, you can specify grid resolutions as fine as 1 mil, any number of pad shapes and sizes, and an unlimited number of nets, connections, or components.

The system supports automatic on-line 45° routing. You can restrict routing and etching to particular areas of the board. Insight routes the horizontal traces first, followed by vertical traces, inserting vias only where needed. The Insight router is a module within the Telesis EDA-3000 PCB Design Application software, which costs $17,500. The EDA-3000 package includes schematic design, automatic placement, and the Insight router.

THE EMBEDDED INTELLIGENCE

Tektronix, on the other hand, has proven that AI techniques needn't be apparent to the end user to be useful. Instead of exposing gate-array design engineers to a foundry's reference database, Tektronix uses a rule-based expert system to conciliate each designer's goals with the foundry's manufacturing constraints. Merlyn-G, Tektronix's automated gate-array layout system, can literally apply any of hundreds of processing options to modify its design strategy when laying out a difficult gate array. Accordingly, it takes a significant amount of expertise to decide which Merlyn-G processing option to apply for the best design.

To provide this expertise, Tektronix developed software modules called TurnChips to tune the Merlyn-G controls for placement and routing (without intervention from the designer) while simultaneously producing a physical design that is functionally equivalent to one reproducible at the foundry's design center. Each foundry-endorsed TurnChip uniquely addresses a particular style of gate array. And it is in the development of each TurnChip that Tektronix's patented AI techniques are used.

TurnChip modules rely on rules concerning tuning decisions and control options to control the execution of programs during the layout process. However, when Tektronix tried to implement the design rules via an interpreter, the system spent more time deciding which program to execute than it took to run the program. So the company developed (and subsequently patented) a compiler for the rule base, which lets the TurnChip provide rapid batch processing by branching from one Merlyn-G program to another.

For engineers who know their equipment better than anyone else, an expert system shell called IN-ATE from Automated Reasoning Corp offers a simple way to generate a customized automatic test system for fault diagnosis. You define the unit under test (UUT) to the system by drawing a schematic diagram of the equipment with a mouse. The schematic includes a hierarchical structure and modular substructure of the UUT, connections between modules of the UUT, and accessible test points. IN-ATE saves the symbolic description of the UUT in a Lisp-format ASCII text file. No other information is required by IN-ATE; the system can automatically learn any other information it requires, or you can add it incrementally.

IN-ATE helps you develop the system's rules via an editor that uses a series of on-screen, menu-like dialog boxes. At each step in the construction of a rule, the editor checks for consistency and correctness. The rules encompass heuristics for fault diagnosis, such as the probability of failure at a particular testpoint, the cost of performing a test, and preconditions that must exist before testing is recommended. Using deductive reasoning and logic modeling, IN-ATE can automatically generate any missing rules. The system also builds and maintains a statistical reliability database based on inductive reasoning relating to failure probabilities.

THE LEARNING ALGORITHM

Even more impressive are the IN-ATE learning capabilities. A learning algorithm lets IN-ATE re-adjust its reliability database and its rules to account for reported UUT failures. The system also uses nonmonotonic logic to refocus and reorganize its internal beliefs about the fault location. IN-ATE can even interactively guide a novice user through each step of a troubleshooting session.

IN-ATE has been used by Northrop Corp for automated testing of avionics on its F-5 fighter. The system also provides IEEE-488 bus support for computer-controllable test instruments. Procedural attachments let IN-ATE automatically program and activate test equipment. IN-ATE comes in several configurations for use with the Macintosh, VAX, and TI's Explorer workstation; price varies from $5000 to more than $30,000.

The specialized test stations (one version costs between $350,000 and $450,000) sold by Teradyne for VLSI in-circuit testing also incorporate an expert system to accurately isolate faults. The expert system analyzes test data using a set of diagnostic rules and knowledge about circuit topology. The system also includes the tests performed by the test station in its diagnostic analysis, including any relationships among nodes that are directly and indirectly involved in each test. If a failure so requires, the system will perform additional tests to clarify a failure condition. Once the system completes all the tests, the expert system eliminates spurious results and generates a repair message for each fault. The system can also produce a history of device failures and generate statistical fault summaries by node and by component.

Teradyne has used AI in its diagnostic systems for the past 10 years, but with the increasing complexity of analog circuits and the large number of active nodes that a test station confronts during a VLSI in-circuit test, the use of AI becomes critical. Each true fault that occurs on a board can create several other apparent faults. Successful device isolation isn't always possible on a heavily populated board. As a result, Teradyne has developed a hierarchical organization of rules that govern the diagnostic process using procedural, production, and table-based schemes to minimize diagnostic time and the amount of memory used to execute the diagnosis. The Teradyne system uses both forward chaining and backward reasoning to arrive at a suitable fault diagnosis.

An overview of Teradyne's diagnostic system and the interactions among the various components in the system are illustrated in Fig 1. Notice that once the knowledge base obtains the test results from the test analyzer, the inference engine has access to information from every area in the system. Note also that the inference engine can also provide data to certain databases in the system.

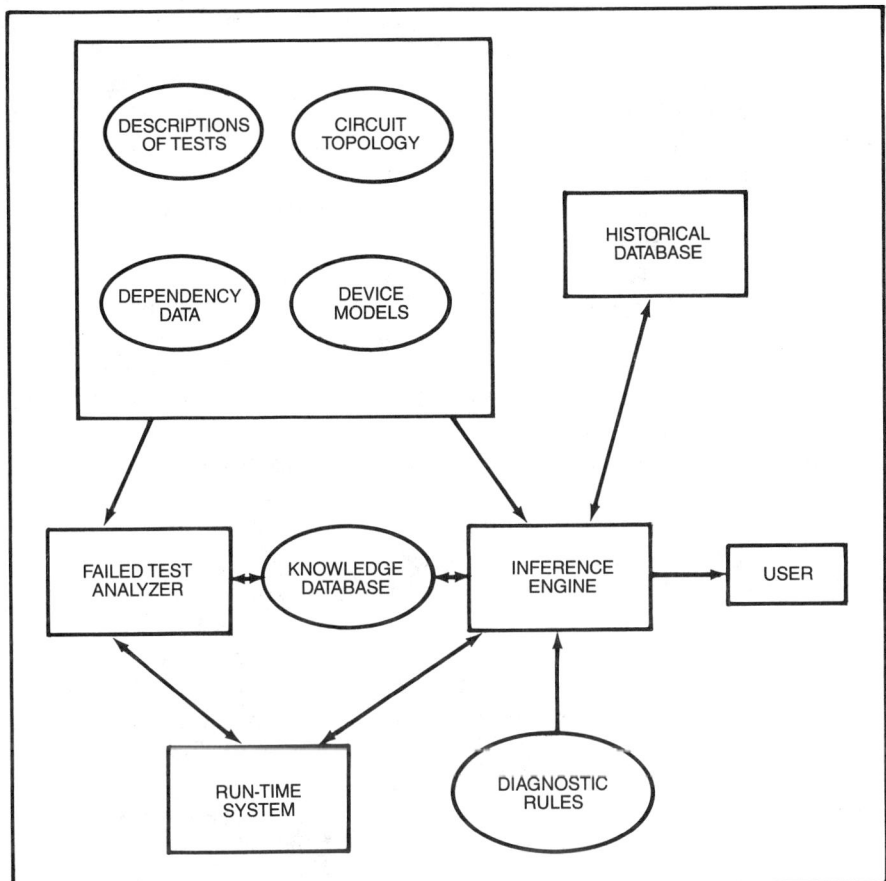

Figure 1. *A diagnostic expert system from Teradyne incorporates a number of interactions among the various system components. Note that the inference engine has access to information throughout the system and can also provide data to certain databases in the system.*

As soon as a fault has been identified, the system modifies the knowledge database so that the inference engine can use the new data to evaluate the remaining faults.

Although most expert systems designed for electronic engineering fall neatly into the traditional CAE/CAD formats, one of the oldest and most useful expert systems solves theoretical and applied mathematical problems. Originally developed in Lisp in the late 1960s at MIT, Macsyma represents over 100 man-years and more than 100,000 lines of source code. More than 600 technical papers have referenced Macsyma, which is described as both a programming language and an interactive system.

Since 1982, Symbolics has supported and enhanced Macsyma for use on four manufacturers' computers: DEC's VAX 11/7XX, Sun Microsystems' Sun-2 and Sun-3, Masscomp's MC5500, and Symbolics's 3600 Series. Price ranges from $6000 to $60,000.

Macsyma uses a library of symbolic math techniques to solve differential equations; to compute Laplace and inverse Laplace transformations; to factor polynomials; to simplify expressions; to compute Poisson series; and to generate Fortran statements from Macsyma expressions. It also uses the techniques to compute definite and indefinite integrals symbolically; to manipulate matrices, vectors, and tensors; to compute indefinite summations; to plot curves and surfaces; and to translate programs into Lisp for later compilation. And Macsyma provides answers in symbolic form, as a graph, or in Fortran code. Useful applications for EEs include VLSI design, analysis of electromagnetic fields and interference, and control theory.

But if you're faced with IC design choices that are of a more fundamental nature, you might want to consider a consultation with VLSI Technology's Design Assistant. This expert system helps you weigh the numerous tradeoffs that must be resolved before designing any complex IC. For example, Design Assistant can compare gate arrays versus cell-based implementation or custom design. It can evaluate partitioning alternatives and the maximum number of elements that you should squeeze into one chip. It even offers suggestions about the type of packaging you should use. Other interesting functions include estimation of chip size and power consumption.

Design Assistant performs these analyses even before you have completed your design. Or you can use the system to provide "what if" analysis before you actually begin your design. To facilitate system conception and to help document your alternatives, the system lets you enter block diagrams with data that describes your design. You can also enter data as gate count, a list of TTL parts, or a captured net list. Naturally, the accuracy of Design Assistant's recommendations improves with the accuracy of the design details.

Currently available only in VLSI Technology design centers, the Design Assistant will be released for use on VAX, Micro VAX, Apollo, Elxsi, Ridge, HP 9000 Series 320, and Sun computers as an integrated part of VLSI Technology's IC design system in the second quarter of 1987. Pricing for the Design Assistant module will be $25,000. The VLSI IC design software varies in price from $20,000 to $140,000.

Notably, not every useful CAE/CAD package that is advertised as an expert system actually uses traditional AI techniques. AI and expert systems are the latest buzzwords for fashionable software, and many companies may stretch traditional AI definitions to gain attention for their products.

Logic Automation provides extensive board-level logic simulation via software modules called Smart-Models that encompass the behavior and characteristics of more than 1000 components—including memories, μPs, PLDs, and ALUs. Although device simulation is commonly used in IC design, board- and system-level designs are typically developed by breadboarding the circuit and then fine-tuning the design during multiple prototype phases. However, the complexity of many board-level devices is increasing to such a degree that simulation has become a necessary tool for designing sophisticated pc boards.

SmartModels are written at the behavioral level, giving them speed and functionality advantages over models developed by other companies using a gate-level approach. Some of the latest models offered by the company include Intel's 80386, IBM's CMAC—a complex number multiplier/accumulator, and TRW's TVC341 4-port memory. Prices for the company's models range from $400 to $5000, with special packages of multiple models costing as much as $25,000.

SmartModels include a symbolic hardware troubleshooting mechanism called Symbolic Hardware Debugging to identify design errors that can escape even prototype testing. As noted by Robert Hunter, marketing vice president, "The I/O protocols of the IBM CMAC, for example, are so complex that it will be virtually impossible to design with it successfully without simulation."

Logic Automation uses experienced IC designers to develop these models. These designers determine from data books the kinds of problems a systems engineer is likely to encounter when using a part. They then write specific checks for these problems into the models to aid in design analysis. When an error occurs during simulation, the model issues a message stating the location, time, and nature of the problem.

It is therefore the Symbolic Hardware Debugging that encompasses the bulk of the expertise drawn upon for what Logic Automation refers to as "Knowledge Based Systems Analysis." The heading of a sidebard in the company's brochure even states that SmartModels actually turn your simulator into an expert system.

Yet, the company's current products do not actually provide inferential reasoning capability. Nor do they incorporate any type of AI-based machine-learning techniques. However, the sophistication of the design analysis offered by Symbolic Hardware Debugging does make them useful—though not AI-based—products.

Suggested Readings

Freeman, E. 1986. Autorouters use sophisticated algorithms to lay out complex, multilayer pc boards. *EDN*. August 7: 67.

Joseph, R. 1984. *An Expert Systems Approach to Completing Partially Routed Printed Circuit Boards*. EE/CS MIT.

Kernoff, A. 1986. *Who's Who in Artificial Intelligence.* Tom Schwartz Associates. Mountain View, CA.

Mosely, J. D. 1986. Expert systems as engineering tools will broaden productivity/creativity options. *EDN.* November 13: 91.

Wang, E. 1987. Artificial intelligence for competitive advantage. *AI Expert.* 2(1): 5.

Reprinted from *EDN*, February 19, 1987. © 1988 Cahners Publishing Company, a Division of Reed Publishing USA.

55

Designing for Assembly:
A Computer-Based Approach

H. Lee Hales

Early manufacturing involvement, simultaneous engineering, and design for manufacture are all concerned with improved communications between engineering and production. But if this effort is to be more than preaching or using slogans, it must be reduced to a formal procedure than can be routinely applied by design and manufacturing engineers. To be in step with today's directions in engineering, such a procedure should be computer-based.

Designing for manufacture can be guided by a stored checklist that can be called and completed by the appropriate parties at key points in the design cycle. Such lists are common for printed circuit design and also for mechanical assemblies where robotic or automated production equipment will be used. Checklists, however, only look at compliance. They do not readily measure the worth of competing designs or help us to choose the best design. For this, we need a scoring procedure, which exists for a variety of design activities, most notably for assembly oriented analysis of mechanical designs. The Hitachi Assemblability Evaluation method and the Boothroyd Dewhurst Design for Assembly (DFA) method are the two best-known scoring procedures (both are commercially available).

In 1985, I chose the Boothroyd Dewhurst DFA method for a research project at Daisy Systems Corp., Mountain View, CA. The purpose of the project was to explore ways in which the computer could be used in a formal approach to designing for product assembly. DFA was the best starting point, because it was widely used and already available in software form on the IBM-PC. This chapter describes the findings of the Daisy Systems project. The Boothroyd Dewhurst DFA methods are fully described in the Suggested Readings at the end of this chapter. I will describe 10 important ways in which the computer can be used to apply and support these methods. The 10 uses are as follows:

1. data entry and calculations;
2. computer-aided drafting;
3. solids modeling;
4. color graphic results;
5. concurrent processes;
6. networking;
7. archiving;
8. data management;
9. reporting; and
10. automatic initialization.

This list is ordered roughly from lower to higher degrees of sophistication. These computer uses are generic and would support any procedural approach to designing for product assembly.

COMPUTER SUPPORT

Data entry and calculations

The most obvious form of computer support is that embodied in the existing, commercially available software from Boothroyd Dewhurst, Inc. These programs capture responses to a sequence of questions and use the responses to extract the appropriate assembly times from a set of reference tables. These operation times are then multiplied by a labor rate and summed, giving comparative costs and design efficiencies for two or more competing designs (Figures 1 and 2).

Computer-aided drafting

For small-to medium-sized assemblies, the most definitive source of input data is the prototype (or production) version of the design. However, as we have noted, an important goal today is to drive the analytical process toward the earliest stages of design, before prototypes are available. In this context, the

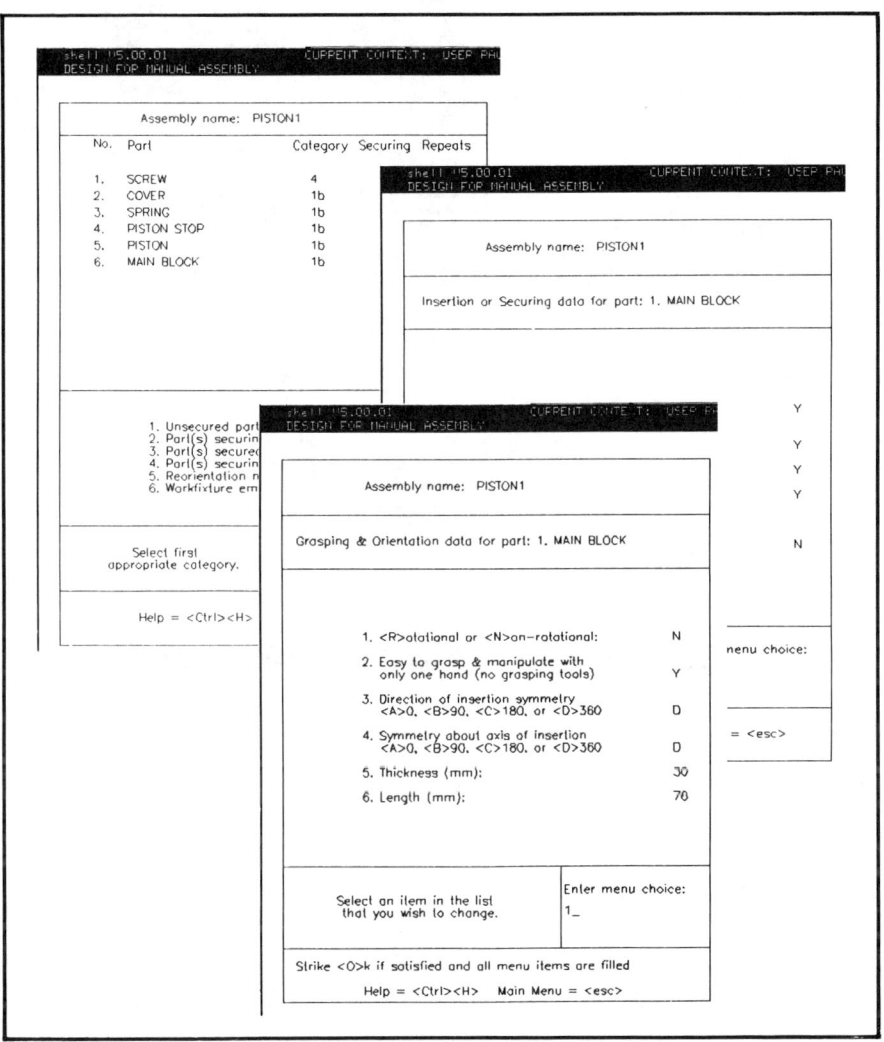

Figure 1.

most definitive sources of input are engineering drawings. An exploded-view (isometric) assembly drawing is typically the most useful document for visualizing the assembly sequence. But, in standard engineering practice, such drawings are not dimensioned. Therefore, we need recourse to the relevant piece part drawings as well.

Shell M5.00.01   CURRENT CONTE T: USER PAULB DEMO
DESIGN FOR MANUAL ASSEMBLY

ID	RP	HC	TH	IC	TI	TA	CA	NM	MANUAL—BENCH ASSEMBLY
Part or Oper'n No.	Manual Handling Code	Manual Insertion Code	Operation Time RP•(TH+TI)	Figures for Min. Parts					Name of Assembly:
	No. of Repeats	Handling Time per Part (s)	Insertion Time per Part (s)	Operation Cost TA•OP					Piston 1b
									Name of Part or Operation
2	1	10	1.5	02	2.5	4.00	1 60	1	Piston
3	1	10	1.5	00	1.5	3.00	1 20	1	Piston Stop
4	1	05	1.84	00	1.5	3.34	1.34	• 1	Spring
5	1	23	2.38	08	6.5	8.86	3.54	0	Cover
6	2	11	1.8	39	8.0	19.60	7.84	0	Screw

Lab. rate 36 $/hour   Totals— 42.25 16.9   4   Efficiency = 29 %

Strike <I>nsert, <D>elete, <C>hange, <L>abor, <O>k

Shell M5.00.01   CURRENT CONTE T: USER PAULB DEMO
DESIGN FOR MANUAL ASSEMBLY

ID	RP	HC	TH	IC	TI	TA	CA	NM	MANUAL—BENCH ASSEMBLY
Part or Oper'n No.	Manual Handling Code	Manual Insertion Code	Operation Time RP•(TH+TI)	Figures for Min. Parts					Name of Assembly:
	No. of Repeats	Handling Time per Part (s)	Insertion Time per Part (s)	Operation Cost TA•OP					Piston 2b
									Name of Part or Operation
1	1	30	1.95	00	1.5	3.45	1.38	1	Main block
2	1	10	1.5	00	1.5	3.00	1 20	1	Piston
3	1	05	1.84	00	1.5	3.34	1.34	1	Spring
4	1	10	1.50	30	2.0	3.50	1.40	1	Cover and stop

Lab. rate 36 $/hour   Totals— 68.29 5.32   4   Efficiency = 90 %

Strike <I>nsert, <D>elete, <C>hange, <L>abor, <O>k

Figure 2.

Today, an increasing number of drawings are being produced with computer-aided design (CAD) systems (figure 3). A useful tool would give simultaneous access to both CAD-generated drawings and the data entry and calculation routines previously described (figure 4). This integrated computer support overcomes the inconvenience of separate software, on separate systems, perhaps in separate locations—all of which inhibit casual use of the techniques by conceptual designers.

The drawings in figures 3 and 4 were produced on a two-dimensional (2-D) CAD system. A three-dimensional (3-D) system is better if the "back side" of the assembly or its piece parts contained relevant information. In a 3-D system, one can change viewpoints to see any side of an object. In a 2-D system, appropriate orthographic views must be assembled and a new isometric must be projected.

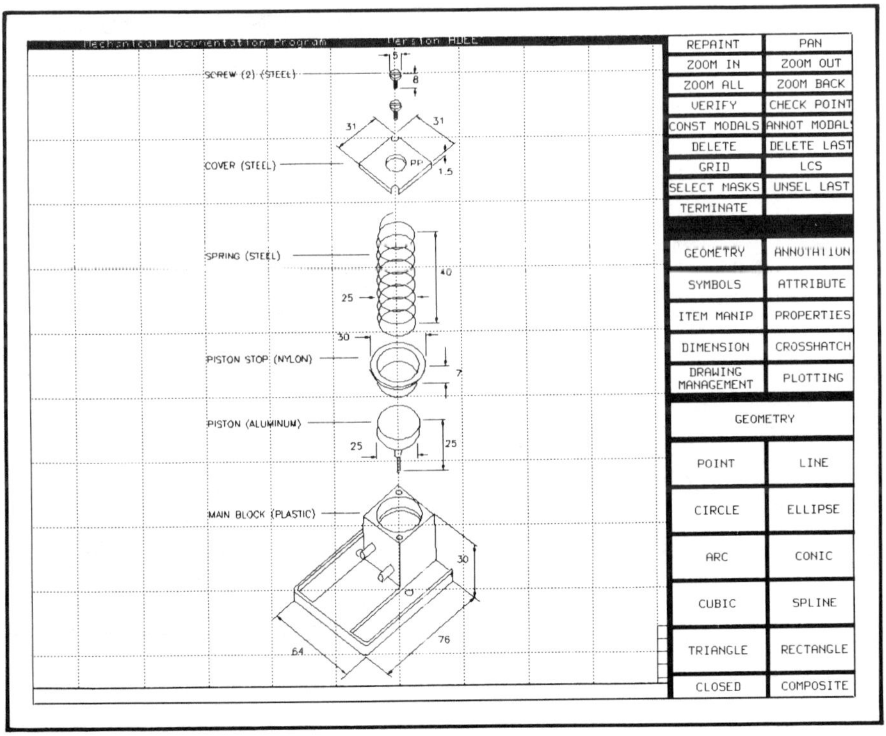

Figure 3.

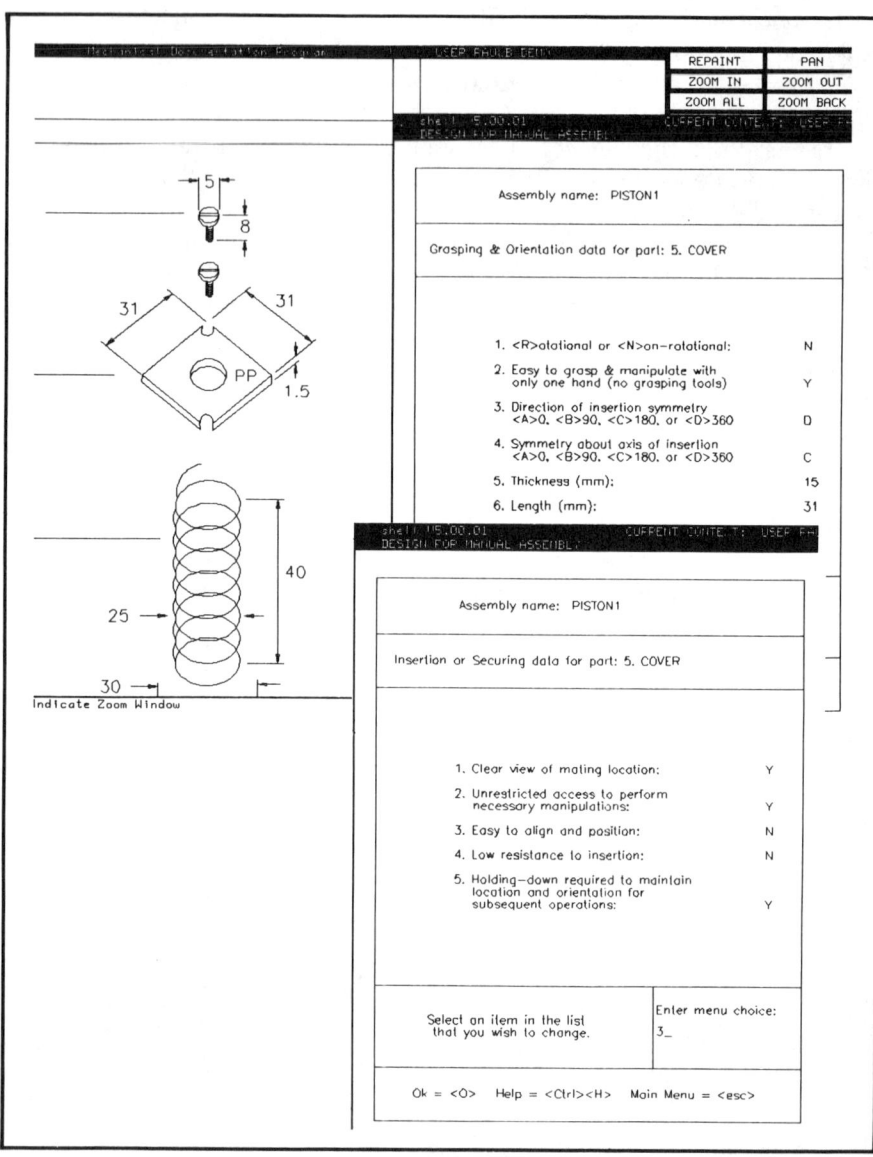

Figure 4.

Solids modeling

Solids modeling is the emerging technology of choice for 3-D computer-based design. A model can be seen from any side, and because it is fully defined mathematically, it can also be sectioned to provide complete cut-away views. Hidden-line removal is often more effective than that provided on conventional 3-D CAD (wireframe) systems. Solids modeling is of special value in assembly analysis because it can graphically display interferences between objects in a scene. This is typically not possible with conventional 3-D systems.

Modeling also provides automatic calculation of mass properties, such as weight and center of gravity. Where design changes for assembly purposes may affect target weights and centers of gravity, the use of a modeler will expedite the necessary calculations for any proposed change.

Color graphic results

The results of a Boothroyd and Dewhurst analysis are currently displayed in tabular form. Although this is an essential form of output, it does not display results on the assembly drawing or solid model. And, in large assemblies, tabular results can be tedious to interpret. One useful measure is to use simple bar charts and color codes to highlight analytical results. The most obvious application is to make a bar charge of operations cost or time, by piece part, to reveal those parts that contribute most to total assembly cost (figure 5). If a color code is added to this chart—red for high (relative) cost; yellow for medium; green for low—the color can be used as an instruction, coloring or shading each piece part in our CAD assembly drawing or solid model. Coloring each part according to its relative assembly dramatically presents results to all members of the design team.

Such coloring can only be achieved if the CAD drawing and the solid model have been defined at the piece part level and match exactly the parts list used in design-for-assembly analysis. Most CAD drawings are simply artwork—collections of lines, arcs, and text. These lines and arcs do not "know" that they represent part no. 1243 and that they should be colored red, for example, to indicate high cost. For this to occur, the CAD user must first perform some high-level operations that join a collection of lines and arcs into a single entity, with the name of "part no. 1234." Once this has been done, a simple procedural interface can be written to scan the design-for-assembly cost table, extract the proper color code, and redraw the part in its appropriate color. This process is easy to establish in some CAD systems, but may be impossible in others, especially if automatic coloring is the goal. Solids modelers can make the process easier, because parts are completely defined, mathematical entities—not artwork collections of lines and arcs. This is yet another, although minor, reason to favor modeling in design-for-assembly analysis.

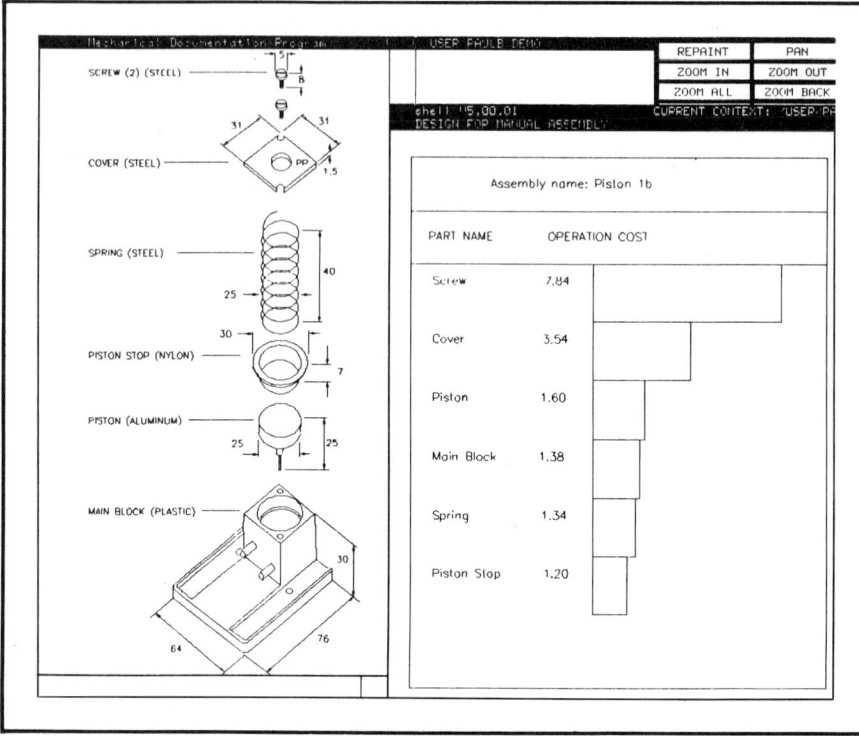

Figure 5.

Concurrent processes

Using CAD or solids modeling assumes that our computer can support at least two concurrent processes—the CAD or modeling function and the design-for-assembly analysis. These processes need to be viewed and used interchangeably through windows that can be opened and closed, made larger or smaller, or stacked and shuffled, such as that done with a sheaf of drawings and a stack of forms. In practice, a 19-inch, high resolution graphics display is required. Even then, "real estate" problems occur. The windowing software must be very fast and flexible or response time becomes tedious and inhibits the analytical process.

Networking

So far, I have descibed the computer support needed at an individual user's workstation. Networking builds on these capabilities and makes it possible to obtain input and disseminate results among a team of engineers—provided, of course, that each is attached to the net. In a typical project environment, one network node contains the project files and is tightly administered to control updates and changes. A variety of users—design engineers, manufacturing engineers, procurement specialists—can then have access to the design-for-assembly analysis, moving it or copying it to their local station for inputs, reviews, or changes.

Terminal emulation, in conjunction with networking and windowing, makes it possible to reach out, electronically, beyond the immediate realm of drawings, models, and design-for-assembly routines. As shown in figure 6, a window could be opened and a piece part material cost could be obtained during the design-for-assembly analysis. The material cost might reside in a purchasing system at another corporate location.

Archiving

Just as drawings that describe a product design are archived, a structured way of archiving the assembly studies that go with them, at least during the development and pilot production period, is needed.

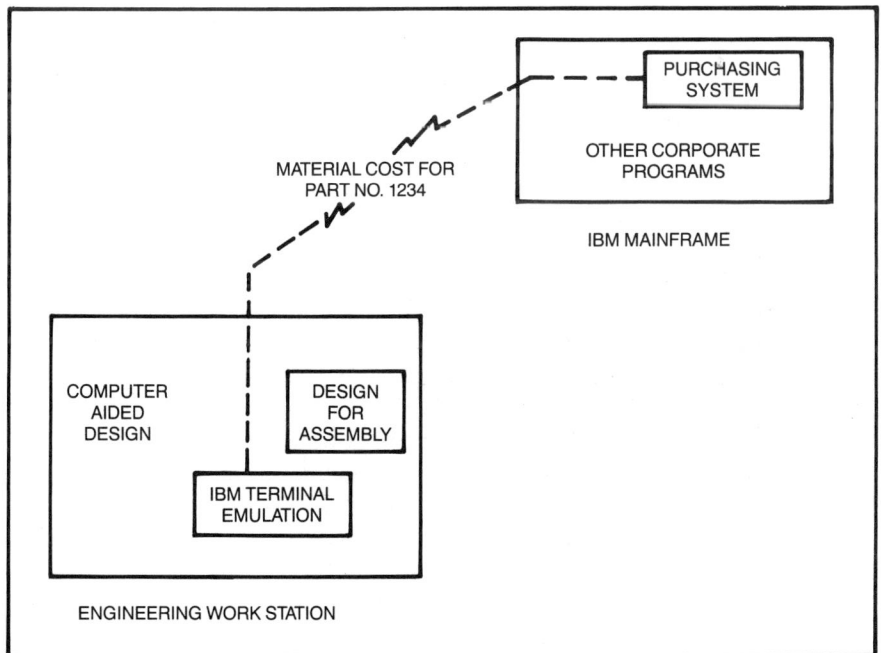

Figure 6.

Archiving and readily retrieving assembly studies is particularly important when the assemblies and piece parts being analyzed belong to a product family. In this situation, a design change adopted in one product or option may have an impact on others. A good archiving system would flag this condition. The results of one assembly analysis could be used elsewhere in the product family.

Two or three competing designs can be developed simultaneously, before one is chosen for production. Typically, these designs begin with a number of common assemblies for which analyses can be shared. As each design evolves, variations emerge, and the designs migrate toward less commonality. During this process, a good archiving scheme can keep track of revisions and changes to designs and their corresponding assembly analyses.

At production times, final designs are often configured from selected versions of earlier subassemblies and parts. In this context, the final results of a design-for-assembly analysis might also be "configured" from the appropriate analyses of each subassembly.

Data management

The need for data management goes hand-in-hand with a good archiving scheme. Design-for-assembly analyses are subject to many changing assumptions and parameters. Moreover, many assemblies contain parts that are shared with other assemblies and products. Design changes cannot be made in isolation without considering impact on other parts and assemblies.

Data management addresses these issues by keeping track of decisions and analyses, their assumptions, parameters, and implications. The central elements of a data management scheme are as follows:

1. *Labor costs.* Design decisions are sensitive to labor-cost assumptions. These can vary among potential assembly locations. In a large manufacturing organization the number of locational alternatives for fabricating and assembling can rapidly multiply the number of analyses that are required for a definitive design decision. Some way to manage the resulting data and show the sensitivity of design decisions to the labor cost issue would be useful.
2. *Equipment costs.* These costs vary in much the same way as labor and compound the analytical problem previously described.
3. *Time standards.* The current Boothroyd Dewhurst software provides a set of time standards for parts handling and insertion. In practice, some industrial engineers choose to substitute their own time standards. At the very least they may want to test results for sensitivity to some or all of the time standards used. Time standards, in practice, may also be a function of assembly location. Some way to manage the substitution and variation of time standards would also be useful.

4. *Ripple effects.* Perhaps the most important element of a data management scheme is including bill-of-material or where-used information. This is important when changes are proposed to parts that have already been designed or are in production. We need to know about all mating, touching, and surrounding parts that might be affected by a design-for-assembly decision. Where product families or modular products are involved, the ripple effects of a change may span several products and assemblies beyond the scope of the initial design analysis.

Commercially available data base management tools are suited to these data management tasks. Ideally, such tools would be available as part of a networked, integrated environment—one that includes the CAD, solids modeling, color graphics, concurrent processing, and archiving already described.

Reporting

Effective reports require more than simple cost tables or even color graphic charts. To "sell" a design change, a formal, written presentation, buttressed by the appropriate drawings, charts, and tables, is needed. The ideal tool for this purpose is a word-processor that can "scissor" or clip graphics from CAD and tables from assembly analysis and insert them as needed into a larger document. Output can then be obtained from a letter-quality printer. This class of computer system goes by several names: technical publications, technical documentation, desk top publishing, and text-graphics merge, to name just a few. The underlying characteristic of such systems is affectionately known as WYSIWYG, which stands for "What You See Is What You Get." The term is apt, for unlike a typical word processor or text editor, a good engineering documentation tool shows the user, page-by-page, exactly what his output will look like before he prints it. All control characters and instructions are kept "off-page." For our purposes, the WYSIWYG capability is best implemented as a concurrent process, coresident on the engineer's work station, along with his design-for-assembly calculations, CAD, and/or solids modeling.

Automatic initialization

Given a list of parts and a set of solid models that represent those parts, much of the information needed by the Boothroyd and Dewhurst methods could be automatically initiated. Part names and dimensions—even symmetry information—are contained in the modeling files. Unfortunately, there is a major impediment to the automatic capture and use of this data. To obtain useful results, the analyst must often ignore certain features of the piece parts under study, specifically those features that have no bearing on handling and insertion during the assembly process. Writing software that will correctly ignore features in a solid model is very difficult. Even determining the thickness or length of an irregular part can be a problem. The data extraction routine must determine if a maximum dimension, an average dimension, the most-prevalent dimension along a given cross-section, etc., are to be obtained.

In some instances, using unwarranted detail is harmless and causes no difference in analytical results. In others, however, it corrupts the assessment.

In practice, the keyboard entry of part names, dimensions, and symmetry turns out to be a rather small part of the total analytical effort. Given the technical difficulty of automation, it seems better left to manual intervention at least for now. The payoff seems far greater from implementing the computer supports already discussed. With continued study, automatic initialization may become realistic at some future date.

CONCLUSION

The direction in engineering today is toward doing more and more early design tasks on networks of work station computers. The ultimate goal of this computing effort is to eliminate or reduce the need for early prototypes and, in general, to speed up the design cycle. Issues such as manufacturability and ease of assembly must be raised early in the design process. For this reason, they must be addressed in computer-based form and integrated into the engineer's normal computing environment. The ideas summarized here can help to serve as a useful guide for those working to achieve this integration.

Suggested Readings

Boothroyd, G. and P. Dewhurst. 1983. *Design for Assembly: A Designer's Handbook.* G. Boothroyd and P. Dewhurst, Inc. Wakefield, RI.
Boothroyd G. and P. Dewhurst. 1984. *Design for Assembly.* (Reprint Series) Penton/IPC Education Division. Cleveland, OH.

Reprinted from *1986 International Electronics Assembly Conference Proceedings.*

H. Lee Hales is National Director of Computer-integrated Manufacturing at Coopers and Lybrand, Duluth, Georgia.

56

Tapping CAD Data
For PC Board Panelization

Lisa D. White

Within the world of electronics system design, computer-integrated manufacturing (CIM) ties together the three distinct areas of the product development cycle: design and analysis (CAE), PCB layout (CAD), and manufacturing and test (CAM). The automation of each area evolved separately and, consequently, each uses hardware and software specific to its needs. Hence, CAE, CAD and CAM are all "islands of automation," consisting of and separated by dissimilar computer-aided tools.

A few CAE and CAD workstations have been developed to bridge that information gap through the implementation of an integrated database, though from the perspective of CAM, the provinces of CAE and CAD have continued to look like unapproachable islands. A workstation "panel editor," however, can bridge the CAE/CAD database to CAM and provide the ability to create manufacturable printed circuit panel artwork from one or more single copies of a PCB.

The data available from the panel editor that are crucial to the various islands of automation within CAM are accessed through a data query language facility and through a host of direct machine interfaces. The panel editor places the solutions to manufacturing problems in the manufacturing environment. Manufacturers can create and modify their own panel artwork without having to rely on the CAE/CAD groups. Similarly, the database and tools for configuring and programming fabrication, assembly and test equipment are available in the workstation.

By strengthening communication between CAE/CAD and CAM and by facilitating fabrication, assembly and test tool preparation, the overall design cycle is shortened. When data are accessed and transferred directly without the use of complicated post-processors or batch programs, data integrity is maintained and manufacturability is ensured.

PANEL EDITOR

The two main components of a particular manufacturing workstation built by Cadnetix are the panel editor and a post-processing capability that is closely linked with the database. Together, they address the areas of PCB fabrication, assembly and test. The panel editor allows the creation of manufacturable panel artwork from single images of PCB artwork, as delivered from a CAD workstation.

To further streamline the PCB fabrication process, data for artwork, solder stencils, and solder masks can be generated at the workstation for realization by a photoplotter. Direct interfaces from the panel editor to popular drill and profiling machines are also available.

After PCB fabrication data are generated, assembly data for single PCBs or full panels can be extracted through the manufacturing workstation. Similarly, physical and logical data required by ATE for assembly of test fixtures and programming of test equipment are also available.

The panel editor allows the user to construct a multiple-image panel for the fabrication and assembly of one or more PCBs. Multilayer PCB images are analyzed and modified through the panel editor's 24 trace layers and 24 drafting layers. Another 12 layers are dedicated to silk screen, assembly, title (logos, etc.), board and package outline pad and via manipulation.

With the panel editor providing the ability to modify the artwork, the PCB manufacturer has the freedom to accept PCB artwork that has been designed anywhere—in or out of house. The manufacturer can implement PCB layouts that were designed without their unique manufacturing process anomalies and tooling incompatibilities in mind, or those that require minor design-rule changes. The implementation occurs without requiring the intervention of the PCB layout designer. All the necessary changes can be made within the realm of manufacturing.

After accepting data from a PCB layout either in the system's CAD workstation format or in Gerber photoplotter format, the panel editor generates its own drawing of a PCB. The drawing can be automatically scaled upon input to the panel editor—important to documentation generation where detailing is required.

The user creates a multiple image panel using step-and-repeat, rotate, and mirror functions, and may elect to construct the images with more than one board type to facilitate assembly of related boards, as in mother/daughter assemblies.

Alternatively, the manufacturer may choose to begin the panel creation process even before the board layout is completed. PCB outlines are among the first parameters to be defined in the design cycle and a PCB outline is sufficient to plan the use of panel real estate. Once the layout is finished, the outlines can be automatically updated with the additional data.

The panel editor also provides the manufacturer with the capability of analyzing the board placements with respect to copper distribution on the panel. Copper area calculation is used to give an approximate indication of both the amount of copper etching represented by the displayed graphics and the overall ratio of copper area to total block image area. With such a capability, manufacturers can ensure the most even distribution of copper across the panel for ease of manufacturability.

In order to monitor manufacturing processes, the user can incorporate test coupons anywhere on the panel which can later be dissected for the analysis of fabrication parameters such as through-hole plating and electrical isolation. Also, for known process extremes or deficiencies, thieving bars can be placed on the panel to ensure that the PCBs are fabricated properly.

In the particular workstation cited, board artwork can be modified to accommodate manufacturing rules or to compensate for process anomalies. For example, if a manufacturer notes from the PCB artwork that a few pads and traces would likely short together during wave soldering due to underetching, or if the traces and pads left unaltered would violate the manufacturer's minimum distance requirements, one option is to compensate by changing the artwork.

Changes are easily accomplished. When the user wishes to modify discrete images, a single copy of a board is selected from the panel and viewed using the hierarchical editing capability. The user changes the image as necessary—pad shaving, increasing or decreasing trace widths, and editing copper areas on the board, for example. When all modifications are entered, the PCB single copy is returned to the panel database and all like-copies are automatically updated with the same modifications.

The power in such ease-of-modification lies in the fact that the manufacturer no longer has to rely on layout designers (who may or may not reside in-house) to make such compensations. Thus, the manufacturing workstation not only makes the manufacturer more autonomous, it shortens the product development cycle.

PROFILING AND DATA QUERY LANGUAGE

Depanelization of boards can happen immediately after panel fabrication or later in the production process, after the panel has been loaded with components. The timing depends in large part on the degree of automation of the production line. Depanelization can be done manually, or it can be done automatically with routers, laser, waterjet and shearing tools for board removal. Each tool requires the board outline and cut-out definitions that are part of the panel editor database.

It is common to do depanelization by combining manual and automatic methods in a route-and-tab process. In this method a board outline is almost completely routed, save for a few retaining tabs. The automatic router requires the definition of the portion of the board outline that is to be routed and the portion left behind for the tabs. Later, final depanelization is accomplished by manually removing the retaining tabs.

The panel editor's database also contains physical data needed to program, configure, and prepare fabrication and assembly tools such as photoplotters, NC drillers, profile routers, auto-insertion tools, pick-and-place tools, punch and reinsert tools, board handlers and axial sequencers. The manufacturing environment also includes automatic test equipment (ATE) in the forms of bare-board, in-circuit and functional testers. Some of the same data necessary for manufacturing and assembly tools are required for building fixtures (test heads) for ATE.

Data query language allows the user to directly select and filter data to suit the various manufacturing and test tools. The resulting data are in an ADCII format which can be viewed and edited on-screen or sent to an output device. Such a language should use the same commands as IBM's Structured Query Language, now a de facto standard in relational database manipulation. English-like syntax makes data query language easy to learn and remember.

Reprinted from *Electronic Packaging and Production*, March, 1987. © 1987

Part 6

Integration of Product and Process Design in High Technology Equipment Production

57

Integration of Product and Process Design in High Technology Equipment Production

Arvind Ballakur
Michael K. Pratt

Considerable attention has been given to the significant impact of product design on manufacturability (DFM) and, more specifically, on ease of product assembly (DFA) (Day 1986; Daetz 1987; Whitney 1987; Boothroyd and Dewhurst 1982; Wallich 1987; Kosugi 1985). Most knowledge in this area is based on an empirical understanding. This impact is more managable by applying appropriate design guidelines and design rating schemas.

The term 'design for X' (DFX), where X is any process which product design can affect, suggests that similar relationships exist, many possibly outside of the scope of manufacturing (installation, repair, etc.) (Daetz 1987; Wallich 1987; Burt and Soukup 1985), that must be managed as well. However, managing these many relationships simultaneously by simply applying guidelines or rating schema that separately focus on specific topics is difficult.

This chapter presents a generic approach to model the coupling between product design and associated processes. This methodology is intended to enhance the understanding and analysis of the fundamental relationships between product design and associated processes, which may grant basic knowledge about these relationships. This methodology can both strengthen the application of current design guidelines and lead to identifying additional relationships beyond the scope of current DFX understanding. This chapter steps through a general description of the methodology and gives a simple example of its application to electronics products.

RELATED WORK

Integrating product design and manufacturing processes has been addressed extensively. We will review several examples found in the current literature.

Douglas Daetz of Hewlett-Packard, in a recent paper (1987), discusses the effect of product design on product quality and product cost. He stresses the critical need for focusing on quality and manufacturability during development. He recommends six product design measures for achieving these goals:

1. designing product and process concurrently;
2. measuring and striving for assembly simplicity;
3. minimizing the number of parts;
4. minimizing the number of part numbers;
5. maximizing the percentage of preferred parts; and
6. minimizing the number of vendors.

Daetz also quotes recent studies from General Electric (GE) which reveal that for typical products in which material costs are 65%–80% of the total product cost, the design directly impacts 75%–90% of the manufacturing costs (1987).

Styslinger and Melkanoff (1985) describe a case study at Hughes Aircraft where they applied the concept of group technology to circuit card assembly. This study is useful in the present context of product-process integration because of the depth in which they describe their understanding of the circuit card assembly process to implement a flexible coding system for more than 70,000 circuit card assemblies in 1,200 different configurations.

In an article entitled "Manufacturing and Design: A Symbiosis," Daniel Whitney of Charles Stark Draper Laboratories (1987) describes the need for design and manufacturing engineers to work together as a team to create the product and its assembly system more or less at the same time. The team should also study the interaction among the choice of components, the method of manufacturing, and the sequence of assembly. They must jointly decide on the path that yields the lowest cost, the easiest inventory management, the greatest manufacturing flexibility, and the widest options for redesign.

The aforementioned cases are examples of integrating product design with manufacturing processes. Several basic approaches to managing this integration have been recommended. These approaches include the following:

1. organizational changes to include manufacturing engineers as part of the design team (Whitney 1987; Wallich 1987);

2. providing manufacturing feedback (Daetz 1987; Styslinger and Melkanoff 1985; Whitney 1987; Wallich 1987);
3. early manufacturing involvement (Whitney 1987; Wallich 1987);
4. quantitative rating schemes (Day 1986; Daetz 1987; Whitney 1987; Boothroyd and Dewhurst 1982; Andreason et al. 1983; Schreiber 1985; Domas and Helander 1984; Wallich 1987);
5. creating design guidelines (Universal Instruments Corporation 1985); and
6. using statistical design techniques (Toguchi and WU 1980; Kacker 1985; Sullivan 1987).

These referenced works represent significant progress in understanding coupling between a product's design and many of the manufacturing processes that are impacted. Such an understanding leads to appropriate guidelines for designers. As mentioned, this chapter attempts to develop a generic "systems engineering" model that will facilitate a more specific analysis of the nature of product/process coupling. This model should also prove useful in examining processes beyond those in manufacturing, such as sales, distribution, repair, etc., as well.

METHODOLOGY

The methodology is comprised of several steps. Product and process parameters are characterized in the first two steps. This is followed by a definition step in which the product-process interaction space (herein referred to as the product-process solution space [PPSS] is established. Elements of the PPSS capture the relationship between the corresponding product and process parameters. The elements are rank-ordered based on their relative impact in terms of global costs, sensitivity to changes, etc. The high impact areas are studied in depth to understand their inherent relationships, which can lead to forming specific technology groups. These groups can then be exploited with traditional GT classification and coding methods (Swain and Hewitt 1985, Styslinger and Melkenoff 1985; Hyer and Wemmerlov 1984; Hyer and Wemmerlov 1985; Houtzeel 1982; Hyde 1981) to arrive at standardized solutions, faster process plans, higher design reuse, and shorter procurement intervals.

Step 1: Product Characterization

The first step of the methodology is to characterize the product in such a way that it can be decoupled into its fundamental elements. Since, in most cases, it is impossible to identify dimensions of characterization that are completely orthogonal, it is important to capture the relationships between product characteristics where cross-coupling is known.

For example, a conceptual framework composed of a product's architecture, physical hardware, and application as three fundamental dimensions may prove useful for characterizing many electronics products at the highest level. The product's architectural characteristics could be further decomposed into bus characteristics, instruction set, file structure, operating system, etc. The physical hardware characteristics could be further decomposed into cabinet/ enclosure, shelf, circuit packs/cards, and interconnection cabling. Adding further detail, a cabinet could, in turn, be decoupled into side plates, back plates, front door, and so on. A similar decomposition for application characteristics can be developed.

Obvious cross-coupling would occur, for example, between the parameters of a circuit pack/card (length and width dimensions) and the parameters of the cabinet (side plate dimensions, footprint) and must be identified where known.

Step 2: Process Characterization

A similar exercise to characterize the fundamental processes associated with a given product must also be performed. The traditional focus has been on those processes associated with the manufacture of a product for obvious reasons. However, if the process scope is expanded to include the many other processes that have a substantial impact on the market success and profitability of a product would be useful.

For example, the various processes involved in realizing electronics products can be represented as shown in Figure 1. To physically build the product, manufacturing operations are required. To know what to build and what components to order requires forecasting and customer order—entry processes. For delivering the product to the customer, extensive coordination and handling are required to ensure that all elements of the product arrive in time from different warehouses/stocking locations. The product must then be installed and maintained on the customer's premises. Finally, the payments from all customers must be collected. All of these processes can be affected in varying degrees by the characteristics of the product's design.

Step 3: Solution Space Definition

The first two steps characterized the product and the process. The next step is to understand and define the boundaries of PPSS. This can be conceptually modeled as a two-dimensional matrix of product and process characterisitcs (figure 2). Each element of the matrix represents the impact/interaction of the corresponding product and process characteristics. The PPSS is, therefore, defined as that space where the relationship between a product's characteristics (e.g., physical parameters) and a process' characteristics (e.g., nature of material handling required) are stated explicitly.

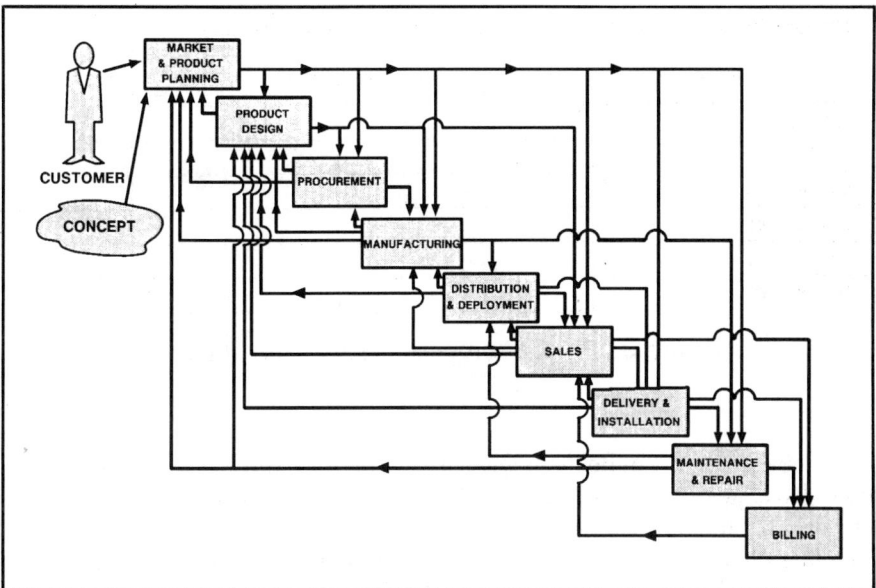

Figure 1. *Product realization process characterization.*

Product Characteristics	Processes					
	Market & Product Planning	Product & Process Introduction	Procurement	Volume Manfg.	Distrbn. & Deplymt.	Sales
ARCHITECTURAL						
APPLICATION						
PHYSICAL						

Figure 2. *Product-process solution space.*

Step 4: Space Partitioning

In most cases, the PPSS is a very large, sparse matrix. It may prove useful to divide the PPSS into mutually exclusive subspaces of principle interest for separate analysis (figure 3). Most of the DFA and DFM efforts can be viewed as subspaces of the total PPSS, for example.

Product Characteristics	Volume Manufacturing				
(Physical)	Receiving	Assembly	Test	Material Handling	Packaging
Cabinet					
Shelf					
Circuit Pack		X			
Component					
Cabling					
Backplane					

Figure 3. *Product-process solution sub-space.*

A global analysis can be done when all the product-process subspace interactions have been analyzed. As previously mentioned, because a completely orthogonal, totally decoupled representation is usually not possible, cross-coupling effects must be considered. For example, a circuit pack component insertion machine must be adjusted for the printed wiring board thickness and component height. The nature of coupling of the corresponding product-process relationship can be expressed as height (setting) of insertion head equal to the sum of the printed wiring board thickness and the maximum component height.

Step 5: Impact Analysis

Product-process relationships can be analyzed at a macrolevel using the PPSS matrix, or subspace matrices, described earlier. In this step, the various interactions are classified. The purpose of this step is to give priority to and narrow the scope of analysis for further study. A number of different criteria can be used to classify the product-process interactions.

Degree of Sensitivity

In certain instances, a process parameter (e.g., machine fixture setting) may be highly sensitive to a small change in a specific product paramerter (e.g., part thickness). Similarly a product parameter (e.g., electrical connectivity) may be highly sensitive to a change in process parameters (e.g., soldering temperature). At the other extreme, certain product and process parameters may not have any interactions at all. For example, cabinet height does not have any interaction with circuit pack component insertion processes.

Therefore, degree of sensitivity may be used as a criterion for classifying the nature of a product-process interaction, and the procedure for this classification is as follows:

1. Determine the sensitivity of each product (or process) to a small change in the given process (or product) parameter.
2. Classify the nature of the impact into different categories: high, medium, low, none. This categorizes the interactions based on the degree of sensitivity of the corresponding product and process parameters.

Nature of Sensitivity

For product simplification and standardization, the nature of the product-process sensitivity must be determined. That is, it is not only important to understand the qualitative degree of sensitivity, but also to determine whether the process can accommodate a range of design parameter values, or if a discrete value must be chosen. This is an assessment of process accommodation with minimal impact of cost, time, or quality, not of tolerance to process variation once the desired parameters have been selected (Taguchi and Wu 1980). For instance, current process capabilities may be capable of accommodating a finite range of variation in a specific product parameter, such as material handling process capabilities and circuit pack dimensions, or product ordering process capabilities and number of product options available.

Cost Impact

Cost is an important criterion for classifying product-process interactions. Analyzing cost of an interaction involves understanding the direct and indirect effects of the interaction and then determining the magnitude of each effect on total costs. For example, if an electronics product incorporates a printed backplane as opposed to a wire-wrapped backplane, then the economic impact of the two alternative designs on the manufacturing process can be quantified in terms of captial, assembly, test, and material costs.

Step 6: Relationship Characterization

Determining the relative impact of the product-process interactions, as described in the previous step, helps in giving the interactions priority and characterizing only those relationships that have high impact. This step involves conducting a detailed study of the significant product-process interactions and determining the best choice of specific parameters (range or discrete) which satisfy design needs and process capabilities. For example, a given component insertion machine may be capable of accommodating circuit pack dimensions ranging from 7 to 12 inches in either length or width with minimal impact, but can only insert components with 0.5-inch centers.

Step 7: Technology Group Formation

The product-process relationships can now be used to support the concept of "technology groups." One way of describing a technology group is a set of products, or product elements, having a common set of design parameters such that they require similar processing steps and process parameter settings. For example, knowing that the range of travel of an automatic insertion machine is related to the length and width of a circuit pack, a technology group can be defined which identifies specific circuit pack dimensions that match the available insertion process capabilities. Specifically, circuit packs that are 8 by 12 inches in dimension might comprise a technology group for typical minicomputer products, and the corresponding process parameters can, therefore, be determined and exploited with minimal effort in subsequent designs.

The determination of specific technology groups is shaped by applying the methodology through the "a priori" relationships derived in *Step 6*. This differs from traditional approaches in applying group technology, where either features of existing products are analyzed to form technology groups or the product classification procedures are *ad hoc* (Hyer and Wemmerlov 1985; Houtzeel 1982; Hyde 1981). Technology groups thus established can lead directly to design standardization and structured process planning.

APPLICATION TO CIRCUIT PACK DESIGN/ASSEMBLY

The methodology described previously can be applied to integrate products and processes from a very broad, high level analysis to a detailed microanalysis. In this section, an application of the methodology to integrated circuit pack design with circuit pack assembly is presented. The example described in this section is only to illustrate how the methodology can be applied and should not be treated as complete in description.

Step 1: Product Characterization

As previously discussed, many typical electronics products such as a minicomputer, for example, exploit a card on board (COB) scheme. In this scheme, various functions are realized on individual circuit packs and integrated into a system by interconnection via a backplane. A circuit pack contains electronics components mounted on it, and one of its edges has an interconnection device. The shelf, which holds the board, has nesting slots for the circuit packs. They guide the circuit packs such that the interconnection device on the pack plugs onto the backplane.

For this example, a partial decomposition of the physical characteristics of this typical minicomputer could be an enclosure/cabinet, shelf, circuit pack, component, cabling and backplane.

Step 2: Process Characterization

Processes ranging from sales and marketing to installation and repair are typically employed in the minicomputer market. If, for example, only those processes in manufacturing are considered, they might be partially characterized as receiving, material handling, assembly, test, repair, and packaging.

Step 3: Solution Space Definition

The PPSS for this partial example would be defined as the matrix of product and process characteristics previously described (figure 3). This matrix is the conceptual model that will be used to identify areas of significant coupling between product design and associated processes in subsequent steps.

Step 4: Space Partitioning

As this example partially characterizes only physical product characteristics and manufacturing processes, it is already a subset of the total PPSS. For illustration, the matrix element of circuit pack product characteristics and manufacturing assembly processes, as marked in Figure 3 with an "X," will be further decomposed as a subspace of significant interest in this example. Potential cross-coupling should be identified when subspaces are selected for closer investigation. For example, component height would be coupled with circuit pack thickness in determining the automatic insertion head height setting.

Step 5: Impact Analysis

In this step, the relationships between specific product and process parameters are investigated. Figure 4 illustrates a few parameters of interest for this example. In this example, an "X" is used to denote elements of the subspace where coupling is likely, and further investigation is warranted. For those elements so designated, it might be determined that they all comprise coupling with high impact, as denoted by the "H" in parenthesis. Further, it is assumed that the process accommodation can be identified, as denoted by the "R" and "D," also in parenthesis.

		Impact on Processes			
No.	Product Parameter	Auto-Insertion Fixture Setting	Insertion Head Travel Adjustment	Solder Temperature	Solder Height	Insertion Head Center Space
1	Circuit pack width	X(H/R) *				
2	Circuit pack length	X(H/R) *				
3	PWB thickness		X(H/R)	X(H/R)**	X(H/R)	
4	PWB base material			X(H/R)**		
5	Component center space					X(H/D)

H: High Impact, R: Range, D: Discrete, *, ** : Elements with Cross-coupling

Figure 4. *Impact analysis—example.*

Also identified in Figure 4 are additional instances of cross-coupling, as denoted by asterisks.

Step 6: Relationship Characterization

Once the product and process parameters which are highly coupled is known, and the nature of that coupling is captured, then specific values (range or discrete) which are best suited to the design requirements and process capabilities can be determined. For this example, those values shown in Figure 5 might illustrate the range of circuit pack widths and lengths best matched with the auto insertion fixture capabilities. Choices of circuit pack dimensions within this range will have a manageable impact on the assembly process.

Auto-Insertion Fixture	CP Width		CP Length	
	Minimum	Maximum	Minimum	Maximum
Standard - A	3.000"	4.000"	6.000"	7.500"
Optional - B	4.500"	5.600"	8.000"	9.000"
Optional - C	6.000"	6.500"	9.500"	10.000"

Figure 5. *Relationship characterization—example.*

Step 7: Technology Group Formation

Now that the relationship between the characteristics of a product's design and the associated processes has been captured, specific product and process parameters can be chosen to comprise selected technology groups. Figure 6, for example, illustrates specific choices that could be made. These values of circuit pack dimensions fall within the ranges previously determined to capture the best match between design needs and process capabilities. These specific values may have been chosen because of purchasing economies or previous use in existing designs, for example. Because the process parameters required to accommodate these particular product parameters are already known, they can be easily captured with typical GT coding schemes and exploited in future designs.

Technology Group	CP Width	CP Length	Auto-Insertion Fixture
1	3.513″ 4.000″ 4.000″	7.000″ 7.000″ 7.500″	Standard Fixture A
2	5.000″ 5.513″	8.000″ 9.000″	Optional Fixture B
3	6.000″ 6.000″	9.500″ 9.800″	Optional Fixture C

Figure 6. Technology group formation—example.

SUMMARY

This chapter presented a methodology that enables the user to develop a generic model of the relationship between a product's design and associated processes. This model can aid in determining critical aspects of product-process coupling for a particular product. Identifying specific parameters best suited to meet design requirements and process capabilities can then be pursued. Although the generic model is applicable to any product, a partial application of the methodology to a typical electronics product is developed.

Suggested Readings

Andreasen, M. M., S. Kahler, and T. Lund. 1983. *Design for Assembly*. IFS Publications Ltd., and Springer-Verlag.

Ballakur, A., and M. K. Pratt. 1987. Integration of product and process design in high technology equipment production. *Third IEEE/CHMT International Electronic Manufacturing Technology Symposium Proceeding*. Los Angeles, CA. October. pp. 220.

Boothroyd, G., and P. Dewhurst. 1982. *Design for Assembly Handbook*. University of Massachusetts. Department of Mechanical Engineering. Amherst, MA.

Burt, D. N., and W. R. Soukup. 1985. Purchasing's role in new product development. *Harvard Business Review*.

Daetz, D. 1987. The effect of product design on product quality and product cost. *Quality Progress*.

Day, T. J. 1986. Producibility-design considerations for printed circuit board assemblies. *Proceedings of 1986 International Electronics Assembly Conference*. September. Institute of Industrial Engineers. Norcross, Georgia.

Domas, K. and M. Helander. 1984. Manual versus robotic assembly: some implications of product design. *Proceedings of the Human Factors Society, 28th Annual Meeting*.

Houtzeel, A. 1982. *Integrating CAD/CAM Through Group Technology.* Organization for Industrial Research. Waltham, MA.

Hyer, N. L., and U. Wemmerlov. 1984. Group technology and productivity. *Harvard Business Review*.

Hyer, N. L., and U. Wemmerlov. 1985. Group technology oriented coding systems: structures, applications and implementation. *Production and Inventory Management*.

Hyde, W. F. 1981. *Improving Productivity by Classification, Coding, and Data Base Standardization*. Marcel Dekker, Inc. New York.

Kackar, R. 1985. Off-Line quality control, parameter design, and the Taguchi Method. *Journal of Quality Technology*.

Kosugi, Y. 1985. Product redesign approach to realize assembly by robot. *Proceedings of 1985 International Electronics Assembly Conference*. Institute of Industrial Engineers. Norcross, Georgia.

Schreiber, R. R. 1985. Design for assembly. *Robotics Today*. 7(3).

Styslinger, T. P., and M. A. Melkanoff. 1985. Group technology for electronics assembly. *SME FMS for Electronics Assembly*. February. Society of Manufacturing Engineers. Dearborn, Michigan.

Swain, R. W., and W. C. Hewitt. 1985. Group technology for electronic assembly. *Proceedings of 1985 International Electronics Assembly Conference*. Institute of Industrial Engineers. Norcross, Georgia.

Sullivan, L. P. 1987. The power of Taguchi methods. *Quality Progress.*
Taguchi, G., and Y. Wu. 1980. *Introduction to Off-Line Quality Control.* Central Japan Quality Control Association. Nagoya, Japan.
Wallich, P. 1987. How and when to make tradeoffs. *IEEE Spectrum.*
Whitney, D. E. 1987. Manufacturing and Design: A Symbiosis. *IEEE Spectrum.*
Universal Instruments Corporation 1985. *Design Guidelines for Leaded Component Insertion.* Universal Instruments Corporation. Binghampton, NY.

This chapter is an enhanced version of the paper by Arvind Ballakur (Associate Editor) and Michael K. Pratt (AT&T Bell Laboratories) which appeared in the *1987 IEEE International Electronic Manufacturing Technology Symposium Proceedings.*

Michael K. Pratt is a Supervisor of Product Realization Operations Engineering Group at AT&T Bell Laboratories, Holmdel, NJ.

Part 7

Group Technology Classification and Coding for Electronic Components: Development and Applications

58

Group Technology Classification and Coding for Electronic Components: Development and Applications

Han P. Bao

INTRODUCTION

Group Technology and its Partnership With Computer-Aided Design/ Computer-Aided Manufacturing (CAD/CAM)

Group technology (GT) takes advantage of what could be called "the doctrine of sameness." As defined by Ham (1975), GT is the "realization that many problems are similar, and that, by grouping similar problems, a single solution can be found to a set of problems, thus saving time and effort." Therefore, the similarity of parts makes most design and manufacturing functions much easier.

The principle of GT leads to developing component classification and coding (C&C), which could provide the common denominator for most CAD/CAM applications through all sectors of manufacturing, including electronics manufacturing. The dependency of CAD/CAM applications on a C&C system is illustrated in Figure 1.

Recognizing that applying GT through classification and coding is not limited to the physical is important. Abstractions, such as information, can also be characterized through GT (Shaffer 1981). As a result, GT concepts can be useful in many industrial functions other than manufacturing, such as design and management. Some examples of GT applications are design data retrieval,

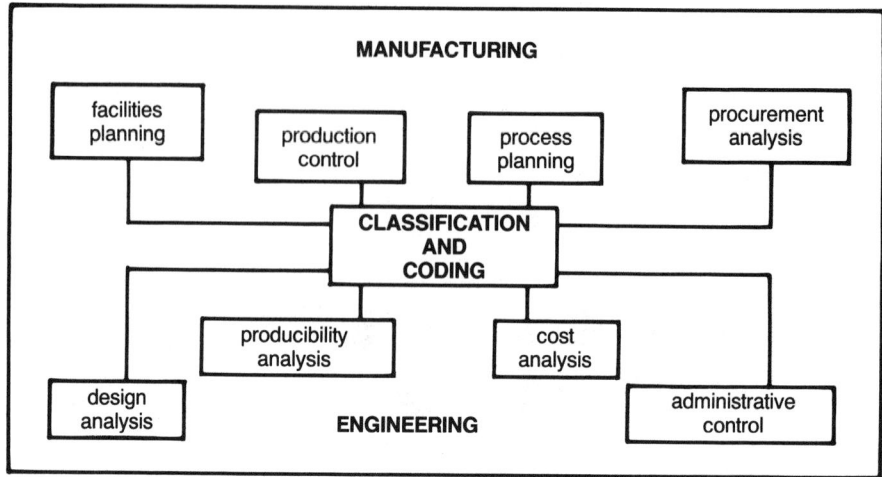

Figure 1. *Classification and coding facilitates CAD/CAM activities (Barnes 1976).*

component standardization, reliability of estimates, effective machine operations, productivity, accurate costing, customer service, and reduction in planning effort, paper work, set-up time, down time, work-in-process, finished part stock, and overall costs.

Group Technology in Electronics Manufacturing

The electronics industry can be classified into several segments according to product lines. Although there is a wide range of electronics products—from basic components, such as resistors and capacitors, to complicated telecommunication equipment, and from consumer products to military ones—one easy way to categorize the industry is to separate basic electronics components manufacturing from end-product electronics equipment. Electronics equipment may be further divided into two main streams—electro-mechanical units and electrical units. Regardless of the type of units involved, printed curcuit board (PCB) assemblies probably represent the most significant activity in the industry and are the subject of discussion here. Therefore, the scope of work discussed in this chapter is limited to the production of PCB assembly.

Applying an appropriate electronics component classification and coding system can provide real solutions to the following problems:

1. *Design function.* The design of a PCB must meet many functional requirements. In the early development stage, one of the designer's first tasks is to

select the components to make up the bill of material (BOM). The components must fit the physical and circuit needs. Other factors, such as operating environment and relevant standards, may also need consideration. To select the right components to be used in the circuit, the designer may have to search through suppliers' catalogs, match these components to their existing part numbers, or assign new part numbers to new items. This is time-consuming and may even result in duplicated part numbers for components that have been used in the past. Apart from this problem, the designer may not be able to specify the best components because of the tendency to use only relatively few familiar catalogs.

With an appropriate C&C system, the designer will be able to retrieve design information from the data base quickly and easily. Furthermore, the system can increase his productivity, reduce his frustration, lower material costs, and reduce unnecessary inventory through part standardization. Another problem in the design area is the difficulty for a designer to get involved with the assembly process. Often manufacturing plants are geographically separated from their design groups. Processing difficulties due to designs and specifications often arise because of the lack of communication. Also, the designer may have difficulty in knowing what effects specific design decisions will have on manufacturing costs because of the lack of specific knowledge or feedback from manufacturing. The manufacturability of a PCB assembly can be determined with the help from a C&C system. Through coding of the manufacturing-related attributes of the components and other production information, an indicator of relative manufacturing cost, called manufacturability rating, can be obtained before the PCB assembly is completely designed. This rating can serve as a design tool to contain production cost.

2. *Manufacturing function.* Process planning is typically a major manufacturing task. Because there are many ways to make a PCB assembly, different process planners have different opinions about what makes the best plan. Unfortunately, many plans can be suboptimal, hence uneconomical. Computer-aided process planning (CAPP) can help the planners achieve good process plans consistently. Apart from process rationalization and standardization, it can increase productivity of process planners, reduce turnaround time, and be incorporated with other application programs, such as cost estimating, work standardization, and others, to automate many of the time-consuming manufacturing support functions. To develop CAPP and some other application programs as mentioned previously, individual families have to be identified. Again, an appropriate C&C system is the best vehicle.

Overhead costs in the industry are also a problem. Much of these costs are caused by in-process inventories. C&C also can help cut overhead costs significantly. With C&C, operation cells may be formed for easy handling of product varieties. Jobs may be group-scheduled and fixtures and tooling may be designed for part families. All these activities can simplify production control and scheduling while cutting overhead costs associated with work-in-process (WIP) inventories, material handling, set-ups, and tooling.

Automation is becoming an important practice in electronics manufacturing. The design concepts of a computer-controlled cart appear to be the best approach to realize an automated factory for electronics assembly (Peeble 1983). This system uses the carts to move magazines of PCBs throughout the factory. A C&C can help to develop an automated electronics manufacturing factory. Tracking PCBs throughout the factory can be achieved by reading a label containing GT-based codes. Thus, not only are the PCBs tracked, but also process plans can be generated automatically.

3. *Materials management function.* This function covers many activities and has quite an extensive data requirement. Some examples of activities are demand forecast, receiving and ordering, gross and net material requirement, inventory planning and control, materials purchasing, and suppliers sources. The electronics industry has a unique problem because it involves a lot more varieties of basic materials than most industries and most of the materials are supplied from outside. The problem is further complicated by the large number and specialization of suppliers. A C&C system can promote component standardization by eliminating unnecessary part numbers in part families and can also help identify the primary suppliers. If primary suppliers are not available, the system can suggest alternative sources. Other benefits of C&C are the easy updating of information in the data base, the ability to determine more accurate costs for new products, and improved tool cost-planning control.

4. *CAD/CAM integration.* In the past, CAD and CAM often use different data bases, and CAM engineers often control many parameters that are totally outside the scope of the CAD program. C&C can ease integrating CAD and CAM by providing manufacturing information to the designers while they work on their design, thus avoiding costly designs.

From the previous discussion, there is clearly a definite need for a viable classification and coding system of electronics parts for the industry. There are C&C systems in existence today for electronics manufacturing activities. But these systems were developed in-house for very specific applications. A generic classification and coding system is yet to be developed to serve the important task of producing PCB assemblies. The work discussed in this chapter is about a prototype C&C system of electronics components generally found on a PCB and the potential impact of such a system on an integrated design-manufacturing-management operation will be demonstrated conslusively through the various applications programs included in this chapter.

FUNDAMENTALS OF BMCODE

The conceptual design of BMCODE was influenced by the coding structure of DCLASS from Brigham Young University (Allen 1979). DCLASS was built on a number of coding schemes—each scheme designed for a specific application. For example, there is a part-family scheme, a materials scheme, a process scheme, a fabrication tool scheme, and an equipment scheme. Application programs were written to use one or more of these schemes. However, DCLASS was meant for a machined or mechanical part only, and therefore, cannot be applicable to products such as a PCB assembly. Nevertheless, the concept of segregating the different bodies of information is sound because it leads to more compact and easier applications. Therefore, the concept was adopted as a guideline for the BMCODE system.

In its present form, BMCODE has the following four coding schemes:

1. part-family classification;
2. part-assembly attribute;
3. board layout; and
4. process and equipment.

The information contained in each scheme is shown in Table 1.

Details of each scheme will be given later. Following the details, a prototype data base will be shown that contains, in appropriate files the various codes to be presented. The last section of this chapter will illustrate this data base in various applications.

Finally, to gain acceptance of the system, the classification procedure follows standard practices and vocabulary of the electronics industry very closely. Whenever applicable, the classification (and coding) methods follow existing standards, such as those of the American National Standard Institution (ANSI), the Electronics Industry Association (EIA), the Institute for Interconnecting and Packaging of Electronic Circuits (IPC), the Institute of Electrical and Electronic Engineers (IEEE), and the military standards. Popular catalogs are also used as guidelines and they include the Electronics Engineers Master catalog, the DATA books, the Electronic Design Goldbooks, and the Electronics Buyers Guide. Popular textbooks in manufacturing engineering are also used.

CODING SCHEMES IN BMCODE

The potential impact of BMCODE on electronics manufacturing has been generally discussed previously and summarized in Table 2. For each of the five coding schemes in BMCODE, the coding structure is discussed first, followed by a small sample of coding tables. The entire coding set will appear in a separate handbook, which will tentatively be published in the near future.

TABLE 1 – Information Contained in the Electronics C&C System

CLASSIFICATION AND CODING SCHEME	Application	Type	Construction	Electrical Specification	Package Style	Dimension and/or Weight	Standard Requirements	Special Manufacturing Requirements	Special Handling Requirements	Placement of Components on PCB	PCB Characteristics	Use of Special Devices on Assembly	Assembly Process	Assembly Equipment	Supplier
Part Family Classification	●	●	●	●	●	●	●								
Part Assembly Attribute				●	●		●	●							
PCB Assembly Layout										●	●	●			
Process and Equipment/Method													●	●	
Supplier															●

For this book, the following coding tables have been included:

1. as examples of part-family coding—
   resistors
   capacitors;
2. as examples of part-assembly coding—
   entire coding set of part assembly attributes;
3. as examples of board layout coding—
   entire coding set of board layout; and
4. as examples of process and equipment—
   entire coding set of process and equipment

TABLE 2 – Application of the Electronics C&C System and Their Information Requirements.

System Application	Information														
	Application	Type	Construction	Electrical Specification	Package Style	Dimension and/or Weight	Standard Requirements	Special Manufacturing Requirements	Special Handling Requirements	Placement of Components on PCB	PCB Characteristics	Use of Special Devices on Assembly	Assembly Process	Assembly Equipment	Supplier
Design and Retrieval	●	●	●	●	●	●	●								
Part Selection and Substitution	●	●	●	●	●	●	●								●
Part Standardization	●	●	●	●	●	●	●								
Manufacturability Rating						●	●	●	●	●	●	●			
Process and Equipment Selection						●	●	●	●	●	●	●	●	●	
Automated Process Planning						●	●	●	●	●	●	●	●	●	
Labor Cost Estimation						●	●	●	●	●	●	●	●	●	
Material Cost Estimation	●	●	●	●	●	●	●								
Purchasing and Inventory Control	●	●	●	●	●	●	●								●
Maintenance Scheduling									●					●	
Equipment Depreciation and Replacement														●	
Elimination of Unnecessary Equipment													●	●	
Capacity planning													●	●	

Part-Family Classification Scheme

The code structure of each part type in the part-family classification scheme is shown in Figure 2.

The first character of the code is always an alphabet that is the ANSI standard symbol for the part type. The rest of the code is a string of numerics. The length and interpretation of the code depend on the type of the part.

The part-family code is essentially a semipolycode with the last three digits being polycodes and the preceding digits being monocodes. The three polycode digits represent, respectively, the minimum operating temperature, maximum operating temperature, and the industry recognition agency. The monocode portion has a variable length, depending on the type of part involved, and contains information on general use, functional characteristics, construction, ratings, and packaging style.

There are nine types of electronics components classified in this scheme, namely:

1. resistors;
2. capacitors;
3. inductors;
4. diodes;
5. transistors;
6. integrated circuits;
7. relays;
8. transformers; and
9. connectors.

Together they represent more than 90% of all components found on a PCB. Thus, a classification and coding system that covers these nine component

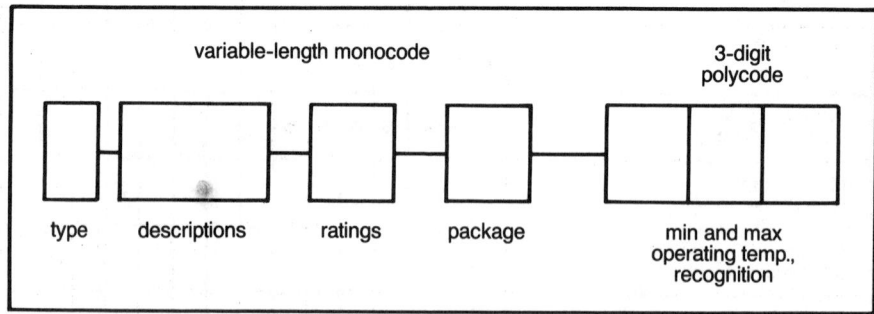

Figure 2. *Code structure of the part-family classification scheme.*

types should be adequate in providing the necessary information for the applications listed in Table 2.

The part-family classification scheme is described with a series of charts. Charts F1 and F2 apply to all categories of components listed previously. The actual format and the interpretation of the code of a component type are given in the chart labeled "code description" on the first page for each type of component. This chart also indicates which other chart to look at for any particular digit in the code string. All other charts for the same component type provide detailed classification and coding for appropriate digit or digits. The heading of any of these charts will show the type of component, the description of the information coded, the code format with an indicator to the location of the coding digit or digits, and which chart to refer to after that chart.

The following standards have been observed in the development of this scheme:

General classification (component type): ANSI Y32.2-1975, IEEE 1-69, IEEE 200-75.

Resistors: EIA RS-155-B-70, EIA RS-172-B-75, EIA RS-196-A-70, EIA RS-229-A-65, EIA RS-322-65, EIA RS-333-67, EIA RS-344-68, EIA RS-345-68, EIA RS-360-68, EIA RS-396-71, EIA RS-452-78, EIA RS-460-80, MIL-R-39032, MIL-R-22, MIL-R-39023, MIL-R-94.

Capacitors: EIA RS-153-B-72, EIA RS-164-A-67, EIA RS-198-B-71, EIA RS-218-67, EIA RS-335-A-72, EIA RS-376-70, EIA RS-377-70, EIA RS-401-73, EIA RS-495-82, MIL-C-20, MIL-C-10950, MIL-C-23269, MIL-C-39022, MIL-C-19978, MIL-C-27287, MIL-C-39006, MIL-C-39003, MIL-C-39018, MIL-C-81, MIL-C-92, MIL-C-14409.

Inductors: EIA RS-175-56, EIA RS-181-57, EIA RS-197-A-73, MIL-T-27, MIL-C-15305, MIL-C-83446.

Diodes: IEEE 216-60, EIA JEDEC 77-81, EIA JEDEC 95-76, EIA JEDEC 99-77, EIA RS-397-72, EIA RS-482-81, MIL-S-19500.

Transistors: IEEE 216-60, EIA JEDEC 74-69, EIA JEDEC 77-81, EIA JEDEC 93-75, EIA JEDEC 95-76, EIA JEDEC 99-77, MIL-S-19500.

Integrated circuits: IEEE 274-66, EIA JEDEC 216-60, EIA JEDEC 77-81, EIA JEDEC 95-76, EIA JEDEC 99-77, EIA RS-428-75, MIL-M-38510, MIL-M-55565, MIL-HDBK-175.

Relays: EIA RS-436-77, EIA RS-443-79, EIA RS-473-81, MIL-R-5757, MIL-R-6106, MIL-R-83726.

Transformers: IEEE 264-77, IEEE 295-69, IEEE 390-75, IEEE 391-76, IEEE C57.12.80-78, EIA RS-180-58, MIL-T-27, MIL-T-21038.

Connectors: IEEE 287-68, EIA RS-380-A-78, EIA-RS-429-76, IPC FC-217-82, IPC 2.2-81, IPC 2.3-81, IPC 2.4-81, IPC 2.5-81, IPC 2.6-81, IPC 2.7-81, IPC 2.8-81, IPC 2.9-81, IPC 2.10-81, IPC 2.11-81, MIL-C-28754, MIL-C-39012.

Part-Assembly Attribute Scheme

The part-assembly attribute scheme is developed specifically for the production function of electronics manufacturing. It involves those features or attributes of the part that affect its assembly process.

Thus, it contains several types of information, such as the following:

1. range of assembly processes (e.g., insertion, soldering, etc.);

2. alternative methods available for each process (e.g., hand insertion, automatic insertion, etc.); and

3. requirements and restrictions of each method for both handling and processing, such as time, special equipment, etc.).

Several part attributes are required to encode the information described previously. One of the most important attributes is the part basic shape (e.g., axial, radial, DIP, etc.). Each basic shape may have many special features superimposed on it (e.g., arrangement of leads, use of screws). Apart from this, other attributes represent dimensions, special requirements, and considerations, such as polarity requirement, heat, moisture or static sensitivity, circuit connection, use of heat sinks, etc.

There are three major divisions in the part-assembly attribute classifications —namely, through-hole mount, nonstandard mount, and surface mount. Only the first two divisions are classified and shown here. The surface mount category will be included in a handbook to be published in the near future.

The part-assembly attribute code has a fixed length, semipolycode structure. Its general format is shown in Chart M1. The first eight digits are monocodes and the last six are polycodes. The code begins with the letter M (for manufacturing) to identify the coding scheme. The rest of the code is numeric. The second digit, labeled "general classification," refers to the type of mounting requirement—it may be through-hole or surface mount. The description of digits three to six depend on the value of digit two. For example, if a part is through-hole mounted, these digits will indicate the package type and information regarding that particular package, such as lead arrangement, lead dimensions and pitch, etc. However, if the part is nonstandard-mounted, these digits will indicate a specific mounting method and the corresponding manufacturability of the component in assembly. Digits seven and eight describe the ranges of body dimensions that, again, depend on digit two and perhaps digit three. Digits nine through 14, which form the polycode segment of the code, concern information such as sensitivity to heat, moisture, static discharge, polarity, conformal coating, circuit connection, and stability due to the weight and size of the part.

PCB Layout

The PCB layout code structure is shown in Chart B1. It is a 15-digit polycode structure. All digits are numeric except for the first one, which is the letter B (for board) to indicate the type of C&C scheme. The information encoded in this scheme includes board size and shape, number of layers, use of multilead holes, special alignment or fixture requirements, special requirements for wave soldering, general component orientation, relative component locations, use of jumpers, multiwired hardnesses or multilayer assemblies, solder-side components, piggy-back mounted parts, and large or heavy components, such as heat sinks, buss bars, etc.

Process and Equipment

The purpose of this scheme is to form families of related processes to aid in selecting production methods or equipment, computer-aided process planning, labor cost estimation, and facilities management functions, such as capacity planning, equipment maintenance scheduling, and equipment depreciation and replacement.

Chart P1 shows the format of the assembly process and equipment classification code. It has a monocode structure that consists of seven alphanumeric characters arranged in four fields.

The first field identifies the assembly process. The first digit in this first field represents a general division of assembly processes. The second digit represents a specific process within this division. For example, the codes "30," "31," and "32" may indicate hand soldering, drag soldering, and wave soldering, respectively.

The next field is a one-character alpha code indicating an equipment family capable of performing the process specified in the process code.

The third field has three digits to uniquely identify the manufacturer. This code and the following model code must be developed in-house by the user. The first digit of the manufacturer code is an alphabetic character representing the first character of the manufacturer's name. The next two digits are numeric codes used to uniquely identify a given manufacturer.

The final field in the code string is a one-character alphabetic code to identify a particular model number.

PART FAMILY

Chart F1: Component Types

Code Location: Digit 1
Next Digit: (see chart below)

Resistor	(R)
	Chart R1
Capacitor	(C)
	Chart C1
Inductor	(L)
	Chart L1
Diode	(V)
	Chart VI
Component Type Transistor	(Q)
	Chart Q1
IC	(I)
	Chart I1
Relay	(K)
	Chart K1
Transformer	(T)
	Chart T1
Connector	(J)
	Chart J1

PART FAMILY

Chart F2: General Attributes

Code Location: Last Three Digits
Next Digit: none

DIGIT	ATTRIBUTE	Code					
		0	1	2	3	4	5
Second from Last	Min. Operating Temp. ('C)	not applicable	>0	-1 to -25	-26 to -40	-41 to -55	$<$-55
Next to Last	Max. Operating Temp. ('C)	not applicable	<70	71 to 85	85 to 100	101 to 125	>125
Last	Recognitions	none	UL or IEC equiva-lence	QPL (milit.)	UL or IEC equiva-lence and QPL	other	

UL - Underwriters Laboratory
IEP - International Electrotechnical Commission
QPL - Qualified Product List (Department of Defense)

RESISTOR

Chart R1: Code Description

CHART	DESCRIPTION	DIGIT LOCATION	CODE FORMAT
			R-NNNN-NNNNN-NN-NNN
--	Resistor	1	
R2	General Type (N=0-1)	2	
R3, R4	Application (N=0-6)	3	
R5	Construction (N=0-9)	4	
R6	Number of Turns (N=0-2)	5	
R7, R8	Total Resistance (N=0-9)	6	
R7, R8	Tolerance (N=0-8)	7	
R7, R8	Rated Wattage (N=0-8)	8	
R7, R8	Max. Voltage (N=0-8)	9	
R7, R8	Temperature Coeff. (N=0-9)	10	
R9, R10	Packaging (NN=00-85)	11-12	
F2	Min. Operating Temp. (N=0-5)	13	
F2	Max. Operating Temp. (N=0-5)	14	
F2	Recognition (N=0-4)	15	

RESISTOR

Chart R2: General Type

R-NNNN-NNNNN-NN-NNN

Code Location: Digit 2
Next Digit: (see chart below)

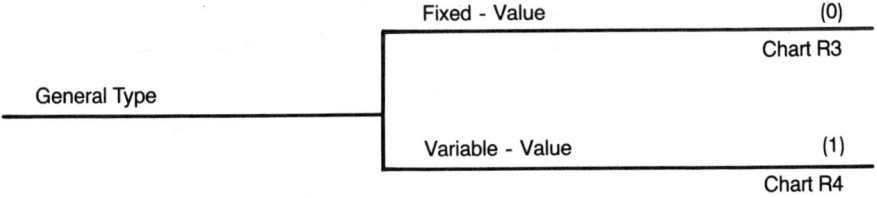

General Type

Fixed - Value (0) — Chart R3

Variable - Value (1) — Chart R4

RESISTOR

Chart R3: Application

R-ONNN-NNNNN-NN-NNN

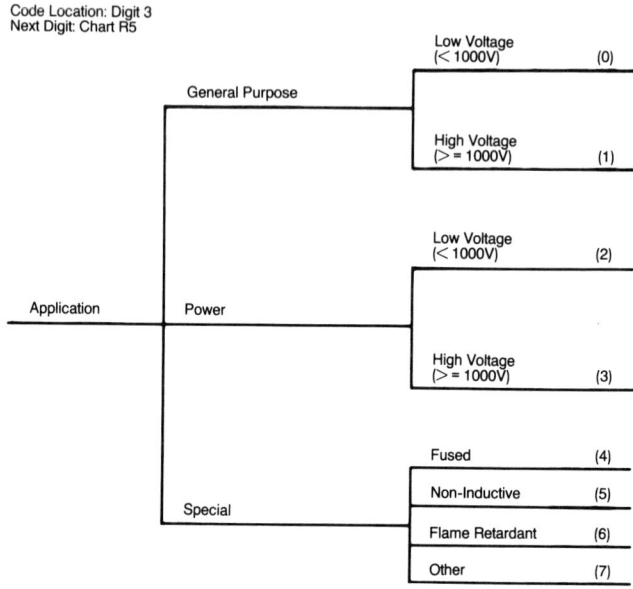

RESISTOR

Chart R4: Application

R-1NNN-NNNNN-NN-NNN

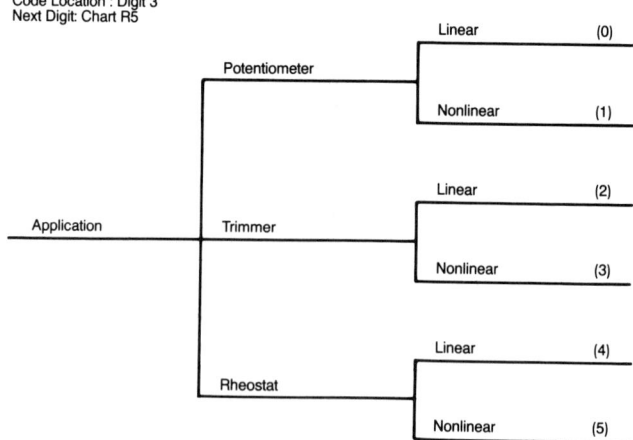

RESISTOR

CHART R5: CONSTRUCTION

R-NNNN-NNNNN-NN-NNN

Code Location: Digit 4
Next Digit: Chart R6

	Wire Wound		(0)
		Carbon	(1)
	Film	Metalized	(2)
		Other	(3)
	Carbon Composite		(4)
Construction	Cermet		(5)
	Ceramic		(6)
	Conductive Plastic		(7)
	Chip		(8)
	Other		(9)

RESISTOR

CHART R6: Number of Turns

R-NNNN-NNNNN-NN-NNN

Code Location: Digit 5
Next Digit: Chart R8 or R9

	Not Applicable	(0)
Number of Turns	Single	(1)
	Multiple	(2)

RESISTOR

Chart R7: Ratings (Fixed Value Resistors)

R-0NNN-NNNNN-NN-NNN

Code Location: Digits 6-10
Next Digit: Chart R9

DIGIT	ATTRIBUTE	CODE									
		0	1	2	3	4	5	6	7	8	9
6	Resistance (Ohm)	<1	>1- 10	>10- 50	>50- 100	>100- 500	>500 - 1K	>1K - 50K	>50K- 100K	>100K - 1M	>1M
7	Tolerance (+/- %)	<0.1	0.25	0.50	1	2	5	10	20	>20	
8	Rated Wattage	<0.1	>0.1- 0.5	>0.5- 1	>1- 5	>5- 10	>0 - 50	>50 - 100	>100 - 500	>500	
9	Maximum Voltage	<100	101- 200	201- 300	301- 500	501- 1000	1001- 2000	2001- 3000	3001- 5000	>5000	
10	Temperature Coefficient (+/- ppm/°C)	not spec.	<1	>1- 5	>5- 25	>25 50	>50- 100	>100- 200	>200- 500	>500- 800	>800

RESISTOR

Chart R8: (Variable - Value Resistors)

R-0NNN-NNNNN-NN-NNN

Code Location: Digits 6-10
Next Digit: Chart R10

DIGIT	ATTRIBUTE	CODE									
		0	1	2	3	4	5	6	7	8	9
6	Resistance (Ohm)	<100	>100- 500	>500- 1K	>1K- 10K	>10K- 50K	>50K- 100K	>100K- 500K	>500K- 1M	>1M- 10M	>10M
7	Tolerance (+/- %)	<0.1	0.25	0.50	1	2	5	10	20	>20	
8	Rated Wattage	<1	>1- 5	>5- 10	>10- 50	>50- 100	>100- 500	>500- 1000	>1000		
9	Maximum Voltage	<100	101- 200	201 300	301- 500	501- 1000	1001- 2000	2001- 3000	3001- 5000	>5000	
10	Temperature Coefficient (+/- ppm/°C)	NS	<1	>1- 5	>5- 25	>25 50	>50- 100	>100- 200	>200- 500	>500- 800	>800

RESISTOR

Chart R9: Package Style (Fixed Value Resistors)

R-ONNN-NNNNN-NN-NNN

Code Location: Digits 11 and 12
Next Digit: Chart F2

	Axial Lead	Molded	(00)
		Comformally Coated	(01)
		Hermatically Sealed	(02)
	Single-end Axial Lead		(10)
	Tubular	Axial Lead	(20)
		Ferrule Terminal	(21)
		Radial Tab	(22)
	Cemet Embedded		(30)
Package Style	Flat Strip	Standard	(40)
		Stacked	(41)
	DIP		(50)
	SIP		(60)
	Aluminum Case	Chasis Mount	(70)
		Chasic Foodthrough	(71)
		Chasis Standoff	(72)
	High Power	Tub Type	(80)
		Tub Type Disk	(81)
		Edison Base	(82)
		Edgewound Bare Ribbon	(83)
		Appliance Motor Control	(84)
		Corrugated Edgewound Ribbon	(85)

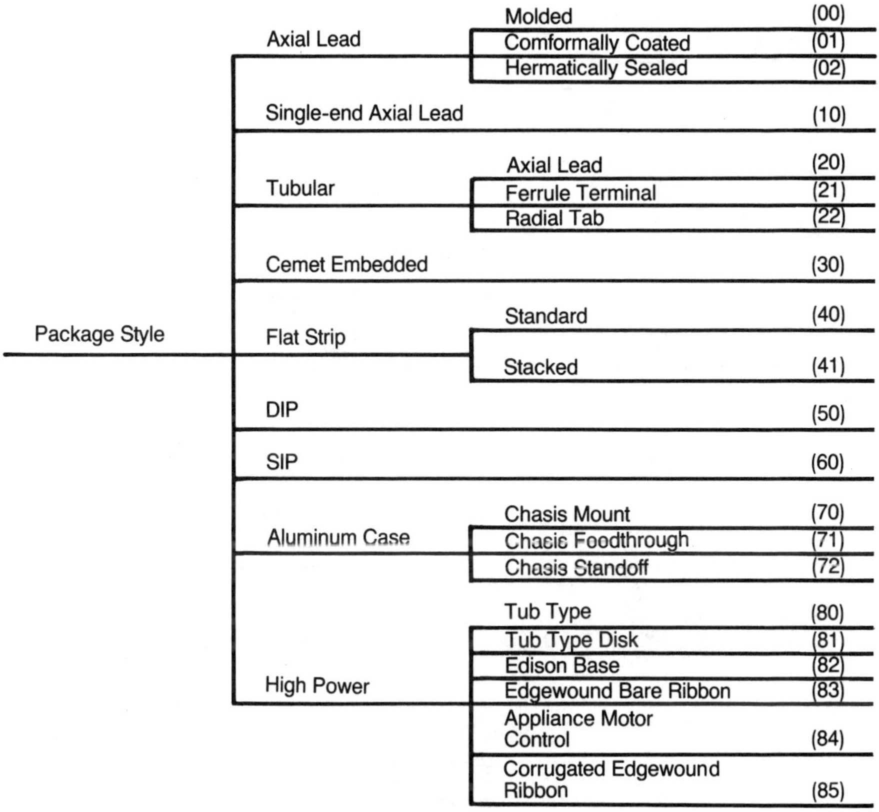

RESISTOR

Chart R10: Package Style: (Variable Value Resistors)

R-1NNN-NNNNN-NN-NNN

Code Location: Digits 11 and 12
Next Digit: Chart F2

		CODE					
DIGIT	ATTRIBUTE	0	1	2	3	4	5
11	Shape	Round	Square	Rectangular			
12	Terminal	Flexible Insulated Wire Lead	PC Pin-Base Mount	PC Pin-Edge Mount Top Adjust	PC Pin-Edge Mount Side Adjust	PC Pin-Staggered	Other

CAPACITOR

Chart C1: Application and Code Structure

Code Location: Digit 2
Next Digit: (see chart below)

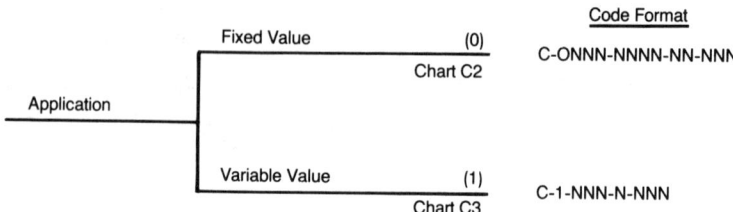

Code Format

Fixed Value   (0)

Chart C2

C-ONNN-NNNN-NN-NNN

Application

Variable Value   (1)

Chart C3

C-1-NNN-N-NNN

CAPACITOR

Chart C2: Code Description (Fixed Value Capacitor)

CHART	DESCRIPTION	DIGIT LOCATION	CODE FORMAT
			C—ONNN-NNNN-NN-NNN
--	Capacitor	1	
C1	Application: Fixed Value	2	
C4	Type and Construction	3-4	
C5	Polarity (N = 0.1)	5	
C5	Capacitance (N = 0-9)	6	
C5	Tolerance (N = 0-7)	7	
C5	DC-rated Voltage (N = 0-9)	8	
C5	Dissipating Factor (N = 0-9)	9	
C6	Packaging	10-11	
F2	Min. Operating Temp. (N = 0-5)	12	
F2	Max. Operating Temp. (N = 0-5)	13	
F2	Recognition (N = 0-4)	14	

CAPACITOR

Chart C3: Code Description (Variable Value Capacitor)

CHART	DESCRIPTION	DIGIT LOCATION	CODE FORMAT
			C—1-NNN-N-NNN
--	Capacitor	1	
C1	Application: Variable Value	2	
C7	Nominal Max. Capacitance (N=0-9)	3	
C7	DC-rated Voltage (N = 0-9)	4	
C7	Dissipating Factor (N = 0-9)	5	
C8	Packaging	6	
F2	Min. Operating Temp. (N = 0-5)	7	
F2	Max. Operating Temp. (N = 0-5)	8	
F2	Recognition (N = 0-4)	9	

CAPACITOR

Chart C4: Type cr Construction (Fixed-Value Capacitor)

C-0NNN-NNNN-NN-NNN

Code Location: Digits 3 and 4
Next Digit: Chart C5

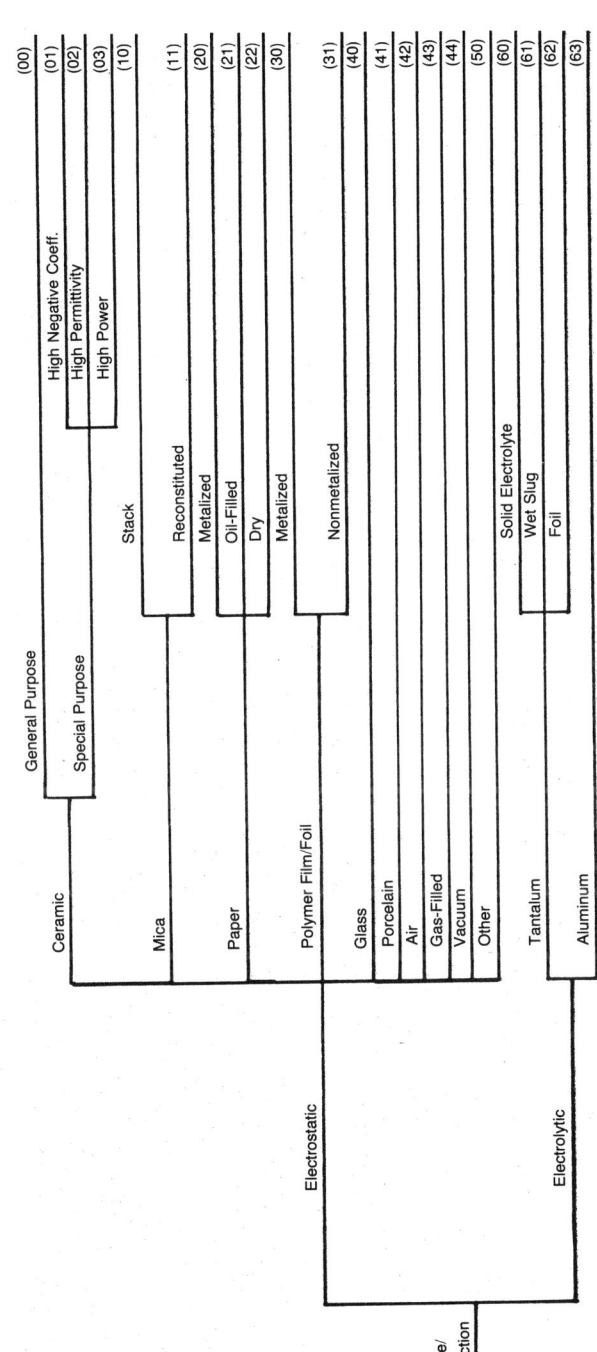

CAPACITOR
Chart C5: Polarity and Ratings (Fixed-value Capacitor)

C-0NNN-NNNN-NN-NNNN

Code Location: Digits 5 to 9
Next Digit: Chart C6

DIGIT	ATTRIBUTE	0	1	2	3	4	5	6	7	8	9
							CODE				
5	Polarity	Polar	Non-polar								
6	Capacitance	<= 10 pF	11 - 100 pF	101 - 1,000 pF	1,001 - 10,000 pF	10,001 - 100,000 pF	100,001 pF to 1 uF	>1 to 10 uF	>10 to 100 uF	>100 to 1,000 uF	>1,000 uF
7	Tolerance (+/- %)	0.5	1	2	3	5	10	20	>20		
8	DC Rated Volt.	<= 6	7 - 50	51 - 100	101 - 200	201 - 400	401 - 1,000	1,001 - 2,500	2,501 - 5,000	5,001 - 10,000	>10,000
9	Dissipating Factor at 1 kHz. 25°C	Not specified	<= 0.001	>0.001 to 0.005	>0.005 to 0.01	>0.01 to 0.05	>0.05 to 1.0	>0.10 to 0.25	>0.25 to 0.50	>0.50 to 2.0	>0.20

CAPACITOR

Chart C6: Packaging (Fixed Value Capacitor)

C-ONNN-NNNN-NN-NNN

Code Location: Digits 10 and 11
Next Digit: Chart F2

Packaging			
	Axial		(00)
	Radial Lead	Tubular	(10)
		Ceramic Disk	(11)
		Dipped	(12)
		Molded Body	(13)
		Can	(14)
	Radial Lug Can	PC Mounted	(20)
		Twisted Mounted	(21)
	Feedthrough	Standard	(30)
		Solder Eyelet	(31)
		Miniature Threaded Bushing	(32)
		Solder Eyelet Hollow	(33)
	Standoff		(40)
	Chip		(50)
	Oil-Filled Case		(60)
	Button		(70)
	Hermatically Sealed		(80)
	Other		(90)

CAPACITOR

Chart C7: Ratings (Variable-value Capacitor)

C-1-NNN-N-NNN

Code Location: Digits 3 to 5
Next Digit: Chart C8

DIGIT	ATTRIBUTE	CODE									
		0	1	2	3	4	5	6	7	8	9
3	Nominal Max.	<= 5	>5 to 10	>10 to 15	>15 to 20	>20 to 30	>30 to 50	>50 to 150	>150 to 500	>500 to 1,000	>1,000
4	DC-Rated Voltage	<= 25	26 - 50	51 - 100	101 - 200	201 - 400	401 - 1,000	1,001 - 2,501	2,501 - 5,000	5,001 - 10,000	10,000
5	Dissipating Factor at 1 kMz. 25°C	Not Specified	<= 0.001	>0.001 - 0.005	>0.005 - 0.01	>0.01 - 0.05	>0.05 - 0.10	>0.10 - 0.25	>0.25 - 0.50	>0.50 - 2.0	>2.0

CAPACITOR

Chart C8: Packaging (Variable Value Capacitor)

C-1-NNN-N-NNN

Code Location: Digit 6
Next Digit: Chart F2

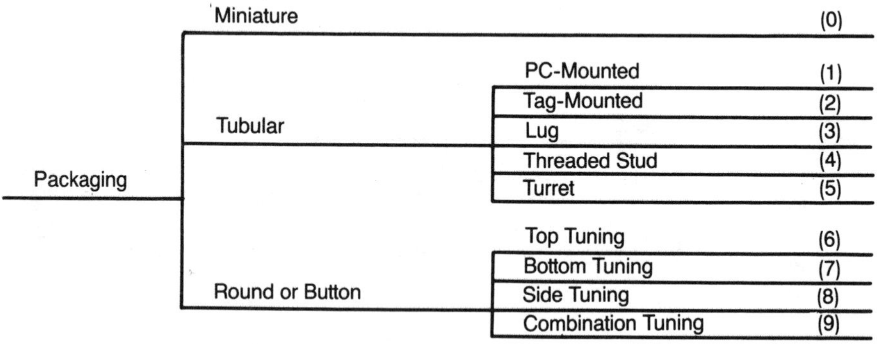

	Miniature		(0)
Packaging	Tubular	PC-Mounted	(1)
		Tag-Mounted	(2)
		Lug	(3)
		Threaded Stud	(4)
		Turret	(5)
	Round or Button	Top Tuning	(6)
		Bottom Tuning	(7)
		Side Tuning	(8)
		Combination Tuning	(9)

PART ASSEMBLY ATTRIBUTE

Chart M1: Code Description

CHART	DESCRIPTION	CODE LOCATION	CODE FORMAT
			M—N-NNNN-NN-NNN-NNN
--	Manufacturing (Assembly)	1	
M2	Mounting Method	2	
M5-18	Attributes for Specific Types	3-6	
M5-18	Body Dimensions	7-8	
M4	Polarity Requirements	9	
M4	Heat/Moisture Sensitivity	10	
M4	Static Sensitivity	11	
M4	Confirmal Coating Requirement	12	
M4	Large Part or Use of Heat Sink	13	
M4	Circuit Connection	14	

PART ASSEMBLY ATTRIBUTE

Chart M2: Mounting Method

M-N-NNNN-NN-NNN-NNN

Code Location: Digit 2
Next Digit: (see chart below)

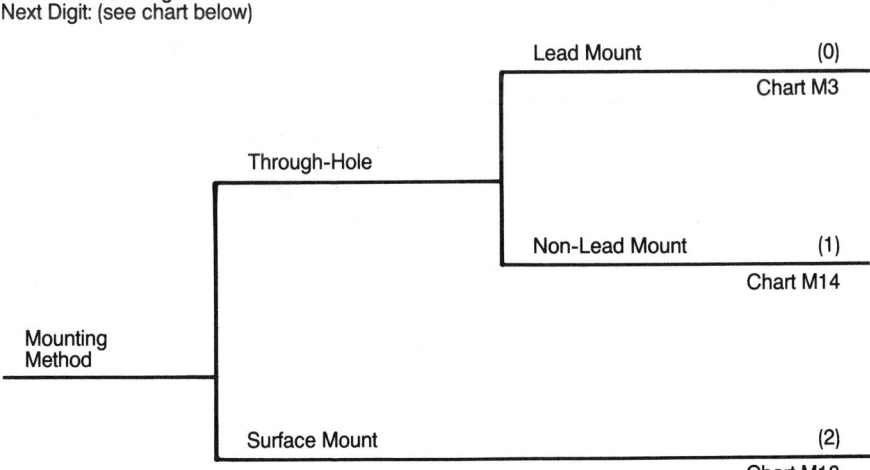

PART ASSEMBLY ATTRIBUTE

Chart M3: Lead Mount Package Style

M-O-NNN-NN-NNN—NNN

Code Location: Digit 3
Next Digit: (see chart below)

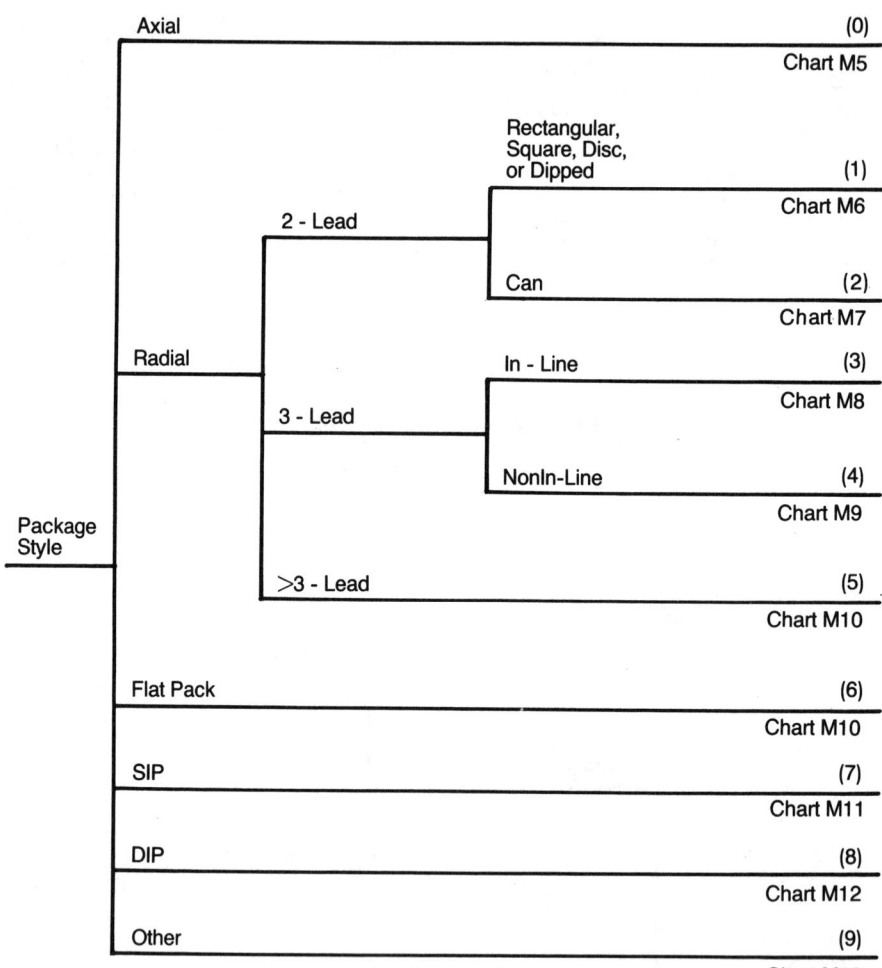

PART ASSEMBLY ATTRIBUTE

Chart M4: Generic Attributes

M-N-NNNN-NN-NNN-NNN

Code Location: Digits 9-14
Next Digit: None

DIGIT	ATTRIBUTE	CODE		
		0	1	2
9	Polarity Require-ment	No	Yes	
10	Heat- or Moisture-Sensitive	No	Yes, only post solder assembly required	Yes, special off-line cleaning required
11	Static Sensitivity	No	Damage threshold >350 V	Damage threshold >350 V
12	Conformal Coating Requirement	No	Yes, No masking required	Yes, masking required
13	Large Part (>0.75 cu. inch)	No	Yes	
14	Circuit Connection	Mount-ing leads	Quick disconnect	Hardwire

PART ASSEMBLY ATTRIBUTE
Chart M5: Dimensions (Axial)

M-0-0NNN-NN-NNN-NNN

Code Location: Digits 4-8
Next Digit: Chart M4

DIGIT	ATTRIBUTE	CODE									
		0	1	2	3	4	5	6	7	8	9
4	Lead Diameter (inch)	<0.015	>0.015 -0.025	>0.025 >0.032	>0.032 >0.037	>0.037 -0.051	---	---	---	---	>0.051
5,6	Nonsignificant	All									
7	Body Length (inch)	---	<0.024	>0.024 -0.590	>0.590 -1.12	>1.12 -1.15	---	---	---	---	>1.15
8	Body Diameter (inch)	---	<0.080	>0.080 -0.250	>0.250 -0.312	>0.312 -0.375	>0.375 -0.412	---	---	>0.417 -0.500	>0.500

PART ASSEMBLY ATTRIBUTE
Chart M6: Dimensions (Rectangular, 2 - Lead Radial)

M-0-1NNN-NN-NNN-NNN

Code Location: Digits 4-8
Next Digit: Chart M4

DIGIT	ATTRIBUTE	CODE									
		0	1	2	3	4	5	6	7	8	9
4	Lead Diameter	>0.046	<0.046								
5	Lead Span (inch)	<0.165	0.166 -0.228								
6	Nonsignificant	All									
7	Body Thickness (inch)	---	<0.080	>0.080 -0.319 (square -body) >0.080 -0.216 (other)	---	---	---	---	---	>0.319 -0.500 (square body) >0.216 -0.500 (other)	>0.500
8	Body Height (inch)	---	<0.240	>0.240 -0.590	---	---	---	---	---	>0.590 -0.126	>1.26

PART ASSEMBLY ATTRIBUTE
Chart M7: Dimensions (Can, 2 - Lead Radial)

M-0-2NNN-NN-NNN-NNN

Code Location: Digits 4-8
Next Digit: Chart M4

DIGIT	ATTRIBUTE	CODE									
		0	1	2	3	4	5	6	7	8	9
4	Lead Diameter	>0.046	<0.046								
5	Lead Span (inch)	<0.165	0.166 -0.228	---	---	---	---	---	---	---	>0.228
6	Nonsignificant	All									
7	Body Diameter (inch)	---	<0.080	>0.080 -0.354	---	---	---	---	---	>0.354 -0.500	>0.500
8	Body Height (inch)	---	<0.240	>0.240 -0.590	---	---	---	---	---	>0.590 -1.26	>1.26

PART ASSEMBLY ATTRIBUTE
Chart M8: Dimensions (3 - Lead, In-line Radial)

M-0-2NNN-NN-NNN-NNN

Code Location: Digits 4-8
Next Digit: Chart M4

DIGIT	ATTRIBUTE	CODE									
		0	1	2	3	4	5	6	7	8	9
4	Lead Diameter (inch)	<0.018	>0.018 -0.027	---	---	---	---	---	---	---	>0.027
5	Lead Span (inch)	<0.094	>0.094 -0.114	---	---	---	---	---	---	---	>0.114
6	Nonsignificant	All									
7	Body Diameter (inch)	---	<0.080	>0.080 -0.236	---	---	---	---	---	>0.236 -0.500	>0.500
8	Body Height (inch)	<0.124	>0.124 -0.240	>0.240 -0.274	---	---	---	---	---	>0.472 -0.590	>0.590

PART ASSEMBLY ATTRIBUTE
Chart M9: Dimensions (3 - Lead, Non-in-line Radial)

M-0-4NNN-NN-NNN-NNN

Code Location: Digits 4-8
Next Digit: Chart M4

DIGIT	ATTRIBUTE	CODE		
		0	8	9
4, 5, 6	Nonsignificant	All		
7	Body Width or Diameter (inch)	<0.08	>0.08 -0.50	>0.50
8	Body Height (inch)	<0.24	>0.24 -0.59	>0.59

PART ASSEMBLY ATTRIBUTE
Chart M10: Dimensions (>3 - Lead, Radial and Flat Pack)

5
M-0-(:)NNN-NN-NNN-NNN
6

Code Location: Digits 4-8
Next Digit: Chart M4

DIGIT	ATTRIBUTE	CODE									
		0	1	2	3	4	5	6	7	8	9
4	No. of leads	4-5	6-7	8-9	10-13	14-17	18-21	22-28	>028		
5, 6	Nonsignificant	All									
7	Body Width or Dia. (inch)	<0.08	---	---	---	---	---	---	---	>0.08 -0.50	>0.50
8	Body Height (inch)	<0.24	---	---	---	---	---	---	---	>0.84 -0.59	>0.59

PART ASSEMBLY ATTRIBUTE
Chart M11: Dimensions (SIP Package)

M-0-7NNN-NN-NNN-NNN

Code Location: Digits 4-8
Next Digit: Chart M4

DIGIT 4	DIGIT 5	DIGIT 6	DIGIT 7	DIGIT 8
# of leads	lead pitch -inch	non-significant	overall thick. -inch	overall leng. -inch
(code)	(code)	(code)	(code)	(code)

DIGIT 4	DIGIT 5	DIGIT 6	DIGIT 7	DIGIT 8
			<0.080 (0)	<0.180 (0)
	0.100 (1)	All (0)	>0.080-0.090 (1)	>0.180-0.190 (1)
			>0.090 (8)	>0.190 (8)
			<0.100 (0)	<0.218 (0)
2 (1)	0.125 (2)	All (0)	>0.100-0.110 (1)	>0.210-0.234 (1)
			>0.110 (8)	>0.234 (8)
	Other (0)	All (0)	Any (0)	Any (0)
			<0.080 (0)	<0.218 (0)
	0.100 (1)	All (0)	>0.080-0.090 (1)	>0.218-.234 (1)
			>0.090 (8)	>0.234 (8)
			<0.100 (0)	<0.474 (0)
4 (2)	0.125 (2)	All (0)	>0.100-0.110 (1)	>0.474-0.484 (1)
			>0.110 (8)	>0.484 (8)
	Other (0)	All (0)	Any (0)	Any (0)

PART ASSEMBLY ATTRIBUTE
Chart M11 Cont.

DIGIT 4	DIGIT 5	DIGIT 6	DIGIT 7	DIGIT 8
# of leads (code)	lead pitch -inch (code)	non-significant (code)	overall thick. -inch (code)	overall leng. -inch (code)

DIGIT 4	DIGIT 5	DIGIT 6	DIGIT 7		DIGIT 8	
	0.100 (1)	All (0)	<0.080	(0)	<0.574	(0)
			>0.080-0.090	(1)	>0.574-0.584	(1)
			>0.090	(8)	>0.584	(8)
6 (3)	0.125 (2)	All (0)	<0.100	(0)	<0.724	(0)
			>0.100-0.110	(1)	>0.724-0.734	(1)
			>0.110	(8)	>0.734	(8)
	Other (0)	All (0)	Any	(0)	Any	(0)
	0.100 (1)	All (0)	<0.080	(0)	<0.774	(0)
			>0.080-0.090	(1)	>0.774-.784	(1)
			>0.090	(8)	>0.784	(8)
8 (4)	0.125 (1)	All (0)	<0.100	(0)	<0.974	(0)
			>0.100-0.110	(1)	>0.974-0.984	(1)
			>0.110	(8)	>0.984	(8)
	Other (0)	All (0)	Any	(0)	Any	(0)
	0.100 (1)	All (0)	<0.080	(0)	<0.974	(0)
			>0.080-0.090	(1)	>0.974-0.984	(1)
10 (5)			>0.090	(8)	>0.984	(8)
	Other (0)	All (0)	Any	(0)	Any	(0)
>10 (0)	Any (0)	All (0)	Any	(0)	Any	(0)

PART ASSEMBLY ATTRIBUTE

Chart M12: Dimensions (Dip Package)

M-0-8NNN-NN-NNN-NNN

Code Location: Digits 4-8
Next Digit: Chart M4

DIGIT 4 \# of leads	DIGIT 5 lead span -inch	DIGIT 6 non- significant	DIGIT 7 body length -inch	DIGIT 8 non- significant

	0.300 (1)	All (0)	<0.093 (0) ≥0.093 - 0.098 (1) >0.098 (8)	All (0)
2 (1)	Other (0)	All (0)	Any (0)	All (0)
	0.300 (1)	All (0)	<0.193 (0) ≥0.193 - 0.198 (1) >0.198 (8)	All (0)
4 (2)	Other (0)	All (0)	Any (0)	All (0)
	0.300 (1)	All (0)	<0.300 (0) ≥0.300 - 1.25 (1) >1.25 (8)	All (0)
6-20 (3)	Other (0)	All (0)	Any (0)	All (0)
	0.400 (2)	All (0)	<1.10 (0) ≥1.10 - 2.05 (1) >2.05 (8)	All (0)
22-40 (4)	0.600 (3)	All (0)	<1.10 (0) ≥1.10 - 2.05 (1) >2.05 (8)	All (0)
	Other (O)	All (0)	Any (0)	All (0)
>40 (0)	Any (O)	All (0)	Any (0)	All (0)

PART ASSEMBLY ATTRIBUTE

Chart M13: Nonsignificant Code
(Unclassified lead mounted package)

M-0-9NNN-NN-NNN-NNN

Code Location: Digits 4-8
Next Digit: Chart M4

DIGIT	ATTRIBUTE	CODE
		0
4 to 8	Nonsignificant	all

PART ASSEMBLY ATTRIBUTE

Chart M14: Mounting Access (Nonlead Mount)

M-1-NNNN-NN-NNN-NNN

Code Location: Digit 3
Next Digit: (see chart below)

Access Method

Part not secured immediately when mounted or added
- Part and associated tool (including hand) can easily reach desired location (0) — Chart M15
- Part and associated tool cannot easily reach desired location due to obstructed access or restricted vision (1) — Chart M15
- Part and associated tool cannot easily reach desired location due to obstructed access and restricted vision (2) — Chart M15

Part secured immediately when mounted or added
- Part and associated tool (including hand) can easily reach desired location (3) — Chart M17
- Part and associated tool cannot easily reach desired location due to obstructed access or restricted vision (4) — Chart M17
- Part and associated tool cannot easily reach desired location due to obstructed access and restricted vision (5) — Chart M17

PART ASSEMBLY ATTRIBUTE

Chart M15: Positioning

0
M-1-(1)NNN-NN-NNN-NNN
2

Code Location: Digit 4
Next Digit: Chart M16

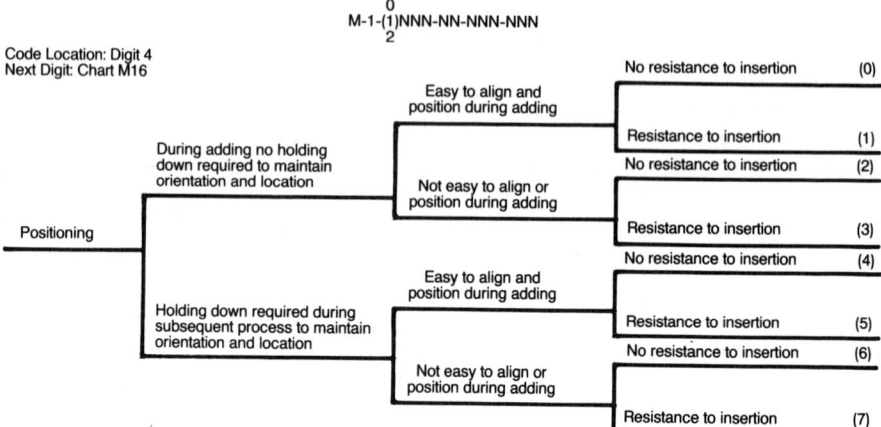

PART ASSEMBLY ATTRIBUTE

Chart M16: Fastening Process

0
M-1-(1)NNN-NN-NNN-NNN
2

Code Location: Digit 5
Next Location: Chart M18

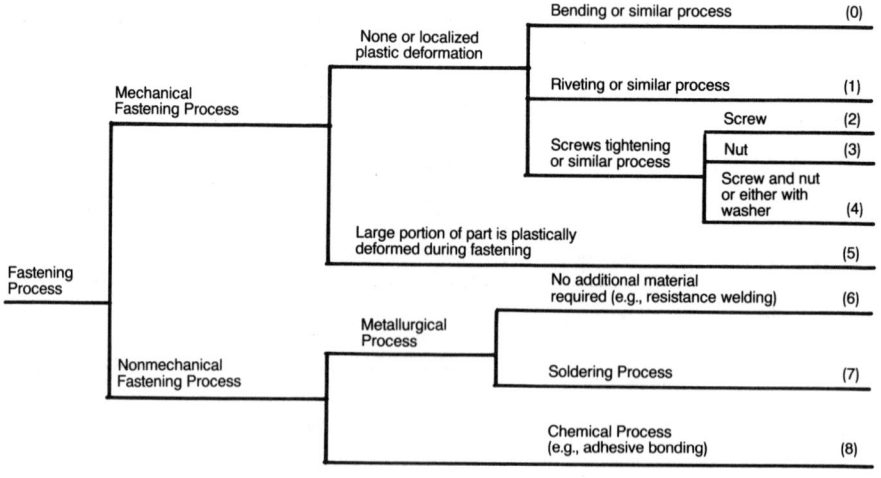

PART ASSEMBLY ATTRIBUTE

Chart M17: Fastening Process and Positioning

$$\overset{3}{\underset{5}{\text{M-1-(4)NNN-NN-NNN-NNN}}}$$

Code Location: Digits 4-5
Next Digit: Chart M18

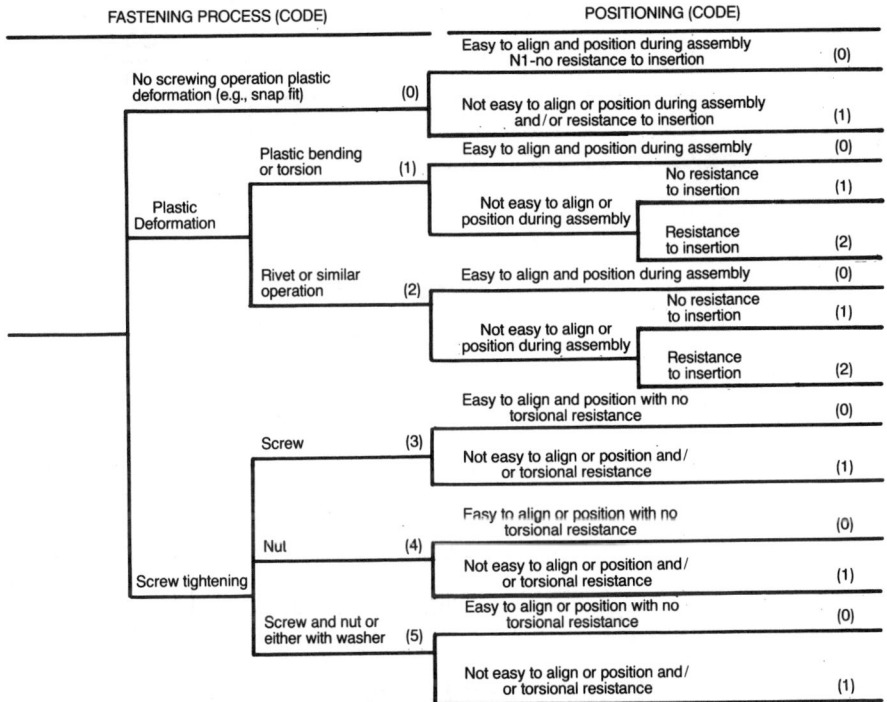

PART ASSEMBLY ATTRIBUTE

Chart M18: Dimensions (Nonlead-Mounted Package)

$$\overset{0}{\underset{5}{\text{M-1-(:)NNN-NN-NNN-NNN}}}$$

Code Location: Digits 6-8
Next Digit: Chart M4

		CODE									
DIGIT	ATTRIBUTE	0	1	2	3	4	5	6	7	8	9
6	Number of Securing Points	>9	1	2	3	4	5	6	7	8	9
7	Body Width of Diameter (inch)	>0.24	>0.24 -0.50	---	---	---	>0.50				
8	Body Height (inch)	<0.08	>0.08 -0.50	---	---	---	---	>0.50			

PART ASSEMBLY ATTRIBUTE

Chart M19: Nonsignificant Code (Surface-mount Part)

M-2-NNNN-NN-NNN-NNN

Code Location: Digits 3-14
Next Digit: None

DIGIT	ATTRIBUTE	CODE
		0
3-14	Nonsignificant	All

PCB ASSEMBLY LAYOUT

Chart B1: Code Description

CHART	DESCRIPTION	DIGIT LOCATION	CODE FORMAT

B-NNN-NNN-NNN-NNN-NN

CHART	DESCRIPTION	DIGIT LOCATION
--	PCB Assembly Layout	1
B2	PCB Size and Shape	2
B2	Number of Layers of PCB	3
B2	No. of Holes w/More Than One Lead	4
B2	No. of Piggyback Mounted Parts	5
B2	Special Alignment or Fixturing Requirement of PCB	6
B2	PCB Edge Clearance	7
B3	No. of PCB Edges Masked for Wave Soldering	8
B3	No. of Tooling Holes Sealed Before Wave Soldering	9
B3	No. of Jumpers	10
B3	Use of Multiwired Harnesses or Multilayer Assemblies	11
B4	Orientation of Axial, Radial, DIP, or SIP Packages	12
B4	Use of Solder-Sided Components	13
B4	No. of Bus Bars and Single Unit Heat Sinks	14
B4	Use of Multiple-Devise Heat Sinks	15

PCB ASSEMBLY LAYOUT

Chart B2: Classification and Coding

B-NNN-NNN-NNN-NNN-NN

Code Location: Digits 2 to 7
Next Digit: Chart B3

DIGIT	ATTRIBUTE	CODE				
		0	1	2	3	4
2	PCB Size and Shape	rectangular and smaller than 8x8 in.	rectangular and between 8x8 in. and 9x9 in.	rectangular and between 9x9 in. and 9x16 in.	non-rectangular and not larger than 9x16 in.	larger than 9x16 in.
3	No. of layers	single	double	3 to 4	>4	
4	No. of holes w/more than one lead	0	1 - 2	3	4 - 6	>6
5	No. of piggy-back mounted parts	0	1	2	>2	
6	Special alignment or fixturing requirement of PCB	no	yes			
7	PCB edge clearance	no component closer to edge than 3/8 in.	some components closer to edge than 3/8 in.			

PCB ASSEMBLY LAYOUT

Chart B3: Classification and Coding

B-NNN-NNN-NNN-NNN-NN

Code Location: Digits 8 to 11
Next Digit: Chart B4

DIGIT	ATTRIBUTE	CODE				
		0	1	2	3	4
8	No. of PCB Edges Masked for Wave Soldering	0	1	>1		
9	No. of Tooling Holes Sealed Before Wave Soldering	0	1 - 3	>3		
10	No. of Jumpers	0	1-3, none requiring tack	4-6, none requiring tack	1-6, some requiring tack	>6
11	Use of Multi wired harnesses or Multilayered Assemblies.	none	some			

PCB ASSEMBLY LAYOUT

Chart B4: Classification and Coding

B-NNN-NNN-NNN-NNN-NN

Code Location: Digits 12 to 15
Next Digit: none

DIGIT	ATTRIBUTE	CODE				
		0	1	2	3	4
12	Orientation of Axial, Radial, DIP and SIP	all oriented in common direction	one component category w/90' mixed orientation	two components categories or more w/90' mixed orientation		
13	Use of Solder-side Component	none	some. none requiring specific alignment	some. some requiring specific alignment		
14	No. of Bus Bars and Single Unit Heat Sinks	0	1	2-4	5-6	>6
15	Use of Multi Device Heat Sinks	none	some			

PROCESS AND EQUIPMENT/METHOD

Chart P1 : Code Description

CHART	DESCRIPTION	DIGIT LOCATION	CODE FORMAT
			NN-A-ANN-A
P2	Process Division	1	
P3-P9	Process Group	2	
--	Equipment Type for Process	3	
--	First Letter of Name of Equipment Manufacturer	4	
--	Manufacturer Code	5-6	
--	Manufacturer Model Number Code	7	

N - Numeric Code
A - Alphabetic Code

PROCESS AND EQUIPMENT/METHOD

Chart P2: Process Division

NN-A-ANN-A

Code Location: Digit 1
Next Digit: (see chart below)

	Preparatory	(0)
		Chart P3
	Mounting	(1)
		Chart P4
	Pre-soldering	(2)
		Chart P5
Process Division	Soldering	(3)
		Chart P6
	Cleaning	(4)
		Chart P7
	Testing	(5)
		Chart P8
	Conformal Coating	(6)
		Chart P9

PROCESS AND EQUIPMENT/METHOD

Chart P3: Process Group (Preparatory Process)

NN-A-ANN-A

Code Location: Digit 2
Next Digit: none

	Lead Forming	(0)
Preparatory Process		
	Component Sequencing	(1)

PROCESS AND EQUIPMENT/METHOD

Chart P4: Process Group (Mounting Process)

NN-A-ANN-A

Code Location: Digit 2
Next Digit: none

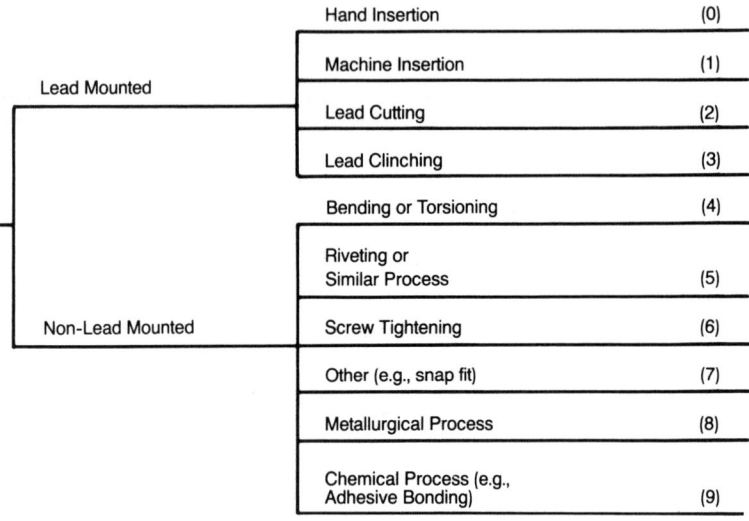

	Hand Insertion	(0)
	Machine Insertion	(1)
Lead Mounted	Lead Cutting	(2)
	Lead Clinching	(3)
	Bending or Torsioning	(4)
	Riveting or Similar Process	(5)
Non-Lead Mounted	Screw Tightening	(6)
	Other (e.g., snap fit)	(7)
	Metallurgical Process	(8)
	Chemical Process (e.g., Adhesive Bonding)	(9)

PROCESS AND EQUIPMENT/METHOD

Chart P5: Process Group Presoldering Process

NN-A-ANN-A

Code Location: Digit 2
Next Digit: none

	Skin Packing		(0)
	Masking		(1)
		Spray	(2)
		Brush	(3)
Presoldering Process	Fluxing	Dip	(4)
		Foam	(5)
		Wave	(6)
	Preheating		(7)

PROCESS AND EQUIPMENT/METHOD

Chart P6: Process Group (Soldering Process)

NN-A-ANN-A

Code Location: Digit 2
Next Digit: none

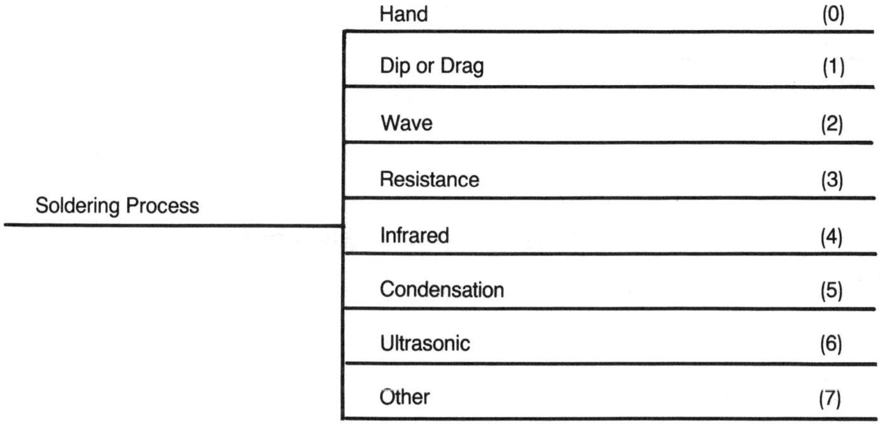

Soldering Process		
	Hand	(0)
	Dip or Drag	(1)
	Wave	(2)
	Resistance	(3)
	Infrared	(4)
	Condensation	(5)
	Ultrasonic	(6)
	Other	(7)

PROCESS AND EQUIPMENT/METHOD

Chart P7: Process Group (Cleaning Process)

NN-A-ANN-A

Code Location: Digit 2
Next Digit: none

Cleaning Process			
	Flux Removal	Neutralizing	(0)
		Alkaline Cleaning	(1)
		Organic-Solvent Cleaning	(2)
	Spray Rinsing		(3)
	Drying		(4)

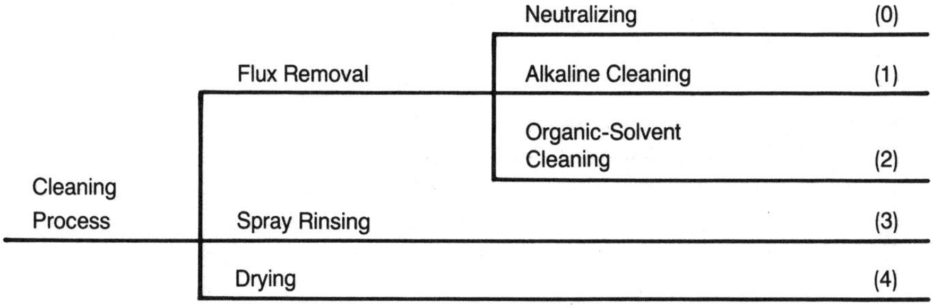

PROCESS AND EQUIPMENT/METHOD

Chart P8: Process Group (Testing Process)

NN-A-ANN-A

Code Location: Digit 2
Next Digit: none

Testing Process	In-Circuit Testing	(0)
	Functional Testing	(1)

PROCESS AND EQUIPMENT/METHOD

Chart P9: Process Group (Conformal Coating Process)

NN-A-ANN-A

Code Location: Digit 2
Next Digit: none

Conformal Coating	Spray	(0)
	Vacuum Deposit	(1)
	Dip	(2)

THE BMCODE DATA BASE

The Conceptual Data Model

The design of the data base follows the method suggested by McFadden and Hoffer (1985), which is summarized in Figure 3.

The data base development starts with a user requirement analysis, then proceeds by translating the user requirements into a conceptual design, an implementation design, and a physical design, respectively.

The user requirements analysis stage identifies user needs for data and results in view descriptions, called user views, and data element descriptions. The user views are discussed later.

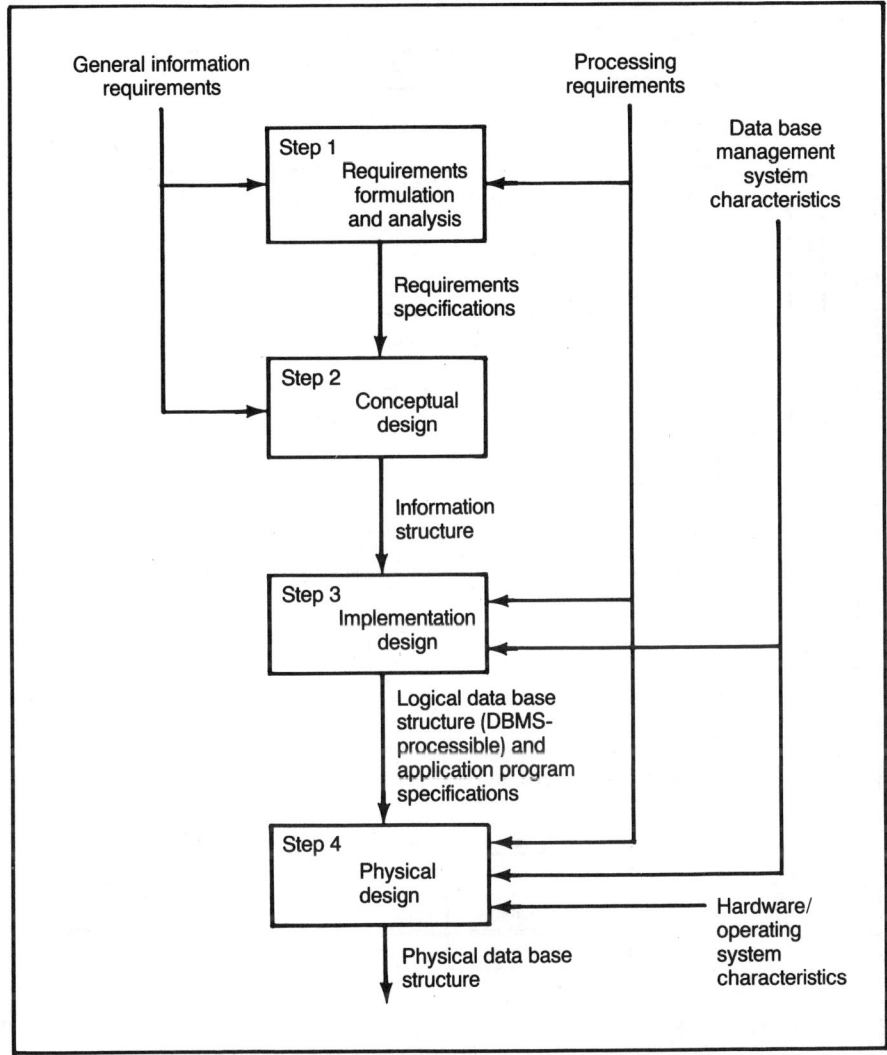

Figure 3. *Data base design procedure (McFadden and Hoffer 1985).*

The conceptual design is expressed in third normal form relations with the logical access maps shown with each user view. A third normal form relation is a relation in which each nonkey attribute is fully dependent on the primary key, and there are no transitive or hidden dependencies. A logical access map is a diagram showing the sequence of logical accesses to conceptual data base records or relations. The overall conceptual data model is summarized in Figure 4.

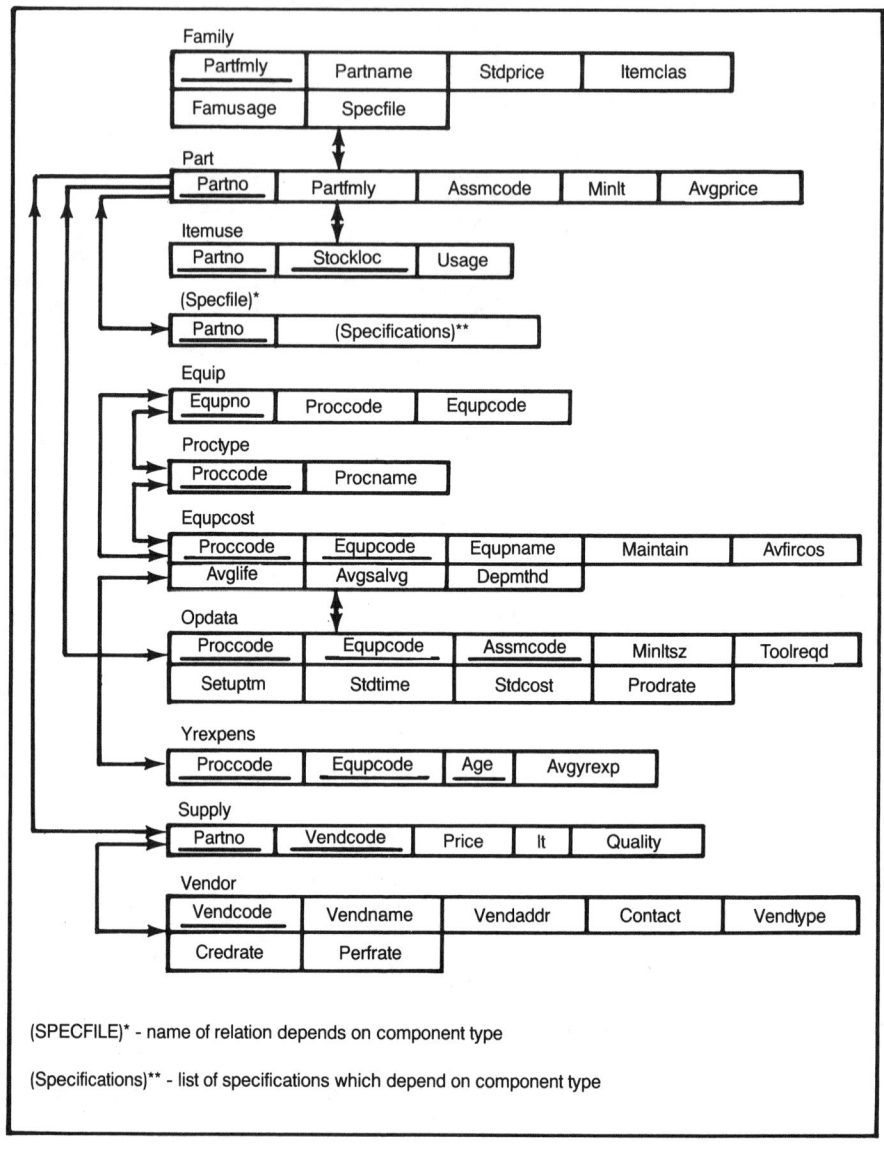

Figure 4. *The conceptual data model of the prototype data base.*

The implementation design is the mapping of the conceptual data model into an internal data model or schema that can be processed by the data base management system (DBMS). Figure 5 shows an example of the result of this design step.

The physical design stage is concerned with designing stored record formats, selecting access methods, and deciding on physical factors, such as record blocking. It is also concerned with data base security, integrity, back-up, and recovery. Because these features are already built in the DBMS, and because the DBMS has been predetermined, there is nothing more to be done in this step.

Relation: Part
* * * * * * * * *

Last mod: 2/01/87   read password: none
schema: Elect   modify password: none

name	type	length	key
Partnum	text	8 characters	yes
Partfamly	text	20 characters	
Assmcode	text	20 characters	
Minlt	int	1	
Avrgprice	real	1	

Current number of rows = 284

Figure 5. *"Part" Relation in the Data Schema.*

The Data Schema

A data base has been developed to demonstrate some applications of the electronics classification and coding system. The Relational Information Management System Version 5 was used on a DEC VAX computer operating under VMS to generate this data base. As implied by its name, this DBMS uses a relational data structure.

Because the purpose of this data base was only for demonstration, its scope was rather limited and will not cover all aspects of electronics manufacturing. Nevertheless, it was designed to be as close to the real environment as possible within the intended scope.

The data schema of this prototype data base is listed in Table 3.

TABLE 3 – The Data Schemes of the Data Base 'ELECT'

RELATION: PART
LAST MOD. 85/11/13   READ PASSWORD: NONE
SCHEMA: ELECT   MODIFY PASSWORD: NONE

NAME	TYPE	LENGTH	KEY
PARTNUM	TEXT	8 CHARACTERS	
PARTFMLY	TEXT	20 CHARACTERS	
ASSMCODE	TEXT	20 CHARACTERS	
MINLT	INT	1	
AVGPRICE	REAL	1	

CURRENT NUMBER OF ROWS = 240

RELATION: FAMILY
LAST MOD: 85/11/13   READ PASSWORD: NONE
SCHEMA: ELECT   MODIFY PASSWORD: NONE

NAME	TYPE	LENGTH	KEY
PARTFMLY	TEXT	20 CHARACTERS	
PARTNAME	TEXT	20 CHARACTERS	
STDPRICE	REAL	1	
ITEMCLAS	TEXT	1 CHARACTERS	
FAMUSAGE	INT	1	
SPECFILE	TEXT	8 CHARACTERS	

CURRENT NUMBER OF ROWS = 195

RELATION: ITEMUSE
LAST MOD: 85/10/28   READPASSWORD: NONE
SCHEMA: ELECT   MODIFY PASSWORD: NONE

NAME	TYPE	LENGTH	KEY
PARTNUM	TEXT	8 CHARACTERS	
STOCKLOC	TEXT	8 CHARACTERS	
USAGE	INT	1	

CURRENT NUMBER OF ROWS = 65

RELATION: SUPPLY
LAST MOD: 85/11/13   READPASSWORD: NONE
SCHEMA: ELECT   MODIFY PASSWORD: NONE

NAME	TYPE	LENGTH	KEY
PARTNUM	TEXT	8 CHARACTERS	
VENDCODE	TEXT	8 CHARACTERS	
PRICE	REAL	1	
dLT	INT	1	
QUALITY	REAL	1	
EQUPCODE	TEXT	5 CHARACTERS	
EQUPNAME	TEXT	20 CHARACTERS	
MAINTAIN	TEXT	8 CHARACTERS	
AVEIRCOS	INT	1	
AVGLIFE	INT	1	
AVGSALVG	INT	1	
DEPMETHD	TEXT	15 CHARACTERS	

CURRENT NUMBER OF ROWS = 0

RELATION: YRLYEXP
LAST MOD: 85/07/09   READ PASSWORD: NONE
SCHEMA: ELECT   MODIFY PASSWORD: NONE

NAME	TYPE	LENGTH	KEY
PROCCODE	TEXT	2 CHARACTERS	
EQUPCODE	TEXT	5 CHARACTERS	
AGE	INT	1	
AVGYREXP	INT	1	

CURRENT NUMBER OF ROWS = 0

RELATION: OPDATA
LAST MOD: 85/11/08   READ PASSWORD: NONE
SCHEMA: ELECT   MODIFY PASSWORD: NONE

NAME	TYPE	LENGTH	KEY
PROCCODE	TEXT	2 CHARACTERS	

CURRENT NUMBER OF ROWS = 740

RELATION: VENDOR
LAST MOD: 85/10/28   READ PASSWORD: NONE
SCHEMA: ELECT   MODIFY PASSWORD: NONE

NAME	TYPE	LENGTH	KEY
VENCODE	TEXT	8 CHARACTERS	
VENDNAME	TEXT	20 CHARACTERS	
STREET	TEXT	20 CHARACTERS	
CITY	TEXT	10 CHARACTERS	
STATE	TEXT	2 CHARACTERS	
ZIP	TEXT	5 CHARACTERS	
PHONE	TEXT	10 CHARACTERS	
CONTACT	TEXT	10 CHARACTERS	
VENDTYPE	TEXT	1 CHARACTERS	
CREDRATE	TEXT	1 CHARACTERS	
PERERATE	TEXT	1 CHARACTERS	

CURRENT NUMBER OF ROWS = 0

RELATION: EQUIP
LAST MOD: 85/11/08   READ PASSWORD: NONE
SCHEMA: ELECT   MODIFY PASSWORD: NONE

NAME	TYPE	LENGTH	KEY
EQUIPNO	TEXT	8 CHARACTERS	
PROCCODE	TEXT	2 CHARACTERS	
EQUPCODE	TEXT	5 CHARACTERS	

CURRENT NUMBER OF ROWS = 0

RELATION: PROCTYPE
LASTMOD: 85/07/09   READ PASSWORD: NONE
SCHEMA: ELECT   MODIFY PASSWORD: NONE

NAME	TYPE	LENGTH	KEY
PROCCODE	TEXT	2 CHARACTERS	
PROCNAME	TEXT	20 CHARACTERS	

CURRENT NUMBER OR ROWS = 0

RELATION: EQUPCOST
LAST MOD: 85/07/09   READ PASSWORD: NONE
SCHEMA: ELECT   MODIFY PASSWORD: NONE

NAME	TYPE	LENGTH	KEY
PROCCODE	TEXT	2 CHARACTERS	

Use of Data Base

As mentioned previously, the user views identify specific needs for data or information contained in the data base. A typical user view contains the following elements:

1. *Objective of the user view.* Some examples may be part selection and standardization, part substitution, equipment selection, and part-family usage.
2. *Description of the objective,* including specific data request.
3. A list of data elements (information entities in a relation file) that are relevant to this objective and their relationships (i.e., one-to-one, one-to-many, many-to-many, many-to-one).
4. A logical access map. This map will show the location of each data element in the relation to the data base.
5. The query language for achieving the goal(s) set forth in this user view.

Examples of user views are given to perform part selection and standardization, part substitution, equipment selection, and part-family usage.

The name of the user view and its description is straightforward and does not require much explanation. The data, elements required for this task are Partfmly, Partname, stdprice, avgprice, etc. as listed. Partfmly is the GT code of this part; Partname is its generic name—for example, low voltage power resistor: stdprice is the average of all parts within the same family, that is, parts with the same GT code; Partno is the unique part ID: for example, R03; Avrgprice is the average of all prices quoted (by different manufacturers) for the same specific part; Specfile is the name of the relation-containing data elements that are very specific to this part, for example, its exact resistance, tolerance, rated voltage etc.; Minlt is the minimum lead-time, and the other data elements are relatively easy to understand because of their names.

The relationships between the data elements are indicated by connecting lines with single or multiple arrow heads. A single arrow head means a one-to-one relationship, a double arrow head means a one-to-many relationship. The logical access map provides a means to access the various data elements contained in many different relations.

The query language—in this case using the Relational Information Management (RIM™*) data base system—is the command that allows the user to perform many of the tasks involved in this user view. Thus, to determine the GT code for a part with given Part ID, the user would type "SELECT PARTFMLY FROM PART WHERE PARTNUM EQS" (Part Id). Other tasks may be performed with the RIM query commands as listed.

---

* Developed by Boeing Commercial Aircraft Company under U.S. Government Contract NAS 1-14700.

User View # 1: Part Selection and Standardization

DESCRIPTION. In selecting a part for a design, a part-family code, which codes design requirements, is used to retrieve a list of part numbers that belong to the part family. Relevant information on those parts is also obtained. A suitable part is then selected by the designer, if it exists. Otherwise, a new part will have to be selected from other sources. Part standardization is achieved by this selection procedure because a new part will be used only if its equivalent is not available in the data base. Also, some parts that have similar properties may be eliminated from the inventory and the data base.

PRIMARY USER: Designer

#	Name	#	Name	#	Name	#	Name
1	Partfmly	2	Partno	36	Price	31	Vandaddr
2	Partname	40	Specfile	37	Lt		
4	Stdprice	5	Minlt	38	Quality	*	
21	Avgprice	29	Vendcode	30	Vendname		Specs.

User View:

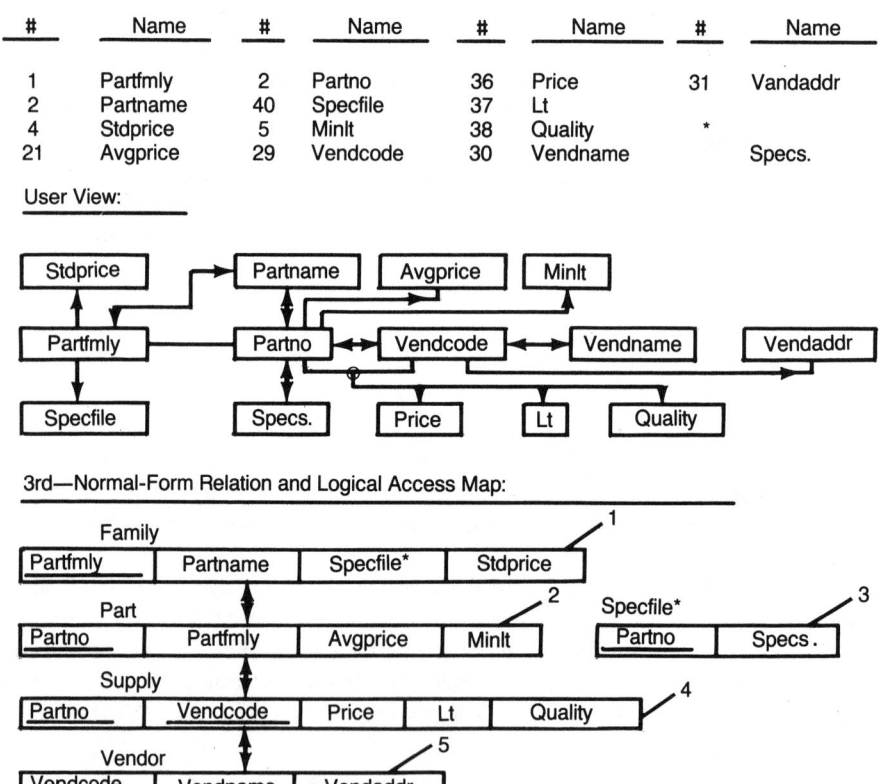

3rd—Normal-Form Relation and Logical Access Map:

DATA ELEMENT:

RIM QUERIES

SELECT PARTFMLY FROM PART WHERE PARTNUM EQS (part no.)
SELECT SPECFILE FROM FAMILY WHERE PARTFMLY EQS (values of
  PARTFMLY obtained above)
PROJECT SUBPART FROM PART USING PARTNO WHERE PARTFMLY
  EQS (value of PARTFMLY obtained above)
JOIN SUBPART USING PARTNO FORMING SUBSPEC
SEL ALL FROM SUBSPEC
INTERSECT SUBPART WITH SUPPLY FORMING SUBVEND USING
  ALL
SET ALL FROM SUBVEND

User View # 2: Part Substitution

DESCRIPTION. Parts in the data base that have similar properties to the part
to substitute are identified through the part-family part code of that part. Their
specifications are reported to select substitute. However, if such a part is not
available, its potential suppliers can be identified as the suppliers of any part in
the part family.

PRIMARY USER: Designer

DATA ELEMENT:

RIM QUERIES

SELECT ALL FROM FAMILY WHERE PARTFMLY EQS (part family)
PROJECT SUBPART FROM PART USING ALL WHERE PARTFMLY EQS
  +(part family)
JOIN SUBPART USING PARTNO WITH (value of SPECFILE obtained above)
  FORMIG PARTINFO
SELECT ALL FROM PARTINFO

#	Name	#	Name	#	Name	#	Name
1	Partfmly	29	Vendcode	36	Price	*	Specs.
2	Partno	30	Vendname	37	Lt	5	Minlt
40	Specfile	31	Vendaddr	38	Quality	21	Avgprice

User View:

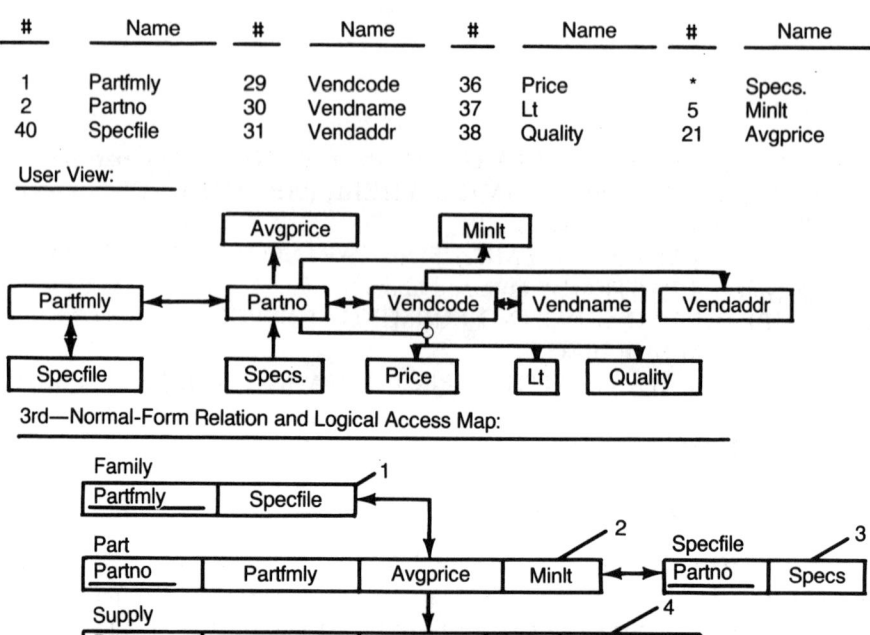

3rd—Normal-Form Relation and Logical Access Map:

User View # 3: Equipment Selection

Description. All equipment that can be used on a given part for an assembly process are listed with their operational information.

PRIMARY USER: Process Planner

DATA ELEMENT:

RIM QUERIES
SELECT ASSMCODE FROM PART WHERE PARTNO EQS (part no.)
SELECT PROCNAME FROM PROCTYPE WHERE PROCCODE EQS (proc. code)
PROJECT PROCEQUP FROM EQUPCOST USING PROCCODE
　　EQUPCODE + EQUPNAME WHERE PROCCODE EQS (proc. code)

#	Name	#	Name	#	Name	#	Name
2	Partno	8	Procname	11	Minlots	14	Stdtime
6	Assmcode	9	Equpcode	12	Toolreqd	15	Stdcost
7	Proccode	10	Equpname	13	Setuptm	17	Prodrate

User View:

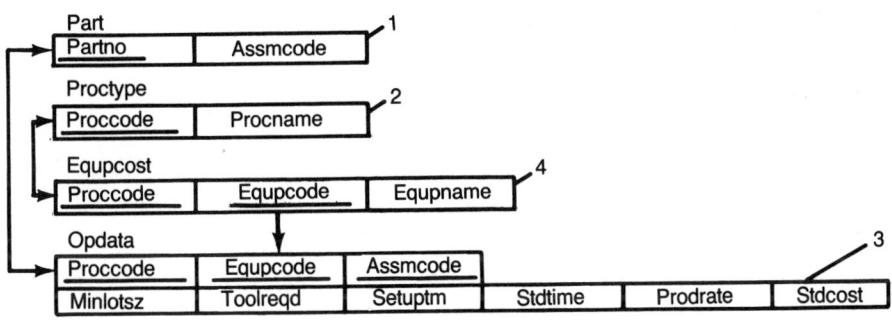

3rd—Normal-Form Relation and Logical Access Map:

PROJECT OPERATN FROM OPDATA USING ALL WHERE PROCCODE
   EQS (proc. code) AND EQUPCODE EQS (equip. code)
INTERSECT PROCEQUP WITH OPERATN FORMING RESULT
SELECT ALL FROM RESULT

User View # 4: Part family usage

DESCRIPTION. The total year-to-date usage of all the parts in a part family is calculated. The result is then stored in the data base for applications such as ABC analysis and material usage forecasting.

PRIMARY USER: Materials management

DATA ELEMENT:

#	Name	#	Name	#	Name	#	Name
1	Partfmly	28	Usage				
2	Partno	39	Famusage				
27	Stockloc						

User View:

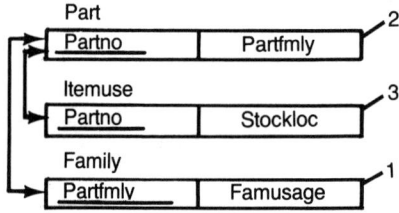

3rd—Normal-Form Relation and Logical Access Map:

Part
| Partno | Partfmly | 2

Itemuse
| Partno | Stockloc | 3

Family
| Partfmly | Famusage | 1

CONCLUSION

The electronics classification and coding system contained in this chapter represents a conceptual attempt to organize and document electronics part characteristics and other management information that influence CAD/CAM activities in PCB assembly operations. The system needs to be implemented in a variety of actual manufacturing environments, tested, and subsequently refined. Because the system is generic in scope, it may need customization to meet the needs of each individual user. Tailoring the system may be achieved by pruning or adding classification trees or attributes.

The work discussed here is by no means complete. The next system expansion should be involved with components other than the nine types of components mentioned, in particular the surface mount devices which are being used more and more in the electronics industries. Also, the manufacturing process of the bare board must be part of the system. As mentioned previously a handbook with the entire and expanded version of the BMCODE system is forthcoming.

Besides the applications discussed so far, another benefit that this C&C system may offer is the development of standard times and cost information. The part assembly attribute scheme was designed in such a way that each classification of an attribute would reflect some special restrictions, requirements, and/or time needed to perform an assembly process. If a part has been coded using the part assembly attribute scheme, the means exists to map that part to a standard time and/or standard cost for using a given assembly method. An assembly process and equipment/method code may also be used to represent the assembly method for this mapping purpose.

Another potential benefit is the foundation for developing automated process planning. The system has been designed to incorporate information related to the assembly of an electronics component to a PCB. A large amount of data required for process planning may be derived from various codes in the system. Standard approaches applied in artificial intelligence, such as hierarchical problem decomposition, tree knowledge representation, and heuristic search, can be applied to the PCB assembly. The codes developed here can be used as interfaces or pointers to the appropriate segments of the industrial-sized data base. Artificial intelligence tasks can, therefore, be conceived to effectively access the data base through these codes and solve the process planning problem.

Acknowledgement

I thank Dr. Manop Reodecha, Lecturer in industrial engineering at the Chulalongkorn University, Thailand, for most of the development work reported in this chapter.

Suggested Readings

Allen, D. K. 1979. D-CLASS: Transportable CAM Data Base Structure. CAM Software Laboratory. Brigham Young University. Utah.

Bao, H., and M. Reodecha. 1986. An Approach to Appraising the Manufacturability of a Printed Circuit Board Assembly. Proceedings of First International Conference on Product Design for Assembly. Newport, R.I.

Barnes, R. D. 1976. Group Technology Concepts Relative to the CAM-I Automated Process Planning (CAPP) System. Proceedings of Executive Seminar on Coding, Classification, and Group Technology for Automated Planning. St Davis. p. 136.

Ham, I. 1975. Introduction to Group Technology. Proceedings of CAM-I's Classification and Coding Workshop. CAM-I. Arlington, TX.

McFadden, F. R., and J. A. Hoffer. 1985. Data Base Management. The Benjamin/Cummings Publishing Company, Inc. Menlo Park, CA.

Peeble, J. F. 1983. Advancements in Electronic Assembly – The Automated Factory. SME Technical paper EE83-122. Society of Manufacturing Engineers. Dearborn, MI.

Shaffer, G. 1981. "GT via automated process planning. *American Machinist.* 124. pp. 112-122.

Han P. Bao is Associate Professor of Industrial Engineering at the University of Missouri-Columbia, Columbia, Missouri.

Glossary

ABC CLASSIFICATION—Classification of the items in an inventory in decreasing order of annual dollar volume or other criteria. This array is then split into three classes, called A, B, and C. Class A contains the items with the highest annual dollar volume and receives the most attention. The medium Class B receives less attention, and Class C, which contains the low-dollar volume items, is controlled routinely. The ABC principle is, that effort saved through relaxed controls on low-value items will be applied to reduce inventories of high-value items. Syn: distribution by value.

ABSORPTION COSTING—An inventory evaluation in which variable costs and a portion of the fixed costs are assigned to each unit of production. The fixed costs are usually allocated to units of output on the basis of direct labor hours, machine hours, or material costs. When a large portion of the unit cost in an absorption system is allocated, inaccuracies may develop particularly in comparisons between products.

ACCUMULATION BIN—Where a product is assembled, this is usually a physical location used to accumulate all of the components that go into the assembly before sending the assembly order out to the assembly floor. Syn: kitting area, assembly bin

ACTUAL COST SYSTEM—A cost system which historically collects costs as they are applied to the production, and allocates indirect costs based upon their specific costs and achieved volume.

AGGREGATE INVENTORY MANAGEMENT—Specifically planning the overall levels of inventory that will be required and making sure that the individual replenishment techniques execute this overall policy.

ALLOCATION—1. In an MRP system, an allocated item is one for which a picking order has been released to the stock room but not yet sent out of the stock room. It is an "uncashed" stock room requisition. 2. A process used to distribute material in short supply.

ASSEMBLY—A group of subassemblies and/or parts which are put together; the total unit constitutes a major subdivision of the final product. When two or more components or subassemblies are put together by the application of labor or machine hours, it is called an assembly. An assembly may be an end item or a component of a higher level assembly.

ASSEMBLY LEADTIME—The time that normally elapses between the time a work order is issued to the assembly floor and its receipt into stock or shipping.

ASSEMBLY ORDER—A manufacturing order to an assembly or blending department authorizing it to put components together into an assembly or blend. Syn: blend order.

AUTOMATION—The substitution of machine work for human physical and mental work, or the use of machines for work not otherwise able to be accomplished, entailing a less continuous interaction with men than previous equipment used for similar tasks.

AVAILABLE INVENTORY—The on-hand balance minus allocations, reservations, backorders and (usually) quantities held for quality problems.

AVAILABLE TO PROMISE—The uncommitted portion of a company's inventory or planned production. This figure is frequently calculated from the Master Production Schedule and is maintained as a tool for order promising.

BACKFLUSH—The deduction from inventory of the component parts used in an assembly or subassembly by exploding the bill of materials by the production count of assemblies produced. Syn: post-deduct inventory transaction processing.

BACKLOG—All of the customer orders booked, i.e. received but not yet shipped. Sometimes referred to as "open orders" or the "order board."

BACKORDER—An unfilled customer order or commitment. It is an immediate (or past due) demand against an item whose inventory is insufficient to satisfy the demand.

BACKWARD SCHEDULING—A scheduling technique where the schedule is computed starting with the due date for the order and working backward to determine the required start date. This can generate negative times, thereby identifying where time must be made up.

BALANCING OPERATIONS—In repetitive just-in-time production, trying to match actual output cycle times of all operations to the cycle time of use for parts as required by final assembly, and eventually as required by the market.

BILL OF MATERIAL—A listing of all the sub-assemblies, parts and raw materials that go into a parent assembly showing the quantity of each required to make an assembly. There are a variety of formats of Bill of Material, including Single Level bill of material, Indented bill of material, Modular (Planning) bill of material, Transient bill of material, Matrix bill of material, Costed bill of material, etc.

BLANKET ORDER—A long-term commitment to a vendor for material against which short-term releases will be generated to satisfy requirements.

BOTTLENECK—A facility, function, department, etc., that impedes production—for example, a machine or work center where jobs arrive at a faster rate than they leave.

BUCKETED SYSTEM—An MRP, DRP or other time-phased system in which all time-phased data are accumulated into time periods or "buckets." If the period of accumulation would be one week, then the system would be said to have weekly buckets.

BUCKETLESS SYSTEM—An MRP, DRP or other time-phased system in which all time-phased data are processed, stored and displayed using dated records rather than defined time periods or "buckets."

BUFFER—A storage area in the computer where data are held temporarily until the computer can process it.

BUSINESS PLAN—A statement of income projections, costs and profits usually accompanied by budgets and a projected balance sheet as well as a cash flow (source and application of funds) statement. It is usually stated in terms of dollars only. The business plan and the production plan, although frequently stated in different terms, should be in agreement with each other.

CAD/CAM—The integration of computer aided design and computer aided manufacturing to achieve automation from design through manufacturing.

CAPACITY—1. In a general sense, refers to an aggregated volume of workload. It is a separate concept from priority. 2. The highest reasonable output rate which can be achieved with the current product specifications, product mix, work force, plant and equipment.

CAPACITY REQUIREMENTS PLANNING (CRP)—The function of establishing, measuring, and adjusting limits or levels of capacity. The term capacity requirements planning in this context is the process of determining how much labor and machine resources are required to accomplish the tasks of production. Open shop orders, and planned orders in the MRP system, are input to CRP which "translates" these orders into hours of work by work center by time period.

CARRYING COST—Cost of carrying inventory, usually defined as a percent of the dollar value of inventory per unit of time (generally one year). Depends mainly on cost of capital invested as well as the costs of maintaining the inventory such as, taxes and insurance, obsolescence, spoilage, and space occupied. Such costs vary from 20-35% annually, depending on type of industry.

CAUSE AND EFFECT DIAGRAM—A precise statement of a problem or phenomenon with a branching diagram leading from the statement to the known potential causes. Syn: fishbone diagram, Ishikawa diagram.

CENTRALIZED DISPATCHING—Organization of the dispatching function into one central location. This often involves the use of data collection devices for communication between the centralized dispatching function, which usually reports to the Production Control Department, and the shop manufacturing departments.

COMPONENT—An inclusive term used to identify a raw material, ingredient, part or subassembly that goes into a higher level assembly, compound or other item. May also include packaging materials for finished items.

CONTROL CHART—A statistical device usually used for the study and control of repetitive processes. It is designed to reveal the randomness or non-randomness of deviations from a mean or control value, usually by plotting these.

COST CENTER—The smallest segment of an organization for which costs are collected. The criteria in defining cost centers are that the cost be significant and the area of responsibility be clearly defined. A cost center may not be identical to a work center. Normally, it would encompass more than one work center.

COST OF CAPITAL—Refers to the inputted cost of maintaining a dollar of capital invested for a certain period, normally one year. This cost is normally expressed as a percentage and may be based upon factors such as the average expected return on alternative investments and current bank interest rate for borrowing.

CRITICAL PATH METHOD (CPM)—A network planning technique used for planning and controlling the activities in a project. By showing each of these activities and their associated times, the "critical path" can be determined. The critical path identifies those elements that actually constrain the total time for the project.

CRITICAL RATIO—A dispatching rule which calculates a priority index number by dividing the time to due date remaining by the expected elapsed time to finish the job. Typically ratios of less than 1.0 are behind, ratios greater than 1.0 are ahead, and a ratio of 1.0 is on schedule.

CUMULATIVE LEAD TIME—The longest length of time involved to accomplish the activity in question. For any item planned through MRP it is found by reviewing each bill of material path below the item, and whichever path adds up to the greatest number defines cumulative material lead time. Syn: aggregate lead time, stacked lead time, composite lead time, critical path lead time.

CUMULATIVE MANUFACTURING LEAD TIME—The composite lead time when all purchased items are assumed to be in stock.

CYCLE COUNTING—A physical inventory-taking technique where inventory is counted on a periodic schedule rather than once a year. For example, a cycle inventory count may be taken when an item reaches its reorder point, when new stock is received, or on a regular basis usually more frequently for high-value fast-moving items and less frequently for low-value or slow-moving items. Most effective cycle counting systems require the counting of a certain number of items every work day.

CYCLE TIME—1. In industrial engineering, the time between completion of two discrete units of production. For example, the cycle time of motors assembled at a rate of 120 per hour would be 30 seconds, or one every half minute. 2. In materials management, the length of time from when material enters a production facility until it exits. Syn: throughput time.

DATA BASE—A data file philosophy designed to establish the independence of computer programs from data file. Redundancy is minimized and data elements can be added to, or deleted from, the file designs without necessitating changes to existing computer programs.

DATA BASE MANAGEMENT—A set of rules about file organization and processing, generally contained in complex software, which controls the definition and access of complex, interrelated files which are shared by numerous application systems.

DEMAND—A need for a particular product or component. The demand could come from any number of sources, i.e. customer order, forecast, interplant, branch warehouse, service part, or to manufacturing the next higher level. At the finished goods level, "demand data" are usually different from "sales data" because demand does not necessarily result in sales, i.e. if there is no stock there will be no sale.

DEPENDENT DEMAND—Demand is considered dependent when it is directly related to or derived from the demand for other items or end products. Such demands are therefore calculated, and need not and should not be forecast. A given inventory item may have both dependent and independent demand at any given time. (cf. independent demand.)

DETERMINISTIC MODELS—Models where no uncertainty is included. Examples include inventory models without safety stock considerations.

DIRECT COSTS—Variable costs which can be directly attributed to a particular job or operation. (cf. variable costs.)

DIRECT-DEDUCT INVENTORY TRANSACTION PROCESSING—A method of doing bookkeeping which decreases the book (computer) inventory of an item as material is issued from stock, and increases the book inventory as material is received into stock. The key concept here is that the book record is updated coincident with the movement of material out of or into stock. As a result, the book record is a representation of what is physically in stock. (cf. pre-deduct inventory transaction processing, post-deduct inventory transaction processing.)

DISBURSEMENT—The issuance of raw material or components from a stores room.

DISCOUNTED CASH FLOW—A method of investment analysis in which future cash flows are converted, or discounted, to their value at present time. The rate of return for an investment is that rate at which the present value of all related cash flow equals zero. (cf. present value.)

DISPATCH LIST—A listing of manufacturing orders in priority sequence according to the dispatching rules. The dispatch list is usually communicated to the manufacturing floor via hard copy or CRT display, and contains detailed information on priority, location, quantity, and the capacity requirements of the manufacturing order by operation. Dispatch lists are normally generated daily and oriented by work center. (Syn: foremen's report).

DISPATCHING—The selecting and sequencing of available jobs to be run at individual work stations and the assignment of these jobs to workers.

DISPATCHING RULE—The logic used to assign priorities to jobs at a work center. (e.g., critical ratio, due date rule, slack time rule.)

DISTRIBUTED PROCESSING—A data processing organizational concept under which computer resources of a company are installed at more than one location with appropriate communication links. Processing is performed at the user's location generally on a mini-computer, and under the user's control and scheduling, as opposed to processing for all users being done on a large, centralized computer system.

DISTRIBUTED SYSTEMS—Refers to computer systems in multiple locations throughout an organization, working in a cooperative fashion, with the system at each location primarily serving the needs of that location but also able to receive and supply information from other systems within the network.

DISTRIBUTION CENTER—A warehouse with finished goods and/or service items. A typical company, for example, might have a manufacturing facility in Philadelphia and distribution centers in Atlanta, Dallas, Los Angeles, San Francisco, and Chicago. The term distribution center is synonymous with the term branch warehouse, although the former has become more commonly used recently. When there is a warehouse that serves a group of satellite warehouses, this is usually called a regional distribution center.

DISTRIBUTION REQUIREMENTS PLANNING—The function of determining the needs to replenish inventory at branch warehouses. Frequently, a time-phased order point approach is used where the planned orders at the branch warehouse level are "exploded" via MRP logic to become gross requirements on the supplying source. In the case of multi-level distribution networks, this explosion process can continue down through the various levels of master warehouse, factory warehouse, etc., and become input to the master production schedule. Demand on the supplying source(s) is recognized as dependent, and standard MRP logic applies.

DISTRIBUTION RESOURCE PLANNING (DRP)—The extension of Distribution Requirements Planning into the planning of the key resources contained in a distribution system: warehouse space, manpower, money, trucks and freight cars, etc. (cf. distribution requirements planning.)

DYNAMIC PROGRAMMING—A method of sequential decision-making in which the result of the decision in each stage affords the best possible means to exploit the expected range of likely (yet unpredictable) outcomes in the following decision-making stages.

ECONOMIC ORDER QUANTITY (EOQ)—A type of fixed order quantity, which determines the amount of an item to be purchased or manufactured at one time. The intent is to minimize the combined costs of acquiring and carrying inventory.

EFFICIENCY—The relationship between the planned resource requirements, such as labor or machine time, for a task(s) and the actual resource time charged to the task(s).

END ITEM—A product sold as a completed item or repair part; any item subject to a customer order or sales forecast. Syn: finished product.

EXPEDITING—The "rushing" of "chasing" of production or purchase orders which are needed in less than the normal lead time. (cf. dispatching.)

EXPONENTIAL DISTRIBUTION—A continuous probability distribution where the probability of occurrence either steadily increases or decreases. The steady increase case (positive exponential distribution) is used to model phenomena such as customer service level versus cost. The steady decrease case (negative exponential distribution) is used to model things such as the weight given to any one time period of demand in exponential smoothing.

EXPONENTIAL SMOOTHING—A type of weighted moving average forecasting technique in which past observations are geometrically discounted according to their age. The heaviest weight is assigned to the most recent datum. The smoothing is termed "exponential" because data points are weighted in accordance with an exponential function of their age. The technique makes use of a smoothing constant to apply to the difference between the most recent forecast and the critical sales datum, which avoids the necessity of carrying historical sales data. The approach can be used for data which exhibit no trend or seasonal patterns or for data with either (or both) trend and seasonality.

EXTERNAL SET-UP TIME—Elements of a set-up procedure performed while the process is in production; the machine is running. (cf. internal set-up time.)

FABRICATION—A term used to distinguish manufacturing operations for components as opposed to assembly operations.

FIFO—First in, first out method of inventory evaluation. The assumption is that oldest inventory (first in) is the first to be used (first out).

FINAL ASSEMBLY—The highest or "zero level" assembled product. Frequently used as a name for the manufacturing department where the product is assembled.

FINAL ASSEMBLY SCHEDULE (FAS)—Also referred to as the "finishing schedule" as it may include other operations than simply the final operation. It is a schedule of end items either to replenish finished goods inventory or to finish the product for a make-to-order product. For make-to-order products, it is prepared after receipt of a customer order, is constrained by the availability of material and capacity, and it schedules the operations required to complete the product from the level where it is stocked for master scheduled) to the end item level.

FIRM PLANNED ORDER—A planned order that can be frozen in quantity and time. The computer is not allowed to automatically change it; this is the responsibility of the planner in charge of the item that is being planned. This technique can aid planners working with MRP systems to respond to material and capacity problems by firming up selected planned orders. Additionally, firm planned orders are the normal method of stating the master production schedule.

FIXED COST—An expenditure that does not vary with the production volume, for example: rent, property tax, salaries of certain personnel.

FIXED INTERVAL REORDER SYSTEM—A periodic reordering system where the time interval between orders is fixed, such as weekly, monthly or quarterly, but the size of the order is not fixed and orders vary according to usage since the last review. This type of inventory control system is employed where it is convenient to examine inventory stocks on a fixed time cycle, such as in warehouse control systems, in systems where orders are placed mechanically, or for handling inventories involving a very large variety of items under some form of clerical control. Also called fixed reorder cycle system.

FLOW SHOP—A shop in which machines and operators handle a standard, usually uninterrupted material flow. The operators tend to perform the same operations for each production run. A Flow Shop is often referred to as a mass production shop, or is said to have a continuous manufacturing layout. The shop layout (arrangement of machines, benches, assembly lines, etc.) is designed to facilitate a product "flow."

FORWARD SCHEDULING—A scheduling technique where the scheduler proceeds from a known start date and computes the completion date for an order usually proceeding from the first operation to the last.

GANTT CHART—The earliest and best known type of control chart especially designed to show graphically the relationship between planned performance and actual performance, named after its originator, Henry L. Gantt. Used for machine loading, where one horizontal line is used to represent capacity and another to represent load against that capacity or for following job progress where one horizontal line represents the production schedule and another parallel line represents the actual progress of the job against the schedule in time. Syn: job progress chart.

GROSS REQUIREMENTS—The total of independent and dependent demand for a part or an assembly prior to the netting of on hand inventory and scheduled receipts.

GROUP CLASSIFICATION CODE—A part of material classification technique which provides for designation of characteristics by successively lower order groups of code. Classification may denote, for example, function, type of material, size, shape, etc. (cf. group technology.)

GROUP TECHNOLOGY—An engineering and manufacturing philosophy which identifies the "sameness" of parts, equipment or processes. It provides for rapid retrieval of existing designs and anticipates a cellular type production equipment layout.

HANDLING COST—The cost involved in handling inventory. In some cases, the handling cost incurred may depend on the size of the inventory. For example, inventories over a fixed maximum level may have to be stored in a nearby warehouse at substantial cost per case of handling and trucking material stored outside, or production in excess of immediate needs of a given product may be specially packed and stored at a substantial extra-handling cost.

HEURISTIC—A form of problem solving where the results or rules have been determined by rule of thumb or intuition instead of by optimization.

IDLE TIME—Time when operators or machines are not producing product because of setup, maintenance, lack of material, tooling. Syn: down time.

INCREMENTAL COST—1. Cost added in the process of finishing a part or assembly, assembling a group of parts or adding part(s) or assembly(s) to a higher level assembly. If the cost of the components of a given assembly equals $5 and the additional cost of assembling the components is $1, then the incremental assembly cost is $1, while the total cost of the finished assembly is $6. 2. Additional cost incurred as a result of a decision selecting a different method of procuring a part, achieving a goal, fulfilling a requirement, etc.

INDENTED BILL OF MATERIAL—A form of multi-level bill of material. It exhibits the highest level subassemblies closest to the left side margin and all the components going into these subassemblies are shown indented to the right of the margin. All subsequent levels of components are indented farther to the right. If a component is used in more than one subassembly within a given product structure, it will appear more than once, under every subassembly in which it is used.

INDEPENDENT DEMAND—Demand for an item is considered independent when such demand is unrelated to the demand for other items. Demand for finished goods, parts required for destructive testing and service parts requirements are some examples of independent demand.

INFINITE LOADING—Showing the work behind work centers in the time periods required regardless of the capacity available to perform this work. The term infinite loading is considered to be obsolete today, although the specific computer programs used to do infinite loading can now be used to perform the technique called capacity requirements planning. Infinite loading was a gross misnomer to start with, implying that a load could be put into a factory regardless of its availability to perform. The poor terminology obscured the fact that it is necessary to generate capacity requirements and compare these with available capacity before trying to adjust requirements to capacity.

INFORMATION—The meaning derived from data which have been arranged and displayed in such a way that they can be related to that which is previously known.

INPUT/OUTPUT CONTROL—A technique for capacity control where actual output from a work center is compared with the planned output developed by CRP. The input is also monitored to see if it corresponds with plans so that work centers will not be expected to generate output when jobs are not available to work on.

INTERMITTENT PRODUCTION—A production system in which the productive units are organized according to function. The jobs pass through the functional departments in lots and each lot may have a different routing.

INTERNAL SET-UP TIME—Elements of a set-up procedure performed while the process is not running.

INTEROPERATION TIME—The time between the completion of an order at one work center and its start at the next.

INTRANSIT LEAD TIME—The time lag between the date of shipment (at supplier shipping point) and the date of receipt (at the customer's dock). Normally customers' orders specify the date by which goods should be at his dock. Consequently this date should be offset by intransit lead time for establishing a ship date for the supplier.

INVENTORY—Items which are in stock point or work-in-process and which serve to decouple successive operations in the process of manufacturing a product and distributing it to the consumer. Inventories may consist of finished goods ready for sale; they may be parts or intermediate items; they may be work-in-process; or they may be raw materials.

INVENTORY TURNOVER—The number of times that an inventory "turns over," or cycles during the year. One way to compute inventory turnover is to divide the average inventory level into the annual cost of sales. For example, if average inventory were three million dollars and cost of sales were twenty-one million dollars, the inventory would be considered to "turn" seven times per year.

INVENTORY VALUATION—The value of the inventory at either its cost or its market value. Because inventory value can change with time, some recognition must be taken of the age distribution of inventory. Therefore, the cost value of inventory, under accounting practice, is usually computed on a first-in-first-out (FIFO), last-in-first-out (LIFO) basis, or a standard cost system to establish the cost of goods sold. (cf. absorption costing, variable costing, retail method.)

JOB SHOP—A functional organization whose departments or work centers are organized around particular types of equipment or operations, such as drilling, forging, spinning, or assembly. Products flow through departments in batches corresponding to individual orders, which may be either stock orders or individual customer orders. Syn: intermittent production.

JUST-IN-TIME—A logistics approach designed to result in minimum inventory by having material arrive at each operation just in time to be used. The implication is that each operation is closely synchronized with the subsequent ones to make that possible. In the narrow sense, just-in-time refers to the movement of material so as to have only the necessary material at the necessary place at the necessary time. In the broad sense, it refers to all the activities of manufacturing which make the just-in-time movement of material possible.

KANBAN—A method of just-in-time production which uses standard containers with a single card attached to each. It is a pull system in which work centers which use parts signal with a card that they wish to withdraw parts from feeding operations. Kanban in Japanese loosely translated means "card," literally "billboard" or "sign."

KIT—The components of an assembly which have been pulled from stock and readied for movement to the assembly area.

KITTING—The process of removing components of an assembly from the stock room and sending them to the assembly floor as a kit of parts. This action may take place automatically whenever a full set of parts is available and/or it may be done only upon authorization by a designated person.

LABOR PRODUCTIVITY—The rate of output of a worker or group of workers, per unit of time, compared to an established standard or rate of output.

LAYOUT—The kits of components ahead of the assembly department waiting to be put together. Syn: staged material.

LEAD TIME—A span of time required to perform an activity. In a production and inventory control context, the activity in question is normally the procurement of materials and/or products either from an outside supplier or from one's own manufacturing facility. The individual components of any given lead time can include some or all of the following: order preparation time, queue time, move or transportation time, receiving and inspection time.

LEAD TIME INVENTORY—This is inventory which is carried on hand during the lead time period in simple inventory systems. The lead time inventory will be equal to forecasted usage during the replenishment lead time. Syn: active inventory.

LEAD TIME OFFSET—A term used in MRP where a planned order receipt in one time period will require the release of that order in some earlier time period based on the lead time for the item. The difference between the due date and the release date is the lead time offset.

LEVEL—Every part or assembly in a product structure is assigned a level code signifying the relative level in which that part or assembly is used within that product structure. Normally the end items are assigned level "0" and the components/subassemblies going into it level "1" and so on. MRP explosion process starts from level "0" and proceeds downwards one level at a time.

LEVEL SCHEDULE—A schedule such that the use of all parts in materials is as evenly distributed over time as possible.

LIFO—Last in, first out method of inventory evaluation. The assumption is that the most recently received (last in) is the first to be used or sold (first out). (cf. FIFO.)

LINEAR PROGRAMMING—Mathematical models for solving linear optimization problems through minimization (or maximization) of a linear function subject to linear constraints. For example, in blending gasoline and other petroleum products, many intermediate distillates may be available. Prices and octane ratings, as well as upper limits on capacities of input materials which can be used to produce various grades of fuel are given. The problem is to blend the various inputs in such a way that: (a) cost will be minimized (profit will be maximized), (b) specified optimum octane ratings will be met, and (c) the need for additional storage capacity will be avoided.

LINE BALANCING—An assembly line process can be divided into elemental tasks, each with a specified time requirement per unit of product and a sequence relationship with the other tasks. Line balancing is the assignment of these tasks to work stations so as to minimize the number of work stations and to minimize the total amount of unassigned time at all stations. Line balancing can also mean a technique for determining the product mix that can be run down an assembly line to provide a fairly consistent flow of work through that assembly line at the planned line rate. For example, if an automotive assembly line happened to be scheduled one day with nothing but convertibles, some workers would be standing idle while others would not be able to keep pace with the line.

LONG-RANGE RESOURCE PLANNING—A planning activity for long-term capacity decisions, based on the production plan and perhaps on even more gross data (e.g., sales per year) beyond the time horizon for the production plan. This activity is to plan long-term capacity needs out to the time period necessary to acquire gross capacity additions such as a major factory expansion.

LOT SIZE—The amount of a particular item that is ordered from the plant or a vendor. Syn: order quantity.

LOT SPLITTING—Dividing a lot into two or more sub-lots and simultaneously processing each sub-lot on identical (or very similar) work centers.

MACHINE UTILIZATION—The percent of time that a machine is running production as opposed to idle time.

MAKE-TO-ORDER PRODUCT—The end item is finished after receipt of a customer order. Frequently long leadtime components are planned prior to the order arriving in order to reduce the delivery time to the customer. Where options or other subassemblies are stocked prior to customer orders arriving, the term "assemble-to-order" is frequently used.

MANUFACTURING LEADTIME—The total time required to manufacture an item. Included here are order preparation time, queue time, set-up time, run time, move time, inspection and put-away time.

MANUFACTURING RESOURCE PLANNING (MRP II)—A method for the effective planning of all resources of a manufacturing company. Ideally, it addresses operational planning in units, financial planning in dollars, and has a simulation capability to answer "what if" questions. It is made up of a variety of functions, each linked together: Business Planning, Production Planning, Master Production Scheduling, Material Requirements Planning, Capacity Requirements Planning and the execution support systems for capacity and material. Output from these systems would be integrated with financial reports such as the business plan, purchase commitment report, shipping budget, inventory projections in dollars, etc. Manufacturing Resource Planning is a direct outgrowth and extension of closed-loop MRP.

MARGINAL COST—The additional out of pocket costs incurred when the level of output of some operation is increased by one unit.

MARGINAL REVENUE—The additional income received when the level of output of some operation is increased by one unit.

MASTER PRODUCTION SCHEDULE (MPS)—For selected items, it is a statement of what the company expects to manufacture. It is the anticipated build schedule for those selected items assigned to the master scheduler. The master scheduler maintains this schedule and, in turn, it becomes a set of planning numbers which "drives" MRP. It represents what the company plans to produce expressed in specific configurations, quantities, and dates. The MPS should not be confused with a sales forecast which represents a statement of demand. The master production schedule must take forecast plus other important considerations (backlog, availability of material, availability of capacity, management policy and goals, etc.) into account prior to determining the best manufacturing strategy. Syn: master schedule.

MATERIAL REQUIREMENTS PLANNING (MRP)—A set of techniques which uses bills of material, inventory data and the master production schedule to calculate requirements for materials. It makes recommendations to release replenishment orders for material. Further, since it is time-phased, it makes recommendations to reschedule open orders when due dates and need dates are not in phase. Originally seen as merely a better way to order inventory, today it is thought of as primarily a scheduling technique, i.e., a method for establishing and maintaining valid due dates on orders.

MEAN—The arithmetic average of a group of values.

MEDIAN—The middle value in a set of measured values when the items are arranged in order of magnitude. If there is no middle value, the median is the average of the two middle values.

MICROWAVE TRANSMISSION—A highly effective method of communication using high frequency radio waves and special equipment. Transmission rates of over 5,000 characters per second are attainable by this method in comparison to approximately 300 characters per second using voice grade channels.

MODULAR BILL (OF MATERIAL)—A type of planning bill which is arranged in product modules or options. Often used in companies where the product has many optional features.

NET REQUIREMENTS—In MRP, the net requirements for a part or an assembly are derived as a result of netting gross requirements against inventory on hand and the scheduled receipts. Net requirements, lot sized and offset for lead time, become planned orders.

NETTING—The process of calculating net requirements.

OPERATING SYSTEM—A group of procedures for operating a computer usually including techniques for scheduling operations within the computer.

PARETO'S LAW—A concept developed by Vilfredo Pareto, an Italian economist, that simply says that a small percentage of a group accounts for the largest fraction of the effort, value, etc. For example, twenty percent of the inventory items comprise eighty percent of the inventory value. Syn: 80/20 rule.

PAST DUE—An order that has not been completed on time. Syn: delinquent.

PEGGED REQUIREMENT—A requirement at a component level that shows the next level parent item and the source of the demand that actually created the requirement.

PERT—Program Evaluation and Review Technique—This is a project planning technique similar to the Critical Path Method, which additionally includes obtaining a pessimistic, most likely, and optimistic time for each activity from which the most likely completion time for the project along the critical path is computed. (cf. critical path scheduling.)

PICKING—The process of withdrawing from stock the components to make the products, or the finished goods to be shipped to a customer.

PIECE PARTS—Consists of individual items in inventory at the simplest level in manufacturing. For example, bolts and washers.

PLANNED ORDER—A suggested order quantity and the due date created by MRP processing, when it encounters net requirements. Planned orders are created by the computer; exist only within the computer; and may be changed or deleted by the computer during subsequent MRP processing if conditions change. Planned orders at one level will be exploded into gross requirements for components at the next lower level. Planned orders also serve as input to capacity requirements planning, along with released orders, to show the total capacity requirements in future time periods.

PLANNING BILL (OF MATERIAL)—An artificial grouping of items, in bill of material format, used to facilitate master scheduling and/or material planning. (cf. common parts bill, modular bill, super bill.)

PRESENT VALUE—The value today of future cash flows. For example, the promise of ten dollars a year from now is worth something less than ten dollars in hand today. (cf. discounted cash flow.)

PROCESS CAPABILITY—The basic physical capability of production equipment and procedures to hold dimensions and other characteristics of products (e.g., electrical resistance) with acceptable bounds *for the process itself.* Not the same as tolerances or specifications required of the produced units themselves.

PROCESS CHART—A graphic representation of events occurring during a series of actions or operations and of information pertaining to those operations. Syn: flow chart.

PROCESS CONTROL—The function of maintaining a process within a given range of capability by feedback, correction, etc.

PROCESS COST SYSTEM—A costing system in which the costs are collected by time period and averaged over all the units produced during the period. This system can be used with either actual or standard costs in the manufacture of a large number of identical units.

PROCUREMENT LEAD TIME—The time required by the buyer to select a supplier, and to place and obtain a commitment for specific quantities of material at specified times. (cf. purchasing lead time.)

PRODUCT MIX—The combination of individual product types and the volume produced that make up the total production volume. Changes in the product mix can mean drastic changes in the manufacturing requirements for certain types of labor and material. (cf. sales mix.)

PRODUCT STRUCTURE—The way components go into a product during its manufacture. A typical product structure would show, for example, raw material being converted into fabricated components, components being put together to make subassemblies, subassemblies going into assemblies, etc.

PULL SYSTEM—1. In distribution, refers to a system for replenishing field warehouse inventories wherein replenishment decisions are made at the field warehouse itself, not at the central warehouse or plant. 2. In production, refers to the production of items only as demanded for use, or to replace those taken for use. 3. In a material control context, refers to the withdrawal of inventory as demanded by the using operations. Material is not issued until a signal comes from the user.

PUSH SYSTEM—1. In distribution, refers to a system for replenishing field warehouse inventories wherein replenishment decision-making is centralized, usually at the manufacturing site or central supply facility. 2. In production, refers to the production of items at times required by a given schedule planned in advance. 3. In a material control context, refers to the issuing of material according to a given schedule and/or issued to a job order at its start time.

QUALITY CIRCLE—A group of about five to twelve people who normally work as a unit and meet frequently for the purpose of seeking and overcoming problems concerning the quality of items produced, process capability or process control. Syn: quality control circle.

QUEUE TIME—The amount of time a job waits at a work center before set-up or work is performed on the job. Queue time is one element of total manufacturing lead time. Increases in queue time result in direct increases to manufacturing lead time.

RATED CAPACITY—Capacity calculated from data such as utilization and efficiency, hours planned to be worked, etc. Syn: theoretical capacity.

RECEIVING—This function includes the physical receipt of material; the inspection of the shipment for conformance with the purchase order (quantity and damage); identification and delivery to destination; and preparing receiving reports.

REORDER QUANTITY—In a fixed order system of inventory control, the fixed quantity which should be ordered each time the available stock (on hand plus on order) falls below the order point. However, in a variable reorder quantity system the amount ordered from time period to time period will vary. Syn: replenishment order quantity.

REPETITIVE MANUFACTURING—Production of discrete units, planned and executed via a schedule, usually at relatively high speeds and volumes. Material tends to move in a sequential flow.

REPLENISHMENT LEAD TIME—The total period of time that elapses from the moment it is determined that a product is to be reordered until the product is back on the shelf available for use.

REQUIREMENTS EXPLOSION—A method of calculating future demand for an item. Future production quantities are multiplied by the quantity in the bill of material. The results represent future demand.

ROUGH-CUT CAPACITY PLANNING—The process of converting the production plan and/or the master production schedule into capacity needs for key resources: manpower, machinery, warehouse space, vendors' capabilities and, in some cases, money. Product load profiles are often used to accomplish this. The purpose of rough-cut capacity planning is to evaluate the plan prior to attempting to implement it. Syn: resource requirements planning.

ROUTING—A document for the manufacture of a particular item, the sequence of operations, transportations, storages, and inspections to be used and usually the standard times applicable, and the machines, equipment, tools, work centers, number of workmen, and materials that are required. Syn: operation list, route sheet, operation sheet or chart, process chart, manufacturing data sheet, bill of operation.

SAFETY CAPACITY—The planning or reserving for excess manpower and equipment above known requirements for unexpected demand. This reserve capacity is in lieu of safety stock. Syn: reserved capacity procurement.

SAFETY STOCK—1. In general, a quantity of stock planned to be in inventory to protect against fluctuations in demand and/or supply. 2. The average amount of stock on hand when a replenishment quantity is received. 3. In the context of Master Production Scheduling, safety stock can refer to additional inventory and/or capacity planned as protection primarily against forecast errors and/or short term changes in the backlog. This investment is often under the control of the master scheduler in terms of where it should be planned. Sometimes referred to as "overplanning" or a "Market Hedge."

SAFETY TIME—In an MRP system, material can be ordered to arrive ahead of the requirement date. The difference between the requirement data and the planned in-stock date is safety time.

SCHEDULED RECEIPTS—Within MRP, open production orders and open purchase orders are considered as "scheduled receipts" on their due date and will be treated as part of available inventory during the netting process for the time period in question. Scheduled receipt dates and/or quantities are not normally altered automatically by the MRP system. Further, scheduled receipts are not exploded into requirements for components as MRP logic assumes that all components required for the manufacture of the item in question have been either allocated or issued to the shop floor.

SEQUENCING—Determining the order in which a manufacturing facility is to process a number of different jobs in order to achieve certain objectives.

SET-UP TIME—The time required for a specific machine, assembly line or work center to convert from the production of one specific item to another. (cf. external set-up-time, internal set-up time, major set-up, minor set-up.)

SHIPPING—This activity provides facilities for the outgoing shipment of parts, products, and components. Packaging, marking, weighing, and loading for shipment is part of this activity.

SHIPPING LEAD TIME—The number of working days normally required for goods in transit between a shipping and receiving point, plus acceptance time in days at the receiving point. (cf. transit time.)

SHOP FLOOR CONTROL—A system for utilizing data from the shop floor as well as data processing files to maintain and communicate status information on shop orders (manufacturing orders) and work centers. The major subfunctions of shop floor control are: 1. assigning priority of each shop order. 2. maintaining work-in-process quantity information. 3. conveying shop order status information to the office. 4. providing actual output data for capacity control purposes. 5. providing quantity by location by shop order for work-in-process inventory and accounting purposes. 6. providing measurement of efficiency, utilization and productivity of manpower and machines. (cf. closed loop MRP).

SIMULATION—The technique of utilizing representative or artificial data to reproduce in a model various conditions that are likely to occur in the actual performance of a system. Frequently used to test the behavior of a system under different operating policies. (cf. Monte Carlo technique, model.)

SINGLE-LEVEL BACKFLUSH—A form of backflush which covers only the parts used in an assembly or subassembly by exploding the bill of materials by the production count of assemblies produced. (cf. backflush, superflush.)

SINGLE-LEVEL BILL OF MATERIAL—A single-level bill shows only those components that are directly used in an upper-level item. It does not show any relationships more than the one level down.

SINGLE-LEVEL WHERE USED—Single-level where used for a component lists each assembly in which that component is directly used and in what quantity. This information is usually made available through the technique known as "implosion."

STAGING—Pulling of the material requirements for an order from inventory before the material is required. This action is taken as a protection from inaccurate inventory records, but leads to increased problems in inventory records and availability.

STANDARD ALLOWANCE—The established or accepted amount by which the normal time for an operation is increased within an area, plant, or industry to compensate for the usual amount of fatigue and/or personal and/or unavoidable delays.

STANDARD COSTS—The normal expected costs of an operation, process, or product including labor, material, and overhead charges, computed on the basis of past performance costs, estimates, or work measurement.

STANDARDIZATION—The process of designing and/or reviewing products to establish and use standard specifications for them and/or components. One of the goals of standardization is to reduce the number of items involved.

STOCHASTIC MODELS—Models where uncertainty is explicitly considered in the analysis.

STOCK—Stored products or service parts ready for sale as distinguished from stores which are usually components or raw materials.

STOCKKEEPING UNIT (SKU)—An item at a particular geographic location. For example, a product stocked at six different distribution centers would represent six SKU's, plus perhaps another for the plant at which it was manufactured.

SUB ASSEMBLY—An assembly which is used at a higher level to make up another assembly going into the automobile engine which is an assembly. Syn: intermediate. (cf. component.)

SYNCHRONIZED PRODUCTION—A term sometimes used to mean repetitive just-in-time production.

THROUGHPUT—The total volume of production through a facility (machine, work center, department, plant, or network of plants).

TIME FENCE—A policy or guideline established to note where various restrictions or changes in operating procedures take place. For example, changes to the master production schedule can be accomplished easily beyond the cumulative lead time whereas changes inside the cumulative lead time becomes increasingly more difficult to a point where changes should be resisted. Time fences can be used to define these points.

TWO-BIN SYSTEM—A type of fixed order system in which inventory is carried in two bins. A replenishment quantity is ordered when the first bin is empty. When the material is received, the serve bin is refilled and the excess is put into the working bin. This term is also used loosely to describe any fixed order system even when physical "bins" do not exist. (cf. fixed order system.)

TWO-LEVEL MPS—A master scheduling approach wherein a super bill (of material) is master scheduled along with selected key options, features and attachments.

VALUE ANALYSIS—The systematic use of techniques which serve to identify required function, establish a value for that function, and finally to provide that function at the lowest overall cost. This approach focuses on the functions of an item rather than the methods of producing the present product design.

VARIABLE COSTS—An operating cost that varies directly with production volume; for example, materials consumed, power, direct labor, sales commissions. (cf. fixed cost.)

WORK IN PROCESS—Product in various stages of completion throughout the plant including raw material that has been released for initial processing and completely processed material awaiting final inspection and acceptance as finished product or shipment to a customer. Many accounting systems also include semi-finished stock and components in this category. Syn: in-process inventory. (cf. movement inventory.)

YIELD—The ratio of usable output from a process to the materials of value input to the process. Yield is usually expressed as a percentage and may be in terms of total input or of a specific raw material.

ZERO INVENTORIES—A philosophy of manufacturing based on planned elimination of all waste and consistent improvement of productivity. It encompasses the successful execution of all manufacturing activities required to produce a final product, from design engineering to delivery and including all stages of conversion from raw material onward. The primary elements of zero inventories are to have *only* the required inventory when needed; to improve quality to zero defects; to reduce lead times by reducing set-up times, queue lengths, and lot sizes; and to incrementally revise the operations themselves to accomplish these things at minimum cost. In the broad sense it applies to all forms of manufacturing, job shop and process as well as repetitive. Syn: just-in-time production, stockless production.

Reprinted with permission, the American Production and Inventory Control Society, Inc. APICS *Dictionary*, Fifth Edition, 1984.

Bibliography

ASSOCIATIONS

American Electronics Association
5201 Great America Parkway
Santa Clara, CA 95054
(408) 987-4200
A trade organization serving the electronics industry.

American Production and Inventory Control Society, Inc.
500 West Annandale Road
Falls Church, VA
(703) 237-8344
A professional society for production and inventory control.

American Society for Quality Control
230 West Wells Street
Milwaukee WI 53203
(414) 272-8575
*A technical professional society which advances the theory
and practices of quality control.*

Electronics Industries Association
2001 Eye Street
Washington, D.C. 20006
(202) 457-4900
*A national trade organization representing the full spectrum
of electronic manufacturers in the United States.*

Human Factors Society
P.O. Box 1369
Santa Monica, CA 90406
(213) 394-1811
*An interdisciplinary organization of professional people
involved in the human factors field.*

Institute of Electrical and Electronic Engineers
820 Second Avenue
New York, NY 10017
(212) 644-7966
*A transnational professional society of engineers and scientists
in electrical engineering, electronics, and allied fields.*

Institute of Industrial Engineers
Electronics Industry Division
25 Technology Park/Atlanta
P.O. Box 6150
Norcross, GA 30091-6150
(404) 449-0460
*An international society devoted to productivity improvement
through the application of industrial engineering techniques.*

Institute for Interconnecting and Packaging Electronic Circuits
3451 Church Street
Evanston, IL 60203
(312) 677-2850
*An international trade organization of member companies representing
manufacturers and users of electronic interconnection devices.*

Society of Manufacturing Engineers
One SME Drive
P.O. Box 930
Dearborn, MI 48121
(313) 271-1500
*A technical society which follows and assesses developments
in manufacturing; publishing and distributing information.*

DICTIONARIES, DIRECTORIES AND HANDBOOKS

Electronic Industry Telephone Directory
Harris Publishing Company
2057 Aurora Road
Twinsburg, OH 44087-1999

Electronics Market Data Book (Yearly)
Electronic Foreign Trade (12 issues/year)
Electronic Market Trends (12 issues/year)
*Sources: A Directory of Electronics Information Agencies Outside
the United States*
Electronics Industries Association
2001 I Street
Washington, D.C. 20006

IEEE Standard Dictionary of Electrical and Electronics Terms
ANSI/IEEE Std 100-1984
Institute of Electrical and Electronic Engineers, Inc.
New York, NY 10017
and
Wiley-Interscience, a division of John Wiley & Sons, Inc.

Electronics Engineer's Handbook
McGraw-Hill Book Company
New York, NY 10011

APICS Dictionary, Fifth Edition
American Production and Inventory Control Society
500 West Annadale Road
Falls Church, VA 22046

MAGAZINES AND JOURNALS

Circuits Manufacturing (Monthly)
(Magazine for Manufacturing Engineers and Managers)
Miller Freeman Publications, Inc.
500 Howard Street
San Francisco, CA 94105

Assembly Engineering (Monthly)
(Design and Manufacturing Technology for Better Production)
Hitchcock Publishing Company
25 W 550 Geneva Road
Wheaton, IL 60188

EDN (38 times/year)
(Electronic Technology for Engineers and Engineering Managers)
Cahners Publishing Company
275 Washington Street
Newton, MA 02158-1630

Electronic Engineering Times
(Industry Newspaper for Engineers and Technology Management)
CMP Publications, Inc.
600 Community Drive
Manhasset, NY 11030

Electronic Packaging and Production (Monthly)
Cahners Publishing Company
275 Washington Street
Newton, MA 02158-1630

Information Week (Weekly)
CMP Publications, Inc.
600 Community Drive
Manhasset, NY 11030

IEE SPECTRUM (Monthly)
Institute of Electrical and Electronics Engineers, Inc.
New York, NY 10017

Electronics (Weekly)
McGraw-Hill, Inc.
1221 Avenue of the Americas
New York, NY 10020

Electronic News (Weekly)
Fairchild Publications
7 E 12th Street
New York, NY 10003

ABOUT THE EDITORS

Johnson A. Edosomwan is currently Program Manager of New Products at the IBM Americas Group Headquarters in Mt. Pleasant, New York; he also serves as Adjunct Professor of Industrial Engineering at the New York Polytechnic University. Dr. Edosomwan received his Doctor of Science in Engineering Administration and Economics from George Washington University; Professional Engineering Degree in Industrial Engineering from Columbia University; MSIE and BSIE degrees from the University of Miami. His consulting experience involves work with companies such as Arthur Young and Touche Ross. Dr. Edosomwan is author/editor of seven books. His numerous contributions and achievements as an engineer, educator and consultant in the engineering profession has earned him reference in such publications as *Who's Who of Intellectuals, Men of Achievement in the World*, and *The International Directory of Distinguished Leadership*. Dr. Edosomwan is a member of Alpha Pi Mu, Tau Beta Pi, Omicron Delta Kappa and Sigma Xi honor societies. He currently serves as a Director of the Society for Integrated Manufacturing (SIM) of the Institute of Industrial Engineers and he is a past Director of IIE's Electronics Industry Division. Dr. Edosomwan is the recipient of IIE's 1988 Outstanding Young Industrial Engineer Award.

Arvind Ballakur is a member of the Technical Staff in the Operations Engineering Department at the AT&T Bell Laboratories in Holmdel, New Jersey. He is responsible for coordinating JIT/pull manufacturing in a new manufacturing line at a AT&T factory in Massachusetts. Dr. Ballakur's career at AT&T includes work on parts proliferation, integrating products and processes, computer integrated manufacturing architecture, and MRP-II/JIT integration. He received his Ph.D in Industrial Engineering from the University of Wisconsin at Madison. He also holds a Certificate in Production and Inventory Management (CPIM) from the American Production and Inventory Control Society (APICS). Dr. Ballakur is an active member of the Institute of Industrial Engineers, and he is currently Director-Elect of the Electronics Industry Division. His work with IIE also involves organizing workshops and conferences on Productivity and Technology Management. Dr. Ballakur has published in several professional journals including the *International Journal of Production Research*, the *Journal of Manufacturing Systems*, and *Computers and Industrial Engineering*.